W0107573

Ganzha · Mayr · Vorozhtsov (Eds.)

Computer Algebra in Scientific Computing

Springer-Verlag Berlin Heidelberg GmbH

Victor G. Ganzha   Ernst W. Mayr
Evgenii V. Vorozhtsov (Eds.)

Computer Algebra in Scientific Computing

CASC 2001

Proceedings of the Fourth International Workshop on Computer Algebra in Scientific Computing, Konstanz, Sept. 22-26, 2001

Springer

Editors

Victor G. Ganzha
Ernst W. Mayr

Institut für Informatik
Technische Universität München
80290 München, Germany

e-mail: ganzha@informatik.tu-muenchen.de
mayr@in.tum.de

Evgenii V. Vorozhtsov

Institute of Theoretical and Applied Mathematics
Russian Academy of Sciences
Novosibirsk 630090, Russia

email: vorozh@itam.nsc.ru

Mathematics Subject Classification (2000): 68Q40, 65M06, 13M10, 12Y05, 34A25, 20C40, 34D20, 68T35

Die Deutsche Bibliothek - CIP-Einheitsaufnahme

Computer algebra in scientific computing : proceedings of the Fourth International Workshop on Computer Algebra in Scientific Computing, Konstanz, Sept. 22 - 26, 2001 / CASC 2001. Victor G. Ganzha ... (ed.). - Berlin ; Heidelberg ; New York ; Barcelona ; Hong Kong ; London ; Milan ; Paris ; Singapor ; Tokyo : Springer, 2001

ISBN 978-3-642-62684-5 ISBN 978-3-642-56666-0 (eBook)
DOI 10.1007/978-3-642-56666-0

http://www.springer.de

Originally published by Springer-Verlag Berlin Heidelberg New York in 2001
Softcover reprint of the hardcover 1st edition 2001

Cover design: *design & production*, Heidelberg
Production: LE-TEX Jelonek, Schmidt & Vöckler GbR, Leipzig
Typesetting by the authors using a Springer-TEX Makro Package
SPIN 10844008 46/3142YL-5 4 3 2 1 0

Preface

CASC 2001 continues a tradition — started in 1998 — of international conferences on the latest advances in the application of computer algebra systems (CASs) to the solution of various problems in scientific computing. The three earlier conferences in this sequence, CASC'98, CASC'99, and CASC 2000, were held, respectively, in St. Petersburg, Russia, in Munich, Germany, and in Samarkand, Uzbekistan, and proved to be very successful.

We have to thank the program committee, listed overleaf, for a tremendous job in soliciting and providing reviews for the submitted papers. There were more than three reviews per submission on average. The result of this job is reflected in the present volume, which contains revised versions of the accepted papers. The collection of papers included in the proceedings covers various topics of computer algebra methods, algorithms and software applied to scientific computing.

In particular, five papers are devoted to the implementation of the analysis of involutive systems with the aid of CASs. The specific examples include new efficient algorithms for the computation of Janet bases for monomial ideals, involutive division, involutive reduction method, etc.

A number of papers deal with application of CASs for obtaining and validating new exact solutions to initial and boundary value problems for partial differential equations in mathematical physics. Several papers show how CASs can be used to obtain analytic solutions of initial and boundary value problems for ordinary differential equations and for studying their properties.

Several papers present the application of CASs to the solution of differential geometry tasks. A number of papers deals with group and Lie symmetry analysis as applied, in particular, to equations governing plane motions of a viscous heat-conducting gas.

There are also papers devoted to problems that are typical CA applications: polynomial ideals, polynomial algebra, and quantifier elimination.

A number of papers is aimed in a new interesting direction for the development of applied CAS packages: the integration of object-oriented programming in CAS environments with the capabilities of Java.

A novel feature of this conference is an enhanced emphasis on engineering applications of computer algebra. In particular, such applied problems are considered as the modelling of shape memory metal alloys, the stability of a satellite with a solar sail, reliability problems in aerospace systems, automatic motion planning of an automobile or aircraft within moving traffic, detection of all singular positions of planar mechanisms, etc.

The invited lecture by R. Maeder shows in detail how MATHEMATICA routines can be moved to and efficiently be implemented in a parallel environment.

The CASC 2001 workshop was supported financially by a generous grant from the Deutsche Forschungsgemeinschaft (DFG) and Visual Analysis AG. We are grateful to W. Meixner for his technical help in the preparation of the camera ready manuscript for this volume. We also thank the publisher, Springer Verlag, for their support in preparing these proceedings.

Our particular thanks are due to the members of the local organizing committee in Konstanz who have ably handled local arrangements in this particularly pleasant location in the very South of Germany.

Munich, July 2001

V.G. Ganzha
E.W. Mayr
E.V. Vorozhtsov

Workshop Organization

CASC 2001 was organized jointly by the Technische Universität München, the FH Weingarten-Ravensburg, and the FH Konstanz.

Workshop Chairs

Vladimir Gerdt (JINR, Dubna)
Ernst W. Mayr (TU München)

Program Commitee

Alkiviadis G. Akkritas (Volos)
Gerd Baumann (Munich)
Stefan Braun (Munich)
Hans-Joachim Bungartz (Augsburg)
Andreas Dolzmann (Passau)
Victor Edneral (Moscow)
Victor Ganzha (Munich, co-chair)
Simon Gray (Ashland)
Evgenii Grebenikov (Moscow)
Jaime Gutierrez (Santander)
Ilias Kotsireas (London, Ontario)
Robert Kragler (Weingarten)
Richard Liska (Prag)
Michal Mnuk (Munich)
Werner Seiler (Mannheim)
Akhmadjon Soleev (Samarkand)
Stanley Steinberg (Albuquerque)
Nikolay Vassiliev (St. Petersburg)
Gilles Villard (Grenoble)
Evgenii Vorozhtsov (Novosibirsk, co-chair)
Andreas Weber (Tubingen)
Franz Winkler (Linz)
Christoph Zenger (Munich)

Organizing Committee

Werner Meixner (Munich, chair)
Annelies Schmidt (Munich, secretary)
Elke-Dagmar Heinrich (Konstanz)
Robert Kragler (Weingarten, chair)

Electronic Address

WWW site: `http://wwwmayr.in.tum.de/CASC2001`

Table of Contents

Jets. A Maple-Package for Formal Differential Geometry

Mohamed Barakat

Lehrstuhl B für Mathematik, RWTH-Aachen, Templergraben 64
D-52062 Aachen, Germany
mohamed.barakat@post.rwth-aachen.de
http://wwwb.math.rwth-aachen.de

Abstract. The Maple-package `jets` was first designed to be an extension of the package `desolv`. In the current stage it became an independent package going beyond symmetries to handle different aspects of formal differential geometry, including some important parts of the variational bicomplex. We demonstrate this by computing the set of all Hamiltonian structures of a order at most 3, which are compatible with D_x. This set includes among others the famous KdV-operator $D_{xxx} + \frac{2}{3}uD_x + \frac{1}{3}u_x$.

1 Introduction

The Maple-package `jets`, originally an extension of the package `desolv`[1] adding to it the facility of computing generalized symmetries of differential equations, is at the current stage an independent package going beyond symmetries to handle different aspects of what I. M. Gel'fand, in his 1970 address to the International Congress in Nice, called "formal differential geometry". Important parts of the variational bicomplex, as playing a crucial role in the formal theory, are implemented in `jets`. Most of the implementation of the variational aspects in `jets`, such as variational symmetries, higher Euler operators, homotopy operators and conservation laws, was done by Gehrt Hartjen as part of his diploma thesis [Har]. As dual to functional forms and the vertical derivative also functional multi-vectors and the Nijenhuis-Schouten bracket are also implemented in `jets`, enabling one to handle Hamiltonian systems of evolution equations and nonlinear integrable systems. The package adds to Maple the important feature of dealing with jet calculus, a thing which is still missing in modern computer algebra systems. Almost every formula appearing in [Olv] can now be computed using `jets`.

2 Hamiltonian Structures and the Nijenhuis-Schouten Bracket

As mentioned in the abstract, the aim of this paper is to demonstrate a nontrivial application of the package `jets` by computing the set of all Hamiltonian

[1] `desolv` was written by Khai Vu and Colin McIntosh. `jets` still uses `desolv` to solve linear PDE systems.

structures of a order at most 3, which are compatible with D_x. This is done in section 3. To this end we define the notion of functional multi-vectors, Hamiltonian structures and the Nijenhuis-Schouten bracket. The notions used in sequel are standard and can be found in [Olv]. Further details are found in [Bar].

Let $E \to M$ be a fibred manifold in p independent variables $(x^i) = (x^1, \ldots, x^p)$ and q dependent variables $(u^\alpha) = (u^1, \ldots, u^q)$. By $J_\infty(E) \to M$ we denote the infinite jet bundle having the jet variables (x^i, u^α_J) as coordinates, where J is an arbitrary multi-index. By $\mathcal{A}$ we denote the space of differential expressions over E, i.e. smooth real-valued functions of finitely many arbitrary jet variables. By $\mathcal{V}^1$ we denote the space of evolutionary vector fields, or equivalently the space of characteristics over a jet bundle. This space can be identified with the Cartesian power $\mathcal{A}^q$. Further we define locally $\mathcal{F}^0 := \mathcal{A}/\mathsf{Div}(\mathcal{A}^p)$ and call it the space of functionals[2]. By $\mathcal{F}^1$ we denote the $\mathcal{F}^0$-dual space of $\mathcal{V}^1$. We can also identify it with $\mathcal{A}^q$. Further let $\mathcal{F}^n$ (resp. $\mathcal{V}^n$) denote the space of functional n-forms (resp. n-vectors).

We first note the following two basic formulas. The first one relates the prolongation of an evolutionary vector field and the Fréchet derivative

$$\mathsf{pr}\,\mathbf{v}_Q(L) = \mathsf{D}_L Q, \tag{1}$$

where $Q = (Q^1, \ldots, Q^q)^{\mathrm{tr}}$ is a characteristic, $\mathbf{v}_Q = Q^\alpha \frac{\partial}{\partial u^\alpha}$ and evolutionary vector field, $\mathsf{pr}\,\mathbf{v}_Q = D_J Q^\alpha \frac{\partial}{\partial u^\alpha_J}$ (prolongation formula) and $\mathsf{D}_L = (\frac{\partial L}{\partial u^1_J}\mathsf{D}_J, \ldots, \frac{\partial L}{\partial u^q_J}\mathsf{D}_J)$ (Fréchet derivative). The proof follows immediately from the prolongation formula and the definition of the Fréchet derivative. The second formula is the standard Leibniz rule

$$\mathsf{pr}\,\mathbf{v}(L \cdot P) = \mathsf{pr}\,\mathbf{v}L \cdot P + L \cdot \mathsf{pr}\,\mathbf{v}P \tag{2}$$

where $\mathbf{v}$ is a generalized vector field and L, P are arbitrary differential expression.

We still need the following lemma.

Lemma 1. *For a differential operator $\mathcal{D} = P^J D_J$ ($P^J \in \mathcal{A}$) and differential function $T \in \mathcal{A}$, we have the following Leibniz rule:*

$$\mathsf{pr}\,\mathbf{v}_Q(\mathcal{D}T) = \mathsf{pr}\,\mathbf{v}_Q(\mathcal{D})T + \mathcal{D}\mathsf{pr}\,\mathbf{v}_Q(T), \tag{3}$$

or equivalently by (1)

$$\mathsf{D}_{\mathcal{D}T}(Q) = \mathsf{pr}\,\mathbf{v}_Q(\mathcal{D})T + \mathcal{D}\mathsf{D}_T Q. \tag{4}$$

Proof. [Olv], Formula (5.38). □

Definition 1 (Adjoint operator). *The formal adjoint operator of a matrix differential operator $\mathcal{D} = (P^J_{\alpha\beta} D_J)$ is defined by*

$$\mathcal{D}^* = ((-1)^{|J|} D_J P^J_{\beta\alpha}).$$

[2] $\mathsf{Div}P = D_i P^i$, where $P = (P^1, \ldots, P^p)$ and $D_i = D_{x^i}$

Definition 2 (Euler operator). *For $L \in \mathcal{A}$ the operator*

$$\mathsf{E}(L) := \mathsf{D}_L^*(1) \tag{5}$$

is called the EULER *operator.*

Lemma 2 ([Olv], Formula (4.15)). *A Lagrangian $L \in \mathcal{A}$ transforms infinitesimally according to the rule*

$$\mathcal{L}_{\mathbf{v}} L = \mathsf{pr}\,\mathbf{v} L + L \mathsf{Div}(\xi), \tag{6}$$

where $\mathbf{v} = \xi^i \frac{\partial}{\partial x^i} + \eta^\alpha \frac{\partial}{\partial u^\alpha}$ is a generalized vector field[3].

PROOF. [Olv], Theorem 4.12. □

Corollary 1 (Lie derivative of functionals). *For a Lagrangian L viewed as an element of $\mathcal{F}^0$, i.e. as a functional 0-form, the Lie derivative $\mathcal{L}_{\mathbf{v}}$ satisfies*

$$\mathcal{L}_{\mathbf{v}} L = \mathsf{pr}\,\mathbf{v}_Q L = \mathsf{E}(L) \cdot Q. \tag{7}$$

PROOF. The following are identities between functionals. For a generalized vector field $\mathbf{v}$ with characteristic Q

$$\begin{aligned}
\mathcal{L}_{\mathbf{v}} L &\overset{(6)}{=} \mathsf{pr}\,\mathbf{v} L + L \mathsf{Div}(\xi) \\
&= \mathsf{pr}\,\mathbf{v}_Q L + \xi^i D_i L + L D_i \xi^i \\
&= \mathsf{pr}\,\mathbf{v}_Q L + \mathsf{Div}(L\xi) \\
&= \mathsf{pr}\,\mathbf{v}_Q L \\
&\overset{(1)}{=} 1 \cdot \mathsf{D}_L(Q) \\
&= \mathsf{D}_L^*(1) \cdot Q \\
&\overset{(5)}{=} \mathsf{E}(L) \cdot Q
\end{aligned}$$

□

Definition 3 (Lie derivative of vector fields). *Let $\mathbf{v}$ be a generalized vector field and R a characteristic, i.e. $R \in \mathcal{V}^1$. Define the Lie derivative of R with respect to $\mathbf{v}$ by*

$$\begin{aligned}
\mathcal{L}_{\mathbf{v}}(R) &= \mathsf{pr}\,\mathbf{v}_Q R - \mathsf{pr}\,\mathbf{v}_R Q \\
&\overset{(1)}{=} \mathsf{pr}\,\mathbf{v}_Q R - \mathsf{D}_Q R,
\end{aligned} \tag{8}$$

where Q is the characteristic of $\mathbf{v}$.

Proposition 1 (Lie derivative of functional 1-forms). *For the Lie derivative of a source form $\Delta \in \mathcal{F}^1$ the following two statements are equivalent:*

[3] [Olv] proves this for point vector fields only. The above Lie derivative coincides with the notion of projected Lie derivative $\mathcal{L}_{\mathbf{v}}^{\sharp}$ introduced in [And], Chapter 3.

(i) Δ *transforms infinitesimally according to*

$$\mathcal{L}_{\mathbf{v}_Q}\Delta = \mathrm{pr}\,\mathbf{v}_Q\Delta + \mathrm{D}_Q^*\Delta. \tag{9}$$

(ii) $\mathcal{L}_{\mathbf{v}_Q}$ *satisfies the following* Leibniz rule *for an arbitrary characteristic* R

$$\mathcal{L}_{\mathbf{v}_Q}(\Delta \cdot R) = \mathcal{L}_{\mathbf{v}_Q}\Delta \cdot R + \Delta \cdot \mathcal{L}_{\mathbf{v}_Q}R. \tag{10}$$

This is an identity of functionals, i.e. the left and right hand sides are equal up to local divergence.

Proof. Both directions follow from the following equalities:

$$\begin{aligned}
&\mathrm{E}(\mathcal{L}_{\mathbf{v}_Q}(\Delta \cdot R)) - \mathrm{E}(\Delta \cdot \mathcal{L}_{\mathbf{v}_Q}R)\\
&\overset{(7)}{=} \mathrm{E}(\mathrm{pr}\,\mathbf{v}_Q(\Delta \cdot R)) - \mathrm{E}(\Delta \cdot \mathcal{L}_{\mathbf{v}_Q}R)\\
&\overset{(2),(8)}{=} \mathrm{E}(\mathrm{pr}\,\mathbf{v}_Q\Delta \cdot R + \Delta \cdot \mathrm{pr}\,\mathbf{v}_Q R) - \mathrm{E}(\Delta \cdot (\mathrm{pr}\,\mathbf{v}_Q R - \mathrm{D}_Q R))\\
&= \mathrm{E}(\mathrm{pr}\,\mathbf{v}_Q\Delta \cdot R + \Delta \cdot \mathrm{D}_Q R)\\
&= \mathrm{E}((\mathrm{pr}\,\mathbf{v}_Q\Delta + \mathrm{D}_Q^*\Delta) \cdot R).
\end{aligned}$$

□

Remark 1. The identity of functionals

$$\mathcal{L}_{\mathbf{v}_Q}\Delta \cdot R = \mathrm{pr}\,\mathbf{v}_Q\Delta \cdot R + \Delta \cdot \mathrm{pr}\,\mathbf{v}_R Q, \tag{11}$$

which is part of the proof, appears as formula (4.2) in [GDo2].

Lemma 3. *The following identity holds for a general* $\mathcal{K} : \mathcal{F}^1 \to \mathcal{V}^1$

$$(\mathrm{pr}\,\mathbf{v}_{.}(\mathcal{K})\Delta)^*\Sigma = (\mathrm{pr}\,\mathbf{v}_{.}(\mathcal{K}^*)\Sigma)^*\Delta \tag{12}$$

Proof. For an arbitrary characteristic S

$$\begin{aligned}
&\mathrm{E}(S \cdot ((\mathrm{pr}\,\mathbf{v}_{.}(\mathcal{K})\Delta)^*\Sigma - (\mathrm{pr}\,\mathbf{v}_{.}(\mathcal{K}^*)\Sigma)^*\Delta))\\
&= \mathrm{E}(\mathrm{pr}\,\mathbf{v}_S(\mathcal{K})\Delta \cdot \Sigma - \mathrm{pr}\,\mathbf{v}_S(\mathcal{K}^*)\Sigma \cdot \Delta)\\
&= \mathrm{E}(\mathrm{pr}\,\mathbf{v}_S(\mathcal{K})\Delta \cdot \Sigma - \Sigma \cdot \mathrm{pr}\,\mathbf{v}_S(\mathcal{K})\Delta)\\
&= 0.
\end{aligned}$$

□

Definition 4 (Nijenhuis-Schouten bracket). *For* $\mathcal{D}, \mathcal{E} \in \mathcal{V}^2$ *the Nijenhuis-Schouten bracket* $[\mathcal{D}, \mathcal{E}] : \mathcal{F}^1 \times \mathcal{F}^1 \times \mathcal{F}^1 \to \mathcal{F}^0$ *is defined as follows:*

$$[\mathcal{D}, \mathcal{E}](\Delta_1, \Delta_2, \Delta_3) := \mathcal{L}_{\mathcal{D}\Delta_1}\Delta_2 \cdot \mathcal{E}\Delta_3 + \mathcal{L}_{\mathcal{E}\Delta_1}\Delta_2 \cdot \mathcal{D}\Delta_3 + (\text{cycle}), \tag{13}$$

where the word (cycle) means summation over all cyclic permutations of the indices $1, 2, 3$. $\mathcal{D}$ *and* $\mathcal{E}$ *are viewed as differential operators from* $\mathcal{F}^1$ *into* $\mathcal{V}^1$.

This definition is a generalisation of the classical Nijenhuis-Schouten bracket from differential geometry, which is one of its advantages. It appears in [GDo2], Formula (3.3). Nevertheless there are two major *drawbacks* of this definition. The first one is that the right hand side is a functional, so it has no normal form. This means that checking the vanishing of the bracket or extracting conditions for its vanishing is not a direct procedure. The second one is that one needs more than total differentials of the Δ_i's, meaning that we cannot compute with general Δ_i's, complicating the check of vanishing of the bracket. Besides, from this definition we do not see that the bracket of two 2-vectors is a $(3,0)$-tensor, *even* a 3-vector. In the following we want to make use of the freedom of adding divergences to circumvent these drawbacks. The following formula cures both drawbacks.

Proposition 2 (Nijenhuis-Schouten bracket). *For $\mathcal{D},\mathcal{E}\in\mathcal{V}^2$ the following formula is an equivalent definition of the Nijenhuis-Schouten bracket $[\mathcal{D},\mathcal{E}]$*

$$
\begin{aligned}
[\mathcal{D},\mathcal{E}](\Delta) = {} & \operatorname{pr}\mathbf{v}_{\mathcal{D}\Delta}(\mathcal{E}) - \operatorname{pr}\mathbf{v}_{\mathcal{D}\cdot}(\mathcal{E})\Delta + (\operatorname{pr}\mathbf{v}_{\mathcal{D}\cdot}(\mathcal{E})\Delta)^* + \\
& \operatorname{pr}\mathbf{v}_{\mathcal{E}\Delta}(\mathcal{D}) \doteq \operatorname{pr}\mathbf{v}_{\mathcal{E}\cdot}(\mathcal{D})\Delta + (\operatorname{pr}\mathbf{v}_{\mathcal{E}\cdot}(\mathcal{D})\Delta)^*.
\end{aligned} \tag{14}
$$

PROOF.

$$
\begin{aligned}
& \mathsf{E}([\mathcal{D},\mathcal{E}](\Delta_1,\Delta_2,\Delta_3)) \\
= {} & \mathsf{E}(\mathcal{L}_{\mathcal{D}\Delta_1}\Delta_2\cdot\mathcal{E}\Delta_3) + \mathsf{E}(\mathcal{L}_{\mathcal{E}\Delta_1}\Delta_2\cdot\mathcal{D}\Delta_3) + (\text{cycle}) \\
\overset{(11)}{=} {} & \mathsf{E}(\operatorname{pr}\mathbf{v}_{\mathcal{D}\Delta_1}\Delta_2\cdot\mathcal{E}\Delta_3) + \mathsf{E}(\Delta_2\cdot\operatorname{pr}\mathbf{v}_{\mathcal{E}\Delta_3}(\mathcal{D}\Delta_1)) + \\
& \mathsf{E}(\operatorname{pr}\mathbf{v}_{\mathcal{E}\Delta_1}\Delta_2\cdot\mathcal{D}\Delta_3) + \mathsf{E}(\Delta_2\cdot\operatorname{pr}\mathbf{v}_{\mathcal{D}\Delta_3}(\mathcal{E}\Delta_1)) + \\
& (\text{cycle}) \\
\overset{(3)}{=} {} & \mathsf{E}(\operatorname{pr}\mathbf{v}_{\mathcal{D}\Delta_1}\Delta_2\cdot\mathcal{E}\Delta_3) + \mathsf{E}(\Delta_2\cdot\operatorname{pr}\mathbf{v}_{\mathcal{E}\Delta_3}(\mathcal{D})\Delta_1) + \mathsf{E}(\Delta_2\cdot\mathcal{D}\operatorname{pr}\mathbf{v}_{\mathcal{E}\Delta_3}\Delta_1) + \\
& \mathsf{E}(\operatorname{pr}\mathbf{v}_{\mathcal{E}\Delta_1}\Delta_2\cdot\mathcal{D}\Delta_3) + \mathsf{E}(\Delta_2\cdot\operatorname{pr}\mathbf{v}_{\mathcal{D}\Delta_3}(\mathcal{E})\Delta_1) + \mathsf{E}(\Delta_2\cdot\mathcal{E}\operatorname{pr}\mathbf{v}_{\mathcal{D}\Delta_3}\Delta_1) + \\
& (\text{cycle}) \\
= {} & \mathsf{E}(\operatorname{pr}\mathbf{v}_{\mathcal{D}\Delta_1}\Delta_2\cdot\mathcal{E}\Delta_3) + \mathsf{E}(\Delta_2\cdot\operatorname{pr}\mathbf{v}_{\mathcal{E}\Delta_3}(\mathcal{D})\Delta_1) - \mathsf{E}(\mathcal{D}\Delta_2\cdot\operatorname{pr}\mathbf{v}_{\mathcal{E}\Delta_3}\Delta_1) + \\
& \mathsf{E}(\operatorname{pr}\mathbf{v}_{\mathcal{E}\Delta_1}\Delta_2\cdot\mathcal{D}\Delta_3) + \mathsf{E}(\Delta_2\cdot\operatorname{pr}\mathbf{v}_{\mathcal{D}\Delta_3}(\mathcal{E})\Delta_1) - \mathsf{E}(\mathcal{E}\Delta_2\cdot\operatorname{pr}\mathbf{v}_{\mathcal{D}\Delta_3}\Delta_1) + \\
& (\text{cycle}) \\
= {} & \mathsf{E}(\operatorname{pr}\mathbf{v}_{\mathcal{D}\Delta_1}\Delta_2\cdot\mathcal{E}\Delta_3) + \mathsf{E}(\Delta_2\cdot\operatorname{pr}\mathbf{v}_{\mathcal{E}\Delta_3}(\mathcal{D})\Delta_1) - \mathsf{E}(\operatorname{pr}\mathbf{v}_{\mathcal{E}\Delta_3}\Delta_1\cdot\mathcal{D}\Delta_2) + \\
& \mathsf{E}(\operatorname{pr}\mathbf{v}_{\mathcal{E}\Delta_1}\Delta_2\cdot\mathcal{D}\Delta_3) + \mathsf{E}(\Delta_2\cdot\operatorname{pr}\mathbf{v}_{\mathcal{D}\Delta_3}(\mathcal{E})\Delta_1) - \mathsf{E}(\operatorname{pr}\mathbf{v}_{\mathcal{D}\Delta_3}\Delta_1\cdot\mathcal{E}\Delta_2) + \\
& (\text{cycle}) \\
\overset{(\text{cycle})}{=} {} & \mathsf{E}(\Delta_3\cdot\operatorname{pr}\mathbf{v}_{\mathcal{D}\Delta_1}(\mathcal{E})\Delta_2 + \Delta_1\cdot\operatorname{pr}\mathbf{v}_{\mathcal{D}\Delta_2}(\mathcal{E})\Delta_3 + \Delta_2\cdot\operatorname{pr}\mathbf{v}_{\mathcal{D}\Delta_3}(\mathcal{E})\Delta_1 \\
& +\Delta_3\cdot\operatorname{pr}\mathbf{v}_{\mathcal{E}\Delta_1}(\mathcal{D})\Delta_2 + \Delta_1\cdot\operatorname{pr}\mathbf{v}_{\mathcal{E}\Delta_2}(\mathcal{D})\Delta_3 + \Delta_2\cdot\operatorname{pr}\mathbf{v}_{\mathcal{E}\Delta_3}(\mathcal{D})\Delta_1) \\
= {} & \mathsf{E}(\Delta_3\cdot\operatorname{pr}\mathbf{v}_{\mathcal{D}\Delta_1}(\mathcal{E})\Delta_2 - \operatorname{pr}\mathbf{v}_{\mathcal{D}\Delta_2}(\mathcal{E})\Delta_1\cdot\Delta_3 + (\operatorname{pr}\mathbf{v}_{\mathcal{D}\cdot}(\mathcal{E})\Delta_1)^*\Delta_2\cdot\Delta_3 \\
& +\Delta_3\cdot\operatorname{pr}\mathbf{v}_{\mathcal{E}\Delta_1}(\mathcal{D})\Delta_2 - \operatorname{pr}\mathbf{v}_{\mathcal{E}\Delta_2}(\mathcal{D})\Delta_1\cdot\Delta_3 + (\operatorname{pr}\mathbf{v}_{\mathcal{E}\cdot}(\mathcal{D})\Delta_1)^*\Delta_2\cdot\Delta_3) \\
= {} & \mathsf{E}(\Delta_3\cdot(\operatorname{pr}\mathbf{v}_{\mathcal{D}\Delta_1}(\mathcal{E}) - \operatorname{pr}\mathbf{v}_{\mathcal{D}\cdot}(\mathcal{E})\Delta_1 + (\operatorname{pr}\mathbf{v}_{\mathcal{D}\cdot}(\mathcal{E})\Delta_1)^* \\
& \quad +\operatorname{pr}\mathbf{v}_{\mathcal{E}\Delta_1}(\mathcal{D}) - \operatorname{pr}\mathbf{v}_{\mathcal{E}\cdot}(\mathcal{D})\Delta_1 + (\operatorname{pr}\mathbf{v}_{\mathcal{E}\cdot}(\mathcal{D})\Delta_1)^*)\Delta_2).
\end{aligned}
$$

□

Remark 2. The right hand side of the formula

$$\begin{aligned}&[\mathcal{D},\mathcal{E}](\Delta_1,\Delta_2,\Delta_3) \\ &= \Delta_3\cdot \mathsf{pr}\,\mathbf{v}_{\mathcal{D}\Delta_1}(\mathcal{E})\Delta_2 + \Delta_1\cdot \mathsf{pr}\,\mathbf{v}_{\mathcal{D}\Delta_2}(\mathcal{E})\Delta_3 + \Delta_2\cdot \mathsf{pr}\,\mathbf{v}_{\mathcal{D}\Delta_3}(\mathcal{E})\Delta_1 \\ &\quad +\Delta_3\cdot \mathsf{pr}\,\mathbf{v}_{\mathcal{E}\Delta_1}(\mathcal{D})\Delta_2 + \Delta_1\cdot \mathsf{pr}\,\mathbf{v}_{\mathcal{E}\Delta_2}(\mathcal{D})\Delta_3 + \Delta_2\cdot \mathsf{pr}\,\mathbf{v}_{\mathcal{E}\Delta_3}(\mathcal{D})\Delta_1,\end{aligned} \tag{15}$$

which is part of the proof, appears as formula (7.30) in [Olv]. This formula is an identity of functionals. This definition still has the first drawback, that trivial functionals do not in general vanish identically, but only up to local divergence. The second drawback is eliminated and one can see the $(3,0)$-tensoriality of the expression. But due to the first drawback it still *not* completely easy to see that this expression is in fact a 3-vector. If we instead use Proposition 2 to define the bracket, these properties follow immediately:

Lemma 4. *The Nijenhuis-Schouten bracket satisfies the following properties:*

(i) $[\mathcal{D},\mathcal{E}]$ *is a* 3*-vector, i.e. is totally skew-adjoint:*
  (a) $[\mathcal{D},\mathcal{E}](\Delta)$ *is a total differential operator in the source form* Δ.
  (b) $[\mathcal{D},\mathcal{E}](\Delta)$ *is skew-adjoint.*
  (c) $[\mathcal{D},\mathcal{E}](\Delta)\Sigma = -[\mathcal{D},\mathcal{E}](\Sigma)\Delta$ *is skew-adjoint.*
(ii) $[\mathcal{D},\mathcal{E}] = [\mathcal{E},\mathcal{D}]$.

PROOF. (i.b) follows immediately from the skew-adjointness of $\mathcal{D},\mathcal{E}$ and for (i.c) we further need to notice that $\mathsf{pr}\,\mathbf{v}_{\mathcal{D}\cdot}(\mathcal{E})\Delta = (\mathsf{pr}\,\mathbf{v}_{\cdot}(\mathcal{E})\Delta)\mathcal{D}$ and (12) for functional bi-vectors i.e. skew-adjoint operators $\mathcal{K}:\mathcal{F}^1\to\mathcal{V}^1$. □

Definition 5 (Poisson bracket). *Let* $\mathcal{D}:\mathcal{F}^1\to\mathcal{V}^1$ *be a differential operator. The Poisson bracket of two functionals* L,P *is defined by*

$$\{L,P\} = \mathsf{E}(L)\cdot\mathcal{D}\mathsf{E}(P), \tag{16}$$

which is again a functional.

Definition 6 (Hamiltonian structure). *A differential operator* $\mathcal{D}:\mathcal{F}^1\to\mathcal{V}^1$ *is called Hamiltonian if its Poisson bracket (16) is skew-symmetric*

$$\{L,P\} = -\{P,L\}, \tag{17}$$

and satisfies the Jacobi identity

$$\{\{L,P\},R\} + \{\{P,R\},L\} + \{\{R,P\},L\} = 0, \tag{18}$$

for all functionals L,P,R. *These are identities between functionals.*

Proposition 3. *A differential operator* $\mathcal{D}$ *is Hamiltonian, if and only if* $\mathcal{D}$ *is a* 2*-vector satisfying* $[\mathcal{D},\mathcal{D}]=0$.

PROOF. First we note that if we replace $\mathcal{E}$ by $\mathcal{D}$ in the right hand side of (15), then, up to a factor, we obtain (7.11) in [Olv]. The rest is done by [Olv] Propositions 7.3, 7.4. □

Definition 7 (Hamiltonian equations). *Let $K \in \mathcal{V}^1$ and $u_t = K$ a system of evolution equations. We say the evolution equation is Hamiltonian, if there exists a Hamiltonian structure $\mathcal{D}$ and a functional H, such that*

$$K = \mathcal{D}\mathsf{E}(H). \tag{19}$$

Definition 8 (Bi-Hamiltonian structure). *Let $u_t = \mathcal{E}\mathsf{E}(H_0) = \mathcal{D}\mathsf{E}(H_1)$ be a system with two Hamiltonian structures. The system is called bi-Hamiltonian, if $[\mathcal{D}, \mathcal{E}] = 0$.*

Remark 3. Bi-Hamiltonian systems with a nondegenerate $\mathcal{D}$ possess a recursion operator $\mathcal{R} = \mathcal{E}\mathcal{D}^{-1}$ generating an infinite family of Hamiltonian symmetries, which by Noether's theorem give rise to an infinite family of conservation laws. This is typical for an integrable system. Details are found in [Olv] and [Bar].

Example 1. The KdV equation $u_t = u_{xxx} + uu_x$ has a bi-Hamiltonian structure with $\mathcal{D} = D_x$, $H_1 = \frac{1}{6}u^3 - \frac{1}{2}u_x^2$, $\mathcal{E} = D_{xxx} + \frac{2}{3}uD_x + \frac{1}{3}u_x$ and $H_0 = \frac{1}{2}u^2$.

3 Example

In this section we use `jets` to compute the set of all Hamiltonian structures of order three and jets up to order one, compatible with $\mathcal{D} = D_x$, obtaining candidates for nonlinear integrable systems.

```
> restart;
```

Loading the package:

```
> read`maple/lib/desolv`: with(jets):
```

Defining the list of independent and dependent variables ($p = q = 1$):

```
> ivar:=[x]; dvar:=[u]; var:=op(alljets(1,ivar,dvar));
```

$$ivar := [x]$$

$$dvar := [u]$$

$$var := x, u, u_x$$

Defining the operator $\mathcal{D}$:

```
> DD:=[[1,[x]]];
```

$$DD := [[1, [x]]]$$

Defining the general operator $\mathcal{E} : \mathcal{F}^1 \to \mathcal{V}^1$ of order at most 3, depending on jet variables of order at most 1, on which we impose several conditions:

```
> SMP:=[[Qxxx(var),[x,x,x]],[Qxx(var),[x,x]],
[Qx(var),[x]],[Q(var),[]]];
```

$$SMP := [[Qxxx(x, u, u_x), [x, x, x]], [Qxx(x, u, u_x), [x, x]], [Qx(x, u, u_x), [x]], [Q(x, u, u_x), []]]$$

The first condition is the compatibility of $\mathcal{D}$ and $\mathcal{E}$, i.e. they must commute (Definition 8). This is a linear condition for $\mathcal{E}$:

```
> BRA:=nsbra3(DD,SMP,ivar,dvar):
```

Here we extract the linear conditions, for the resulting functional 3-vector to vanish. We use the fact that $[\mathcal{D}, \mathcal{E}](T)$ is a total differential operator in $T = (T^{u^1}, \ldots, T^{u^q})$:

```
> CND1:=getcond(map(a->a[1],BRA),map(a->a[1],SMP),
ivar,[op(dvar),T.(op(dvar))]):
```

'jsolve' is a wrapper function which uses the package `desolv` to solve the given system of linear PDEs:

```
> jsolve(CND1): RES1:=subs(CND1[4],%):
```

The third list is the general solution, and the fourth list is the list of all functions and constants appearing in the general solution. The first empty list means that `desolv` succeeded to completely solve the system:

```
> RES1;
```

$$[[], [], [Qxxx(x, u, u_x) = F\_3(x), Qxx(x, u, u_x) = F\_4(x), Qx(x, u, u_x) = F\_5(x, u), Q(x, u, u_x) = \frac{1}{2}(\frac{\partial}{\partial x} F\_5(x, u)) + \frac{1}{4} F\_8(x) + \frac{1}{2} u_x (\frac{\partial}{\partial u} F\_5(x, u))], [F\_5(x, u), F\_4(x), F\_3(x), F\_8(x)]]$$

Define the intermediate $\mathcal{E}$, i.e. the general $\mathcal{E}$ that satisfies the linear condition $[\mathcal{D}, \mathcal{E}] = 0$:

```
> SMP1:=convert(subs(RES1[3],SMP),D);
```

$$SMP1 := [[F\_3(x), [x, x, x]], [F\_4(x), [x, x]], [F\_5(x, u), [x]], [\frac{1}{2} D_1(F\_5)(x, u) + \frac{1}{4} F\_8(x) + \frac{1}{2} u_x D_2(F\_5)(x, u), []]]$$

$\mathcal{E}$ is skew-adjoint. This condition also produces a linear system of PDEs:

```
> sadj(SMP1,ivar,dvar):
CND2:=getcond(map(a->a[1],%),RES1[4],ivar,dvar);
```

$$CND2 := [F\_4(x) - \frac{3}{2}(\frac{\partial}{\partial x} F\_3(x)), -\frac{3}{2}(\frac{\partial^2}{\partial x^2} F\_3(x)) + (\frac{\partial}{\partial x} F\_4(x)),$$
$$\frac{1}{4} F\_8(x) - \frac{1}{2}(D^{(3)})(F\_3)(x) + \frac{1}{2}(D^{(2)})(F\_4)(x)],$$
$$[F\_5(x,u), F\_4(x), F\_3(x), F\_8(x)], [x,u], [x = x, u = u]$$

Find the solution and redefine the intermediate $\mathcal{E}$:

```
> RES2:=jsolve(CND2):
SMP2:=convert(esubs(RES2[3],SMP1),D);
```

$$SMP2 := [[F\_3(x), [x,x,x]], [\frac{3}{2} D(F\_3)(x), [x,x]], [F\_5(x,u), [x]],$$
$$[\frac{1}{2} D_1(F\_5)(x,u) - \frac{1}{4}(D^{(3)})(F\_3)(x) + \frac{1}{2} u_x D_2(F\_5)(x,u), []]]$$

For $\mathcal{E}$ to be Hamiltonian, $\mathcal{E}$ must commute with itself (Definition 3):

```
> nsbra3(SMP2,SMP2,ivar,dvar):
CND3:=getcond(map(a->a[1],%),RES2[4],
ivar,[op(dvar),T.(op(dvar))]):
```

The resulting conditions form a nonlinear system of PDEs:

```
> CND3;
```

$$[2\,\%1\,F\_3(x), 2\,\%4\,F\_3(x) - \%5, -\frac{3}{2}\%5 + 3\,\%4\,F\_3(x),$$
$$-\frac{3}{2} D_2(F\_5)(x,u)\,(D^{(2)})(F\_3)(x) + \frac{3}{2}\%4\,D(F\_3)(x) + 3\,F\_3(x)\,\%6,$$
$$3\,\%1\,F\_3(x), 3\,F\_3(x)\,\%3, 6\,F\_3(x)\,\%2 + \frac{3}{2}\%1\,D(F\_3)(x),$$
$$3\,\%1\,F\_3(x), \frac{3}{2} D(F\_3)(x)\,\%3 + 3\,F\_3(x)\,D_{1,2,2,2}(F\_5)(x,u),$$
$$\frac{3}{2}\%1\,D(F\_3)(x) + 3\,F\_3(x)\,\%2, -3\,\%1\,F\_3(x),$$
$$3\,D(F\_3)(x)\,\%2 + 3\,F\_3(x)\,D_{1,1,2,2}(F\_5)(x,u), \%1\,F\_3(x),$$
$$3\,F\_3(x)\,\%3, F\_3(x)\,D_{2,2,2,2}(F\_5)(x,u), F\_3(x)\,D_{1,1,1,2}(F\_5)(x,u)$$
$$+\frac{3}{2} D(F\_3)(x)\,\%6 - \frac{1}{2} D_2(F\_5)(x,u)\,(D^{(3)})(F\_3)(x),$$
$$\frac{3}{2}\%5 - 3\,\%4\,F\_3(x), -3\,F\_3(x)\,\%3, -\%1\,F\_3(x),$$
$$-F\_3(x)\,D_{2,2,2,2}(F\_5)(x,u), -3\,\%1\,F\_3(x),$$
$$-3\,F\_3(x)\,\%2 - \frac{3}{2}\%1\,D(F\_3)(x), -3\,F\_3(x)\,\%3,$$
$$-3\,D(F\_3)(x)\,\%2 - 3\,F\_3(x)\,D_{1,1,2,2}(F\_5)(x,u),$$

$\frac{1}{2}D_2(F\_5)(x,u)\,(D^{(3)})(F\_3)(x) - F\_3(x)\,D_{1,1,1,2}(F\_5)(x,u) - \frac{3}{2}D(F\_3)(x)\,\%6,$
$-\frac{3}{2}\%4\,D(F\_3)(x) - 3\,F\_3(x)\,\%6 + \frac{3}{2}D_2(F\_5)(x,u)\,(D^{(2)})(F\_3)(x),$
$\%5 - 2\,\%4\,F\_3(x), -\frac{3}{2}D(F\_3)(x)\,\%3 - 3\,F\_3(x)\,D_{1,2,2,2}(F\_5)(x,u),$
$-6\,F\_3(x)\,\%2 - \frac{3}{2}\%1\,D(F\_3)(x), -2\,\%1\,F\_3(x)],$
$[F\_3(x), F\_5(x,u)], [u,x], [x = x, u = u, Tu = Tu, u[x] = u_x,$
$Tu[x,x,x] = Tu_{x,x,x}, Tu[x,x] = Tu_{x,x}, u[x,x] = u_{x,x}, u[x,x,x] = u_{x,x,x},$
$Tu[x] = Tu_x]$
$\%1 := D_{2,2}(F\_5)(x,u)$
$\%2 := D_{1,2,2}(F\_5)(x,u)$
$\%3 := D_{2,2,2}(F\_5)(x,u)$
$\%4 := D_{1,2}(F\_5)(x,u)$
$\%5 := D_2(F\_5)(x,u)\,D(F\_3)(x)$
$\%6 := D_{1,1,2}(F\_5)(x,u)$

Because we assume $F\_3(x) \neq 0$, the first equation in CND3 yields:

```
>  SUBS:=SMP2[3,1]=G_1(x)*u+G_2(x);
```

$$SUBS := F\_5(x,u) = G\_1(x)\,u + G\_2(x)$$

Reinserting the new $F\_5(x)$ in CND3:

```
>  esubs(SUBS,CND3[1]);
```

$$[0, 2\,D(G\_1)(x)\,F\_3(x) - G\_1(x)\,D(F\_3)(x),$$
$$-\frac{3}{2}G\_1(x)\,D(F\_3)(x) + 3\,D(G\_1)(x)\,F\_3(x),$$
$$-\frac{3}{2}G\_1(x)\,(D^{(2)})(F\_3)(x) + \frac{3}{2}D(G\_1)(x)\,D(F\_3)(x)$$
$$+\,3\,F\_3(x)\,(D^{(2)})(G\_1)(x), 0,0,0,0,0,0,0,0,0,0,0,$$
$$F\_3(x)\,(D^{(3)})(G\_1)(x) + \frac{3}{2}D(F\_3)(x)\,(D^{(2)})(G\_1)(x)$$
$$-\,\frac{1}{2}G\_1(x)\,(D^{(3)})(F\_3)(x),$$
$$\frac{3}{2}G\_1(x)\,D(F\_3)(x) - 3\,D(G\_1)(x)\,F\_3(x), 0,0,0,0,0,0,0,$$
$$\frac{1}{2}G\_1(x)\,(D^{(3)})(F\_3)(x) - F\_3(x)\,(D^{(3)})(G\_1)(x)$$
$$-\,\frac{3}{2}D(F\_3)(x)\,(D^{(2)})(G\_1)(x), -\frac{3}{2}D(G\_1)(x)\,D(F\_3)(x)$$
$$-\,3\,F\_3(x)\,(D^{(2)})(G\_1)(x) + \frac{3}{2}G\_1(x)\,(D^{(2)})(F\_3)(x),$$
$$G\_1(x)\,D(F\_3)(x) - 2\,D(G\_1)(x)\,F\_3(x), 0,0,0]$$

The first nonzero equation:

```
> eqn:=2*D(G_1)(x)*F_3(x)-G_1(x)*D(F_3)(x);
```

$$eqn := 2\,D(G\_1)(x)\,F\_3(x) - G\_1(x)\,D(F\_3)(x)$$

Solve with MAPLE's internal 'dsolve' command:

```
> sol:=dsolve(eqn,F_3(x));
```

$$sol := F\_3(x) = \_C1\,G\_1(x)^2$$

The solution of the first equation satisfies all other equations!

```
> simplify(esubs([SUBS,sol],CND3[1]));
```

$$[0,0,0,0,0,0,0,0,0,0,0,0,0,0,0,0,0,0,0,0,0,0,0,0,0,0,0,0,0]$$

Define the final $\mathcal{E}$, i.e. the most general Hamiltonian operator of order at most 3, depending on jet variables of order at most 1:

```
> symp:=esubs([SUBS,sol],SMP2);
```

$$symp := [[\_C1\,G\_1(x)^2,\ [x,x,x]], [3\,\_C1\,G\_1(x)\,D(G\_1)(x), [x,x]], [G\_1(x)\,u + G\_2(x), [x]], [\frac{1}{2}\,D(G\_1)(x)\,u + \frac{1}{2}\,D(G\_2)(x) - \frac{3}{2}\,\_C1\,D(G\_1)(x)\,(D^{(2)})(G\_1)(x) - \frac{1}{2}\,\_C1\,G\_1(x)\,(D^{(3)})(G\_1)(x) + \frac{1}{2}\,u_x\,G\_1(x), [\,]]]$$

Check skew-adjointness:

```
> sadj(symp,ivar,dvar);
```

$$0$$

Check compatibility:

```
> nsbra3(DD,symp,ivar,dvar);
```

$$0$$

Check the Hamiltonian condition:

```
> nsbra3(symp,symp,ivar,dvar);
```

$$0$$

Example 1 ([GDo1]):

```
> EX1:=gcollect(
esubs([_C1=0,G_1(x)=2*C[2],G_2(x)=2*C[1]],symp),ivar);
```

$$EX1 := [[2\,C_2\,u + 2\,C_1, [x]], [u_x\,C_2, [\,]]]$$

```
> CEX1:=gcollect(
esubs([_C1=1/(4*D[2]^2),G_1(x)=2*D[2]],symp),ivar);
```

$$CEX1 := [[1, [x,x,x]], [2\,D_2\,u + G\_2(x), [x]], [\frac{1}{2}\,D(G\_2)(x) + u_x\,D_2, [\,]]]$$

The last two operators form a Hamiltonian pair (C_1,C_2,D_2 constants):

```
> nsbra3(EX1,CEX1,ivar,dvar);
```

$$0$$

Example 2 (KdV):

```
> gcollect(
esubs([_C1=9/4,G_1(x)=2/3,G_2(x)=f(x)],symp),ivar);
```

$$[[1,[x,x,x]],[\frac{2}{3}u+f(x),[x]],[\frac{1}{2}D(f)(x)+\frac{1}{3}u_x,[\,]]]$$

For $f(x)=0$ we obtain the KdV-operator:

```
> KdV:=[[1,[x,x,x]],[2/3*u,[x]],[1/3*u[x],[]]];
```

$$KdV := [[1,[x,x,x]],[\frac{2}{3}u,[x]],[\frac{1}{3}u_x,[\,]]]$$

If we choose $G(x) = const$ then $\mathcal{E}$ commutes even with the more general operator $(2C_2u+2C_1)D_x+C_2u_x$:

```
> nsbra3(EX1,esubs(G_1(x)=D[2],symp),ivar,dvar);
```

$$0$$

References

[And] Ian M. Anderson *The Variational Bicomplex.* (to appear).

[Bar] Mohamed Barakat *Functional Spaces. A Direct Approach.* Ph.D. thesis (under review).

[GDo1] I. M. Gel'fand and I. Ya. Dorfman. *Hamiltonian operators and algebraic structures related to them.* Func. Anal. Appl. 13 (1979), 248-262.

[GDo2] I. M. Gel'fand and I. Ya. Dorfman. *Hamiltonian operators and infinite-dimensional Lie algebras.* Func. Anal. Appl. 15 (1981), 173-187.

[Har] Gehrt Hartjen *Variational calculus and conservations laws with* Maple. Diploma thesis (under review).

[Olv] Peter J. Olver. *Applications of Lie Groups to differential Equations. 2nd Edition.* 1998, Springer-Verlag.

Computing Stratifications of Quotients of Finite Groups and an Application to Shape Memory Alloys

Thomas Bayer

Institut für Informatik
Technische Universität München
80290 München, Germany
bayert@in.tum.de
www: http://www14.in.tum.de/personen/bayert

Abstract. When modeling shape memory alloys one has to construct a free energy $\Phi(F,\theta)$ which has to assume global minima on various strata of the orbit space of the symmetry group G depending on the temperature θ. We provide an algorithm for decomposing a G−variety X (G finite) and its orbit space $Z = X/G$ in different strata, each corresponding to a different orbit type. An application to the construction of Φ for Nickel-Titanium alloys is presented.

1 Introduction

A body consisting of shape memory alloy has the property that, after being deformed at room temperature, it can be brought to its original shape by heating over a critical temperature θ_c. This corresponds to a phase transition from the low-temperature phases (martensic phases) to the high-temperature phase (austensic phase). The phase transition from high-temperature to low temperature corresponds to a symmetry breaking of the system, as pointed out by Abud and Sartori, cf. [1].

More precisely, let G be the symmetry group of the underlying alloy. One has to construct the free energy $\Phi(C,\theta)$ where C is the left Cauchy-Green strain tensor and θ denotes the temperature, s.t. Φ assumes minima on different strata (subsets of the same symmetry type) depending on the temperature. In addition, Φ must be invariant w.r.t. G. Symmetry breaking can be understood as moving from points x to points y s.t. $G_x \supset G_y$, i.e., moving from points with more symmetries to points with fewer symmetries. Hence a description of the strata in terms of equations is a major simplification in the construction of Φ.

We provide an algorithmic solution for the construction of equations of the semistrata and an application to Nickel-Titanium (NiTi) alloys, as done in [9]. Note that the strata can be described by means of the equations of the semistrata, in particular, they are Zariski open sets. All computations have been done in the computer algebra system SINGULAR 2-0-0, cf. [4].

2 Shape Memory Alloys and Group Theory

In this section we provide the necessary background from shape memory alloys, the connection to representation theory and tools from invariant theory.

Let $\mathbf{K}$ be a field of characteristic 0, let G be a finite group and $\rho : G \to GL_n(\mathbf{K})$ be a linear representation. A polynomial $f \in \mathbf{K}[x_1, x_2, \ldots, x_n]$ is **invariant** w.r.t. G if $f(\rho(g)^{-1} \cdot \mathbf{x}) = f(\mathbf{x})$ for all $g \in G$. The ring $\mathbf{K}[x_1, x_2, \ldots, x_n]^G$, consisting of all invariant polynomials w.r.t. G, is called the **invariant ring** of G (ρ will be omitted). By Hilbert's Finiteness Theorem, the invariant ring is finitely generated as a $\mathbf{K}$–algebra. Let $h_1, h_2, \ldots, h_m$ be homogeneous algebra generators of $\mathbf{K}[x_1, x_2, \ldots, x_n]^G$, called **fundamental invariants** (i.e., each invariant polynomial is a polynomial in $h_1, h_2, \ldots, h_m$). Then the projection

$$\begin{aligned} \pi : \mathbf{K}^n &\longrightarrow \mathbf{K}^n/G \subseteq \mathbf{K}^m \\ \mathbf{x} &\longmapsto (h_1(\mathbf{x}), h_2(\mathbf{x}), \ldots, h_m(\mathbf{x})) \end{aligned}$$

of $\mathbf{K}^n$ onto its orbit space $\mathbf{K}^n/G$ is called the **Hilbert map**. Note that π maps closed sets to closed sets and that the fibers of π are the orbits of G, thus they are finite and $\deg(\pi) = |G|$. For constructive invariant theory we refer, e.g., to [5] or [7].

In the sequel we omit the application of ρ, i.e., we assume $G \subset GL_n(\mathbf{K})$. The **orbit** of $x \in \mathbf{K}^n$ is denoted by $G(x) := \{g \cdot x : g \in G\}$, the **orbit length** by $\omega(x) := |G(x)|$ and the stabilizer of x by $G_x := \{g \in G : g \cdot x = x\}$. The orbit of a set $X \subset \mathbf{K}^n$ is $G(X) := \{g \cdot x : x \in X, g \in G\}$. We call $\mathbf{K}^n$ the **representation space** of G and the union of all orbits (with the induced topology) the **orbit space** X/G of G. Note that the orbit space is an algebraic variety. A variety $X \subseteq \mathbf{K}^n$ is called a G**–variety** if $G(x) \subseteq X$ for all $x \in X$. For $H \subset G$ and $X \subset \mathbf{K}^n$ we define $X_H := \{x \in X : H(x) = \{x\}\}$. Since the action is linear, X_H equals the intersection $X \cap \mathbf{K}^n_H$. If H is maximal w.r.t. to $\mathbf{K}^n_H$, i.e., $\mathbf{K}^n_H = \mathbf{K}^n_{H'} \Rightarrow H' \subseteq H$ then H is called an **isotropy subgroup**. The **orbit type** of H is the conjugacy class $[H] := \{gHg^{-1} : g \in G\}$ and the orbit type of x it is the orbit type of its stabilizer, i.e., $[x] := [G_x]$. For $g \in G_x$ and $\sigma \in G$ we have $\sigma g \sigma^{-1} \sigma x = \sigma g x = \sigma x$, and therefore two isotropy subgroups H_1, H_2 are conjugated (in the same orbit type) iff $G(\mathbf{K}^n_{H_1}) = G(\mathbf{K}^n_{H_2})$. To each orbit type we associate the **semi-stratum** $S_{[x]} = \cup_{H \in [x]} \mathbf{K}^n_H$ and the **stratum** $S_{([x])} = \{y \in \mathbf{K}^n : G_y \in [x]\}$. Note that the strata form a disjoint partition of the representation space. The relation between symmetry breaking and orbit types was pointed out by Abud and Sartori, cf. [1]. They also proved that the lattice of orbit types is finite if G is a compact Lie-group.

We now briefly sketch how group theory arises in modeling shape memory alloys. Since time dependence of u and θ are not important, we omit t in the notation. Let $\Omega \subset \mathbf{R}^n$ be a configuration and $u : \Omega \to \mathbf{R}^n$ an injective orientation preserving map modeling the movement of the configuration. If u is differentiable at x then the determinant of the derivative $F = \nabla u$ at x is positive, i.e., $F : \Omega \to GL_n(\mathbf{R})^+$. We are interested in the free energy Φ (also called potential in this context) depending on the strain tensor F and temperature θ, i.e.,

$$\begin{aligned} \Phi : GL_n(\mathbf{R})^+ \times \mathbf{R}^+ &\longrightarrow \mathbf{R}^+ \\ (F,\theta) &\longmapsto \Phi(F,\theta). \end{aligned}$$

The free energy must be independent of the position of the observer (frame invariance), i.e., $\Phi(S\cdot F,\theta) = \Phi(F,\theta)$ for all $S \in SO_n(\mathbf{R})$. Hence Φ depends only on the *left Cauchy-Green strain tensor* $C = \sqrt{F * F}$. In our case $C \in Sym_n(\mathbf{R})^+ = \{M \in GL_n(\mathbf{R})^+ : M \text{ is symmetric}\}$. One is interested in modeling phase transitions from high- to low temperature. Let $G \subset GL_n(\mathbf{R})$ be the symmetry group of the high-temperature phase which acts on the space of left Cauchy-Green strain tensors and θ_c be the critical temperature. One obtains that Φ must be constant on the $G-$orbits, that the orbit type $[G]$ corresponds to the high-temperature phase and that there exists a subgroup H s.t. the orbit type $[H]$ corresponds to the martensic phase. Hence Φ must assume a minimum on the stratum $S_{([G])}$ for $\theta > \theta_c$ and a minimum on the stratum $S_{([H])}$ for $\theta < \theta_c$. By computing equations describing these strata (in terms of differences) we provide a major simplification for the construction of Φ. For a much more accurate description we refer to [9].

3 Algorithmic Stratification

We show how to move from the group theoretical description of the strata and orbit types from the previous section to an ideal theoretic description of the semistrata, which can be obtained by means of computer algebra, and present the corresponding algorithms. Once equations of the semistrata are given, the strata can be described as difference sets of semistrata.

In the sequel let $\mathbf{K}$ be an algebraically closed field of characteristic 0. Let $G \subset GL_n(\mathbf{K})$ be a finite group and $\pi : \mathbf{K}^n \to \mathbf{K}^n/G$ be the Hilbert map (the projection on the quotient). The group action of G on a $G-$variety X is given by $\psi = (\psi_1, \psi_2, \ldots, \psi_n)$, each $\psi_i \in \mathbf{K}[s_1, s_2, \ldots, s_k, t_1, t_2, \ldots, t_n]$, i.e.,

$$\begin{aligned} \psi : G \times X &\longrightarrow X \\ (g,x) &\longmapsto \psi(g,x). \end{aligned}$$

By $\mathcal{I}(X) := \{f \in \mathbf{K}[t_1, t_2, \ldots, t_n] : f(x) = 0 \text{ for all } x \in X\}$ and by $\mathcal{V}(I) := \{x \in \mathbf{K}^n : f(x) = 0 \text{ for } f \in I\}$ we denote the **ideal** resp. **variety** associated to a variety resp. ideal. A subset $U \subseteq \mathbf{K}^n$ is closed in the Zariski topology iff $U = \mathcal{V}(I)$ for some ideal $I \subseteq \mathbf{K}[t_1, t_2, \ldots, t_n]$. A morphism $\phi : X \to Y$ has a **singularity** at $x_0 \in X$ iff the rank of the Jacobian matrix $(\frac{\partial \phi}{\partial x})(x_0)$ is not maximal. The set of singular points of ϕ is denoted by $sing(\phi)$. In order to compute a stratification we start by decomposing X into subsets of different maximal orbit lengths.

Definition 1. *For a set $X \subseteq \mathbf{K}^n$ and $d \in \mathbf{N}$ we define $X_d := \{x \in X : \omega(x) \leq G\}$ and $X_{(d)} := X_d - X_{d-1}$.*

The sets $X_{d_1} \subset X_{d_2} \subset \ldots \subset X_{d_r} = X_{|G|}$ where $d_1, d_2, \ldots, d_r$ represent all different orbit lengths of G play a major role in in computing a stratification of X. These sets have important properties.

Proposition 1. *(a) For a variety $X \subseteq \mathbf{K}^n$ the set X_d is Zariski-closed.*
(b) For an isotropy subgroup $H \subseteq G$ the set of points having orbit length equal to $|G/H|$ is Zariski dense in $\mathbf{K}^n_H$.

Proof. (a) Note that $\omega(x) = |G/G_x|$. For a divisor d of $|G|$ we have $X_d = \{x \in X : |\psi^{-1}(x)| \geq \frac{G}{d}\}$ which is closed by upper continuity of the fiber dimension. If $d' < d$ is not a divisor of $|G|$ and there is no divisor in between, then $X_{d'} = X_d$.
(b) By definition, there exists $x \in \mathbf{K}^n_H$ s.t. $\omega(x) = |G/H| = d$. Since $\mathbf{K}^n_H$ is a subvectorspace, it is irreducible and closed in the Zariski topology and by part (a), $\mathbf{K}^n_{d-1}$ is a proper closed subset. Hence the set $\mathbf{K}^n_H - \mathbf{K}^n_{d-1}$ is Zariski open. □

We call the length of a general point in X_H the **generic orbit length** of X_H, denoted by $\omega(X)$. We are interested in a decomposition of the set $X_{|G|-1}$ which equals $X_{|G|-1} = X \cap \mathbf{K}^n_{|G|-1}$. Note that $\mathbf{K}^n_{|G|-1}$ is a finite union of subspaces.

Proposition 2. *(a) Let $h_1, h_2, \ldots, h_m$ be fundamental invariants of G and $\pi = (h_1, h_2, \ldots, h_m)$ be the associated Hilbert map. We have $sing(\pi) = \mathbf{K}^n_{|G|-1}$.*
(b) Let $I_1 \cap I_2 \cap \ldots \cap I_k$ be an irredundant primary decomposition of the ideal $\mathcal{I}(sing(\pi))$. Then the varieties $\mathcal{V}(I_1), \mathcal{V}(I_2), \ldots, \mathcal{V}(I_k)$ are in $1:1$ correspondence to the minimal isotropy subgroups of G.

Proof. (a) Note that the Jacobian matrix $(\frac{\partial \pi}{\partial x})$ is an $m \times n$ matrix and that $n \leq m$ since $\mathbf{K}[x_1, x_2, \ldots, x_n]$ is a finite extension of $\mathbf{K}[x_1, x_2, \ldots, x_n]^G$. Hence π is singular at x_0 iff $rank((\frac{\partial \pi}{\partial x})(x_0)) < n$ iff $\ker(\frac{\partial \pi}{\partial x})(x_0) \neq \{0\}$ since $n \leq m$. From $\pi(x_0) = \pi(g \cdot x_0)$ for all $g \in G$ we obtain $(\frac{\partial \pi}{\partial x})(x_0) = (\frac{\partial \pi}{\partial x})(g \cdot x_0) \cdot g$ and $g \cdot x_0 = x_0$ implies that $(\frac{\partial \pi}{\partial x})(x_0)(id_n - g) = 0$. Let $x_0 \in X_{|G|-1}, v \in \mathbf{K}^n$ and $id \neq g \in G$ s.t. $g \cdot x_0 = x_0$ and $g \cdot v \neq v$. Hence the element $(id - g)v$ is in the kernel of $(\frac{\partial \pi}{\partial x})(x_0)$. For the converse inclusion note that x is a singularity of π implies $|\pi^{-1}(\pi(x))| < \deg(\pi) = |G|$.
(b) Let $H_1, H_2, \ldots, H_r$ denote all minimal nontrivial isotropy subgroups of G. By definition and part (a), $\mathbf{K}^n_{|G|-1} = \cup_{i=1}^r \mathbf{K}^n_{H_i} = \cup_{i=1}^k \mathcal{V}(I_i)$ and each $\mathcal{V}(I_i)$ is a linear subspace of finite dimension. Since the decomposition of I is irredundant and $\mathbf{K}$ is infinite, the claim follows. □

By computing all combinations of intersections of the ideals of an irredundant primary decomposition of $\mathcal{I}(sing(\phi))$ we obtain precisely the vectorspaces associated to isotropy subgroups of G.

Proposition 3. *Let $G \subset GL_n(\mathbf{K})$ be a faithful representation of a cristallographical group (G is finite), $H_1, H_2, \ldots, H_r \subseteq G$ be the minimal isotropy subgroups and $X \subseteq \mathbf{K}^n$ be a G−variety. For $1 \leq d \leq |G| - 1$ let $X_d = \cup_{i=1}^k X_d^{(i)}$ be* *the irredundant decomposition of X_d in irreducible varieties. For $1 \leq i \leq k$ there exists indices $j_1, j_2, \dots, j_{m_i}, 1 \leq j_l \leq r, 1 \leq l \leq m_i$, s.t.*

$$X_d^{(i)} = \bigcap_{l=1}^{m_i} X_{H_{j_l}}.$$

Proof. In the sequel we denote $\mathbf{K}^n$ by V and show that for each subspace $V_d^{(i)}$ of an irredundant decomposition of V_d there exist subgroups H_{j_j} s.t. $V_d^{(i)} = \bigcap_{j=1}^{m_i} V_{H_{j_l}}$. This yields the decomposition for each $X_d^{(i)}$ since $X_d = X \cap \mathbf{K}_d^n$.

We proceed by induction on d. For $d = |G|-1$, each subspace $V_d^{(i)}$ corresponds to a minimal isotropy subgroup by Proposition 2(b). Let $d < |G|$ be s.t. $V_d^{(i)} \subset V_{d+1}^{(i)}$ (otherwise there is nothing to prove), let $H_{j_1}, H_{j_2}, \dots H_{j_{m_i}}$ be isotropy groups required for $V_{d+1}^{(i)}$ and dentote by H the group generated by the element of these subgroups. Let $x \in V_{d+1}^{(i)}$ be s.t. $\omega(x) < \omega(V_{d+1}^{(i)})$,i.e., $x \in V_d^{(i)}$. Note that $G_x \supset H = G_{V_{d+1}^{(i)}}$. Since G is homomorphic to a cristallographical group, there exists $g \in G_x - H$ s.t. $g(y) \neq y$ for some $y \in V_{d+1}^{(i)}$ and g is contained in a minimal isotropy group $H^{(x)}$ where $s \neq j_l$ for $1 \leq l \leq m_i$ (cf., e.g., [6]). Hence x is contained in $V_H \cap V_{H^{(x)}} \subseteq V_d$. It follows that for each $x \in V_d$ we obtain a subgroup $H^{(x)}$ as above, and

$$V_d = \bigcup_{x \in V_d} (V_{d+1} \cap \mathbf{K}^n_{H^{(x)}}).$$

Since there are only finitely many subgroups of G the union is finite and yields the desired decomposition of V_d. □

Hence by computing all possible intersections we obtain a list of subspaces s.t. the strata are unions of elements from this list. Now the strata may be constructed by collecting the subspaces having the same generic orbit length and then computing the orbit of each such subspace. Since the orbit of a fixedpoint space of an isotroy subgroup correspond to an orbit type, we obtain the desired description of the strata. It remains to compute the generic orbit length and the orbit of a subspace, i.e., its image under the orbit map ψ. The latter can be computed as follows (cf., e.g., [8]).

Proposition 4. *Let $I = \langle f_1, f_2, \dots, f_k \rangle$ be the ideal of a variety X and let $\psi : \mathbf{K}^n \to \mathbf{K}^m$ be a morphism given by polynomials $\psi_1, \psi_2, \dots, \psi_m \in \mathbf{K}[x_1, x_2, \dots, x_n]$. Let $\{h_1, h_2, \dots, h_r\}$ be a Gröbner basis of the ideal $\langle f_1, f_2, \dots, f_k, \psi_1 - y_1, \psi_2 - y_2, \dots \psi_m - y_m \rangle$ w.r.t. a term order $<$ s.t. $x_i < y_j$. Then the ideal of the Zariski closure of X is given by $\langle h_i : h_i \in \mathbf{K}[y_1, y_2, \dots, y_m], 1 \leq i \leq r \rangle$.*

By using the concept of generic points we can compute the orbit length of the points of the Zariski dense subset of X_H, mentioned in Proposition 1(b).

Proposition 5. *Let $V = \mathbf{K}^n_H$ be a subspace of dimension k associated to the isotropy group H and let $\phi = (\phi_1, \phi_2, \ldots, \phi_n), \phi_i \in \mathbf{K}[a_1, a_2, \ldots, a_k]$, be a parameterization of V. For the image J of the ideal $I = \langle t_i - \phi_i : 1 \leq i \leq n \rangle \subset \mathbf{K}(a_1, a_2, \ldots, a_k)[t_1, t_2, \ldots, t_n]$ under the orbit map ψ we have*

$$\omega(V) = \dim_{\mathbf{K}(a_1, a_2, \ldots, a_k)} \mathbf{K}(a_1, a_2, \ldots, a_k)[t_1, t_2, \ldots, t_n]/J.$$

Proof. The computation of the image of $\mathcal{V}(I)$ under the orbit map corresponds to the computation of the image of the generic point ξ of V under the orbit map ψ. Hence J is the ideal of the orbit of the generic point of V, i.e., there exists a Zariski open dense set $U \subset \mathbf{K}^n_H$ s.t. $\omega(\xi) = \omega(u)$ for $u \in U$. By Proposition 1(b), $\omega(\xi) = \omega(V)$. □

We describe the main algorithm for computing a stratification and the algorithm for determining the generic orbit length. All standard algorithms (Gröbner bases, dimension, normal-form, primary decomposition, radical, Jacobian, minors) are available in `SINGULAR` (either built in or contained in libraries). For a description of the algorithms from commutative algebra we refer, e.g., to [8]. Almost all of them rely on Gröbner bases (cf. [3]). We omit implementation details like change of baserings, membership test, etc.

Algorithm 1 *Semistrata(G, ψ, X)*
In: *finite group $G \subset GL_n(\mathbf{K})$ as a list of matrices, ψ a list of polynomials in $\mathbf{K}[s_1, s_2, \ldots, s_k, t_1, t_2, \ldots, t_n]$ defining the action of G, algebraic variety X.*
Out: *list of equations for X_d...*
begin
I_G = defining ideal of G.
$\pi = (h_1, h_2, \ldots, h_r)$; // algebra generators of $\mathbf{K}[t_1, t_2, \ldots, t_n]^G$;
$I = minors(\det(jacobian(\pi)), n)$; // n-minors of the Jacobian
$decomp$ = primary decomposition of $\sqrt{I}$; // radical is sufficient
$allIntersections$ = all possible unions of ideal of decomp;
$allVS$ = list of all ideals from $allIntersections$, each ideal appearing only once;
$orbitLength = \{GenericOrbitLength(I_G, \psi, allVS[i]) : 1 \leq i \leq |allVS|\}$;
for $i = 1$ **to** $|orbitLength|$ **do**
    $collectedSpaces = \{V : V \in allVS, \omega(V) = orbitLength[i]\}$;
    for *each $V \in collectSpaces[i]$* **do**
        $orbitV = \psi(G, V)$; // orbit of V
        if *$orbitV \notin Semistrata[i]$ then $Semistrata[i] = Semistrata[i] \cup orbitV$;*
    end-for;
end-for;
return(*Semistrata*);
end

The correctness of *Semistrata* follows from Proposition 2, Proposition 3 and from the correctness of *GenericOrbitLength*. All other mentioned algorithms are either described in [8], or are straightforward to implement.

Algorithm 2 *GenericOrbitLength*(I_G, ψ, I_V)
In: *ideal* I_G *of a finite group* G, *ideal* I_V *of a vectorspace* V.
Out: *cardinality of the generic orbit*
Note: *Basering is* $\mathbf{K}(a_1, a_2, \ldots, a_k)[s_1, s_2, \ldots, s_k, t_1, t_2, \ldots, t_n, Y_1, Y_2, \ldots, Y_n]$.
begin
$I = std(I_V)$; // *Gröbner Basis of V*
$c = 0$;
for $i = 1$ **to** n **do**
  if $\deg(NormalForm(t_i, I)) > 0$ *then* **begin**
    $c := c + 1$;
    $I = std(I \cup \{t_i - a_c\})$;
  end-if
end-for
$I = I \cup \{\psi_i - Y_i : 1 \leq i \leq n\}$;
$J = std(I) \cap \mathbf{K}(a_1, a_2, \ldots, a_k)[Y_1, Y_2, \ldots, Y_n]$;
return$(\dim_{\mathbf{K}(a_1, a_2, \ldots, a_k)}(\mathbf{K}(a_1, a_2, \ldots, a_k)[Y_1, Y_2, \ldots, Y_n]/J))$;
end

In order to determine the generic orbit set of a subspace, the algorithm *GenericOrbitLength* computes the image of the generic point of the subspace w.r.t. the orbit map. The correctness of the *GenericOrbitLength* follows from Proposition 5.

Remark 1. Since the Hilbert map π is surjective and closed, we obtain the stratification of the orbit space by computing the images of the stratification of the representation space, i.e, let $H_1, H_2, \ldots, H_k$ be all nonconjugated isotropy subgroups of G. Then, as disjoint union, we have

$$\mathbf{K}^n/G = \bigcup_{i=1}^{k} \pi(\mathbf{K}^n_{([H_i])}).$$

4 Application

We consider an application to Nickel-Titanium alloys, as done by J. Zimmer in his PhD thesis, cf. [9]. For the computations we have used the computer algebra system `SINGULAR`, version 2-0-0, cf. [4] and the libraries `finvar.lib`, cf. [5], `primdec.lib`, cf. [4] and `rinvar.lib`, cf. [2].

Following Zimmer, the symmetry group for the high-temperature phase is the octahedron group G, given in the 3−dimensional representation generated by

$$g_1 = \begin{pmatrix} 0 & -1 & 0 \\ 1 & 0 & 0 \\ 0 & 0 & 1 \end{pmatrix}, g_2 = \begin{pmatrix} 0 & 0 & 1 \\ 1 & -1 & 0 \\ 1 & 0 & 0 \end{pmatrix}.$$

The left Cauchy-Green strain tensor C is contained in the 6−dimensional space $Sym_3(\mathbf{R})^+$. We obtain an induced 6−dimensional representation on the representation space $\mathbf{R}^6$

$$\rho : G \longrightarrow GL_6(\mathbf{R})$$

where

$$\rho(g_1) = \begin{pmatrix} 0 & 1 & 0 & 0 & 0 & 0 \\ 1 & 0 & 0 & 0 & 0 & 0 \\ 0 & 0 & 1 & 0 & 0 & 0 \\ 0 & 0 & 0 & -1 & 0 & 0 \\ 0 & 0 & 0 & 0 & 0 & -1 \\ 0 & 0 & 0 & 0 & 1 & 0 \end{pmatrix}, \rho(g_2) = \begin{pmatrix} 0 & 0 & 1 & 0 & 0 & 0 \\ 0 & 1 & 0 & 0 & 0 & 0 \\ 1 & 0 & 0 & 0 & 0 & 0 \\ 0 & 0 & 0 & 0 & -1 & 0 \\ 0 & 0 & 0 & -1 & 0 & 0 \\ 0 & 0 & 0 & 0 & 0 & 1 \end{pmatrix}.$$

The free energy $\Phi(C,\theta)$ has to assume minima on different strata depending on the temperature θ, and, since Φ is constant on the orbits of G, Φ is a function in the invariants of G (see below).

A minimal set of homogeneous algebra generators for $\mathbf{R}[t_1, t_2, \ldots, t_6]^G$ is given by the 9 invariants

$$\begin{aligned}
h_1 &= t_1 + t_2 + t_3, \\
h_2 &= t_4^2 + t_5^2 + t_6^2 \\
h_3 &= t_1^2 + \frac{1}{2}t_1t_2 + t_2^2 + \frac{1}{2}t_1t_3 + \frac{1}{2}t_2t_3 + t_3^2, \\
h_4 &= t_4t_5t_6, \\
h_5 &= t_1t_4^2 + t_2t_4^2 - \frac{1}{2}t_3t_4^2 - \frac{1}{2}t_1t_5^2 + t_2t_5^2 + t_3t_5^2 + t_1t_6^2 - \frac{1}{2}t_2t_6^2 + t_3t_6^2, \\
h_6 &= t_1^3 + t_2^3 + t_3^3, \\
h_7 &= t_4^4 + \frac{1}{2}t_4^2t_5^2 + t_5^4 + \frac{1}{2}t_4^2t_6^2 + \frac{1}{2}t_5^2t_6^2 + t_6^4, \\
h_8 &= t_1^2t_4^2 + t_2^2t_4^2 + t_1t_3t_4^2 + t_2t_3t_4^2 + t_1t_2t_5^2 + t_2^2t_5^2 + t_1t_3t_5^2 + \\
&\quad t_3^2t_5^2 + t_1^2t_6^2 + t_1t_2t_6^2 + t_2t_3t_6^2 + t_3^2t_6^2, \\
h_9 &= t_3t_4^4 + t_1t_5^4 + t_2t_6^4.
\end{aligned}$$

The primary decomposition of the defining ideal I of $sing(\pi)$ is given by

$$\begin{aligned}
I = \ & \langle t_6, t_5\rangle \cap \langle t_6, t_4\rangle \cap \langle t_5, t_4\rangle \cap \langle t_4 - t_5, t_1 - t_3\rangle \cap \langle t_4 + t_5, t_1 - t_3\rangle \cap \\
& \langle t_5 - t_6, t_1 - t_2\rangle \cap \langle t_5 + t_6, t_1 - t_2\rangle \cap \langle t_4 + t_6, t_2 - t_3\rangle \cap \langle t_4 - t_6, t_2 - t_3\rangle\,.
\end{aligned}$$

Hence there are 9 minimal isotropy subgroups, each of order 2 (cf. the table containing ω). By computing all possible unions of the above ideals and eliminating double elements we obtain a list of 21 ideals corresponding to 21 subspaces given by

$V_1 = \{(a,b,c,d,0,0)\}$	$V_8 = \{(a,b,b,c,d,c)\}$	$V_{15} = \{(a,b,b,0,0,0)\}$
$V_2 = \{(a,b,c,0,d,0)\}$	$V_9 = \{(a,b,b,c,d,-c)\}$	$V_{16} = \{(a,a,b,0,0,0)\}$
$V_3 = \{(a,b,c,0,0,d)\}$	$V_{10} = \{(a,b,c,0,0,0)\}$	$V_{17} = \{(a,a,a,b,-b,b)\}$
$V_4 = \{(a,b,a,c,-c,d)\}$	$V_{11} = \{(a,a,b,c,0,0)\}$	$V_{18} = \{(a,a,a,b,-b,-b)\}$
$V_5 = \{(a,b,a,c,c,d)\}$	$V_{12} = \{(a,b,b,0,c,0)\}$	$V_{19} = \{(a,a,a,b,b,-b)\}$
$V_6 = \{(a,a,b,c,d,-d)\}$	$V_{13} = \{(a,b,a,0,0,c)\}$	$V_{20} = \{(a,a,a,b,b,b)\}$
$V_7 = \{(a,a,b,c,d,d)\}$	$V_{14} = \{(a,b,a,0,0,0)\}$	$V_{21} = \{(a,a,a,0,0,0)\}$

where a, b, c, d are variables ranging over $\mathbf{R}$. The 9 minimal isotropy subgroups correspond to the vectorspaces $V_1, V_2, \ldots, V_9$. An application of *GenericOrbitLength* to each subspace yields that there are orbits of length $1, 3, 4, 6, 12$ respectively. The subspaces having the same generic orbit length are collected as follows.

ω	Dim.	Vectorspaces
12	4	$V_1, V_2, V_3, V_4, V_5, V_6, V_7, V_8, V_9$
6	3	$V_{10}, V_{11}, V_{12}, V_{13}$
4	2	$V_{17}, V_{18}, V_{19}, V_{20}$
3	2	V_{14}, V_{15}, V_{16}
1	1	V_{21}

By computing the orbits of $V_1, V_2, \ldots, V_{21}$ and eliminating double elements we obtain 7 semistrata ($S_{d,i}$ denotes the $i-$th semistratum having $\omega(S_{d,i}) = d$), namely

$$\begin{aligned} S_{12,1} &= V_1 \cup V_2 \cup V_3, \\ S_{12,2} &= V_4 \cup V_5 \cup V_6, \cup V_7 \cup V_8 \cup V_9, \\ S_{6,1} &= V_{10}, \\ S_{6,1} &= V_{11} \cup V_{12} \cup V_{13}, \\ S_4 &= V_{17} \cup V_{18} \cup V_{19} \cup V_{20}, \\ S_3 &= V_{14} \cup V_{15} \cup V_{16}, \\ S_1 &= V_{21} \end{aligned}$$

Note that $X_d = S_d$ or $X_d = S_{d,1} \cup S_{d,2}$. The semistrata $S_{(d)}$ and $S_{(d),i}$ are Zariski open sets and given by the difference $S_d - X_{d-1}$ and $S_{d,i} - X_{d-1}$ respectively. The 7 different symmetry types and their orbit types are given in the table below. The association of orbit types to strata were done by hand, using Chapter 5 of [9]. Note that by using `ZeroSet (rinvar.lib)` one can obtain the matrices of each isotropy subgroup of G. By D_i we denote the $i-$th Dihedral group and by $H^{(i)}$ we denote multiple occurrences of nonconjugated representations of H in G.

Orbit type	$[G]$	$[D_4]$	$[D_3]$	$[D_2^{(1)}]$	$[D_2^{(2)}]$	$[\mathbf{Z}_2^{(1)}]$	$[\mathbf{Z}_2^{(2)}]$
Stratum	$S_{(1)}$	$S_{(3)}$	$S_{(4)}$	$S_{(6,1)}$	$S_{(6,2)}$	$S_{(12,1)}$	$S_{(12,2)}$
Dim.	1	2	2	3	3	4	4
ω	1	3	4	6	6	12	12

By construction of the model (cf. [9]), we know that for $\theta > \theta_c$, $\Phi(F, \theta)$ has to assume a minimum on $S_{(1)}$, the stratum associated to $[G]$, and a minimum on $S_{(3)}$, the stratum associated to $[D_4]$, for $\theta < \theta_c$. For a further simplification, one considers Φ as a function on the orbit space. Note that the orbit space of G, acting on $\mathbf{C}^6$, is the $6-$dimensional variety $\mathbf{C}^6/G$, here embedded in $\mathbf{C}^9$, whose ideal can be generated by 13 equations, where the degrees range from 5 to 10. The ideals of the images of the semistrata S_1 and S_3 in $\mathbf{C}^9$ are given by

$$\begin{aligned}\mathcal{I}(\pi(S_1)) &= \langle y_9, y_8, y_7, y_5, y_4, y_2, 4y_3^2 - 9y_1y_6, 2y_1y_3 - 9y_6, y_1^2 - 2y_3\rangle \\ \mathcal{I}(\pi(S_3)) &= \langle y_9, y_8, y_7, y_5, y_4, y_2, 29y_1^6 - 144y_1^4y_3 + 192y_1^2y_3^2 + 90y_1^3y_6 - 32y_3^3 \\ &\quad -216y_1y_3y_6 + 81y_6^2\rangle\end{aligned}$$

as ideals in the coordinate ring $\mathbf{C}[y_1, y_2, \ldots, y_9]$ of $\mathbf{C}^9$. For our purposes we only consider solutions in $\mathbf{R}^9$ and split $\Phi = \Phi_1 + \Phi_2$ in two parts s.t. Φ_1 assumes minima on $S_{(1)}$ and $S_{(3)}$ (depending on θ) and Φ_2 assures that Φ does not assume minima on all other strata. We conclude that Φ_1 is a function in the variables y_1, y_3, y_6 (and θ of course). Instead of constructing a function in the 9 invariants which has to assume minima on a 1 dimensional subspace and the union of three 2–dimensional subspaces in $\mathbf{R}^6$ it suffices to construct Φ as a function of y_1, y_3, y_6 according to the restrictions stated above. It turns out (cf. [9]) that $\Phi = \Phi_1$ is sufficient and a free energy is given by

$$\begin{aligned}\Phi(y_1, y_3, \theta) = y_3 + \frac{1}{3}y_1^2 &+ \left(\theta - \theta_c + \frac{1}{4}\right)\left(\frac{4}{3}y_3 - \frac{2}{3}y_1^2\right) \\ &- \frac{1}{2}\left(\frac{4}{3}y_3 - \frac{2}{3}y_1^2\right)^2 + \frac{1}{3}\left(\frac{4}{3}y_3 - \frac{2}{3}y_1^2\right)^3 \\ &+ \frac{\sqrt{2}}{2}(\theta - \theta_c)\left(\frac{4}{3}y_3 - \frac{2}{3}y_1^2\right)^{\frac{3}{2}}.\end{aligned}$$

To obtain the free energy on the representation space $\mathbf{R}^6$, one has to substitute h_1 for y_1 and h_3 for y_3.

Finally, we present the runtime of the calculation done with SINGULAR V2-0-0 on a 1 GHz Athlon with 768MB. By *Number* we denote the number of elements of the result, i.e., for *Invariants, Orbit space* and *Group equations* it is the number of resulting polynomials, for *decomposition* the number of primary ideals, for *All vectorspaces* the number of vectorspaces, and for *Strata* the number of strata. The runtime time is measured in seconds.

#	Step	Time	Number
1	Group equations	1.71	104
2	Invariants	1323.63	9
3	Decomposition of $sing(\pi)$	4.19	9
4	All vectorspaces	12.90	21
5	GenericOrbitLength	4.31	5
6	Strata	2.90	7
7	Image of S_1, S_3	0.38	–
8	Orbit space (not nec.)	114.87	
	Time of step $1-7$	1350.02	–

Note that computation of the equations for the orbit space is not necessary, so it does not contribute to the total running time.

5 Conclusion and Future Work

We have presented constructive tools, implemented in the computer algebra system SINGULAR, for the description of the strata in terms of equations, given the the crystallographic (finite) group G. The main difficulty is still to find out the strata where Φ has to assume minima, and, of course, the construction of the function Φ which assumes the minima on the 'right' strata. It seems to be interesting for physicists to extend the presented approach to compact Lie groups, which appear, e.g., in the determination of minima of Higgs potentials.

Acknowledgements

The author is very grateful to Johannes Zimmer from technical university munich for showing me the connection of shape memory alloys to invariant theory and for the numerous useful discussions.

References

1. Abud, M., Sartori, G., *The Geometry of Spontaneous Symmetry Breaking.* Annales of Physics, 150 (1998).
2. Bayer, T. *Computation of moduli spaces of semiquasihomogeneous singularities and an implementation in* SINGULAR. Diploma thesis, Fachbereich Mathematik, Universität Kaiserslautern (2000).
3. Buchberger, B., *Gröbner Bases - an Algorithmic Method in Polynomial Ideal Theory.* In : Bose, N.K.(ed) : Multidimensional System Theory. D. Reidel, Dordrecht 1985 (pp184 - 232).
4. Greuel, G.-M., Pfister, G., and Schönemann, H., Singular Reference Manual. In *Reports On Computer Algebra*, number 12. Centre for Computer Algebra, University of Kaiserslautern, May 1997. http://www.singular.uni-kl.de
5. Heydtmann, A. E., *Generating Invariant Rings over Finite Groups.* Diploma thesis, University of Saarbrücken(1997).
6. Scheja, G., Storch, U., *Lehrbuch der Algebra - Teil 2.* B.G. Teubner, Stuttgart (1988).
7. Sturmfels, B., *Algorithms in Invariant Theory.* Springer Verlag Wien New York (1993).
8. Vasconcelos, W., *Computational Methods in Commutative Algebra and Algebraic Geometry.* Algorithms and Computations in Mathematics 2, Springer Verlag Berlin Heidelberg New York (1998).
9. Zimmer, J., *Mathematische Modellierung und Analyse von Formgedächtnislegierungen in mehreren Raumdimensionen.* PhD. Thesis, Technische Universität München (2000).

A MuPAD Library for Differential Equations*

Jay Belanger[1], Marcus Hausdorf[2], and Werner M. Seiler[2]

[1] Division of Mathematics and Computer Science, Truman State University, Kirksville, Missouri 63501, USA; belanger@truman.edu

[2] Lehrstuhl für Mathematik I, Universität Mannheim, 68131 Mannheim, Germany; {hausdorf,werner.seiler}@math.uni-mannheim.de

Abstract. We present an overview of the *MuPAD* library *DETools* for the analysis of differential equations. It has been developed within an object-oriented environment for the efficient representation of differential functions. Currently, the main ingredients of the library are a fairly general package for Lie symmetry analysis and a algebraic-geometric completion package for overdetermined systems.

1 Introduction

Differential equations are of central importance in all applied sciences. Whenever a continuous process is modelled mathematically, chances are high that the model is built upon them. Thus it is not surprising that almost any computer algebra system provides functions for analysing differential equations. Within *MuPAD*, this is largely the task of the library *DETools*.

This library is still in a rather early stage of its development and only parts of the planned functionality are already realised in the upcoming new release of *MuPAD* (2.0). We describe briefly the main components of the library in its current state: a fairly general package for Lie symmetry analysis and a algebraic-geometric completion package for overdetermined systems. In addition, there exist some miscellaneous functions, especially for the symbolic and numeric analysis of mechanical systems.

Of course, *MuPAD* is not the first computer algebra system to provide such packages; there exist so many symmetry and completion packages that it is impossible to list them here. A good overview is contained in the survey article [14]. As some of these packages have been developed for many years, they are considerably larger and offer more functionality as our library in its present state. On the other hand, our library uses at several places new ideas both on the theoretical side and on the level of implementation.

DETools is developed within an object-oriented environment for differential functions realised with the `domains` package of *MuPAD* [5]. This environment contains a number of categories and domains for efficiently representing different types of differential functions (linear, polynomial, general expressions). It

* This work has been supported by Deutsche Forschungsgemeinschaft, Landesgraduiertenförderung Baden-Württemberg and INTAS grant 99-1222.

permits the easy interaction between different packages of the library, as the exchange of data causes no problems.

As many users of *MuPAD* are not familiar with domains, the library has been written in such a way that no knowledge of the environment is required for using its basic functionality. Only for advanced applications it is of advantage to directly use the domains. The main reason for embedding the library in the environment is the possibility of generic implementations. For example most differential equations appearing in applications have at most polynomial non-linearities. The *DETools* library will automatically perform all computations with such equations in polynomial arithmetics which is much faster than working with general expressions.

This article is written in a rather informal style. No details on the underlying mathematics are given; they can be found in the cited references. Instead emphasis is put on presenting some concrete examples. The article is organised as follows. The next section reviews our object-oriented environment for differential equations. The following two sections discuss first the Lie symmetry package and then the completion package. The final section gives an overview of some miscellaneous functions contained in the library.

2 An Object-Oriented Environment

Object-oriented programming in computer algebra tries to mimic a categorical approach to mathematics: each object belongs to a computational domain the properties of which are determined by its category. So the general properties of our basic objects, *differential variables* and *functions*, are specified in the two categories `Cat::DifferentialVariable` and `Cat::DifferentialFunction`. Concrete representations are implemented in a series of domains within these categories. Each user may extend the environment by developing own domains. As long as they satisfy the specifications of the mentioned categories, they will interact without problems with all other domains. However, for most purposes the domains already provided should suffice.

2.1 Differential Variables

`Cat::DifferentialVariable`. We distinguish three different types of differential variables. The *dependent variables* represent the unknown functions in a differential equation; their arguments are the *independent variables*; and of course there appear *derivatives* of the dependent variables with respect to the independent ones. In order to have a consistent scheme of denoting these variables, we assume that the independent and the dependent variables are given in the form of ordered lists. Then each can be addressed by its *index* giving its position in the respective list. Derivatives are additionally characterised by a *multi* or a *repeated index*: a list showing with respect to which independent variables the differentiation takes place. For example,

$$\frac{\partial^5 u(x,y)}{\partial x^4 \partial y} \quad \text{yields} \quad \begin{cases} \texttt{[4,1]} & \text{as multi index,} \\ \texttt{[1,1,1,1,2]} & \text{as repeated index.} \end{cases}$$

These three ingredients constitute the specification of any differential variable in `Cat::DifferentialVariable`. The category provides a number of operations on the variables, most importantly differentiation. One must distinguish between differentiation of a dependent variable or derivative with respect to independent variables (henceforth called *total differentiation*) and *partial differentiation* of one variable with respect to another; in the latter case only 0 and 1 are possible results. Other methods include the determination of the *order* and *class* (the lowest occurring independent variable in a differentiation) of a variable, manipulations of multi indices and orderings.

`Dom::DifferentialVariable`. In order to speed up some of the computations, the representation of an element from this domain has, in addition to the three ingredients described above, two further slots in which the order and class of a variable are stored explicitly. Their values are determined every time a variable is created or changed. The form of the output is set according to how the names of the variables are given at the creation of the domain: if this is done by providing two lists with the names of the independent resp. dependent variables, repeated index notation with the indices substituted by the corresponding variable names is chosen; otherwise, if two indexed identifiers are given, multi index notation is used. An important task of `Dom::DifferentialVariable` is to convert to and from other representations for differential variables, e. g. in the `diff` or `D` notation of *MuPAD* and thus to provide a convenient interface. It reimplements many methods from `Cat::DifferentialVariable` in a more efficient way now that the concrete representation of the objects is known.

Example 1. We create two different domains of differential variables; the first one uses repeated indices, the second one multi indices. Both contain two independent variables and one dependent variable.

MuPAD

```
>> DV  := Dom::DifferentialVariable([x,t], u):
>> IDV := Dom::DifferentialVariable(x[2],u[1]):
>> DV(diff(u(x,t),x,t)):DV(u([x,t]));
>> IDV(D([1,2],u)):IDV(u([1,1]));
```

Output

```
u([x, t])
u([1, 1])
```

The last two lines show several methods to generate the derivative u_{xt} resp. $u_{x_1 x_2}$ in both domains. Note that for improved readability the output is always in a condensed notation which may also be used for input. •

`Dom::RestrictedDifferentialVariable.` This domain is used for imposing restrictions on an already existing domain of differential variables. At creation, it takes as arguments this domain and a criterion which must be fulfilled by the variables belonging to the new domain. This can be for example a condition on the order or the type of the variables. The domain is mainly used in connection with linear differential functions or polynomials for specifying the type of the coefficients (see below).

2.2 Differential Functions

`Cat::DifferentialFunction.` The category of differential functions over a domain of differential variables `DV` inherits the methods from `Cat::SemiGroup`, i. e. addition of elements is defined. If multiplication is possible,[1] too, it should belong to `Cat::PartialDifferentialRing`. The category provides default implementations for a variety of operations on differential functions and infers only the methods `_plus`, `diff`, `subs` (substitution of differential variables), `solve`, `has` (checking whether a certain expression occurs in a function) and `indets` (set of all indeterminates of a function). Among the methods already implemented are the following ones. `leader` returns the leading differential variable[2]; `class` and `order` give the respective values of this leading derivative. `diffSubs` substitutes a differential variable and all its derivatives; `eval` allows to substitute general expressions for the dependent variables, which is useful for entering a solution ansatz into an equation. `totalDiff` performs total differentiations. `autosimplify` and `autoreduce` provide simplification of a list of differential function without resp. with consideration of differential consequences.

`Dom::DifferentialExpression.` This domain "lifts" the standard *MuPAD* expressions to `Cat::DifferentialFunction`. Thus besides the methods mentioned there many *MuPAD* functions like `simplify` or `normal` are lifted, too. The representation consists of two slots: the first contains the actual expression, the second a set of all occurring differential variables. Note, however, that for efficiency reasons this set is in general only a superset of the actual variables: if, as a result of an arithmetic operation, a variable disappears, it is not automatically eliminated from the set.

`Dom::LinearDifferentialFunction.` This domain is used for linear differential functions. Many operations like autoreduction and autosimplification can be performed much more efficiently (and completely algorithmically) for linear systems, so that the default implementations in `Cat::DifferentialFunction` are overwritten. The functions are linear in all differential variables not appearing in the given coefficient ring. Thus one can also represent quasi-linear functions

[1] This is probably the case in all domains except those representing linear functions where obviously multiplication would lead to something non-linear.

[2] The default ordering of differential variables is graded reverse lexicographic.

with this domain. However, it may then happen that some operations are not possible, as the result is no longer quasi-linear.

`Dom::DifferentialPolynomial.` *MuPAD* possesses a very fast polynomial arithmetic directly implemented in the kernel. The goal of this domain is to exploit this functionality and to enable faster manipulation of expressions polynomial in differential variables. The problem one encounters, however, is that the list of variables is in general infinite. In order to benefit from the internal polynomial arithmetic the variable lists of the polynomials to be added or multiplied must coincide. A too frequent adaption of the lists considerably slows down computations. `Dom::DifferentialPolynomial` follows the strategy to always work with lists containing *all* variables up to a fixed order. Therefore, a differential polynomial has three components: the actual polynomial, the set of occurring differential variables (as for `Dom::DifferentialExpression`) and the order up to which the polynomial contains differential variables. The variable lists are adapted only *before* two polynomials are added or multiplied (if their orders do not coincide); this is done directly on the internal list representation.

Example 2. We now create some domains of differential functions using the domain `DV` of the previous example.

MuPAD

```
>> DE := Dom::DifferentialExpression(DV):
>> RDV := Dom::RestrictedDifferentialVariable(DV,
>>                                    Types={"Indep"}):
>> RDE := Dom::DifferentialExpression(RDV):
>> DP := Dom::DifferentialPolynomial(DV, RDE):
>> LDF := Dom::LinearDifferentialFunction(Vars=[[x,y,z],u],
>>                               Rest=[Types={"Indep"}]):
>> LDF::coeffRing;
```

Output

```
Dom::DifferentialExpression(Dom::RestrictedDifferentialVariable
   (Dom::DifferentialVariable([x, y, z], u), Types = {Indep}))
```

`DE` represents arbitrary expressions in all differential variables contained in `DV`. The expressions in the domain `RDE` may dependent only on the independent variables in `DV`. The domain `DP` contains differential polynomials over `DV` with arbitrary functions of the independent variables as coefficients. The last but one line shows a shortcut in creating a domain of linear differential functions `LDF`: instead of first explicitly creating a domain of differential variables and then a coefficient domain, one may specify these information with some options and the corresponding domains are automatically generated. The output shows the generated coefficient ring.

In order to demonstrate the efficiency of the polynomial arithmetics, we compare timings[3] of additions in `Dom::DifferentialPolynomial` and `Dom::DifferentialExpression` by creating two differential polynomials, `dp1` and `dp2`, and converting them to equivalent differential expressions.

MuPAD
```
>> dp1 := DP((u([x,x])^2*u([x])-3*x*t^2*u([t,t])^2)^5):
>> dp2 := DP((u*u([x,t])-x^2*u([x])^3*u([t,t])+u([x])^2)^4):
>> de1 := DE(dp1): de2 := DE(dp2):
>> time(dp1+dp2 $ i=1..1000);
>> time(de1+de2 $ i=1..1000);
```
Output
```
170

420
```

The addition in `Dom::DifferentialPolynomial` is more than twice as fast as in `Dom::DifferentialExpression` (0.17 ms vs. 0.42 ms). •

2.3 Interface

Many *MuPAD* users are not familiar with domains or do not want to work with them. Indeed, for a quick calculation it is rather cumbersome, if one must first generate a number of domains and then explicitly create some elements in these etc. For this reason the *DETools* library has been written in such a way that all packages and functions within it can be used without explicit reference to domain elements.

Almost all functions of the *DETools* library operate in the following way. If they are passed as arguments elements of one of the above described domains, the computations will take place in this domain. Thus advanced users can always prescribe a special domain of computation. If differential functions or equations are passed as standard *MuPAD* expression, an automatic coercion into an appropriate domain is tried. In order to enhance efficiency, it is always attempted to coerce into a domain with as much structure as possible. Five different domains are tried in the following order:

1. linear functions with constant coefficients
2. linear functions with variable coefficients
3. polynomial functions with constant coefficients
4. polynomial functions with variable coefficients
5. differential expressions

Of course, these conversion attempts require some computation time. However, in longer calculations, as they typically occur in symmetry or completion

[3] All timings were obtained on a Linux-PC with Pentium III processor (733MHz) and 512MB memory.

theory, this time is well spent, if all subsequent arithmetical or differential operations are accelerated. In smaller examples the time spent on conversions does not matter at all.

3 The Lie Symmetry Package

Lie symmetry analysis [15] in one of the most important techniques to study differential equations. There exist many computer algebra packages for aiding symmetry computations; a survey can be found in [14]. Currently, the *DETools* library provides functions for setting up the determining systems for several types of symmetry generators: `detSys` for Lie point symmetries, `genDetSys` for generalised symmetries, and `ncDetSys` for a very general form of non-classical symmetries. For the analysis of the determining systems, the completion package described in the next section can be used. For lack of space we will discuss in the sequel only `detSys` and `ncDetSys`, but the properties of `genDetSys` are very similar.

3.1 Infinitesimal Generators and Prolongation

Basic objects in Lie symmetry analysis are vector fields as symmetry generators. For the analysis of differential equations these vector fields must be prolonged to fields operating on jet bundles. We have developed a special domain, `Dom::JetVectorField`, for representing such fields. It provides three main operations: one can *prolong* and *project* vector fields to higher resp. lower order jet bundles and one can *apply* a field to a differential function.

Some other useful methods are implemented, too. `ansatz` returns a generic ansatz for symmetry analysis; the names of the coefficient functions must be given as argument. `characteristic` computes the characteristic of an infinitesimal generator needed for non-classical symmetries. `evolutionaryForm` transforms a vector field into an equivalent one in evolutionary form, i. e. one where all components in the directions of the independent variables vanish. For the prolongation of evolutionary fields, which are important for generalised symmetries, the command `evolProlong` exists.

Most of these operations are straightforward and comparatively cheap. Only the prolongation requires a costly recursive determination of the coefficients. But most differential equations are sparse; so if one applies the prolonged vector field to an equation, only a small subset of all coefficients are effectively needed. Rather surprisingly, most symmetry packages described in the literature do not seem to exploit this simple fact. The `prolong` method of `Dom::JetVectorField` permits this. It accepts as a second argument either an integer, in which case the result is the full prolongation up to that order, or a list of differential variables, in this case only the coefficients in those directions are determined.

Example 3. This strategy of only determining the really needed coefficients can lead to drastic savings in computation time. As a typical example of an evolution equation consider the *Cahn-Hilliard equation*

$$u_t = uu_{xx} + u_x^2 - u_{xxxx} \,. \tag{1}$$

We create a domain of jet vector fields and a suitable ansatz $\xi\partial_x + \tau\partial_t + \eta\partial_u$. Then we compute once the full prolongation up to fourth order and once only the necessary coefficients.

MuPAD

```
>> JVF := Dom::JetVectorField(DE):
>> ans := JVF::ansatz([xi,tau,eta]):
>> time(JVF::prolong(ans,4));
>> vars := map([u([t]),u([x]),u([x,x]),u([x,x,x,x])], DV):
>> time(JVF::prolong(ans, vars));
```

Output

```
820
210
```

As one can see, the second call is about four times faster, as it does not determine any coefficient in direction of a mixed derivative $u_{t\cdots x\cdots}$. In higher-dimensional examples the difference can be even larger. •

3.2 Lie Point Symmetries

The computation of the determining system is carried out in the classical way: the prolonged ansatz is applied to the equations; the result is simplified by entering the solved original equations and then regarded as a polynomial in the derivatives; finally, the vanishing of the extracted coefficients yields the determining system. By default, a general ansatz is chosen for the infinitesimal generator. But it is also possible to work with a user defined ansatz via the option `Ansatz`. The names of the occurring unknown functions must be specified with the option `Params`.

The output domain for the determining system can be controlled by the option `Expr`. If its value is `FALSE`, the output equations are contained in a domain of linear differential functions. This is the default, if the input equations belong to one of the domains describe above (i. e. the user explicitly works with domains). If `Expr` has the value `NoDiff` (which is the default otherwise), the output consists of standard *MuPAD* expression; however, differential variables are represented in condensed notation. This is similar to `Expr=TRUE` where also pure expressions are returned but derivatives are represented in the full `diff` notation (see the example below).

With the option `AutoReduced=TRUE`, the determining system is automatically simplified. Finally, an idea first mentioned in [4] has been implemented via the option `Selected`. One may specify a sublist of the input equations; the prolonged ansatz is then applied only to these. This strategy can be fairly useful for tackling very large systems step by step.

Example 4. The determining system for the Cahn-Hilliard-equation (1) can be computed with the command

```
MuPAD
>> detools::detSys(u([t])=u*u([x,x])+u([x])^2-u([x,x,x,x]),
>>    [x,t], u, Expr=NoDiff, Autoreduced=TRUE)
```

```
Output
--               PHI1
|  XI1([x]) + ----, XI2([x]), PHI1([x]), XI1([t]), XI2([t])
--               2 u
     2 PHI1                                              PHI1 --
   + ------, PHI1([t]), XI1([u]), XI2([u]), PHI1([u]) - ----  |
       u                                                  u   --
```

Here no user defined domains are involved in either input or output. The result consists of standard *MuPAD* expressions. But for improved readability the condensed notation `XI2([x])` is used instead of the usual notation `diff(XI2(x,t,u),x)` etc. •

3.3 Non-Classical Symmetries

Non-classical symmetries have been originally introduced by Bluman and Cole [2]. However, they and subsequently most of the authors who applied non-classical symmetries considered only what one might call the "first level". They require that the *invariant surface condition* for the used ansatz is compatible with the original differential equation, i. e. all integrability conditions are identically satisfied. However, for the existence of symmetry reductions it suffices that the invariant surface condition is consistent with the original equation, i. e. they possess common solutions. This was shown independently by Pucci and Saccomandi [18] and one of the present authors [21].

It is in principle possible to compute *any* solution of a given system of differential equations via a non-classical symmetry reduction. In practice, this seems hardly feasible, as at each level of non-classical symmetries the determining system becomes smaller, i. e. less overdetermined, but the equations in it are getting more and more complex. Furthermore, the determining systems for non-classical symmetries are non-linear. Thus it is usually impossible to compute their general solution; one must find from other considerations a useful ansatz for the symmetry generators.

Our symmetry package seems to be the first one that is able to set up the determining systems also for non-classical symmetries of "higher level". The command `ncdetsys` proceeds mostly in the same way as `detSys` does; the difference is that the invariant surface condition derived from the used ansatz (by the method `characteristic` from `Dom::JetVectorField`) is added to the input equations. This necessitates some further slight modifications, since the resulting determining equations are no longer linear. Non-classical symmetries of a "higher level" can be computed with the option `Steps`. The idea is to iterate the process of applying the ansatz to the combined system and restricting to the solution manifold by substituting the equations. With `Steps=1` (the default), we obtain the Bluman-Cole method; a higher number yields the above mentioned higher level symmetries.

Example 5. Bluman and Cole introduced their method for the heat equation. The relevant determining system can be computed in *MuPAD* as follows.

MuPAD

```
>> detools::ncDetSys(u([t])=u([x,x]), [x,t], u,
>>              Ansatz=xi(x,t,u)*D(x)+D(t)+eta(x,t,u)*D(u),
>>              Params=[xi,eta], Autoreduced=TRUE, Expr=NoDiff)
```

Output

```
--                                                 xi([t])
|  - eta([t]) + eta([x, x]) - 2 eta xi([x]),  -------
--                                                    2

    xi([x, x])
  - ---------- + eta([x, x]) + xi xi([x]) - eta xi([u]),
        2

                                                            --
  xi([u, u]), - 2 xi([x, u]) + eta([u, u]) + 2 xi xi([u])   |
                                                            --
```

As one can see, the determining system is quasi-linear. This example also demonstrates how one can prescribe an ansatz for the symmetry generators. In this case we have set the component into t-direction to 1. In order to obtain all non-classical symmetries at the first level one would have to redo this computation without the `D(t)` in the ansatz. •

4 Completion to Involution

If one studies under- or overdetermined systems of differential equations, an important step is the so-called *completion* of the system. This comprises in particular the construction of all hidden integrability conditions of the system. A typical application is the analysis of determining systems in Lie symmetry theory (see above); but completion is also crucial for the numerical analysis [13, 22, 24] of such systems, for constrained mechanical systems [25] or for control theory (see examples below) to name just a few fields.

There exist many different approaches to completion but no rigorous definition of what a completed system is. Depending on the approach chosen the result of a completion may depend on additional ingredients like a ranking of derivatives or an involutive division. The completion algorithm implemented in *DETools* is rooted in the formal geometric theory of differential equations [16,

20] but also uses methods from commutative algebra, namely involutive bases [3, 8, 23]. For theoretical details we must refer to [12] which also contains many references to alternative approaches.

We should stress that while our completion algorithm can be used for the analysis of determining systems, it is not the most efficient choice for this particular task. In the analysis of determining systems, the main emphasis is on constructing as fast as possible as many integrability conditions as possible in the hope that they may facilitate the explicit integration of the system. In contrast, our algorithm aims at obtaining intrinsic geometric information, as it represents an efficient realisation of the Cartan-Kuranishi completion.

Former implementations of the Cartan-Kuranishi algorithm [19] working purely geometrical performed many redundant computations. Our new algorithm avoids these using some algebraic concepts. An open question is its relative efficiency compared with approaches based on differential forms [9].

4.1 The Algorithm

The general framework for our algorithm is the *Cartan-Kuranishi completion* in formal theory. A system of differential equations of order q is interpreted as a fibred submanifold $\mathcal{R}_q$ of a jet bundle $J_q\mathcal{E}$. Two natural operations are prolongation to higher order jet bundles and projection to lower order bundles which may produce integrability conditions. The algorithm completes (under some mild regularity assumptions) any system to an *involutive* one. Involution does not only imply *formal integrability*, i. e. the absence of hidden integrability conditions, but also poses some combinatorial conditions on the *symbol* $\mathcal{M}_q$ (the highest order part of the linearised system). As Fig. 1 shows the algorithm can be formulated in a completely intrinsic manner, i. e. without reference to any particular system of coordinates.

Algorithm: *Cartan-Kuranishi completion*
Input: differential equation $\mathcal{R}_q$
Output: involutive differential equation $\mathcal{R}^{(s)}_{q+r}$

$r \leftarrow 0,\ s \leftarrow 0$
repeat
  while $\mathcal{M}^{(s)}_{q+r}$ is not involutive **do**
    $r \leftarrow r+1$
  involutive $\leftarrow \left(\mathcal{R}^{(s+1)}_{q+r} = \mathcal{R}^{(s)}_{q+r}\right)$
  if not(*involutive*) **then** $s \leftarrow s+1$
until *involutive*

Fig. 1. Cartan-Kuranishi completion of differential equations

The implemented algorithm proceeds exactly as outlined above which has the advantage over, say, purely differential algebraic techniques that all intrinsic geometric information is retained. For efficiency reasons, the tests of a symbol resp. a system for involution, however, rely heavily on the algebraic theory and a special representation of the system. In general, one can describe the main idea as to partition for each equation the set of independent variables into *multiplicative* and *non-multiplicative* ones. In each step, only the prolongations by non-multiplicative variables are effectively computed. Then they are reduced by multiplicative prolongations determined only when needed. In an involutive system all non-multiplicative prolongations can be reduced to zero by the multiplicative ones.

4.2 The Completion Command

`detools::complete` expects as its first argument the list of equations to be completed. Currently, it can handle only linear equations or equations with polynomial non-linearities. The completion process can be influenced by a number of options. `Output` controls the amount of information printed during the completion; this can range from nothing to all details of every reduction step. If the option `FractionFree=TRUE`, divisions by coefficients are avoided; otherwise the coefficient of the leading derivative is always normalised to one. In order to avoid problems with δ-singular coordinate systems, one can give some positive value i to the option `Random`. Then a linear transformation of the independent coordinates is performed with a random matrix whose entries are integers between $-i$ and i.

Example 6. We consider the linear system $u_{zz} - yu_{xx} = 0$, $u_{yy} = 0$, a classical example due to Janet.

MuPAD

```
>> sys := map([u([z,z])-y*u([x,x]), u([y,y])], LDF):
>> detools::complete(sys, Output=3);
```

Output

```
[u([z, z]) - y u([x, x]), u([y, y]), u([y, y, z]), u([x, x, y]),
   u([x, x, y, z]), u([x, x, x, x]), u([x, x, x, x, z])]
```

Besides the final result shown above, the `complete` function with the given value of the option `Output` produces a number of messages reporting on the progress of the computation. As this output is rather lengthy, we show it only in an edited form in Fig. 2.

As one can see, the output permits us to reconstruct the steps of the intrinsic Cartan-Kuranishi completion. In this particular case it produces the following sequence of differential equations

$$\mathcal{R}_2 \to \mathcal{R}_3 \to \mathcal{R}_3^{(1)} \to \mathcal{R}_4^{(1)} \to \mathcal{R}_4^{(2)} \to \mathcal{R}_5^{(2)} . \tag{2}$$

Current system: $\mathcal{R}_2^{(0)}$; Dimension: 8 $\mathcal{M}_2^{(0)}$ not involutive; Dimension: 4	$\mathcal{M}_4^{(1)}$ involutive; Dimension: 2 $\mathcal{R}_4^{(1)}$ not involutive
Current system: $\mathcal{R}_3^{(0)}$; Dimension: 12 $\mathcal{M}_3^{(0)}$ involutive; Dimension: 4 $\mathcal{R}_3^{(0)}$ not involutive	Current system: $\mathcal{R}_4^{(2)}$; Dimension: 12 $\mathcal{M}_4^{(2)}$ not involutive; Dimension: 1
	Current system: $\mathcal{R}_5^{(2)}$; Dimension: 12
Current system: $\mathcal{R}_3^{(1)}$; Dimension: 11 $\mathcal{M}_3^{(1)}$ not involutive; Dimension: 3	$\mathcal{M}_5^{(2)}$ involutive; Dimension: 0 $\mathcal{R}_5^{(2)}$ involutive
Current system: $\mathcal{R}_4^{(1)}$; Dimension: 13	Final system: $\mathcal{R}_5^{(2)}$; Dimension: 12

Fig. 2. Output produced during the completion of the Janet example

As the final step did not produce any integrability conditions, $\mathcal{R}_4^{(2)}$ is a formally integrable but not involutive system. •

4.3 Compatibility Conditions

For systems of linear differential equations, the computation of compatibility conditions corresponds to determining a list of generators for the syzygy module of the system. The right hand side of each equation is set to a dependent variable of a new domain of differential variables (with the independent variables being the same as for the input system); all prolongations and reductions in the completion process are also simultaneously applied to these right hand sides. If the left hand side of an equation is reduced to zero, the right hand side yields a compatibility condition; thus all relations between the equations have become apparent when the completion is finished.

If this is repeated, i. e. one computes the compatibility conditions of the compatibility conditions and so on, the process terminates and yields a kind of finite free resolution, called the *Janet sequence.* One may also go in the other direction and try to find for a given system another system the compatibility of which is exactly the first system. This is referred to as computing a *potential* or a *parametrisation.* These concepts are important in control theory [26]. For checking whether a parametrisation exists, and in case that it does, for computing it, we use the method described in [17].

Example 7. Computation of compatibility conditions can be invoked with the option `Compatibility=TRUE`. For the above mentioned methods involving the Janet sequence, the commands `janet` and `parametrise` exist. In this example, we work with matrices of linear differential operations instead of systems of linear differential equations; the commands `makeMatrix` and `makeSystem` convert between these two representations.

One example of a Janet sequence are the well known operators *grad*, *rot* and *div*. Here, each operator describes the compatibility conditions of the previous one. In *MuPAD*, this looks as follows:

MuPAD

```
>> LDO := Dom::LinearDifferentialOperator([x,y,z]):
>> MLDO := Dom::Matrix(LDO):
>> grad_op := MLDO([[D(x)],[D(y)],[D(z)]]):
>> detools::janet(grad_op);
```

Output

```
-- +-      -+  +-                         -+                          --
|  | D(x) |   |  D(y), -D(x),    0      |                            |
|  |      |   |                          |  +-                   -+  |
|  | D(y) |,  |  D(z),    0,    -D(x)   |,  | D(z), D(y), D(x) |  |
|  |      |   |                          |  +-                   -+  |
|  | D(z) |   |    0,    D(z), -D(y)    |                            |
-- +-      -+  +-                         -+                          --
```

A more interesting example where it is shown that the linearised Einstein equations do not possess a potential can be found in [10]. •

5 Miscellaneous Functions

Besides the two packages discussed above, the *DETools* library contains a some further functions. However, most of them are in a less developed state. We discuss below a solver for partial differential equations and a very useful utility function. Before, we briefly mention some functions for the numerical and symbolic analysis of mechanical systems. For lack of space we cannot give a more detailed exposition, as this would require a separate paper.

The analysis of mechanical systems starts with setting up the equations of motion. The *DETools* library contains functions for generating *Euler-Lagrange equations* and *Hamiltonian equations* for finite- and infinite-dimensional systems. *Constrained systems* can already in principle be treated with the help of the completion package but improved routines for different constraint formalisms are in preparation. The equations of motion can be generated in such a way that they can directly be numerically solved with the functions of the *MuPAD* `numeric` library. There also exist several functions of the visualisation of the output either as trajectories or phase plots.

Of independent interest for the numerical analysis of differential equations are some functions for the generation and analysis of numerical methods. For example, it is possible to compute the coefficients for finite difference methods with arbitrary stencils following an algorithm of Fornberg [6, 7]. For ordinary differential equation the so-called *method of modified equations* [1] has been implemented. It constructs a perturbation of the given equation which better approximates the numerical solution of the original equation obtained with some numerical scheme.

5.1 A Solver for Partial Differential Equations

In many cases, the important properties of the solutions of differential equations are found by analysing the equations rather than the solutions themselves. In some cases, however, it may be useful to have an explicit solution to examine. `detools::pdesolve` can be used to solve certain partial differential equations. The equation can be given in ordinary `diff` notation, or in condensed form, and the function to be solved for must be specified. With no optional arguments, `detools::pdesolve` will return a general solution; initial conditions can be provided to get a particular solution.

Example 8. The initial conditions are given as the values of the variables on an initial surface. Since this example has three independent variables, the initial conditions are given using two parameters. The parameters must be given as a separate list.

MuPAD
```
>> detools::pdesolve(u = u([x]) + x*u([y]) - 2*u([z]),
>>     u(x,y,z), [u=S,x=T,y=S+T,z=S-T],[S,T]);
```
Output
```
                                2 1/2
u = exp(x - (4 x - 2 y + 2 z + x )    )
                                     2 1/2
   (2 x + z - (4 x - 2 y + 2 z + x )    )
```

The differential equation was converted into a linear differential function, and then the method `detools::charSolve` was called applying the classical method of characteristics. This is equivalent to the following.

MuPAD
```
>> de := LDF(u - u([x]) - x*u([y]) + 2*u([z])):
>> detools::charSolve(de,[u=S,x=T,y=S+T,z=S-T],[S,T]);
```

Here the user directly calls `charSolve`. This is possible for most solution methods known to `pdesolve`. •

As an alternative to finding a closed form solution, if `pdesolve` is given the option `"Series"`, it will find a power series solution. To do this, *MuPAD* must be able to solve the equation for an expression of the form $\partial^n u/\partial t^n$, where u is the dependent variable, t is one of the independent variables, and n is the order of the equation. Any initial conditions in this case can be specified by a list whose first element is an equation `t = ` *value*, specifying an initial surface, and whose second element is another list, consisting of the values of $[u, \partial u/\partial t, \ldots, \partial^{n-1}u/\partial t^{n-1}]$ on the initial surface. Other optional arguments are the centre and the order of the resulting series.

Example 9. We compute a series solution around the point $(1,2)$ of the differential equation $yu_{xx} - u_y = x$ for the initial conditions $u(1,y) = y$ and $u_x(1,y) = y + 2$.

MuPAD

```
>> detools::pdesolve(y*u([x,x]) - u([y]) = x, u(x,y),
>>                   [1,2], 3, [x=1,[y,y+2]], "Series");
```

Output

```
                                          2
2 + 4 (x - 1) + (y - 2) + 1/2 (x - 1)  + (x - 1) (y - 2) +
   O([x - 1, y - 2], 3)
```

As sufficiently many initial conditions are prescribed, no undetermined parameters appear in the solution. Without initial data the general power series solution is computed. •

5.2 The "Derivative Tree"

In computations with differential equations one relatively often encounters the following problem; thus it is rather surprising that to our knowledge nobody has published a solution so far. Assume that we are given a function $f(x,y)$ and we need f_{xx} and f_{xy}. Obviously, the most efficient way to compute these two derivatives is to first compute f_x. If one started with computing f_y, one would need one differentiation more. This example is trivial but assume that we need six different derivatives up to fourth order of a function depending on five variables, then it is less clear how one can find an optimal way to compute these derivatives with the least number of differentiations.

The *DETools* library contains a function `derList2Tree` taking as argument a list of multi indices representing the needed derivatives and returning a tree-like structure representing a way to compute these derivatives. It is based on a simple heuristic procedure,[4] since an exact algorithm seems hard to find and would perhaps incur more overhead than time saved. In all examples we have studied so far, it has produced the optimal solution.

We define for a multi index $\mu = (\mu_1, \ldots, \mu_n)$ its norm $||\mu|| = \sum_{i=1}^{n} |\mu_i|$. Using the distance function

$$d(\mu,\nu) = \begin{cases} 0 & \text{if } ||\mu - \nu|| = ||\mu|| - ||\nu||, \\ ||\mu - \nu|| & \text{else,} \end{cases} \tag{3}$$

we proceed for an input set $\mathcal{S}$ of multi indices as follows:

1. Compute the multi index α with $\alpha_i = \min_{\mu \in \mathcal{S}} \mu_i$ and subtract it from all multi indices in $\mathcal{S}$; if the zero index occurs now in $\mathcal{S}$, it is removed and the position of the corresponding multi index to be computed in the original list is inserted.

[4] We are indebted to Sven Helmer (Universität Mannheim) for this heuristics.

2. Determine those $\mu, \nu \in \mathcal{S}$ for which $d(\mu, \nu)$ is maximal.
3. Partition $\mathcal{S}$ into two lists: $\mathcal{S}_\mu$ contains all multi indices lying closer (with respect to the distance function d) to μ, $\mathcal{S}_\nu$ all those closer to ν.
4. Apply the above procedure recursively to $\mathcal{S}_\mu$ and $\mathcal{S}_\nu$.

Example 10. Consider the result of the call

MuPAD

```
>> detools::derList2Tree([[0,4],[2,3],[3,2],[2,1],[4,0]])
```

Output

```
[[[[0, 3], [[[0, 1], 1], [[2, 0], 2]]],
   [[2, 0], [[[[2, 0], 5], [[0, 1], 4, [1, 1], 3]]]]]]
```

It can be interpreted as follows. Go to `[0,3]` and branch: the first multi index of the argument list is reached at `[0,4]`; then proceed to `[2,3]` and return the second index. Start again at `[0,0]` and go to `[2,0]` this time, then branch: first to `[4,0]` (the fifth index) and second to `[2,1]` (the fourth) and `[3,2]` (the third). The total (and also minimal) number of differentiations in this example is 12. •

The function `derList2Tree` will be rarely called interactively, but a number of functions we have already mentioned internally use it. This includes the methods `eval` and `diffSubs` in any domain from `Cat::DifferentialFunction`. In particular, the fast prolongation of symmetry generators (see Example 3) is based on the "derivative tree".

References

1. M.O. Ahmed and R.M. Corless. The method of modified equations in Maple. Technical report, Dept. of Applied Mathematics, University of Western Ontario, London (Canada), 1997.
2. G.W. Bluman and J.D. Cole. The general similarity solution of the heat equation. *J. Math. Mech.*, 18:1025–1042, 1969.
3. J. Calmet, M. Hausdorf, and W.M. Seiler. A constructive introduction to involution. In *Proc. Int. Symp. Applications of Computer Algebra – ISACA '2000.* World Scientific, Singapore, to appear.
4. B. Champagne, W. Hereman, and P. Winternitz. The computer calculation of Lie point symmetries of large systems of differential equations. *Comp. Phys. Comm.*, 66:319–340, 1991.
5. K. Drescher. Axioms, categories and domains. Automath Technical Report No. 1, Universität Paderborn, 1996.
6. B. Fornberg. Generation of finite difference formulas on arbitrary space grids. *Math. Comp.*, 51:699–701, 1988.
7. B. Fornberg. Calculation of weights in finite difference methods. *SIAM Rev.*, 40:685–691, 1998.
8. V.P. Gerdt and Yu.A. Blinkov. Involutive bases of polynomial ideals. *Math. Comp. Simul.*, 45:519–542, 1998.

9. D.H. Hartley and R.W. Tucker. A constructive implementation of the Cartan-Kähler theory of exterior differential systems. *J. Symb. Comp.*, 12:655–667, 1991.
10. M. Hausdorf. *Geometrisch-algebraische Vervollständigung allgemeiner Systeme von Differentialgleichungen.* Master's Thesis, Universität Karlsruhe, 2000.
11. M. Hausdorf and W.M. Seiler. Differential equations in MuPAD I: An object-oriented environment. Internal Report 2000-17, Universität Karlsruhe, Fakultät für Informatik, 2000.
12. M. Hausdorf and W.M. Seiler. An efficient algebraic algorithm for the geometric completion to involution. Preprint Universität Mannheim, 2000.
13. M. Hausdorf and W.M. Seiler. Perturbation versus differentiation indices. This proceedings.
14. W. Hereman. Symbolic software for Lie symmetry analysis. In N.H. Ibragimov, editor, *CRC Handbook of Lie Group Analysis of Differential Equations, Volume 3: New Trends in Theoretical Development and Computational Methods*, chapter 13. CRC Press, Boca Raton, Florida, 1995.
15. P.J. Olver. *Applications of Lie Groups to Differential Equations.* Graduate Texts in Mathematics 107. Springer-Verlag, New York, 1986.
16. J.F. Pommaret. *Systems of Partial Differential Equations and Lie Pseudogroups.* Gordon & Breach, London, 1978.
17. J. F. Pommaret and A. Quadrat. Generalized Bezout Identity. *Appl. Alg. Eng. Comm. Comp.*, 9:91-116, 1998.
18. E. Pucci and G. Saccomandi. On the weak symmetry groups of partial differential equations. *J. Math. Anal. Appl.*, 163:588–598, 1992.
19. J. Schü, W.M. Seiler, and J. Calmet. Algorithmic methods for Lie pseudogroups. In N. Ibragimov, M. Torrisi, and A. Valenti, editors, *Proc. Modern Group Analysis: Advanced Analytical and Computational Methods in Mathematical Physics*, pages 337–344. Kluwer, Dordrecht, 1993.
20. W.M. Seiler. *Analysis and Application of the Formal Theory of Partial Differential Equations.* PhD thesis, School of Physics and Materials, Lancaster University, 1994.
21. W.M. Seiler. Involution and symmetry reductions. *Math. Comp. Model.*, 25:63–73, 1997.
22. W.M. Seiler. Indices and solvability for general systems of differential equations. In V.G. Ghanza, E.W. Mayr, and E.V. Vorozhtsov, editors, *Computer Algebra in Scientific Computing — CASC '99*, pages 365–385. Springer-Verlag, Berlin, 1999.
23. W.M. Seiler. A combinatorial approach to involution and δ-regularity. Preprint Universität Mannheim, 2000.
24. W.M. Seiler. Completion to involution and semi-discretisations. *Appl. Num. Math.*, to appear, 2001.
25. W.M. Seiler and R.W. Tucker. Involution and constrained dynamics I: The Dirac approach. *J. Phys. A*, 28:4431–4451, 1995.
26. E. Zerz. *Topics in Multidimensional Linear Systems Theory.* Lecture Notes in Control and Information Sciences 256, Springer-Verlag, London, 2000.

Algebraic Identification Algorithm and Application to Dynamical Systems

Farida Benmakrouha[1], Christiane Hespel[1], Gérard Jacob[2], and Edouard Monnier[1]

[1] INSA-IRISA 20 Avenue Buttes de Coësmes, 35043 Rennes Cedex, France.
[2] LIFL (UPRESA 8022 CNRS) Université Lille I,
59655 Villeneuve d'Ascq Cedex, France.

Abstract. We present an algorithm for the identification of any unknown dynamical system and a partial validation of the produced model. The method consists in computing a dynamical system (B_k) whose behaviour approximates the exact system one's, "up to order k", when knowing only a finite set of input/output data. The approximated dynamical system provided is a bilinear system depending on some parameters. In order to preserve the generic feature of the computation, the algorithm produces a pair (bilinear system B, constraint C on parameters), where B is produced under constraint C. A computational tool is provided, in the form of Maple package DYNAMICMODEL. The model validation is a central problem in system identification [2]. It consists generally in a test that falsifies or not falsifies the model, using a validation data set. For this continuous and symbolic model, we propose an overestimation of the mistake resulting from the systems (B_k) at order k and we show the convergence towards the exact output system. We propose also a majoration, for a certain class $(\mathcal{DP})$, of the gap between two consecutive outputs.

1 Introduction

In spite of the theoretical interest of the identifiabilty problem [11, 12], the model identification process is very important in practice. Many authors have dealt with parameters identification of a given parametric model. These models are suitable when they are phenomenological (knowledge-based), or when they have to satisfy some physical laws. On the contrary, the behavioural models do not require any knowledge of the process, which is considered as a "black-box". According to the second point of view, we intent here to identify an unknown nonlinear functional considered as a black-box, only known by its input/output behaviour. We make the following hypothesis: the unknown functional is an "analytical causal functional" in the sense of M.Fliess [3]. In other words, it can be encoded by a noncommutative generating series, which is nothing but a noncommutative transfer function. We present the identification process in section 2, the bilinear approximation in section 3, the convergence study in section 4 and 5, and finally examples of simulations in section 7.

1.1 Symbolic Encoding

1.1.1 Word Encoding

$\mathcal{Z} = \{z_0, \cdots, z_m\}$ being a finite alphabet, a word $w = z_{i_1}, \cdots, z_{i_l}$ is built by concatenating some letters of $\mathcal{Z}$. We denote by ε the empty word, by $\mathcal{Z}^*$ the set of the words w. A noncommutative series s on the variables $\{z_0, \cdots, z_m\}$ is defined as a formal sum

$$s = \sum_{w \in \mathcal{Z}^*} \langle s \mid w \rangle w \ .$$

In the case of m inputs defined by the vector $a(t) = (a_j(t))_{j=1..m}$, the associated alphabet $\mathcal{Z} = \{z_0, \cdots, z_m\}$, ($z_0$ associated to the drift $a_0(t) \equiv 1$) is called "encoding alphabet".
Then the iterated integrals are encoded by the words. For instance,

$$w = z_1 z_1 z_0 \rightarrow \int_0^t \left[\int_0^{s_1} \left[\int_0^{s_2} a_1(s_3) ds_3 \right] a_1(s_2) ds_2 \right] ds_1 \ .$$

$$\begin{cases} \textit{autonomous part (drift):} & z_0 \mapsto 1 \\ \textit{first control input:} & z_1 \mapsto a_1(t) \end{cases} .$$

1.1.2 Causal Functional and Generating Series

- Definitions
The generating series [3],

$$G = \sum_{w \in \mathcal{Z}^*} \langle G \mid w \rangle w \ . \tag{1}$$

is indexed by the words on the encoding alphabet $\mathcal{Z}$. The coefficients of the Chen series $\mathcal{C}_{\mathbf{a}}(t)$ are recursively defined by:
$\langle \mathcal{C}_{\mathbf{a}}(t) \mid \varepsilon \rangle = 1$, and for any $w = u z_j$, $u \in \mathcal{Z}^\star$, $z_j \in \mathcal{Z}$,

$$\langle \mathcal{C}_{\mathbf{a}}(t) \mid u z_j \rangle = \int_0^t \langle \mathcal{C}_{\mathbf{a}}(\tau) \mid u \rangle \, a_j(\tau) d\tau \ . \tag{2}$$

An alternative definition is given by the differential equation:

$$\frac{d}{dt} \mathcal{C}_{\mathbf{a}}(t) = \mathcal{C}_{\mathbf{a}}(t)[z_0 + \sum_{j=1}^{m} a_j(t) z_j], \qquad \mathcal{C}_{\mathbf{a}}(0) = 1 \ . \tag{3}$$

The system output is obtained by replacing in G the word w by the input contribution:

$$y(t) = \sum_{w \in Z^*} \langle G \mid w \rangle \langle \mathcal{C}_{\mathbf{a}}(t) \mid w \rangle \ . \tag{4}$$

- System and Inputs Contributions
  The system contribution is described by the generating series G and the input contribution by the Chen series $\mathcal{C}_{\mathbf{a}}(t)$:
  - $\langle G \mid w\rangle$ is the *geometric contribution* (independent of the input).
  - $\langle \mathcal{C}_{\mathbf{a}}(t) \mid w\rangle$, is the *input contribution* (independent of the system). The input contributions are given by the iterated integrals.
- Autonomous Part
  Autonomous behaviour of the system (when $a_1(t) = \cdots = a_m(t) = 0$) needs to use a special input $a_0(t) \equiv 1$ and to encode it by a specific letter z_0.

1.1.3 Case of One Input $a(t)$ with Drift

For simplicity sake, we consider now the case of one input $a(t)$ with drift (autonomous part). All the algorithms developped in this paper are adaptable in the case of several inputs.

1.2 Special Cases

1.2.1 State Affine Systems

$$(\Sigma)\begin{cases} \dot{\mathrm{q}}(\mathrm{t}) = g_0(q) + \sum_{i=1}^{m} a_i(t) g_i(q) \\ y \quad = h(q) \ . \end{cases} \tag{5}$$

with

$$g_i = \sum_{s=1}^{N} g_i^s \frac{\partial}{\partial q^s} \quad \forall i \ \ 0 \le i \le m \ ,$$

and

- state vector $q \in \mathcal{V}$, analytical variety,
- $\{g_i\}$ vector fields on $\mathcal{V}$,
- scalar observation function $h : \mathcal{V} \to I\!R$.

In this case, the coefficient $\langle G \mid w\rangle$ for every $w = z_{i_1} \cdots z_{i_k}$, of the input/output functional is given by:

$$\langle G \mid z_{i_1} z_{i_2} \cdots z_{i_k}\rangle = [f_{i_1} \circ f_{i_2} \circ \cdots \circ f_{i_k} \circ h(q)]_0 \ . \tag{6}$$

1.2.2 The Special Case of the Class $(\mathcal{DP})$: Examples of Duffing and Pendulum Equations.

We choose some well known examples in order to test our identification algorithm. Let us remark that our identification algorithm holds for any dynamical system. Consider the class $(\mathcal{DP})$ included in the class of affine state systems:

$$y^{(2)}(t) + f(y^{(1)}(t)) + g(y(t)) = a(t) \ . \tag{7}$$

where $a(t)$ is the input function
f (resp g) is an unknown function of $y^{(1)}$(resp. y).
In this case, equation (7) can be written:

$$(\Sigma) \qquad \begin{cases} \dot{q} = f_0(q) + f_1(q) a(t) \\ y(t) = q_1(t) \ . \end{cases}$$

- a(t) is the real input,
- $(q_1(t), q_2(t))$ is the current state,
- $f_0 = F\frac{\partial}{\partial q_2} + q_2\frac{\partial}{\partial q_1}$, with $F(q) = -f(q_2) - g(q_1)$,
- $f_1 = \frac{\partial}{\partial q_2}$.

1. Duffing Equation (used in non linear mechanics)
$$y^{(2)} + \alpha y^{(1)} + \beta^2 y + \gamma y^3 = a_1(t) \ .$$
We set $F_1 = -\alpha q_2 - \beta^2 q_1 - \gamma q_1^3$, (Σ_1) being the system.
2. Pendulum Equation
$$y^{(2)} + k_1 y^{(1)} - k_2 sin(y) = a_1(t) \ .$$
We set $F_2 = -k_1 q_2 - k_2 sin(q_1)$, (Σ_2) being the system.

2 Identification of the Generating Series up to Order *k*

Our aim consists in computing the coefficients of the generating series for every words whose length is $\leq k$. First we compute the iterated derivatives of the output in terms of the derivatives of the input.

2.1 Iterated Derivatives of the Output

Let $a = a(0)$, $\dot{a} = \dot{a}(0)$, $\ddot{a} = \ddot{a}(0)$, $\cdots$, *(where a is a differential letter)*.
The iterated derivatives of the output may be computed by derivating the equation (4) and by using the equations (2,3)

$$\begin{array}{rl} y(0) = & \langle G \mid \varepsilon \rangle \\ \dot{y}(0) = & \langle G \mid z_0 \rangle + a \, \langle G \mid z_1 \rangle \\ \ddot{y}(0) = & \langle G \mid z_0^2 \rangle + a \, \langle G \mid z_0 z_1 + z_1 z_0 \rangle + a^2 \, \langle G \mid z_1^2 \rangle + \dot{a} \, \langle G \mid z_1 \rangle \\ \frac{d^3}{dt^3} y(0) = & \langle G \mid z_0^3 \rangle + a \, \langle G \mid z_0^2 z_1 + z_0 z_1 z_0 + z_1 z_0^2 \rangle \\ & + a^2 \, \langle G \mid z_0 z_1^2 + z_1 z_0 z_1 + z_1^2 z_0 \rangle + a^3 \, \langle G \mid z_1^3 \rangle \\ & + \dot{a} \, \langle G \mid 2 z_0 z_1 + z_1 z_0 \rangle + \ddot{a} \, \langle G \mid z_1 \rangle + a\dot{a} \, \langle G \mid 3 z_1^2 \rangle \\ \cdots = & \cdots \end{array}$$

Now we denote by $G(n)_{i_1,\cdots,i_q}^{e_1,\cdots,e_q}$, the coefficient of $(a^{(i_1)})^{e_1} \cdots (a^{(i_q)})^{e_q}$ in the equation whose left member is $y^{(n)}(t)$. The relation between the iterated derivatives of the input and the iterated derivatives of the ouput is more generally, :

$$y^{(n)}(0) = \sum_{\sum_j e_j(1+i_j) \leq n} (a^{(i_1)})^{e_1} \cdots (a^{(i_q)})^{e_q} \ G(n)_{i_1,\cdots,i_q}^{e_1,\cdots,e_q} \ , \qquad (8)$$

where $G(n)_{i_1,\cdots,i_q}^{e_1,\cdots,e_q}$ are time independant noncommutative coefficients.
(e.g. $G(3)_0^3 = \langle G \mid z_1^3 \rangle$ and $G(3)_{0,1}^{1,1} = \langle G \mid 3z_1^2 \rangle$).

Our solution consists of two steps: to compute the contribution $G(n)_{i_1,\cdots,i_q}^{e_1,\cdots,e_q}$ of any multiderivative $(a^{(i_1)})^{e_1}\cdots(a^{(i_q)})^{e_q}$ and to split up each word w appearing in $G(n)_{i_1,\cdots,i_q}^{e_1,\cdots,e_q}$. The identification up to order 3 will require to provide the output derivatives up to order 5.

2.2 First Step: Identify the $G(n)_{i_1,\cdots,i_q}^{e_1,\cdots,e_q}$

We choose the input polynomial: $a(t) = \sum_{i=0}^{k} \frac{c_i}{i!} t^i$. The generating series is defined on the alphabet $\mathcal{Z} = \{z_0, z_1\}$.

The method consists in defining some order on the input multiderivatives (indexing the columns) and in choosing suitable values of the inputs and their derivatives (indexing the rows). Then we obtain a determinant which can be written as a Vandermonde determinants product, by Gauss transformations. Example for $n = 3$, table 1:

Table 1. First step

	1	a	a^2	a^3	$a^{(1)}$	$aa^{(1)}$
$a(0) \equiv c_{01}$	1	c_{01}	c_{01}^2	c_{01}^3	0	0
$a(0) \equiv c_{02}$	1	c_{02}	c_{02}^2	c_{02}^3	0	0
$a(0) \equiv c_{03}$	1	c_{03}	c_{03}^2	c_{03}^3	0	0
$a(0) \equiv c_{04}$	1	c_{04}	c_{04}^2	c_{04}^3	0	0
$a^{(1)}(0) \equiv c_{11}$	1	0	0	0	c_{11}	0
$a(0) \equiv c_{01},\ a^{(1)}(0) \equiv c_{12}$	1	c_{01}	c_{01}^2	c_{01}^3	c_{12}	$c_{01}c_{12}$

2.2.1 Method

This subsection is detailed in [7]. The problem consists in identifying the coefficients $G(n)_{i_1,\cdots,i_q}^{e_1,\cdots,e_q}$, by dealing with a single equation of the system, for some panel of inputs c_i and corresponding outputs $y^{(n)}(0)$.

The method is recursive (for the number q of different derivation orders).

2.2.2 What $G(n)_{(i_j)}^{(e_j)}$ and What Panel Input/Output Are Useful for the Identification up to Order k?

In order to compute the input multiderivatives coefficients, we try to attain two objects: compute a minimal set by selecting the useful coefficients and economize the inputs by using several times the sames values.

We replace now the letter z_0 by 0 and the letter z_1 by 1.

– What $G(n)_{(i_j)}^{(e_j)}$ are Useful for the Identification?

The words concerned by this identification have a length $\leq k$. Anticipating next step, the coefficients of these words can be identified when knowing some $(G(n)_{(i_j)}^{(e_j)})_{j \in J}$, bijectively associated with these words.
For $k = 3$, the tables 2,3 give the correspondences between every word w, its differential monomial m and the order n of derivation of y.

$$n = weight(monomial) + depth(word) \ ,$$

$$depth(w) = |w|_{z_0} \ ,$$

$$weight((a^{(i_1)})^{e_1} \cdots (a^{(i_q)})^{e_q}) = \sum_{j=1}^{q} (i_j + 1)e_j \ .$$

Using these tables, we can know what $G(n)_{(i_j)}^{(e_j)}$ are useful and in what linear equation they appear.

Table 2. Second step, $|w| \leq 2$

w	ϵ	1	0	10	01	00	11
m	1	a_1	a_0	$a_1 a_0$	$a_1^{(1)} a_0$	a_0^2	$a_1^2 a_0$
n	0	1	1	2	3	2	2
G	$G(0)_0^0$	$G(1)_0^1$	$G(1)_0^0$	$G(2)_0^1$	$G(3)_1^1$	$G(2)_0^0$	$G(2)_0^2$

Table 3. Second step, $|w| = 3$

w	000	010	001	100	011	110	101	111
m	a_0^3	$a_1^{(1)} a_0^2$	$a_1^{(2)} a_0^2$	$a_1 a_0^2$	$(a_1^{(1)})^2 a_0$	$a_1^2 a_0$	$a_1 a_1^{(1)} a_0$	a_1^3
n	3	4	5	3	5	3	4	3
G	$G(3)_0^0$	$G(4)_1^1$	$G(5)_2^1$	$G(3)_0^1$	$G(5)_1^2$	$G(3)_0^2$	$G(4)_{0,1}^{1,1}$	$G(3)_0^3$

Unfortunately, other $G(n)_{(i_j)}^{(e_j)}$ than those of this tabular, are required in order to build some Vandermonde determinant, using to compute them.
For instance, the computation of $G(5)_1^2$ requires the computation of $G(5)_1^1$. The computation of $G(4)_{0,1}^{1,1}$ requires the computation of $G(4)_0^{i \leq 4}$, $G(4)_1^1$, $G(4)_1^2$, $G(4)_{0,1}^{2,1}$, in order to construct the necessary Vandermonde determinant.
In brief, after selecting the $G(n)_{(i_j)}^{(e_j)}$, $j \in J$ appearing in the tabular, we have to choose the $G(n)_{(i_l)}^{(p_l)}$, $l \in L \subset J$.

Finally, we have to compute every $G(n)^{(e_j)}_{(i_j)}$, except those which do not appear in the Vandermonde concerned, and those previously computed (e.g. $G(2)^1_1 = G(1)^1_0$).

- What Way for Economizing the Inputs?

When the algorithm needs an input value in order to identify the $G(n)^{(e_j)}_{(i_j)}$, a value previously provided can be used, if it does not cancel the corresponding Vandermonde determinant. This is possible particularly for different derivation orders. For instance, $a_1(t) \equiv 0$ is suitable for identifying $G(p)^0_0, \quad 0 \leq p \leq 3$. In the same way, $a(t) \equiv 1$ is suitable for identifying $G(1)^1_0$ and for providing the first input value for the identification of $G(2)^1_0, G(2)^2_0$ et $G(3)^1_0, G(3)^2_0, G(3)^3_0$.

- What Way for Computing the Derivatives Values of the Corresponding Output?

The matter does not consists in measuring the derivatives values of the output for the selected inputs, but in computing them in order to provide them exact for the series identification.
The examples of Duffing (Σ_1 , $F = F_1$) and pendulum (Σ_2 , $F = F_2$) equations provide:

$$\begin{aligned} y(0) &= q_1(0) \\ y^{(1)}(0) &= q_2(0) \\ y^{(p)}(0) &= F^{(p-2)}(0) + a^{(p-2)}(0), \quad p \geq 2 \ . \end{aligned} \tag{9}$$

The values of the iterated derivatives of the output can then be computed in terms of the input values.

2.2.3 Algorithm

This algorithm is generic: the computation is symbolic.

- The implementation of the first step requires the production of all the monomials $(a^{(i_1)})^{e_1} \cdots (a^{(i_q)})^{e_q}$ verifying the mathematic property:

$$\sum_j e_j(1 + i_j) \leq n \ . \tag{10}$$

  Instead of systematically generating every monomial and then controlling the property (10), we choose a method providing the only relevant monomials respecting (10).
  In this algorithm, a differential monomial is presented by 2 lists:
  1. The list *lo* equal to the list of the derivation orders.
  2. The list *le* of the associated exponents.

  Its weight is $\sum_i (lo_i + 1) * le_i$ and its length l is the length of *lo* or *le*.
  The concept of reference monomial is used in the retained method. A reference monomial is a monomial such that its exponents are all equal to 1.

Consequently, a reference monomial can be presented without ambiguity, by its list of derivation orders lo. Then the set of reference monomials

$$S = \{RM\} = \{a^{(i_1)} \cdots a^{(i_q)}\}$$

is a seed for computing families of monomials defined by the same lo and by a weight $w \geq weight(RM)_{lo}$.
We compute the set $S_{L,W}$ of reference monomials $RM_{L,W}$, whose length is L and whose weight is W. We note

$$RM_L = [i_1, ..., i_L]$$

a reference monomial whose length is L.
The computing of $S_{L,W}$ is made recursively.

- case $L = 1$. We immediatly obtain:

$$weight(RM_1) = W = i_1 + 1 \ .$$

  And we construct

$$S_{1,W} = \{[W-1]\} \ .$$

- Induction hypothesis : We assume that for every length $L < N$ we know how to compute the set $S_{L,W}$ of reference monomials $RM_{L,W}$, whose length is L and whose weight is:

$$W = \sum_{j=1}^{L} (i_j + 1) \ .$$

- case $L = N$. We want to construct

$$S_{N,W} = \{RM_N \mid weight(RM_N) = W\} \ .$$

  We can rewrite

$$W = \sum_{j=1}^{N-1} (i_j + 1) + i_N + 1 \ . \tag{11}$$

  We note that the i_j appearing in $\sum_{j=1}^{N-1} (i_j + 1)$ compose a reference monomial RM_{N-1}. We know how to construct it according to induction hypothesis.
  The weight of this RM_{N-1} is $W - (i_N + 1)$.
  Effectively we have to construct several RM_{N-1}, one for each possible value for i_N. But what are these possible values? From (11) we have

$$i_N = W - \sum_{j=1}^{N-1} (i_j + 1) - 1 \ .$$

* The greatest value G for i_N is reached for the least value of the weight of RM_{N-1}. In this case:

$$G = W - \frac{N(N-1)}{2} - 1 \ .$$

* The least value P for i_N is reached for

$$P = \lceil N + \frac{W - \frac{N(N+1)}{2}}{N} \rceil \ .$$

Then, for every integer $i_N \in [P, G]$ we compute the reference monomials $RM_{N-1} = [i_1, ..., i_{N-1}]$ such that $weight(RM_{N-1}) = W - (i_N + 1)$.

2.3 Second step

In the second step, we compute the coefficients $\langle G|w\rangle$ from the coefficients $G(n)_{i_1,\cdots,i_q}^{e_1,\cdots,e_q}$ computed in the first step. These last coefficients represent the contribution of G on a family of noncommutative polynomials g_μ. The problem consists in inversing the relations expressing the polynomials g_μ in terms of the words w i.e. in computing the words w as linear combinations of the g_μ.

This part can be solved by a recursive Gauss elimination process. It is performed with the help of a combinatorial bijection between words of length l on $\{z_0, z_1\}$ and differential monomials of length $= l - depth(word)$.
Since the polynomials g_μ appearing in the equations (12) are homogeneous in $\{z_0, z_1\}$, it remains to separate the words equal except for the letters order. The problem consists in identifying the coefficients in G of the words whose length is l, containing l_1 occurrences of z_1, by using:

$$(y^{(n)}(0) = \sum_\mu a^\mu \langle G|g_\mu\rangle)_{n\in I} \ , \tag{12}$$

by denoting $a^\mu = (a^{(i_1)})^{e_1} \cdots (a^{(i_q)})^{e_q}$.

2.3.1 Method

This subsection is detailed in [8]. The method involves three steps:

1. Search for the Useful Words
2. Search for the Associated Differential Monomials (by a combinatorial bijection)
3. Search for the Solution of the Useful Linear Equations System, by a Recursive Gauss Elimination Process.

Let us explain the method with an example: $l = |w| = 4, l_1 = |w|_{z_1} = 2$

– Search for the Useful Words: 1100, 1010, 0110, 1001, 0101, 0011

- Search for the Associated Differential monomials:

  Every selected word is coded by a pair (i_1, i_2), where i_1 and i_2 express the shifting of the first and the second 1 with respect to the starting point where these two 1 are wedged in the left.

$$\begin{aligned}
1100 &\longrightarrow (i_1 = 0, i_2 = 0) \longrightarrow {a_1}^2{a_0}^2 \\
1010 &\longrightarrow (i_1 = 0, i_2 = 1) \longrightarrow a_1{a_1}^{(1)}{a_0}^2 \\
0110 &\longrightarrow (i_1 = 1, i_2 = 1) \longrightarrow ({a_1}^{(1)})^2{a_0}^2 \\
1001 &\longrightarrow (i_1 = 0, i_2 = 2) \longrightarrow a_1{a_1}^{(2)}{a_0}^2 \\
0101 &\longrightarrow (i_1 = 1, i_2 = 2) \longrightarrow {a_1}^{(1)}{a_1}^{(2)}{a_0}^2 \\
0011 &\longrightarrow (i_1 = 2, i_2 = 2) \longrightarrow ({a_1}^{(2)})^2{a_0}^2 \ .
\end{aligned}$$

  A correspondence is then established between the word and the differential monomial.
- Solution of the Selected Linear Equations System and Identification of the coefficients in G of the words whose length is l, containing l_1 occurrences of z_1:
  We have to solve, for every length $l \leq k$ and for every number $l_1 \leq l$ of z_1 occurrences, each selected linear equations system.
  The terms appearing in these systems determinants are given by:

$$\begin{aligned}
y^{(n)}(0) = \sum_{k=0}^{n} \sum_{i_j} \langle G | z_{i_1} \cdots z_{i_k} \rangle & \\
\sum_{\rho_1 + \cdots + \rho_k = n-k} \binom{\rho_1}{\rho_1} \binom{\rho_1 + \rho_2 + 1}{\rho_2} \cdots & \qquad (13) \\
\binom{\rho_1 + \cdots \rho_k + k - 1}{\rho_k} a_{i_1}^{(\rho_1)}(0) \cdots a_{i_k}^{(\rho_k)}(0) \ . &
\end{aligned}$$

The generating series G of the unknown system is then identified up to order k.

2.3.2 Algorithm

This algorithm is generic: the computation is symbolic.
For this second step, the algorithm immediatly follows the previous method. The only point we want to emphasize is the genericity of the identification solution. With this aim in view, we execute some substitutions at the end of the step 2.

3 Approximation by an Associated Bilinear System (B_k)

By using the two previous steps, we can identify the coefficients of the generating series G up to order k. In this section, the aim consists in obtaining a best approximant by computing a non commutative Padé type-approximant which coincides with G up to order k. In other words, we compute "a bilinear approximant" of G.

3.1 Method

This subsection is detailed in [4, 5].
We provide a minimal rank bilinear system (B_k). The method consists in extending the Hankel matrix $H(G)_{\leq k}$ of G (restricted to the words whose length is $\leq k$), while preserving the linear dependence relations standing in $H(G)_{\leq k}$ and while maintaining its Hankel matrix specifications. We then obtain an Hankel matrix $H(G_k)$, where G_k is a rational series approximating G up to order k.
It remains to reconstruct the state equations of the associated bilinear system (B_k), according to the informations appearing in $H(G_k)$.

3.2 Algorithm

This part needs a discussion about the parameters $q_1(0), q_2(0)$, in order to provide a generic algorithm. Maple does not allow to practise this kind of programming.
Instead of constructing a bilinear system, our algorithm produces, a list of pairs (bilinear system, constraint on parameters).
We begin filling the Hankel matrix for $\mid w \mid \leq k = 3$. The words are ordered by increasing length, and by lexicographical order for a given length.
The procedure *creathankel* admits as inputs the coefficients (from step 2) and constructs the Hankel matrix as specified above. Let us note that these coefficients are not some numerical values but are parameterized by $q_1(0)$ and $q_2(0)$.
We want to extract free columns, processing from left to right. If a column can be expressed as linear combination of previous columns, we have to maintain the linear relation by expanding suitably the columns. The difficulty is that this relation depends on the parameters values.
Unfortunatly, Maple system on which we develop our package, does not support any discussion. So, for the linear system solution with parameters, the answer provided by Maple may be false for some particular values of parameters. Therefore in the process deciding if a column C is a linear combination of some previous columns, we have to examine the rank of the matrix A obtained by adding to the basis vectors, the vector issued from the column C.
In order to replace the procedure *rank* of *linalg* package we must define a new procedure *genrank* generating a pair list, each pair associating a possible rank with a constraint on the parameters. The rank of a matrix M can be expressed as the number of non-zero rows of triangular matrix M_g resulting from Gauss elimination on M. And then, in order to replace the non-generic procedure *gausselim* of *linalg* package, we must define a new procedure *gengausselim* providing a pair list, each pair associating a possible matrix M_g with a constraint on the parameters.
At last, by using procedures *gengausselim* and *genrank*, the extraction of an associated bilinear system is supported by a procedure *genbilin* . The constraint process requires some procedures. Pratically, whenever a new constraint appears, we add it to the set of the previous constraints, by conjunction. It would remain to study the satisfaction of this constraint set.

4 Convergence of the Systems (B_k)

We propose a partial validation of the model by computing the mistake due to approximation of the unknown system by the (B_k). We show that we may adjust the required precision by increasing the order k. In this estimation, we consider that experimental data are exactly known.

4.1 Difference between the Outputs $\overline{y}_p(t)$ of (B_p) and $\overline{y}_q(t)$ of (B_q)

We study the convergence of the systems (B_k) towards the system (Σ) by computing the difference between the outputs $\overline{y_p}(t)$ and $\overline{y_q}(t)$ of the systems provided by the identification method at order p and q.

- We show that systems (B_k) outputs converge uniformly on interval [0,T].
- We propose a majoration of the gap $\mid \overline{y}_p(t) - \overline{y}_q(t) \mid \forall p, q \geq \nu$:

$$\mid \overline{y}_p(t) - \overline{y}_q(t) \mid \leq (2 \mid M \mid) \frac{(|M| T(m+1)\Delta)^{\nu}}{\nu!} (\frac{P_\nu}{1 - P_\nu}) \ .$$

where Δ,$\mid M \mid$,P_ν are constants independent of p,q.

Proof. We have according to [5]

$$\langle G_p \mid w \rangle = \langle G_q \mid w \rangle \quad \forall w, \mid w \mid \leq k, \quad \forall k \leq p \leq q \ .$$

that gives :

$$\mid \overline{y_p}(t) - \overline{y_q}(t) \mid = \mid \sum_{|w| \geq k+1} (\langle G_p \mid w \rangle - \langle G_q \mid w \rangle) \int_0^t \delta w \mid \ . \tag{14}$$

Moreover, according to [3], we have $\mid \int_0^t \delta w \mid \leq \frac{A_t^l t^l}{l!}$ where

$$A_t = \sup(1, \max_{0 \leq \tau \leq t, 1 \leq i \leq m} \mid a_i(\tau) \mid) \ .$$

We assume that $A_t \leq \Delta$. Then we have:

$$\mid \int_0^t \delta w \mid \leq \frac{\Delta^l t^l}{l!} \ .$$

And

$$\mid \langle G_p \mid w \rangle - \langle G_q \mid w \rangle \mid \leq 2 \mid M_{q,p} \mid^{l+1} \ .$$

where, since every G_r (rational) can be written $\langle G_r \mid w \rangle = \lambda_r \mu_r(w) \gamma_r$ and λ_r is the unit vector

$$\mid M_{q,p} \mid = \sup_{z_i \in Z} (\| \mu_p(z_i) \|, \| \mu_q(z_i) \|, \| \gamma_q \|) \quad for \quad w = z_{i_1} ... z_{i_l} \ .$$

By assuming $| M_{q,p} | \leq | M |, \quad \forall p, q$, we have

$$| \overline{y}_p(t) - \overline{y}_q(t) | \leq \sum_{l \geq k+1} \frac{2 | M |^{l+1} (m+1)^l T^l \Delta^l}{l!} \quad \forall t \in [0, T] \ .$$

This series of general term $u_l = \frac{(|M| T(m+1)\Delta)^l}{l!}$ converges if $0 < \frac{u_{l+1}}{u_l} = P_l < 1$. Let ν the smallest integer such that $0 < P_\nu < 1$ i.e. $| M | T(m+1)\Delta < \nu + 1$ we obtain $\forall p, q \geq k \geq \nu$

$$| \overline{y}_p(t) - \overline{y}_q(t) | \leq (2 | M |) u_k \sum_{r \geq 1} P_k^r) \ .$$

We have $\forall t \in [0, T], \quad \forall p, q \geq k$

$$| \overline{y}_p(t) - \overline{y}_q(t) | \leq (2 | M |) u_k (\frac{P_k}{1 - P_k}) \rightarrow 0 \quad when \quad k \rightarrow +\infty \ ,$$

and

$$| \overline{y}_p(t) - \overline{y}_q(t) | \leq (2 | M |) u_\nu (\frac{P_\nu}{1 - P_\nu}) \ .$$

□

4.2 Overestimation of the Mistake $| \overline{\mathbf{y}}_\mathbf{p}(\mathbf{t}) - \mathbf{y}(\mathbf{t}) |$

We show that the limit of systems (B_k) outputs is in fact the exact system output, under the hypothesis that considered functions are analytic. So, the previous majoration holds, by passage to the limit

$$| \overline{y}_p(t) - y(t) | \leq (2 | M |) \frac{(|M| T(m+1)\Delta)^\nu}{\nu!} (\frac{P_\nu}{1 - P_\nu}) \quad \forall t \in [0, T] \ .$$

Proof. We show that the limit z(t) is such that

$$z^{(j)}(0) = y^{(j)}(0) \quad \forall j \geq 0 \ ,$$

where $y(t)$ is the exact output.

We assume that $z^{(i)}(0) \neq y^{(i)}(0)$ for some $i \geq 0$. Let i the smallest integer such that $\overline{y}_p^{(i)}(0) \neq z^{(i)}(0) \ \forall p \geq i$. The gap is :

$$E(\overline{y}_p, z) \geq \frac{t^i}{i!} \, || \, \overline{y}_p^{(i)}(0) - z^{(i)}(0)) \, | - | \, (\epsilon_p(t) - \epsilon(t)) \, || \ ,$$

where $\epsilon_p(t)$, (resp. $\epsilon(t)$) is the Taylor remainder of $\overline{y}_p$, (resp. $z(t)$) at order i. From $\overline{y}_p^{(i)}(0) \neq z^{(i)}(0)$, we can say that $\exists \epsilon$ so that $| \overline{y}_p^{(i)}(0) - z^{(i)}(0) | \geq \epsilon$. Moreover, $\forall \epsilon_1, \exists T_p$,

$$| \epsilon_p(t) - \epsilon(t) | < \epsilon_1 \quad \forall t \in [0, T_p] \ .$$

We choose $\epsilon_1 < \epsilon$. Finally, we have :

$$| \overline{y_p}(t) - z(t) | \geq \frac{t^i}{i!}(\epsilon - \epsilon_1) \quad \forall t \in [0, T_p], \ \forall p \geq i \ .$$

The Taylor expansions coinciding up to order i, we can overestimate the mistake $E(\overline{y}_{i+1}, \overline{y}_i)$.

$$\forall \epsilon_2, \quad \exists T_i, \quad \forall t \in [0, T_i], \ | \overline{y}_i(t) - \overline{y}_{i+1}(t) | \leq \quad \frac{t^i}{i!}\epsilon_2 \ .$$

If we take $\epsilon_2 = \epsilon - \epsilon_1$ and $T_{i,p} = min(T_p, T_i)$, we have :

$$\forall t \ \in [0, T_{i,p}], \quad \forall p \geq i, \quad | z(t) - \overline{y_p}(t) | \geq | \overline{y}_i(t) - \overline{y}_{i+1}(t) | \ ,$$

that contradicts the fact $z(t) = \lim_{p \to +\infty} \overline{y}_p(t)$. □

5 Majoration of $| \overline{\mathbf{y}}_2(\mathbf{t}) - \overline{\mathbf{y}}_1(\mathbf{t}) |$

The previous majoration gives information about convergence speed. But, we can't compute effectively this bound, because the majoration is, in general case, too large. So, we propose, in this section, a fine and computable majoration. In the case of a system of the class $(\mathcal{DP})$, we find a majoration of $| \overline{y}_2(t) - \overline{y}_1(t) |$, where $\overline{y}_2$ (resp $\overline{y}_1$) is the output provided by identification at order 2 (resp 1). In this way, we can measure the impact of initial conditions and determine intervals [0 T] where approximation is "acceptable". We show this in Table4.

In equation (7) defining the class $(\mathcal{DP})$, we denote $F(q(0))$ by F and $q_1(0)$ (resp. $q_2(0)$) by q_1 (resp. q_2). We assume now that $q_1 \neq 0$, and the rank of $\begin{bmatrix} q_1 & q_2 \\ q_2 & F \end{bmatrix}$ is 2.

Matrix μ_2 (resp μ_1) representing bilinear systems (B_2) (resp(B_1)) is given by :

$$\mu_2(z_0) = \begin{bmatrix} 0 & 1 \\ \frac{F}{q_1} & 0 \end{bmatrix} \ ,$$

$$\mu_2(z_1) = \begin{bmatrix} 0 & 0 \\ \frac{1}{q_1} & 0 \end{bmatrix} \ ,$$

$$\mu_1(z_0) = \frac{q_2}{q_1} \ , \qquad\qquad \mu_1(z_1) = 0 \ .$$

5.1 Computation of $\sum_{|\mathbf{w}|=\mathbf{l}} \langle \mathbf{G_2} | \mathbf{w} \rangle, \forall \mathbf{l} \geq \mathbf{2}$

$\langle G_2 | w \rangle = \lambda \mu(z_{i_l}) ... \mu(z_{i_1}) \gamma$ where :

- λ is the unit vector [1 , 0],
- γ is the input vector $[q_1, q_2]^T$,
- $w = z_{i_1} ... z_{i_l}$.

We show by induction that

$$\sum_{|w|=l} \langle G_2|w\rangle = \lambda[A,B]^T \ ,$$

where

- l=2n, $n \geq 1$, $A \ = \ \frac{F^n}{q_1^{n-1}} + \frac{1}{q_1^{n-1}} + \sum_{i=1}^{n-1} \binom{n}{i} \frac{F^i}{q_1^{n-1}}$,
  $B \ = \ \frac{\alpha^n q_2}{q_1^n} + \frac{q_2}{q_1^n} + \sum_{i=1}^{n-1} \binom{n}{i} \frac{F^i q_2}{q_1^n}$,
- l = 2n +1, $n \geq 1$, $A \ = \ \frac{F^n q_2}{q_1^n} + \frac{q_2}{q_1^n} + \sum_{i=1}^{n-1} \binom{n}{i} \frac{F^i q_2}{q_1^n}$,
  $B \ = \ \frac{F^{n+1}}{q_1^n} + \frac{1}{q_1^n} + \sum_{i=1}^{n} \binom{n+1}{i} \frac{F^i}{q_1^n}$.

Proof. **Basis :**

1. For $| w | = 3$, it is easy to verify that $A = \frac{Fq_2}{q_1} + \frac{q_2}{q_1}$, $B = \frac{F^2}{q_1} + \frac{1}{q_1} + \frac{2F}{q_1}$
2. For $| w | = 2$, it is easy to verify that $A = F + 1$, $B = \frac{Fq_2}{q_1} + \frac{q_2}{q_1}$.

Induction hypothesis: Holds for w, $| w | = p < l$.
Induction step : Let w, $| w |= l$, w = $z_i w'$ with $| w' |< l$, and $z_i = z_0$ or $z_i = z_1$.

We note that if $\sum_{|w'|=l-1} \langle G_2|w'\rangle \ = \ \lambda[A,B]^T$ then

$$\sum_{|w|=l} \langle G_2|w\rangle \ = \ \lambda[A',B']^T \ = \ \lambda(\mu(z_0)[A,B]^T + \mu(z_1)[A,B]^T) \ .$$

That gives :

$$\sum_{|w|=l} \langle G_2|w\rangle \ = \ \lambda([B, \frac{\alpha A}{q_1} + \frac{A}{q_1}]^T) \ ,$$

i.e.
$A' \ = \ B$, $B' \ = \ \frac{FA}{q_1} + \frac{A}{q_1}$.

- **Let l = 2n+2**

$$A' = \frac{F^{n+1}}{q_1^n} + \frac{1}{q_1^n} + \sum_{i=1}^{n} \binom{n+1}{i} \frac{F^i}{q_1^n} \ ,$$

$$B' = \frac{F^n q_2}{q_1^{n+1}} + \frac{q_2}{q_1^{n+1}} + \sum_{i=1}^{n-1} \binom{n}{i} \frac{F^i q_2}{q_1^{n+1}} + B'' \ ,$$

$$B'' = \frac{F^{n+1} q_2}{q_1^{n+1}} + \frac{F q_2}{q_1^{n+1}} + \sum_{i=1}^{n-1} \binom{n}{i} \frac{F^{i+1} q_2}{q_1^{n+1}} \ .$$

Lastly,

$$B' = \frac{F^{n+1} q_2}{q_1^{n+1}} + \frac{q_2}{q_1^{n+1}} + \frac{F^n q_2}{q_1^{n+1}} + B''' \ ,$$

$$B''' = \sum_{i=1}^{n-1} (\binom{n}{i} + \binom{n}{i-1}) \frac{F^i q_2}{q_1^{n+1}} + \binom{n}{n-1} \frac{F^n q_2}{q_1^{n+1}} .$$

That gives :

$$B' = \frac{F^{n+1} q_2}{q_1^{n+1}} + \frac{q_2}{q_1^{n+1}} + \sum_{i=1}^{n} \binom{n+1}{i} \frac{F^i q_2}{q_1^{n+1}} .$$

- **Let l=2n+1**
  $A' = \frac{F^n q_2}{q_1^n} + \frac{q_2}{q_1^n} + \sum_{i=1}^{n-1} \binom{n}{i} \frac{F^i q_2}{q_1^n}$.
  By a similar proof, we show that $B' = \frac{F^{n+1}}{q_1^n} + \frac{1}{q_1^n} + \sum_{i=1}^{n} \binom{n+1}{i} \frac{F^i}{q_1^n}$.

□

We have lastly :

$$\sum_{|w|=l} | \langle G_2|w\rangle | \leq (F+1)^{[l/2]} \frac{q_{1,2}}{q_1^{[l/2]}} ,$$

where $q_{1,2} = max(q_1, q_2)$.

5.2 Computation of $\sum_{|\mathbf{w}|=\mathbf{l}} \langle \mathbf{G_1}|\mathbf{w}\rangle, \forall \mathbf{l} \geq 2$

$$\sum_{|\mathbf{w}|=\mathbf{l}} \langle \mathbf{G_1}|\mathbf{w}\rangle = \frac{\mathbf{q_2^l q_1}}{\mathbf{q_1^l}} .$$

5.3 Majoration of $| \overline{\mathbf{y}}_2(\mathbf{t}) - \overline{\mathbf{y}}_1(\mathbf{t}) |$

$$| \overline{y}_2(t) - \overline{y}_1(t) | \leq S_1 + S_2 ,$$

$$S_1 = \sum_{l \geq 2} |F+1|^{[l/2]} \frac{q_{1,2}}{q_1^{[l/2]}} \frac{\Delta^l T^l}{l!} ,$$

$$S_2 = \sum_{l \geq 2} \frac{q_2^l}{q_1^l} \frac{q_1 \Delta^l T^l}{l!} .$$

Like previously, we set ν_1 and ν_2 the smallest integers such that :

$$P'_{\nu_1} = \frac{| F+1 |^{1/2} q_1^{-1/2} \Delta T}{\nu_1 + 1} < 1 ,$$

$$P''_{\nu_2} = \frac{\frac{q_2}{q_1} \Delta T}{\nu_2 + 1} < 1 .$$

We have :

$$S_1 \leq \sum_{l=2}^{\nu_1 - 1} |F+1|^{[l/2]} \frac{q_{1,2}}{q_1^{[l/2]}} \frac{\Delta^l T^l}{l!} + u'_{\nu_1} \frac{1}{1 - P'_{\nu_1}} ,$$

$$S_2 \leq \sum_{l=2}^{\nu_2 - 1} \frac{q_2^l}{q_1^l} \frac{q_1 \Delta^l T^l}{l!} + u''_{\nu_2} \frac{1}{1 - P''_{\nu_2}} .$$

6 Conclusion

Several black-box modelling of the nonlinear dynamical systems have be provided and validated [9, 10].
Like many others, our black-box method is limited by the inputs number, otherwise a combinatory explosion is possible.
The originality of our method consists in several points. First our method works in continuous time, instead of discrete time. Secondly, our method can adjust itself to the required precision by computing the generating series to upper order. Thirdly, our computation is completely generic, by using constraints on parameters. Lastly, our method allows to compute effectively the gap.
Instead of computing a local identification , we can compute a global identification for $t \in [0, V]$ for V bounded, by a fusion of local identifications on every interval $[t_i, t_i + s]$ of $[0, V]$. For instance, for a given step s and a given order k, we can compute the gap on $[0, V]$. An application restricted to numerical domain has been developed for the problem of insulin infusion[1].

7 Applications and Simulations

7.1 Applications

The Duffing equation is given by (7). We fix $\alpha = \beta = \gamma = 1$, $q_2 = 1$, $a(t) = sin(t) + cos(t)$,$\nu 1 = \nu_2 = 2$. This table gives, for different initial values y(0), the bound computed in section 5, on interval [0 T]. We see that, for a mistake about 6%, the interval [0 T] is divided by 4 when the initial value is 10 times greater.

Table 4. bound $|\overline{y}_2 - \overline{y_1}|$ for Duff Equ.

Init Val	bound $\mid \overline{y}_2 - \overline{y_1} \mid$	T
$y(0) = 1$	0.41	0.4
$y(0) = 10$	0.72	0.20
$y(0) = 100$	6.67	0.05

7.2 Simulations

We present simulations for several inputs (polynomial, oscillating), and for several orders ($k = 2$, $k = 5$). We fix

$$\begin{aligned} a_1(t) &= 200sin(50t) \\ a_2(t) &= t^7 - t^4 - 17 \\ a_3(t) &= 220sin(60t) - (220/7)t \end{aligned}$$

We can make the following remarks:

- For any input, the interval of coincidence for (Σ) and (B_k) outputs is increasing with k.
- The interval of coincidence depends on the given input: For an oscillating input, the coincidence interval for (Σ) and (B_k) outputs can concern only the first oscillation, what can constrain this interval to be very small.

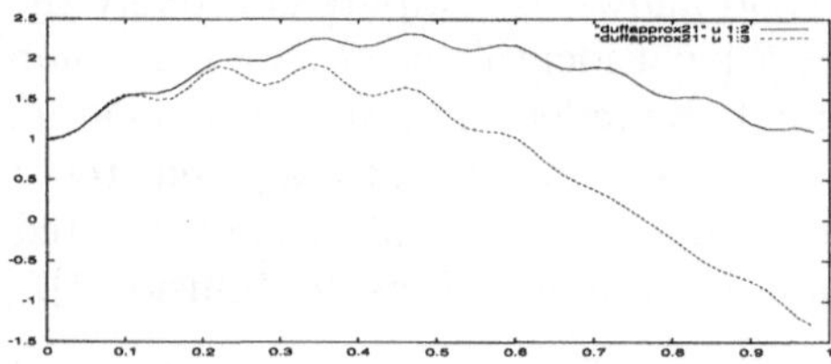

Fig. 1. Duffing equation (k=2,a(t) = a_1(t))

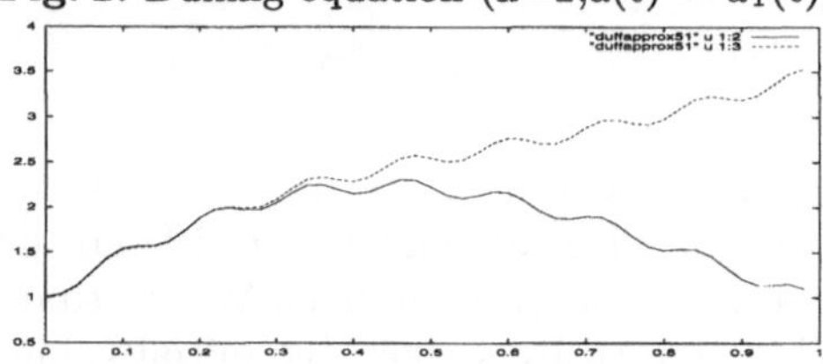

Fig. 2. Duffing equation(k=5,a(t) = a_1(t))

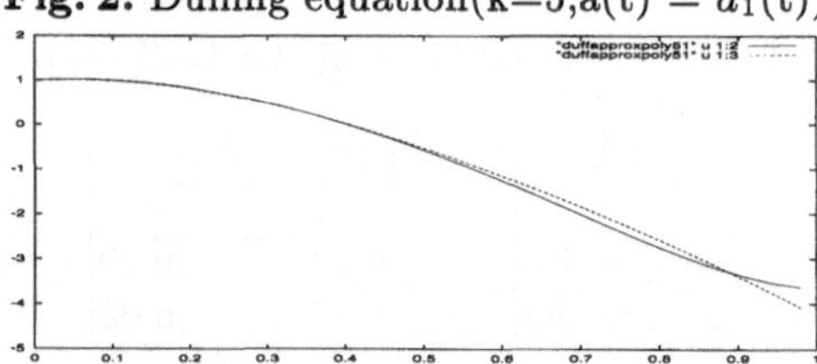

Fig. 3. Duffing equation(k=5,a(t) = a_2(t))

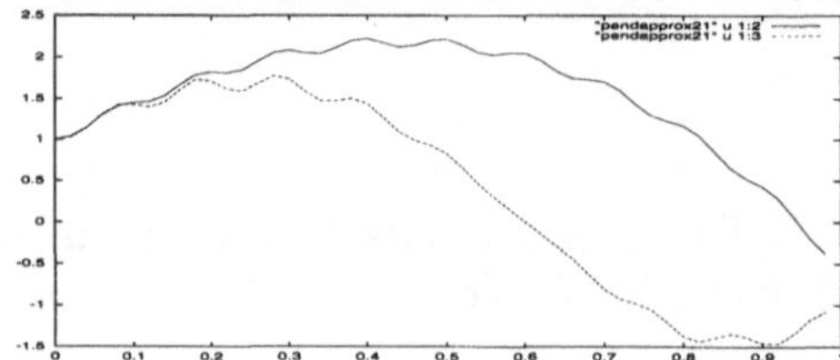

Fig. 4. Pend equation(k=2,a(t) = a_3(t))

References

1. Benmakrouha F., Foursov M., Hespel C., Hespel J.P., Jacob G., Monnier E., *Algebraic Identification: application to insulin infusion*, ISGIID'2000, Evian, September 14-15 2000.
2. Jung L., *Model Validation and Model Error Modeling*, Technical Report, August 1999.
3. Fliess M., *Fonctionnelles causales non linéaires et indéterminées non commutatives*, Bull. Soc. Math. France 109, pp.3-40, 1981.
4. Hespel C., *Approximation de séries formelles par des séries rationnelles*, RAIRO Inform.Théor., vol.18, n° 3, pp. 241-258, 1984.
5. Hespel C., Jacob G., *Approximation of nonlinear dynamic systems by rational series*, Theoret. Comput. Sciences, 79, pp.151-162, 1991.
6. Hespel C., *Iterated derivatives of a non linear dynamic system and Faà di Bruno formula*, Mathematics and Computers in Simulation, vol. 42, pp. 641-657, 1996.
7. Hespel C., Jacob G., *First steps towards exact algebraic identification*, Discrete Math., vol. 180, pp. 211-219, 1998.
8. Hespel C., Jacob G., *On Algebraic Identification of Causal Functionals*, Discrete Math., vol.225, pp. 173-191, 2000.
9. Sjoberg J., Zhang Q., Ljung L., Benveniste A., Delyon B., Glorennec P.Y., et al., *Non linear Black-box modelling in System Identification: a Unified Overview*, Automatica, vol 31, 12, pp. 1691-1724, 1995.
10. Judisky A., Hjalmarsson H., Benveniste A., Delyon B., Ljung L., Sjoberg J., Zhang Q., *Non linear Black-box modelling in System Identification: Mathematical fundations*, Automatica, vol 31, 12, pp. 1725-1750, 1995.
11. Ollivier F., *Le problème de lidentifiabilité structurelle globale : étude théorique, méthodes effectives et bornes de complexité*, Thèse de l'Ecole Polytechnique, 1990.
12. Walter E., *Identifiability of State Space Models*, Lecture Notes in Biomathematics, n° 46, Springer-Verlag, 1982.

Cooperation between a Dynamic Geometry Environment and a Computer Algebra System for Geometric Discovery

Francisco Botana[1] and José L. Valcarce[2]

[1]Departamento de Matemática Aplicada, Universidad de Vigo
Campus A Xunqueira, 36005 Pontevedra, Spain, fbotana@uvigo.es
[2]IES Pontepedriña, 15704 Santiago, Spain, jvalcarce@edu.xunta.es

Abstract. The use of computer algebra is usually considered beneficial for mechanised reasoning in mathematical domains. We describe a program, in the application domain of geometric discovery, which supports this claim. When interfacing a standard dynamic geometry environment and Mathematica, we enhance the educational uses of geometric problem solving environments through the symbolic capabilities of computer algebra software.

Keywords: problem–solving environments, symbolic–numeric interface, dynamic geometry, automatic discovery in geometry, Groebner basis method.

1 Introduction

Several approaches to automatic theorem proving in geometry using constructive methods were proposed in the 1980s. Systems based on Groebner bases [9], algebraic variety dimension [2], or Wu's method [22, 3] were much more successful than earlier proposals based on purely logical or axiomatic approaches. The reported work deals with an extension of these approaches to geometric discovery. It describes a successful cooperation between computer algebra systems and dynamic geometry environments, filling a gap recently highlighted [19].

Discovery is a Windows graphical user interface for automatic discovery in Euclidean geometry. It provides an intelligent, interactive environment for doing geometry. Discovery links REX [20], a dynamic geometry environment written in Visual Prolog [15], and Mathematica, a well–known computer algebra system. Lugares [21] is another program specialized in the graphing of geometric loci. Lugares, which will be described in a later paper, links REX and CoCoA [4] and shares some algorithms with Discovery.

The dynamic geometry component of Discovery offers the usual characteristics of such software [8, 13, 18]: dynamic transformations, dynamic measurements, free dragging, animation and graphic locus generation [5, p. 246]. This system offers a new approach. The user experiments by drawing a construction and realizes that a property is true for just some instances of the figure. The user can state this property as true and ask Discovery what the necessary conditions for it are. In other words, Discovery deals, just in a graphic environment, with arbitrary geometry statements aiming to find complementary hypotheses so that the statements will become true. This proposal was recently done by Recio and co–workers in [17, 16]. The work of Kapur and Mundy [11] is a solid precedent about automatic discovery. A brief comment of Chou about using his prover for theorem discovery [3, p. 275] should also be remarked. Our main concern when

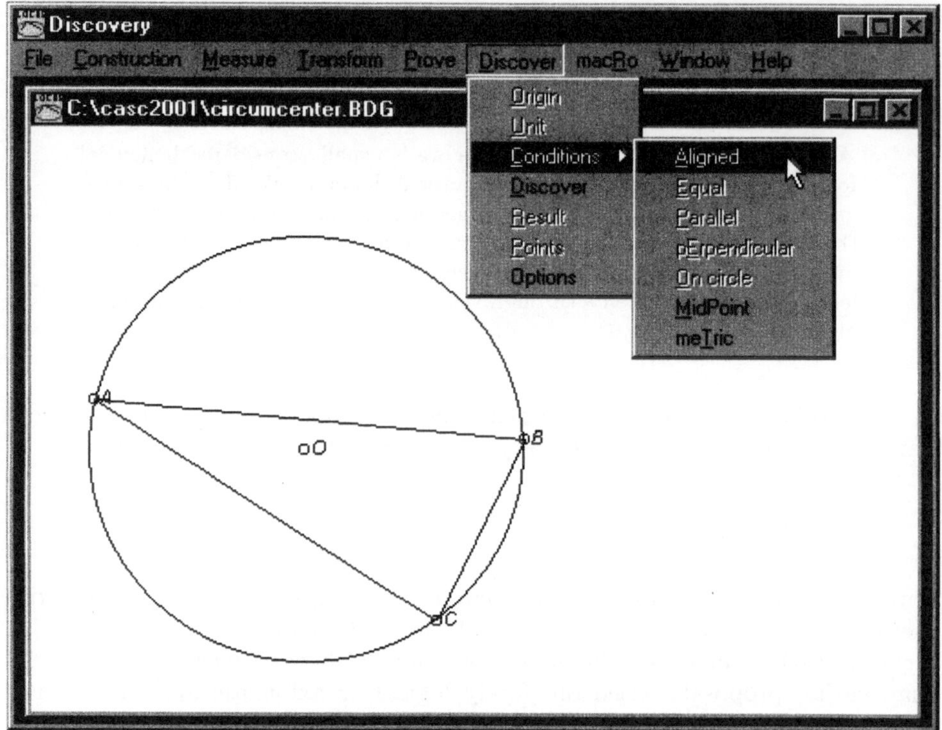

Fig. 1. Declaring the alignment of A, B and O

designing Discovery was simplicity, since its primary use is an educational one: as a standard dynamic geometry environment, easy diagram construction and manipulation are feasible. Furthermore, the Mathematica implementation of the Groebner basis algorithm allows us an approach to automated theorem proving method based on Kapur's work [9, 10]. We stress the emphasis on the discovery side: imposing some extra conditions to the construction, the program returns linguistic sentences meaning the necessary conditions so that the first ones will be true. In this way, the user is not aware of the fact that an automatic prover is working: the system acts as a black box.

For the sake of illustration, let us consider this simple construction: a triangle ABC and its circumcenter O (Fig. 1). After drawing the figure, the user drags a vertex, say A, and observes that the points A, B and O are sometimes aligned.

Using the Discover option, where the alignment of these points is declared, the system returns the sentence

CB is perpendicular to CA

as a necessary condition for the alignment.

As far as we know, there are just a few computer programs mixing a complete dynamic geometry environment with a prover. Cinderella [18] uses a randomized proving technique [12] to deduce properties about a given construction. This approach, which resembles the Cabri [13] property checker but rigorously based, makes it hard or even impossible to use the system for the discovery in the way described above. Geometry Expert (GEX) [6] is "a program for dynamic diagram drawing and automated geometry theorem proving and discovering" [5, p. 246]. It exhibits an outstanding implementation of many prove methods but, in our opinion, it lacks a single approach to automatic discovery.

2 An Illustration of the Interface

The objects that dynamic geometry environments use are easily described by a declarative language, such as Prolog. In REX, the geometric objects, their properties and relationships are declared in a Prolog facts database; the construction list. Each construction consists of a set of object definitions. The definition of an object has the following structure:

$$object(\#, object\_type(\#1, \#2, ...), hidden?, label, show\_label?, selected?)$$

where

- # is the number of the object in the construction list,
- object_type is a terminal element in the grammar of REX,
- $\#n$ is the number of one of the object's parents in the construction list or two coordinates if object_type is a free point,
- label is a text variable, and
- hidden?, show_label? and selected? are boolean variables.

A text–based input and output of the construction list is available in the program through a History window (Fig. 2).

When trying to discover any properties about a construction, the system reads the construction list and returns its properties and the involved points. In our case these properties are

- Midpoint(A_7, B, C)
- Midpoint(A_9, A, B)

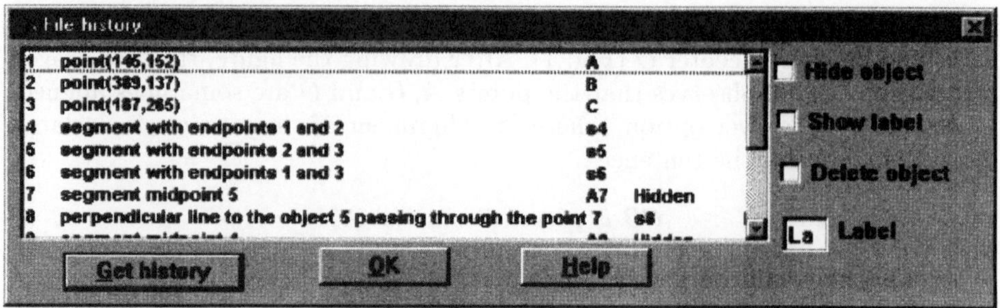

Fig. 2. The History window for the triangle and its circumcircle

- Perpendicular(BC, OA_7)
- Perpendicular(AB, OA_9)

(where the points A_7 and A_9 are hidden in the construction), and, following the convention of naming a free coordinate u and a bounded one x, the symbolic coordinates of points are:

$$A(u_1, u_2), B(u_3, u_4), C(u_5, u_6), A_7(x_1, x_2), A_9(x_3, x_4), O(x_5, x_6)$$

With the alignment of points A, B, O imposed by the user, Aligned(A, B, O), all the discovery elements are given to Mathematica (Fig. 3).

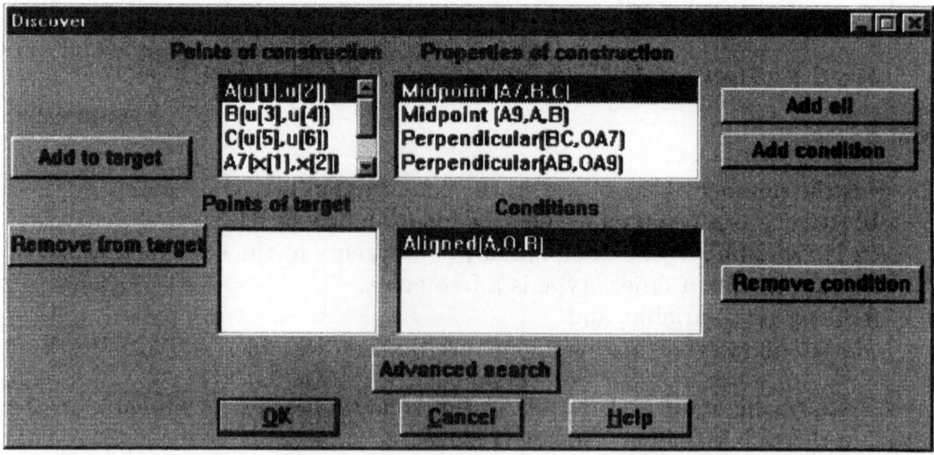

Fig. 3. The Discover dialog window

The hypotheses and the condition(s) are translated into polynomial equations over variables u's and x's

$$\begin{aligned} x_1 &= (u_3 + u_5)/2 \\ x_2 &= (u_4 + u_6)/2 \\ x_3 &= (u_1 + u_3)/2 \\ x_4 &= (u_2 + u_4)/2 \\ (u_6 - u_4)(x_6 - x_2) &= (u_3 - u_5)(x_5 - x_1) \\ (u_4 - u_2)(x_6 - x_4) &= (u_1 - u_3)(x_5 - x_3) \\ (x_6 - u_2)(u_3 - x_5) &= (u_4 - x_6)(x_5 - u_1) \end{aligned}$$

The bounded variables are eliminated from the ideal of (*hypotheses, condition(s)*). The Mathematica function used by Discovery is GroebnerBasis, setting the MonomialOrder option to DegreeLexicographic. (With this order, empirical findings report computation times faster than with the pure Lexicographic one [1]).

The vanishing of every element in the elimination ideal

$$(hypotheses, condition(s)) \cap K[free\ variables]$$

is a necessary condition for the theorem *hypotheses* $\Rightarrow$ *condition(s)* to be held. In this case the elimination ideal has only one generator, which factorizes to

$$(u_1^2+u_2^2-2u_1u_3+u_3^2-2u_2u_4+u_4^2)(u_1u_3+u_2u_4-u_1u_5-u_3u_5+u_5^2-u_2u_6-u_4u_6+u_6^2)$$

The first factor is a degenerated condition ($A = B$), and the second one is the perpendicularity of lines AC and BC. These factors are returned by the program as necessary conditions for the collinearity of points A, B, O (Fig. 4).

The user could have added some construction points on the Discover dialog window (Fig. 3). In that case, the system generates a database of linguistic sentences–equations pairs about geometric properties involving these points. For example, given the points A, B, C and the *aligned* filter (an option in the Advanced search window, not shown in Fig. 3), the pair ("The points A, B, C are aligned", $(u_4 - u_6) * (u_2 - u_5) - (u_2 - u_6) * (u_3 - u_5) = 0$) would be added to the database. The size of the database is controlled through these filters. After having tried to match every factor of the necessary conditions with each equation in the database, the program returns the linguistic sentences, only in the case that some match has been obtained (Fig. 5). The degenerated condition

$$u_1^2 + u_2^2 - 2u_1u_3 + u_3^2 - 2u_2u_4 + u_4^2 = 0 (= (u_1 - u_3)^2 + (u_2 - u_4)^2)$$

does not match with the corresponding database equations

$$u_1 = u_3, u_2 = u_4$$

because the used method considers complex zeros and the formula

$$(u_1 - u_3)^2 + (u_2 - u_4)^2 = 0 \Rightarrow u_1 = u_3 \text{ AND } u_2 = u_4$$

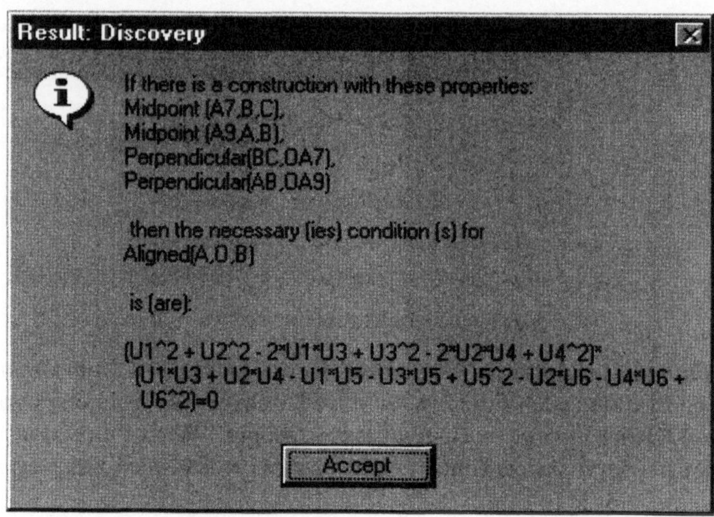

Fig. 4. The equation describing the necessary conditions

is a theorem if u_1, u_2, u_3, u_4 are assumed to range over the real numbers, but it is not so if the range is an extension of the real field, as it happens here.

Once we have got the non–degenerated conditions necessary for the theorem to be held, it is a common place to check its general truth. Discovery uses Wu's method [22] to confirm that in any triangle, rectangle in C, the vertices A, B and the circumcenter are aligned. The reason why we use Wu's approach instead of following the Groebner basis method is their computational cost: doubly exponential, in the worst case [23], for Groebner while somewhat better for Wu.

Although the method of discovery here described is not complete [17, p. 76], it works for a considerable amount of Euclidean constructions, justifying the usefulness of Discovery as a didactical tool and, even, the automating of tedious calculations in scientific work.

3 Dynamic Geometry Systems and Geometric Discovery

In this section we will have a deeper look at the way that current dynamic geometry environments deal with discovery, showing the inherent limitations to numerical approaches. As said above, only three dynamic geometry programs, Cabri, Cinderella and GEX have a way to detect geometric properties different from the visual one.

Cabri incorporates a property checker that can decide about collinearity, parallelism, perpendicularity, equidistance and membership. Although its authors have not published the method it uses, it seems to be based on the testing of the actual positions of the elements on the screen. So, if you carefully draw two horizontal independent lines, the checker will assert they are parallel. This

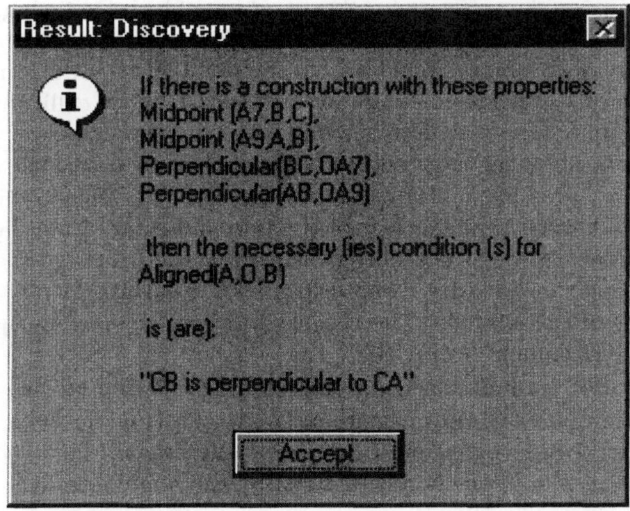

Fig. 5. A sentence describing one necessary condition

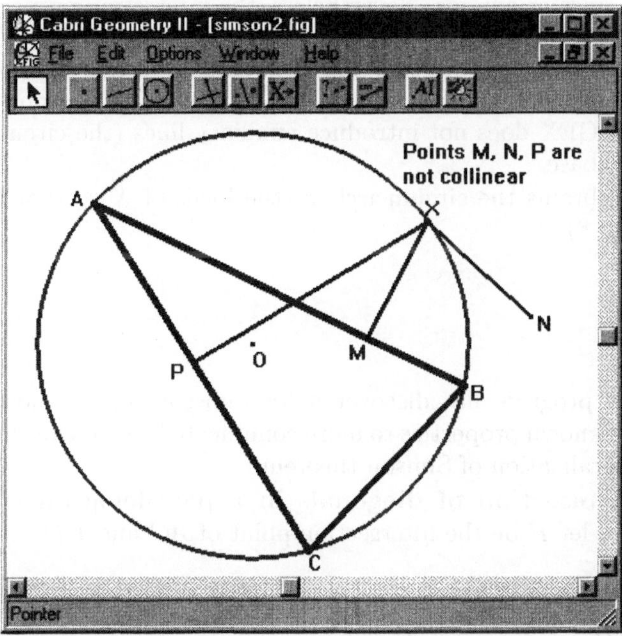

Fig. 6. Cabri does not help on searching the locus

numerical–based behavior, which a priori cannot be arguable, becomes an undesirable feature when using the software for the searching of loci. On trying to find the Simson line of a triangle, let us inscribe a triangle in a circle, select a point on it and finally draw perpendiculars from the point to each side of the triangle. The three feet of the perpendiculars are, as stated by Simson, collinear. This kind of visual proof, a paradigmatic example of a dynamic geometry use, cannot be made without knowing the result in advance. If you change the construction slightly, (that is, drawing a triangle and a point on its plane and finding the locus of points X such that the feet of the perpendiculars from X to the sides are collinear), the user will not obtain help from the property checker. It will not state the point collinearity when dragging X to arbitrary positions, except for the vertices of the triangle. The checker will even reject the conjecture that the locus is the circumcircle (Fig 6).

Cinderella uses a randomized proving technique based on Schwartz–Zippel theorem. In the case of Simson theorem, Cinderella (Fig. 7) behaves as Cabri: if the circumcircle and a point on it are previously drawn, the automatic prover declares the third foot is on the line defined by the remaining two. However, we get no statement when making the alternative construction.

Although the third program, GEX, is not specially well–suited for discovery, it has a deductive database method for proof, which can also be used for discovery. It uses forward chaining (i.e., reasoning from the hypotheses to the conclusion) to find all the properties of a geometric configuration that can be deduced from a set of rules or axioms. Nevertheless, the interesting properties are usually hidden among a lot of irrelevant statements [14]. Furthermore, GEX can not cope with discovery as stated in this paper (finding necessary conditions for a statement to become true). In our example, the locus of X in Simson theorem is not found because GEX does not introduce auxiliary lines (the circumcircle) in the geometric database.

Discovery easily obtains the circumcircle as the locus of X so that the feet will be collinear (Fig. 8)

4 Examples

The results that the program has discovered (or rediscovered, in most cases) range from easy well–known properties to more complicated Euclidean theorems, such as a recent generalization of Simson theorem.

Example 1. **The bisection of diagonals in a parallelogram**. Given a quadrilateral $ABCD$, let P be the intersection point of AC and BD.
Imposed conditions:
  Midpoint(P, A, C)
  Midpoint(P, B, D)
Necessary conditions discovered:
  AB is parallel to CD, and
  AD is parallel to BC

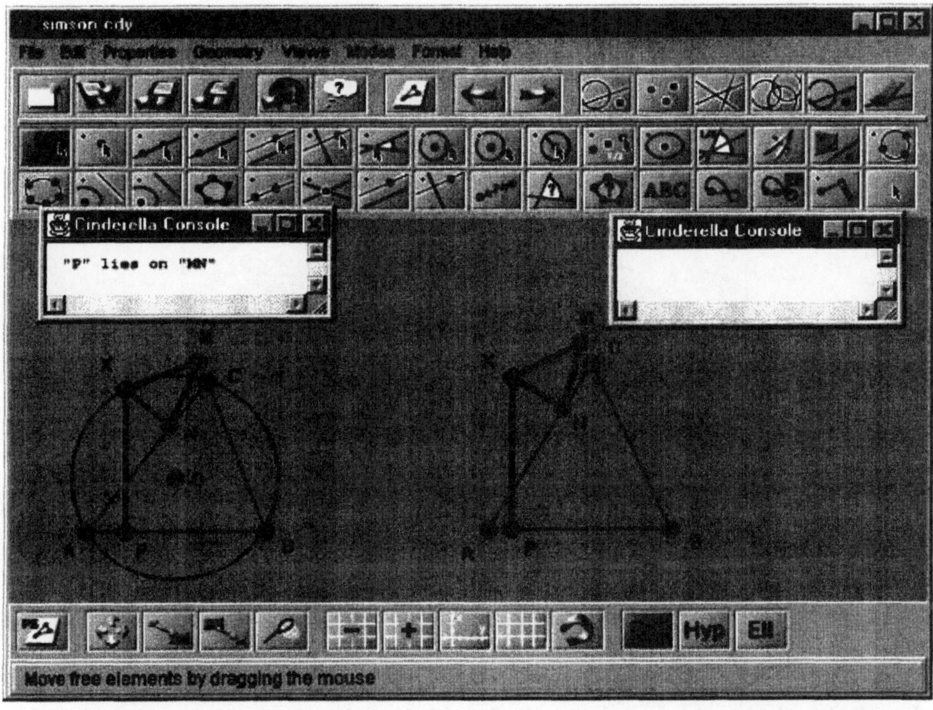

Fig. 7. Cinderella checks Simson theorem, but it does not discover it

Example 2. **The alignment of the symmetrical points of a given point with respect to the sides of a triangle.** Given a triangle ABC and an arbitrary point X, let X', X'' and X''' be the symmetrical images of X with respect to lines AB, AC and BC.
Imposed conditions:
Aligned(X', X'', X''')
Necessary conditions discovered:
The points A, B, C are aligned, or
The locus of X is a circle passing through A, B, C

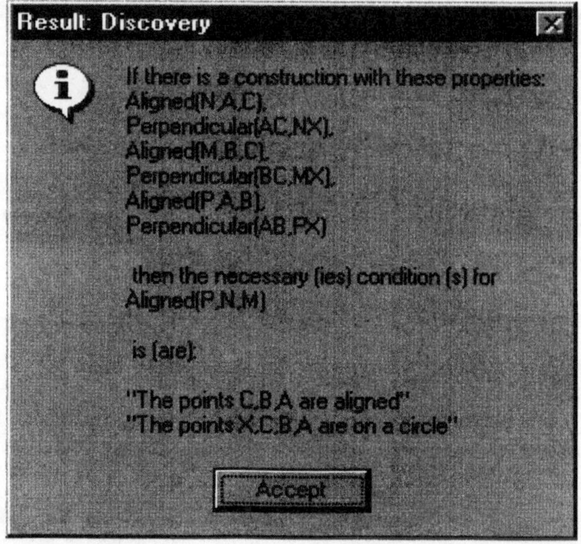

Fig. 8. The equation describing the necessary conditions

Example 3. **Steiner's generalization of Simson theorem.** Given a triangle ABC and an arbitrary point X, let M, N, P be the orthogonal projections of X on the sides of the triangle.
Imposed conditions:
Oriented_Area(M, N, P)=k.
Necessary conditions discovered:
The locus of X is a circle passing through A, B, C (if $k = 0$)
The locus of X is a circle (if $k \neq 0$)

Note that if $k \neq 0$, no reference to the circumcircle is made. The program has not enough intelligence to infer that the center of the circle is the circumcenter.

Example 4. **Guzmán generalization of Simson–Steiner theorems** [7]. Given a triangle ABC,an arbitrary point X, and three projection directions, not all three equal, nor parallel to the sides of the triangle. Let M, N, P be the projections along these directions on the sides.

Imposed conditions:

Oriented_Area(M, N, P)=k.

Necessary conditions discovered:

The locus of X is a conic

If no point is added in the Discover window, the system returns a polynomial condition with 154 terms, which is the equation of a conic in the coordinates of X.

5 Conclusions

According to current trends in mathematical education, getting "mathematical power" is a basic goal for students. Computer algebra and dynamic geometry systems have been useful tools, in spite of their independent evolvement. This paper describes a computer program that fills the gap between both fields in the domain of elementary geometric discovery. It mixes techniques from both fields (computational conmutative algebra and algebraic geometry, and dynamic geometry design) implying a step forward in the interactive exploration of geometric properties. We hope that this step will be just the first of Professor Recio's dream [16, p. 70]: "Ya existen hoy calculadoras de bolsillo que incluyen programas geométricos y simbólicos... ¡sólo falta que, mañana, se relacionen entre sí!" (Today there exist pocket calculators that include geometric and symbolic programs... keeping each other in touch tomorrow is only missing!).

References

1. Buchberger, B.: Groebner Bases: an Algorithmic Method in Polynomial Ideal Theory. In N.K. Bose, *Multidimensional systems theory*, Reidel, Dordrecht, 184-232 (1985)
2. Carrá, G., Gallo, G.: A procedure to prove geometrical statements. In L. Huguet, A. Poli, *Proc. 5th Int. Conf. on Applied Algebra, Algebraic Algorithms and Error-Correcting Codes*, LNCS, 356, 141–150. Springer, Berlin (1989)
3. Chou, S. C.: Proving Elementary Geometry Theorems Using Wu's Algorithm. *Automated Theorem Proving: After 25 years*, Contemporary Mathematics, AMS, 29, 243–286 (1984)
4. Capani, A., Niesi, G.,L. Robbiano: CoCoA, a system for doing Computations in Commutative Algebra. Available via anonymous ftp from: cocoa.dima.unige.it
5. Gao, X. S.: Automated Geometry Diagram Construction and Engineering Geometry. In *Proc. ADG'98*, Beijing, 232–257, Springer (1998)
6. Gao, X. S., Zhang, J. Z., Chou, S. C.: *Geometry Expert.* Nine Chapters Publ. Taiwan (1998)
7. Guzmán, M.: An extension of the Wallace–Simson theorem: projecting in arbitrary directions. *American Mathematical Monthly*, 106(6), 574–580 (1999)
8. Jackiw, N.: *The Geometer's Sketchpad.* Key Curriculum Press, Berkeley (1997)
9. Kapur, D.: Geometry theorem proving using Hilbert's Nullstellensatz. In *Proc. SYMSAC'86*, Waterloo, 202–208, ACM Press (1986)
10. Kapur, D.: A refutational approach to geometry theorem proving. *Artificial Intelligence*, 37, 61–94 (1988)

11. Kapur, D., Mundy, J. L.: Wu's method and its application to perspective viewing. In D. Kapur, J. L. Mundy, *Geometric Reasoning.* The MIT Press, Cambridge (1989)
12. Kortenkamp, U.: Foundations of dynamic geometry. Ph. D. Thesis, ETH, Zurich (1999)
13. Laborde, J. M., Bellemain, F.: *Cabri Geometry II.* Texas Instruments, Dallas (1998)
14. Nevins, A. J.: Plane Geometry Theorem Proving Using Forward Chaining. *Artificial Intelligence* 6, 1-23 (1975)
15. Prolog Development Center.
16. Recio, T.: *Cálculo simbólico y geométrico.* Síntesis, Madrid (1998) (in Spanish)
17. Recio, T., Vélez, M. P.: Automatic discovery of theorems in elementary geometry. *Journal of Automated Reasoning,* 23, 63–82 (1999)
18. Richter–Gebert, J., Kortenkamp, U.: *The Interactive Geometry Software Cinderella.* Springer, Berlin (1999)
19. Roanes, E.: Boosting the Geometrical Possibilities of Dynamic Geometry Systems and Computer Algebra Systems through Cooperation. Plenary talk at *5th Int. Conf. on Technology for Mathematics Teaching.* Klagenfurt, Austria (2001) (to appear)
20. Valcarce, J. L., Botana, F.: Developing a Dynamic Geometry Environment with Prolog. Technical Report, University of Vigo, Pontevedra (1999)
21. Valcarce, J. L., Botana, F.: Lugares. Manual de Referencia. Technical Report, University of Vigo, Pontevedra (2001)
22. Wu, W. T.: *Mechanical Theorem Proving in Geometries.* Springer, Viena (1994)
23. Yap, C. K.: A new lower bound for conmutative Thue systems, with applications. *Journal of Symbolic Computation,* 12, 1–28 (1991)

On the Stability of Steady Motions of Solar-Sail Satellite

Larissa Bourlakova

Institute of Systems Dynamics and Control Theory SB RAS,
134, Lermontov str., Irkutsk, 664033, Russia
irteg@icc.ru

Abstract. A satellite with solar sail moving along a circular heliocentric orbit is considered. A complete set of steady motions for such a system has been obtained, and Lyapunov's method has been employed to investigate their stability. The sufficient conditions are compared with the necessary ones. To the end of solving the problem, the capabilities of the software "Stability" for symbolic computations have been used [1].

1 Introduction

Many publications have been devoted to the problem of motion of an artificial satellite with solar sail (see [2],[3] as well as the bibliography in these monographs). In [4] some steady motions of a satellite with solar sail are considered in a restricted statement when the satellite's mass center moves evenly along a circular heliocentric orbit. Some results of investigation of such motions have been considered, and the respective periodic solutions close to the stationary ones constructed. This work unfortunately has some shortcomings, which have motivated us to consider this problem again.

In solving the problem, the potentialities of the software "Stability" [1] have been employed. This software is intended for symbolic and numeric-symbolic investigation of interconnected mechanical systems. It has been designed on the basis of the computer algebra software system "Mathematica".

The notation used in [4] as well as the choice of generalized coordinates have been retained and used in the present paper. Similarly to [4], we assume that a dynamically symmetric satellite is equipped with a weightless sail the normal to which is constantly oriented to the sun. Let O be the satellite's mass center, and O_1 the sail geometric center (pressure center).

For the purpose of description of satellite's orientation let us introduce two right Cartesian coordinate systems: the orbital one $O\xi\eta\zeta$, whose axis $O\zeta$ is oriented from the sun, the axis $O\eta$ is oriented normally to the orbital plane (in the direction of the orbital angular velocity $\boldsymbol{\omega_0}$); and the system $OXYZ$ of principal central axes of inertia for the body (A, B, C are the inertia moments), furthermore, the axis OX is the symmetry axis ($B = C$), which is directed along the direction $\boldsymbol{OO_1}$. Let us choose the following Euler angles in the capacity of generalized coordinates: θ is the precession angle measured from the axis $O\zeta$ in the orbital plane, φ is the nutation angle measured from the orbital plane, ψ is the angle of eigen rotation. The directional cosines are given in the table.

Table

	X	Y	Z
ξ	$\sin\theta\cos\varphi$	$\sin\varphi\sin\theta\cos\psi - \cos\theta\sin\psi$	$-\sin\varphi\sin\theta\sin\psi - \cos\psi\cos\theta$
η	$-\sin\varphi$	$\cos\varphi\cos\psi$	$-\cos\varphi\sin\psi$
ζ	$\cos\varphi\cos\theta$	$\sin\varphi\cos\theta\cos\psi + \sin\theta\sin\psi$	$\sin\theta\cos\psi - \sin\varphi\cos\theta\sin\psi$

2 Equations of steady motions

The Lagrangian $L = T_2 + T_1 + T_0 + U$ for the mechanical system under consideration represented in terms of selected generalized coordinates writes [4]:

$$\begin{aligned} L = {} & A(\dot\psi^2 - 2\dot\psi(\dot\theta + \omega_0)\sin\varphi + (\dot\theta+\omega_0)^2\sin^2\varphi)/2 \\ & + B(\dot\varphi^2 + (\dot\theta+\omega_0)^2\cos^2\varphi)/2 + 3/2(B-A)\omega_0^2\cos^2\theta\cos^2\varphi \\ & + Psl\cos\theta\cos\varphi, \end{aligned} \tag{1}$$

where T_i is some part of kinetic energy, which is the ith form of generalized velocities, U is the force function of both Newton gravitation and solar pressure, P is the modulus of solar pressure, s is the sail surface, l is the length of $\boldsymbol{OO_1}$.

As obvious from (1), the time and the coordinate ψ are cyclic, hence the problem has the first integrals:

$$H = T_2 - T_0 - U = h \quad \text{the Jacoby-type energy integral;}$$

$$V = \frac{\partial L}{\partial \dot\psi} = A(\dot\psi - (\dot\theta + \omega_0)\sin\varphi) = Ap_0 \quad \text{the cyclic integral.}$$

Let us use the Routh-Lyapunov's theorem [5] for the purpose of the investigation of steady motions. Construct a linear bundle of integrals

$$\begin{aligned} K = H - \lambda V = {} & (A/2)(\dot\psi^2 - 2\dot\psi\dot\theta\sin\varphi \\ & + (\dot\theta^2 - \omega_0^2)\sin^2\varphi - 2\lambda(\dot\psi - (\dot\theta+\omega_0)\sin\varphi)) + (B/2)(\dot\varphi^2 + (\dot\theta^2 - \omega_0^2)\cos^2\varphi) \\ & - (3/2)(B-A)\omega_0^2\cos^2\theta\cos^2\varphi - Psl\cos\theta\cos\varphi. \end{aligned} \tag{2}$$

It is well known that in the conservative systems with cyclic integrals, the collection of necessary conditions of stationarity of the integral (2) with respect to velocities

$$\frac{\partial K}{\partial \dot\varphi} = B\dot\varphi = 0, \quad \frac{\partial K}{\partial \dot\psi} = A(\dot\psi - \dot\theta\sin\varphi - \lambda) = 0,$$

$$\frac{\partial K}{\partial \dot\theta} = A(\dot\theta\sin^2\varphi + \lambda\sin\varphi - \dot\psi\sin\varphi) + B\dot\theta\cos^2\varphi = 0$$

has the unique solution: $\dot{\theta}=0$, $\dot{\varphi}=0$ $\dot{\psi}=\lambda$. Note that from now on $\cos\varphi\neq 0$, otherwise T_2 degenerates.

Having substituted the solutions obtained for the velocities into the conditions of stationarity of integral (2) with respect to the coordinates and having eliminated (with the aid of the integral V) the cyclic velocity, we obtain the equations for constant values of positional coordinates in steady motions:

$$\sin\theta\cos\varphi\left[3\omega_0^2(B-A)\cos\theta\cos\varphi+Psl\right] = 0 \quad (\cos\varphi\neq 0\,); \tag{3}$$

$$Ap_0\omega_0\cos\varphi+B\omega_0^2\sin\varphi\cos\varphi+\left[3\omega_0^2(B-A)\cos\theta\cos\varphi+Psl\right]\cos\theta\sin\varphi=0. \tag{4}$$

Let us obtain the solution for (3)–(4).

2.1. Let in (3) $\sin\theta=0$. Then there exist the solutions:

$$\underline{\text{Solution:}}\qquad \theta_0=0\ (\pi)\ ,\ \varphi_0=0\ (\pi)\quad \text{for}\ \ p_0=0; \tag{5}$$

$$\underline{\text{Solution:}}\qquad \theta_0=\pi(0)\ ,\ \varphi_0=0(\pi)\quad \text{for}\ \ p_0=0. \tag{6}$$

The variants in brackets define equal spatial positions for the axis of the dynamic symmetry. In solution (5), the symmetry axis OX is directed from the gravitation center, whereas in (6) it is directed to the gravitation center (see the Table).

Solution: for $p_0=0$

$$\theta_0=0\ (\pi)\ ,\ \cos\varphi_0=\frac{Psl}{\omega_0^2(3A-4B)}\qquad\left(\frac{Psl}{\omega_0^2(4B-3A)}\right). \tag{7}$$

The condition for the existence of above solutions writes: $(Psl)^2<\omega_0^4(4B-3A)^2$. The both of the solutions suggest the same value of $\cos(\widehat{\zeta,x})$.

Solution: $\theta_0=0\ (\pi), \sin\varphi=\sin\varphi_0$, where $\sin\varphi_0$ satisfies the equation

$$\left((4B-3A)\sin\varphi_0+Ap_0\omega\right)^2\left(1-\sin^2\varphi_0\right)-P^2s^2l^2\sin^2\varphi_0=0. \tag{8}$$

This equation has two real solutions when $a_1=(4B-3A)^2\omega_0^4-A^2p_0^2\omega^2-P^2s^2l^2<0$, and four real solutions when $a_1\geq 0$.
Having solved this equation with the aid of the system of computer algebra, we have obtained:

$$\sin\varphi_0=-\frac{b}{2a}\pm\frac{m_3}{2\sqrt{3}a}\pm\frac{1}{2\sqrt{3}a}\sqrt{2b+4a^2-4c^2-m_2-\frac{6\sqrt{3}(a^2+c^2)b}{m_3}}\,,$$

here

$$m_3=\sqrt{2a^2+b^2-2c^2+m_2},\qquad m_2=-\frac{\left(\sqrt[3]{4}(a^2-b^2-c^2)^2+\sqrt[3]{m_1{}^2}\right)}{\sqrt[3]{2m_1}}\,,$$

$$m_1 = 2\left(a^6 - 3a^4(b^2+c^2) - (b^2+c^2)^3 + 3a^2(b^4 - 16b^2c^2 + c^4)\right) + \sqrt{m_0},$$
$$m_0 = -4(-a^2+b^2+c^2)^6$$
$$+ 4\left(a^6 - 3a^4(b^2+c^2) - (b^2+c^2)^3 + 3a^2(b^4 - 16b^2c^2 + c^4)\right)^2,$$
$$a = (4B - 3A)\omega_0^2\ ,\quad b = Ap_0\omega_0\ ,$$
$$c = \pm Psl \ \text{(the sign shown below is for } \cos\theta_0 = -1).$$

Remark. In case of linear equations there may appear subsidiary solutions. For the purpose of sorting the solutions obtained it is advisable to substitute the solutions into the initial equations, but in virtue of complexity (cumbersome character) of expressions incurred, the simplifying operations are not always efficient. So, there is the need to verify "reliability" of the solutions by some other aids.
The values of the angles φ_0 must be chosen so that $\sin\varphi_0 \cos\varphi_0 < 0$. If $c = 0$ (there is an orbited satellite without any sail), then this solution transforms into some "conic" precession [2]:

$$\sin\theta_0 = 0, \quad \sin\varphi_0 = \frac{Ap_0}{(3A - 4B)\omega_0}.$$

2.2. Let in (3) $\left[3\omega_0^2(B-A)\cos\theta\cos\varphi + Psl\right] = 0$. Then, when solving it in one system with (4), we have the following solutions.

<u>Solution:</u> $$\sin\varphi_0 = -\frac{Ap_0}{B\omega_0},\ \cos\theta_0 = \pm\frac{PslB}{3\omega_0(A-B)\sqrt{B^2\omega_0^2 - A^2p_0^2}} \quad (9)$$

The condition of existence of the solutions (9) is $(Ap_0)^2 < (B\omega_0)^2$.

<u>Solution:</u> $$\varphi_0 = 0, \quad \cos\theta_0 = \frac{Psl}{3\omega_0^2(A-B)} \quad \text{for } p_0 = 0. \quad (10)$$

The condition of existence of the solution (10) is $(Psl)^2 < 9\omega_0^4(A-B)^2$.
In [4] the solutions (5),(6),(7),(9) were presented. Our solution (9) refines the solution (9) found in [4].

Introduce the notations for undisturbed motion:

$$\theta = \theta_0 + x_1,\ \varphi = \varphi_0 + x_2,\ \dot\psi = \lambda + \dot x_3.$$

Then in the neighbourhood of the steady-state motions obtained, the equations of the 1st approximation write:

$$\begin{aligned} Bcos^2\varphi_0\ddot x_1 &= (Ap_0\cos\varphi_0 + 2B\omega_0\sin\varphi_0\cos\varphi_0)\dot x_2 \\ &+ (3(B-A)\omega_0^2\sin 2\theta_0\sin\varphi_0\cos\varphi_0 + Psl\sin\theta_0\sin\varphi_0)x_2 \\ &- 3(B-A)\omega_0^2\cos^2\varphi_0\cos 2\theta_0 + Psl\cos\varphi_0\cos\theta_0)x_1; \\ B\ddot x_2 &= -(Ap_0\cos\varphi_0 + 2B\omega_0\sin\varphi_0\cos\varphi_0)\dot x_1 \\ &- (B\omega_0^2\cos 2\varphi_0 - Ap_0\omega_0\sin\varphi_0 + 3(B-A)\omega_0^2\cos 2\varphi_0\cos^2\theta_0 \\ &+ Psl\cos\varphi_0\cos\theta_0)x_2 \\ &+ (3(B-A)\omega_0^2\sin 2\theta_0\sin\varphi_0\cos\varphi_0 + Psl\sin\theta_0\sin\varphi_0)x_1. \end{aligned} \quad (11)$$

Here $\dot{x}_i, \ddot{x}_i$ $(i = 1, 2)$ are deviations of the positional velocities as well as accelerations in disturbed motion from their values in steady motion. Equations (11) describe the motion of a linear system, in which both potential and gyroscopic forces act:

$$M\ddot{x} + 2G\dot{x} + Kx = 0, \quad M = M^T\ , \ G = -G^T\ , \ K = K^T \tag{12}$$

The characteristic equation for the system (11) has the form:

$$\begin{aligned}
&n^4 B^2 \cos^2 \varphi_0 + n^2 \cos \varphi_0 (BPsl \cos \theta_0 + BPsl \cos^2 \varphi_0 \cos \theta_0 \\
&+ B\omega_0^2 \cos^3 \varphi_0 (4B - 3A - (A - B) \cos 2\theta_0) \\
&+ \cos \varphi_0 (A^2 p_0^2 + 3AB\omega_0 p_0 \sin_0 + 3B^2 \omega_0^2 sin^2 \varphi_0 + 3(A - B)B\omega_0^2 \sin^2 \theta_0)) \\
&+ 1/4(2Psl\omega_0^2 \cos^3 \varphi_0 \cos \theta_0 (5B - 3A - 9(A - B) \cos 2\theta_0) \\
&+ 6(A - B)\omega_0^4 \cos^4 \varphi_0 \cos 2\theta_0 (3A - 5B + 3(A - B) \cos 2\theta_0) \\
&+ Psl\omega_0 \cos \varphi_0 \sin \varphi_0 \cos \theta_0 (-4Ap_0 \\
&- 2\omega_0(-15A + 17B + 9(A - B) \cos 2\theta_0) \sin \varphi_0) \\
&- 4(Psl)^2 \sin^2 \varphi_0 \sin^2 \theta_0 + 4 \cos^2 \varphi_0 (-9(A - B)^2 \omega_0^4 \cos^4 \theta_0 \sin^2 \varphi_0 \\
&+ \cos^2 \theta_0 ((Psl)^2 + 3A(A - B)\omega_0^3 p_0 \sin \varphi_0 + 3/2(A - B)\omega_0^4 (-9A + 11B \\
&+ 9(A - B) \cos 2\theta_0 \sin^2 \varphi_0 - 3(A - B)\omega_0^3 \sin \varphi_0 (Ap_0 + B\omega_0 \sin \varphi_0 \sin^2 \theta_0)) = 0.
\end{aligned}$$

3 Stability of Steady Motions

In accordance with Routh-Lyapunov's theorem [5], sufficient stability conditions of selected steady-state motions with respect to the variables $\theta, \varphi, \dot{\theta}, \ \dot{\varphi}\ , \ \dot{\psi}$ (the cyclic variable ψ is unstable) may be obtained as the conditions of signdefiniteness for the 2nd variation $\delta^2 K$ of the integral (2) in the neighbourhood of the steady motion for the constant value of the integral V:

$$\delta V = \dot{x}_3 - \sin \varphi_0 \dot{x}_1 - \omega_0 \cos \varphi_0 x_2 + O(x_2, \dot{x}_1) = 0. \tag{13}$$

Having eliminated $\dot{x}_3$ from $\delta^2 K$ with the aid of (13), we have:

$$\begin{aligned}
2\delta^2 \tilde{K} = &B \cos^2 \varphi_0 \dot{x}_1^2 + B\dot{x}_2^2 + \cos \varphi_0 \left(3(B - A)\omega_0^2 \cos \varphi_0 \cos 2\theta_0 + Psl \cos \theta_0\right) x_1^2 \\
&+ \left(B\omega_0^2 \cos 2\varphi_0 - Ap_0\omega \sin \varphi_0 + 3(B - A)\omega_0^2 \cos^2 \theta_0 \cos 2\varphi_0\right. \\
&+ \left. Psl \cos \theta_0 \cos \varphi_0\right) x_2^2 \\
&- 2\left(3(B - A)\omega_0^2 \cos \varphi_0 \sin 2\theta_0 \sin \varphi_0 + Psl \sin \theta_0 \sin \varphi_0\right) x_1 x_2 + \dots .
\end{aligned} \tag{14}$$

The function (14) is signdefinite when the form $W(x_1, x_2)$ in $\delta^2 \tilde{K}$ is positive definite. The conditions obtained will be the sufficient conditions of secular stability. Let us consider this on an example of steady motions.

Solution (5). As far as this solution is concerned, the form

$$2\delta^2 \tilde{K} = B\dot{x}_1^2 + B\dot{x}_2^2 + (3(B - A)\omega_0^2 + Psl)x_1^2 + \ ((4B - 3A)\omega_0^2 \ + Psl)x_2^2$$

is positive definite, and the motion is stable if $Psl > 3(A-B)\omega_0^2$. This requirement is also the necessary (without any boundary) condition of stability, what follows from equations of the first approximation.

For the solution (6) the Lyapunov function

$$2\delta^2\tilde{K} = B\dot{x}_1^2 + B\dot{x}_2^2 + (3(B-A)\omega_0^2 - Psl)x_1^2 + ((4B-3A)\omega_0^2 - Psl)x_2^2$$

is positive definite, and hence the motion is stable when $Psl < 3(B-A)\omega_0^2$. For both of these solutions the conditions do not coincide with those in [4], but transform into the conditions of [3] if there is no solar pressure. Note also that the solution (6) can be obtained from (5) when negative sign is assumed for l.

In the case of Solution (7) the form (14) is signvariable:

$$2\delta^2\tilde{K} = B\cos^2\varphi_0\dot{x}_1^2 + B\dot{x}_2^2 - B\omega_0^2\cos^2\varphi_0 x_1^2 - (4B-3A)\omega_0^2\sin^2\varphi_0 x_2^2.$$

There is no secular stability. So, let us consider the equations of the first approximation (11):

$$\begin{aligned} B\cos^2\varphi_0\ddot{x}_1 &= 2B\omega_0\sin\varphi_0\cos\varphi_0\dot{x}_2 + B\omega_0^2\cos^2\varphi_0; \\ B\ddot{x}_2 &= -2B\omega_0\sin\varphi_0\cos\varphi_0\dot{x}_1 + (4B-3A)\omega_0^2\sin^2\varphi_0. \end{aligned}$$

The characteristic equation $Bn^4 + \omega_0^2(3A\sin^2\varphi_0 - B)n^2 + \omega_0^4(4B-3A)\sin^2\varphi_0 = 0$ has the roots with positive real parts, and consequently the motion is unstable when at least one of the inequalities fails to hold:

$$4B-3A > 0, \quad 3A\sin^2\varphi_0 - B > 2\omega_0^2\sqrt{B(4B-3A)\sin^2\varphi_0}.$$

For example, the motion is unstable when $3A > 4B$. Another sufficient instability condition can be obtained from the theorem [6] : if $-K + GM^{-1}G > 0$ (see 12) then the motion is unstable (for a linear system). In the case under scrutiny, this condition suggests the inequality $B > A$. If the body parameters are such that $A > B > 3A/4$ then it is advisable to consider the possibility of gyroscopic stabilization of the linear system.

Solution (8). The Lyapunov function (14) for this solution writes:

$$\begin{aligned} 2\delta^2\tilde{K} = {} & B\cos^2\varphi_0\dot{x}_1^2 + B\dot{x}_2^2 + (3(B-A)\omega_0^2\cos^2\varphi_0 \pm Psl\cos\varphi_0)x_1^2 \\ & + (\pm Psl\cos\varphi_0 + (4B-3A)\omega_0^2\cos 2\varphi_0 - Ap_0\omega_0\sin\varphi_0)x_2^2. \end{aligned}$$

Due to conditions of existence of the solution (8), the coefficient for x_2^2 transforms:

$$a_2 = a_3 + B\omega_0^2\cos^2\varphi_0 \pm Psl\sin{\varphi_0}^2/\cos\varphi_0 \ ,$$

where $a_3 = 3(B-A)\omega_0^2\cos{\varphi_0}^2 \pm Psl\cos\varphi_0$. The quadratic form under scrutiny is positive definite when $a_2 > 0$, $a_3 > 0$. These conditions hold, for example, for $\cos\theta_0\cos\varphi_0 > 0,\ B > A$.

The requirement $B > A$ is the sufficient condition of secular stability of "conic" precession of a symmetric rigid body on a circular orbit [2]. When there is a sail, for the protruded satellite there appears the possibility of stabilization of "conic" precession by potential forces.

The characteristic equation for the 1st approximation system

$$B^2 n^4 \cos^2\varphi_0 + \cos\varphi_0(-Psl + 3(A-B){\omega_0}^2\cos\varphi_0)(-Psl\cos\varphi_0$$
$$+\omega_0((3A-4B)\omega_0\cos 2\varphi_0 + Ap_0\sin\varphi_0))$$
$$+1/2n^2\cos\varphi_0((-9AB{\omega_0}^2 + 12B^2{\omega_0}^2 + 2A^2p_0^2)\cos\varphi_0$$
$$+B(Psl\cos 2\varphi_0 - (3A-2B){\omega_0}^2\cos 3\varphi_0 + 3(c + A\omega_0 p_0\sin 2\varphi_0))) = 0$$

has the solutions:

$$n^2 = \tfrac{1}{4B^2}\big(\sec\varphi_0\big(-(-9AB{\omega_0}^2 + 12B^2{\omega_0}^2 + 2A^2p_0^2)\cos\varphi_0$$
$$-B(Psl\cos 2\varphi_0 - (3A-2B){\omega_0}^2\cos 3\varphi_0 + 3(Psl + A\omega_0 p_0\sin 2\varphi_0))\pm$$
$$\surd[-16B^2\cos\varphi_0(-Psl + 3(A-B){\omega_0}^2\cos\varphi_0)$$
$$(-Psl\cos\varphi_0 + \omega_0((3A-4B)\omega_0\cos 2\varphi_0 + Ap_0\sin\varphi_0))$$
$$+((-9AB{\omega_0}^2 + 12B^2{\omega_0}^2 + 2A^2p_0^2)\cos\varphi_0 + B(Psl\cos 2\varphi_0 - (3A$$
$$-2B){\omega_0}^2\cos 3\varphi_0 + 3(Psl + A\omega_0 p\sin 2\varphi_0)))^2))\big]\ .$$

The motion can be stable only in case when the roots of the characteristic equation are imaginary. Consequently, the necessary conditions of motion (8) stability write:

$$a_4 = \cos\varphi_0(-Psl + 3(A-B){\omega_0}^2\cos\varphi_0)(-Psl\cos\varphi_0 + \omega_0((3A-4B)\omega_0\cos 2\varphi_0$$
$$+Ap_0\sin\varphi_0)) \geq 0;$$

$$\cos\varphi_0((-9AB{\omega_0}^2 + 12B^2{\omega_0}^2 + 2A^2p_0^2)\cos\varphi_0$$
$$+B(Psl\cos 2\varphi_0 - (3A-2B){\omega_0}^2\cos 3\varphi_0$$
$$+3(c + A\omega_0 p_0\sin 2\varphi_0))) > 4B\sqrt{a_4 cos^2\varphi_0}\ ;$$

Solution (9). The motion is stable with respect to the variables $\theta, \varphi, \dot{\theta}, \dot{\varphi}, \dot{\psi}$ when for the form

$$2\delta^2\tilde{K} = B\cos^2\varphi_0\dot{x}_1^2 + B\dot{x}_2^2 + 3(A-B)\omega_0^2\cos^2\varphi_0\sin^2\theta_0 x_1^2$$
$$+\left(B\omega_0^2\cos^2\varphi_0 - 3(B-A)\omega_0^2\cos^2\theta_0\sin^2\varphi_0\ \right)x_2^2$$
$$-6(B-A)\omega_0^2\cos\varphi_0\cos\theta_0\sin\theta_0\sin\varphi_0 x_1 x_2$$

the Sylvester conditions $A > B$, $3B(A-B)\omega_0^4\cos^4\varphi_0\sin^2\theta_0 > 0$ hold (the second condition is satisfied due to the first one). Consequently, $A > B$ is the sufficient condition of stability of motion (9), and furthermore, this is the condition of secular stability. The result obtained differs from that of [4], but when there is no solar pressure, it coincides with the stability condition of "hyperboloidal" satellite's precession [2].

Under the conditions of existence of the solution (9), the 1st approximation differential equations (11) may be written in the following form:

$$B\cos^2\varphi_0\ddot{x}_1 = 2B\omega_0\cos\varphi_0\sin\varphi_0\dot{x}_2 + 3(B-A)\omega_0^2\cos^2\varphi_0\sin^2\theta_0 x_1$$
$$+3(B-A)\omega_0^2\cos\varphi_0\sin\varphi_0\cos\theta_0\sin\theta_0 x_2,$$
$$B\ddot{x}_2 = -2B\omega_0\cos\varphi_0\sin\varphi_0\dot{x}_1 + 3(B-A)\omega_0^2\cos\varphi_0\sin\varphi_0\cos\theta_0\sin\theta_0 x_1$$
$$-(B\omega_0^2\cos^2\varphi_0 - 3(B-A)\omega_0^2\sin^2\varphi_0\cos^2\theta_0 x_1.$$

The characteristic equation

$$B^2n^4\cos^2\varphi_0 + Bn^2\omega_0^2\cos^2\varphi_0(B+3(B-A)(\cos^2\theta_0\sin^2\varphi_0+\sin^2\theta_0))$$
$$+3(A-B)B\omega_0^2\cos^4\varphi_0\sin^2\theta_0 = 0$$

for $B > A$ has the roots with a positive real part. Consequently, the condition $A > B$ is both necessary and sufficient (without any boundary) condition of stability of motion (9).

Solution (10) represents a particular case of (9), and the stability condition of this motion holds: $A > B$.

4 Conclusions

Within the frame of the model of solar sail satellite there is a possibility to obtain additional (with respect to the satellite without any sail) orientations in steady motions of the satellite. Stability conditions of the motions, which coincide with the motions of the satellite without any sail, are not restricted. Stabilization of steady motions by potential forces of light pressure is possible for some configurations of solar sail satellites.

References

1. Banshchikov, A., Bourlakova, L.: Information and Research System "Stability". Informations RAS. The theory and control systems. (Izvestia RAN. Teoria i sistemi upravlenia) **2** (1996) 13–20
2. Beletsky, V.V.: Motion of a Satellite with Respect to the Mass Center in the Gravitation Field, Izdatelstvo MGU, Moscow (1975) (in Russian)
3. Belketsky, V.V.: Essays on Motion of Space Bodies, Nauka Publ., Moscow (1977) (in Russian)
4. Al Nadgar, M.Yu.: Periodic motions of artificial solar satellite with a sail. Cosmic Research(Kosmicheskie Issledovaniya) **XXVI** (1988) 956–958
5. Lyapunov, A.M.: On permanent screw motions of a rigid body in liquids. Collection of Works. Vol. 1, Izdatelstvo Academy Nauk, Moscow (1954) 276–320 (in Russian)
6. Hagedorn, P.: Über die Instabilität konservativer Systeme mit gyroskopischen Kräften. Arch. Rat. Mech. Anal. **58** (1975) 1–9

Application of Computer Algebra for Investigation of Group Properties of the Navier-Stokes Equations for Compressible Viscous Heat-Conducting Gas *

Vasiliy V. Bublik

Institute of Theoretical and Applied Mechanics, Russian Academy of Sciences, Siberian Branch, Institutskaya str., 4/1, Novosibirsk 630090, Russia
bublik@itam.nsc.ru

Abstract. In the paper the equations describing plane motion of a viscous heat-conducting perfect gas with polytropic equation of state are investigated. On the basis of methods of the group analysis of differential equations [1] some classes of exact solutions are constructed, namely, invariant and partially invariant. We use essentially the computer algebra systems for finding the symmetry group and for constructing the solution.

1 Introduction

During the development of new numeric methods and computer code packages for solving mathematical physics problems an important stage is the stage of testing formulae, algorithms and their program realizations on exact solutions. One of techniques for obtaining the exact solutions are the methods of group analysis of differential equations [1]. With the aid of these methods one can obtain the classes of invariant, partially invariant and differentially invariant solutions. In this work we consider the partially invariant solutions.

The construction of partially invariant solutions entails significant computing difficulties caused by the fact that for obtaining such solutions the completion to involution of systems of partial differential equations is required. The algorithms, which enable one to conduct such an analysis in finite number of operations, are well known [2–4]. However, in practice the execution of needed calculations by hand for more or less complicated systems of equations often is beyond one's power because of a huge amount of analytical calculations. The computer algebra systems such as *Mathematica*, *MACSYMA*, *Maple*, *REDUCE*, *AXIOM*, and *MuPAD*, are extremely helpful in such computations. For the first time the science school of N. N. Yanenko applied the systems of computer algebra for the completion analyses [5–8].

* This research was financially supported by INTAS (grant 99-1222) and the Russian Foundation for Basic Research (projects 99-01-00515 and 01-01-06171).

Until now, a large number of programs and software packages for compatibility research of systems of differential equations have been developed worldwide and are being developed at present. However, practice has shown that the more universal is the program, the narrower is the range of problems to which it is applicable. It is related to scantiness of computer resources (even at the present stage of evolution of computer facilities). Therefore, for intricate and practically important problems of mathematical physics it is necessary to search for the individual approaches for finding the solutions of overdetermined systems of differential equations.

First of all it is necessary to refuse a completely automatic execution of the compatibility research programs. Necessary condition for obtaining a result in such problems is an essential usage of a single-step dialog mode for performance of the programs. Practically in this case the computer algebra systems are used only for chores (differentiations, substitutions and other manipulations with expressions). A problem of selecting the following operation rests completely with man. Thus, for a successful solution of the problem it is necessary to actively apply not only properly the compatibility theory of differential equations, but also many other branches of mathematics, mechanics and physics.

As the involutivity conditions of systems in their pure form are often impossible to write down, one has to abandon the solution of this problem in favour of a problem of construction of the solution of the system under study. Furthermore, for applications it is often more important to have not a system of differential equations in involution but its general solution. In practice it looks approximately so. At first, steps of one of known algorithms of completion to involution are realized for system under consideration. Then at some stage either an integration of some equations or other simplifications are carried out, for example, with some physical properties. After that the research is continued. The physical properties of model are actively used during research (for example, positiveness of some parameters). It helps essentially to lower the number of cases at the completion of systems to involution: there is no need to examine those cases, which certainly do not have physical meaning. Also the usage of group properties helps to simplify significantly the research. Firstly, the group properties of original model allow fixing a part of integration constants, which reduces the number of unknown parameters. Secondly, it is necessary to pay attention to a problem of reduction during the construction of partially invariant solutions. For example, after partial integration it is possible to reveal that the solution will reduce to invariant one by a certain selection of constants (or functions of integration). It is natural that there is no need in investigating these cases up to their end, as this solution can be obtained in a simpler way.

2 Model Description

The system of equations governing plane motion of a viscous heat-conducting prefect gas with a polytropic equation of state is considered:

$$\rho(u_t + uu_x + vu_y) = -p_x + (\lambda(u_x + v_y))_x + (2\mu u_x)_x + (\mu(u_y + v_x))_y \ , \quad (1)$$

$$\rho(v_t + uv_x + vv_y) = -p_y + (\lambda(u_x + v_y))_y + (\mu(u_y + v_x))_x + (2\mu v_y)_y \ , \tag{2}$$

$$\rho_t + (u\rho)_x + (v\rho)_y = 0 \ , \tag{3}$$

$$\begin{aligned} c_V\rho(T_t + uT_x + vT_y) + p(u_x + v_y) &= (\varkappa T_x)_x + (\varkappa T_y)_y \\ &+\lambda(u_x + v_y)^2 + \mu\left(2u_x^2 + 2v_y^2 + (u_y + v_x)^2\right) \ . \end{aligned} \tag{4}$$

Here u and v are the components of velocity vector, ρ is the density, T is the temperature, $p = R\rho T$ is the pressure, $\varepsilon = c_V T$ is the internal energy, R is the gas constant, c_V is the specific heat of gas at constant volume, $\mu = m_0 T^\omega$ is the first viscosity coefficient, $\lambda = l_0 T^\omega$ is the second viscosity coefficient, $\varkappa = \varkappa_0 T^\omega$ is the heat conduction coefficient. In our research we will essentially allow for the following conditions having physical meaning:

$$\rho > 0 \ , \quad T > 0 \ , \quad 3\lambda + 2\mu \geq 0 \ , \quad \mu > 0 \ , \quad \varkappa \geq 0 \ , \quad R > 0 \ , \quad c_V > 0 \ .$$

In case $3\lambda + 2\mu = 0$ the group admitted by (1)–(4) was found in [9] for unsteady motions and in [10] for steady flows. For the arbitrary λ and μ the admissible group was found for three-dimensional motions in [11].

The principal Lie algebra admitted by (1)–(4) is Lie algebra L_8 with basis

$$\begin{gathered} X_1 = \partial_x \ , \quad X_2 = \partial_y \ , \quad X_3 = t\partial_x + \partial_u \ , \quad X_4 = t\partial_y + \partial_v \ , \\ X_5 = y\partial_x - x\partial_y + v\partial_u - u\partial_v \ , \quad X_6 = \partial_t \ , \quad X_7 = t\partial_t + x\partial_x + y\partial_y - \rho\partial_\rho \ , \\ X_8 = x\partial_x + y\partial_y + u\partial_u + v\partial_v + 2(\omega - 1)\rho\partial_\rho + 2T\partial_T \ . \end{gathered}$$

The optimal system of subalgebras of the Lie algebra L_8 is constructed in [12]. In this paper we will dwell on the consideration of three-dimensional subalgebras, which are presented in an optimal system by 46 series of dissimilar subalgebras.

3 Invariant Solutions of Rank 0

The first class of the solutions, the simplest one for the construction and research, which may be constructed on the basis of three-dimensional subalgebras, are invariant solutions of rank 0. It has been proposed recently to name such solutions the "simple" solutions [13]. The "simplicity" of these solutions is related to the fact that they are described by finite algebraic equations, which enable a more full research of their properties. However, in spite of their simplicity, such solutions were purposfully almost not studied. For the system of equations of axisymmetric motions of a viscous heat-conducting perfect gas with polytropic equation of state and power-law dependence of viscosity and heat conduction coefficients on temperature the description of all invariant solutions of rank 0 was done first in [9].

Among the three-dimensional subalgebras, the necessary condition for existence of invariant solution is satisfied by 37 subalgebras. All of these subalgebras have been investigated for possibility of physical solutions existence. Here, the physical solution means the (local) solution with positive values of density and temperature. This concept of the physical solution is invariant with respect to transformations admitted by (1)–(4). The requirement of physical solution may impose severe restrictions on the values of some parameters. So, for example, on the basis of some subalgebras it is generally impossible to construct the physical solutions. Some other subalgebras may give such solutions only at some fixed values of parameter ω. These limitations result in that all invariant solutions may be united in 15 classes, among which there are spiral motions, radial scattering, straight-line motions, stratified motions, decelerating gas flow. These solutions have been described in [14].

4 Partially Invariant Solutions Reducible to Invariant Solutions

Second class of solutions, which may be constructed on the basis of three-dimensional subalgebras, are regular partially invariant solutions of rank 1 and defect 1. The construction of invariant solutions is much easier than the construction of partially invariant solutions. Therefore, the criteria, which enable one to eliminate in advance the partially invariant solutions reducible to invariant solutions, help us to focus our efforts on the construction of the non-reducible solutions. The criterion for reducibility of arbitrary partially invariant manifold to invariant manifold was proved in [15]. Unfortunately, this criterion is not always convenient for practical use. Therefore, the problem of proving the particular theorems giving sufficient conditions is topical. We investigate in the present work the problem of reducibility of regular partially invariant solutions of rank 1 and defect 1 for equations of plane motion of viscous heat-conducting perfect gas. A similar problem for axisymmetric gas motion has been investigated in [16].

Theorem. *If the universal invariant of the subgroup $H \subset G_8$ can be chosen in the form*

$$J = (\xi(t,x,y),\ \ A(t,x,y)u + B(t,x,y)v + C(t,x,y),\ \ D(t,x,y)\rho,\ \ E(t,x,y)T)\ ,$$

where ξ, A, B, C, D, and E are some functions, then the corresponding regular partially invariant H-solution of rank 1 and defect 1 of (1)–(4) is reducible to invariant solution.

All invariant solutions of rank 1, to which in this case the reduction takes place, are described in [12]. The condition of the above Theorem is satisfied by 9 series of subalgebras. The general scheme of proving the theorem coincides with that of [16].

5 Regular Partially Invariant Solutions of Rank 1 and Defect 1 Not Reducible to Invariant Solutions

Among three-dimensional subalgebras, the solutions have remained which were not considered in sections 3 and 4. On the basis of these subalgebras one may try to find the regular partially invariant solutions of rank 1 and defect 1 not reducible to invariant solutions. The complete study of such solutions demands using the theories of completion analysis of systems of partial differential equations. For all subalgebras such an analysis was not completely conducted, therefore, we shall restrict our consideration here to the illustration of one example of the irreducible solution.

Consider the subalgebra $\{X_3, X_4, X_5 + \alpha X_8\}$ with $\alpha \neq 0$. When $\omega \neq 1$ the invariants of corresponding subgroup are

$$t \ , \quad \left((ut-x)\sin\frac{\ln\rho}{2\alpha(\omega-1)} + (Vt-y)\cos\frac{\ln\rho}{2\alpha(\omega-1)}\right)\rho^{1/(2-2\omega)} \ ,$$
$$\left((ut-x)\cos\frac{\ln\rho}{2\alpha(\omega-1)} - (Vt-y)\sin\frac{\ln\rho}{2\alpha(\omega-1)}\right)\rho^{1/(2-2\omega)} \ , \quad T\rho^{1/(1-\omega)} \ .$$

We shall search for the solution in the form

$$u = \frac{1}{t}\rho^{1/(2\omega-2)}\left(\varphi(t)\sin\frac{\ln\rho}{2\alpha(\omega-1)} + \psi(t)\cos\frac{\ln\rho}{2\alpha(\omega-1)}\right) + \frac{x}{t} \equiv U(t,x,\rho) \ ,$$
$$v = \frac{1}{t}\rho^{1/(2\omega-2)}\left(\varphi(t)\cos\frac{\ln\rho}{2\alpha(\omega-1)} - \psi(t)\sin\frac{\ln\rho}{2\alpha(\omega-1)}\right) + \frac{y}{t} \equiv V(t,y,\rho) \ ,$$
$$T = T_1(t)\rho^{1/(\omega-1)} \ , \quad \rho = \rho(t,x,y) \ .$$

For functions U and V we have:

$$\frac{\partial U}{\partial \rho} = \frac{t(\alpha U + V) - (\alpha x + y)}{2(\omega-1)t\rho} \ , \quad \frac{\partial V}{\partial \rho} = \frac{t(\alpha V - U) - (\alpha y - x)}{2(\omega-1)t\rho} \ .$$

The substitution of solution representation into (1)–(4) gives us a system for determining the functions $\rho(t,x,y)$, $\varphi(t)$, $\psi(t)$, and $T_1(t)$. As a full compatibility analysis of this system is rather laborious we shall restrict our consideration to particular case. Let us namely consider the motions with homogeneous deformation. In this case we impose the following conditions on the velocity components:

$$u_{xx} = u_{xy} = u_{yy} = v_{xx} = v_{xy} = v_{yy} = 0 \ . \tag{5}$$

Conditions (5) may easily be rewritten into a system of second order equations for function ρ:

$$A\rho\rho_{xx} + B\rho_x^2 = 0 \ , \quad A\rho\rho_{xy} + B\rho_x\rho_y = 0 \ , \quad A\rho\rho_{yy} + B\rho_y^2 = 0 \ ,$$
$$C\rho\rho_{xx} + D\rho_x^2 = 0 \ , \quad C\rho\rho_{xy} + D\rho_x\rho_y = 0 \ , \quad C\rho\rho_{yy} + D\rho_y^2 = 0 \ ,$$
$$A = (\alpha U + V)t - (\alpha x + y) \ , \quad C = (\alpha V - U)t - (\alpha y - x) \ ,$$
$$B = \frac{(\alpha^2 - 2\alpha(\omega-1) - 1)(Ut-x) + 2(\alpha-\omega+1)(Vt-y)}{2(\omega-1)} \ ,$$
$$D = \frac{(\alpha^2 - 2\alpha(\omega-1) - 1)(Vt-y) - 2(\alpha-\omega+1)(Ut-x)}{2(\omega-1)} \ .$$

The compatibility conditions of this system are as follows:

$$\left((Ut-x)^2+(Vt-y)^2\right)\rho_x=0\ ,\quad \left((Ut-x)^2+(Vt-y)^2\right)\rho_y=0\ .$$

For $\rho_x=\rho_y=0$ we have:

$$u=u_1(t)+x/t\ ,\quad v=v_1(t)+y/t\ ,\quad \rho=\rho(t)\ ,\quad T=T(t)\ ,$$

i.e. the solution is invariant with respect to the subgroup $\{X_3,X_4\}$.

Consider the case $u=x/t$, $v=y/t$, $\rho=\rho(t,x,y)$, $\rho_x^2+\rho_y^2\neq 0$. Equations (1)–(3) are reduced to

$$\omega\left(2(l_0+m_0)T_1^{\omega-1}-Rt\right)=0\ ,\quad T\rho_t+x\rho_x+y\rho_y+2\rho=0\ .$$

When $\omega=0$ the solution of (1)–(4) is reconstructed from the solution $\Psi(\xi,\eta)$, and $T_1(t)$ of the system

$$\Psi_{\xi\xi}+\Psi_{\eta\eta}=A\ ,\quad \frac{c_V(t^2T_1)'-4(l_0+m_0)}{t^2T_1}=Ak_0-\frac{2R}{t}$$

by using the following formulae:

$$u=\xi\ ,\quad v=\eta\ ,\quad \rho=\frac{1}{t^2\Psi}\ ,\quad T=t^2T_1\Psi\ ,\quad \xi=\frac{x}{t}\ ,\quad \eta=\frac{y}{t}\ .$$

The solution will not reduce to invariant solution when $\Psi_\xi^2+\Psi_\eta^2\neq 0$.

When $\omega\neq 0$ the solution of (1)–(4) having physical meaning does not exist.

Thus, for the system (1)–(4) we obtain the partially invariant solutions which do not reduce to invariant solutions. These solutions are reconstructed from the solutions of the Poisson equation. The initial- and boundary-value problems for (1)–(4) may easily be rewritten into a boundary-value problem for the Poisson equation. The velocity specification on the domain boundary should be matched to the solution. The Dirichlet problem corresponds to the specification of temperature on the domain boundary. The Neumann problem corresponds to the specification of the heat flux on the domain boundary. Mixed problems are also possible. The elliptic equations in the domains with curvilinear boundary can now be solved numerically with very high accuracy [17].

References

1. Ovsiannikov, L. V.: *Group Analysis of Differential Equations*, Nauka, Moscow (1978) (English transl. published by Academic Press, New York (1982))
2. Cartan, E.: *Les Systèmes Différentielles Extérieurs et Leurs Applications Géometriques*, Hermann, Paris (1945)
3. Kuranishi, M.: *Lectures on Involutive Systems of Partial Differential Equations*, Publ. Soc. Math., Sao Paulo (1967)
4. Pommaret, J. F.: *Systems of Partial Differential Equations and Lie Pseudogroups*, Gordon & Breach, New York (1978)

5. Shurygin, V. A., Yanenko N. N.: About realization on electronic computing machine of algebraic-differential algorithms. *Problems of Cybernetics* **1** (1961) 33–43 (in Russian)
6. Arais, E. A., Shapeev, V. P., Yanenko, N. N.: Realization of Cartan's method of exterior differential forms on an electronic Computer. *Doklady Akademii Nauk SSSR* **214(4)** (1974) 737–738 (English translation in *Sov. Math. Dokl.* **15(1)** (1974) 203–205).
7. Ganzha, V. G., Meleshko, S. V., Murzin, F. A., Shapeev, V. P., Yanenko, N. N.: Computer implementation of algorithm of compatibility of systems of partial differential equations. *Doklady Akademii Nauk SSSR* **261(5)** (1981) 1044–1046 (English translation in *Sov. Math. Dokl.*)
8. Sidorov, A. F., Shapeev, V. P., Yanenko, N. N.: *Method of Differential Constraints and Its Applications in Gas Dynamics*, Nauka, Novosibirsk (1984) (in Russian)
9. Bublik, V. V.: Group classification of the two-dimensional equations of motion of a viscous heat conducting perfect gas. *Prikl. Mekh. Tekhn. Fiz.* **37(2)** (1996) 27–34 (English translation in *Appl. Mech. and Techn. Phys.* by Plenum Publishing Corp.)
10. Meleshko, S. V.: Group classification of two-dimensional stable viscous gas equations. *Int. J. Nonlin. Mech.* **34(3)** (1998) 449–456
11. Bublik, V. V.: Group classification of the Navier-Stokes equations for compressible viscous heat-conducting gas. In: *Computer Algebra in Scientific Computing/ CASC 2000*, V.G. Ganzha, E.W. Mayr and E.V. Vorozhtsov (Eds.), Springer-Verlag, Berlin (2000) 61–67
12. Bublik, V. V.: Invariant solutions of rank 1 of equations for plane motions of a viscous heat-conducting perfect gas. *Prikl. Mekh. Tekhn. Fiz.* **38(3)** (1997) 26–31 (English translation in *Appl. Mech. and Techn. Phys.* by Plenum Publishing Corp.)
13. Ovsiannikov, L. V.: On "simple" solutions of equations for dynamics of polytropic gas. *Prikl. Mekh. Tekhn. Fiz.* **40(2)** (1996) 5–12. (English translation in *Appl. Mech. and Techn. Phys.* by Plenum Publishing Corp.)
14. Bublik, V. V.: "Simple" solutions of equations of dwo-dimensional motions of a viscous heat-conducting perfect gas. *Dinamika Sploshnoi Sredy*, Novosibirsk **116** (2000) 123–127 (in Russian)
15. Bytev, V. O.: To a problem of reduction. *Dinamika Sploshnoi Sredy*, Novosibirsk **5** (1970) 146–148 (in Russian)
16. Bublik, V. V.: Exact solutions of equations for axisymmetric motions of a viscous heat-conducting perfect gas described by systems of ordinary differential equations. *Prikl. Mekh. Tekhn. Fiz.* **40(5)** (1997) 51–54 (English translation in *Appl. Mech. and Techn. Phys.* by Plenum Publishing Corp.)
17. Shapeev, A. V., Shapeev, V. P.: Difference scheme of improved approximation for solving the elliptic equations in domain with curvilinear boundary. *Zhurnal Vychislitel'noi Matematiki i Matematicheskoi Fiziki* **40(2)** (2000) 223–232 (in Russian)

Mathematica and Nilpotent Lie Superalgebras

L.M. Camacho, J.R. Gómez, R.M. Navarro, and I. Rodríguez

Dpto. Matemática Aplicada I. Univ. Sevilla. Avda. Reina Mercedes S.N. 41012 Sevilla (Spain). E-mail: lcamacho@us.es - jrgomez@us.es
Dpto. de Matemáticas. Univ. de Extremadura. Avda. de la Universidad S.N. 10071 Cáceres (Spain). E-mail: rnavarro@unex.es
Dpto. Matemáticas. Escuela Politécnica Superior, Universidad de Huelva, Huelva (Spain). Email: rodgar@uhu.es

Abstract. The aim of this work is to study a family of Lie superalgebras which generalize Heisenberg Lie algebras. We prove the existence of a special basis for these superalgebras with arbitrary dimension of even part and dimension of odd part up to three. By using the software *Mathematica 4.0* we classify these superalgebras for arbitrary dimension of even part and dimension of odd part up to two.

1 Introduction

During the last few decades the theory of Lie superalgebras (or, as they also called, $\mathbf{Z}_2$-graded Lie algebras) has seen a remarkable evolution, both in mathematics and physics. The historical background is referred to the review by Corwin, Ne' eman and Sternberg [3] which presents the subject as it was known in 1974. The first comprehensive description of the mathematical theory of Lie superalgebras is due to Kac [6] in 1977, who establishes the classification of all finite-dimensional simple Lie superalgebras over an algebraically closed field of characteristic zero. Other authors have studied semi-simple Lie superalgebras and their cohomology [7], [8].

However, nilpotent Lie superalgebras are practically unknown, so far. In [10] we can find the definition and Engel's theorem for Lie superalgebras. Filiform Lie superalgebras has been studied [4] recently. These are a special case of superalgebras with even part a filiform Lie algebra. This paper is placed within this context. Heisenberg algebra plays a fundamental role in quantum mechanics [1], [9], thus we generalize this concept to the theory of Lie superalgebras. We are going to define Heisenberg Lie superalgebras as Lie superalgebras with even part a Heisenberg algebra.

We will not presuppose any knowledge of the theory of Lie superalgebras, however, we assume that the reader is familiar with the standard theory of Lie algebras

2 Preliminary

The vector space V is said to be $\mathbf{Z}_2$−graded if it admits a decomposition in direct sum $V = V_0 \oplus V_1$. An element X of V is called *homogeneous of degree*

α ($deg(X) = d(X) = \alpha$), $\alpha \in \mathbf{Z}_2$, if it is an element of V_α. In particular, the elements of V_0 (resp. V_1) are also called even (resp. odd).

Let $V = V_0 \oplus V_1$ and $W = W_0 \oplus W_1$ two graded vector spaces. A linear mapping $f : V \longrightarrow W$ is said to be homogeneous of degree $\alpha \in \mathbf{Z}_2$, if $f(V_\beta) \subset W_{\alpha+\beta(mod\ 2)}$ for all $\beta \in \mathbf{Z}_2$. The mapping f is called a *homomorphism* of the $\mathbf{Z}_2$-graded vector space V into the $\mathbf{Z}_2$-graded vector space W if f is homogeneous of degree 0. It is now evident how we define an isomorphism or an automorphism of $\mathbf{Z}_2$-graded vector spaces.

The first definition of Lie superalgebra can be found in [10].

Definition 1. *A Lie superalgebra is a* $\mathbf{Z}_2$*-graded vector space,* $\mathfrak{g} = \mathfrak{g}_{\overline{0}} \oplus \mathfrak{g}_{\overline{1}}$, *endowed with a bracket product [,] verifying:*

1. $[\mathfrak{g}_\alpha, \mathfrak{g}_\beta] \subset \mathfrak{g}_{\alpha+\beta(mod2)}$ $\qquad \forall \alpha, \beta \in \mathbf{Z}_2$.
2. $[X,Y] = -(-1)^{\alpha\cdot\beta}[Y,X]$ $\qquad \forall X \in \mathfrak{g}_\alpha, \forall Y \in \mathfrak{g}_\beta$.
3. $(-1)^{\gamma\cdot\alpha}[X,[Y,Z]] + (-1)^{\alpha\cdot\beta}[Y,[Z,X]] + (-1)^{\beta\cdot\gamma}[Z,[X,Y]] = 0$
   for all $X \in \mathfrak{g}_\alpha, Y \in \mathfrak{g}_\beta, Z \in \mathfrak{g}_\gamma$ *with* $\alpha, \beta, \gamma \in \mathbf{Z}_2$.
   This property is called *graded Jacobi identity* and we will denote it by $J_g(X,Y,Z)$.

In 1995 another definition for Lie superalgebras is established; for more details see [2].

Definition 2. *Let* $\mathfrak{g} = \mathfrak{g}_{\overline{0}} \oplus \mathfrak{g}_{\overline{1}}$ *be a* $\mathbf{Z}_2$*-graded vector space. We call* $\mathfrak{g}$ *a Lie superalgebra if the following conditions are satisfied:*

1. $\mathfrak{g}_{\overline{0}}$ *is a Lie algebra (with Lie bracket or law* μ*).*
2. $\mathfrak{g}_{\overline{1}}$ *is a* $\mathfrak{g}_{\overline{0}}$*−module (with an operation or product denoted by* "." : $\mathfrak{g}_{\overline{0}} \times \mathfrak{g}_{\overline{1}} \longrightarrow \mathfrak{g}_{\overline{1}}$*).*
3. *There exists a bilinear and symmetric mapping* $B : \mathfrak{g}_{\overline{1}} \times \mathfrak{g}_{\overline{1}} \longrightarrow \mathfrak{g}_{\overline{0}}$, *that verifies:*
   3.1. B *is* $\mathfrak{g}_{\overline{0}}$*-invariant, that is:*
   $\mu(X, B(Y_1,Y_2)) = B(X.Y_1,Y_2) + B(Y_1, X.Y_2)$ $\forall X \in \mathfrak{g}_{\overline{0}}; \forall Y_1, Y_2 \in \mathfrak{g}_{\overline{1}}$.
   3.2. $B(Y_1,Y_2).Y_3 + B(Y_2,Y_3).Y_1 + B(Y_3,Y_1).Y_2 = 0$ $\qquad \forall Y_i \in \mathfrak{g}_{\overline{1}}, 1 \le i \le 3$.

Remark 1. Under the conditions of definition 2 we define a new product, [,], over $\mathfrak{g}$ as follows:

$$\begin{array}{lll} [X_1,X_2] = \mu(X_1,X_2) & \text{if} & X_1,X_2 \in \mathfrak{g}_{\overline{0}} \\ [X_1,Y_1] = -[Y_1,X_1] = X_1.Y_1 & \text{if} & X_1 \in \mathfrak{g}_{\overline{0}}, Y_1 \in \mathfrak{g}_{\overline{1}} \\ [Y_1,Y_2] = B(Y_1,Y_2) & \text{if} & Y_1,Y_2 \in \mathfrak{g}_{\overline{1}} \end{array}$$

Thanks to this product [,] it is easy to check that the two above definitions are equivalent, in particular being $\mathfrak{g}_{\overline{1}}$ a $\mathfrak{g}_{\overline{0}}$-module along with the conditions 3.1 and 3.2 lead to graded Jacobi identity.

Throughout this paper, Lie superalgebras will be considered under the conventions of the definition 2 and attending the Remark 1

We say that two Lie superalgebras $\mathfrak{g}$, $\mathfrak{g}'$ are isomorphic if there exists a $\mathbf{Z}_2$-graded vector spaces isomorphism $\Phi : \mathfrak{g} \longrightarrow \mathfrak{g}'$ satisfying $\Phi([X,Y])=[\Phi(X),\Phi(Y)]$ for all $X,Y \in \mathfrak{g}$ (and then ϕ is called an isomorphism of Lie superalgebras). Recall that isomorphism (homomorphism, automorphism) of $\mathbf{Z}_2$-graded vector spaces are always assumed to be consistent with the $\mathbf{Z}_2$-gradations, that is, they are homogeneous linear mappings of degree zero. Particularly, changes of basis are isomorphisms of Lie superalgebras.

The *descending central sequence* of a Lie superalgebra $\mathfrak{g} = \mathfrak{g}_{\bar{0}} \oplus \mathfrak{g}_{\bar{1}}$ is defined by $\mathcal{C}^0(\mathfrak{g}) = \mathfrak{g}$, $\mathcal{C}^{k+1}(\mathfrak{g}) = [\mathcal{C}^k(\mathfrak{g}), \mathfrak{g}]$ for all $k \geq 0$. If $\mathcal{C}^k(\mathfrak{g}) = \{0\}$ for some k, Lie superalgebra is called *nilpotent*. The smallest integer k such as $\mathcal{C}^k(\mathfrak{g}) = \{0\}$ is called the nilindex of $\mathfrak{g}$.

Engel's theorem for Lie algebras and its direct consequences remain valid for Lie superalgebras, and the proof is the same as for Lie algebras [5].

Proposition 1. *A Lie superalgebra $\mathfrak{g}$ is nilpotent if and only if $ad(X)$ is nilpotent for every homogeneous element X of $\mathfrak{g}$.*

Corollary 1. *Let V be a vector space of dimension m and let $\mathfrak{h}$ be a set of nilpotent endomorphisms of V. Then there exists a descending sequence of vector subspaces $V_m, \dots, V_1, V_0$ of V, with dimensions $m, m-1, \dots 0$, respectively, and such that $h(V_{i+1}) \subseteq V_i \ \forall h \in \mathfrak{h} \ i = 0, 1, \dots, m-1$.*

Remark 2. Let $\mathfrak{g} = \mathfrak{g}_{\bar{0}} \oplus \mathfrak{g}_{\bar{1}}$ be a nilpotent Lie superalgebra. If we consider $V = \mathfrak{g}_{\bar{1}}$ (taking $\mathfrak{g}_{\bar{1}}$ as vector space) and $\mathfrak{h}$ the operator *ad* restricted to $\mathfrak{g}_{\bar{0}}$ the conditions of the above corollary are satisfied, and then we have a descending sequence of subspaces $V = V_m \supset \dots \supset V_1 \supset V_0$ of dimensions $m, m-1, \dots 0$, such that $[\mathfrak{g}_{\bar{0}}, V_{i+1}] \subseteq V_i$.

We denote by $\mathcal{L}^{n+m}$ the set of the Lie superalgebras $\mathfrak{g} = \mathfrak{g}_{\bar{0}} \oplus \mathfrak{g}_{\bar{1}}$ with $dim(\mathfrak{g}_{\bar{0}}) = n$ and $dim(\mathfrak{g}_{\bar{1}}) = m$.

By taking an homogeneous basis $\{X_0, X_1, \dots, X_{n-1}, Y_1, \dots, Y_m\}$ in $\mathfrak{g}$ (with $\mathfrak{g} \in \mathcal{L}^{n+m}$), the superalgebra is completely determined by its structure constants, that is, by the set of constants $\{C_{ij}^k, D_{ij}^k, E_{ij}^k\}_{i,j,k}$ that verify

$$[X_i, X_j] = \sum_{k=0}^{n-1} C_{ij}^k X_k \qquad 0 \leq i < j \leq n-1$$

$$[X_i, Y_j] = \sum_{k=1}^{m} D_{ij}^k Y_k \qquad 0 \leq i \leq n-1, 1 \leq j \leq m$$

$$[Y_i, Y_j] = \sum_{k=0}^{n-1} E_{ij}^k X_k \qquad 1 \leq i \leq j \leq m$$

being [,] the pointed bracket of $\mathfrak{g}$.

From the graded Jacobi identity we have the following restrictions for the structure constants:

$$(S)\begin{cases}
\sum_{l=0}^{n-1}(C_{ij}^{l}C_{kl}^{s}+C_{jk}^{l}C_{il}^{s}+C_{ki}^{l}C_{jl}^{s})=0 & 0\le i<j<k\le n-1,\\
 & s=0,\dots,n-1\\
\sum_{l=1}^{m}(D_{jk}^{l}D_{il}^{s}-D_{jl}^{s}D_{ik}^{l})-\sum_{l=0}^{n-1}C_{ij}^{l}D_{lk}^{s}=0 & 0\le i<j\le n-1,\\
 & k=1,\dots,m,\ s=1,\dots,m\\
\sum_{l=1}^{m}(D_{ij}^{l}E_{lk}^{s}+D_{ik}^{l}E_{jl}^{s})-\sum_{l=0}^{n-1}E_{jk}^{l}C_{il}^{s}=0 & i=0,\dots,n-1,\\
 & 1\le j\le k\le m,\ s=0,\dots,n-1\\
\sum_{l=0}^{n-1}(E_{ij}^{l}D_{lk}^{s}+E_{ik}^{l}D_{lj}^{s}+E_{jk}^{l}D_{li}^{s})=0 & 1\le i\le j\le m,0\le k\le n-1\\
 & s=1,\dots,m
\end{cases}$$

One could think of a program of classifying all Lie superalgebras by considering the above equations to be solved for unknown structure constants. This turns out to be a very complicated problem because of the non-linearity of the equations.

In this paper we are going to solve (S) for some special classes of nilpotent Lie superalgebras (which we call Heisenberg) using changes of basis, nilindex, and the software *Mathematica 4.0.*

3 Heisenberg Lie Superalgebras

Let $\mathfrak{g}$ be a nilpotent Lie superalgebra, ($\mathfrak{g}=\mathfrak{g}_{\bar{0}}\oplus\mathfrak{g}_{\bar{1}}$), $\mathfrak{g}$ is a Heisenberg Lie superalgebra (HSA) if $\mathfrak{g}_{\bar{0}}=\mathcal{H}_r$, where $\mathcal{H}_r$ is the Heisenberg Lie algebra of dimension $2r+1$ and law

$$[X_{2i},X_{2i+1}]=X_{2r},\qquad 0\le i\le r-1$$

in a certain basis that we denote by $\{X_0,X_1,\dots,X_{2r}\}$.

Theorem 1. *If $\mathfrak{g}=\mathfrak{g}_{\bar{0}}\oplus\mathfrak{g}_{\bar{1}}$ is a HSA with $dim(\mathfrak{g}_{\bar{0}})=2r+1$ and $dim(\mathfrak{g}_{\bar{1}})=2$, then there exists a suitable homogeneous basis of $\mathfrak{g}$, $\{X_0,X_1,\dots,X_{2r},Y_1,Y_2\}$, with $\{X_0,X_1,\dots,X_{2r}\}$ a basis of $\mathfrak{g}_{\bar{0}}$ and $\{Y_1,Y_2\}$ a basis of $\mathfrak{g}_{\bar{1}}$, such that*

$$\begin{cases}[X_{2i},X_{2i+1}]=X_{2r}, & 0\le i\le r-1\\ [X_0,Y_1]=\epsilon Y_2, & \epsilon\in\{0,1\}\end{cases}$$

Proof. Let $\mathfrak{g}=\mathfrak{g}_{\bar{0}}\oplus\mathfrak{g}_{\bar{1}}$ be a HSA, then $\mathfrak{g}_{\bar{0}}$ is the Heisenberg algebra of dimension $2r+1$. There exists an adapted basis of $\mathfrak{g}_{\bar{0}}$, $\{X_0,X_1,\dots,X_{2r}\}$, with $[X_{2i},X_{2i+1}]=X_{2r}$, $0\le i\le r-1$.

Moreover, we have that $\mathfrak{g}_{\bar{1}}$ is a $\mathfrak{g}_{\bar{0}}$-module with $\mathfrak{g}_{\bar{0}}$ a nilpotent Lie algebra and according to remark 2 there exists a fibration of $\mathfrak{g}_{\bar{1}}$ (taking $\mathfrak{g}_{\bar{1}}$ as vector space), V_2, such that

$$V_0 = 0 \subset V_1 \subset V_2, \qquad dim\,(V_i/V_{i-1}) = 1, \quad \text{and} \quad [\mathfrak{g}_{\bar{0}}, V_i] \subseteq V_{i-1}, \quad 1 \leq i \leq 2$$

Let $\{Y_2\}$ and $\{Y_1, Y_2\}$ be the basis of the spaces V_1 and V_2, respectively. Hence,

$$\begin{cases} [X_i, Y_2] = 0, \\ [X_i, Y_1] = D_{i1}^2 Y_2, \; 0 \leq i \leq 2r \end{cases}$$

Thus, we will consider two cases:

a) If for all i, $0 \leq i \leq 2r$, $D_{i1}^2 = 0$, then, we obtain the basis of theorem for $\epsilon = 0$.

b) If there exists i, $0 \leq i \leq 2r$, such that $D_{i1}^2 \neq 0$, we can suppose that $D_{01}^2 \neq 0$. In other case, if there exists i, $1 \leq i \leq 2r-1$, with $D_{i1}^2 \neq 0$ (if $i = 2r$ by $J_{\mathfrak{g}}(X_0, X_1, Y_1)$ we lead to contradiction) and making a simple basis change we obtain $D_{01}^2 \neq 0$.
Finally, we arrive at the basis of theorem (for $\epsilon = 1$), previous change of scale.

□

Theorem 2. *If $\mathfrak{g} = \mathfrak{g}_{\bar{0}} \oplus \mathfrak{g}_{\bar{1}}$ is a HSA with $dim(\mathfrak{g}_{\bar{0}}) = 2r+1$ and $dim(\mathfrak{g}_{\bar{1}}) = 3$, then there exists a suitable homogeneous basis of $\mathfrak{g}$, $\{X_0, X_1, \ldots, X_{2r}, Y_1, Y_2, Y_3\}$, with $\{X_0, X_1, \ldots, X_{2r}\}$ a basis of $\mathfrak{g}_{\bar{0}}$ and $\{Y_1, Y_2, Y_3\}$ a basis of $\mathfrak{g}_{\bar{1}}$, such that*

$$(*) \begin{cases} [X_{2i}, X_{2i+1}] = X_{2r}, & 0 \leq i \leq r-1 \\ [X_0, Y_1] = \epsilon_1 Y_2 & \\ [X_0, Y_2] = \epsilon_1 \epsilon_2 Y_3, & \epsilon_1, \epsilon_2 \in \{0, 1\} \end{cases}$$

or,

$$(**) \begin{cases} [X_{2i}, X_{2i+1}] = X_{2r}, & 0 \leq i \leq r-1 \\ [X_0, Y_1] = Y_3 & \\ [X_0, Y_2] = 0 & \end{cases}$$

Proof. Analogously to the case of $dim(\mathfrak{g}_{\bar{1}}) = 2$ and using that $[X_{2r}, Y_j] \subset \; < Y_{j+2}, \ldots, Y_m >$ (obtained from $J_{\mathfrak{g}}(X_0, X_1, Y_j)$ and from finite induction for j), we obtain a basis $\{X_0, \ldots, X_{2r}, Y_1, Y_2, Y_3\}$ such that

$$\begin{cases} [X_{2i}, X_{2i+1}] = X_{2r}, & 0 \leq i \leq r-1 \\ [X_j, Y_1] = D_{j,1}^2 Y_2 + D_{j,1}^3 Y_3, & 0 \leq j \leq 2r-1 \\ [X_{2r}, Y_1] = D_{2r,1}^3 Y_3 & \\ [X_j, Y_2] = D_{j,2}^3 Y_3, & 0 \leq j \leq 2r-1 \end{cases}$$

At this point, we are going to consider different cases depending on the nullity of structure constants.

Case 1: $D^3_{j,2} = 0$ for all j, $0 \leq j \leq 2r-1$. We have $D^3_{2r,1} = 0$ from $J_g(X_0, X_1, Y_1)$.

1.1. If $D^3_{j,1} = 0$ for all j, $0 \leq j \leq 2r-1$.

1.1.1. If $D^2_{j,1} = 0$ for all j, $0 \leq j \leq 2r-1$. Then $\mathfrak{g}_{\overline{1}}$ is the trivial $\mathfrak{g}_{\overline{0}}$-module, that is, in the terms of definition 2, the product ".": $\mathfrak{g}_{\overline{0}} \times \mathfrak{g}_{\overline{1}} \longrightarrow \mathfrak{g}_{\overline{1}}$ is identically null. We have obtained the basis $(*)$ with $\epsilon_1 = \epsilon_2 = 0$.

1.1.2. If there exists j, $0 \leq j \leq 2r-1$, such that $D^2_{j,1} \neq 0$ it is possible to obtain the basis $(*)$ with $\epsilon_1 = 1$, $\epsilon_2 = 0$ (by a simple change of basis and a change of scale).

1.2. If there exists j, $0 \leq j \leq 2r-1$, such that $D^3_{j,1} \neq 0$.

1.2.1. If $D^2_{j,1} = 0$ for all j, $0 \leq j \leq 2r-1$, it is the case 1.1.2. (by a simple change of basis).

1.2.2. If there exists j, $0 \leq j \leq 2r-1$, such that $D^2_{j,1} \neq 0$, a simple change of basis permits to obtain the same basis of case 1.1.2.

Case 2: If there exists j, $0 \leq j \leq 2r-1$, such that $D^3_{j,2} \neq 0$.

2.1. If $D^2_{j,1} = 0$ for all j, $0 \leq j \leq 2r-1$, then we have $D^3_{2r,1} = 0$ (from $J_g(X_0, X_1, Y_1)$) and we obtain (by the basis change $Y_1' = Y_2$, $Y_2' = Y_1$ and a change of scale)

$$\begin{cases} [X_0, Y_1] = Y_3 \\ [X_0, Y_2] = D^3_{0,2} Y_3 \end{cases}$$

The basis change $Y_2' = Y_2 - D^3_{0,2} Y_1$ permits to obtain the basis $(**)$

2.2. If there exists j, $0 \leq j \leq 2r-1$, such that $D^2_{j,1} \neq 0$

2.2.1. If there exists i, $0 \leq i \leq 2r-1$ such that $D^3_{i,2} \neq 0$ and $D^2_{i,1} \neq 0$, by certain changes of basis we obtain

$$\begin{cases} [X_0, Y_1] = Y_2 \\ [X_0, Y_2] = Y_3 \end{cases}$$

that corresponds to the basis $(*)$ with $\epsilon_1 = \epsilon_2 = 1$. In this case $\mathfrak{g}_{\overline{1}}$ is called $\mathfrak{g}_{\overline{0}}-$module filiform.

2.2.2. If there exists $(j,k) \in \{(2i, 2i+1), (2i+1, 2i)\}, 0 \leq i \leq r-1$ with $D^3_{j,2} \neq 0$ and $D^2_{k,1} \neq 0$. By basis changes of the form $X_{2i}' = X_{2i} + \lambda X_{2i+1}$ with $\lambda \in \mathbf{C}$ (that type of changes are the only ones that are admissible for Heisenberg algebras) along with other simple basis changes we lead to the above case.

2.2.3. In other cases we can obtain

$$\begin{cases} [X_0, Y_1] = Y_2 \\ [X_0, Y_2] = 0 \end{cases}$$

that is the basis (*) with $\epsilon_1 = 1$ and $\epsilon_2 = 0$.

□

Remark 3. In the case $dim(\mathfrak{g}_{\bar{1}}) = 1$ it is obtained a suitable homogeneous basis of $\mathfrak{g}$, $\{X_0, X_1, \ldots, X_{2r}, Y_1\}$ such that

$$\left\{ [X_{2i}, X_{2i+1}] = X_{2r}, \quad 0 \leq i \leq r-1 \right.$$

4 Computational Processing

Classification problems up to isomorphisms can be reduced to determinate the structure constants of non-isomorphic superalgebras. As the structure constant depend on the basis, it is convenient to find a basis for which one has a maximum number of structure constants equal to zero. We will consider the basis found in Theorem 1.

First, for the classifications of the HSA it is necessary to compute all graded Jacobi identity. It is here where we use this program.

The program can be separated in three parts:

1. We introduce some conditions about $\mathfrak{g}$, that is, $\mathfrak{g}_{\bar{0}}$ is a Lie algebra and $\mathfrak{g}_{\bar{1}}$ is a $\mathfrak{g}_{\bar{0}}$-module. (See definition 2)

```
r1 = 2;
dim = 2 r1 + 1;
dim1 = 2;
basemu0 = Table[x[i], {i, 0, dim}];
basemu1 = Table[y[j], {j, 1, dim1}];

d[i_, j_] := Subscript[D, i, j];
e[i_, j_, k_] := Superscript[Subscript[e, i, j], k];

(*mu0, Lie algebra*)

mu0[0, x_] := 0;
mu0[x_, 0] := 0;
mu0[x_, x_] := 0;
mu0[x_, y_] := Simplify[-mu0[y, x]] /; OrderedQ[{y, x}];
mu0[x_ + y_, z_] := Simplify[mu0[x, z] + mu0[y, z]];
mu0[z_, x_ + y_] := Simplify[mu0[z, x] + mu0[z, y]];
mu0[x_, a_ y_] := a mu0[x, y];
mu0[a_ x_, y_] := a mu0[x, y];
```

```
(*mu1*)

mu1[0, x_] := 0;
mu1[x_, 0] := 0;
mu1[x_, y_] := Simplify[mu1[y, x]] /; OrderedQ[{y, x}];
mu1[x_ + y_, z_] := Simplify[mu1[x, z] + mu1[y, z]];
mu1[z_, x_ + y_] := Simplify[mu1[z, x] + mu1[z, y]];
mu1[x_, a_ y_] := a mu1[x, y];
mu1[a_ x_, y_] := a mu1[x, y];

(*mu2*)

mu2[0, x_] := 0;
mu2[x_, 0] := 0;
mu2[x_, y_] := Simplify[-mu2[y, x]] /; OrderedQ[{y, x}];
mu2[x_ + y_, z_] := Simplify[mu2[x, z] + mu2[y, z]];
mu2[z_, x_ + y_] := Simplify[mu2[z, x] + mu2[z, y]];
mu2[x_, a_ y_] := a mu2[x, y];
mu2[a_ x_, y_] := a mu2[x, y];
```

2. We introduce the expression of the brackets of the possible Lie superalgebras. Still we do not known if the law corresponds or not to a Lie superalgebra.

```
(*The family H_r*)

For[i = 0, i <= r1 - 1, i++,
    mu0[x[2i], x[2i + 1]] = x[2r1]];
mu0[x[i_], x[j_]] := 0;

(*the remain of bracket products*)

For[i = 0, i <= 2r1 - 1, i++,
    mu2[x[i], y[1]] = d[i, 1] y[2]];

If[i > 2r1 - 1, mu2[x[i], y[1]] = 0];

For[i = 0, i <= 2r1, i++,
    mu2[x[i], y[2]] = 0];

For[i = 1, i <= dim1, i++,

  For[j = 1, j <= dim1, j++,
    mu1[y[i], y[j]] =\sum_{k=0}^{k=2r}e[i, j, k] x[k]]];
```

3. We compute the graded Jacobi identity. We obtain the restrictions between the structure constants.

```
    (*Graded Jacobi identity*)

jac0[i_Integer, j_Integer, k_Integer] := Collect[
      mu0[x[i], mu0[x[j], x[k]]] + mu0[x[j], mu0[x[k], x[i]]] +
          mu0[x[k], mu0[x[i], x[j]]], basemu0];

jac1[i_Integer, j_Integer, k_Integer] := Collect[
      mu2[y[i], mu1[y[j], y[k]]] + mu2[y[j], mu1[y[k], y[i]]] +
          mu2[y[k], mu1[y[i], y[j]]], basemu1];

jac2[i_Integer, j_Integer, k_Integer] := Collect[
      mu2[x[i], mu2[x[j], y[k]]] - mu2[mu0[x[i], x[j]], y[k]] -
          mu2[x[j], mu2[x[i], y[k]]], basemu1];

jac3[i_Integer, j_Integer, k_Integer] := Collect[
      mu0[x[i], mu1[y[j], y[k]]] - mu1[mu2[x[i], y[j]], y[k]] -
          mu1[y[j], mu2[x[i], y[k]]], basemu0];

Jacobimu0 := Module[{i, j, k},
      For[i = 0, i <= dim - 3, i++,
        For[j = i + 1, j <= dim - 2, j++,
                For[k = j + 1, k <= dim - 1, k++,

            If[jac0[i, j, k] == 0,
              Print["JAC-0(", i, j, k")->", jac0[i, j, k]], {},
              Print["JAC-0(", i, j, k")->", jac0[i, j, k]]]]]]];


Jacobimu1 := Module[{i, j, k},
      For[i = 1, i <= dim1, i++,
        For[j = 1, j <= dim1, j++,
                For[k = 1, k <= dim1, k++,

            If[jac1[i, j, k] == 0,
              Print["JAC-1(", i, j, k")->", jac1[i, j, k]], {},
              Print["JAC-1(", i, j, k")->", jac1[i, j, k]]]]]]];

Jacobimu0mu2 := Module[{i, j, k},
      For[i = 0, i <= dim - 2, i++,
        For[j = i + 1, j <= dim - 1, j++,
                For[k = 1, k <= dim1, k++,

            If[jac2[i, j, k] == 0,
              Print["JAC-2(", i, j, k")->", jac2[i, j, k]], {},
              Print["JAC-2(", i, j, k")->", jac2[i, j, k]]]]]]];
```

```
JacobimuOmulmu2 := Module[{i, j, k},
      For[i = 0, i <= dim - 1, i++,
        For[j = 1, j <= dim1, j++,
                For[k = 1, k <= dim1, k++,

            If[jac3[i, j, k] == 0,
              Print["JAC-3(", i, j, k")->", jac3[i, j, k]], {},
              Print["JAC-3(", i, j, k")->", jac3[i, j, k]]]]]];

lista0 = Select[
      Flatten[Table[
          Coefficient[jac0[i, j, k], basemu0],
          {i, 0, dim - 3}, {j, i + 1, dim - 2},
          {k, j + 1, dim - 1}]], ! NumberQ[#] &];
lista1 = Select[
      Flatten[Table[
          Coefficient[jac1[i, j, k], basemu1],
          {i, 1, dim1}, {j, 1, dim1},
          {k,1, dim1}]], ! NumberQ[#] &];
lista2 = Select[
      Flatten[Table[
          Coefficient[jac2[i, j, k], basemu1],
          {i, 0, dim - 2}, {j, i + 1,dim - 1},
          {k, 1, dim1}]], ! NumberQ[#] &];
lista3 = Select[
      Flatten[Table[
          Coefficient[jac3[i, j, k], basemu0],
          {i, 0, dim - 1}, {j, 1,dim1},
          {k, 1, dim1}]], ! NumberQ[#] &];
lista = Join[lista0, lista1, lista2, lista3];
ecuaciones = Map[(# == 0) &, lista]
```

Remark 4. The program that we explicit allows to obtain the relations between the structure constants for concrete dimensions. By induction we can compute the relations for Lie superalgebras in arbitrary dimension.

In this paper, we have obtained the classifications for some concrete dimensions with the aid of the program. Then, we have induced the expressions for the general case and finally, we have proved the general result.

5 Application: Classification of HSA with Arbitrary Dimension of Even Part and Dimension of Odd Part up to Two

We present the classification of the HSA with arbitrary dimension of even part and dimension of odd part up to two, with aid of the program.

Theorem 3. *If $\mathfrak{g}$ is a HSA with arbitrary dimension of even part and dimension of odd part up to two, then it is isomorphic to one of the superalgebras, pairwise non-isomorphic, whose laws can be expressed in a suitable basis by*

$\mathfrak{g}_1$:

$$\left\{ [X_{2i}, X_{2i+1}] = X_{2r},\ 0 \le i \le r-1 \right.$$

$\mathfrak{g}_2$:

$$\begin{cases} [X_{2i}, X_{2i+1}] = X_{2r},\ 0 \le i \le r-1 \\ [Y_1, Y_1] = X_{2r} \end{cases}$$

$\mathfrak{g}_3$:

$$\begin{cases} [X_{2i}, X_{2i+1}] = X_{2r},\ 0 \le i \le r-1 \\ [Y_1, Y_2] = X_{2r} \end{cases}$$

$\mathfrak{g}_{2j}$:

$$\begin{cases} [X_{2i}, X_{2i+1}] = X_{2r}, & 0 \le i \le r-1 \\ [X_{2i}, Y_1] = Y_2, & 0 \le i \le r-j+1 \end{cases}$$

with $2 \le j \le r+1$

$\mathfrak{g}_{2j+1}$:

$$\begin{cases} [X_{2i}, X_{2i+1}] = X_{2r}, & 0 \le i \le r-1 \\ [X_{2i}, Y_1] = Y_2, & 0 \le i \le r-j+1 \\ [Y_1, Y_1] = X_{2r} \end{cases}$$

with $2 \le j \le r+1$

$\mathfrak{g}_{2r+j+4}$:

$$\begin{cases} [X_{2i}, X_{2i+1}] = X_{2r} & 0 \le i \le r-1 \\ [X_{2i}, Y_1] = Y_2, & 0 \le i \le j \\ [Y_1, Y_1] = 2X_1 \\ [Y_1, Y_2] = X_{2r} \end{cases}$$

with $0 \le j \le r-1$

Proof. From Theorem 1, we can consider two cases:

Case 1: $\epsilon = 0$.

In this case, we have $[\mathfrak{g}_{\overline{0}}, Y_1] = 0$ ($\mathfrak{g}_{\overline{1}}$ is a trivial $\mathfrak{g}_{\overline{0}}$-module).

It is easy to see that $[Y_i, Y_j] \subset < X_{2r} >$. In other case, if there exists k such that $X_{2k} \in [Y_i, Y_j]$, $0 \le k \le 2r-1$, ($X_k \in [Y_i, Y_j] \Leftrightarrow E_{ij}^k \neq 0$) from $J_g(Y_i, Y_j, X_{k+1})$ or $J_g(Y_i, Y_j, X_{k-1})$ (k even or odd, respectively) we obtain that $X_{2r} = 0$, that is a contradiction.

The family of laws to be classified is:

$$\begin{cases} [X_{2i}, X_{2i+1}] = X_{2r},\ 0 \le i \le r-1 \\ [Y_1, Y_1] = E_{11}^{2r} X_{2r} \\ [Y_1, Y_2] = E_{12}^{2r} X_{2r} \\ [Y_2, Y_2] = E_{22}^{2r} X_{2r} \end{cases}$$

We can suppose that $E_{22}^{2r}=0$. In fact,

- If $E_{11}^{2r}=0$, we make the change $Y_1'=Y_2,\ Y_2'=Y_1$.
- If $E_{11}^{2r}\neq 0$, we make the change $Y_1'=Y_1,\ Y_2'=cY_1+Y_2$, with

$$c=\frac{-E_{12}^{2r}\pm\sqrt{(E_{11}^{2r})^2-E_{11}^{2r}E_{22}^{2r}}}{E_{11}^{2r}}$$

Thus, we obtain the family

$$\begin{cases}[X_{2i},X_{2i+1}]=X_{2r},\ 0\le i\le r-1\\ [Y_1,Y_1]=E_{11}^{2r}X_{2r},\\ [Y_1,Y_2]=E_{12}^{2r}X_{2r}.\end{cases}$$

- If $E_{12}^{2r}=0$, we find the superalgebras $\mathfrak{g}_1$ (if $E_{11}^{2r}=0$) or $\mathfrak{g}_2$ (if $E_{11}^{2r}\neq 0$).
- If $E_{12}^{2r}\neq 0$, we proceed to the change of basis $Y_1'=\dfrac{1}{E_{12}^{2r}}Y_1-\dfrac{E_{11}^{2r}}{2E_{12}^{2r}}Y_2,\ Y_2'=Y_2$ and we obtain $\mathfrak{g}_3$

Case 2: $\epsilon=1$.

- From $J_g(X_0,Y_1,Y_1)$ we have that $[Y_1,Y_2]\subset Im\ ad(X_0)=<X_{2r}>$.
- From $J_g(X_0,Y_1,Y_2)$ we have that $[Y_2,Y_2]=0$.

Then the family of laws to be classified is

$$\begin{cases}[X_{2i},X_{2i+1}]=X_{2r}, & 0\le i\le r-1\\ [X_0,Y_1]=Y_2\\ [X_i,Y_1]=D_{i,1}^2Y_2, & 1\le i\le 2r-1\\ [Y_1,Y_1]=\sum\limits_{k=0}^{2r}E_{11}^kX_k\\ [Y_1,Y_2]=E_{12}^{2r}X_{2r}\end{cases}$$

Next, we compute the graded Jacobi identity (with aid of the program)

- From $J_{\mathfrak{g}}(Y_1,Y_1,Y_1)$ we obtain $E_{11}^0+\sum\limits_{k=1}^{2r-1}E_{11}^kD_{k,1}^2=0$.
- From $J_{\mathfrak{g}}(X_{2i},Y_1,Y_1)$ with $0\le i\le r-1$ we obtain

$$\begin{array}{ll}E_{11}^{2i+1}=2D_{2i,1}^2E_{12}^{2r}, & 1\le i\le r-1\\ E_{11}^1=2E_{12}^{2r}, & i=0\end{array}$$

- From $J_{\mathfrak{g}}(X_{2i+1},Y_1,Y_1)$ with $0\le i\le r-1$ we obtain

$$E_{11}^{2i}=-2D_{2i+1,1}^2E_{12}^{2r},\quad 0\le i\le r-1.$$

We can suppose that $D^2_{2i+1,1} = 0,\ 0 \leq i \leq r-1$. If $D^2_{2i,1} \neq 0$ the following change of basis

$$\begin{cases} X'_{2i} &= \dfrac{1}{D^2_{2i,1}} X_{2i}, \\ X'_{2i+1} &= -D^2_{2i,1} X_{2i+1} + D^2_{2i+1,1} X_{2i}. \end{cases}$$

permits to suppose that $D^2_{2i+1,1} = 0$ and $D^2_{2i,1} = 1$. If $D^2_{2i,1} = 0$ we proceed to the following change

$$\begin{cases} X'_{2i} &= X_{2i+1}, \\ X'_{2i+1} &= X_{2i}. \end{cases}$$

moreover, $X'_{2r} = -X_{2r}$.

Then, we obtain that $E^{2i}_{11} = 0,\ 0 \leq i \leq r-1$ (by means of the restrictions), now the family is

$$\begin{cases} [X_{2i}, X_{2i+1}] = X_{2r}, & 0 \leq i \leq r-1 \\ [X_0, Y_1] = Y_2 & \\ [X_{2i}, Y_1] = D^2_{2i,1} Y_2, & 1 \leq i \leq r-1, \\ [Y_1, Y_1] = 2E^{2r}_{12} \left(X_1 + \displaystyle\sum_{i=1}^{r-1} D^2_{2i,1} X_{2i+1} \right) + E^{2r}_{11} X_{2r} & \\ [Y_1, Y_2] = E^{2r}_{12} X_{2r} & \end{cases}$$

with $D^2_{2i,1} \in \{0, 1\},\ 0 \leq i \leq r-1$.

2.1. If $E^{2r}_{12} = 0$,

$$\begin{cases} [X_{2i}, X_{2i+1}] = X_{2r},\ 0 \leq i \leq r-1, \\ [X_0, Y_1] = Y_2, \\ [X_{2i}, Y_1] = D^2_{2i,1} Y_2,\ 1 \leq i \leq r-1, \\ [Y_1, Y_1] = E^{2r}_{11} X_{2r}. \end{cases}$$

2.1.1. If $D^2_{2i,1} = 1,\ 1 \leq i \leq r-1$ for all i. The family of laws is

$$\begin{cases} [X_{2i}, X_{2i+1}] = X_{2r},\ 0 \leq i \leq r-1, \\ [X_0, Y_1] = Y_2, \\ [X_{2i}, Y_1] = Y_2, \qquad 1 \leq i \leq r-1, \\ [Y_1, Y_1] = E^{2r}_{11} X_{2r}. \end{cases}$$

and we obtain $\mathfrak{g}_4$ (if $E^{2r}_{11} = 0$) or $\mathfrak{g}_5$ (if $E^{2r}_{11} \neq 0$).

2.1.2. If $D^2_{2i,1} = 1$, for all i such that $1 \leq i \leq r-2$, and $D^2_{2r-2,1} = 0$. In the same way, we have $\mathfrak{g}_6$ (if $E^{2r}_{11} = 0$) and $\mathfrak{g}_7$ (if $E^{2r}_{11} \neq 0$).

2.1.j. If $D^2_{2i,1} = 1$, for all i such that $1 \leq i \leq j$, and $D^2_{2i,1} = 0$, for all i such that $j+1 \leq i \leq r-1$. We have $\mathfrak{g}_{2(r-j+1)}$ (if $E^{2r}_{11} = 0$) and $\mathfrak{g}_{2(r-j+1)+1}$ (if $E^{2r}_{11} \neq 0$).

2.1.r. If $D^2_{2i,1} = 0$, for all i such that $1 \leq i \leq r-1$. We obtain $\mathfrak{g}_{2(r+1)}$ (if $E^{2r}_{11} = 0$) and $\mathfrak{g}_{2(r+1)+1}$ (if $E^{2r}_{11} \neq 0$).

2.1.– In other case, we can make a simple change of basis to reduce this case to some of the previous cases.

2.2. If $E_{12}^{2r} \neq 0$. Making this change of basis

$$\begin{cases} X_0' &= X_0, \\ X_1' &= E_{12}^{2r}\left(X_1 + \sum_{i=1}^{r-1} D_{2i,1}^2 X_{2i+1}\right) + \frac{1}{2}E_{11}^{2r}X_{2r}, \\ X_{2i}' &= X_{2i}, & 1 \le i \le r-1, \\ X_{2i+1}' &= E_{12}^{2r}X_{2i+1}, & 1 \le i \le r-1, \\ X_{2r}' &= E_{12}^{2r}X_{2r} \\ Y_j' &= Y_i, & j = 1,2. \end{cases}$$

we lead to

$$\begin{cases} [X_{2i}, X_{2i+1}] = X_{2r}, \ 0 \le i \le r-1, \\ [X_0, Y_1] = Y_2, \\ [X_{2i}, Y_1] = D_{2i,1}^2 Y_2, \ 1 \le i \le r-1, \\ [Y_1, Y_1] = 2X_1, \\ [Y_1, Y_2] = X_{2r}. \end{cases}$$

2.2.1. If $D_{2i,1}^2 = 1$, for all i such that $1 \le i \le r-1$, we obtain $\mathfrak{g}_{3r+3}$.

2.2.j. If $D_{2i,1}^2 = 1$, for all i such that $1 \le i \le j$, and $D_{2i,1}^2 = 0$, for all i such that $j+1 \le i \le r-1$, we obtain $\mathfrak{g}_{2r+j+4}$.

2.2.r. If $D_{2i,1}^2 = 0$, for all i such that $1 \le i \le r-1$, we obtain $\mathfrak{g}_{2r+4}$.

2.1.– In other case, we can make a simple change of basis to reduce this case to some of the previous cases.

The above Lie superalgebras are pairwise non-isomorphic. In fact, it is enough to observe the next table.

Table 1. Resume

Superalgebras	$dim(\mathcal{Z}(\mathfrak{g}))$	$dim(C^1(\mathfrak{g}))$	$dim(Cent_{\mathfrak{g}}(\mathfrak{g}_1))$
$\mathfrak{g}_1$	3	1	
$\mathfrak{g}_2$	2	1	
$\mathfrak{g}_3$	1	1	
$\mathfrak{g}_{2j} \quad 2 \le j \le r+1$	2	2	r+j+1
$\mathfrak{g}_{2j+1} \quad 2 \le j \le r+1$	2	2	r+j
$\mathfrak{g}_{2r+j+4} \quad 0 \le j \le r-1$	1	3	

References

1. G.G.A. Bäuerle, E.A. De Kerf. *Lie Algebras Part 1*. Studies in Mathematical Physics I. Elsevier, 1990.
2. K. Bauwens, L. Le Bruyn. *Some remarks on solvable Lie superalgebras*. Jour. of Pure and App. Alg. 99 (1995) 113-134.
3. L. Corwin, Y. Ne'eman, S. Sternberg. Rev. Mod. Phys. 47 (1975) 573.
4. M. Gilg, *Super-algèbres*. PhD thesis, Mulhouse, 2000.
5. N. Jacobson. *Lie algebras*. Interscience Publishers, Wiley, New York, (1962).
6. V.G. Kac. *Lie Superalgebras*. Advances in Mathematics 26, 8-96 (1977).
7. D.A. Leites. *Lie superalgebras*. JOSMAR, 30,n 6, 1984, 2481-2513.
8. D.A. Leites. *Towards classification of simple Lie superalgebras*. In: Chan L-L., Nahm W. (eds.) Differential geometric methods in theorical physics (Davis, CA, 1988) NATO Adv. Sci. Inst. Ser. B Phys., 245, Plenum, New York, 1990,633-651.
9. D.H. Sattinger, O.L. Weaver. *Lie Groups and Algebras with Applications to Physics, Geometry, and Mechanics*. Springer-Verlag New York Inc., 1986.
10. M. Scheunert. *The Theory of Lie Superálgebras*. Lecture Notes in Math. 716 (1979).

Neighborhoods of an Ordinary Linear Differential Equation

Giuseppa Carrà Ferro and Valentina Marotta

Department of Mathematics and Computer Science
Viale A. Doria, 6 - 95125, Catania, Italy
{carra,marotta}@dmi.unict.it *

Abstract. In this paper the notion of neighborhood of a linear ordinary differential equation with constant coefficients is introduced. The general concept of neighborhood is often used in Scientific Computation where involved data have limited accuracy. In this context we introduce the notion of pseudo-integral of a differential equation. Furthermore we use the notion of differential resultant in order to know the relations between the integrals of a linear ordinary differential equation with constant coefficients and the integrals of a linear ordinary differential equation in its neighborhood.

1 Introduction

Many data are often "uncertain" if either they come from physical measurements and observations or they are the output of numerical computations. Another source of error in data is due to the fact that computers work with the Floating Point arithmetic, that is a finite arithmetic. So a real number is represented approximately because of the limited number of digits. For all these reasons it is growing up the necessity of combining methods in computer algebra and numerical analysis. If we know that a value is not exact we represent it by an interval. For instance, after measuring a physical value $\bar{x}$ we can only conclude that the actual (unknown) value lies within the interval $[\bar{x}-\eta, \bar{x}+\eta]$, where η is the upper bound on the measurement error (depending on the measuring instrument). The notions of neighborhood of a polynomial are discussed in [ST1,ST2] and by one of the authors in [MV1]. Here we extend these concepts to the ordinary linear differential equations with constant coefficients. In many real life situations the coefficients of the differential equation cannot be expressed exactly. For instance we consider a well-known differential equation, which is used in electronic circuits applications:

$$L\frac{d^2y}{dt^2} + R\frac{dy}{dt} + \frac{1}{C}y = 0$$

where $L \geq 0$ is the inductance, $R \geq 0$ is the resistance and $C \geq 0$ is the capacity; for example if $C = 3$, $1/C$ is an irrational value. If we know that the coefficients have limited accuracy, we work with the family of the differential

* Supported by INTAS grant n.99-1222.

operators having coefficients within a given tolerance instead of working with a single differential operator: all these operators form the differential operator neighborhood. We define the concept of pseudo-integral of a differential equation as the integral of a differential equation in its neighborhood and we focus its numerical meaning, when we look for the integrals of the initial differential equations. Furthermore some relations between the differential equation and the differential equations in its neighborhood can be investigated by using the differential resultant.

The paper is organized as follows: Section 2 sketches some preliminary notions about the differential operators and the used notation; in Section 3 we define the differential equation neighborhood; in Section 4 the concept of pseudo-integral and its applications are discussed; in Section 5 we briefly analyze the concept of differential resultant; in Section 6 common solutions of an ordinary linear differential equation with constant coefficients and ordinary linear differential equations in its neighborhood are studied using the differential resultant and its properties. For our experiments we use two computer algebra packages: MAPLE 6 [GMH1] and MATHEMATICA 4.1 [WO1]. The use of MATHEMATICA 4.1 is essential because this package works with interval arithmetic [AC1,HA1].

Finally the authors will use often general concepts and notions of differential algebra, since many results in the paper can be extended in the future to many other cases, for example either to the case of ordinary algebraic differential equations with coefficients in a differential field of functions or to the case of linear partial differential equations with constant coefficients.

2 Preliminaries and Notation

Here we introduce some general concepts and notations in differential algebra. We deal with a commutative ring of functions $\mathbb{K}$ in one independent variable x, that contains the field $\mathbb{Q}$. Every undefined notion is as in [KO1]. A *derivation* on $\mathbb{K}$ is an application $d : \mathbb{K} \to \mathbb{K}$ such that:

1. $d(a+b) = d(a) + d(b) \;\; \forall a, b \in \mathbb{K}$
2. $d(a \cdot b) = ad(b) + bd(a) \;\; \forall a, b \in \mathbb{K}$

$(\mathbb{K}, d)$ is called a differential ring; if $\mathbb{K}$ is a field, then $(\mathbb{K}, d)$ is called a differential field.

Example 1. $\mathbb{K} = C^\infty(\mathbb{R})$, $\mathbb{Q}[x]$, $\mathbb{R}[x]$, $\mathbb{C}[x]$ with $d = \partial/\partial x$ are all examples of differential rings; $\mathbb{K} = \mathbb{Q}(x)$, $\mathbb{C}(x)$, $\mathbb{C}(x, e^{c_i x}\colon c_i \in \mathbb{C},\, i \in I)$ with $d = \partial/\partial x$ are examples of differential fields [GCL1].

In this paper we will assume that $(\mathbb{K}, d)$ is a differential field of functions, i.e. $(\mathbb{K}, d)$ is a differential extension of $\mathbb{C}(x)$ with exponential functions, in other words $(\mathbb{K}, d)$ is a field of elementary functions over $\mathbb{C}$ ([GCL1]). Denote with $C_{\mathbb{K}} = \{\varrho \in \mathbb{K} : \delta(\varrho) = 0\}$ the differential ring of constants of $\mathbb{K}$. If $(\mathbb{K}, d)$ is as above then $C_{\mathbb{K}} = \mathbb{C}$. Let us assume to work with the derivation $d = \partial/\partial x$ and, for brevity, we denote it by δ. Let $\mathbb{K}[\delta]$ be the ring of the ordinary differential operators with coefficients in $\mathbb{K}$. If $D_a^n \in \mathbb{K}[\delta]$, then

$$D_a^n = \sum_{i=0}^{n} a_i \delta^i \tag{1}$$

where $a = (a_0, \ldots, a_n)$, $a_i \in \mathbb{K}$, $a_n \neq 0$ and n is the order of the operator. We have the following rules for operations over $\mathbb{K}[\delta]$: if $D_a^n = \sum_{i=0}^{n} a_i \delta^i$ and $D_b^m = \sum_{j=0}^{m} b_j \delta^j$, then

1. $D_a^n + D_b^m = \sum_{k=0}^{max(m,n)} (a_k + b_k)\delta^k$
2. $D_a^n \cdot D_b^m = \sum_{k=0}^{m+n} c_k \delta^k$ where

$$c_k(a,b) = \sum_{s=max(0,k-n)}^{min(m,k)} \sum_{i=s}^{n} \binom{i}{s} b_{k-s}^{(i-s)} a_i \quad k = 0, \ldots, m+n$$

and $b_{k-s}^{(i-s)}$ is the $(i-s)$-th derivative of b_{k-s}.

If $D_a^n = \delta^r$ (i.e. $n = r$, $a_r = 1$ and $a_i = 0\ \forall i \neq r$), D_c^{r+m} has the following coefficients:

$$c_{r,k}(b) = \sum_{s=max(0,k-r)}^{min(r,k)} \binom{r}{s} b_{k-s}^{(r-s)}$$

for $k = 0, \ldots, m+r$ and $c_{r,k} = 0$ for $k < 0$ and $k > m + r$.

Definition 1. *Let $(\mathbb{K}, d)$ be a differential field. $S = \mathbb{K}\{y\} = \mathbb{K}[\delta^n y : n \in \mathbb{N}_0]$ is the differential ring of the differential polynomials in the differential indeterminate y with coefficients in the differential ring $\mathbb{K}$; we denote $\delta^n y$ by $y^{(n)}$.*

Given $f \in S$ the *order* of f is $ord(f) = \max\{n \in \mathbb{N}_0$: f contains a power product in $y, \delta y, \ldots, \delta^n y$ with nonzero coefficients$\}$. By definition as above, an ordinary differential equation in one dependent variable y with coefficients in $\mathbb{K}$ is an equality $f = 0$ for some differential polynomial $f \in S$.

3 Differential Equation Neighborhood

Let $(\mathbb{K}, \delta)$ be a differential field such that $\mathbb{R} \subseteq C_{\mathbb{K}}$. Let $D_a^n = \sum_{j=0}^{n} a_j \delta^j \in \mathbb{K}[\delta]$ be a differential operator with $a \in \mathbb{R}^{n+1}$.

The *tolerance* e of D_a^n is a non negative vector $e = \{e_0, \ldots, e_n\}$, such that

$$e_j \in \mathbb{R} \quad e_j \geq 0 \quad \text{for } j = 0, \ldots, n$$

Definition 2. *The* **neighborhood** $\mathcal{N}(D_a^n, e)$ *of a differential operator* D_a^n *in* $\mathbb{K}[\delta]$, $a \in \mathbb{R}^{n+1}$, *with tolerance* e *is the family of differential operators* $\tilde{D}_a^n$ *in* $\mathbb{K}[\delta]$, $\tilde{D}_a^n = \sum_{j=0}^{n} \tilde{a}_j \delta^j$, *such that* $\tilde{a}_j$ *satisfies one of the following conditions for each* $j = 0, \ldots, n$*:*

$$\tilde{a}_j = a_j \quad \textit{if } e_j = 0, \textit{ i.e. } a_j \textit{ is exact}$$

$$|\tilde{a}_j - a_j| \leq e_j \quad \textit{otherwise}$$

When $|\tilde{a}_j - a_j| \leq e_j$ we have that the exact value $\tilde{a}_j \in [a_j - e_j, a_j + e_j]$, i.e. $\tilde{a}_j = a_j + \epsilon_j$ such that $|\epsilon_j| \leq e_j$. ϵ_j is called *perturbation* for all j and the generic operator $\tilde{D}^n_a$ in the neighborhood is called *perturbed.*

This definition can be extended to the more general case of linear differential polynomials. Let $f_a = D^n_a(y) + a_{-1} \in S$ be a linear differential polynomial with $a = (a_{-1}, a_0, \ldots, a_n) \in \mathbb{R}^{n+2}$.
The *tolerance* e of f_a is a non negative vector $e = \{e_{-1}, e_0, \ldots, e_n\}$, such that

$$e_j \in \mathbb{R} \quad e_j \geq 0 \quad \text{for } j = -1, 0, \ldots, n$$

Definition 3. *The* **neighborhood** $\mathcal{N}(f_a, e)$ *of a linear differential polynomial* $f_a = D^n_a(y) + a_{-1} \in S$ *with tolerance* e *is the family of linear differential polynomials* $\tilde{f}_a = \tilde{D}^n_a(y) + \tilde{a}_{-1} \in S$ *with* $\tilde{D}^n_a = \sum_{j=0}^{n} \tilde{a}_j \delta^j$, *such that* $\tilde{a}_j$ *satisfies one of the following conditions for each* $j = -1, 0, \ldots, n$:

$$\tilde{a}_j = a_j \quad \textit{if } e_j = 0, \;\; \textit{i.e. } a_j \textit{ is exact}$$

$$|\tilde{a}_j - a_j| \leq e_j \quad \textit{otherwise}$$

ϵ_j is still called *perturbation* for all j and the generic differential polynomial $\tilde{f}_a$ in the neighborhood is called *perturbed.*

4 Pseudo-Integral of a Differential Equation

The treatment of data with limited accuracy is based on the *backward error principle.* An approximate result is meaningful if it is the exact result for a problem within a tolerance neighborhood of the specified problem. According with this definition the notion of integral of a differential equation must be replaced by the notion of pseudo-integral.

Definition 4. *Let* $(\mathbb{K}, \delta)$ *be a differential field such that* $\mathbb{R} \subseteq C_{\mathbb{K}}$. *Let* D^n_a *in* $\mathbb{K}[\delta]$ *with* $a \in \mathbb{R}^{n+1}$ *be a linear ordinary differential operator. A function* $\phi(x) \in \mathbb{K}$ *is a* pseudo-integral *of the homogeneous linear ordinary differential equation* $D^n_a(y) = 0$ *with tolerance* e *if it is an integral of the ordinary linear homogeneous differential equation* $\tilde{D}^n_a(y) = 0$ *for some operator* $\tilde{D}^n_a \in \mathcal{N}(D^n_a, e)$.

This notion can be extended to the more general case of ordinary linear differential equations in one dependent variable y.

Definition 5. *Let* $(\mathbb{K}, \delta)$ *be a differential field such that* $\mathbb{R} \subseteq C_{\mathbb{K}}$. *Let* $f_a = D^n_a(y) + a_{-1} \in S$ *with* $a \in \mathbb{R}^{n+2}$ *be a linear ordinary differential polynomial. A function* $\phi(x) \in \mathbb{K}$ *is a* pseudo-integral *of the linear ordinary differential equation* $f_a = 0$ *with tolerance* e *if it is an integral of some linear differential equation* $\tilde{f}_a = \tilde{D}^n_a(y) + \tilde{a}_{-1} = 0$ *with* $\tilde{f}_a \in \mathcal{N}(f_a, e)$.

Definition 6. *Let* $(\mathbb{K}, \delta)$ *be a differential field such that* $\mathbb{R} \subseteq C_{\mathbb{K}}$*. Let* D_a^n *with* $a \in \mathbb{R}^{n+1}$ *be a linear ordinary differential operator. A* pseudo-integral domain *of the homogeneous ordinary linear differential equation* $D_a^n(y) = 0$ *with tolerance* e *is the set of all pseudo-integrals*

$$\mathcal{I}(D_a^n, e) = \{\phi(x) \in \mathbb{K} : \exists \tilde{D}_a^n \in \mathcal{N}(D_a^n, e) : \tilde{D}_a^n(\phi(x)) = 0\}$$

In similar way we have the following definition.

Definition 7. *Let* $(\mathbb{K}, \delta)$ *be a differential field such that* $\mathbb{R} \subseteq C_{\mathbb{K}}$*. Let* $f_a = D_a^n(y) + \tilde{a}_{-1}$ *with* $a \in \mathbb{R}^{n+2}$ *be a linear ordinary differential polynomial. A* pseudo-integral domain *of the linear differential equation* $f_a = 0$ *with tolerance* e *is the set of all pseudo-integrals*

$$\mathcal{I}(f_a, e) = \{\phi(x) \in \mathbb{K} : \exists \tilde{f}_a \in \mathcal{N}(f_a, e) : \tilde{f}_a(\phi(x)) = 0\}$$

We are interested in investigating the pseudo-integral domain of an ordinary linear differential equation with constant coefficients.

Given a differential operator D_a^n, let $f_a = D_a^n(y) = \sum_{k=0}^n a_k y^{(k)}(x)$ be the related linear homogeneous differential polynomial. Denote with $p_{D_a^n}(t) = \sum_{k=0}^n a_k t^k \in \mathbb{R}[t]$ the characteristic polynomial of f_a.

It is well known in differential equation theory that there is a one to one correspondence between the set of all zeros with their multiplicity of the characteristic polynomial of f_a and a set of distinct integrals of the differential equation $f_a = 0$, that are linear independent over the field of constants. We shall use such algebraic equation in order to solve the differential equation $f_a = 0$.

A value $\lambda \in \mathbb{C}$ is a zero of $p_{D_a^n}(t)$ if and only if $e^{\lambda t}$ is an integral of $f_a = 0$. So we have to compute the zeros of the equation $p_{D_a^n}(t) = 0$. Since the coefficients $(a_0, \ldots, a_n)$ of the operator D_a^n have limited accuracy, then the coefficients of the polynomial $p_{D_a^n}(t)$ have the same accuracy. It follows that $p_{D_a^n}(t)$ represents a family of polynomials named *polynomial neighborhood* [MV1]. For a polynomial neighborhood we can find the pseudozeros, i.e. the zeros of all perturbed polynomials in the neighborhood. Since the characteristic polynomial has degree n, then it has n zeros $\{z_j = \alpha_j + i\beta_j \colon (\alpha_j, \beta_j) \in \mathbb{R}^2, j = 1, \ldots, n\}$, in the field of complex numbers $\mathbb{C}$. When we work in the polynomial neighborhood they can be identified with pairs of n intervals, that are called *pseudozero components.* In terms of integral of the differential equation $f_a = 0$, instead of a function we find a class of functions. For instance suppose that the function $\phi(x) = ce^{\lambda x}$, $c \in \mathbb{R}$ is an integral of $f_a = 0$ coming from a zero $\lambda = \alpha + i\beta$ of $p_{D_a^n}(t)$ when we approximate the real coefficients. The integral $\phi(x)$ becomes the pseudo-integral $ce^{([\lambda_{1\alpha}, \lambda_{2\alpha}] + i[\lambda_{1\beta}, \lambda_{2\beta}])x}$, $\lambda_{1\alpha}, \lambda_{2\alpha}, \lambda_{1\beta}, \lambda_{2\beta} \in \mathbb{R}$, where $[\lambda_{1\alpha}, \lambda_{2\alpha}]$ and $[\lambda_{1\beta}, \lambda_{2\beta}]$ are equal to the intervals obtained by solving $p_{D_a^n}(t) = 0$ and by regarding its coefficients as uncertain, i.e. $[\lambda_{1\alpha}, \lambda_{2\alpha}] + i[\lambda_{1\beta}, \lambda_{2\beta}]$ is a pseudozero of $p_{D_a^n}(t)$.

For example suppose that we find n distinct zeros $\{z_1, \ldots, z_n\}$ of the characteristic polynomial and associate to each z_j the pseudozero $[\lambda_{1j\alpha}, \lambda_{2j\alpha}] + i[\lambda_{1j\beta}, \lambda_{2j\beta}]$. If all pseudozeros are distinct, then

$$i(x) = \sum_{k=0}^{n} c_k e^{([\lambda_{1k\alpha}, \lambda_{2k\alpha}] + i[\lambda_{1k\beta}, \lambda_{2k\beta}])x}$$

is the general pseudo-integral of $f_a = 0$. If the pseudozeros components are not distinct, we can have a zero of $p_{D_a^n}(t)$ with multiplicity h greater than one, i.e. an integral $\phi(x) = e^{\lambda x}(c_0 + c_1 x + \ldots + c_h x^h)$, $c_0, \ldots, c_h \in \mathbb{R}$ and a pseudo-integral $\phi(x) = e^{([\lambda_{1\alpha}, \lambda_{2\alpha}] + i[\lambda_{1\beta}, \lambda_{2\beta}])x}(c_0 + c_1 x + \ldots + c_h x^h)$, $c_0, \ldots, c_h \in \mathbb{R}$.

Let us see some numerical examples.

Example 2. Let $D = \delta^2 + \delta - 2$ be a differential operator and let $f = y'' + y' - 2y = 0$ be the related differential equation. By solving the characteristic polynomial $p_D(t) = t^2 + t - 2$ we find the zeros $\lambda_1 = -2$ and $\lambda_2 = 1$, so the general integral of $f = 0$ is $\phi(x) = c_1 e^{-2x} + c_2 e^x$, $c_1, c_2 \in \mathbb{R}$. Let us suppose that the coefficients of D are not exact and they have a tolerance $e = (10^{-1}, 10^{-1}, 10^{-1})$. The characteristic polynomial neighborhood has roots $[0.90, 1.10]$ and $[-2.25, -1.78]$. It follows that the general pseudo-integral is $i(x) = c_1 e^{[-2.25,-1.78]x} + c_2 e^{[0.90,1.10]x}$, where $e^{[\alpha,\beta]} = \{e^{\gamma x} : \alpha \leq \gamma \leq \beta\}$ by interval theory.

Example 3. Let $D = \delta^2 - 0.3\delta + 0.02$ be a differential operator and let $f = y'' - 0.3y' + 0.02y = 0$ be the related differential equation. The general integral of $f = 0$ is $\phi(x) = c_1 e^{0.1x} + c_2 e^{0.2x}$, $c_1, c_2 \in \mathbb{R}$. Let $e = (0, 10^{-1}, 10^{-2})$ be the tolerance vector; by using interval arithmetic the zeros of the polynomial $p_D(t) = t^2 + (-0.3 + [-10^{-1}, 10^{-1}])t + (0.02 + [-10^{-2}, 10^{-2}])$ are $\lambda_1 = \frac{[0.2,0.4]+\sqrt{[-0.08,0.12]}}{2}$ and $\lambda_2 = \frac{[0.2,0.4]-\sqrt{[-0.08,0.12]}}{2}$, i.e. the discriminant of any characteristic polynomial of a differential equation in the neighborhood can be positive, negative or zero. As a consequence, this differential equation is very sensitive to the change in its coefficients, in fact we can find three possible class of results: real distinct solutions, coincident real solutions and complex solutions. This means that a *bad* approximation in its coefficients can change drastically the solution.

Roughly speaking, it seems that small perturbations in the coefficients of the differential polynomial give small variations in the integrals of the differential equation associated. But it is not always true.

Example 4. Let $D = (\delta - 1)(\delta - 2) \cdots (\delta - 20) = \delta^{20} - 210\delta^{19} + \cdots + 20!$. The characteristic polynomial associated to the differential equation $D(y) = 0$ is a well-known polynomial in numerical analysis, named the *Wilkinson's polynomial.* If we change the coefficient -210 to $-210 + 10^{-7}$, the roots $1, 2, \ldots, 20$ of the polynomial change dramatically (it has a lot of complex roots).

A differential operator as in Example 4 is said to be *sensitive*, which means that the relative error in the result is much greater than the relative error in the data, i.e. it is *ill-conditioned.* Our analysis can be very useful in order to investigate more accurately this class of cases and check the sensitivity of a differential operator.

5 Resultant Theory in the Differential Case

The differential resultant problem for differential operators was first investigate by Ore. The definition of resultant of two ordinary differential operators was given by Berkovich and Tsirulik in 1986 [BT1] and extended to the partial case by one of the authors in 1994 [CF1]. Furthermore properties of subresultants of linear ordinary differential operators were studied by Chardin in 1991, while properties of subresultants of linear ordinary differential polynomials were studied by Li in 1995. Related studies were due by Riquier, Janet, Ritt and Kolchin [JA1,KO1,RQ1,RT1].

5.1 Differential Resultant of Differential Operators

The notion of resultant of two differential operators with constant coefficients coincides with the well known notion of Sylvester resultant of the corresponding characteristic polynomials [MC1].

Definition 8. *Let $D_a^n = \sum_{i=0}^n a_i\delta^i$ and $D_b^m = \sum_{j=0}^m b_j\delta^j$ be two differential operators with constant coefficients. The resultant matrix of D_a^n and D_b^m is the following $(n+m)\times(n+m)$ matrix:*

$$M(D_a^n, D_b^m) = \begin{pmatrix} a_n & \dots & a_0 & 0 & \dots & 0 \\ 0 & a_n & \dots & a_0 & 0 & \dots \\ \vdots & \ddots & \ddots & \dots & \ddots & \dots \\ 0 & \dots & 0 & a_n & \dots & a_0 \\ b_m & \dots & b_0 & 0 & \dots & 0 \\ 0 & b_m & \dots & b_0 & 0 & \dots \\ \vdots & \ddots & \ddots & \dots & \ddots & \dots \\ 0 & \dots & 0 & b_m & \dots & b_0 \end{pmatrix}$$

and the differential resultant $\delta Res(D_a^n, D_b^m)$ is the determinant of the matrix $M(D_a^n, D_b^m)$.

The definition of resultant of k ordinary differential operators with constant and non-constant coefficients and $k \geq 2$ can be found in [BT1].

5.2 Differential Resultant of Differential Polynomials

General Case. By using the Macaulay's definition of algebraic resultant of multivariate polynomials, the definition of resultant of two differential polynomials f_1 and f_2 was given in [CF2].

Definition 9. *Let $f_1, f_2 \in S$ with $ord(f_1) = n$ and $ord(f_2) = m$. The differential resultant of f_1 and f_2 is*

$$\delta Res(f_1, f_2) = Res(\delta^m f_1, \dots, \delta f_1, f_1, \delta^n f_2, \dots, \delta f_2, f_2) \quad (2)$$

In particular given two non-homogeneous ordinary linear differential equations $f_1 = D_a^n(y) + a_{-1} = 0$ and $f_2 = D_b^m(y) + b_{-1} = 0$, $\delta Res(f_1, f_2)$ is defined as the determinant of a $(n+m+2)\times(n+m+2)$ matrix $M(f_1, f_2)$, that is called resultant matrix of f_1 and f_2. The entries of $M(f_1, f_2)$ are the coefficients of the differential polynomials $\{\delta^m f_1, \ldots, \delta f_1, f_1, \delta^n f_2, \ldots, \delta f_2, f_2\}$, i.e.

$$\delta Res(f_1, f_2) = det(M(f_1, f_2)) = \det \begin{pmatrix} a_n & \ldots & a_0 & 0 & 0 & \ldots & 0 \\ 0 & a_n & \ldots & a_0 & 0 & 0 & \ldots \\ \vdots & \ddots & \ddots & \ldots & \ddots & \ldots & \vdots \\ 0 & \ldots & 0 & a_n & \ldots & a_0 & a_{-1} \\ b_m & \ldots & b_0 & 0 & 0 & \ldots & 0 \\ 0 & b_m & \ldots & b_0 & 0 & 0 & \ldots \\ \vdots & \ddots & \ddots & \ldots & \ddots & \ldots & \vdots \\ 0 & \ldots & 0 & b_m & \ldots & b_0 & b_{-1} \end{pmatrix}$$

Homogeneous Case. If the differential polynomials f_1 and f_2 are homogeneous, then by using the Macaulay's definition of algebraic resultant of multivariate homogeneous polynomials we have the following definition.

Definition 10. *Let $f_1 = D_a^n(y)$ and $f_2 = D_b^m(y)$ be in S, with $ord(f_1) = n$ and $ord(f_2) = m$. The differential resultant of f_1 and f_2 is*

$$\delta Res(f_1, f_2) = Res(\delta^{m-1} f_1, \ldots, \delta f_1, f_1, \delta^{n-1} f_2, \ldots, \delta f_2, f_2) \tag{3}$$

This last definition shows the equality between the notions of differential resultant of two ordinary differential operators with constant coefficients, differential resultant of the corresponding ordinary linear homogeneous differential polynomials and the algebraic Sylvester's resultant of the corresponding characteristic polynomials.

6 Using the Resultant for the Neighborhood

In this section we use the differential resultant in order to know the relations between the solutions of an ordinary linear differential equation $f_a = 0$ and the solutions of a generic perturbed equation $\tilde{f}_a = 0$ in its neighborhood. For instance, we can know (often experimentally) one or more solutions of the differential equation although the inaccuracy of its coefficients, thus we can require stronger conditions between f_a and the perturbed $\tilde{f}_a$. These conditions can be defined by means of the differential resultant.

6.1 Homogeneous Case

Let $f_a = D_a^n(y)$ be an ordinary linear differential polynomial in S and let $f_a = 0$ be the corresponding linear homogeneous differential equation. The differential resultant of a differential operator $D_a^n = \sum_{j=0}^n a_j \delta^j$ and a generic perturbed operator $\tilde{D}_a^n = \sum_{j=0}^n \tilde{a}_j \delta^j \in \mathcal{N}(D_a^n, e)$ is computed. The differential resultant is the determinant of the resultant matrix defined in Sect.5.1; in our case we work with the following $(n+n) \times (n+n)$ matrix:

$$\delta Res(D_a^n, \tilde{D}_a^n) = \det \begin{pmatrix} a_n & \dots & a_0 & 0 & \dots & 0 \\ 0 & a_n & \dots & a_0 & 0 & \dots \\ \vdots & \ddots & \ddots & \dots & \ddots & \dots \\ 0 & \dots & 0 & a_n & \dots & a_0 \\ a_n+\epsilon_n & \dots & a_0+\epsilon_0 & 0 & \dots & 0 \\ 0 & a_n+\epsilon_n & \dots & a_0+\epsilon_0 & 0 & \dots \\ \vdots & \ddots & \ddots & \dots & \ddots & \dots \\ 0 & \dots & 0 & a_n+\epsilon_n & \dots & a_0+\epsilon_0 \end{pmatrix}$$

It is well known in linear algebra that the determinant of the matrix does not change (and the resultant too), when we substitute the $(n+1)$-th row with the difference between the $(n+1)$-th row and the first row, the $(n+2)$-th row with the difference between the $(n+2)$-th row and the second row and so on until the $2n$-th row, that we substitute with the difference between the $2n$-th row and the n-th row. We obtain a new matrix:

$$\begin{pmatrix} a_n & \dots & a_0 & 0 & \dots & 0 \\ 0 & a_n & \dots & a_0 & 0 & \dots \\ \vdots & \ddots & \ddots & \dots & \ddots & \dots \\ 0 & \dots & 0 & a_n & \dots & a_0 \\ \epsilon_n & \dots & \epsilon_0 & 0 & \dots & 0 \\ 0 & \epsilon_n & \dots & \epsilon_0 & 0 & \dots \\ \vdots & \ddots & \ddots & \dots & \ddots & \dots \\ 0 & \dots & 0 & \epsilon_n & \dots & \epsilon_0 \end{pmatrix}$$

It follows that the differential resultant between D_a^n and $\tilde{D}_a^n$ is the differential resultant between D_a^n and the differential operator $D_\varepsilon^n = \sum_{j=0}^n \epsilon_j \delta^j$, with $\varepsilon = (\epsilon_0, \dots, \epsilon_n)$.

Given two differential operators D_a^n and $\tilde{D}_a^n$, let $f_a = D_a^n(y) = 0$ and $\tilde{f}_a = \tilde{D}_a^n(y) = 0$ be the corresponding homogeneous differential equations.

The following result holds for linear homogeneous differential polynomials in S and it was given by Berkovich [BT1].

Theorem 1 (Homogeneous Case). *Let D_a^n and D_b^m be two ordinary differential operators in $\mathbb{K}[\delta]$ of order n (respectively of order m). Let r be the rank of the differential resultant matrix of D_a^n and D_b^m. The system $\{D_a^n(y) = 0, D_b^m(y) = 0\}$ has a solution different from zero if and only if $n + m > r$.*

Since we can associate to each linear homogeneous differential equation its characteristic polynomial, we can use the following theorems for the commutative polynomial case:

Theorem 2. *Let R be a unique factorization domain. Let $A(x) = \sum_{i=0}^{n} a_i x^i$ and $B(x) = \sum_{j=0}^{m} b_j x^j$ be primitive polynomials in $R[x]$, then*

$$Res(A, B) = 0 \Leftrightarrow G.C.D.(A, B) \neq 1$$

Theorem 3. *Let R be a unique factorization domain. Let $A(x) = \sum_{i=0}^{n} a_i x^i$ and $B(x) = \sum_{j=0}^{m} b_j x^j$ be primitive polynomials in $R[x]$ and let $D = G.C.D.(A, B)$. If the rank of the resultant matrix is $m + n - k$ then $deg(D) = k$.*

Laydacker's theorem [LA1]:

Theorem 4. *Let R be a unique factorization domain. Let $A(x) = \sum_{i=0}^{n} a_i x^i$ and $B(x) = \sum_{j=0}^{m} b_j x^j$ be polynomials in $R[x]$. Let $M_d(A, B)$ be the matrix obtained by the resultant matrix $M(A, B)$, when we put it in triangular echelon form by using row transformations only. The entries of the last nonzero row of the matrix $M_d(A, B)$ are the coefficients of a G.C.D. of A and B.*

Let us see the meaning of the theorems as above in the differential context. First of all, we observe that given two linear homogeneous differential equations they always have the trivial common solution $y(x) = 0$; furthermore if $\varphi(x)$ is a common integral, $c\varphi(x)$, with c constant, is also an integral. Once the resultant matrix is given, then we can put it in triangular echelon form by using row transformations only. By using the Laidacker's result the entry in the last row and column is the resultant of the differential operators. If the first nonzero row from the bottom is the $(2n - h)$-th, then the entries of such row are the coefficients of the polynomial $q(t)= G.C.D.(p_{D_a^n}(t), p_{\tilde{D}_a^n}(t))$, that has degree h. This last fact means that $D_h = G.C.D.(D_a^n, \tilde{D}_a^n)$ has order h and $q(t)$ is its characteristic polynomial. Finally the system of ordinary linear homogeneous differential equations $\{D_a^n(y) = 0, \tilde{D}_a^n(y) = 0\}$ has as a set of solutions the solutions of the linear ordinary homogeneous differential equation $D_h(y) = 0$.

There are two remarks, that are very useful.

Remark 1. $\delta Res(D_a^n, \tilde{D}_a^n)$ is a homogeneous polynomial of degree $2n$ in the ϵ_j's and a_j's and a homogeneous polynomial of degree n as polynomial in the only ϵ_j's.

If $\delta Res(D_a^n, \tilde{D}_a^n) = 0$ then the system of the two differential equations has, at least, a solution different from the trivial solution zero.

Remark 2. A trivial case $\delta Res(D_a^n, \tilde{D}_a^n) = 0$ occurs when the coefficients of the differential operators are the same up to a nonzero constant. This case occurs when we have the following choice for the perturbations:

$$\epsilon_n \in [-e_n, e_n], \quad \epsilon_j = a_j \frac{\epsilon_n}{a_n} \quad j = 0, \dots, n-1 \tag{4}$$

where $a_n \neq 0$, since $n = ord(f)$. In order to guarantee that for any choice of ϵ_n the corresponding values of each ϵ_j belong to their intervals, we have to compute

$$\tau = \min_{j=0,\dots,n-1} \left\{ \left| \frac{e_j a_n}{a_j} \right| \right\} \quad \text{with} \quad a_j \neq 0$$

Now if $\epsilon_n \in [-\tau, +\tau] \cap [-e_n, e_n]$ by applying (4) we can find all differential operators in the neighborhood, whose coefficients are proportional to the coefficients of D_a^n.

In the general case we have the following result: if f_a and $\tilde{f}_a$ have at least k common integrals, with $1 \leq k \leq n$, that are linear independent over the field of constants, then the entries from the $(2n-k+1)$-th row to the $2n$-th row of the resultant matrix $M_d(D_a^n, \tilde{D}_a^n)$ must vanish. We find a system of $\frac{k(k+1)}{2}$ homogeneous equations in the ϵ_j's. If $k = n$ the only solution is given by (4), i.e. f_a and $\tilde{f}_a$ have a proportional coefficients (this means that they define the same differential equation). Otherwise we find the conditions for the ϵ_j's such that f_a and $\tilde{f}_a$ have k linear independent common integrals.

Example 5. Let $D = \delta^2 - 1$ be a differential operator and let $\tilde{D} = (1+\epsilon_2)\delta^2 + \epsilon_1\delta - 1 + \epsilon_0$ be a generic perturbed in the neighborhood, with $|\epsilon_j| \leq e_j, j = 0, 1, 2$. The resultant matrix is

$$\begin{pmatrix} 1 & 0 & -1 & 0 \\ 0 & 1 & 0 & -1 \\ \epsilon_2+1 & \epsilon_1 & \epsilon_0-1 & 0 \\ 0 & \epsilon_2+1 & \epsilon_1 & \epsilon_0-1 \end{pmatrix} \Rightarrow \text{triangular form} \Rightarrow \begin{pmatrix} 1 & 0 & -1 & 0 \\ 0 & 1 & 0 & -1 \\ 0 & 0 & \epsilon_1 & \epsilon_0+\epsilon_2 \\ 0 & 0 & 0 & \epsilon_1^2-(\epsilon_0+\epsilon_2)^2 \end{pmatrix}$$

$\delta Res(D, \tilde{D}) = \epsilon_1^2 - (\epsilon_0+\epsilon_2)^2$. If $D(y) = y'' - y = 0$ and $\tilde{D} = (1+\epsilon_2)y'' + \epsilon_1 y' + (\epsilon_0 - 1)y = 0$ have a common integral, then $\delta Res(D, \tilde{D}) = \epsilon_1^2 - (\epsilon_0+\epsilon_2)^2 = 0$, i.e. the perturbations have to satisfy $\epsilon_1 - \epsilon_0 - \epsilon_2 = 0$ or $\epsilon_1 + \epsilon_0 + \epsilon_2 = 0$. The differential operators D and $\tilde{D}$ have the same coefficients up to a nonzero constant if $\epsilon_1 = 0$ and $\epsilon_0 + \epsilon_2 = 0$; in this case $D(y) = 0$ and $\tilde{D}(y) = 0$ have the same integrals. If $\epsilon_1 \neq 0$ and the resultant vanishes, then there exists a common solution $ce^{\lambda x}$ of $D(y) = 0$ and $\tilde{D}(y) = 0$, with $\lambda = \frac{-\epsilon_0-\epsilon_2}{\epsilon_1}$ and $c \in \mathbb{R}$.

For instance if $\lambda = 1$, from the above considerations we find that $\{\epsilon_0, \epsilon_1, \epsilon_2\}$ have to satisfy $\{\epsilon_0 \in [-e_0, e_0], \epsilon_2 \in [-e_2, e_2], \epsilon_1 = -\epsilon_0 - \epsilon_2\}$. If we choose the tolerance $e = (10^{-1}, 10^{-1}, 10^{-1})$ and we substitute $\epsilon_0 = 0.01, \epsilon_2 = -0.1, \epsilon_1 = 0.09$ in $\tilde{D}$, we find the differential equation $\tilde{D}(y) = 0.9y'' + 0.09y' - 0.99y = 0$, having solution $y(x) = c_1 e^x + c_2 e^{-\frac{11}{10}x}$.

6.2 Non-Homogeneous Case

In the non-homogeneous case we can use the resultant of two linear ordinary differential polynomials: $f_a = D_a^n(y) + a_{-1}$ and $\tilde{f}_a = \tilde{D}_a^n(y) + \tilde{a}_{-1}$. We want to find conditions in order that the differential equations f_a and $\tilde{f}_a$ have a common solution.

There is a theorem analogous to Theorem 1 in the non-homogeneous case. Let $D_a^n(y) = a_{-1}$ and $D_b^m(y) = b_{-1}$ be two non-homogeneous linear ordinary differential equations with $a_{-1}, b_{-1} \in \mathbb{R}$, $n = ord(D_a^n)$ and $m = ord(D_b^m)$. We can assume that only one among a_{-1} and b_{-1} is different from zero. In fact if this is not the case, we can transform the system of differential equations to this form by substituting the equation $D_a^n(y) = a_{-1}$ with the equation $b_{-1}(D_a^n(y) - a_{-1}) - a_{-1}(D_b^m(y) - b_{-1}) = 0$. Let $\bar{D} = b_{-1}(D_a^n) - a_{-1}(D_b^m)$.

Let $D_{\bar{a}}^{\bar{n}}(y) = \bar{D}(y)$ be the differential polynomial with a vanishing right side, and let $D_{\bar{b}}^{\bar{m}}$ be the differential operator corresponding to $b_{-1}\delta(D_b^m(y) - b_{-1}) - \delta(b_{-1})(D_b^m(y) - b_{-1})$.

We obtain two new homogeneous equations: $D_{\bar{a}}^{\bar{n}}(y) = 0$ and $D_{\bar{b}}^{\bar{m}}(y) = 0$.

Theorem 5 (Non-Homogeneous Case). *Let $D_a^n(y) - a_{-1}$ and $D_b^m(y) - b_{-1}$ be two linear ordinary differential polynomials in $\mathbb{K}\{y\}$ of order n (respectively of order m). Let r be the rank of the differential resultant matrix $M(D_a^n, D_b^m)$ of the corresponding linear ordinary differential operators D_a^n and D_b^m. Let $D_{\bar{a}}^{\bar{n}}$ and $D_{\bar{b}}^{\bar{m}}$ be the ordinary differential operators as above. Let $\bar{r} = rank(M(D_{\bar{a}}^{\bar{n}}, D_{\bar{b}}^{\bar{m}}))$. The system $\{D_a^n(y) = a_{-1},\ D_b^m(y) = b_{-1}\}$ has a solution if and only if $\bar{n} + \bar{m} - \bar{r} > n + m - r$.*

Unfortunately this theorem does not give any procedure in order to find such solution and we need to use other tools.

Given $f_a = \sum_{j=0}^n a_j\delta^j + a_{-1}$ and $\tilde{f}_a = \sum_{j=0}^n \tilde{a}_j\delta^j + \tilde{a}_{-1}$ we construct the resultant matrix as in Sect.5.2. We have the following $(2n+2) \times (2n+2)$ matrix

$$M(f_a, \tilde{f}_a) = \begin{pmatrix} a_n & \dots & a_0 & 0 & 0 & \dots & 0 \\ 0 & a_n & \dots & a_0 & 0 & 0 & \dots \\ \vdots & \ddots & \ddots & \dots & \ddots & \dots & \vdots \\ 0 & \dots & 0 & a_n & \dots & a_0 & a_{-1} \\ a_n + \epsilon_n & \dots & a_0 + \epsilon_0 & 0 & 0 & \dots & 0 \\ 0 & a_n + \epsilon_n & \dots & a_0 + \epsilon_0 & 0 & 0 & \dots \\ \vdots & \ddots & \ddots & \dots & \ddots & \dots & \vdots \\ 0 & \dots & 0 & a_n + \epsilon_n & \dots & a_0 + \epsilon_0 & (a_{-1} + \epsilon_{-1}) \end{pmatrix}$$

As before, we subtract the first row from the $(n+2)$-th, the second row from the $(n+3)$-th, ..., the $(n+1)$-th row from the $(2n+2)$-th, and we obtain a new matrix, that has the same determinant of $M(f_a, \tilde{f}_a)$:

$$\begin{pmatrix} a_n & \dots & a_0 & 0 & 0 & \dots & 0 \\ 0 & a_n & \dots & a_0 & 0 & 0 & \dots \\ \vdots & \ddots & \ddots & \dots & \ddots & \dots & \vdots \\ 0 & \dots & 0 & a_n & \dots & a_0 & a_{-1} \\ \epsilon_n & \dots & \epsilon_0 & 0 & 0 & \dots & 0 \\ 0 & \epsilon_n & \dots & \epsilon_0 & 0 & 0 & \dots \\ \vdots & \ddots & \ddots & \dots & \ddots & \dots & \vdots \\ 0 & \dots & 0 & \epsilon_n & \dots & \epsilon_0 & \epsilon_{-1} \end{pmatrix}$$

Now we put this matrix in triangular form using row transformations only and shifting all the vanishing rows to the bottom. The entry in the last row is the resultant between f_a and $\tilde{f}_a$. If this value is not zero, f_a and $\tilde{f}_a$ have not any common solutions. Instead if $\delta Res(f_a, \tilde{f}_a)$ vanishes we cannot assert that there is a common solution. Here is an example:

Example 6. Let $D = \delta$ be a differential operator and let $f_1 = D(y) = 0$ and $f_2 = D(y) - 1 = 0$ two non-homogeneous differential equations. We have

$$\delta Res(f_1, f_2) = Res(\delta^2 f_1, \delta f_1, \delta f_2, f_2) = Res(\delta^2 y, \delta y, \delta^2 y, \delta y - 1) = 0$$

but f_1 and f_2 have not common solutions.

In this case we need further investigations. We look for the entries in the $(2n-1)$-th row, regarding them as the coefficients of a first order differential equations. If this equation has no solutions, then f_a and $\tilde{f}_a$ have not a common solution, otherwise if such entries are zero we look for the entries in the $(2n-2)$-th row and so on.

Example 7. Let $f = \delta^2(y) - \delta(y) - 1$ be a differential polynomial and let $\tilde{f} = (1+\epsilon_2)\delta^2(y) + (-1+\epsilon_1)\delta(y) + \epsilon_0 y + (-1+\epsilon_{-1})$ be a generic perturbed in the neighborhood, with $|\epsilon_j| \le e_j$, $j = -1, 0, 1, 2$. The resultant matrix is

$$M(f,\tilde{f}) = \begin{pmatrix} 1 & -1 & 1 & 0 & 0 & 0 \\ 0 & 1 & -1 & 1 & 0 & 0 \\ 0 & 0 & 1 & -1 & 1 & -1 \\ \epsilon_2+1 & \epsilon_1-1 & \epsilon_0+1 & 0 & 0 & 0 \\ 0 & \epsilon_2+1 & \epsilon_1-1 & \epsilon_0+1 & 0 & 0 \\ 0 & 0 & \epsilon_2+1 & \epsilon_1-1 & \epsilon_0+1 & \epsilon_{-1}-1 \end{pmatrix}$$

Its triangular form is:

$$\begin{pmatrix} 1 & -1 & 0 & 0 & 0 & 0 \\ 0 & 1 & -1 & 0 & 0 & 0 \\ 0 & 0 & 1 & -1 & 0 & -1 \\ 0 & 0 & 0 & \epsilon_1+\epsilon_2 & \epsilon_0 & \epsilon_2+\epsilon_{-1} \\ 0 & 0 & 0 & 0 & -\epsilon_0^2-\epsilon_0\epsilon_1-\epsilon_0\epsilon_2 & \epsilon_1^2+\epsilon_1\epsilon_2-\epsilon_{-1}\epsilon_0-\epsilon_{-1}\epsilon_1-\epsilon_{-1}\epsilon_2-\epsilon_0\epsilon_2 \\ 0 & 0 & 0 & 0 & 0 & -\epsilon_0^2\epsilon_2-\epsilon_0^2\epsilon_1-\epsilon_0^3 \end{pmatrix}$$

Then $\delta Res(f,\tilde{f}) = \epsilon_0^2(-\epsilon_2-\epsilon_1-\epsilon_0)$. We distinguish two cases:

1. If $\epsilon_0 = 0$, $-\epsilon_0^2-\epsilon_0\epsilon_1-\epsilon_0\epsilon_2 = 0$ and $\epsilon_1^2+\epsilon_1\epsilon_2-\epsilon_{-1}\epsilon_0-\epsilon_{-1}\epsilon_1-\epsilon_{-1}\epsilon_2-\epsilon_0\epsilon_2 \neq 0$, then there are no common solutions;
2. If $\epsilon_0 \neq 0$ then $\delta Res(f,\tilde{f}) = 0$ only when $\epsilon_2 = -\epsilon_0 - \epsilon_1$. In this case $-\epsilon_0^2 - \epsilon_0\epsilon_1 - \epsilon_0\epsilon_2 = 0$ and $\epsilon_1^2 + \epsilon_1\epsilon_2 - \epsilon_{-1}\epsilon_0 - \epsilon_{-1}\epsilon_1 - \epsilon_{-1}\epsilon_2 - \epsilon_0\epsilon_2 = \epsilon_0^2 \neq 0$, i.e. there are no common solutions.

It follows that the only case in which we have common solutions is $\epsilon_0 = 0$ and $\epsilon_1^2+\epsilon_1\epsilon_2-\epsilon_{-1}\epsilon_1-\epsilon_{-1}\epsilon_2 = 0$, i.e. $(\epsilon_1-\epsilon_{-1})(\epsilon_1+\epsilon_2) = 0$. This gives the following relations:

(a). If $\epsilon_1 = \epsilon_{-1}$ and $\epsilon_1 + \epsilon_2 \neq 0$ with $\epsilon_2 \in [-e_2, e_2]$, then the common solution is the integral of the first order differential equation $(\epsilon_1 + \epsilon_2)y' = -(\epsilon_1 + \epsilon_2)$, i.e. $y(x) = -x + c$, $c \in \mathbb{R}$.
(b). If $\epsilon_1 = -\epsilon_2$ and $\epsilon_1 - \epsilon_{-1} \neq 0$ with $\epsilon_{-1} \in [-e_{-1}, e_{-1}]$, then there are no common solutions.
(c). If $\epsilon_1 = -\epsilon_2$ and $\epsilon_1 - \epsilon_{-1} = 0$ with $\epsilon_{-1} \in [-e_{-1}, e_{-1}]$, then the coefficients of f and $\tilde{f}$ are proportional and the corresponding differential equations have the same solutions.

Finally the following example shows some relations between numerical solutions of an ordinary linear differential equations with constant coefficients and the integrals of a perturbed differential equation in the neighborhood with assigned tolerance, that has at least a nontrivial common integral with the given equation.

Example 8. Let $D = \delta^2 - 1$ be the differential operator as in Example 5 and let $\tilde{D} = (1 + \epsilon_2)\delta^2 + \epsilon_1\delta - 1 + \epsilon_0$ be its generic perturbed in the neighborhood with $|\epsilon_j| \leq e_j = 10^{-1}$ for $j = 0, 1, 2$. By Example 5 if $\delta Res(D, \tilde{D}) = \epsilon_1^2 - (\epsilon_0 + \epsilon_2)^2 = 0$, then $D(y) = y'' - y = 0$ and $\tilde{D} = (1 + \epsilon_2)y'' + \epsilon_1 y' + (\epsilon_0 - 1)y = 0$ have at least a nontrivial common integral. If $\epsilon_1 + \epsilon_0 + \epsilon_2 = 0$ and $\epsilon_1 \neq 0$, then $D(y) = 0$ and $\tilde{D}(y) = 0$ have ce^{-x} with c constant as common integrals and the general integral of $\tilde{D}(y) = 0$ is

$$y(x) = c_1 e^{-x} + c_2 e^{\frac{1-\epsilon_0}{1+\epsilon_2}x}$$

Now suppose that we solve numerically the differential equation $D(y) = y'' - y = 0$ with the initial conditions $y(0) = 1$ and $y'(0) = 1$ by using the Eulero's method

$$y_n = y_{n-1} + hy'_{n-1} + h^2 y''_{n-1}$$

and constant stepsize h [SK1]. We find $y_n = (1+h)^n$ by the initial conditions and then we approximate the solution with the function $y(x) = (1+h)^x$. The function $y(x) = (1+h)^x$ is equal to the integral $e^{\frac{1-\epsilon_0}{1+\epsilon_2}x}$ for suitable ϵ_0, ϵ_2 whenever $\frac{1-\epsilon_0}{1+\epsilon_2} = \ln(h+1)$. For instance, if $h = 1$ then we find the function 2^x, which is not in the neighborhood, because it gives values of the perturbations without the accepted tolerance. Instead, if we choose $h = 1.5$, after straightforward manipulations, we find that all the values of the perturbations are within the given tolerance.

References

[AC1] Alefeld, G., Claudio, D. : The basic properties of interval arithmetic, its software realization and some applications. Computer and Structures, **67** (1998) 3–8
[BT1] Berkovich, L.M., Tsirulik, V.G.: Differential Resultants and some of their Applications. Differential Equations, Plenum Publ. Corp., **22** (1986) 750–757
[CF1] Carrà Ferro, G.: A Resultant Theory for Systems of Linear Partial Differential Equations. Lie Groups and their Applications **1** (1994) 47–55
[CF2] Carrà Ferro, G.: A Resultant Theory for the Systems of Two Ordinary Algebraic Differential Equations. Appl. Algebra in Engineering, Communication and Computing) **8** Springer, (1997) 539–560

[GCL1] Geddes, K.O., Czapor, S. R., Labahn, G.: Algorithms for Computer Algebra. Kluwer Academic Publishers, Boston (1992)

[GMH1] Geddes K.O., Monogan M.B., Heal K.M., Labahn G., Vorkoetter S.M. McCarron S.: MAPLE 6, Programming Guide. Waterloo Maple Inc., Canada (2000)

[HA1] Hansen, E.: Global Optimization using Interval Analysis. Marcel Dekker Inc. New York-Basel-Hong Kong (1992)

[JA1] Janet, M.: Sur les sistémes d'équations aux dérivée partielles. J, Math. **3** (1920) 65–151

[KO1] Kolchin, E.R.: Differential Algebra and Algebraic Groups. Academic Press London-New York(1973)

[LA1] Laidacker, M.A.: Another Theorem relating Sylvester's Matrix and the Greatest Common Divisors. Mathematics Magazine **42** (1969) 126–128

[MC1] Macaulay, F.S.: The Algebraic Theory of Modular Systems. Proc. Cambridge Univ. Press Cambridge(1916)

[MV1] Marotta, V.: Resultants and Neighborhoods of a Polynomial. to be presented at SNSC'01 (2001)

[RQ1] Riquier, C.H.: Les systémes d'équations aux dérivée partielles. Gauthier-Villars Paris (1910)

[RT1] Ritt, J.F.: Differential Algebra. AMS Coll. Publ. New York **33**(1950)

[SK1] Skeel, R.D., Keiper, J.B.: Elementary Numerical Computing with MATHEMATICA. McGraw-Hill International Editions, Computer Science Series (1993)

[ST1] Stetter, H.J.: Polynomial with Coefficients of Limited Accuracy. Computer Algebra in Scientific Computing (Eds. V.G.Ganzha, E.W.Mayr, E.V.Vorozhtsov), Springer (1999) 409–430

[ST2] Stetter, H.J.: Nearest Polynomial with a Given Zero, and Similar Problems. SIGSAM Bull. **33** (1999) n. 4, 2–4

[WO1] Wolfram, S.: The Mathematica Book. 4th ed., Wolfram Media, Cambridge University Press (1999)

Invariants of Finite Groups and Involutive Division

C. F. Cid and W. Plesken

Lehrstuhl B für Mathematik, RWTH-Aachen, Germany
plesken@willi.math.rwth-aachen.de

1 Introduction

The invariant ring of a finite matrix group is known to be well behaved for reflection groups and messy in general. Involutive division is a newly discovered tool in commutative algebra and in this note it is applied to the problem of finding a presentation of the ring of invariants of a finite matrix group. The first author has implemented the JANET-algorithm in MAPLE following [GeB 98a] and [GeB 98b], more precisely Gerdt's involutive algorithm for Janet's (involutive) division. The results of this are collected in two MAPLE-packages called INVOLUTIVE and JANET, the first dealing with polynomials and the second with linear partial differential equations. Both of these packages have a collection of other routines serving various purposes. There is also a loose connection with the MAPLE-package JETS by Mohammed Barakat, which deals with symmetries of differential equations, conservation laws etc.. Here we report on our experience with applying the package INVOLUTIVE to questions of invariant theory of finite groups. We outline an algorithm constructing a presentation of the ring of invariants of a finite complex matrix group and representing each invariant in a unique way as an expression in the generators. We also report on the limits with the present MAPLE implementation. As far as the invariant theory of finite groups proper is concerned, there is a MAPLE-package available to perform the tasks discussed here, cf. [Kem 98] or [Kem 99], based on GROEBNER basis techniques and even a very effective implementation in MAGMA. The issue here is more to demonstrate the flexibility of involutive division and the JANET algorithm to these aims. Here we also restrict the discussion to the classical case of fields of characteristic zero, where MOLIEN's series is available.

We wish to thank Vladimir Gerdt for many discussions and his guidance with the implementation, Daniel Robertz for adding some functions to the system, and Mohammed Barakat for contributing some of the functions of his MAPLE package JETS, which turned out to be rather useful and to J.-F. Pommaret, whose book [Pom 94] got us interested in JANET's algorithm.

2 Summary of the Relevant Theory

In this section let $G \leq \mathrm{GL}(n, \mathbb{C})$ be a finite matrix group. Identify the natural G-module $\mathbb{C}^{n\times 1}$ with the $\mathbb{C}$-vector space $\mathbb{C}[x_1, \ldots, x_n]_1$ of homogeneous polynomials of degree 1 in the polynomial ring $\mathbb{C}[x_1, \dots, x_n]$. In this way G acts on $\mathbb{C}[x_1, \dots, x_n]$ by algebra automorphisms. The subring $\mathbb{C}[x_1, \dots, x_n]^G$ of G-fixed points is the object of our interest. The oldest result on this is MOLIEN's formula for the dimensions of the

$$\mathbb{C}[x_1, \dots, x_n]_i^G := \mathbb{C}[x_1, \dots, x_n]_i \cap \mathbb{C}[x_1, \dots, x_n]^G$$

where $\mathbb{C}[x_1, \dots, x_n]_i$ is the $\mathbb{C}$-vector space of homogeneous polynomials of degree i with $i \in \mathbb{Z}_{\geq 0}$. This formula is given by

$$m_G(s) := \sum_{g \in G} \det(I_n - sg)^{-1} = \sum_{i=0}^{\infty} \mathrm{Dim}(\mathbb{C}[x_1, \dots, x_n]_i^G) s^i.$$

Its left hand side, the MOLIEN series can easily be evaluated using characters, as provided for by GAP, cf. [GAP 00]. Of course

$$\mathbb{C}[x_1, \dots, x_n]^G = \bigoplus_{i=0}^{\infty} \mathbb{C}[x_1, \dots, x_n]_i^G$$

is a graded ring, and EMMY NOETHER has shown that it is generated by the invariants of degree $\leq |G|$, cf. [Ben 93], where a relative version of this result is also proved: If $U \leq G$ is a subgroup whose ring of invariants is generated by invariants of degrees $\leq b$, then the ring of G-invariants is generated by invariants of degrees $\leq b|G : U|$. This will turn out to be useful.

Individual invariants can be obtained constructively as images under the projection operator

$$\pi : \mathbb{C}[x_1, \dots, x_n] \to \mathbb{C}[x_1, \dots, x_n]^G : p \mapsto \frac{1}{|G|} \sum_{g \in G} p \circ g.$$

Of course, if $|G|$ gets bigger, one will split up this sum into in iterated sum over transversals in a subgroup chain of G. Details can be found in [Sta 79], [Ben 93], [Stu 93].

3 Presentation for the Ring of Invariants

Keeping the notation of the last section, we now want to present an algorithm for finding a presentation of $\mathbb{C}[x_1, \dots, x_n]^G$. More precisely, we want to outline an algorithm to find generators and unique expressions for each invariant in terms of these generators. Of course, we want to do with less generators than given by NOETHER's theoretical bound, possibly with the minimal number of generators possible, all generators being assumed to be homogeneous.

Problem 1: Given homogeneous invariants $i_1, \dots, i_k \in \mathbb{C}[x_1, \dots, x_n]^G$. Find the generating function for the

$$\mathrm{Dim}(\mathbb{C}[i_1, \dots, i_k] \cap \mathbb{C}[x_1, \dots, x_n]_i).$$

The classical approach to this is as follows: Look at the polynomial ring $\mathbb{C}[I_1, \ldots, I_k]$ and its epimorphisms onto $\mathbb{C}[i_1, \ldots, i_k]$ mapping the indeterminate I_l to the invariant polynomial i_l:

$$\sigma : \mathbb{C}[I_1, \ldots, I_k] \to \mathbb{C}[i_1, \ldots, i_k] : I_l \mapsto i_l.$$

Construct a free resolution of $\mathbb{C}[I_1, \ldots, I_k]$-modules of the kernel $\ker\sigma$ of this map. From general commutative algebra, this resolution is bound to determinate after at most n steps. Assigning appropriate degrees to the I_l (namely $\deg(i_l)$) and to the generators of the free modules, one gets the desired generating function above as an alternating sum of products of certain geometric series multiplied by powers of the variable s, for details see [Ben 93] . Technically this can be done by using JANET's algorithm for linear differential equations translated into polynomial equations by starting out with $x_1, \ldots, x_n, I_1, \ldots, I_k$ as indeterminates, and $I_1 - i_1, \ldots, I_k - i_k$ as relations and to use pure lexicographic order to eliminate the x_i. In this way, one gets generators for the kernel and can proceed from there to construct the free resolution, cf. We have an automatic function doing this. For instance the example 6.6 of $G = \langle -I_3 \rangle \leq \mathrm{GL}(3, \mathbb{C})$ of [Sta 79] runs completely automatically. There are two independent observations to make:

First the good news: with the JANET basis for $\ker\sigma$ it is no longer necessary to construct the free resolution: the generating function can be read off from the JANET basis of $\ker\sigma$.

And the bad news: There seem to be strict limits to this approach, the bottleneck being the performance of the JANET algorithm on the rather big system above involving the x_i and the I_j.

The first point is clarified by the following proposition, demonstrating the nice and clear structure of the JANET-approach.

Proposition 1. *Let $p_1, \ldots, p_r$ be a JANET basis of $\ker\sigma$ in the degree-lexicographical order and let $M(p_l)$ be the subset of $\{I_1, \ldots, I_k\}$ of multiplicative variables for p_l. Assign the degree $d_j := \deg(i_j)$ to I_j and let D_l be the resulting degree for the leading monomial of p_l. Then the HILBERT series of $\mathbb{C}[i_1, \ldots, i_k]$ is given by*

$$\sum_i \mathrm{Dim}(\mathbb{C}[i_1, \ldots, i_k] \cap \mathbb{C}[x_1, \ldots, x_n]_i) s^i = \prod_{j=1}^{n} \frac{1}{1 - s^{d_j}} - \sum_{l=1}^{r} s^{D_l} \prod_{I_j \in M(p_l)}^{n} \frac{1}{1 - s^{d_j}}$$

Proof. The TAYLOR expansion at $s = 0$ of $\prod_{j=1}^{n} \frac{1}{1-s^{d_j}}$ is the generating function for the number of monomials in $\mathbb{C}[I_1, \ldots, I_k]$ according to the degrees. $s^{D_l} \prod_{I_j \in M(p_l)}^{n} \frac{1}{1-s^{d_j}}$ counts the homogeneous generators of $\ker\sigma$, which are multiples of p_l by multiplicative variables. Each $\mathbb{C}$-basis vector of $\ker\sigma$ has a unique representation as a product of some p_l by multiplicative variables. Hence the sum of the $s^{D_l} \prod_{I_j \in M(p_l)}^{n} \frac{1}{1-s^{d_j}}$ has to be subtracted from the complete $\prod_{j=1}^{n} \frac{1}{1-s^{d_j}}$

to obtain the generating function for the monomials which map onto a basis of $\mathbb{C}[i_1, \dots, i_k]$. q. e. d.

Please note, the last result also indicates an alternative of computing the HILBERT series, which is commonly used in the JANET approach, even if the generators are not homogeneous: One simply assigns the degree 1 to each indeterminate I_i and the same formula yields the answer. As pointed out by the referee, this degree-1-version is already in [Ape 98]. Moreover the result as it stands follows also immediately from the observation in [Sei 00] that the JANET-Algorithm yields a STANLEY-decomposition, cf. [Sei 00] for details. At the moment we have not pursued the point further, how to read off a free resolution from the JANET data, cf. [Pom 94] and [Sei 00].

Coming to the second point of slow performance for the elimination of the x_j. This we have overcome by a simple use of linear algebra. Here is the algorithm to produce elements and finally generators of the kernel. This algorithm has been implemented by D. Robertz in MAPLE as part of the package INVOLUTIVE. Recall that we assigned degrees $d_l = \deg(i_l)$ to the generators I_l of $\mathbb{C}[I_1, \dots, I_k]$ thus defining a new grading for the polynomial ring $\mathbb{C}[I_1, \dots, I_k]$. Denote the homogeneous components of degree i of $\mathbb{C}[I_1, \dots, I_k]$ by $\mathbb{C}[I_1, \dots, I_k]_i$.

Algorithm 1 *Input: Homogeneous polynomials* $i_1, \dots, i_k \in \mathbb{C}[x_1, \dots, x_n]$ *and a degree* $d \in \mathbb{N}$.
Output: For each $i, 0 \le i \le d$ *(linearly independent) elements* $b_1^{(i)}, \dots, b_{\delta(i)}^{(i)} \in \mathbb{C}[I_1, \dots, I_k]_i$ *such that*

$$(\sigma(b_1^{(i)}), \dots, \sigma(b_{\delta(i)}^{(i)}))$$

is a $\mathbb{C}$*-basis for*

$$\sigma(\mathbb{C}[I_1, \dots, I_k]_i) = \mathbb{C}[i_1, \dots, i_k]_i$$

and elements $p_1^{(i)}, \dots, p_{\rho(i)}^{(i)} \in \mathbb{C}[I_1, \dots, I_k]_i$ *such that*

$$p_1^{(1)}, \dots, p_{\rho(1)}^{(1)}, \dots p_{\rho(i)}^{(i)}$$

multiplied by the monomials of $\mathbb{C}[I_1, \dots, I_k]$ *of appropriate degrees generate the kernel of* σ *restricted to* $\mathbb{C}[I_1, \dots, I_k]_i$ *as a* $\mathbb{C}$*-vector space.*
Algorithm: Assume that the data for $\mathbb{C}[I_1, \dots, I_k]_j$ *for* $j = 1, \dots, i-1$ *are available already as sequences* $b^{(j)}$ *and* $p^{(j)}$*. Form the set*

$$R_i := \{I_l | d_l = i\} \cup \{b_r^{(j)} b_s^{(i-j)} | 1 \le j \le \frac{i}{2}, 1 \le r \le \rho(j), 1 \le s \le \rho(i-j)\} \\ \subset \mathbb{C}[I_1, \dots, I_k]_i.$$

Select $b_1^{(i)}, \dots, b_{\rho(i)}^{(i)} \in R_i$ *maximal such that* $(\sigma(b_1^{(i)}), \dots, \sigma(b_{\rho(i)}^{(i)}))$ *is* $\mathbb{C}$*-linearly independent in* $\mathbb{C}[i_1, \dots, i_k]_i$*, i. e. is a* $\mathbb{C}$*-basis of* $\mathbb{C}[i_1, \dots, i_k]_i$*. Each of the remaining elements* r *of* R_i *yields an element of the form* $r - \sum a_s(r) b_s^{(i)} \in \ker \sigma$ *with unique* $a_s(r) \in \mathbb{C}$ *as an element of the sequence* $p^{(i)}$.

This rather obvious algorithm serves two purposes: to compute the dimensions of the $\mathbb{C}[i_1, \ldots, i_k]$ and to produce relations, which ultimately will generate $\ker \sigma$. The delicate point of course is the choice of the parameter d, which in general might have to be chosen rather big. It seems however that the case of invariant rings is not so bad behaved. On the other hand, one can easily construct examples, outside the range of invariant theory, where the JANET algorithm with lexicographic elimination order is faster than the above algorithm. The slow performance of the later occurs usually if the invariants are complicated, e. g. more than 2 variables and substantial degrees and many summands.

Summarizing we end up with a presentation which might not contain enough relators. Two situations are possible: One has enough generators. This case is favourable and treated below. Or some generators are missing. Even if the above algorithm does not go far enough to detect this, it is most unlikely that the resulting HILBERT series is equal to the MOLIEN series, e. g. that the missing relators compensate the missing generators.

With the relative version of NOETHER's result on the degrees of generating invariants, cf. Section 2, and the help of the above algorithm one can often get a reasonable bound d in 1 until where one has to check. Hence we are left with an easier problem.

Problem 2 : Given homogeneous invariants $i_1, \ldots, i_k$ generating $\mathbb{C}[x_1, \ldots, x_n]^G$. Find a presentation for $\mathbb{C}[x_1, \ldots, x_n]^G$ in these generators.

Now the setup above can be used to construct relators, use the JANET's algorithm for computing the HILBERT series and thus obtain a proof that the presentation is complete, in case HILBERT and MOLIEN series agree, and to rerun algorithm 1, in case there are coefficients in the HILBERT series which are bigger than the corresponding coefficients in the MOLIEN series. Ultimately this procedure has to come to an end. It can easily be arranged that one gets a minimal set of generators.

4 Examples

As a first example we reproduce a MAPLE-session using the package INVOLUTIVE to find a presentation of the ring of invariants of the matrix group

$$O := \langle \begin{pmatrix} -1 & 0 & 0 & 0 \\ 0 & 1 & 0 & 0 \\ 0 & 0 & 1 & 0 \\ 0 & 0 & 0 & 1 \end{pmatrix}, \begin{pmatrix} 0 & 1 & 0 & 0 \\ 1 & 0 & 0 & 0 \\ 0 & 0 & 1 & 0 \\ 0 & 0 & 0 & 1 \end{pmatrix}, \begin{pmatrix} 0 & 0 & 1 & 0 \\ 0 & 0 & 0 & 1 \\ 1 & 0 & 0 & 0 \\ 0 & 1 & 0 & 0 \end{pmatrix} \rangle$$

and its subgroup G of determinant 1 elements. Here is the MAPLE-session with complete details. The commands are self explanatory in view of the last section.

First group: O:=(C_2 wr C2) wr C_2 of degree 4.

Problem: Find presentation for the ring I(O) of invariants of O.

Note the group O contains a reflection subgroup of index 2 the ring of invariants of which is generated by x^2+y^2,z^2+u^2,x ^2*y^2,z^2*u^2. By the refinement of Noether's Theorem the O-invariants up to degree 8 generate the ring of O-invariants.

```
> restart;
> with(jets): with(Involutive):
```

GAP yields the following Molien series for O:

```
> m0 := (1+s^6) / ((1-s^8)*(1-s^4)^2*(1-s^2));
```

$$mO := \frac{1+s^6}{(1-s^8)\,(1-s^4)^2\,(1-s^2)}$$

```
> taylor(m0, s=0, 20);
```

$$1+s^2+3\,s^4+4\,s^6+8\,s^8+10\,s^{10}+16\,s^{12}+20\,s^{14}+29\,s^{16}+35\,s^{18}+\mathrm{O}(s^{20})$$

```
> var_erz := [[], [p1=x^2+y^2+z^2+u^2], [],
> [p2=x^4+y^4+z^4+u^4, p3=x^2*y^2+z^2*u^2], [],
> [p4=x^6+y^6+z^6+u^6],[], [p5=x^8+y^8+z^8+u^8]];
```

$$\begin{aligned}var\_erz := [[],\ [p1 = x^2+y^2+z^2+u^2],\ [],\\ [p2 = x^4+y^4+z^4+u^4,\ p3 = x^2\,y^2+z^2\,u^2],\ [],\\ [p4 = x^6+y^6+z^6+u^6],\ [],\ [p5 = x^8+y^8+z^8+u^8]]\end{aligned}$$

```
> l := relations(var_erz, 12):
```

$$0;1;0;3;0;4;0;8;0;10;0;16$$

These numbers are the dimensions of the spaces of invariants of degrees 1 to 12 which generated by the products of the invariants p1, .. ,p5. Since these numbers agree with the coefficients in the expansion of the Molien series, we have proved now that the ring of O-invariants is generated by p1, .. ,p5. Clearly none of the generators can be omitted. We also obtain the following relations among the pi's:

```
> l[1];
```

$$\begin{aligned}&[\frac{3}{8}\,p1^2\,p2^2+\frac{1}{2}\,p2\,p5-\frac{1}{3}\,p1\,p2\,p4+\frac{1}{2}\,p1^2\,p3\,p2+\frac{2}{9}\,p1^3\,p4-\frac{1}{6}\,p1^4\,p2\\&-\frac{1}{2}\,p1^2\,p3^2-\frac{1}{4}\,p1^2\,p5-p1\,p3\,p4-\frac{1}{2}\,p2^2\,p3+\frac{1}{72}\,p1^6+p3^3+p3\,p5\\&-\frac{1}{9}\,p4^2-\frac{1}{4}\,p2^3+\frac{1}{2}\,p2\,p3^2]\end{aligned}$$

```
> J:=InvolutiveBasis(l[1], l[2]);
```

$$\begin{aligned}J := [&27\,p1^2\,p2^2+36\,p2\,p5-24\,p1\,p2\,p4+36\,p1^2\,p3\,p2+16\,p1^3\,p4\\&-12\,p1^4\,p2-36\,p1^2\,p3^2-18\,p1^2\,p5-72\,p1\,p3\,p4-36\,p2^2\,p3+p1^6\\&+72\,p3^3+72\,p3\,p5-8\,p4^2-18\,p2^3+36\,p2\,p3^2]\end{aligned}$$

```
> h0 := PolWeightedHilbertSeries(l[3], s);
```

$$hO := \frac{1}{(1-s^2)\,(1-s^4)^2\,(1-s^6)\,(1-s^8)} - \frac{s^{12}}{(1-s^2)\,(1-s^4)^2\,(1-s^6)\,(1-s^8)}$$

```
> simplify(mO-hO);
```

$$0$$

Since the Hilbert series agrees with the Molien series and since we started out with generators, we have proved that the relator above yields a presentation for the ring of invariants of O.

We now proceed to the subgroup G of O consisting of all matrices of determinant 1 in O.

Problem: Find a presentation for the ring I(G) of invariants of G.

GAP yield the following Molien series.

```
> mG := (1+s^6+s^8+s^14) / ((1-s^8)*(1-s^4)^2*(1-s^2));
```

$$mG := \frac{1+s^6+s^8+s^{14}}{(1-s^8)\,(1-s^4)^2\,(1-s^2)}$$

```
> taylor(mG,s=0,28);
```

$$1+s^2+3\,s^4+4\,s^6+9\,s^8+11\,s^{10}+19\,s^{12}+24\,s^{14}+37\,s^{16}+45\,s^{18}+63\,s^{20}$$
$$+76\,s^{22}+101s^{24}+119\,s^{26}+\mathrm{O}(s^{28})$$

Since 1+s^6+s^8+s^14= (1+s^6)(1+s^8) we expect that I(G) viewed as I(O)-module to be free with basis 1 and some G-invariant of degree 8. To find a suitable invariant of degree 8 one factors the Jacobi determinant of p1,p2,p3, p5 to find the G-invariant p6.

```
> var_erz := [[], [p1=x^2+y^2+z^2+u^2], [],
> [p2=x^4+y^4+z^4+u^4, p3=x^2*y^2+z^2*u^2], [],
> [p4=x^6+y^6+z^6+u^6], [], [p5=x^8+y^8+z^8+u^8,
> p6=x*u*y*z*(u-z)*(u+z)*(x-y)*(x+y)]];
```

$var\_erz := [[], [p1 = x^2+y^2+z^2+u^2], [],$
$[p2 = x^4+y^4+z^4+u^4,\ p3 = x^2\,y^2+z^2\,u^2], [], [p4 = x^6+y^6+z^6+u^6], [],$
$[p5 = x^8+y^8+z^8+u^8,\ p6 = x\,u\,y\,z\,(u-z)\,(u+z)\,(x-y)\,(x+y)]]$

```
> l := relations(var_erz, 16):
```

$$0; 1; 0; 3; 0; 4; 0; 9; 0; 11; 0; 19; 0; 24; 0; 37;$$

Comparing these dimensions with the coefficients of the Molien series makes us suspect that p1 to p6 generate I(G). However, this is no proof this time. We also have obtained a list of 21 relations, too long to be reproduced here:

```
> nops(l[1]);
```

$$21$$

```
> JG:=InvolutiveBasis(l[1],l[2]):
> nops(JG);
```

$$24$$

Whereas the coefficients in the original relations (contained in l[1]) were rather small, quite a few coefficients in the Janet basis get rather big (30 digits and more).

```
> hG := PolWeightedHilbertSeries(l[3], s);
```

$$
\begin{aligned}
hG := &\frac{1}{(1-s^2)(1-s^4)^2(1-s^6)(1-s^8)^2} - \frac{s^{16}}{(1-s^4)^2(1-s^6)(1-s^8)^2} \\
&- \frac{s^{12}}{(1-s^2)(1-s^4)^2(1-s^6)(1-s^8)^2} - \frac{s^{20}}{(1-s^4)(1-s^6)(1-s^8)^2} \\
&- \frac{s^{18}}{(1-s^4)^2(1-s^6)(1-s^8)^2} - \frac{s^{24}}{(1-s^6)(1-s^8)^2} \\
&- \frac{s^{24}}{(1-s^4)(1-s^6)(1-s^8)^2} - \frac{s^{22}}{(1-s^4)(1-s^6)(1-s^8)^2} \\
&- \frac{s^{28}}{(1-s^4)(1-s^6)(1-s^8)^2} - \frac{s^{28}}{(1-s^4)^2(1-s^6)(1-s^8)^2} \\
&- \frac{s^{26}}{(1-s^4)^2(1-s^6)(1-s^8)^2} - \frac{s^{28}}{(1-s^6)(1-s^8)^2} \\
&- 2\,\frac{s^{32}}{(1-s^4)(1-s^6)(1-s^8)^2} - \frac{s^{26}}{(1-s^4)(1-s^6)(1-s^8)^2} \\
&- \frac{s^{30}}{(1-s^4)(1-s^6)(1-s^8)^2} - \frac{s^{36}}{(1-s^4)(1-s^6)(1-s^8)^2} \\
&- \frac{s^{34}}{(1-s^4)(1-s^6)(1-s^8)^2} - \frac{s^{40}}{(1-s^4)(1-s^6)(1-s^8)^2} \\
&- \frac{s^{38}}{(1-s^4)(1-s^6)(1-s^8)^2} - \frac{s^{44}}{(1-s^4)(1-s^6)(1-s^8)^2} \\
&- \frac{s^{42}}{(1-s^4)(1-s^6)(1-s^8)^2} - \frac{s^{48}}{(1-s^4)^2(1-s^6)(1-s^8)^2} \\
&- \frac{s^{46}}{(1-s^4)(1-s^6)(1-s^8)^2} - \frac{s^{50}}{(1-s^4)^2(1-s^6)(1-s^8)^2}
\end{aligned}
$$

```
> simplify(mG-hG);
```

$$0$$

Hence we believe that we have a presentation for I(G). But the final proof now goes as follows: Obviously I(O)1 +I(O)p6 is a free I(O)-module of rank 2 contained in I(G). We want to show equality. But this follows from comparing the Molien series mG of I(G) with (1+s^8)mO, which turn out to be equal. Note now we know that I(G) is generated by p1 to p6 and hence we have a presentation for I(G).

The next (rather small) example is to demonstrate how much one can do, if one follows the obvious approach outlined at the beginning of the last section, by getting a JANET basis or involutive basis, as it is called in the polynomial case in the package for the generators x_i and I_j with relators $I_j - i_j$ in the notation of the last section. The group chosen here is

$$G := \langle \begin{pmatrix} 0 & 1 \\ -1 & 0 \end{pmatrix} \rangle \leq H := \langle \begin{pmatrix} 0 & 1 \\ 1 & 0 \end{pmatrix}, \begin{pmatrix} -1 & 0 \\ 0 & 1 \end{pmatrix} \rangle$$

Note, H is a reflection group, which keeps things easy. Here comes the MAPLE-session with comments:

```
> with(jets): with(Involutive):
```

The ring I(G) of invariants of G (isomorphic to C_4) is clearly generated by x^2+y^2, x^2*y^2,x^3*y-y^3*x.

Below we find an involutive basis for the three relations given in L in the lexicographic ordering. Note that q4 comes earlier in this orderring than p2 and p4, corresponding to the H-invariants x^2+y^2, x^2*y^2.

```
> L:=[p2-(x^2+y^2),p4-x^2*y^2, q4-(x^3*y-y^3*x)];
```

$$L := [p2 - x^2 - y^2,\, p4 - x^2\, y^2,\, q4 - x^3\, y + y^3\, x]$$

```
> Lvars:=[x,y,q4,p2,p4];
```

$$Lvars := [x,\, y,\, q4,\, p2,\, p4]$$

```
> B:=InvolutiveBasis(L,Lvars,1);
```

$$\begin{aligned}B := [&q4^2 - p2^2\, p4 + 4\, p4^2,\, y\, q4^2 - y\, p4\, p2^2 + 4\, y\, p4^2,\\ &y^2\, q4^2 - p2^2\, p4\, y^2 + 4\, p4^2\, y^2,\, y^3\, q4^2 - p2^2\, p4\, y^3 + 4\, p4^2\, y^3,\\ &p4 - y^2\, p2 + y^4,\, -p2\, y^3\, q4 + y\, p2^2\, q4 - p2^2\, x\, p4 + 4\, x\, p4^2 - 2\, p4\, y\, q4,\\ &x\, q4 + y^3\, p2 - y\, p2^2 + 2\, y\, p4,\\ &4\, y\, x\, p4 - 2\, y^2\, q4 + p2\, q4 - y\, x\, p2^2,\, -x\, y\, q4 - 2\, y^2\, p4 + p2\, p4,\\ &-2\, y^2\, x\, p4 + y^3\, q4 - y\, p2\, q4 + p2\, x\, p4,\, 2\, x\, p4 - x\, y^2\, p2 - y\, q4,\\ &-x\, y^2\, q4 - 2\, y^3\, p4 + y\, p4\, p2,\, q4 + 2\, y^3\, x - y\, x\, p2,\, p2 - x^2 - y^2]\end{aligned}$$

There are three possibilities for the normal form of a polynomial p in x and y with respect to the involutive basis B: Either the normal form only involves p2 and p4, which is tantamount to p being an H-invariant, or it also involves q4, saying that p is G-invariant but not H-invariant, or it involves some x or y, in which case it is not H-invariant.

```
> PolInvReduce((x^2+(x^2-y^2)^3)^2+(y^2+(x^2-y^2)^3)^2,
> B,Lvars);
```

$$\begin{aligned}&2\, p2^6 - 4\, y^2\, p2^3 - 24\, p2^4\, p4 + 2\, p2^4 + 16\, y^2\, p2\, p4 + 96\, p2^2\, p4^2 - 8\, p2^2\, p4\\ &- 128\, p4^3 + p2^2 - 2\, p4\end{aligned}$$

```
> PolInvReduce((x-y)^3,B,Lvars);
```

$$2\, y^2\, x + 2\, y^3 + x\, p2 - 3\, y\, p2$$

```
> PolInvReduce(x^7*y^3-x^3*y^7,B,Lvars);
```

$$p2\, q4\, p4$$

That the Hilbert series with the degrees given as below takes the simple form 1/(1-s)^2 corresponds to the fact that the ring for which B is an involutive basis is isomorphic to the polynomial ring in x and y.

```
> h:=PolWeightedHilbertSeries([x=1,y=1,q4=4,p2=2,p4=4], s);
```

$$h := \frac{1}{(1-s)^2\,(1-s^4)^2\,(1-s^2)} - \frac{s^8}{(1-s^4)^2\,(1-s^2)} - \frac{s^9}{(1-s^4)^2\,(1-s^2)}$$
$$-\frac{s^{10}}{(1-s^4)^2\,(1-s^2)} - \frac{s^{11}}{(1-s^4)^2\,(1-s^2)}$$
$$-2\,\frac{s^4}{(1-s)\,(1-s^4)^2\,(1-s^2)} - \frac{s^9}{(1-s^4)\,(1-s^2)}$$
$$-\frac{s^5}{(1-s^4)^2\,(1-s^2)} - \frac{s^6}{(1-s^2)\,(1-s^4)} - \frac{s^6}{(1-s^2)\,(1-s^4)^2} - \frac{s^7}{1-s^4}$$
$$-\frac{s^5}{(1-s^4)\,(1-s^2)} - \frac{s^7}{(1-s^4)^2\,(1-s^2)} - \frac{s^2}{(1-s)^2\,(1-s^4)^2\,(1-s^2)}$$

```
> simplify(h);
```

$$\frac{1}{(-1+s)^2}$$

Finally the commad NotHas extracts from B all relations not involving x or y, thus yielding a presentation for I(G) on the generators p2,p4,q4.

```
> R:=NotHas(B,[x,y]);
```

$$R := [q4^2 - p2^2\,p4 + 4\,p4^2]$$

5 Some Refined Techniques

One problem that was not solved satisfactorily in Section 3 was how to prove that one had generators for the ring of invariants in order to conclude from the equality of the HILBERT series and the MOLIEN series that one had a presentation of the ring of invariants. We give some hints in this section, how to use the JANET algorithm directly on the invariants towards this aim. At the same time, this technique can be used to obtain a standard expression of any given invariant in terms of the generators. The theoretical concept used in this section is the fact that the ring $I(G)$ of invariants of a finite complex matrix group $G \leq \mathrm{GL}(n,\mathbb{C})$ is COHEN-MACAULY, cf. [Ben 93] pg. 50. This implies that there exist n homogeneous invariants, $f_1,\dots,f_n \in I(G)$, which form a system of parameters, i. e., which are algebraically independent and have the property that $I(G)$ is a free $\mathbb{C}[f_1,\dots,f_n]$-module of finite rank.

Proposition 2. *Let $f_1,\dots,f_n \in I(G)$ be homogeneous. The following three statements are equivalent.*
1) $f_1,\dots,f_n$ form a set of parameters for $I(G)$.
2) $f_1,\dots,f_n$ form a set of parameters for $\mathbb{C}[x_1,\dots,x_n]$.

3) $\mathbb{C}[x_1, \dots, x_n]/(f_1, \dots, f_n)$ *is a finite dimensional* $\mathbb{C}$ *algebra.*
Moreover, if the cosets of $b_1, \dots, b_t \in \mathbb{C}[x_1, \dots, x_n]$ *form a* $\mathbb{C}$*-basis for the residue class algebra* $\mathbb{C}[x_1, \dots, x_n]/(f_1, \dots, f_n)$*, then* $I(G)$ *is generated by the* f_i *and the* $\pi(b_j)$ *with* π *as defined at the end of Section 2.*

Proof. Obviously all three conditions imply that $f_1, \dots, f_n$ are algebraically independent so that we only have to deal with the other issues. The implication o 2) implies 1) can be taken from the poof of Theorem 4.3.6 in [Ben 93] , where the COHEN-MACAULY property of the ring of invariants is proved. The reversed implication follows form Theorem 4.3.5 in [Ben 93]. That 2) implies 3) is obvious. We shall see how the JANET algorithm can be modified to obtain a constructive proof of the implication 3) to 2), which at the same time yields an algorithm to construct a $\mathbb{C}[f_1, \dots, f_n]$-basis of $\mathbb{C}[x_1, \dots, x_n]$ or with some more effort of $I(G)$ and how to express any given element of $\mathbb{C}[x_1, \dots, x_n]$ resp. $I(G)$ in this basis. In fact, we shall formulate this part of the proof as an algorithm, which is slightly more general than the situation considered here. q. e. d.

Algorithm 2 *Input: Algebraically independent elements*

$$f_1, \dots, f_n \in \mathbb{C}[x_1, \dots, x_n]$$

such that

$$\mathbb{C}[x_1, \dots, x_n]/(f_1, \dots, f_n)$$

is finite dimensional.
Output: A $\mathbb{C}[f_1, \dots, f_n]$*-basis* $(b_1, \dots b_s)$ *of* $\mathbb{C}[x_1, \dots, x_n]$ *and a procedure to express any given element of* $\mathbb{C}[x_1, \dots, x_n]$ *in this basis.*
Algorithm: Perform the usual JANET*-algorithm on* $f_1, \dots, f_n$ *with the usual degree lexicographical ordering with the following variation: Instead of starting with* $f_1, \dots, f_n$*, introduce a symbol* a_i *for each* f_i*, start out with the pairs* $(f_i; a_i)$*, and perform all the operations in both components, e. g. multiplication with* x_j*, addition and subtraction. The operations are done according to the usual rules coming from the first components, so that one ends up with* $(r_j; \sum r_{ji}a_i)$*, where the* $r_j \in \mathbb{C}[x_1, \dots, x_n]$ *form a* JANET*-basis for the ideal generated by* $f_1, \dots, f_n$ *and the* r_{ji} *lie in* $\mathbb{C}[x_1, \dots, x_n]$*.*
The $\mathbb{C}[f_1, \dots, f_n]$*-basis of* $\mathbb{C}[x_1, \dots, x_n]$ *is given by all monomials in* $\mathbb{C}[x_1, \dots, x_n]$*, which do not occur as a leading monomial of some polynomial of* $(f_1, \dots, f_n)$*, i. e. which are not multiples of the leading monomials of the* r_j *in the* JANET *basis.*
Procedure to express a given element h of $\mathbb{C}[x_1, \dots, x_n]$ *in the above basis: Perform involutive division on h with the* $r_j - \sum r_{ji}a_i$ *(instead of the usual* r_j*). In this process one builds up linear combinations of monomials of the* a_i*, the coefficients of which are polynomials in* $x_1, \dots, s_n$*. These coefficients are processed according to the* JANET*-rules, until the process terminates, i. e. only monomials of the constructed basis occur. Now one rewrites the expression as a sum of the basis elements with polynomials in the* a_i *as coefficients.*

Proof. That the algorithm terminates is clear from JANET's algorithm. That the procedure terminates is also clear for the same reason. The procedure shows that the monomials representing a $\mathbb{C}$-basis of $\mathbb{C}[x_1, \dots, x_n]/(f_1, \dots, f_n)$ form a generating set for $\mathbb{C}[x_1, \dots, x_n]$ as $\mathbb{C}[f_1, \dots, f_n]$-module. Tensoring with the $\mathbb{C}[f_1, \dots, f_n]$-module $\mathbb{C}$, where all f_i act on $\mathbb{C}$ as multiplication by 0, shows that the rank of $\mathbb{C}[x_1, \dots, x_n]$ as $\mathbb{C}[f_1, \dots, f_n]$-module is equal to the number of these monomials. It follows they even form a set of free generators, since the matrices over $\mathbb{C}[f_1, \dots, f_n]$ expressing one by the others are quadratic and therefore inverse to each other. q. e. d.

It is clear that the modified JANET algorithm 2 does everything one can hope for from the side of commutative algebra: It decides whether given homogeneous invariants $f_1, \dots, f_n$ form parameters for the invariant ring, by checking the finite dimensionality of algebra $\mathbb{C}[x_1, \dots, x_n]/(f_1, \dots, f_n)$. Once this is established, it expresses each new invariant in the normal form by the procedure of the algorithm and thereby enables one to quickly find the free generators b_i of $I(G)$ as $\mathbb{C}[f_1, \dots, f_n]$-module. So one has

$$I(G) = \bigoplus_i \mathbb{C}[f_1, \dots, f_n] b_i.$$

And finally once the b_i are given in the normal form, it can also by a slight extension of involutive division express each given invariant as a linear combination of the b_i with coefficients in $\mathbb{C}[f_1, \dots, f_n]$. Finally, it therefore is also able to quickly derive a presentation of $I(G)$, simply by expressing the products $b_i b_j$ in this normal form.

References

[Ape 98] J. Apel, The Theory of Involutive Divisions and Apllications to Hilbert Function Computations. J. Symbolic Computation (1998) **25**, 683-704.

[Ben 93] D. J. Benson, Polynomial Invariants of Finite Groups. LMS Lecture Notes 190, Cambridge Univ. Press 1993.

[GAP 00] The GAP Group, GAP — Groups, Algorithms, and Programming, Version 4.2; Aachen, St Andrews, 1999. (http://www-gap.dcs.st-and.ac.uk/gap)

[GeB 98a] V. P. Gerdt, Y. A. Blinkov, Involutive bases of polynomial ideals. Mathem. and Computers in Simulation 45 (1998), 519-541.

[GeB 98b] V. P. Gerdt, Y. A. Blinkov, Minimal involutive bases. Mathem. and Computers in Simulation 45 (1998), 543-560.

[Kem 98] G. Kemper, Computational Invariant Theory. The Curves Seminar at Queen's, Volume XII, in: Queen's Papers in Pure and Applied Math. 114 (1998), 3-26.

[Kem 99] G. Kemper, HILBERT Series and degree bounds in invariant theory. in: B. Heinrich Matzat, Gert-Martin Greuel, Gerhard Hiss, eds, Algorithmic Algebra and Number Theory, Springer-Verlag, Heidelberg, 1999.

[Pom 94] J.- F. Pommaret, Partial Differential Equations and Group Theory. Kluver Academic Publishers 1994.

[Sei 00] W. M. Seiler, A Combinatorial Approach to Involution and δ-Regularity. Preprint University Mannheim (2000).

[Sta 79] R. P. Stanley, Invariants of finite groups. AMS Bulletin vol. 1, No. 3 (1979), 475-511.

[Stu 93] B. Sturmfels, Algorithms in Invariant Theory. Springer 1993.

Symbolic Computation and Boundary Conditions for the Wave Equation

A. S. Deakin* and H. Rasmussen

Department of Applied Mathematics
University of Western Ontario
London, Ontario, Canada, N6A 5B7

Abstract. We develop a procedure using Maple for the estimation of the accuracy of approximate boundary conditions for the wave equation. To solve this equation in an infinite three dimensional domain, a spherical artificial boundary is introduced to restrict the computational domain Ω. To determine the nonreflecting boundary condition on $\partial\Omega$, we start with a finite number of spherical harmonics for the Helmholtz equation. With a precise choice of nodes on the sphere, the theorem on Gauss-Jordan quadrature establishes the discrete orthogonality of the spherical harmonics when summed over these nodes. The nonreflecting boundary condition for the Helmholtz equation follows readily upon solving the exterior Dirichlet problem. The boundary condition for the time dependent wave equation follows directly by taking the inverse Fourier transform of the boundary condition for the Helmholtz equation. The boundary condition has the following properties: only the first derivatives in space and time appear; once the coefficients are updated in a simple way from the previous time step, the boundary condition involves only the nodes at the current time step. We derive using Maple some very precise estimates of the accuracy of these approximate boundary conditions.

1 Introduction

In the past, computer algebra packages, such as Maple and Mathematica, have been used in the analysis and study of partial differential equations. For example, exact solutions are derived in Hereman and Nuseir [11] while symmetries are analyzed in Göktas and Hereman [8]. Another important application is the study and analysis of the accuracy of approximate boundary conditions in infinite domains, and this has not yet attracted much attention. In this paper we will describe such an application to a specific problem, namely, the three dimensional wave equation. While our procedure at the moment is only applied to this specific problem, we believe that the general approach will be applicable to a wider class of problems.

In order to develop a numerical procedure for a linear partial differential equation in an infinite domain we divide the domain into two parts: an inner

* This research was supported by NSERC of Canada.

finite subdomain enclosing the physical region of interest and an outer infinite region. This division is carried out by introducing an artificial boundary B such that the exterior Dirichlet problem can be solved analytically. The computational zone Ω is restricted to the interior of the artificial boundary. Inside Ω there may be nonlinear source terms, and boundaries with appropriate boundary conditions could also be present. Our concern in this paper is the nonreflecting boundary condition that must be imposed on B. The geometry of the problem is shown in Figure 1.

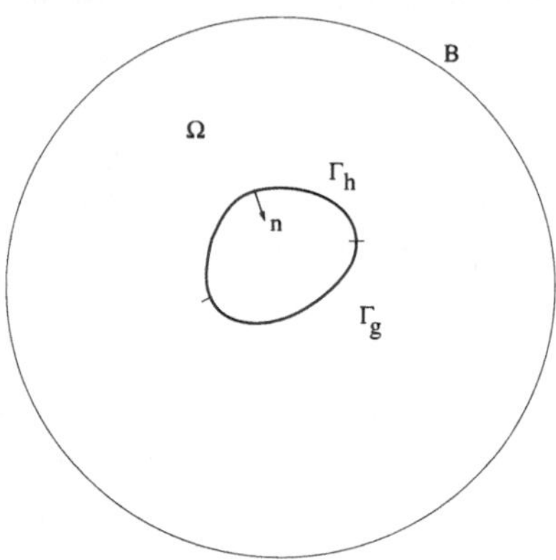

Fig. 1. The basic geometric configuration

In particular we consider, in three space dimensions, the time dependent wave equation and the Helmholtz equation where the artificial boundary B is a sphere of radius a that encloses Ω. In the infinite region exterior to B, we require that $U(x,t)$ satisfy the wave equation with zero initial conditions:

$$\frac{\partial^2}{\partial t^2}U(x,t) - \nabla^2 U(x,t) = 0, \quad |x| > a, \quad t > 0 \tag{1}$$

$$U(x,0) = 0, \qquad U_t(x,0) = 0. \tag{2}$$

where, for the moment, we assume that $U(x,t)$ is given on $|x| = a$. To solve this problem we consider is the reduced wave equation or the Helmholtz equation with the radiation boundary condition at infinity:

$$\nabla^2 u(x,k) + k^2 u(x,k) = 0, \quad |x| > a \tag{3}$$

$$\lim_{r\to\infty} r(u_r - iku) = 0, \quad r = |x|. \tag{4}$$

and $u(x,t)$ is given on $|x| = a$ The solutions $U(x,t)$ and $u(x,k)$ are connected by the Fourier transform and its inverse given by

$$u(x,k) = \int_0^\infty U(x,t)e^{ikt}\,dt, \quad U(x,t) = \frac{1}{2\pi}\int_{-\infty}^\infty u(x,k)e^{-ikt}\,dk \tag{5}$$

There are several recent papers in which nonreflecting boundary conditions are derived for the time dependent wave equation. Grote and Keller [9] derive boundary conditions on B for the wave equation. Two of these boundary conditions involve high r derivatives of U and U_t and the third one contains only the derivatives U_r and U_t. For this last case, Grote and Keller [10] use the finite difference method for three problems to illustrate the accuracy of this boundary condition. In addition, they compare the numerical accuracy of this boundary condition with that of a modified boundary condition and two local boundary conditions based on the ideas of Bayliss and Turkel [2]. Givoli and Cohen [7] also consider the time dependent wave equation in three dimensions; however, their boundary condition uses the Kirchhoff formula where the surface, over which the functions are integrated, is enclosed by the artificial boundary B. This exact boundary condition, although nonlocal in space and time, is limited to a fixed amount of past information. The boundary conditions we analyze in this paper have been presented in [14] and [15].

As discussed by Grote and Keller [10], the numerical results for the time dependent wave equation are very accurate. Based on our experience in applying nonreflecting boundary conditions to potential problems in two dimensions (Deakin and Dryden [3]) and in a three dimensional region between two planes (Deakin and Rasmussen [4]), nonreflecting boundary conditions lead to highly accurate solutions.

In this paper, we start with the Gauss-Jacobi quadrature formulas ([13]) to establish the orthogonality properties of the spherical harmonics when summed over a special set of nodes. Then the exterior Dirichlet problem for the Helmholtz equation (3,4) is solved for a finite number of eigenfunctions (see (7)). The boundary condition for the Helmholtz equation (§3) follows readily and, the Fourier inversion gives the boundary condition for the wave equation (§4). The way in which we estimate the accuracy of our boundary conditions is presented in §5 where we consider a point source in Ω so that we know the exact solution. In (§6) we describe the role that Maple plays in the work described in this paper while in (§7) we discuss the accuracy of the obtained boundary conditions.

We use Maple to accurately compute the roots of two polynomials: the roots of a Legendre polynomial that determines the special set of nodes, and the ratio of spherical Hankel functions is computed using the roots of a generalized hypergeometric function. As a measure of the accuracy of our boundary condition, we compare the derivative of a solution for the point source on the boundary with the derivative as determined by the boundary condition. We use Maple to integrate exactly the integrals in our boundary condition.

2 Formulation

In spherical coordinates, we define nodes as $x_{pq} = (a, \theta_p, \phi_q)$ where $\phi_q = (q - 1)2\pi/M$ $(q = 1, \ldots, M)$ and, for reasons to be stated shortly, θ_p $(p = 1, \ldots, N)$ are the roots of the Legendre polynomial $P_N(\cos\theta) = 0$ $(0 < \theta < \pi)$. The boundary conditions that we derive in §3 and §4 involve a sum over these nodes.

The boundary condition for the Helmholtz equation is determined from the solution of the exterior Dirichlet problem for $|x| > a$. The general eigenfunction expansion that satisfies the radiation condition at infinity is

$$u(x,k) = \sum_{n=0}^{\infty} \sum_{m=-n}^{n} B_{nm} \frac{H^{(1)}_{n+1/2}(kr)\sqrt{a}}{H^{(1)}_{n+1/2}(ka)\sqrt{r}} Y_n^m(\theta,\phi), \tag{6}$$

where $H^{(1)}_{n+1/2}(kr)$ is the Hankel function of the first kind, the spherical harmonic $Y_n^m(\theta,\phi)$ is defined by $e^{im\phi}P_n^m(\cos\theta)$, and $P_n^m(\cos\theta)$ is the associated Legendre function. The Fourier coefficients B_{nm} are uniquely determined once $u(x,k)$ is given on the artificial boundary $|x| = a$. In the sequel, we formulate a discrete version of the Fourier coefficients and, in Appendix B, we present the corresponding approach that involves the usual Fourier coefficients.

Lemma 1. *Suppose that the solution to (3,4) is a linear combination of a finite number of spherical harmonics* $Y_n^m(\theta,\phi)$,

$$u(r,\theta,\phi,k) = \sum_{n=0}^{N-1} \sum_{m=-n}^{n} B_{nm} \frac{H^{(1)}_{n+1/2}(kr)\sqrt{a}}{H^{(1)}_{n+1/2}(ka)\sqrt{r}} Y_n^m(\theta,\phi). \tag{7}$$

Then, given u at the primary nodes, the coefficients in this solution are uniquely determined by the discrete orthogonality properties

$$\sum_q e^{i(m'-m)\phi_q} = \delta_{m'm}M, \quad |m'-m| < M, \quad (M \geq 2N-1), \tag{8}$$

$$\int_{-1}^{1} P_n^m(\mu)P_{n'}^{-m}(\mu)\,d\mu = \sum_{p=1}^{N} \lambda_p P_n^m(\mu_p)P_{n'}^{-m}(\mu_p) = \delta_{n'n}\frac{2(-1)^m}{2n+1}, \tag{9}$$

where $\mu = \cos\theta$, $\mu_p = \cos\theta_p$ *(p=1,...,N) are the zeros of the Legendre polynomial* $P_N(\mu)$, λ_p *is the Christoffel number associated with* μ_p, *and* $\delta_{m'm}$ *is the Kronecker delta (see [13]).*

Applying the orthogonality conditions to (7), the N^2 Fourier coefficients are given by

$$B_{nm} = (-1)^m \frac{2n+1}{2M} \sum_{p,q} \lambda_p Y_n^{-m}(\theta_p,\phi_q)u(x_{pq},k). \tag{10}$$

One expression for the coefficient λ_p is (see [13], p. 47)

$$\lambda_p = \frac{2}{NP_{N-1}(\mu_p)P'_N(\mu_p)} = \frac{2(1-\mu_p^2)}{(NP_{N-1}(\mu_p))^2}. \tag{11}$$

(In [13], we take the orthogonal polynomial $p_n(x)$ to be $(n+1/2)^{1/2}P_n(x)$.) Note that the orthogonality property (8) is unaffected if $\phi_q = \alpha + (q-1)2\pi/M$ where α is any arbitrary number. Thus, without loss of generality, we can take $\alpha = 0$ since a rotation of the coordinate system about the z-axis would remove this term.

3 Boundary Condition for the Helmholtz Equation

The derivative $\partial u/\partial r$ at any point x on the boundary is derived from (7) by differentiating with respect to r and using the addition theorem ([5])

$$P_n(\cos\psi(x, x_{pq})) = \sum_{m=-n}^{n} (-1)^m Y_n^m(\theta, \phi) Y_n^{-m}(\theta_p, \phi_q), \tag{12}$$

where $\psi(x, x_{pq})$ is the angle between the rays from the origin to the points x and x_{pq}. Thus, u and $\partial u/\partial r$ at any point on the artificial boundary are given by

$$u(x, k) = \sum_{p,q} \sum_{n=0}^{N-1} C_{pq}^n(x) u(x_{pq}, k), \tag{13}$$

$$\frac{\partial}{\partial r} u(x, k) = \sum_{p,q} M_{pq}(x) u(x_{pq}, k), \tag{14}$$

where

$$M_{pq}(x) = \sum_{n=0}^{N-1} C_{pq}^n(x) \frac{\frac{\partial}{\partial a}(H_{n+1/2}^{(1)}(ka)a^{-1/2})}{H_{n+1/2}^{(1)}(ka)a^{-1/2}}, \tag{15}$$

$$C_{pq}^n(x) = \lambda_p P_n(\cos\psi(x, x_{pq}))(2n+1)/(2M). \tag{16}$$

An alternate formulation of the boundary condition (14), that is suitable particularly if $|ka| >> 1$, is readily derived by expressing the ratio of Hankel functions in (15) in another form. From the definitions

$$H_{n+1/2}^{(1)}(z) = \sqrt{\frac{2}{\pi z}} e^{i(z-(n+1)\pi/2)} F(n+1, -n; \frac{1}{2iz}), \tag{17}$$

$$F(n+1, -n; z) = \sum_{j=0}^{n} \frac{(n+j)!}{(n-j)!} \frac{(-z)^j}{j!}, \tag{18}$$

where F is the generalized hypergeometric function ${}_2F_0$ ([5]), we have

$$a\frac{\frac{\partial}{\partial a}(H_{n+1/2}^{(1)}(ka)a^{-1/2})}{H_{n+1/2}^{(1)}(ka)a^{-1/2}} + 1 - ika = \frac{k\frac{\partial}{\partial k}F(n+1, -n; 1/(2ika))}{F(n+1, -n; 1/(2ika))} \equiv G_n(ka), \tag{19}$$

and $G_n(ka) = O(1/(ka))$ as $|ka| \to \infty$.

The simplest way to compute $G_n(ka)$ is to determine the zeros of $F(n+1, -n; 1/(2iz))$ which are the zeros of $H_{n+1/2}^{(1)}(z)$. In the complex z-plane for each $n \geq 1$, there are n simple zeros $z_{nj} = a_{nj} - ib_{nj}$ $(j = 1, \ldots, n)$ where $b_{nj} \geq 1$, and these roots lie symmetrically with respect to the imaginary axis. Only for n odd is there a pure imaginary root. Furthermore, for any n and j, all roots z_{nj} are distinct ([1],[5]). From (19), we have $G_n(ka) = \sum_j z_{nj}/(ka - z_{nj})$.

From (13) and (14) $\frac{\partial}{\partial r}(ru) - ikru$ at any point x on the artificial boundary may be expressed as a linear combination

$$\left(\frac{\partial}{\partial r}(ru) - ikru\right)(x,k) = \sum_{p,q}\sum_{n=1}^{N-1} C_{pq}^n(x) G_n(ka) u(x_{pq},k). \tag{20}$$

4 Boundary Condition for the Wave Equation

The inverse Fourier transform for $G_n(ka)$ in (20) is readily evaluated as

$$\sum_j (-iz_{nj}/a)\exp(-iz_{nj}t/a)H(t) \tag{21}$$

where $H(t)$ is 0 for $t < 0$ and 1 for $t > 0$. Upon taking the inverse transform of (20), we have the following result.

Theorem 1. *U and $(\frac{\partial}{\partial r} + \frac{\partial}{\partial t})(rU)$ at any point x on the artificial boundary are expressed in terms of U at the nodes by*

$$U(x,t) = \sum_{p,q}\sum_{n=0}^{N-1} C_{pq}^n(x) U(x_{pq},t), \tag{22}$$

$$\left(\frac{\partial}{\partial r} + \frac{\partial}{\partial t}\right)(rU)(x,t) = -\sum_{p,q}\sum_{n=1}^{N-1} C_{pq}^n(x) \sum_j \frac{iz_{nj}}{a} I_{nj}(x_{pq},t), \tag{23}$$

where

$$I_{nj}(x_{pq},t) = \int_0^t \exp(-iz_{nj}(t-\xi)/a) U(x_{pq},\xi)\, d\xi \tag{24}$$

and the sum in j is over the zeros z_{nj} of the Hankel function $H_{n+1/2}^{(1)}(z)$.

Although I_{nj} involves an integral over U for all times less than t, any numerical application of the boundary condition (23) does not require that at each time step we sum U over all previous time steps. To show this, we note that

$$\frac{d}{dt} I_{nj}(x_{pq},t) = (-iz_{nj}/a) I_{nj}(x_{pq},\tau) + U(x_{pq},t) \tag{25}$$

with $I_{nj}(x_{pq},0) = 0$. Thus we need only update I_{nj} at each time step.

The simple boundary condition $(\frac{\partial}{\partial r} + \frac{\partial}{\partial t})(rU) = 0$ is the first in a sequence of boundary conditions, when applied to rU instead of U, defined by Bayliss and Turkel [2].

5 Accuracy of Boundary Conditions

While it is fairly simple to derive the approximate boundary conditions given by (23), it is far more difficult to estimate the accuracy of this expressions. One way would be to write a numerical procedure for solving the wave equation with condition (23). Then we could solve the interior problem in Ω for situations for which we have exact solutions, such as point sources, and the accuracy could then be estimated by comparing the numerical solution with the exact solution. However, the accuracy of the numerical solution will depend, in addition to the approximations in the boundary condition, on the truncation error in the finite difference or finite element procedure and on the accuracy with which the system of algebraic equations, resulting from the finite difference procedure, is solved. It will be difficult to decide the relative importance of these sources of errors and hence state with any degree of confidence the accuracy of our approximate boundary conditions.

For these reasons we decide to use a different approach. We determine the accuracy of our boundary conditions by using point sources for which we know the solution of the Helmholtz and the time dependent wave equation. From these solutions we can calculate the exact expressions for the normal derivative, U_r at the artificial boundary B. However we can also use our approximate boundary condition to estimate U_r. The difference between these two expressions is an estimate of the accuracy of our procedure. With Maple we can make sure that the only difference between the two expressions is only due to the approximate nature of of the boundary condition.

For the time dependent wave equation, we take the solution of (1) to be

$$W(x,t) = \begin{cases} f(t-s)/(4\pi s) & t > 0 \\ 0 & t \leq 0 \end{cases} \qquad f(t) = \sin(\beta t) \times \begin{cases} g_i(t/\gamma) & 0 < t < \gamma \\ 1 & t \geq \gamma \end{cases} \tag{26}$$

where β and γ are parameters, and $g_i(t)$ is a polynomial of degree $2i$ such that the derivatives of $W(x,t)$ of order i are continuous. We use two cases: $g_1(t) = 2t - t^2$ and $g_3(t) = -10t^6 + 36t^5 - 45t^4 + 20t^3$.

Owing to symmetry, the boundary condition (23) simplifies since $U(x_{pq}, t)$ is independent of q. Summing (12) in q and using the orthogonality property (8), we have

$$\sum_q C^n_{pq}(x) U(x_{pq}, t) = \frac{2n+1}{2} \lambda_p P_n(\mu) P_n(\mu_p) U(x_{p1}, t). \tag{27}$$

From (20) , the boundary condition becomes

$$U_r(x,t) = -\frac{U(x,t)}{a} - U_t(x,t) - \frac{1}{a} \sum_p \int_0^t K_p(a,\mu,t-\xi) U(x_{p1},\xi)\, d\xi \tag{28}$$

where

$$K_p(a,\mu,t) = \sum_j \sum_{n=1}^{N-1} \frac{2n+1}{2} \frac{iz_{nj}}{a} \lambda_p P_n(\mu) P_n(\mu_p) e^{-iz_{nj}t/a}. \tag{29}$$

To measure the accuracy of the boundary condition (28), we compare the normal derivative W_r on B to the derivatives u_r and U_r determined by the boundary conditions. To this end, we compute the norm $E(a, N, t)$ on B:

$$E(a, N, t) = \max_{x \in B} |U_r(x, t) - W_r(x, t)|. \tag{30}$$

U_r is computed from (28) where $U(x, t) = W(x, t)$ and $U_t(x, t) = W_t(x, t)$. We have chosen the functions in (26) so that the integral in (28) can be integrated exactly by Maple.

We have three figures to illustrate the accuracy of the boundary condition (28). The upper bound on the error for all figures can be estimated by taking a "smooth" upper bound to the errors shown. We used the point on B nearest to the point source to compute these curves. At the large downward spikes the largest error occurs at another point on B. Hence the large downward spikes in the figures are much smaller than shown. In all figures, the case $N = 1$ corresponds to the boundary condition, developed by Bayliss and his co-workers, that is described in §4.

6 Application of a Computer Algebra System

It is clear from the complexity of the different expressions described above, that it would be very difficult if not impossible, to carry out all the required analytical manipulations by hand. A computer algebra package is required. We use Maple 7.

Let us consider in more detail the five different subproblems that have to be solved.

- (1) The coordinates x_{pq} on the boundary B are obtained from the zeros of the Legendre polynomials $P_N(\mu)$. For each value of N we used the command *fsolve(P(N,x),x,x=0..1)*. This solves the equation and gives us all roots to arbitrary precision. In most of our applications, we only did this once and stored the roots for later use.
- (2) The roots μ_p are then substituted into equation (11) to given values for the coefficients λ_p.
- (3) The quantities z_{nj} are the roots of the generalized hypergeometric function ${}_2F_0$. From equation (18) we see that we must solve the polynomial equation

$$\sum_{j=0}^{n} \frac{(n+j)!}{(n-j)!} \frac{(-z)^j}{j!} = 0$$

  Here we use the command *fsolve(poly(n,z), z ,complex)* where poly is the left hand side of the equation.
- (4) The expressions for λ_p and x_{pq} are used to calculate the coefficients $C^n_{pq}(x)$, equation (16), for different values of x.

– (5) In order to asses the accuracy of the boundary conditions we evaluate expression (28) at different points on the boundary. The integrals are done analytically while the sum is evaluated in floating point. The results are presented in Figures 2 to 4. The process is quite slow, about two hours of computing time was required on a personal computer, Pentium II 350 MHz. This time could be reduced somewhat by only doing the integration using Maple while the numerical evaluation of the sums could be carried in C++.

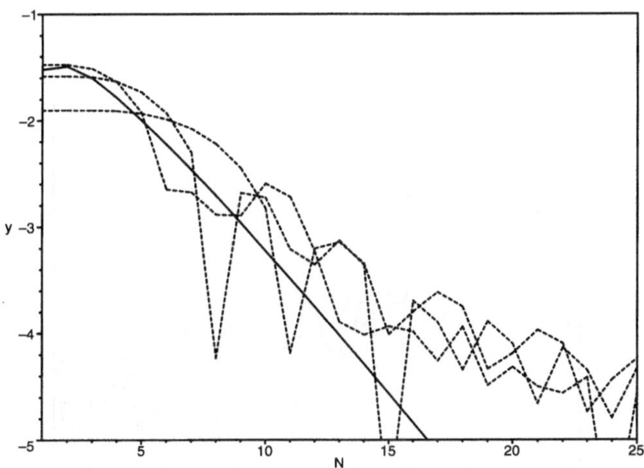

Fig. 2. The error $E(a, N, t) = 10^y$ in (30) is shown vs. N. For $N = 1$, the kernel in (29) is defined to be zero. The third derivatives of $W(x, t)$ in (26) are continuous and the parameters are $\beta = 1$, $\gamma = 1$, and $a = 2$. The three cases $t = 1.5$ (dot), $t = 1.75$ (dot) and $t = 2.0$ (dot) provide an estimate for the upper bound of the error. For $t = 5.0$ (solid) the errors are much smaller

7 Results

The purpose of the paper is to show that our approximate treatment of the far field boundary conditions is reasonably accurate. We show this by plotting some representative results for the error $E(a, N, t)$ as defined by equation (30). Here a is the radius of the artificial boundary B, N is the number of terms in our expansion and t is the time. In Figure 2 we plot the log of this error vs N for three different values of t when the third derivatives of $W(x, t)$ are continuous. We see that the accuracy increases by about two orders of magnitude as N increases from 5 to 20 as N increases. In Figure 3 where it is only the first derivatives which are continuous the error only decreases from approximately

10^{-2} to $10^{-3.5}$ In Figure 2, the maximum error is approximately linear in N; whereas, in Figure 3, itr drops more rapidly initially and then changes more slowly for $N > 10$. Hence, the smoothness of the solution has some effect on the accuracy of the boundary condition. Note that for $t = 5$ where $g_i(t) = 1$ in (26) on B the boundary condition is much more accurate compared with the error for earlier values of t when the disturbance at some points on B is not purely oscillatory. Comparing Figure 3 ($\beta = 1$) and Figure 4 ($\beta = 6$), we see that the boundary condition is less accurate as the frequency β is increased.

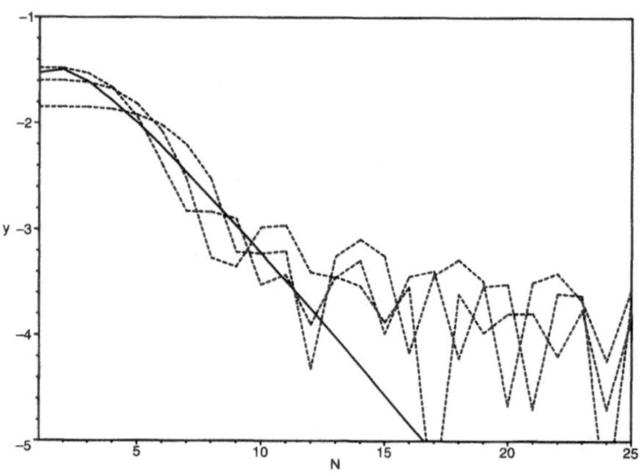

Fig. 3. The same as in Fig. 2 except that the first derivatives of $W(x, t)$ in (26) are continuous

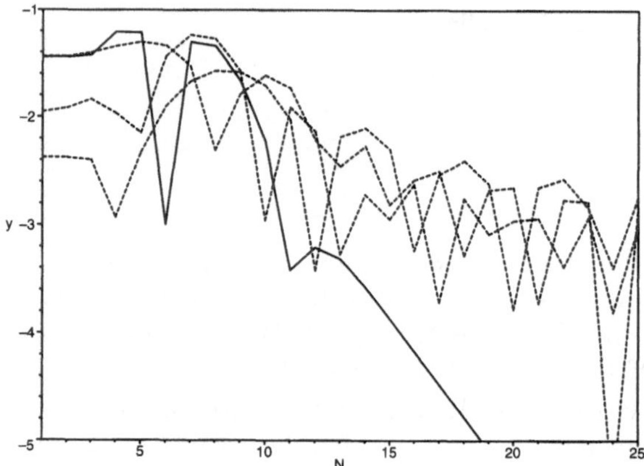

Fig. 4. The same as in Fig. 3 except that $\beta = 6$

These results show that the proposed procedure for calculating an approximate far field boundary condition gives reasonable numerical accuracy when it is applied to a non-trivial test problem for which we have accurate values to compare with. We also show, as expected, that when the number of terms in the expansion is increase we get a considerable increase in accuracy.

Usually when the accuracy of an approximate boundary condition is assessed, the partial differential equation in the inner region is solved numerically. This makes it more difficult to estimate the accuracy of the boundary condition since the calculated results contains both the errors due to the approximate nature of the boundary conditions and also the truncation error associated with the numerical procedure for the partial differential equation. In our approach we only measure the errors due to the boundary condition.

References

1. Abramowitz, M. and Stegun, I.A.: *Handbook of Mathematical Physics*, Dover, New York (1965)
2. Bayliss, A. and Turkel, E.: Radiation boundary conditions for wave-like equations. *Commun. Pure Appl. Math.* **33** (1980), 707–725
3. Deakin, A.S. and Dryden, J.R.: Numerically derived boundary conditions on artificial boundaries. *J. Comp. Appl. Math.* **58** (1995), 1–16
4. Deakin, A.S. and Rasmussen, H.: Sparse boundary conditions on artificial boundaries for three-dimensional potential problems. *J. Comput. Phys.* **129** (1996), 111-120
5. Erdélyi, A. (ed.): *Higher Transcendental Functions*, Bateman manuscript project, Vol. I, II, R. E. Kreiger, Florida (1985)
6. Fiedler, M.: *Special Matrices and Their Applications in Numerical Mathematics*, Martinus Nijhoff, Boston (1986)
7. Givoli, D. and Cohen, D.: Nonreflecting boundary conditions based on Kirchhoff-type formulae. *J. Comput. Phys.* **117** (1995) 102–113
8. Göktas, Ü. and Hereman, W.: Algorithmic computation of higher-order symmetries for nonlinear evolution and lattice equations. *Advances in Computational Mathematics* **11** (1999) 55–80
9. Grote, M.J. and Keller, J.B.: Exact nonreflecting boundary conditions for the time dependent wave equation. *SIAM J. Appl. Math.* **55** (1995) 280–297
10. Grote, M.J. and Keller, J.B.: Nonreflecting boundary conditions for time-dependent scattering. *J. Comput. Phys.* **127** (1996) 52–65
11. Hereman, W. and Nuseir, A.: Symbolic methods to construct exact solutions of nonlinear partial differential equations. *Mathematics and Computers in Simulation* **43** (1997) 13–27
12. Keller, J.B. and Givoli, D.: Exact non-reflecting boundary conditions. *J. Comput. Phys.* **82** (1989) 172–192
13. Szegö, G.: *Orthogonal Polynomials*, AMS Colloq. Publ., Vol. 23, AMS, New York, 1959
14. Sofronov, I.L.: Conditions for complete transparency on a sphere for the three-dimensional wave equation. *Russ. Acad. Sci. Dokl. Math.* **46** (1983) 397–401
15. Sofronov, I.L.: Artificial boundary conditions of absolute transparency for two- and three-dimensional external time-dependent scattering problems. *Euro. J. Appl. Math.* **9** (1998) 561–588

Parametric Systems of Linear Congruences

Andreas Dolzmann and Thomas Sturm

FMI, University of Passsau, Passau, Germany,
{dolzmann,sturm}@uni-passau.de,
http://www.fmi.uni-passau.de/{dolzmann,sturm}

Abstract. Based on an extended quantifier elimination procedure for discretely valued fields, we devise algorithms for solving multivariate systems of linear congruences over the integers. This includes determining integer solutions for sets of moduli which are all power of a fixed prime, uniform p-adic integer solutions for parametric prime power moduli, lifting strategies for these uniform p-adic solutions for given primes, and simultaneous lifting strategies for finite sets of primes. The method is finally extended to arbitrary moduli.

1 Introduction

We devise methods for testing multivariate systems of linear congruences for feasibility. In the positive case we obtain at least one sample solution for the system. Our methods allow to prescribe for each constraint a particular modulus in contrast to having only one modulus for the entire system:

$$a_{11}x_1 + \cdots + a_{1n}x_n \equiv b_1 \bmod \mu_1$$
$$\vdots$$
$$a_{m1}x_1 + \cdots + a_{mn}x_n \equiv b_m \bmod \mu_m,$$

where $a_{ij} \in \mathbb{Z}$. In the easiest case, $\mu_1, \ldots, \mu_m$ are various powers of a fixed prime number p:

$$\mu_1 = p^{k_1}, \quad \ldots, \quad \mu_m = p^{k_m}.$$

We then extended our approach to a parametric p, which stands for an arbitrary prime. Finally, we can apply the methods derived for such a parametric p to the general situation where

$$\mu_1, \quad \ldots, \quad \mu_m \in \{2, 3, 4, \ldots\}.$$

This is an important multivariate generalization of the problem solved by one of the key algorithms in computer algebra: The Chinese Remainder Theorem, which states the feasibility and gives a solution procedure for the special case where the $\mu_1, \ldots, \mu_m$ are pairwise relatively prime, and there is only one variable with coefficient 1, i.e.,

$$n = 1, \quad a_{11} = \cdots = a_{m1} = 1.$$

See Theorem 1 for details.

Our method will reduce the problem of solving such systems to one or several *extended quantifier elimination* problems over the rational numbers with p-adic valuations. The obtained p-adic integer sample solutions are then lifted to the integers. Similar extended quantifier elimination procedures have been successfully applied for constraint solving over the reals [9, 16].

For given fixed prime numbers p there are efficient linear algebra approaches for the discussed problem available [12, 3, 15]. Our work, in contrast, focuses on *uniform* solutions for a *parametric* prime p. It is not at all clear how the linear algebra methods cited above could be extended to the parametric case we are going to discuss here. With our approach about 98 percent of the computation time will be performed uniformly for all primes. The rest can be done at the result output stage without any delay. We shall furthermore prove that this grade of uniformity is the best one can expect.

The complexity of our methods is dominated by the complexity of the extended quantifier elimination procedure. This is single exponential in the number of variables but only polynomial in the number of congruences.

The plan of the paper is as follows: In Section 2 we recall some basic facts about p-adically valued fields, in which we compute our solutions before lifting them to the integers. In Section 3 we describe the connection between p-adic solving and integer solving. In Section 4 we give an overview on our method for the p-adic solving step which is extended quantifier elimination. Quantifier elimination procedures operate on first-order formulas. Section 5 explains how to obtain a suitable input formula for a system of linear congruences. One crucial advantage of quantifier elimination procedures for solving is that they can process parametric input in a very natural way. Section 6 exhibits how to exploit this for linear congruence systems with parametric moduli. We can then obtain p-adic, i.e. unlifted, solutions that are, up to a finite case distinction, uniformly correct for all possible choices of primes. We also demonstrate the theoretical limits for such uniform solving. In Section 7, we explain how and to what extent also the lifting step from our uniform p-adic solutions to integer solutions can be performed uniformly. The methods developed here for the simultaneous lifting for finite sets of primes allow us to finally extend our method to congruence systems to arbitrary, i.e. not necessarily prime, moduli. This is described in Section 8. The conclusions in Section 9 summarize and evaluate our results.

All methods and algorithms discussed here are efficiently implemented within the widespread computer algebra system REDUCE, based on the package REDLOG [7] by the authors. All our computations have been performed using 32 MB Lisp heap on an 800 MHz Athlon PC running Linux.

2 P-adic Valuations

For a given prime p, the p-adic *valuation* on the rational numbers is a map $v_p : \mathbb{Q} \to \mathbb{Z} \cup \{\infty\}$, where

$$v_p(0) = \infty, \quad v_p(r/s) = \max\{\, n \in \mathbb{N} : p^n \mid r \,\} - \max\{\, n \in \mathbb{N} : p^n \mid s \,\}.$$

Such valuations have the following properties: $v(a) = \infty$ if and only if $a = 0$, and

$$v(ab) = v(a) + v(b), \quad v(a+b) \geq \min\{v(a), v(b)\}.$$

It follows that $v(a+b) = \min\{v(a), v(b)\}$ if $v(a) \neq v(b)$. This fact is referred to as the *ultra-metric triangle equality.* Note that for $z \in \mathbb{Z}$ we have $v_p(p^z) = z$, i.e., v_p is onto. Due to a famous theorem by Ostrowski [14] the p-adic valuations are essentially the only maps with these properties.

The elements of non-negative value form a ring, the *valuation ring*

$$\mathbb{Z}_p = \{\, r/s \in \mathbb{Q} : \gcd(r, s) = 1 \text{ and } p \nmid s \,\}.$$

The elements of $\mathbb{Z}_p$ are called the *p-adic integers.* In $\mathbb{Z}_p$ the elements of positive value form a maximal ideal, the *valuation ideal* $p\mathbb{Z}_p$, which is the only maximal ideal in $\mathbb{Z}_p$. The elements of $\mathbb{Z}_p \setminus p\mathbb{Z}_p$ form the multiplicative group of units of $\mathbb{Z}_p$:

$$p\mathbb{Z}_p = \{\, r/s \in \mathbb{Q} : \gcd(r, s) = 1 \text{ and } p \mid r \,\}$$
$$\mathbb{Z}_p \setminus p\mathbb{Z}_p = \{\, r/s \in \mathbb{Q} : \gcd(r, s) = 1,\ p \nmid r \text{ and } p \nmid s \,\}.$$

From the maximality of $p\mathbb{Z}_p$ it follows that $\mathbb{Z}_p/p\mathbb{Z}_p$ is a field, the *residue class field* wrt. v_p. Up to isomorphism, this residue class field is particularly simple:

$$\mathbb{Z}_p/p\mathbb{Z}_p = (\mathbb{Z} + p\mathbb{Z}_p)/p\mathbb{Z}_p \simeq \mathbb{Z}/(\mathbb{Z} \cap p\mathbb{Z}_p) = \mathbb{Z}/p\mathbb{Z}.$$

All ideals in the ring $\mathbb{Z}_p$ of p-adic integers are of the form

$$p^k\mathbb{Z}_p = \{\, x \in \mathbb{Z}_p : v_p(x) \geq k \,\} = \{\, r/s \in \mathbb{Q} : \gcd(r, s) = 1,\ p^k \mid r \,\} \qquad (k \in \mathbb{N}).$$

A valuation can be essentially recovered from its valuation ring. To avoid a two-sorted language, we may thus drop the information about the concrete values by using the language of rings together with *abstract divisibilities.* These divisibilities express ordering relations in the value group $\mathbb{Z}$ by relating rational numbers:

$$x \mid y :\longleftrightarrow v(x) \leq v(y), \quad x \sim y :\longleftrightarrow v(x) = v(y)m \quad x \not\sim y :\longleftrightarrow v(x) \neq v(y).$$

We furthermore add a constant π of value 1 to our language, which is interpreted as the p of our p-adic valuation. Note that our language does not include reciprocals. For convenience, we allow ourselves to identify terms with polynomials in $\mathbb{Z}[\mathbf{x}, \pi]$ where $\mathbf{x} = (x_1, \ldots, x_n)$ are the contained variables, and π is the constant of our language.

3 Solving Congruences

In the previous section we have introduced the valuation rings $\mathbb{Z}_p$ wrt. p-adic valuations on the rational numbers. We have demonstrated that these rings have a particularly nice algebraic structure, which suggests that they admit sophisticated algebraic methods for solving. We are going to focus on such methods in the following section, after here making clear the connection between solving over $\mathbb{Z}_p$ on one hand, and solving over the integers, which we are actually interested in, on the other hand.

The following lemma shows that p-adic solutions can easily be lifted to integer solutions, while integer solution are themselves already p-adic solutions.

Lemma 1. *Let $f_1, \dots, f_m \in \mathbb{Z}[x_1, \dots, x_n]$ be polynomials that are linear in x_1, $\dots$, x_n. Let p be prime, and let $k_1, \dots, k_m \in \mathbb{N}$. Consider the following systems S and S' of congruences over $\mathbb{Z}$ and $\mathbb{Z}_p$, respectively:*

$$S = \{f_j(\mathbf{x}) \equiv 0 \bmod p^{k_j}\mathbb{Z} : 1 \leq j \leq m\}$$
$$S' = \{f_j(\mathbf{x}) \equiv 0 \bmod p^{k_j}\mathbb{Z}_p : 1 \leq j \leq m\}.$$

Then S has a solution $\mathbf{a} \in \mathbb{Z}^n$ iff S' has a solution $\mathbf{a}' \in \mathbb{Z}_p^n$. More precisely, every solution $\mathbf{a} \in \mathbb{Z}^n$ for S is already a solution for S', and every solution $\mathbf{a}' \in \mathbb{Z}_p^n$ for S' can be easily lifted to a solution $\overline{\mathbf{a}'}$ for S in $\mathbb{Z}^n$.

Proof. To begin with, observe that $\mathbb{Z}$ is a subring of $\mathbb{Z}_p$, and that for our p, k_1, $\dots$, k_m, the corresponding ideals $p^{k_j}\mathbb{Z}$ are exactly the restrictions of the ideals $p^{k_j}\mathbb{Z}_p$; more precisely

$$p^{k_j}\mathbb{Z} = p^{k_j}\mathbb{Z}_p \cap \mathbb{Z} \subseteq p^{k_j}\mathbb{Z}_p \quad (1 \leq j \leq m).$$

Let now $\mathbf{a}$ be a solution for S over $\mathbb{Z}$. Then $f_j(\mathbf{a}) \equiv 0 \bmod p^{k_j}\mathbb{Z}$ corresponding to $f_j(\mathbf{a}) \in p^{k_j}\mathbb{Z}$ implies $f_j(\mathbf{a}) \in p^{k_j}\mathbb{Z}_p$, which in turn corresponds to $f_j(\mathbf{a}) \equiv 0 \bmod p^{k_j}\mathbb{Z}_p$.

Let vice versa $\mathbf{a}' = (r_1/s_1, \dots, r_n/s_n) \in \mathbb{Z}_p^n$ be a solution for S'. Let $k = \max\{k_1, \dots, k_m\}$. We restrict our attention to r_1/s_1. This is a p-adic integer, and thus s_1 is relatively prime to p. We compute using the extended Euclidean algorithm a multiplicative inverse $\overline{s_1}$ of s_1 in $\mathbb{Z}/p^k$:

$$1 = \gcd(s_1, p^k) = \overline{s_1}s_1 + xp^k,$$

i.e., $\overline{s_1}s_1 \equiv 1 \bmod p^k\mathbb{Z}$ over $\mathbb{Z}$, and certainly $\overline{s_1}s_1 \equiv 1 \bmod p^k\mathbb{Z}_p$ over $\mathbb{Z}_p$. Moreover, the corresponding congruences obviously hold for all the k_1, $\dots$, k_m. This means that $r_1 s_1 \overline{s_1}/s_1 = r_1\overline{s_1}$ constitutes an integer solution for x_1 that is congruent to r_1/s_1 wrt. all the $p^{k_1}\mathbb{Z}_p$, $\dots$, $p^{k_m}\mathbb{Z}_p$. We set $\overline{a_1'} = r_1\overline{s_1}$ and applying our method to the other coordinates, we obtain a complete integer solution $\overline{\mathbf{a}'}$ for S' over $\mathbb{Z}_p$. According to our initial observations, this $\overline{\mathbf{a}'}$ is obviously a solution for S over $\mathbb{Z}$.

Although we are accustomed to speak of integer solutions, it is also quite natural to consider these solutions as elements in $(\mathbb{Z}/p^k)^n$, where k is the highest power of p in S. Viewed in this way, the solutions obtained by the lifting procedure in the proof of Lemma 1 will in general not be the canonical representatives, i.e., we have to expect to obtain integers $x_i = \overline{a_i'} \geq p^k$. We can however easily derive from any such solution another solution, which is a canonical representative for an element of $(\mathbb{Z}/p^k)^n$. The following Lemma shows how.

Lemma 2. *Let S be a system of linear congruence as in Lemma 1, and let $\mathbf{a} \in \mathbb{Z}^n$ be an integer solution for S. Let $k = \max\{k_1, \ldots, k_m\}$. Then all elements of $\mathbf{a} + (p^k\mathbb{Z})^n$ are integer solutions of S. In particular there is one solution $\mathbf{c} = (c_1, \ldots, c_n) \in \mathbf{a} + (p^k\mathbb{Z})^n$ with $0 \leq c_i < p^k$ for $i \in \{1, \ldots, n\}$.*

Proof. Our $\mathbf{a} = (a_1, \ldots, a_n)$ solves S if and only if it solves for $j \in \{1, \ldots, m\}$ the equation $f_j = 0$ over $\mathbb{Z}/p^{k_j}\mathbb{Z}$. Let $\mathbf{b} \in (p^k\mathbb{Z})^n$. Then $\mathbf{b} = 0$ in $(\mathbb{Z}/p^{k_j}\mathbb{Z})^n$, and thus $\mathbf{a}+\mathbf{b} = \mathbf{a}$ is a solution of $f_j = 0$ over $\mathbb{Z}/p^{k_j}\mathbb{Z}$. Accordingly, for $i \in \{1, \ldots, n\}$ we can obtain c_i by division with positive remainder of a_i by p^k.

It is not hard to see that $\mathbf{a} + (p^k\mathbb{Z})^n$ in Lemma 2 does not describe the complete solution space of S. As an example consider the system

$$5x_1 + 7x_2 + 1 \equiv 0 \bmod 11.$$

Here $(5, 1)$ is a solution, but so is also $(-3, 2) \notin (5, 1) + (11\mathbb{Z})^2$.

4 Extended Quantifier Elimination

For solving our linear constraints, we use an effective linear quantifier elimination procedure based on *virtual substitution of test points*. Based on ideas of Ferrante and Rackoff [11] for decision problems, virtual substitution methods for quantifier elimination date back to a theoretical paper by Weispfenning [19]. Corresponding methods over the reals have been successfully used for solving problems from numerous areas in science and engineering [9].

For eliminating the quantifiers from an input formula

$$\varphi(u_1, \ldots, u_m) \equiv \mathsf{Q}_1 x_1 \ldots \mathsf{Q}_n x_n \psi(u_1, \ldots, u_m, x_1, \ldots, x_n)$$

where $\mathsf{Q}_i \in \{\exists, \forall\}$, the elimination starts with the innermost quantifier regarding the other quantified variables within ψ as extra parameters. Universal quantifiers are handled by means of the equivalence $\forall x \psi \longleftrightarrow \neg\exists x \neg\psi$. We may thus restrict our attention to a formula of the form

$$\varphi^*(u_1, \ldots, u_k) \equiv \exists x \psi^*(u_1, \ldots, u_k, x),$$

where the u_{m+1}, …, u_k are actually x_i quantified from further outside. The idea is now to find a finite *elimination set* E of terms in u_1, …, u_k such that

$$\exists x \psi^*(u_1, \ldots, u_k, x) \equiv \bigvee_{t \in E} \psi^*[x/t](u_1, \ldots, u_k).$$

That is, the above disjunction is a quantifier-free equivalent for φ^*. Note that it is not necessary to perform any transformation on the boolean structure of ψ^*. The elimination method is single exponential in the number of quantified variables, and double exponential in the number of quantifier blocks. It has turned out suitable for parallelization [5].

By keeping track of the terms t substituted during the elimination process, we obtain instead of a quantifier-free equivalent $\bigvee_{i=1}^{k} \psi^*[x/t_i]$ a guarded expression [6]

$$\begin{bmatrix} \psi^*[x/t_1] & x = t_1 \\ \vdots & \vdots \\ \psi^*[x/t_k] & x = t_k \end{bmatrix}$$

including satisfying sample points. This process of *extended quantifier elimination* can also be repeated for several existential quantifiers. The result then is a set of *conditions* each associated with an *answer* for each eliminated variable obtained by back-substitution.

The construction of elimination sets for linear formulas in valued fields has been described by the second author [17]. Before, Weispfenning had given elimination sets for special cases of valued fields including the case of p-adic valuations [19]. Necessary simplification strategies and implementation issues have been discussed in [8].

The existence of a quantifier elimination procedure for the general case including non-linear formulas has been shown independently by Ax and Kochen [1] and Ershov [10]. The first explicit procedure has been given by Cohen [4]. Considerable progress has been made by Macintyre [13] turning to a more reasonable language including root predicates in analogy to the reals. This has been made explicit by Weispfenning [18].

5 Solving by Extended Quantifier Elimination

We consider for $f_1, \ldots, f_m \in \mathbb{Z}[x_1, \ldots, x_n]$ linear, p prime, and $k_1, \ldots, k_m \in \mathbb{N}$, a system of congruences

$$S = \{f_1(\mathbf{x}) \equiv 0 \bmod p^{k_1}\mathbb{Z}, \ \ldots, \ f_m(\mathbf{x}) \equiv 0 \bmod p^{k_m}\mathbb{Z}\}.$$

According to Lemma 1 it suffices to solve instead the corresponding system

$$S' = \{f_1(\mathbf{x}) \equiv 0 \bmod p^{k_1}\mathbb{Z}_p, \ \ldots, \ f_m(\mathbf{x}) \equiv 0 \bmod p^{k_m}\mathbb{Z}_p\}.$$

over $\mathbb{Z}_p$. The solvability of this new system S' can be expressed by a first order formula as follows:

$$\Phi(S') \equiv \exists x_1 \cdots \exists x_n \Big(\bigwedge_{i=1}^{n} 1 \mid x_i \wedge \bigwedge_{j=1}^{m} p^{k_j} \mid f_j(x_1, \ldots, x_n) \Big).$$

Here, the first conjunction restricts the x_i to be in the valuation ring $\mathbb{Z}_p$. Extended quantifier elimination applied to this formula will decide feasibility and, in the positive case, yield one sample solution. Our notion of *solving* thus resembles the standard notion used in constraint solving. Recall from Lemma 2 that in our situation such a sample solution even describes an infinite subset of the solution space.

Algorithm 1 (Solving Integer Congruences)
Input: A system

$$S = \{ f_j(\mathbf{x}) \equiv 0 \bmod p^{k_j}\mathbb{Z} : 1 \leq j \leq m \}.$$

of congruences over $\mathbb{Z}$ with $f_1, \ldots, f_m \in \mathbb{Z}[x_1, \ldots, x_n]$ linear, p prime, and $k_1, \ldots, k_m \in \mathbb{N}$.
Output: "infeasible," or a sample solution $\mathbf{c} = (c_1, \ldots, c_n)$ over $\mathbb{Z}$ for S with $0 \leq c_i < p^{\max\{k_1,\ldots,k_m\}}$ for $i \in \{1, \ldots, n\}$.

1. Change from S to S' according to Lemma 1
2. Generate from S' the first-order formula $\Phi(S')$ as described above
3. Apply extended quantifier elimination to $\Phi(S')$
4. (a) if the elimination result is false then return "infeasible"
   (b) else lift the solution $\mathbf{a}' \in \mathbb{Z}_p^n$ to a solution $\mathbf{a} \in \mathbb{Z}^n$ according to Lemma 1
5. Apply Lemma 2 to derive from $\mathbf{a}$ a solution

$$\mathbf{c} = (c_1, \ldots, c_n),$$

   where $0 \leq c_i < p^{\max\{k_1,\ldots,k_m\}}$ for $i \in \{1, \ldots, n\}$.

Proof. The correctness follows from Lemma 1, Lemma 2, the definition of extended quantifier elimination, and the correspondence between S' and $\Phi(S')$ discussed above.

In our algorithm, the quantifier elimination step (3) constitutes due to the particular form of $\Phi(S')$ an extreme special case of p-adic quantifier elimination. From the elimination point of view, the crucial syntactic feature of $\Phi(S')$ is that all the x_i occur only on the right hand sides of the abstract divisibilities. That is, we only impose lower bounds on the values of these x_i but no upper bounds.

Restating the elimination procedure given in [17] for this special case in our congruence framework, the p-adic solutions will be determined as follows.

Definition 1. *For a congruence $\gamma = f(\mathbf{x}) \equiv 0 \bmod I$ we denote by $\gamma^{(=)}$ the corresponding equation $f(\mathbf{x}) = 0$. This naturally extends to the notion of a system $S^{(=)}$ of equations corresponding to a system S of congruences.*

Algorithm 2 (Deciding p-adic Congruences)
Input: A system

$$S' = \{ f_j(\mathbf{x}) \equiv 0 \bmod p^{k_j}\mathbb{Z}_p : 1 \leq j \leq m \}$$

of congruences over $\mathbb{Z}$ with $f_1, \ldots, f_m \in \mathbb{Z}[x_1, \ldots, x_n]$ linear, p prime.
Output: "infeasible," or a sample solution $\mathbf{a}' \in \mathbb{Z}_p^n$ for S'.

BEGIN
  $\mathcal{S} := \{(S', \emptyset)\}$
  $\mathcal{S}' := \emptyset$
  for each variable $x \in \{x_1, \ldots, x_n\}$ **do**

```
        for each (S, σ) ∈ 𝒮 do
            for each constraint γ in S do
                if γ contains x then
                    a := solution in ℚ wrt. x for γ^(=)
                    S_a := S with a plugged in for x
                    𝒮' := 𝒮' ∪ {(S_a, σ ∪ {x = a})}
                fi
            od
            S_0 := S with 0 plugged in for x
            𝒮' := 𝒮' ∪ {(S_0, σ ∪ {x = 0})}
        od
        𝒮 := 𝒮'
        𝒮' := ∅
    od
    if there is ({0 ≡ 0, ..., 0 ≡ 0}, σ) in 𝒮 then
        return σ
    else
        return "infeasible"
    fi
END
```

Proof. This is a straightforward consequence of Corollary 8.5 in [17] applied to $\Phi(S')$.

Example 1. We apply our implementation of Algorithm 1 to the following randomly generated system S of congruences:

$$\begin{aligned} 70x_1 + 6x_3 + 89x_4 + 7x_6 + 30 &\equiv 0 \bmod 103^{10} \\ 87x_1 + 93x_2 + 78x_3 + 73x_4 + 53 &\equiv 0 \bmod 103^9 \\ 87x_2 + 41x_5 + 3 &\equiv 0 \bmod 103^3 \\ 12x_2 + 37x_3 + 69x_4 + 15x_6 + 53 &\equiv 0 \bmod 103^3 \\ 75x_1 + 90x_3 + 65x_4 + 14x_5 + 41 &\equiv 0 \bmod 103 \\ 91x_5 + 96x_6 + 55 &\equiv 0 \bmod 103^2. \end{aligned}$$

Extended quantifier elimination applied to $\Phi(S')$ yields the following sample solution over $\mathbb{Z}_{103}$ for S':

$$\begin{aligned} x_1 &= \frac{1120921235}{6450196079}, & x_2 &= -\frac{2555928514}{19350588237}, \\ x_3 &= -\frac{2265478209}{6450196079}, & x_4 &= -\frac{2512869252}{6450196079}, \\ x_5 &= \frac{1335886309}{6450196079}, & x_6 &= -\frac{4961733734}{6450196079}. \end{aligned}$$

Formally, with the naming conventions of Lemma 1, we have found

$$\mathbf{a}' = \left(\frac{1120921235}{6450196079}, -\frac{2555928514}{19350588237}, \ldots, -\frac{4961733734}{6450196079}\right) \in \mathbb{Z}_{103}^6$$

solving S'. After lifting this results in the following corresponding sample solution for the original system S over $\mathbb{Z}$:

$$\begin{aligned} x_1 &= 18804386104945290509, & x_2 &= 8303843175527713857, \\ x_3 &= 6309069755640464645 6 & x_4 &= 83696580514895056415, \\ x_5 &= 9382637398778301034 4 & x_6 &= 13364656665295088119 2. \end{aligned}$$

This formally corresponds to

$$\overline{\mathbf{a}'} = (18804386104945290509, \dots, 13364656665295088119 2) \in \mathbb{Z}^6.$$

The total computation time is 2.3 s, which is almost completely spent for the extended quantifier elimination step. All other steps, in particular the lifting, take less than the accuracy of the system clock, which is 0.01 s.

6 Parametric Moduli

So far, we have considered integer congruence systems with prime power moduli for a fixed prime p. Algorithm 2 suggests that the first-order framework of quantifier elimination is not necessary for solving this problem. The entire elimination procedure can easily be described in terms of manipulating lists of congruence systems. This changes when turning to more general questions. The first more general problem we are going to discuss here, is solving our congruence systems uniformly for a parametric prime p.

Let us take a look at our Algorithm 1 wrt. this generalization:

1. The p-adic system S' can be generated as before now containing parametric ideals $p^{k_j}\mathbb{Z}_p$.
2. The first-order formula $\Phi(S')$ now contains the constant π of our language denoting the parametric p.
3. The extended quantifier elimination is now not a decision procedure. Notice that variable-free atomic formulas cannot be decided. For our generalized Algorithm 2 this means that we drop the final **if** statement but return the extended quantifier elimination result. The conditions in this result will contain two types of atomic conditions:
   (a) Positive conditions on p resulting from the substitution into the congruences.
   (b) Negative conditions on p, which are guarding conditions introduced with substitution for excluding zero denominators.

4/5. The lifting step depends on the concrete choice for p, and has to be considered separate from the p-adic solution phase. The p-adic solution provided by the generalized Algorithm 2 will thus be the final output of our generalized Algorithm 1.

Example 2. We recompute our Example 1 replacing the base 103 of the moduli by a parametric p. We then obtain after 7.43 s the following solution, which is uniform over $\mathbb{Z}_p$ for the valid moduli $p \notin \{3, 6450196079\}$:

$$\big[3 \sim 1 \wedge 6450196079 \sim 1 \quad \mathbf{x} = \mathbf{a}'\big],$$

where the p-adic integer solution $\mathbf{a}'$ happens to be identical to that for the case $p = 103$ in Example 1. Our $\mathbf{a}'$ can be lifted e.g. for $p = 103$ within less then 0.01 s to the integer solution $\overline{\mathbf{a}'} \in \mathbb{Z}^6$ we know from Example 1.

Notice that we have found in the above example a uniform p-adic solution, subject to a guarding condition that straightforwardly states that the system is unsolvable for $p \in \{3, 6450196079\}$.

The remainder of this section is devoted to studying what kind of results concerning uniformity and explicitness we may expect wrt. the stated problem on one hand, and our particular approach to it on the other hand.

Example 3. Consider the following system of two congruences:

$$3x_1 + 5x_2 \equiv 1 \bmod p$$
$$5x_1 + 3x_2 \equiv 1 \bmod p.$$

Our elimination procedure yields the following result distinguishing two cases:

$$\begin{bmatrix} 2 \not\sim 1 \; \{x_1 = \frac{1}{3}, x_2 = 0\} \\ 2 \sim 1 \; \{x_1 = \frac{1}{8}, x_2 = \frac{1}{8}\} \end{bmatrix}.$$

With the result in the example, we would for $p = 2$ be only allowed to lift the first solution, while for all other primes p only the second one is valid. Since we have to know p for lifting anyway, it is easy to automatically detect the valid branch. Anyway, the question arises whether there exists also a uniform solution, which we would consider an intermediate result of better quality. This is in fact not the case here as we going to exhibit in the sequel.

To begin with, note that by inspection of our elimination procedure, we know that our sample solutions will always be numbers not involving the constant $\pi = p$ of our language.

Lemma 3. *A rational number a is a p-adic integer for all primes p if and only if it is an integer. In other words,*

$$\bigcap_{p \text{ prime}} \mathbb{Z}_p = \mathbb{Z}.$$

Proof. Let $a = n/d \in \bigcap_p \mathbb{Z}_p$ be reduced to lowest terms. Then $p \nmid d$ for all primes p. It follows that $d = 1$ and thus $a \in \mathbb{Z}$. Let conversely $a = a/1 \in \mathbb{Z}$, and let p be prime. Then $p \nmid 1$ and thus $a \in \mathbb{Z}_p$.

Let now S be a system of congruences with parametric modulus base p over $\mathbb{Z}$. Denote by S' the corresponding system over $\mathbb{Z}_p$ as made precise in Lemma 1, and by $S^{(=)}$ the corresponding system of equations according to Definition 1.

We learn from Lemma 3 above that a uniform p-adic integer solution for S' is in fact a uniform integer solution for S' and thus also for S. By choosing p sufficiently large, it is not hard to see that this uniform integer solution even solves the corresponding system $S^{(=)}$ of linear equations. Conversely, any integer solution for $S^{(=)}$ is obviously a uniform integer solution for S and thus for S'. The following proposition states this observation more concisely:

Proposition 1 *Let S be a system of linear congruences over $\mathbb{Z}$, let S' be the corresponding system over $\mathbb{Z}_p$, and let $S^{(=)}$ be the corresponding system of linear equations over $\mathbb{Z}$. Assume we determine a (uniform) solution* **a** *for one of these three systems. Then* **a** *is up to some natural homomorphism also a (uniform) solution for the other two systems.*

Consider now the system of equations over $\mathbb{Z}$ corresponding to the congruence system in Example 3:

$$3x_1 + 5x_2 = 1$$
$$5x_1 + 3x_2 = 1.$$

Subtracting the first equation from the second one, we obtain the consequence $x_1 = x_2$. A solution $x_1 = x_2 = a \in \mathbb{Z}$ would thus have to satisfy $8a = 3a + 5a = 1$, which is obviously impossible. Proposition 1 now tells us that there is no uniform solution for the original congruence system, neither over $\mathbb{Z}$ nor over $\mathbb{Z}_p$. Concerning the case distinction, our solution in Example 3 is optimal.

Our procedure is however not optimal in general. In the following example we miss finding a uniform solution, although there exists one.

Example 4. Consider the system consisting of the sole congruence

$$5x_1 + 7x_2 + 1 \equiv 0 \bmod p.$$

Application of our elimination procedure yields that this is solvable for all primes p, giving two guarded sample solutions. The first one holds uniformly, except for $p = 5$, while the second one holds uniformly except for $p = 7$:

$$\begin{bmatrix} 5 \sim 1 \; \{x_1 = -\frac{1}{5}, \; x_2 = 0\} \\ 7 \sim 1 \; \{x_1 = 0, \quad x_2 = -\frac{1}{7}\} \end{bmatrix}.$$

Here $x_1 = -3$ and $x_2 = 2$ solves the corresponding system of equations over $\mathbb{Z}$, and thus constitutes a uniform solution for all p.

In the non-parametric case we could obviously easily obtain either "true" or "false" for each guarding condition, and then pick a "true" solution. In the parametric case here, we have seen conditions of the form $p \sim 1$ and $p \not\sim 1$ for primes p.

In fact, every variable-free formula over our language is simplified to "true," "false," or a formula of one of the forms

$$p_1 \not\sim 1 \vee \cdots \vee p_k \not\sim 1, \quad p_1 \sim 1 \wedge \cdots \wedge p_k \sim 1,$$

where $p_1 < \cdots < p_k$ prime. The first formula states $p \in \{p_1, \ldots, p_k\}$, while the second one states $p \notin \{p_1, \ldots, p_k\}$. We observe, as a consequence, that any parametric congruence system of our considered form is of one of the following four types:

1. generally feasible for all primes p,

2. generally infeasible for all primes p,
3. feasible for finitely many p,
4. feasible for all but finitely many p.

In particular, there is no such system for which there exists a partition $P_1 \dot{\cup} P_2$ of all primes into infinite sets P_1 and P_2, such that the system is feasible for all $p \in P_1$ but infeasible for all $p \in P_2$.

Guarding conditions of the form $p_1 \not\sim 1 \vee \cdots \vee p_k \not\sim 1$ restricting p to a finite set of primes are typically introduced because the congruence system degenerates for these primes. For instance $p_1, \ldots, p_k$ may be the prime factors of a certain coefficient, which becomes zero then, which in turn leads to a special solution that does not work for other primes. Guards of the form $p_1 \sim 1 \wedge \cdots \wedge p_k \sim 1$, in contrast, exclude, as a rule, the prime factors of the denominators of the associated solution.

Arriving from an arbitrary variable-free formula over our language, which possibly contains the constant π, at one of the four forms described above is by no means trivial. It requires a large arsenal of sophisticated simplification strategies. The part of our simplifier that is of general interest for quantifier elimination over discretely valued fields has been described in detail in [8]. Further special-purpose simplification algorithms have been added for the particular project discussed here.

7 Simultaneous Lifting

With the parametric setup of the previous section it is possible to lift the p-adic solutions simultaneously for finitely many primes p. The crucial tool for this is the well-known *Chinese Remainder Theorem* (CRT) [2].

Theorem 1 (Chinese Remaindering). *Let $r_1, \ldots, r_k, m_1, \ldots, m_k \in \mathbb{Z}$, where the m_i are relatively prime. We are interested in the system*

$$S = \{\, x \equiv r_i \bmod m_i \mid 1 \le i \le k \,\}$$

of congruences. For $1 \le i \le k$ set

$$n_i = \prod_{\substack{1 \le j \le k \\ j \ne i}} m_i.$$

Then $\gcd(n_i, m_i) = 1$, and the extended Euclidean algorithm yields a linear combination $1 = s_i n_i + t_i m_i$. Now

$$a = \sum_{j=1}^{k} n_j s_j r_j$$

is a solution to the system S. The set of all solutions is $a + m\mathbb{Z}$, where $m = m_1 \cdots m_k$.

7.1 Simultaneous Branch Lifting

Consider now for a congruence system S with symbolic p a solution branch

$$\left(\gamma, \{x_1 = \tfrac{r_1}{s_1}, \ldots, x_n = \tfrac{r_n}{s_n}\}\right).$$

Let k be the highest power of p in S, and let $P = \{p_1, \ldots, p_l\}$ be a finite set of primes satisfying γ. We are going to apply the Chinese Remainder Theorem for obtaining an integer solution that is simultaneously correct for all the $p_1, \ldots, p_l$, by solving for each of the r_i/s_i the following system of integer congruences:

$$\begin{aligned} y &\equiv 1 \bmod p_1^k \\ &\vdots \\ y &\equiv 1 \bmod p_l^k \\ y &\equiv 0 \bmod s_i. \end{aligned}$$

All the k-th powers of the various p_j are obviously pairwise relatively prime, and since the p_j satisfy γ, they are also relatively prime to s_i. The first l congruences allow us to multiply r_i/s_i with our solution a for y without changing its residue class modulo any of the ideals $p^k\mathbb{Z}_p$ for the various $p_1, \ldots, p_l$, and that $a \not\equiv 0$. The last congruence makes sure that a will be a multiple of s_i, such that $r_i a/s_i \in \mathbb{Z}$.

Example 5. We simultaneously lift the uniform result for $p \notin \{3, 6450196079\}$ of our Example 2.

(i) For the ten primes $\{2, 5, 7, 11, 13, 17, 19, 23, 29, 31\}$, we obtain after 0.01 s a uniform integer solution where $x_1, \ldots, x_6$ have either 93 or 94 digits each.
(ii) For the first 100 primes $\{2, \ldots, 547\}$ different from 3, we obtain after 0.79 s a uniform integer solution where $x_1, \ldots, x_6$ have 2192 digits each.
(iii) For the first 500 primes $\{2, \ldots, 3581\}$ different from 3, we obtain after 31.5 s a uniform integer solution where $x_1, \ldots, x_6$ have either 15228 or 15229 digits each.

7.2 Simultaneous Solution Lifting

Simultaneous branch lifting is applicable only in cases where the list of target primes matches one particular solution branch. This will be in general not be the case as we have demonstrated for our Example 3 concerning, e.g., the primes 2, 3. Nevertheless, we can always find simultaneous integer solutions for a given finite set of primes, provided, of course, the system is solvable for every single prime.

Consider a finite set $P = \{p_1, \ldots, p_l\}$ of primes such that our system S is solvable for each $p \in P$. That is, for each $p_i \in P$ we have a p_i-adic integer solution branch

$$\left(\gamma^{(i)}, \{x_1^{(i)} = a_1^{(i)'}, \ldots, x_n^{(i)} = a_n^{(i)'}\}\right)$$

with a p_i satisfying $\gamma^{(i)}$, and we can independently lift all the solutions for the various p_i, arriving at corresponding integer solutions

$$L_1 = \{x_1^{(1)} = a_1^{(1)}, \dots, x_n^{(1)} = a_n^{(1)}\}$$
$$\vdots$$
$$L_l = \{x_1^{(l)} = a_1^{(l)}, \dots, x_n^{(l)} = a_n^{(l)}\}$$

for $p_1, \dots, p_l$, respectively. It is easy to see that we can equivalently replace all the $a_1^{(1)}, \dots, a_1^{(l)}$ by the solution a_1 of the following system, where k is the highest power of p in the original system S:

$$a_1 \equiv a_1^{(1)} \bmod p_1^k$$
$$\vdots$$
$$a_1 \equiv a_1^{(l)} \bmod p_l^k.$$

This system is definitely solvable by Chinese remaindering. In the same way, we independently find suitable $a_2, \dots, a_n$. such that $x_1 = a_1, \dots, x_n = a_n$ simultaneously solves S for all $p \in P$.

Example 6. We simultaneously lift the result in Example 3 for various finite sets of primes:

(i) For the first ten primes $\{2, 3, 5, 7, 11, 13, 17, 19, 23, 29\}$ we obtain after less than 0.01 s the uniform solution $x_1 = 404355827$, $x_2 = 3639202442$.
(ii) For the first 100 primes $\{2, \dots, 541\}$ we obtain a uniform solution after 0.01 s, where both x_1 and x_2 have 220 digits.
(iii) For the first 500 primes $\{2, \dots, 3571\}$ we obtain a uniform solution after 0.15 s, where both x_1 and x_2 have 1520 digits.

In general, there will be solution branches that match for several primes $P' \subseteq P$, such that we can lift these branches by simultaneous branch lifting. That is, we combine both our approaches.

7.3 Infinite Sets of Primes

Both our approaches allow us to lift simultaneously only for a finite number of primes. Simultaneous lifting of a non-integer solution for an infinite number of primes is in fact impossible as the following lemma shows.

Lemma 4. *Let r/s be a p-adic integer wrt. an infinite set P of primes. If r/s can be simultaneously lifted to an integer for an infinite subset $P' \subseteq P$, then r/s is already an integer.*

Proof. Let the lifting factor $\overline{s} \equiv 1 \bmod p$ for all $p \in P'$. Since P' is infinite, there is $p_0 \in P'$ with $p_0 > \overline{s}$, and from $\overline{s} \equiv 1 \bmod p_0$, i.e., lifting does not change r/s.

8 Arbitrary Moduli

So far we have only considered linear congruence systems modulo powers of one fixed possibly parametric prime modulus. We are now going to extend our ideas to general moduli, where we restrict to the non-parametric case. Consider a system

$$S = \{f_1(\mathbf{x}) \equiv 0 \bmod \mu_1,\ \ldots,\ f_m(\mathbf{x}) \equiv 0 \bmod \mu_m\},$$

where $f_1, \ldots, f_m \in \mathbb{Z}[x_1, \ldots, x_n]$ are polynomials that are linear in $x_1, \ldots, x_n$, and $\mu_1, \ldots, \mu_m \in \mathbb{N}$.

The key observation is that each of the $\mu_1, \ldots, \mu_m$ factors into finitely many prime powers, and that $\mathbf{a} \in \mathbb{Z}^n$ satisfies a given congruence modulo a product of prime powers if and only if it does so for all the single prime powers simultaneously:

$$\bigwedge_{i=1}^{l} f(\mathbf{a}) \equiv 0 \bmod p_i^{k_i} \iff f(\mathbf{a}) \equiv 0 \bmod \prod_{i=1}^{l} p_i^{k_i}.$$

So, we have learned all necessary techniques for solving this more general problem already in the previous section. The following algorithm explains how to organize the computation:

Algorithm 3 (Solving with Arbitrary Moduli)
Input: A system

$$S = \{f_1(\mathbf{x}) \equiv 0 \bmod \mu_1,\ \ldots,\ f_m(\mathbf{x}) \equiv 0 \bmod \mu_m\}.$$

of congruences over $\mathbb{Z}$ with $f_1, \ldots, f_m \in \mathbb{Z}[x_1, \ldots, x_n]$ linear, $\mu_1, \ldots, \mu_m \in \mathbb{N}$.
Output: "infeasible," or a sample solution $\mathbf{a} \in \mathbb{Z}^n$ for S.

```
BEGIN
    P := the prime factors of μ_1, ..., μ_m
    for each p ∈ P do
        T := ∅
        k^(p) := 0
        for j := 1 : m do
            k := the power of p in μ_j
            if k > 0 then
                T := T ∪ {f_j(x) ≡ 0 mod p^k}
                k^(p) := max(k^(p), k)
            fi
        od
        apply Algorithm 1 to T
        if T is feasible then
            a^(p) := an integer solution for T
        else
            return "infeasible"
        fi
```

od
for $i := 1 : n$ **do**
$C := \{\, x_i \equiv a_i^{(p)} \bmod p^{k(p)} \mid p \in P \,\}$
$a_i :=$ an integer solution for x_i by CRT
od
return $(a_1, \dots, a_n)$
END

Example 7. We apply our implementation of Algorithm 3 to the following system S of congruences derived from the randomly generated Example 1:

$$\begin{aligned}
70x_1 + 6x_3 + 89x_4 + 7x_6 + 30 &\equiv 0 \bmod 280\\
87x_1 + 93x_2 + 78x_3 + 73x_4 + 53 &\equiv 0 \bmod 5665\\
87x_2 + 41x_5 + 3 &\equiv 0 \bmod 110\\
12x_2 + 37x_3 + 69x_4 + 15x_6 + 53 &\equiv 0 \bmod 1545\\
75x_1 + 90x_3 + 65x_4 + 14x_5 + 41 &\equiv 0 \bmod 3125\\
91x_5 + 96x_6 + 55 &\equiv 0 \bmod 1925.
\end{aligned}$$

The moduli here factorize as follows:

$$\begin{aligned}
280 &= 2^3 \cdot 5 \cdot 7\\
5665 &= 5 \cdot 11 \cdot 103\\
110 &= 2 \cdot 5 \cdot 11\\
1545 &= 3 \cdot 5 \cdot 103\\
3125 &= 5^5\\
1925 &= 5^2 \cdot 7 \cdot 11.
\end{aligned}$$

We obtain after 2.29 s the following solution:

$$\begin{aligned}
x_1 &= 2873631250, & x_2 &= 3339537828,\\
x_3 &= 289265341729, & x_4 &= 422862329737,\\
x_5 &= 255144121, & x_6 &= 112853162929.
\end{aligned}$$

Notice that our algorithm is based on solution lifting in contrast to branch lifting. In extreme special cases a combination with branch lifting might be more efficient. This is the case when there are many prime factors occurring with equal powers in all of the moduli. One would then solve the system parametrically for this distribution of prime powers.

9 Conclusions

Based on an extended quantifier elimination procedure for p-adically valued fields, we have devised algorithms for solving multivariate linear systems of congruences. Our methods generally split into two parts: First, finding solutions in suitable rings of p-adic integers $\mathbb{Z}_p$. Second, lifting these solutions to the integers $\mathbb{Z}$. The first part is computationally hard, while the second one is straightforward and efficient. For the special case, where each modulus is some power of a fixed prime, the computationally hard first part can be performed uniformly for all primes. This is the crucial advantage of our approach in contrast to well-known linear algebra methods for concrete fixed primes. For this uniform case, we have developed two methods for making the lifting step also as uniform as theoretically possible. These methods can be finally reused for extending our approach to the general case of arbitrary, i.e. not necessarily prime power, moduli. This general case is a considerable generalization of the problem solved by the Chinese Remainder Theorem.

References

[1] James Ax and Simon Kochen. Diophantine problems over local fields. *Annals of Mathematics*, 83:437–456, 1966. Part III.

[2] Thomas Becker, Volker Weispfenning, and Heinz Kredel. *Gröbner Bases, a Computational Approach to Commutative Algebra*, volume 141 of *Graduate Texts in Mathematics*. Springer, New York, 1993.

[3] Johannes Buchmann and Stefan Neis. Algorithms for linear algebra problems over principal ideal rings. Techincal Report TI-7/96, Technische Hochschule Darmstadt, Fachbereich Informatik, D-64283 Darmstadt, Germany, November 1996.

[4] Paul J. Cohen. Decision procedures for real and p-adic fields. *Communications in Pure and Applied Logic*, 25:213–231, 1969.

[5] Andreas Dolzmann, Oliver Gloor, and Thomas Sturm. Approaches to parallel quantifier elimination. In Oliver Gloor, editor, *Proceedings of the 1998 International Symposium on Symbolic and Algebraic Computation (ISSAC 98)*, pages 88–95, Rostock, Germany, August 1998. ACM, ACM Press, New York, 1998.

[6] Andreas Dolzmann and Thomas Sturm. Guarded expressions in practice. In Wolfgang W. Küchlin, editor, *Proceedings of the 1997 International Symposium on Symbolic and Algebraic Computation (ISSAC 97)*, pages 376–383, Maui, HI, July 1997. ACM, ACM Press, New York, 1997.

[7] Andreas Dolzmann and Thomas Sturm. Redlog: Computer algebra meets computer logic. *ACM SIGSAM Bulletin*, 31(2):2–9, June 1997.

[8] Andreas Dolzmann and Thomas Sturm. P-adic constraint solving. In Sam Dooley, editor, *Proceedings of the 1999 International Symposium on Symbolic and Algebraic Computation (ISSAC 99), Vancouver, BC*, pages 151–158. ACM Press, New York, NY, July 1999.

[9] Andreas Dolzmann, Thomas Sturm, and Volker Weispfenning. Real quantifier elimination in practice. In B. H. Matzat, G.-M. Greuel, and G. Hiss, editors, *Algorithmic Algebra and Number Theory*, pages 221–247. Springer, Berlin, 1998.

[10] Juri L. Ershov. On elementary theories of local fields. *Algebra i Logika Sem.*, 4(2):5–30, 1965.

[11] Jeanne Ferrante and Charles W. Rackoff. *The Computational Complexity of Logical Theories.* Number 718 in Lecture Notes in Mathematics. Springer-Verlag, Berlin, 1979.

[12] John A. Howell. Spans in the module $(\mathbb{Z}_m)^s$. *Linear and Multilinear Algebra*, 19(1):67–77, 1986.

[13] Angus Macintyre. On definable subsets of p-adic fields. *Journal of Symbolic Logic*, 41(3):605–610, September 1976.

[14] Alexander Ostrowski. Über einige Lösungen der Funktionalgleichung $\varphi(x)\cdot\varphi(y) = \varphi(xy)$. *Acta Mathematica*, 41:271–284, 1918.

[15] A. Storjohann and T. Mulders. Fast algorithms for linear algebra modulo N^*. In Gianfranco Bilardi, Giuseppe F. Italiano, Andreas Pietracaprina, and Geppino Pucci, editors, *Algorithms — ESA '98*, volume 1461 of *Lecture Notes in Computer Science*, pages 139–150, Berlin, 1998. Springer.

[16] Thomas Sturm. Reasoning over networks by symbolic methods. *Applicable Algebra in Engineering, Communication and Computing*, 10(1):79–96, September 1999.

[17] Thomas Sturm. Linear problems in valued fields. *Journal of Symbolic Computation*, 30(2):207–219, August 2000.

[18] Volker Weispfenning. Quantifier elimination and decision procedures for valued fields. In G. H. Müller and M. M. Richter, editors, *Models and Sets (Aachen, 1983)*, volume 1103 of *Lecture Notes in Mathematics*, pages 419–472. Springer-Verlag, Berlin, Heidelberg, 1984.

[19] Volker Weispfenning. The complexity of linear problems in fields. *Journal of Symbolic Computation*, 5(1&2):3–27, February–April 1988.

Bifurcation Analysis of Low Resonant Case of the Generalized Henon - Heiles System

Victor F. Edneral

Institute for Nuclear Physics of Moscow State University,
Vorobievi Gori, Moscow, 119899, Russia
edneral@theory.sinp.msu.ru

Abstract. The paper describes the computer algebra application of the normal form method to bifurcation analysis of a low resonant case of the generalized Henon - Heiles system. A behavior of all local families of periodic solutions in system parameters is determined. Corresponding approximated solutions were checked by a comparison with the numerical solutions of the system.

Keywords: Resonant normal form; Dynamical systems; Henon–Heiles system; Local periodic families; Computer algebra.

1 Introduction

Normal form methods use a nonlinear change of variables to transform a nonlinear system of ordinary differential equations to a simpler form. In this paper we use an algorithm based on an approach developed by A.D. Bruno [1] for computing the resonant normal form. An important advantage of this approach is its algorithmic simplicity: there are direct recurrence formulas for coefficients of the transformation and of the transformed system, and thus the storage of large intermediate results is not necessary. This approach does not require solving any intermediate systems and there are no restrictions on low resonance cases.

2 The Generalized Henon–Heiles System

The generalized Henon–Heiles system is a couple of second order differential equations:

$$\begin{aligned} &\ddot{x} + l_1 \cdot x + 2 \cdot d \cdot x \cdot y = 0\,, \\ &\ddot{y} + l_2 \cdot y + d \cdot x^2 - c \cdot y^2 = 0\,. \end{aligned} \tag{1}$$

This system is known to be integrable [3] when:

1. $l_1 = l_2, \quad c/d = -1;$
2. $c/d = -6;$
3. $16 l_1 = l_2, \quad c/d = -16.$

Below we discuss the case with the same pure imaginary eigenvalues of the linear part of (1). I.e. we suppose that $l_1 = l_2 \neq 0$. By changing the time variable we choose $l_1 = l_2 = 1$ and thus rewrite (1) to the "low resonant form".

Take into account the fact that if $d = 0$ the system can be integrated in an analytical form because equations for x and y will be independent:

$$\ddot{x} + x = 0\,, \quad \ddot{y} + y - c \cdot y^2 = 0\,.$$

The first equation above has an exponential solution and the second one has a solution in terms of elliptic functions [9], example 6.10:

$$\begin{array}{ll} x(t) = C_1 \cdot \exp(it) + C_2 \cdot \exp(-it), & C_1, C_2 \in \mathbb{C}\,, \\ t = \int \frac{dt}{\sqrt{\frac{2}{3} \cdot c \cdot y^3 - y^2 + C_3}}, & C_3 \in \mathbb{R}\,. \end{array} \tag{2}$$

Thus, let us assume that $d \neq 0$. Then with changing $d \cdot x \to x$, $d \cdot y \to y$ and $c/d \to c$ we obtain (1) in the form:

$$\begin{array}{l} \ddot{x} + x + 2 \cdot x \cdot y = 0\,, \\ \ddot{y} + y + x^2 - c \cdot y^2 = 0\,, \end{array} \tag{3}$$

with the Hamiltonian:

$$h = \frac{1}{2}[(\dot{x})^2 + (\dot{y})^2 + x^2 + y^2] + x^2 y - \frac{c}{3} y^3\,. \tag{4}$$

A linear change of variables:

$$\begin{array}{l} x = y_1 + y_2\,,\; \dot{x} = \; -i\;(y_1 - y_2) \\ y = y_3 + y_4\,,\; \dot{y} = -i\;(y_3 - y_4)\,, \end{array} \tag{5}$$

transforms (3) to the form required by the method:

$$\begin{array}{l} \dot{y}_1 = -i\,y_1 - i\;(y_1 + y_2)\;(y_3 + y_4)\,, \\ \dot{y}_2 = \quad i\,y_2 + i\;(y_1 + y_2)\;(y_3 + y_4)\,, \\ \dot{y}_3 = -i\,y_3 - \frac{i}{2}\;[(y_1 + y_2)^2 - c(y_3 + y_4)^2]\,, \\ \dot{y}_4 = \quad i\,y_4 + \frac{i}{2}\;[(y_1 + y_2)^2 - c(y_3 + y_4)^2]\,, \end{array} \tag{6}$$

The eigenvalues of this system are two pairs of complex conjugate imaginary units: $\mathbf{\Lambda} = (-i, i, -i, i)$ [1]. So this is a deeply resonant problem, i.e. the most difficult (and interesting) type of a problem.

[1] Remark that in paper [6] the other order of variables is used. The corresponding vector of eigenvalues is $\mathbf{\Lambda} = (-i, -i, i, i)$ there. It corresponds to interchanging $y_2 \leftrightarrow y_3$ for agreement with notation of the present paper.

3 A Normal Form for the Generalized Henon–Heiles System

For a pure resonance of the generalized Henon–Heiles system the ratios of all pairs of eigenvalues are ± 1 and the normal form for (6) is:

$$\dot{z}_k = z_k G_k \stackrel{\text{def}}{=} \lambda_k z_k + z_k \sum_{\substack{q_k \geq -1, \\ q_1, \ldots, q_{k-1}, q_{k+1}, \ldots, q_4 \geq 0, \\ q_1 + q_3 = q_2 + q_4 > 0 \\ 0 \leq p \leq q_1 + q_2 + q_3 + q_4}} g_{k,q_1,q_2,q_3,q_4,p}\, z_1^{q_1} z_2^{q_2} z_3^{q_3} z_4^{q_4} c^p, \quad k = 1, \ldots, 4. \quad (7)$$

$G_k = G_k(\mathbf{z}, c)$ are series in $z_1, \ldots, z_4$ and polynomials in c, $g_{k,q_1,q_2,q_3,q_4,p}$ are numeric coefficients which can be calculated by the LISP based program NORT [4].

The set $\mathcal{A}$ in a phase space of system (7) is defined by the system of equations [1], [7]:

$$\begin{aligned} &\mathcal{A} = \{z_1, z_2, z_3, z_4 : \\ &\lambda_k z_k \omega = \lambda_k z_k + z_k \sum g_{k,q_1,q_2,q_3,q_4,c}\, z_1^{q_1} z_2^{q_2} z_3^{q_3} z_4^{q_4} c^p\}, \quad k = 1, \ldots, 4, \end{aligned} \quad (8)$$

where ω is a series in $z_1, \ldots, z_4$ and polynomials in c. ω does not depend on index k along the set $\mathcal{A}$.

Searching for *all* local periodic families of solutions of system (6) is equivalent to searching for the set $\mathcal{A}$ which contains all of them. It is important that along the set $\mathcal{A}$ formal series in formulae above have a basis which consists of convergent power series in z_k variables [2] [1]. So the families of periodic solutions of (8) can be expressed (approximated) in terms of convergent series.

Equations (8) can be recast (by eliminating ω which is non zero for non-trivial solutions) in the form:

$$\begin{aligned} P_1 &\stackrel{\text{def}}{=} z_1 z_2 \cdot [G_1(\mathbf{z}, c) + G_2(\mathbf{z}, c)] = 0, \\ P_2 &\stackrel{\text{def}}{=} z_3 z_4 \cdot [G_3(\mathbf{z}, c) + G_4(\mathbf{z}, c)] = 0, \\ P_3 &\stackrel{\text{def}}{=} z_1 z_4 \cdot [G_1(\mathbf{z}, c) + G_4(\mathbf{z}, c)] = 0, \\ P_4 &\stackrel{\text{def}}{=} z_2 z_3 \cdot [G_2(\mathbf{z}, c) + G_3(\mathbf{z}, c)] = 0. \end{aligned} \quad (9)$$

Of course no more than 3 equations above are independent, but the form (9) is symmetric.

Because of (8) all families of periodic solutions of (7) have the form:

$$z_j = a_j \exp(-i\omega t), \quad z_{j+1} = a_{j+1} \exp(i\omega t), \quad j = 1, 3. \quad (10)$$

The a_j above are integration constants and ω is time independent and plays the role of frequency. It depends on constants c and a_j only.

[2] The parameter c is not supposed to be small.

With (10) we can rewrite (9) as an algebraic problem of solving a system of equations over the ring of formal power series in a_k:

$$\begin{aligned} P_1 &\stackrel{\text{def}}{=} a_1 a_2 \cdot [G_1(a_1,\dots,a_4,c) + G_2(a_1,\dots,a_4,c)] = 0\,,\\ P_2 &\stackrel{\text{def}}{=} a_3 a_4 \cdot [G_3(a_1,\dots,a_4,c) + G_4(a_1,\dots,a_4,c)] = 0\,,\\ P_3 &\stackrel{\text{def}}{=} a_1 a_4 \cdot [G_1(a_1,\dots,a_4,c) + G_4(a_1,\dots,a_4,c)] = 0\,,\\ P_4 &\stackrel{\text{def}}{=} a_2 a_3 \cdot [G_2(a_1,\dots,a_4,c) + G_3(a_1,\dots,a_4,c)] = 0\,. \end{aligned} \tag{11}$$

Searching for all local families of periodic solutions of system (7) is equivalent to determining all solutions of system (11).

The original system (3) is real, thus real solutions of (7) satisfy the reality conditions:

$$\begin{aligned} &z_{j+1} = \overline{z_j}, \quad j = 1,3\\ &\text{or}\\ &a_{j+1} = \overline{a_j}, \quad j = 1,3 \end{aligned} \tag{12}$$

so we can fix (by neglecting a trivial time shift) $a_2 = a_1$ as pure real.

Because system 3 is even in time all families of solutions arise by couples.

4 Calculation of Results

By using our program NORT for system (6) we have calculated the normalizing transformation and normal form till the 10th order in z_k (i.e. till 12th in a_k for P_k series). We calculated periodic solutions and compared them with corresponding numerical solutions at different values of parameters. Below we discuss the bifurcation picture and the phase portrait of system (3) following this analysis [3].

It is proved in [2] that for any reversible resonant system of the 4th order, both the series P_1 and P_2 have the same factor:

$$\begin{aligned} P_1(\mathbf{z}) &= (z_1^r z_4^s - z_2^r z_3^s) \cdot Q_1(\mathbf{z})\,,\\ P_2(\mathbf{z}) &= (z_1^r z_4^s - z_2^r z_3^s) \cdot Q_2(\mathbf{z})\,, \end{aligned}$$

where r and s are smallest positive integers which satisfy the equation $\lambda_1 \cdot r = \lambda_3 \cdot s$.

For our case $s = r = 1$, but the system has an additional symmetry and we can find by factoring that the P_1 and P_2 series calculated by the NORT program have a more complicated factor:

$$\begin{aligned} P_1 = {} & 2i\cdot (a_1^2 a_4^2 - a_2^2 a_3^2)\cdot\\ & \cdot[1 + \tfrac{1}{6}c + +(\tfrac{133}{18} - \tfrac{95}{108}c - c^2)a_1 a_2 +\\ & +(\tfrac{4}{9} - \tfrac{172}{27}c + \tfrac{23}{18}c^2 + \tfrac{89}{108}c^3)a_3 a_4 + O(a^4)]\,,\\ P_2 = {} & -2i\cdot (a_1^2 a_4^2 - a_2^2 a_3^2)\cdot\\ & \cdot[1 + \tfrac{1}{6}c + +(\tfrac{13}{2} - \tfrac{95}{108}c - \tfrac{1}{9}c^2)a_1 a_2 +\\ & +(\tfrac{20}{9} - \tfrac{148}{27}c - \tfrac{1}{2}c^2 - \tfrac{7}{108}c^3)a_3 a_4 + O(a^4)]\,, \end{aligned} \tag{13}$$

[3] All calculations below were carried out for the mechanical energy $h = 1/12$.

where (and in some places below) we adduce for simplicity first terms of the calculated series only.

Recall that we are interested only in local solutions, i.e. in such solutions of (11) which can include the stationary point $a_k = 0$, $k = 1, \dots, 4$ as a particular case.

Thus, instead of the first pair of equations in (11), we have the equation:

$$a_1^2 a_4^2 - a_2^2 a_3^2 = 0\,,$$

because the brackets in (13) cannot add any nontrivial local solution as they contain constant terms at $c \neq -6$, and at $c = -6$ these both brackets are proportional to $a_1 a_2 + 4 a_3 a_4$, which is a sum of squares of modules (see (12)).

So, if we now fix $a_1 = a_2 = a$ as a pure real, then the first couple of equations (13) has two solutions:

1. $a_3 = a_4 = b$ has a pure real value;
2. $a_3 = -a_4$ has a pure imaginary value. Let $a_3 = ib$.

4.1 Case of Pure Real a_3

In this case the second pair of equations (13) gives a single equation:

$$\begin{aligned} P_3 = -P_4 = \alpha_1 \cdot a \cdot b \cdot [a^2 - (c+2)b^2] \cdot [c - 1 - (\tfrac{2431}{180} - \tfrac{29}{90}c + \tfrac{217}{60}c^2)a^2 - \\ -(\tfrac{67}{90} - \tfrac{1289}{180}c + \tfrac{233}{45}c^2 - \tfrac{157}{36}c^3)b^2 + O(a^4) + O(b^4) + O(a^2b^2)] = 0\,. \end{aligned} \tag{14}$$

α_1 here is a nonzero numerical constant. Let us discuss the families of periodic solutions which correspond to zeroing each factor of the product above.

A couple of families of periodic solutions which corresponds to $a = 0$ exists at any values of c and lies in the plane $x = 0$. This is a family with a single internal parameter. We choose the mechanical energy h from (4) as this parameter. At $c = 1$ this family corresponds to family 5 of the classic Henon–Heiles system [8], see Fig. 1 and paper [6]. In Fig. 1, the intersections of periodic solutions of the Henon–Heiles system with Surface Of Section (SOS) [8], which is defined by equations SOS $= \{x = 0,\ \dot{x} = \dot{x}(x, y, \dot{y}, h) > 0\}$, are displayed in coordinates y, $\dot{y}$. The periodic solutions of families 5 lie entirely in the plane $x = \dot{x} = 0$. For this case there is an analytical solution of type 2 in elliptic functions.

The families which correspond to $b = 0$ also exist at any values of c. They look like family 4 of the Henon – Heiles system (Fig. 1). The corresponding intersection flows slowly from left to right at increasing c. The frequency of these periodic families is:

$$\begin{aligned} \omega_4 = 1 &- 5\rho/3 + \rho^2(-281 + 504c)/108 + \\ &+\rho^3(-13913 + 645024c - 323488c^2)/19440 + \\ &+\rho^4(33903721 + 134318856c - 137045376c^2 + 59393664c^3)/699840 \\ &+\rho^5(103971857615 + 172223295216c - 402212367472c^2 + \\ &+294216077568c^3 - 105272265984c^4)/220449600 + O(\rho^6)\,, \end{aligned}$$

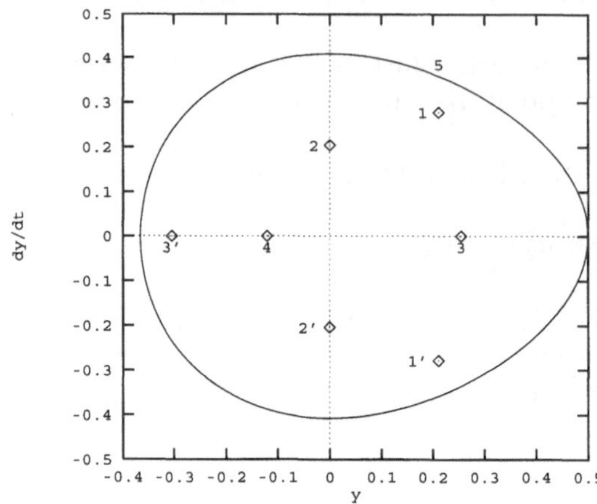

Fig. 1. Intersections of periodic solutions of the Henon–Heiles system ($c = 1$) with the surface of section (SOS) at the energy level $h = \frac{1}{12}$

where:

$$\begin{aligned}\rho = {} & h/2 + 7h^2/72 + h^3(-1099 - 1184c)/3456+\\ & +h^4(-1830221 - 1187297c + 208324c^2)/777600+\\ & +h^5(-64702312929 - 10993351152c + 19241890208c^2-\\ & -4382098944c^3)/6718464000 + O(h^6)\,.\end{aligned}$$

These are one-parametric families.

Let us define a relative error as the maximum of relative difference between the tabulated series (of tenth order) $x(t)_{app}, y_{app}(t), \dot{x}_{app}(t), \dot{y}_{app}(t)$ and the corresponding numerical solution during one period:

$$f_{err} \stackrel{\text{def}}{=} \sup_{t\in[0,2\pi/\omega_i]}$$

$$\sqrt{\frac{(x_{num}(t)-x_{app}(t))^2+(y_{num}(t)-y_{app}(t))^2+(\dot{x}_{num}(t)-\dot{x}_{app}(t))^2+(\dot{y}_{num}(t)-\dot{y}_{app}(t))^2}{x^2_{num}(t)+y^2_{num}(t)+\dot{x}^2_{num}(t)+\dot{y}^2_{num}(t)}}$$

The relative error of the approximation of this case is about 6% at $c = -1$; 0.3% at $c = 1$ and 0.23% at $c = 2$.

The next case is $a^2 = (c+2)\cdot b^2$. At $c = 1$ it corresponds to 2 and 2' families in fig. 1. The families are real if $c \geq -2$ only. It is one parametric case also. With increasing c it slowly flows to the abscissa axis. At $c = -2$ it coincides with curve 5. The frequency is:

$$\begin{aligned}\omega_2 = {} & 1 - 20h_1/3 - 380h_1^2/3 - 878960h_1^3/243-\\ & -88121780h_1^4/729 - 3203319760h_1^5/729 + O(h_1^6)\end{aligned}$$

where $h_1 = h/(2(3+c))$. The relative error is about 4% at $c = 0$; 0.4% at $c = 1$ and 0.06% at $c = 2$.

Zeroing the last brackets in (14) is possible near the point $c = 1$, when the corresponding distance $c-1$ is about a square of amplitudes a or b, or, say, about energy h. So we can suppose that $c = 1 + \varepsilon \cdot h$ and the corresponding solution exists at least at ε lower or about 1. This case agrees with families 1 and 1′ in Fig. 1. The intersection with SOS flows to the abscissa axis and from right to left with increasing ε. These are one parametric families. The frequency is:

$$\omega_1 = 1 - 25\varepsilon h/252 + O(h^2)\,.$$

The relative error here is 0.09% at $\varepsilon = -1$; 0.09% at $\varepsilon = 1$ and 0.2% at $\varepsilon = 2$. A small value of error may say about possibility of spreading this case into wider domain of the parameter ε.

4.2 Case of Pure Imaginary a_3

Let $a_3 = ib$. In this case the second pair of equations (13) also gives a single equation:

$$\begin{aligned} P_3 = P_4 = \alpha_2 \cdot a \cdot b \cdot (c+1)\cdot \\ \cdot[7a^2 - (5c+2)b^2 + O(a^4) + O(b^4) + O(a^2b^2)] = 0\,. \end{aligned} \tag{15}$$

α_2 here is a nonzero numerical constant. Let us one more discuss families of periodic solutions, which correspond to zeroing factors of the product (15).

Cases $a = 0$ and $b = 0$ have been discussed. Case $c = -1$ is a specific one. It is an essential two parametric case, but it is better to choose parameters from a physical meaning. Let us introduce the energy as one of parameters. We will suppose that the amplitudes a and b connected with the energy h in this way:

$$a^2 + b^2 = p(h), \quad a^2 = \beta \cdot p(h), \quad \beta \in [0,1],,$$

where $p(h)$ is the series:

$$\begin{aligned} p(h) = h/2 + 7h^2/72 + 85h^3/3456 - \\ -2173h^4/3888 - 240679781h^5/53747712 + O(h^6)\,. \end{aligned}$$

This parameterization is good because it displays that frequency depends on $p(h)$, i.e. on energy only:

$$\begin{aligned} \omega_{-1} = 1 - 5p(h)/3 - 785p^2(h)/108 - 65495p^3(h)/1296 - \\ -59370835p^4(h)/139968 - 4971155135p^5(h)/1259712 + O(p^6(h))\,. \end{aligned}$$

At the beginning of the interval ($\beta = 0$) a behavior of these periodic families looks like solution 5 in fig. 1. Then the solutions cross the SOS look like 3 and 3′ on the segment $\beta \in (0,1)$ and at the end of the interval ($\beta = 1$) these 3 and 3′ join each other in the point which corresponds to solution 4. Relative error here equals 6% at $\beta = 0$; 5% at $\beta = 0.5$ and again 6% at $\beta = 1$.

The case when the last bracket in (15) is equal to zero is close to the last case of the previous subsection. It is one-parametric case and it is real only when $c \geq -\frac{2}{5}$. The case corresponds to families 3 and 3′ in the Figure. Frequency is:

$$\begin{aligned} \omega_3 = 1 + 2/3h(1+6c)/(9+5c) + \\ +1/3h^2(-135 - 1051c - 2694c^2 - 1503c^3 - \\ -105c^4)/(729 + 1215c + 675c^2 + 125c^3) + O(h^3) \end{aligned}$$

The families start from a shape of 5 at $c = -2/5$, and then go along abscissa to the origin likewise couple 3, 3′. Relative errors equal $6 \cdot 10^{-6}\%$ at $c = -2/5$; $6 \cdot 10^{-5}\%$ at $c = 0$; $4 \cdot 10^{-3}\%$ at $c = 1$ and 0.08% at $c = 1.5$.

There is also a couple of two-parametric local families of complex periodic solutions at zero energy h with frequencies equal to ± 1. Such solutions were discussed in [5], [6].

5 Conclusions

In the normal form method a bifurcation analysis for local families of periodic solutions was carried out for the low resonant case of the generalized Henon–Heiles system. We have found 6 pairs of real families of local periodic solutions. In terms of (3) and Figure 1 they are:

1. a couple of families looks like solutions 5 in the Figure, with one internal parameter. These families exist at any value of external parameter c;
2. a couple of families look like solutions 4 in the Figure, with one internal parameter. The families exist at any value of external parameter c;
3. a couple of families looks like solutions 2 and 2′ in the Figure, with one internal parameter. The couple exists at the value of external parameter $c \geq -2$;
4. a couple of families looks like solutions 1 and 1' in the Figure, with one internal parameter. The couple exists at least near the value of external parameter $c = 1$;
5. a couple of families looks like solutions 3 and 3′ in the Figure, with two internal parameters. The couple exists at the single value of external parameter $c = -1$;
6. a couple of families looks like number 3 and 3′ in the Figure, with one internal parameter. The couple exists at the value of external parameter $c \geq -2/5$.

There are two local two-parametric complex families of periodic solutions at the energy $h = 0$ also.

Acknowledgements

The author is very grateful to Prof. R. I. Bogdanov and Dr. S. Yu. Vernov for interesting discussions and important advices.

References

1. Bruno, A. D.: *Local Method in Nonlinear Differential Equations*, Springer-Verlag, Berlin (1989)
2. Bruno, A. D., Soleev, A.: *Local analysis of a singularity of a reversible ODE system. Complicated cases*; Preprint # 47 of the Keldysh Institute of Applied Mathematics of RAS, Moscow (1995) (in Russian)

3. Chang, Y. F., Tabor, M. and Weiss, J.: Analytic structure of the Henon-Heiles Hamiltonian in integrable and nonintegrable regimes. *J. Math. Phys.* **23** (1982) 531–538
4. Edneral, V. F.: Computer generation of normalizing transformation for systems of Nonlinear ODE. In: *Proceedings of the 1993 International Symposium on Symbolic and Algebraic Computation (Kiev, Ukraine, July 6-8, 1993)*, M. Bronstein (ed.), ACM Press, New York (1993) 14–19
5. Edneral, V. F.: Complex periodic solutions of autonomous ODE systems with analytical right-hand sides near an equilibrium point. *Fundamentalnaya i prikladnaya matematika* **1** (1995) #2 393–398 (in Russian)
6. Edneral, V. F.: A symbolic approximation of periodic solutions of the Henon–Heiles system by the normal form method. *J.Mathematics and Computers in Simulation* **45** (1998) 445–463
7. Edneral, V. F.: About normal form method. In: *Proc. Second Workshop on Computer Algebra in Scientific Computing (CASC'99, Munich, Germany, 1999)*, Ganzha et al. (eds.), Springer (1999) 51–66
8. Henon, M., Heiles, C.: The applicability of the third integral of motion: Some numerical experiments. *Astronomical J.* **69** (1964) 73–79
9. Kamke, E.: *Differentialgleichungen. I*, Verbesserte Auflage, Leipzig (1959)

An Involutive Reduction Method to Find Invariant Solutions for Partial Differential Equations

Joachim Engelmann[1] and Gerd Baumann[1,2]

[1] University of Ulm, Department of Mathematical Physics, Albert-Einstein-Allee 11, 89069 Ulm, Germany,
joachim.engelmann@physik.uni-ulm.de, gerd.baumann@physik.uni-ulm.de,
[2] Visual Analysis AG, Neumarkter Str. 87, 81673 München, Germany,
gerd.baumann@visualanalysis.com

Abstract. A standard approach to solve partial diffrential equations is the construction of invariant solutions. These solutions have to fulfill an additional equation called the invariant surface condition. This condition represents the invariance of the equation under a symmetry transformation. To solve the coupled system of the differential equation and its invariant surface condition we used a *Mathematica*–package which combines involutive and heuristic methods to simplify and solve this coupled system. The procedure is presented by some examples.

1 Introduction

Invariant solutions are important from a physical point of view. For example consider solutions invariant under time translation, Galileian transformations or travelling wave solutions such as the famous solitons. In any of these cases the initial partial differential equation is reduced by so-called similarity variables. They consist of combinations of the independent variables of the differential equation. A characteristic example resulting from scaling symmetries is the variable $\zeta = \frac{x}{\sqrt{t}}$ related to the diffusion equation.

To find such similarity variables and the corresponding invariant solutions, we first have to have some kind of symmetry transformation [1]. Hereby, a symmetry transformation is a transformation which leaves the form of the given differential equation invariant. Solving an additional differential equation, called the invariant surface condition, describing the invariance property under the infinitesimal transformation, results in a system of invariants depending on the independent and dependent variables of the differential equation under consideration. The derivation of the symmetry transformations can be carried out automatically by applying computer algebra programs such as *MathLie* [1] or others. An overview of existing computer algebra packages treating symmetry analysis is given in [2]. Inserting the invariants in the initial differential equation leads to a reduction in the number of independent variables. This reduction procedure is standard for partial differential equations and can be found for example in [3], [4] or in [5,6].

This method is also known as the method of characteristics.
However in a remark in [6] a hint for another way of reduction was given. This alternative approach consists of adding the integrability conditions to the original equations. The result is a combined system of partial differential equations incorporating the invariant surface condition of the system. Unlike the nonclassical method, where a symmetry analysis is performed on this coupled system, in [6] the system is directly reduced using elimination techniques, with no symmetry background at all. With this proceeding a reduction of the initial differential equation is possible.
We implemented this procedure in an algorithmic way to find solutions of partial differential equations invariant under a symmetry transformation. Unlike other computer algebra packages, like `CRACK` [7] or `PDESolve` [1], which aim at "just" solving a system of overdetermined polynomially nonlinear partial differential equations, or `rif` [8] or `diffalg` [9], which "just" perform a reduction of such a system to a simplified form, our resulting procedure implemented in *Mathematica* is able to simplify *and* solve systems as mentioned above in one step, as long as the heuristic solver is able to solve the corresponding differential equations. That means that by starting the calculation simplifications by integrability conditions and solutions of partial differential equations by heuristics are done. This is achived in the following way.
First the involutive reduction is performed. This algorithm is based upon the standard steps for a reduction/completion algorithm such as discussed in [10–15]. Furthermore it is capable of considering different cases for arbitrary functions or constants appearing in the equations. The program distinguishes automatically between different conditions the arbitrary functions have to satisfy. After these simplifications the system under consideration enters a heuristic solver. This solver is in fact an adjustment of the *MathLie* function PDESolve [1] to our needs.
These two steps, namely simplification and solution, are arranged in a loop which is repeated over and over again until the system changes no more. That means the user has not to do anything like he would have to if there are two completely different procedures. In this case the notation might has to be changed, the system may have to be put in another form, the appearing constants of integration might have to be renamed etc. This is not the case. The simplifier and the solver are built in such a way that each one's result is the input of the other procedure. Only in this way an automatic combination is possible.
This interplay of involutive methods to simplify and the ad-hoc methods to solve equations of a system of coupled partial differential equations leads to simplifications which may not be possible by either method alone in just one run. In this way both procedures enrich itself and further algorithmic reductions are possible.
In the following, we demonstrate both methods with illustrative examples. In part 2 we compare the method of invariants against the involutive method on the standard diffusion equation. We derive two solutions, an Error-function solution and a solution incorporating the Airy-functions Ai and Bi. In section 3

we turn to a nonlinear diffusion equation to explain once more how the involutive solver works. Finally, in part 4 we mention some more results obtained with the involutive solver, namely a solution for the standard diffusion equation depending on the hypergeometric function ${}_2F_2$ and solutions for nonlinear diffusion equations where the diffusion "constant" is a rational function of the independent variables.

2 Methodology Applied to the Diffusion Equation

The standard diffusion equation in 1+1 dimensions is given by

$$u_t = u_{x,x}, \tag{1}$$

where lower indices represent partial differentiations with respect to time t and spatial coordinate x. In the following we discuss the invariance of (1) under two infinitesimal transformations leading to two different kinds of solutions. Hereby we compare the standard procedure, the method of invariants or the characteristic method, with the one used in the program.

2.1 Error-Function Solution

The standard diffusion equation allows among other transformations a scaling with the infinitesimal generator

$$\boldsymbol{v} = x\partial_x + 2t\partial_t,$$

so a function $u = f(x,t)$ is invariant under this transformation if it satisfies the first order partial differential equation

$$xu_x + 2tu_t = 0, \tag{2}$$

called the invariant surface condition. To reduce equation (1) with the condition (2) we start with the method of invariants.

Method of Invariants As mentioned above we first look for solutions of (2). This equation has two invariants which we will denote here by ζ and f. Using *Mathematica*, we get

$$\zeta = \frac{x}{\sqrt{t}},$$
$$f = const.$$

Thus we conclude that an invariant solution of (1) is given by

$$u = f\left(\frac{x}{\sqrt{t}}\right).$$

Inserting this result in the original equation (1) we observe that f has to satisfy

$$\zeta f'(\zeta) + 2f''(\zeta) = 0. \tag{3}$$

We mention that the reduced equation is free of our original variables because only ζ appears. This is not always the case as demonstrated in [6].
Again using *Mathematica* to solve (3), we find

$$f = C_1 + C_2\sqrt{\pi}\,\mathrm{erf}\left(\frac{\zeta}{2}\right),$$

with erf denoting the Error-function (see e. g. [16]). Combining the results we get the invariant solution

$$u(x,t) = C_1 + C_2\sqrt{\pi}\,\mathrm{erf}\left(\frac{x}{2\sqrt{t}}\right). \tag{4}$$

More details of the method of invariants are discussed in [3, 4].

Involutive Method Calling our involutive solver InvolutivePDESolve on the combined system (1,2) we get the same result, but completely automatic and in a direct way. When tracing the intermediate results to take a look inside the "black box" InvolutivePDESolve, we see that the involution algorithm reduces (1) with (2). First the involutive simplifier is called. Since the integrability conditions of the resulting system are all satisfied the involutive algorithm ends with the system

$$\begin{aligned} xu_x + 2tu_{x,x} &= 0, \\ xu_x + 2tu_t &= 0. \end{aligned} \tag{5}$$

The following solution step first tries to solve monomial equations, that means equations consisting of just one term. Since (5) does not contain such equations the solution algorithm enters the next step. If an equation is a polynomial in one variable the coefficients of the various powers of this variable are identically zero. Again the system does not change, just like in the next step, which differentiates equations with respect to variables appearing polynomially.
Next, the system enters a solver for ordinary differential equations. Hereby, the first equation of (5) has the solution

$$u = f_3(t) + \sqrt{\pi t}\,\mathrm{erf}\left(\frac{x}{2\sqrt{t}}\right) f_2(t)$$

with the new arbitrary functions $f_2(t)$ and $f_3(t)$. Inserting this in (5), we get

$$\sqrt{\pi t}\,\mathrm{erf}\left(\frac{x}{2\sqrt{t}}\right) f_2(t) + 2t\sqrt{\pi t}\,\mathrm{erf}\left(\frac{x}{2\sqrt{t}}\right) f_2'(t) + 2tf_3'(t) = 0 \tag{6}$$

as the only equation which has to be fulfilled. With this the solver ends and we again enter the involutive part of the algorithm. Since f_2 and f_3 only depend on t differentiation with respect to x and back-insertion of the result in (6) delivers

$$f_3'(t) = 0. \tag{7}$$

The reduction step inserts this in the system. We get

$$f_2(t) + 2tf_2'(t) = 0.$$

With these results the involutive part ends. Entering the solution algorithm the monomial solver integrates (7), so f_3 is a constant. The ODE-solver then gives

$$f_2(t) = \frac{C}{\sqrt{t}}.$$

Combining these results the algorithm InvolutivePDESolve ends with

$$u = C_1 + C_2\sqrt{\pi}\,\mathrm{erf}\left(\frac{x}{2\sqrt{t}}\right).$$

We see that this is exactly the result obtained by the method of invariants.

2.2 Airy Solution

The standard diffusion equation is also invariant under the infinitesimal generator

$$\boldsymbol{v} = \partial_t - 2t\partial_x + xu\partial_u.$$

Looking for invariant solutions under this transformation, we get the additional equation

$$xu + 2tu_x - u_t = 0. \tag{8}$$

Combining equations (1,8) and following the steps in which changes happen we call the involutive solver.

In the first part of the involutive step equation (1) is reduced by (8) to

$$-xu - 2tu_x + u_{x,x}.$$

The following steps, the computation and reduction of the integrability conditions do not change anything. So the involutive procedure delivers the system

$$\begin{aligned} -xu - 2tu_x + u_{x,x} &= 0, \\ -xu - 2tu_x + u_t &= 0. \end{aligned} \tag{9}$$

Entering the solution step the algorithm first looks for monomial equations. Since there are none the program checks for polynomial variables. Again the system does not change under this operation as well as under a differentiation. Next (9) enters the solver for ordinary differential equations. In this step the first equation is solved to give

$$u = e^{tx}(\mathrm{Ai}(x+t^2)f_2(t) + \mathrm{Bi}(x+t^2)f_3(t))$$

with new free functions f_2 and f_3. Hereby Ai and Bi denote Airy-functions [16]. Inserting this in (9) we get an identity and the relation

$$-\mathrm{Ai}(x+t^2)f_2'(t) - \mathrm{Bi}(x+t^2)f_3'(t) + 2t^2\mathrm{Ai}(x+t^2)f_2(t) + 2t^2\mathrm{Bi}(x+t^2)f_3(t) = 0. \quad (10)$$

Since f_2 and f_3 only depend on t, differentiation of this condition with respect to x and inserting this in (10) delivers the following ODEs:

$$f_2'(t) - 2t^2 f_2(t) = 0,$$
$$f_3'(t) - 2t^2 f_3(t) = 0,$$

with which the involutive part ends. These two equations are then integrated by the ad-hoc solver to

$$f_2(t) = C_1 e^{\frac{2}{3}t^2},$$
$$f_3(t) = C_2 e^{\frac{2}{3}t^2}.$$

With these solutions the involutive solver ends with the invariant solution

$$u = e^{\frac{2}{3}t^2+tx}(C_1\mathrm{Ai}(x+t^2) + C_2\mathrm{Bi}(x+t^2)).$$

Again this solution is obtained completely automatic.

3 A Nonlinear Diffusion Equation

As a second example we consider the nonlinear diffusion equation

$$u_t = u_x^2 + u u_{x,x}. \quad (11)$$

This equation is invariant under the infinitesimal generator

$$\boldsymbol{v} = x\partial_x + 2u\partial_u,$$

meaning that the corresponding invariant surface condition

$$2u - xu_x = 0 \quad (12)$$

has to be satisfied. The method of invariants just works the same way as for the standard diffusion equation. We just state the result:

$$u = \frac{x^2}{C - 6t} \quad (13)$$

where C is a constant.
To treat the problem of finding invariant solutions with the involutive method we again discuss those steps in which important changes occur. First the combined system (11,12) enters the involution step. The first part of the involutive algorithm reduces the system with itself. In our example (12) is inserted in (11) to give

$$\begin{aligned} -2u + xu_x &= 0, \\ -6u^2 + x^2 u_t &= 0. \end{aligned} \tag{14}$$

The following completion step does not change anything which is also true for the computation and reduction of the integrability conditions. So the involutive procedure ends with (14).
Now the system enters the heuristic solver. The first step of the solver searches for monomial equations. Since there are none and also no simplifications are possible the system enters the differentiator. Hereby the first equation of (14) is differentiated two times with respect to x leading to the equation

$$xu_{x,x,x} = 0.$$

The following monomial solver provides

$$u = f_1(t) + xf_2(t) + x^2 f_3(t).$$

Inserting this in (14) gives

$$\begin{aligned} xf_3(t) + 2x^2 f_4(t) - 2f_2(t) - 2xf_3(t) - 2x^2 f_4(t) &= 0, \\ -6(f_2(t) + xf_3(t) + x^2 f_4(t))^2 + x^2 f_2'(t) + x^3 f_3'(t) + x^4 f_4'(t) &= 0. \end{aligned}$$

Since each equation has to be satisfied for all values of x the polynomial simplifier delivers

$$\begin{aligned} f_2(t) &= 0, \\ f_3(t) &= 0, \\ -6f_3(t)^2 - 12f_2(t)f_4(t) + f_2'(t) &= 0, \\ -12f_3(t)f_4(t) + f_3'(t) &= 0, \\ -6f_4(t)^2 + f_4'(t) &= 0 \end{aligned}$$

which results in

$$u = f_4(t)x^2, \tag{15}$$

where f_4 has to solve the ODE

$$-6f_4(t)^2 + f_4'(t) = 0. \tag{16}$$

The single remaining condition (16) now enters the solver for ordinary differential equations. The result is

$$f_4(t) = \frac{1}{C_1 - 6t}.$$

Inserting this in (15) we get one solution of (11):

$$u = \frac{x^2}{C - 6t}.$$

With this result the involutive solver stops. This is exactly the same solution as (13) derived by the method of invariants, but in a much more convenient way. It is done automatically!

4 Additional Solutions for other Nonlinear Diffusion Equations

Using the same technique as described before we also calculated the following solutions to some nonlinear diffusion equations:

- The equation

$$u_t = (K(u)u_x)_x \quad \text{with} \quad K(u) = C_1(C_2 + uC_3)^{\frac{C_4}{C_3}}$$

  is invariant under the transformation given by the generator

$$\boldsymbol{v} = (2C_5 + xC_4)\partial_x + (2uC_3 + 2C_2)\partial_u.$$

  The corresponding solution reads

$$u(x,t) = -\frac{C_2}{C_3} + C_6 \left(\frac{(2C_5 + xC_4)^2}{C_7 - 4tC_3C_4 - 2tC_4^2} \right)^{\frac{C_3}{C_4}}.$$

- The following nonlinear diffusion equation

$$u_t = (K(u)u_x)_x \quad \text{with} \quad K(u) = e^{C_1(u - C_2)}$$

  allows the infinitesimal transformation

$$\boldsymbol{v} = (C_4 + \frac{x}{2}C_1C_3)\partial_x + C_3\partial_u.$$

  The solution invariant under this transformation is

$$u(x,t) = C_2 + \log \left(\frac{(2C_4 + xC_1C_3)^2}{C_1 - 2tC_1^2C_3^2} \right)^{\frac{1}{C_1}}.$$

- The standard diffusion equation with diffusion constant K

$$u_t = Ku_{x,x}$$

  is invariant with respect to the generator

$$\boldsymbol{v} = \left(C_4 + \frac{C_3}{2}x \right) \partial_x + (C_2 + tC_3)\partial_t + C_1\partial_u.$$

The solution to this generator is given by

$$u(x,t) = \log (C_2 + tC_3)^{\frac{C_1}{C_3}} - 2C_5\sqrt{\frac{\pi}{KC_3}}\,\operatorname{erf}\left(\frac{2C_4 + xC_3}{2\sqrt{KC_3(C_2+tC_3)}}\right) + C_1\frac{(2C_4 + xC_3)^2}{2KC_3^2(C_2+tC_3)}\,{}_2\mathrm{F}_2\left(1,1;\frac{3}{2},2;-\frac{(2C_4+xC_3)^2}{4KC_3\,(C_2+tC_3)}\right).$$

5 Conclusion

We introduced an implementation of an involutive solver in *Mathematica* capable of a distinction between different cases for arbitrary functions or constants. This procedure combines an involutive simplifier with an heuristic solver. These two parts are combined algorithmically, which means that they act upon a system over and over again. The result of one of the functions is the input of the other and vice versa. We demonstrated that this solver is capable of delivering invariant solutions for partial differential equations.
Further research is needed to see if this procedure is as powerful as the method of invariants to solve partial differential equations. Also, a combination of both methods may be a more efficient tool than just one of them.
The computation and the use of integrability conditions to reduce or simplify the combined system of the differential equation and the invariant surface condition of a given transformation may help in finding invariants. Furthermore it may be possible that the result of the involutive solver is much more suitable to get invariants than the original system, just think of the ability to distinguish between different cases for optional free functions or constants which may simplify the reduced invariant surface condition. This distinction may also reduce problems when trying to find invariants for the invariant surface condition.

References

1. Baumann, G.: Symmetry Analysis of Differential Equations with *Mathematica*. Springer, New York, 2000
2. Hereman, W.: Symbolic software for Lie symmetry analysis. CRC Handbook of Lie Group Analysis of Differential Equations, Volume3: New trends in Theoretical Developments and Computational Methods. Editor: Ibragimov, N. H. CRC Press, Boca Raton, 1996
3. Bluman, G. W., Kumei, S.: Symmetries and Differential Equations. Springer, New York, 1989
4. Olver, P. J.: Applications of Lie Groups to Differential Equations. Springer, New York, 1993
5. Olver, P. J., Rosenau, P.: The construction of special solutions to partial differential equations. Phys. Lett. **114 A** (1986) 107–112
6. Olver, P. J., Rosenau, P.: Group-invariant solutions of differential equations. SIAM J. Appl. Math. **47** (1987) 263–278

7. Wolf, T., Brand, A.: The Computer Algebra Package `CRACK` for investigating PDEs. Proceedings of ERCIM, Partial Diffential Equations and Group Theory. Editors: Johnson, J. H., McKee, S., Vella, A. Bonn, 1992
8. Reid, G. J., Wittkopf, A. D. and Boulton, A.: Reduction of systems of nonlinear partial differential equations to simplified involutive forms. Eur. J. of Appl. Math. **7** (1996) 604 - 635.
9. Boulier, F., Lazard, D., Ollivier, F., Petitot, M.: Computing representations for radicals of finitely generated differential ideals. J. Symb. Comp. **11** 1-45
10. Schwarz, F.: An algorithm for determining the size of symmetry groups. Computing **49** (1992) 95–115
11. Reid, G. J.: A triangularization algorithm which determines the Lie symmetry algebra of any system of PDEs. J. Phys. A: Math. Gen. **23** (1990) L853–L859
12. Reid, G. J.: Algorithms for reducing system of PDESs to standard form, determining the dimension of its solution space and calculating its Taylor series solution. Euro. Jnl. of Appl. Math. **2** (1991) 293–318
13. Reid, G. J.: Finding abstract Lie symmetry algebras of differential equations without integrating determining equations. Euro. Jnl. of Appl. Math. **2** (1991) 319–340
14. Riquier, C.: Les systèmes d'équations aux dérivées partielles. Gauthier-Villars, Paris, 1910
15. Janet, M.: Leçons sur les systèmes d'équations aux dérivées partielles. Gauthier-Villars, Paris, 1929
16. Abramowitz, M., Stegun, I. A.:Handbook of Mathematical Functions. Dover Publications, Dover 1970

Recurrence Functions and Numerical Characteristics of Graphs *

Gani E. Ergashev and Ulugbek H. Narzullaev

Department of Mathematics, Samarkand State University, Samarkand, 703004, Uzbekistan

iipp@samarkand.uz and soleev@samarkand.uz

Abstract. We present a general method for the description of the structural meaning of the canonical representation coefficients of a wide class of the recurrence functions of graphs. The combinatorial objects, the so-called (j, k)-placements, are used. The algorithm, which can be realized on PC and allows to reveal some relations between the numerical characteristics of graphs, is resulted.

1 Introduction

Some complex discrete objects have the following properties: the characteristic of complex object can be computed if the characteristics of simpler objects are known, which are obtained from the given ones by some operation of the disassembly.

For example, this may take place in the process of running a program for PC, in the computation of the electrical links (method of the single movement and short circuit), in the process of solving many combinatorial problems of choosing the optimal variant, etc.

The methods of the computation of number of colourings proposed by G.D. Birkhof [1, 2], H. Witney [4, 5], V. Tutte [6] and A.A. Zykov [7, 8] in different forms and times belong to early applications of this method, in case where the role of the discrete object is played by graph.

In previous works on investigation of numerical characteristics of the discrete object, the recurrence method was used just to compute the defined numerical characteristics. But the wide responsibilities of the changing operation of the disassembly and specific functions with purpose of finding different characteristics of the object and relations between them give rise to the statement of the main problems of the more complete studies of those functions of graphs, which allow such a recurrence computation by some operation of disassembly.

Let $G = (X, U)$ be an arbitrary class of the finite graphs or multi-graphs, where X is the set of vertices, and U is the set of graphs. With respect to conditions of the problem we define the operator $\Gamma : G \longrightarrow G$ (often called

* This work was supported by State Committee for Science and Technology of the Rebublic of Uzbekistan, grant No. 4110.

“disassembly of graph”) taking each graph (multi-graph) $L \in G$ to the set of graphs $L_1, \dots, L_n$, which are simpler than L.

Definition. *It is said that graph L' is simpler than L if there exist such $L = L_1, L_2, \dots, L_n = L'$ that $L_2 \in \Gamma L_1, L_3 \in \Gamma L_2, \cdots, L_n \in \Gamma L_{n-1}$.*

Here we assume that by finite number of applications of the operator Γ we can come to the graphs

$$L_0^1, L_0^2, \cdots, L_0^k,$$

to which we cannot apply the operator Γ, because after a finite number of applications we obtain the graphs which are called "best simple" ones (often the role of these graphs is played by the graphs for which the investigating information is completely determined) and about which we have complete information.

Let $\{\Phi_i^0\}$ be the set of such symbols that to each graph L_i^0 the symbol $\{\Phi_i^0\}$ corresponds. Let $\{\alpha_i\}$ be the countable set of symbols, which are different from all $\{\Phi_i^0\}$ and K be a free associative ring generated by $\{\alpha_i\} \bigcup \{\Phi_i^0\}$.

Let us determine the function $\Phi(L)$ on K corresponding to the disassembly $\Gamma L = \{L_1, \cdots, L_n\}$ in the following manner:

$$\Phi(L) = \sum_{j=1}^{n} \alpha_j \Phi(L_j), \qquad \Phi(L_i^0) = \Phi_i^0.$$

It is evident that the values of $\Phi(L)$ lie in ring K and are the polynomials in the elements, which are generators of ring K.

The natural restriction imposed on the function $\Phi(L)$ is as follows: $\Phi(L)$ does not depend on the order of disassembly, i.e. $\Phi(L)$ takes the same values on isomorphic graphs.

This restriction attracts some factorization of the ring K, i.e. the system of correlations equating some of its elements to each other; correlations of such kind are called conditions of monotonicity.

The second problem consists in the determination of the minimal information about graph L, which allows to find the value of $\Phi(L)$ without analyzing the graph.

For determination of the minimal information delivered by function $\Phi(L)$, the general expression of this function is transformed to the canonical form by using the monotonicity conditions. As a result, $\Phi(L)$ will be expressed by a polynomial in the elements, which generate the ring K.

The most difficult problem is to find out the information (by means of graph construction) carried by the function given through recurrence.

In this way the combinatorial objects $(i,\ j,\ k)$ are introduced, the placements, which enable us to solve the problem mentioned above for rather a wide class of recurrence functions of graphs. In particular, some relations between numerical characteristics of the graphs and multi-graphs are deduced by this method.

2 The Disassembly Function Related to the Number of the Complete Subgraphs

As illustration of the theory described above let us consider the function $\Phi(L)$ which is defined in the following manner: Let K be a free associative ring generated by the generators $\alpha, \omega, \gamma, 1$. Let us consider in K the following reccurence function of the ordinary graph (finite, without loop and without divisible edges):

$$\Phi(L) = \alpha\Phi(L_\alpha) + \gamma\Phi(L_\alpha) + \omega\Phi(L_\omega) + 1 \tag{1}$$

with the initial condition

$$\Phi(\varepsilon_n) = 0, \tag{2}$$

where L is a graph from $G = (X, U)$ and U is the set of the edges of the graph linearly ordered as in natural manner $1, 2, \ldots, m$, and ab is the edge of least number; L_α is the graph obtained from L by removal of ab (without removal of the vertices); L_γ is the graph obtained from L_α by identification of the vertices a and b and removal of all edges, which are neighbouring with vertices a and b and which are not the edges of the triangle inclusive the edge ab; L_ω is the graph obtained from L_α by removal of all edges which are neighbouring with vertices a and b and which are not the edges of the triangle inclusive the edge ab; ε_n is the empty (without edges) graphs with n vertices.

The next theorem holds.

Theorem 1. The function $\Phi(L)$ is single-valued if and only if the following relations hold:

$$\begin{gathered}(\alpha\gamma - \gamma)\alpha = 0, \ \ (\alpha\omega - \omega)\alpha = 0, \ \ (\alpha\gamma - \gamma\alpha)\alpha = 0, \\ (\alpha\omega - \omega\alpha)\alpha = 0, \ (\omega\gamma - \gamma\omega)\alpha = 0, \ \ 2\omega(\gamma + \omega) = 0, \\ \alpha\gamma - \gamma\alpha + \alpha\omega - \omega\alpha = 0, \ \ \alpha\gamma - \gamma + \alpha\omega - \omega = 0\end{gathered} \tag{3}$$

This theorem can be proved in the same manner as the theorems in [10, 11].

Theorem 2. The funcfion $\Phi(L)$ is reducible to the following canonical form:

$$\Phi(L) = \sum_{i=0}^{m-1} \alpha^i + \left[\sum_{j\geq 1} a_j(L)\gamma^j + \sum_{j\geq 1} b_j(L)\omega^j\right]\alpha + \sum_{j,k\geq 1} \delta_j^k(L)\omega^j(\omega+\gamma)^k, \tag{4}$$

where $m = m(L)$ is the number of the edges of $L, and \delta_j^k(L)$ is egual to 0 or 1.

Proof. This statement is proved by the method of mathematical induction on the number $m(L)$ of the edges of the graph.

First note that Theorem 1 implies the following equalities:

$$\omega\gamma^j\alpha = \sum_{r=0}^{j-1}(-1)^r\delta_j^r(L)C_j^r\omega^{r+1}(\omega+\gamma)^{j-r} + (-1)^j\omega^{j+1}\alpha,$$

$$\gamma\omega^j\alpha = \omega^j(\omega+\gamma) - \omega^{j+1}\alpha,\ \ \alpha\delta_j^k(L)\omega^j(\omega+\gamma)^k = \delta_j^k(L)\omega^j(\omega+\gamma)^k,$$
$$\gamma\delta_j^k(L)\omega^j(\omega+\gamma)^k = \delta_j^k(L)\omega^j(\omega+\gamma)^{k+1} - \delta_j^k(L)\omega^{j+1}(\omega+\gamma)^k.$$

Since the graphs L_α, L_ω and L_γ have less edges than L then

$$\Phi(L) = \alpha\Phi(L_\alpha) + \omega\Phi(L_\omega) + \gamma\Phi(L_\gamma) + 1 =$$

$$\alpha\left\{\sum_{i=0}^{m(L_\alpha)-1}\alpha^i + [\sum_{j\geq 1} a_j(L_\alpha)\gamma^j + b_j(L_\alpha)\omega^j]\alpha + \sum_{j,k\geq 1}\delta_j^k(L_\alpha)\omega^j(\omega+k)^k\right\} +$$

$$\gamma\left\{\sum_{i=0}^{m(L_\gamma)-1}\alpha^i + [\sum_{j\geq 1} a_j(L_\gamma)\gamma^j + \sum_{j\geq 1} b_j(L_\gamma)\omega^j]\alpha + \sum_{j,k\geq 1}\delta_j^k(L_\gamma)\omega^j(\omega+\gamma)^k\right\} +$$

$$\omega\left\{\sum_{i=0}^{m(L_\omega)-1}\alpha^i + [\sum_{j\geq 1} a_j(L_\omega)\gamma^j + \sum_{j\geq 1} b_j(L_\omega)\omega^j]\alpha + \sum_{j,k\geq 1}\delta_j^k(L_\omega)\omega^j(\omega+\gamma)^k\right\} + 1$$

$$= [\sum_{i=0}^{m(L_\alpha)-1}\alpha^i + 1] + [\sum_{j\geq 1} a_j(L_\alpha)\gamma^j + \sum_{j\geq 1} b_j(L_\alpha)\omega^j]\alpha + \sum_{j,k\geq 1}\delta_j^k(L_\alpha)\omega^j(\omega+\gamma)^k +$$

$$[m(L_\gamma - 1)]\gamma\alpha + \gamma + [\sum_{j\geq 1} a_j(L_\gamma)\gamma^{j+1} - \sum_{j\geq 1} b_j(L_\gamma)\omega^{j+1}]\alpha +$$

$$\sum_{j\geq 1} b_j(L_\gamma)\omega^j(\omega+\gamma) + \sum_{j,k\geq 1}\delta_j^k(L_\gamma)\omega^j(\omega+\gamma)^{k+1} - \sum_{j,k\geq 1}\delta_j^k(L_\gamma)\omega^{j+1}(\omega+\gamma)^k +$$

$$[m(L_\omega - 1) - 1]\omega\alpha + \omega + [\sum_{j\geq 1}(-1)^j a_j(L_\omega)\omega^{j+1} + \sum_{j\geq 1} b_j(L_\omega)\omega^{j+1}]\alpha +$$

$$\sum_{j\geq 1}\sum_{k=0}^{j-1} a_j(L_\omega)(-1)^k\delta_j^k(L_\omega)\omega^{k+1}(\gamma+\omega)^{j-k} + \sum_{j,k\geq 1}\delta_j^k(L_\omega)\omega^{j+1}(\omega+\gamma)^k =$$

$$\sum_{i=0}^{m(L)-1}\alpha^i + \left\{\sum_{j\geq 1}[a_j(L_\alpha) + a_{j-1}(L_\gamma)]\gamma^j + m(L_\gamma)\gamma\right\}\alpha +$$

$$\left\{\sum_{j\geq 1}[b_j(L_\alpha) - b_{j-1}(L_\gamma) + b_{j-1}(L_\omega) + (-1)^{j-1}a_{j-1})(L_\omega)]\omega^j + m(L_\omega)\omega\right\}\alpha +$$

$$\sum_{j,k\geq 1}\delta_j^k(L_\alpha)\omega^j(\omega+\gamma)^k + \sum_{j,k\geq 1}\delta_j^k(L_\gamma)\omega^j(\omega+\gamma)^{k+1} -$$

$$\sum_{j,k\geq 1}\delta_j^k(L_\gamma)\omega^{j+1}(\omega+\gamma)^k + \sum_{j\geq 1}\sum_{k=0}^{j-1}(-1)^k\delta_j^k(L_\omega)c_j^k\omega^{k+1}(\omega+\gamma)^{j-k} +$$

$$\sum_{j,k\geq 1}\delta_j^k(L_\omega)\omega^{j+1}(\omega+\gamma)^k+\sum_{j\geq 1}b_j(L_\gamma)\omega^j(\omega+\gamma)=\sum_{i=0}^{m(L)-1}\alpha^i$$

$$+\left[\sum_{j\geq 1}a_j(L)\gamma^j+\sum_{j\geq 1}b_j(L)\omega^j\right]\alpha+\sum_{j,k\geq 1}\delta_j^k(L)\omega^j(\omega+\gamma)^k.$$

In order to elucidate the meaning of the coefficients $a_j(L)$ assume $\omega=0,\alpha=1$. This assumption does not give additional restrictions on the generator γ (the conditions of unambiguity (3) are satisfied identically).

Then $\Phi(L)=m(L)+\sum_{j\geq 1}a_j(L)\gamma^j$. Let us consider [10]

$$c_{j,k}(L)=\sum_{r=0}^{j}(-1)^r\tilde{\varphi}(j-r,k+r+1)(L).$$

Then if $k=0$ we obtain:

$$a_j(L)=c_{j,0}(L)=\sum_{r=0}^{j}(-1)^r\tilde{\varphi}(j-r,r+1)(L),$$

where $\tilde{\varphi}^{(j,k)}(L)$ is the number of the equivalence classes of actual (j,k)-placements of the graph L.

Let $\alpha=1$ and $\gamma=-\omega$ (the conditions (3) are again satisfied identically). Then

$$\Phi(L)=m(L)+\sum_{j\geq 1}(-1)^j a_j(L)\omega^j+\sum_{j\geq 1}b_j(L)\omega^j \tag{5}$$

Denote by $r_j(L)$ the number of the complete subgraphs, with j vertices, of the graph L. It is obvious that

$$r_j(L)=r_j(L_\alpha)+r_{j-1}(L_\omega)-r_{j-1}(L_\gamma)(j\geq 3),\quad r_2(L)=r_2(L_\alpha)+1.$$

Therefore, the polynomial $R(L)=\sum_{j\geq 2}r_j(L)\omega^{j-2}$ satisfies the following recurrence relations:

$$R(L)=R(L_\alpha)+\omega R(L_\omega)-\omega R(L_\gamma)+1,\quad R(\varepsilon_n)=0.$$

Whence $R(L)=\Phi(L;\alpha=1,\gamma=-\omega)$, i. e.

$$\sum_{j\geq 2}r_j(L)\omega^{j-2}=m(L)+\sum_{j\geq 1}(-1)^j a_j(L)\omega^j+\sum_{j\geq 1}b_j(L)\omega^j.$$

Then equating the coefficients in both sides of equation, we obtain:

$$r_2(L)=m(L),b_j(L)=r_{j+2}(L)-\sum_{r=0}^{j}(-1)^{j+2}\tilde{\varphi}^{(j-r,r+1)}(L).$$

Now, in order to illustrate our algorithm of computing the recurrence functions and numerical characteristics of graphs, we give the computing of the polynomials $\Phi(L)$ and $R(L)$ for the following graph:

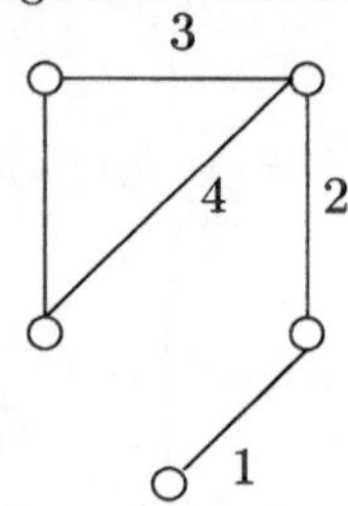

In order to simplify notation let us denote

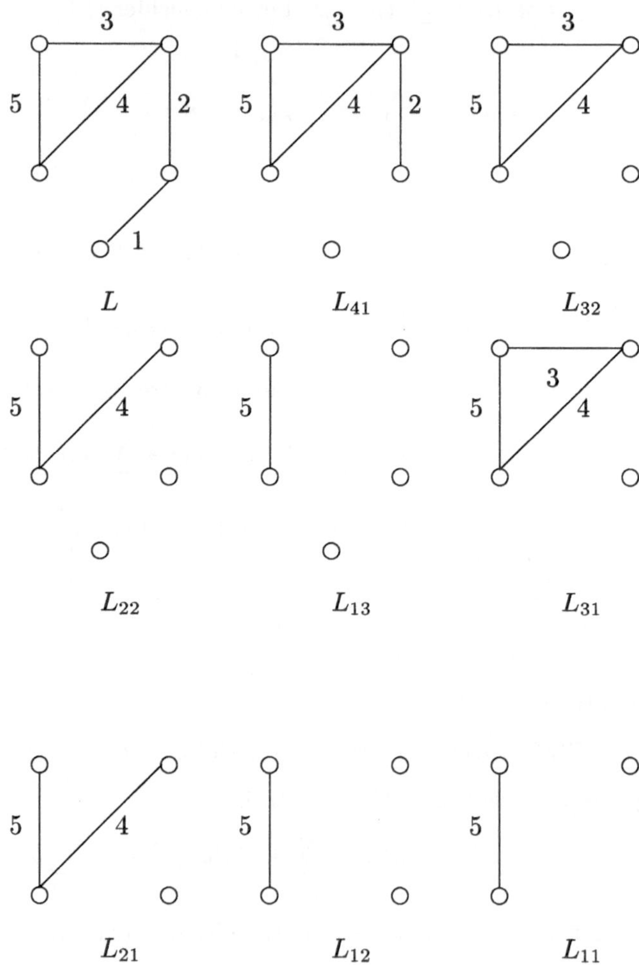

Calculating the function $\Phi(L)$, at the first step we obtain the graphs: $L_\alpha = L_{41}$, $L_\gamma = L_{31}$, $L_\omega = L_{32}$.

At the second step we obtain three graphs from each one: $(L_\alpha)_\alpha = L_{32}$, $(L_\alpha)_\gamma = L_{12}$, $(L_\alpha)_\omega = L_{13}$ etc.

As a result we obtain the decomposition:

$$\begin{aligned}\Phi(L) &= \alpha\Phi(L_{41}) + \gamma\Phi(L_{31}) + \omega\Phi(L_{32}) + 1\\ &= \alpha^2(L_{31}) + \alpha\gamma\Phi(L_{12}) + \alpha\omega\Phi(L_{13}) + \alpha + \gamma\alpha\Phi(L_{21}) + \gamma^2\Phi(L_{11})\\ &\quad + \gamma\omega\Phi(L_{21}) + \gamma + \omega\alpha\Phi(L_{21}) + \omega\gamma\Phi(L_{12}) + \omega^2\Phi(L_{21}) + \omega + 1\end{aligned}$$

$$\begin{aligned}&= \alpha^3\Phi(L_{22}) + \alpha^2\gamma\Phi(L_{12}) + \alpha^2\omega\Phi(L_{22}) + \alpha^2 + \alpha\gamma\alpha\Phi(\varepsilon_4) + \alpha\gamma^2\Phi(\varepsilon_3)\\ &+ \alpha\gamma\omega\Phi(\varepsilon_4) + \alpha\gamma + \alpha\omega\alpha\Phi(\varepsilon_5) + \alpha\omega\gamma\Phi(\varepsilon_4) + \alpha\omega^2\Phi(\varepsilon_5) + \alpha\omega + \alpha\\ &+ \gamma\alpha^2\Phi(L_{12}) + \gamma\alpha\gamma\Phi(\varepsilon_3) + \gamma\alpha\omega\Phi(\varepsilon_4) + \gamma\alpha + \gamma^2\alpha\Phi(\varepsilon_3) + \gamma^3\Phi(\varepsilon_2)\\ &+ \gamma^2\omega\Phi(\varepsilon_3) + \gamma^2 + \gamma\omega\alpha\Phi(L_{12}) + \gamma\omega\gamma\Phi(\varepsilon_3) + \gamma\omega^2\Phi(\varepsilon_4) + \gamma\omega + \gamma\\ &+ \omega\alpha^2\Phi(L_{13}) + \omega\alpha\gamma\Phi(\varepsilon_4) + \omega\alpha\omega\Phi(\varepsilon_5) + \omega\alpha + \omega\gamma\alpha\Phi(\varepsilon_4) + \omega\gamma^2\Phi(\varepsilon_3)\\ &+ \omega\gamma\omega\Phi(\varepsilon_4) + \omega\gamma + \omega^2\alpha\Phi(L_{13}) + \omega^2\gamma\Phi(\varepsilon_4) + \omega^3\Phi(\varepsilon_5) + \omega^2 + \omega + 1\end{aligned}$$

$$\begin{aligned}&= \alpha^4\Phi(L_{13}) + \alpha^3\gamma\Phi(\varepsilon_4) + \alpha^3\omega\Phi(\varepsilon_5) + \alpha^3 + \alpha^2\gamma\alpha\Phi(\varepsilon_4) + \alpha^2\gamma^2\Phi(\varepsilon_3)\\ &+ \alpha^2\gamma\omega\Phi(\varepsilon_4) + \alpha^2\gamma + \alpha^2\omega\alpha\Phi(L_{13}) + \alpha^2\omega\gamma\Phi(\varepsilon_4) + \alpha^2\omega^2\Phi(\varepsilon_5) + \alpha^2\omega + \alpha^2\\ &+ \alpha\gamma + \alpha\omega + \alpha + \gamma\alpha^3\Phi(\varepsilon_4) + \gamma\alpha^2\gamma\Phi(\varepsilon_3) + \gamma\alpha^2\omega\Phi(\varepsilon_4) + \gamma\alpha^2 + \gamma\alpha + \gamma^2\\ &+ \gamma\omega\alpha^2\Phi(\varepsilon_4) + \gamma\omega\alpha\gamma\Phi(\varepsilon_3) + \gamma\omega\alpha\omega\Phi(\varepsilon_4) + \gamma\omega\alpha + \gamma\omega + \gamma + \omega\alpha^3\Phi(\varepsilon_4)\\ &+ \omega\alpha^2\gamma\Phi(\varepsilon_3) + \omega\alpha^2\omega\Phi(\varepsilon_4) + \omega\alpha^2 + \omega\alpha + \omega\gamma + \omega^2\alpha^2\Phi(\varepsilon_5)\\ &+ \omega^2\alpha\gamma\Phi(\varepsilon_4) + \omega^2\alpha\omega\Phi(\varepsilon_5) + \omega^2\alpha + \omega^2 + \omega + 1\end{aligned}$$

$$\begin{aligned}&= \alpha^5\Phi(\varepsilon_5) + \alpha^4\gamma\Phi(\varepsilon_4) + \alpha^4\omega\Phi(\varepsilon_5) + \alpha^4 + \alpha^3 + \alpha^2\gamma + \alpha^2\omega\alpha^2\Phi(\varepsilon_5)\\ &+ \alpha^2\omega\alpha\gamma\Phi(\varepsilon_4) + \alpha^2\omega\alpha\omega\Phi(\varepsilon_5) + \alpha^2\omega\alpha + \alpha^2\omega + \alpha^2 + \alpha\gamma + \alpha\omega + \alpha + \gamma\alpha^2\\ &+ \gamma\alpha + \gamma^2 + \gamma\omega\alpha + \gamma\omega + \gamma + \omega\alpha^2 + \omega\alpha + \omega\gamma + \omega^2\alpha + \omega^2 + \omega + 1\end{aligned}$$

$$\begin{aligned}&= \alpha^4 + \alpha^3 + \alpha^2\gamma + \alpha^2\omega\alpha + \alpha^2\omega + \alpha^2 + \alpha\gamma + \alpha\omega + \alpha\\ &+ \gamma\alpha^2 + \gamma\alpha + \gamma^2 + \gamma\omega\alpha + \gamma\omega + \gamma + \omega\alpha^2 + \omega\alpha + \omega\gamma + \omega^2\alpha + \omega^2 + \omega + 1\end{aligned}$$

$$\begin{aligned}&= \alpha^4 + \alpha^3 + \alpha^2 + \alpha + 1 + \alpha^2(\gamma + \omega) + \omega\alpha + \alpha(\gamma + \omega)\\ &+ \gamma\alpha + \gamma\alpha + (\gamma + \omega)^2 + \omega\gamma\alpha + \gamma + \omega + \omega\alpha + \omega\alpha + \omega^2\alpha\end{aligned}$$

$$\begin{aligned}&= \alpha^4 + \alpha^3 + \alpha^2 + \alpha + 1 + (\gamma + \omega)\alpha + \omega\alpha + (\gamma + \omega)\alpha\\ &+ \gamma\alpha + \gamma\alpha + (\gamma + \omega)^2\alpha + \omega(\gamma + \omega)\alpha - \omega^2\alpha + (\gamma + \omega)\alpha + \omega\alpha + \omega\alpha + \omega^2\alpha\end{aligned}$$

$$\begin{aligned}&= \alpha^4 + \alpha^3 + \alpha^2 + \alpha + 1 + (\gamma + \omega + \omega + \gamma + \omega + \gamma + \gamma\\ &+ \gamma^2 - \omega^2 - \omega^2 + \gamma + \omega + \omega + \omega + \omega^2)\alpha + \omega(\gamma + \omega)\\ &= \alpha^4 + \alpha^3 + \alpha^2 + \alpha + 1 + (\gamma^2 + 5\gamma - \omega^2 + 6\omega)\alpha + \omega(\gamma + \omega).\end{aligned}$$

$$R(L) = \omega + 5.$$

Indeed the graph L has one triangle (complete subgraph with three vertices) and five edges (complete subgraphs with two vertices).

References

1. Birkhof, G. D.: A determinant formula for the number of ways of coloring a map. *Ann. of Math.* **14** (1912) 42–46
2. Birkhof, G. D.: On the number of ways of coloring a map. *Prog. Edinburg Math. Soc.* **2** (1930) 83–91
3. Whitney, H.: The coloring of graphs. *Ann of Math.* **33** (1932) 668–718
4. Whitney, H.: A logical expansion in mathematics. *Bull. Amer. Math. Soc.* **38** (1932) 572–579
5. Whitney, H.: A set topological invariants for gfaphs. *Amer. J. of Math.* **55** (1938) 231–235
6. Tutte, W.T.: A contribution to the theory of chromatic polinomials. *Canadian J. of Math.* **6** (1954) 80–91
7. Zykov, A. A.: Functions of edges and vertices and distributive properties of graphs. *DAN SSSR* **139** (1961) 787–790 (in Russian)
8. Zykov, A. A.: Recursively calculable functions of graphs. Theory of graphs and its applications. In: *Proc. Sympos. held in Smolenice in June 1963* (1963) 99–105
9. Ergashev, G. E.: Structural meaning of some reccurence functions of the graphs. *Applied Mathematics and Programming*, Kishinev **7** (1972) 140–157
10. Ergashev, G. E.: Investigation of some reccurence function of graphs. In: *Problems of Algebra, Number Theory, Differential and Integral Equations*, Samarkand, SSU (1973) 115–123
11. Ergashev, G. E.: Structural meaning of some reccurence functions of the multigraphs. *J.Kibernetika*, Kiev, No. 5 (1979) 15–20
12. Ergashev, G. E.: The numerical characteristics of the graphs with edges of two types. In: *Abstracts of the International Conference "Nonclassic and Ill-Posed Problems of Mathematical Physics and Analysis"*, Samarkand (2000) 95

A New Combinatorial Algorithm for Large Markov Chains (Extended Abstract)

Anna Gambin[1]* and Piotr Pokarowski[2]

[1] Institute of Informatics, email: aniag@mimuw.edu.pl
[2] Institute of Applied Mathematics, email: pokar@mimuw.edu.pl
Warsaw University Banacha 2, 02-097 Warsaw, Poland
tel: +48 22 55 444 06 fax: +48 22 55 444 00

Abstract. We consider large Markov chains which posses specific decomposable structure, the so called *Nearly Completely Decomposable Chains* (NCD chains). A new theoretical approach for approximate computations of NCD chains has been recently introduced in [11, 12]. The method of *forest expansions* gives raise to aggregation algorithms, which approximate effectively the characteristics of Markov chain. The algorithms are based on grouping the states of a Markov chain in such a way that the probability of changing the state inside the group is of greater order of magnitude than interactions between groups. In [5] the algorithm approximating stationary distribution (described by transposed system $\mathbf{L}^T\mathbf{x} = \mathbf{b}$, where $\mathbf{L}$ is derived from transition matrix) was presented; in this paper we illustrate that combinatorial aggregation in the case of non-transposed system (defining e.g. *mean hitting time*) is not less effective. This novel approach allows us to treat both types of problems in the unified manner. To our knowledge for the first time an aggregation scheme was used to calculate Markov chain characteristics other than stationary distribution.

Keywords: Markov chains, decomposability, aggregation, mean hitting time.

1 Introduction

It is often possible to represent the behavior of a physical system by describing all the different states which it can occupy and by indicating how it moves from one state to another in time. If the future evolution of the system depends only on its current state, the system may be represented by a Markov process. When the state space is discrete, the term "Markov Chain" is employed.

Markov chains are used so frequently in various areas of computer science and in other disciplines that it is not difficult to justify their importance. Making use of a suitable defined Markov chain which model the system of interest one

* This work was partially supported by the KBN grant 8 T11C 039 15

is able to locate bottlenecks in communication network; to assess the benefit of increasing the number of CPUs in multiprocessor systems and to quantify the effect of scheduling algorithms on throughput [16]. Markov modeling, in particular reliability modeling, is used to estimate the mean time to failure of components in systems as diverse as software system and aerospace systems.

We propose a new combinatorial aggregation approach which is applicable when the state space of Markov chain is large enough so that the direct methods (e.g. Gauss elimination) are inefficient. It is based on lumping together closely related states. This approach is attractive in that case, because the structure of transition matrix implies that standard iterative methods converge very slowly (see discussion in [16, 17] and the references therein).

Formally, a Markov chain with the state space $S = \{1, \ldots, s\}$ is defined as a sequence $X = (X_t)_{t \in \mathbb{N}}$ of random variables such that the probability of $X_i = j$ depends only on X_{i-1} (for details, we refer to [8, 10]). Thorough this paper, we assume that Markov chains are given by a probability transition matrices — the only difference between these two ways of introducing a Markov chain is that given X, we have a distinguished initial distribution X_0, which is not given in a probability transition matrix.

It was observed that many facts are valid simultaneously for both discrete and continuous time Markov chains. To deal with them at the same time we use, following [11], a **laplacian matrix**, i.e. matrix $\mathbf{L} = (l_{ij})_{i,j=1}^{s}$, $l_{ij} \in \mathbb{R}$ satisfying $l_{ii} = -\sum_{j:\, j \neq i} l_{ij}$ for $i = 1, \ldots, s$. Denote by $\mathbf{I}$ the identity matrix of size s. Let $\mathbf{P}$ be the **transition probability matrix** of a Markov chain. The (i, j)-th element of $\mathbf{P}$, denoted $p_{i,j}$, is the one-step transition probability of going from state i to state j. Matrix $\mathbf{L} = \mathbf{I} - \mathbf{P}$ is a laplacian matrix induced by $\mathbf{P}$. From now on we assume that the Markov chain is introduced by the **Markov chain laplacian matrix** i. e., laplacian matrix with non-positive real off-diagonal entries.

Many characteristics of Markov chains are solutions of systems of linear equations in one of the following form:

$$\mathbf{L}(R|R)\mathbf{x} = \mathbf{b} \tag{1}$$

$$\mathbf{L}^T(R|R)\mathbf{x} = \mathbf{b}, \tag{2}$$

where R is a subset of states, $\mathbf{b}$ is a nonnegative vector of size $s - |R|$ and $\mathbf{L}(R|R)$ is a submatrix of $\mathbf{L}$ resulting from deletion of rows and columns indexed by R; $\mathbf{L}^T$ denotes transposition. The most elegant way to deal with (1) and (2) is to find the analytical formulas for the solution of the system. Unfortunately, it is usually impossible and the only way is to solve the problem numerically [16]. Problems arise from the computational point of view because of the large number of states which systems may occupy. It is not uncommon for thousands of states to be generated even for simple applications. On the other hand these Markov chains are often sparse and possess specific structure.

Motivation and related research. A lot of research has been done concerning the numerical solutions of some linear equations that occur when one studies Markov chains (see for example [2, 16]). Almost all methods for solving a system of linear equations are adapted into this context: iterative and direct methods, projection techniques and the concept of preconditioning (see [17]). Some of them turn out to be quite efficient, others have serious drawbacks. The applicability of a method depends strongly on the structure of the Markov chain considered (see discussion in [3]). In this paper we focus on *nearly uncoupled* or *nearly completely decomposable* Markov chains (see [1, 16]). Such chains often arise in queueing network analysis, large scale economic modeling and computer systems performance evaluation. The state space of these chains can be naturally divided into groups of states such that transitions between states belonging to different groups are significantly less likely than transitions between states within the same group.

As a example consider a NCD Markov chain defined by the following matrix:

$$\mathbf{L} = \begin{pmatrix} \mathbf{L}_{11} & \mathbf{L}_{12} & \dots & \mathbf{L}_{1N} \\ \mathbf{L}_{21} & \mathbf{L}_{22} & \dots & \mathbf{L}_{2N} \\ \vdots & \vdots & \ddots & \vdots \\ \mathbf{L}_{N1} & \mathbf{L}_{N2} & \dots & \mathbf{L}_{NN} \end{pmatrix}$$

where $\mathbf{L}_{11}, \mathbf{L}_{22}, \dots, \mathbf{L}_{NN}$ are square subblocks. Assume that $\mathbf{L}$ is of the form:

$$\mathbf{L} = \mathrm{diag}(\mathbf{L}_{11}, \mathbf{L}_{22}, \dots, \mathbf{L}_{NN}) + \mathbf{E}$$

where component $\mathbf{E}$ incorporates all off-diagonal blocks. In the sequel we assume that the submatrices $\mathbf{L}_{ii}$ are of moderate size, hence we apply the direct method (a modification of the standard Gaussian elimination called GTH after Grassmann-Taksar-Heyman [6]) to solve the systems of the form $\mathbf{L}_{ii}\mathbf{x} = \mathbf{b}$, arising from decomposition of large NCD Markov chain. In Section 3 we discuss the possibility to speedup the computation by choosing some iterative method instead.

For solving NCD Markov chains the methods known as *aggregation* and *iterative aggregation/disaggregation* are the most suitable. Our combinatorial aggregation approach can be seen as a generalization of existing aggregation algorithms (c. f. [9]). In [4, 5] we summarize the most important advantages over previous methods. Here we emphasize the following aspects:

- All known aggregation methods considered in the literature are designed only to solve the problem of stationary distribution, and other characteristics of Markov chain are neglected. In contrast to this, we propose the procedure approximating the **mean hitting time**. Similar algorithms can be obtained for other characteristics being the solution of (1). For the first time the aggregation approach is successful in computing Markov chain characteristics which are solutions of **non-transposed** systems of equations.
- Comparison with GTH procedure shows the height precision level of approximation computed by our algorithms.

Structure of the paper. In Section 2 a mathematical theory behind the algorithms is sketched (see [11, 12] for more detailed treatment). The following section contains the description of the algorithm and the discussion of its complexity. In Section 4 we report results of numerical experiments we have performed — they are very promising and clearly justify the applicability of our approach for NCD Markov chains. Final remarks and some open problems can be found in the last section.

2 Mathematical Preliminaries

This section is devoted to mathematical preliminaries crucial for aggregation algorithms. We define here the Markov chain characteristics we are interested in. It is argued that to compute them, one needs to solve certain systems of linear equations. Then, we recall several known results enabling to express the solution of systems of linear equations by means of weighted directed forests (in the underlying graph). Having such *forests expansion* of a characteristics under consideration, we are looking for an effective procedure to compute it. To this end the concept of *powerly perturbed Markov chain* [11] is used, as in this case there exist recursive formulas enabling to evaluate effectively the forest expansions. This procedure yields an approximation of characteristics considered.

Let $G = (S, E)$ be a directed graph with the set of vertices (called also states) $S = \{1, 2, \ldots s\}$, for $s \geq 1$. The classification of states in a graph follows the Markov chain terminology (c.f. [8, 10] for a detailed treatment of Markov chain theory). A **strong component** of G is any maximal subgraph C of G, with the property that for any two states i, j of C there exists a path from i to j. A strong component is **absorbing** if it has no outgoing edges. Strong absorbing components are also called **closed classes** in the sequel. An acyclic subgraph $f = (S, E_f)$ of G containing all its vertices, in which any state has out-degree at most 1 is called **a directed spanning forest**. A set of states $R \subseteq S$ with no outgoing edges in E_f forms a **root** of a forest. When the root is singleton we talk about **directed spanning tree**. We write shortly forest (tree) instead of directed spanning forest (tree).

Let $\mathcal{F}(R)$ denote the set of all forests in G having the root R (a forest is identified with the set of its edges). For fixed $i \notin R$ and $j \in R$, $\mathcal{F}_{ij}(R) \subseteq \mathcal{F}(R)$ denotes the set of all forests with the root R, containing a path from i to j.

Now we assume G to be an underlying graph of Markov chain induced by a square real matrix A of size s (i. e. there are edges between all pairs (i, j) with $a_{ij} \neq 0$). We define the **(multiplicative) weight** of a forest $f = (S, E_f)$ and the weight of a set $\mathcal{F}$ of forests as follows:

$$w(f) = \prod_{(i,j)\in E_f} (-a_{ij}), \qquad w(\mathcal{F}) = \sum_{f\in\mathcal{F}} w(f).$$

Let $R \subseteq S$, $i, j \notin R$, and $k \in R$; Markov chain characterisctics we consider in the sequel are defined as follows:

$$
\begin{aligned}
m_i(R) &:= \mathbf{E}_i\tau_R \text{ — the } \textbf{mean hitting time} \text{ of } R,\\
\mu_{ij}(R) &:= \mathbf{E}_i\left[\sum_{0\le t<\tau_R} \mathbf{1}(X_t=j)\right] \qquad \text{where} \qquad \mathbf{1}(\phi) := \begin{cases} 1 & \text{if } \phi \text{ is true,}\\ 0 & \text{otherwise.}\end{cases}\\
&\qquad \text{— the } \textbf{mean number of visits before absorption}\\
p_{ik}(R) &:= \mathbf{Pr}_i\{X_{\tau_R}=k\}\\
&\qquad \text{— the } \textbf{probability distribution in the hitting time} \text{ of } R,\\
\mathbf{Pr}_i(A) &:= \mathbf{Pr}(A|X_0=i),\\
\tau_R &:= \min\{t\ge 0 : X_t\in R\} \text{ — the hitting time of the set } R,\\
\mathbf{E}_iA &:= \mathbf{E}(A|X_0=i) \text{ — expectation conditioned by starting from state } i.
\end{aligned}
$$

The **mean hitting time** of R can be calculated by solving following system of linear equation of the form (1) ($\mathbf{m} = (m_i(R))_{i\in S\setminus R}$):

$$\mathbf{L}(R|R)\mathbf{m} = \mathbf{e}; \tag{3}$$

where $\mathbf{e} = (11\dots1)^T$. Two other characteristics can be also presented as solutions of non-transposed systems of the form (1).

Famous Markov Chain Tree Theorem [14] characterizes stationary distribution by a rational function of directed forest weights called the **forest expansion**. A result of [11] extends MCT Theorem to the characteristics introduced above:

$$
\begin{aligned}
m_i(R) &= \frac{\sum_{j\notin R} w(\mathcal{F}_{ij}(R\cup\{j\}))}{w(\mathcal{F}(R))};\\
\mu_{ij}(R) &= \frac{w(\mathcal{F}_{ij}(R\cup\{j\}))}{w(\mathcal{F}(R))};\\
p_{ik}(R) &= \frac{w(\mathcal{F}_{ik}(R))}{w(\mathcal{F}(R))}.
\end{aligned}
$$

In Section 3 we present the combinatorial aggregation algorithms approximating vector $\mathbf{m} = m_i(R)$. The analogous constructions for $\mathbf{p} = p_{ik}(R)$ and $\boldsymbol{\mu} = \mu_{ij}(R)$ is discussed.

The concept of powerly perturbed Markov chains [11] is crucial for effective approximation of the solution of system $\mathbf{L}(R|R)\mathbf{x} = \mathbf{b}$ ($\mathbf{L}^T(R|R)\mathbf{x} = \mathbf{b}$). It subsumes all previously known generalizations of NCD Markov chains, aiming in expressing several different orders of magnitude of interaction strength (see for example [7]).

For given functions $A, B : \mathbb{R}\to\mathbb{R}$, the notation $A(\varepsilon)\sim B(\varepsilon)$ means that: $\lim_{\varepsilon\to 0}\frac{A(\varepsilon)}{B(\varepsilon)} = 1$. We also set $A(\varepsilon)\sim 0$, if there exists $\varepsilon_1\neq 0$ such that for any $\varepsilon\in(-\varepsilon_1,\varepsilon_1)$, $A(\varepsilon)=0$.

A family $\{\mathbf{L}(\varepsilon) = (l_{ij}(\varepsilon))_{i,j=1}^{s},\ \varepsilon \in (0, \varepsilon_1)\}$ of laplacian matrices of size $s \times s$ is a **powerly perturbed** Markov chain, if there exist matrices $\boldsymbol{\Delta} = (\delta_{ij})_{i,j \in S}$, and $\mathbf{D} = (d_{ij})_{i,j \in S}$, $\delta_{ij} \geq 0$ and $d_{ij} \in \mathbb{R}$, for $i, j \in S$, such that the asymptotic behavior of laplacians $\mathbf{L}(\varepsilon)$ is determined by $\boldsymbol{\Delta}$ and $\mathbf{D}$ as follows:

$$-l_{ij}(\varepsilon) \sim \delta_{ij} \varepsilon^{d_{ij}}. \tag{4}$$

Powerly perturbed nonnegative vector is defined as the family $\{\mathbf{b}(\varepsilon),\ \varepsilon \in (0, \varepsilon_1)\}$ of nonnegative vectors of size u, such that for some vectors $\boldsymbol{\zeta} = (\zeta_i)_{i=1}^{u}$ and $\mathbf{z} = (z_i)_{i=1}^{u}$, with $\zeta_i \geq 0$, $z_i \in \mathbb{R}$, for $i = 1, \ldots, u$, the following holds:

$$b_i(\varepsilon) \sim \zeta_i \varepsilon^{z_i}. \tag{5}$$

Consider the following graph induced by matrix $\mathbf{D}$ (we take into account asymptotically nonzero entries):

$$G^*(\mathbf{D}) = (S, \{(i,j) \in S \times S : \ \delta_{ij} \neq 0\}).$$

For an arbitrary forest f and a nonempty set $\mathcal{F}$ of forests in $G^*(\mathbf{D})$ we study parameters:

(i) $\left\{ \begin{aligned} d(f) &:= \sum_{(i,j) \in f} d_{ij} \\ \delta(f) &:= \prod_{(i,j) \in f} \delta_{ij} \end{aligned} \right\}$ an asymptotic weight of the forest f.

(ii) $\left\{ \begin{aligned} d(\mathcal{F}) &:= \min_{f \in \mathcal{F}} d(f) \\ \delta(\mathcal{F}) &:= \sum_{f \in \mathcal{F}:\, d(f) = d(\mathcal{F})} \delta(f) \end{aligned} \right\}$ an asymptotic weight of the set of forests F.

We describe the asymptotics of solutions of systems $\mathbf{L}(R|R)\mathbf{x} = \mathbf{b}$ and $\mathbf{L}^T(R|R)\mathbf{x} = \mathbf{b}$, related to powerly perturbed Markov chains, in terms of directed forests expansions (i.e. the rational function of forests' weights). The following theorem, says that a solution of a system of linear equations, for a perturbed chain, can be treated as a perturbed vector.

Theorem 1 ([11]). *Let matrices Δ and $\mathbf{D}$ be such that (4) above holds, for a powerly perturbed Markov chain $\{\mathbf{L}(\varepsilon), \varepsilon < \varepsilon_1\}$; let $R \subseteq S$, where S is a set of states. Moreover let vectors ζ and $\mathbf{z}$ of size $u := s - |R|$ be such that (5) holds, for a powerly perturbed vector $\mathbf{b}$. Suppose that there exist a forest with the root R in $G^*(\mathbf{D})$. Then the following hold:*

(1) the solution $\mathbf{x}(\varepsilon) = (x_i(\varepsilon))_{i \in S \setminus R}$ of the system $\mathbf{L}^T(R|R)(\varepsilon)\mathbf{x}(\varepsilon) = \mathbf{b}(\varepsilon)$ satisfies for $i \in S \setminus R$ the relation $x_i(\varepsilon) \sim \eta_i \varepsilon^{h_i}$, for some constants η_i, h_i;
(2) the solution $\mathbf{x}(\varepsilon) = (x_i(\varepsilon))_{i \in S \setminus R}$ of the system $\mathbf{L}(R|R)(\varepsilon)\mathbf{x}(\varepsilon) = \mathbf{b}(\varepsilon)$ satisfies for $i \in S \setminus R$ the relation $x_i(\varepsilon) \sim \beta_i \varepsilon^{b_i}$, for some constants β_i, b_i.

This theorem can be seen as a generalization of the Markov Chain Tree Theorem in three respects. First, powerly perturbed Markov chains are considered. Second, a general class of problems is taken into account. And third, part *(2)* deals with non-transposed matrices. From the proof of this theorem we can deduce the asymptotic forest expansions for Markov chain characteristics in terms of parameters $d(\mathcal{F})$ and $\delta(\mathcal{F})$. Both cases *(1)* and *(2)*, although described similarly, differ substantially in difficulty(cf. [4, 5]). — this is visible in different structures of algorithms. While in *(1)* it is sufficient to consider spanning trees rooted in some state i, in *(2)* it is necessary to take into account spanning forests rooted in $R \cup \{j\}$.

Unfortunately, as we will see all obtained expressions for asymptotic coefficients (i.e. constants η, h, β, b from Theorem above) are computationally non-tractable, at least directly, because of their exponential length. In the next section we discuss the aggregation approach, yielding effective and accurate procedures for computing the asymptotic coefficients and approximate values of the interesting characteristics of NCD Markov chain.

3 Aggregation Algorithm for Mean Hitting Time

The mean hitting time ($\mathbf{m}$) is an important characteristics of the Markov chain defined by the system of linear equations with a non-transposed matrix. It is characterized by a subset of states $R \subseteq S$ and the task is to approximately calculate the expected number of steps before reaching this set.

Applying Theorem 1 we derive $m_i(R)(\varepsilon) \sim \beta_{iR}\varepsilon^{b_{iR}}$, where asymptotic coefficients $\boldsymbol{\beta} = (\beta_{iR})$ and $\mathbf{b} = (b_{iR})$ are defined as follows:

$$a_i := \min_{j \notin R}[d(\mathcal{F}_{ij}(R \cup \{j\}))], \qquad b_{iR} := a_i - d(\mathcal{F}(R)),$$

$$\beta_{iR} := \frac{\sum_{j \notin R: d(\mathcal{F}_{ij}(R \cup \{j\}))=a_i} \delta(\mathcal{F}_{ij}(R \cup \{j\}))}{\delta(\mathcal{F}(R))}.$$

We present the algorithm which computes asymptotic coefficients $\boldsymbol{\beta}$ and $\mathbf{b}$ — they can be used to approximate $\mathbf{m}$ (c. f. Section 5). The main idea of the algorithm is to reduce the size of state-space of a Markov chain by lumping together closely related states. This process is repeated in the consecutive phases of aggregation; during each phase graphs induced by matrix $\mathbf{D}$ is considered. The algorithm groups states in each closed class of the graph and solves the system of linear equations restricted to this class. Smaller size, hence tractable, systems of equations can be solved by a direct method. Before passing to a next phase, an aggregation procedure is performed, lumping all states in each closed class into a new, aggregated state.

The algorithm consists of two phases. The first one runs the aggregation scheme, similarly as for transposed case (c.f. [4, 5]); however, the aggregation can leave several closed classes of the original graph not lumped together. Then, the second phase calculates coefficients b_{iR} and β_{iR} for closed classes resulted from first phase (see the description of Algorithm2). Computed values are then propagated throughout each class.

Correctness of the algorithm. The interesting observation is that the task of computing exponents b_{iR} (h_i) is of quite different nature than the task of computing the coefficients β_{iR} (η_i). While the former can be performed using purely combinatorial methods (hence precisely), the latter uses a procedure of solving a system of linear equations, exposed to numerical errors. Although calculating coefficients β_{iR} is of crucial importance, in the sequel we concentrate on b_{iR} and h_i. We overview here the process of aggregation and state some facts on **shortest forests**, useful in computing exponents h_i (b_{iR}). Note that, the coefficients h_i and η_i play an auxiliary role in the algorithm — they are necessary during the aggregation phase and alow to construct an aggregated underlying graph.

Consider the graph $G := G^*(\mathbf{D})$ and its subgraph G_{min}, consisting of the **shortest** edges outgoing from each vertex, i.e., for each vertex i, of those d_{ij} which are equal to

$$m(i) := \min_j d_{ij}. \tag{6}$$

Shortest edges correspond to the largest probability of moving from state i to j. Recall that $\mathbf{D} = \{d_{ij}\}$. In a single step of the aggregation process, the graph G is replaced by another graph $G' = aggr(G)$. Vertices of G' are closed classes I of G_{min} together with transient states in G_{min}. Edges (I, J) in G' are weighted by d_{IJ} defined by the following formula:

$$d_{IJ} := \min_{i\in I, j\in J}(d_{ij} + h(i|I)) \quad \text{where} \quad h(i|I) := \max_{k\in I} m(k) - m(i). \tag{7}$$

Values $h(i|I)$ are computed in aggregation precedure — they correspond to coefficients h_i in a graph induced by a closed class I. Lemma 1 below justifies such an aggregation scheme in order to calculate h_i. Let i denote any state of G such that there exists some tree rooted in i ($\mathcal{F}(i) \neq \emptyset$). By a **shortest tree** rooted in i we mean any tree rooted in i such that $d(f)$ is minimal, i.e., $d(f) = \min_{f'\in\mathcal{F}(i)} d(f') = d(\mathcal{F}(i))$.

Lemma 1 ([11]). *Let f be a shortest tree in G, rooted in i, and let I be a closed class in G_{min} containing i, i.e. $i \in I$. Let f_I be a shortest tree in the subgraph of G_{min} induced by I, rooted in i. Moreover, let f' be a shortest tree in G', rooted in I. The following holds (for simplicity, we apply here notation $d(\_)$ to graph G' as well):*

$$d(f) = d(f_I) + d(f').$$

For a non-transposed case, we need a similar fact for forests.

Lemma 2 ([11]). *Let R be a subset of states of G s.t. $\mathcal{F}(R)$ is nonempty. Let $I \neq J$ be closed classes or transient states in G_{min} s.t. R and $I \cup J$ are disjoint. Fix a state $i \in I$. Let f_j be a shortest tree in the subgraph of G_{min} induced by J, rooted in j. The following holds (similarly as before, we extend here notation $d(\_)$ and $\mathcal{F}\_(\_)$ to graph G'):*

$$\min_{j\in J} d(\mathcal{F}_{ij}(R \cup \{j\})) = \min_{j\in J} d(f_j) + d(\mathcal{F}_{IJ}(R \cup \{J\})).$$

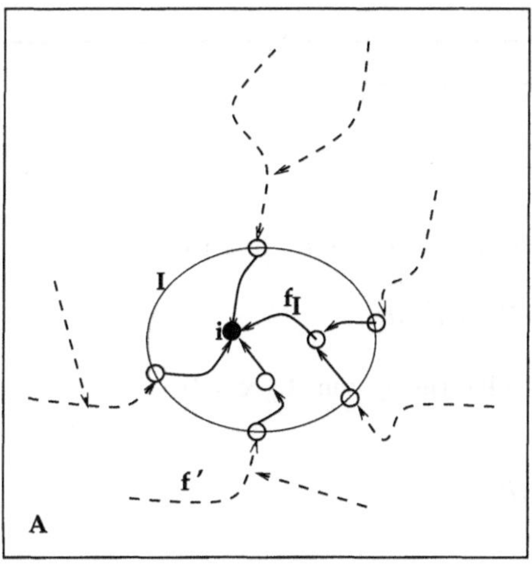

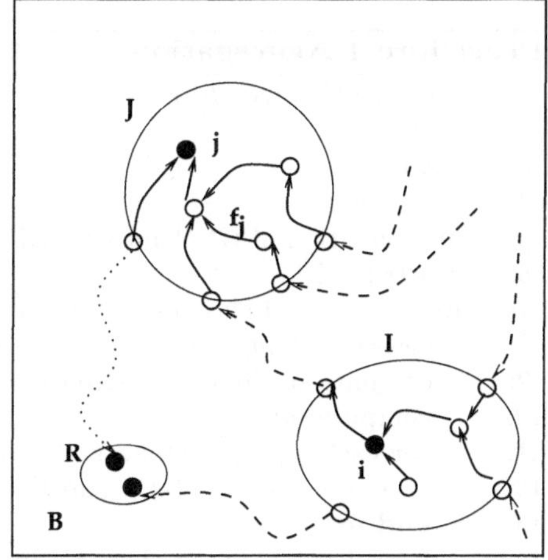

Fig. 1. A: shortest tree rooted in i from Lemma 1, B: shortest forest rooted in $R \cup j$ from Lemma 2

We consider the following aggregation process, which gives rise to the sequence of graphs $G^i = (S^i, E^i)$, for $i = 0, 1, \ldots$; starting from $i = 1$, the superscript i enumerates consecutive phases of Procedure 1. Initially, define the graph G^0 which differs slightly from that in transposed case. Namely, all states from R are replaced by a new state u_R; it has no outgoing edges and for any $i \notin R$, an edge from i to u_R is weighted by $\min_{j \in R} d_{ij}$ (this means that we are interested in reaching *any* of states of R). Moreover, we add one more state u^*, and set $d_{iu^*} := 0$, $\delta_{iu^*} := 1$ for all $i \notin R \cup \{u_R\}$. Intuitively, this new state corresponds to the right-hand side vector $\mathbf{e}$ in equation (3). Notice that edges leading to u^* are ignored, during the construction of a new set of aggregated states (steps 5 and 6 in Procedure 1). However they are updated in step 18 according to the same rule as other edges (cf. formula (7)). For $k = 1, 2, \ldots$, we define inductively $G'_k = aggr(G_{k-1})$. Now, as a new graph G_k we take $(G'_k)_{min}$, whose states are the same as in G'_k and whose edges are the shortest edges in G'_k.

Denote by $\eta(i|I^k)$ and $h(i|I^k)$ coefficients η_i and h_i computed in the subgraph restricted to some closed class I^k. Following this convention, $\eta^k(I^{k-1}|I^k)$ and $h^k(I^{k-1}|I^k)$ correspond to the η and h coefficient for the aggregated state I^{k-1} computed during the k-th phase for the subgraph of G^k restricted to a closed class I^k.

From Lemmas 1 and 2 one can derive following recursive relation:

$$h(i|I^k) = h(i|I^{k-1}) + h^k(I^{k-1}|I^k), \quad \eta(i|I^k) = \eta(i|I^{k-1})\eta^k(I^{k-1}|I^k). \tag{8}$$

Coefficients $h^k(I^{k-1}|I^k)$ can be computed using the value of $m(I^k)$ (step 12): $h^k(I^{k-1}|I^k) = \max_{I \subseteq I^k} m(I) - m(I^{k-1})$. Using (8) we can justify the aggregation scheme: in a given iteration we have only to consider all vertices I aggregated during the previous step, which belong to the closed class I^k. Procedure 1 results

Procedure 1 Aggregation

1: construct $G^0 = (S^0, E^0)$
2: $k := 0$
3: **repeat**
4:   $k := k + 1$
5:   find partition of G^{k-1} into closed classes (ignore edges leading to u^*)
6:   construct S^k
7:   **for** each closed class in S^k, say I^k, s.t. $I^k \neq u^R$ **do**
8:     construct laplacian $\mathbf{L}_k$
9:     compute stationary distribution i. e. solve the system $\mathbf{L}_k^T \mathbf{x} = \mathbf{b}$
10:     compute $m(I^k)$
11:     **for** each aggregated state I^{k-1} in I^k **do**
12:       compute $\eta^k(I^{k-1}|I^k)$ and $h^k(I^{k-1}|I^k)$
13:     **end for**
14:     **for** all neighbors of class I^k **do**
15:       determine shortest edges
16:     **end for**
17:   **end for**
18:   construct new set of aggregated edges E^k (update edges leading to u^*)
19:   $G^k := (S^k, E^k)$
20: **until** G^k has the same number of states as G_{k-1}

in aggregated graph $G^k = (S^k, E^k)$ having the set of closed classes as vertices ($S^k = \{I^k, J^k \dots\}$) and aggregated edges ($d_{I^k J^k} \in E^k$). Aiming in computing coefficients $\mathbf{b}$ and β (i. e. approximating $\mathbf{m}$) we run the second phase of Algorithm 2. Recall that $m(I^k)$ denote the weight of the shortest edge leaving I^k.

Finally, we need to explain how to compute effectively $\eta^k(I^{k-1}|I^k)$ in step 12 of Procedure 1 and β_{iR} in step 18 of Algorithm 2. Recall that we have assumed that the Markov chain under consideration possesses a specific block structure, namely the sizes of all closed classes I^k are small compared with the size of the whole state space. It opens the possibility of using direct methods for solving systems of the form $\mathbf{L}_k \mathbf{x} = \mathbf{b}$ ($\mathbf{L}_k^T \mathbf{x} = \mathbf{b}$), independently inside each class I^k. For the solution the following holds: $x_k(\varepsilon) \sim \beta_{kR} \varepsilon^{b_{kR}}$ ($x_k(\varepsilon) \sim \eta^k \varepsilon^{h^k}$). Now having already computed b_{kR} (h^k) and putting some fixed ε_0, the corresponding coefficients η^k are derived as the solution of the equation: $\beta_{I^k R} = x_{I^k R} \varepsilon_0^{-b_{I^k R}}$ ($\eta^k(I^{k-1}|I^k) = x_{I^{k-1}} \varepsilon_0^{-h^k(I^{k-1}|I^k)}$).

Complexity issues. The upper bound on time cost of combinatorial aggregation is $\mathcal{O}(n^3)$, where n is the number of states. The upper bound on memory needed is $\mathcal{O}(n^2)$. However, the cost of algorithm depends strongly on the structure of a Markov chain under consideration. In [4] we study in detail some important cases. The main conclusion is that if we can profit from a specific structure of a matrix (e.g. if we choose an appropriate ε), time $\mathcal{O}(n^2)$ is sufficient. Moreover, when a matrix is sparse, i.e. the number of edges m is significantly smaller

Algorithm 2 Calculate asymptotic coefficients b_{iR} and β_{iR} i.e. approximate the mean hitting time $m_i = \beta_{iR}\varepsilon^{b_{iR}}$

run **Procedure 1**
for each closed class in S^k, say I^k, s.t. $I^k \neq u^R$ **do**
    compute $b_{I^k R} := d_{I^k u^*} - m(I^k)$
end for
repeat
    remove from S^k classes I^k having minimal $b_{I^k R}$ value, denote the set of removed classes by M
    for each $I^k \in M$ and any state $i \in S^0$ belonging to I^k **do**
        $b_{iR} := b_{I^k R}$
    end for
    for each I^k remaining in S^k **do**
        let $n(I^k) := \min_{J^k \in M}(d_{I^k J^k} - m(I^k) + b_{J^k R})$
        compute $b_{I^k R} := \min(b_{I^k R}, n(I^k))$
    end for
until the set $S^k = \{u_R\}$
construct laplacian $\mathbf{L}$, taking S^k as the set of states
solve the system $\mathbf{L}_{\{u_R\},\{u_R\}}\mathbf{x} = \mathbf{b}^*$, where $\mathbf{b}^*_{I^k}$ equals the probability of moving from I^k to u^* in G_k
for each $I^k \in S^k$ different than u_R and each state $i \in S^0$ belonging to I^k **do**
    compute $\beta_{iR} := x_{I^k}\varepsilon_0^{-b_{I^k R}}$
end for

that $\mathcal{O}(n^2)$, the algorithm uses only $\mathcal{O}(n+m)$ space. This is crucial since matrices appearing in applications are often sparse and it is not rare that $m = \Theta(n)$. These estimates strongly motivate further studies of the issue of establishing an appropriate value of parameter ε, which has a significant impact on time and space cost.

In our analysis we have assumed that all closed classes (i.e. diagonal blocks of NCD Markov chain) are treated by a precise direct method (GTH). One can achieve better bound by choosing an iterative method to solve the subsystems on the desired level of accuracy.

Consider a laplacian with n states and m edges (i.e. m is the number of nonzero entries in the probability transition matrix). Assume that in a step of the aggregation process, the underlying graph is divided into k closed classes of size $n_1, n_2, \ldots, n_k$, respectively (i.e. $n_1 + n_2 + \ldots + n_k = |S|$, transient state are singleton closed classes).

Theorem 2. *The time and space costs of a single phase of aggregation are as follows:*

$$T = \mathcal{O}(n + m + \sum_{i=1}^{k} {n_i}^3), \quad S = \mathcal{O}(n + m + \max_{1 \le i \le k} {n_i}^2).$$

The cost of the Algorithm 2 is equal to the total cost of all aggregation phases increased by $\mathcal{O}(c^3)$, where c is the number of closed classes when the aggregation processes stops; the space cost is also increased by $\mathcal{O}(c^2)$.

4 Numerical Experiments

We describe here the outcomes obtained when combinatorial aggregation algorithms are used to compute stationary distribution ($\boldsymbol{\pi}$) and mean hitting time ($\mathbf{m}$) of several Markov models. Stationary distribution (non-transposed case) [5], is concerned in order to demonstrate that computing the mean hitting time is of the same difficulty. We concentrate mainly on small examples which appear frequently in the literature (cf. [1, 16]). Although the sizes of the matrices considered are quite small, it is still instructive to examine the effect of using our algorithms in such cases. In contrast to this, last problem investigated by us is a real life example and had been extensively studied by many authors (see e. g. [3]).

For each example considered in the sequel we compute the relative error of the solution (denoted by $\boldsymbol{\pi}^*$ and $\mathbf{m}^*$) compared with outcome of the GTH procedure (vectors: $\boldsymbol{\pi}$ and $\mathbf{m}$, respectively). Errors are computed w.r.t. GTH algorithm, despite that it is also exposed itself to numerical errors. GTH is suitable for this purpose as it has an *a priori* error estimation [11].

Two measures of algorithms accuracy correspond to a mean number of correct most significant digits and are given by the following formulas (n is the size of the matrix):

$$\mathbf{Prec}_1 := -\frac{1}{n}\sum_{i=1}^{n}\log_{10}\frac{|\pi_i^* - \pi_i|}{|\pi_i|} + \log_{10} 5,$$

$$\mathbf{Prec}_2 := -\log_{10}\frac{||\boldsymbol{\pi}^* - \boldsymbol{\pi}||_2}{||\boldsymbol{\pi}||_2} + \log_{10} 5.$$

In the case of combinatorial aggregation we discuss also the aggregation process: the value of ε, the number of phases and the number of aggregated states in each phase.

For readability, instead of laplacians we show probability transition matrices (zero entries are marked by dots).

Courtois matrix. The first problem considered in this section is the 8×8 Courtois matrix studied already in [1] as an example of NCD Markov chain. The parameter ε is chosen to be 0.001. During computing stationary distribution first phase of aggregation results in three aggregated states, $I = \{1, 2, 3\}$, $J = \{4, 5\}$ and $K = \{6, 7, 8\}$, which are aggregated in the following step into one closed class.

$$\begin{bmatrix} .85 & . & .149 & .0009 & . & .00005 & . & .00005 \\ .1 & .65 & .249 & . & .0009 & .00005 & . & .00005 \\ .1 & .8 & .0996 & .0003 & . & . & .0001 & . \\ . & .0004 & . & .7 & .2995 & . & .0001 & . \\ .0005 & . & .0004 & .399 & .6 & .0001 & . & . \\ . & .00005 & . & . & .00005 & .6 & .2499 & .15 \\ .00003 & . & .00003 & .00004 & . & .1 & .8 & .0999 \\ . & .00005 & . & . & .00005 & .1999 & .25 & .55 \end{bmatrix}$$

Courtois matrix with two additional states. We have modified the Courtois matrix by adding two **asymptotically transient** states. It is worth noting that this matrix does not satisfy NCD conditions, as not all off-diagonal elements have small values, but it is still feasible for our aggregation algorithms. The parameter ε is chosen to be 0.001. In the first step of aggregation we obtain three closed classes (similarly as before) and two transient states 9 and 10. Second step of aggregation results in one closed class lumping together all states.

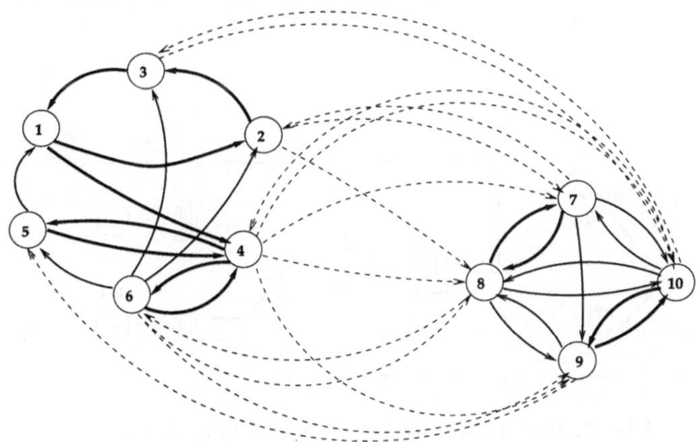

Fig. 2. Aggregation in Stewart matrix

Stewart matrix. This example is proposed by Stewart [16]. Note the impact of different ε values on accuracy of the result.

$$\left[\begin{array}{cccccccccc}
. & .5 & . & .5 & . & . & . & . & . & . \\
. & . & .999994 & . & . & . & .000006 & . & . & . \\
.999995 & . & . & . & . & . & . & . & . & .000005 \\
. & . & . & .019999 & .44 & .44 & .000003 & .000002 & .000003 & .000002 \\
.0001 & . & . & .99 & .0099 & . & . & . & . & . \\
. & .0002 & .0002 & .99 & .00959 & . & . & .000005 & .000005 & . \\
. & .000001 & . & . & . & . & .49 & .49 & .001 & .00099 \\
. & .000003 & . & . & . & .000007 & .49999 & .49 & .006 & .004 \\
. & . & . & . & .000001 & . & . & .009999 & . & .99 \\
. & . & .000005 & .000005 & . & . & .00999 & .01 & .98 & .
\end{array}\right]$$

The following array summarizes results of experiments. Recall that precision measures are accurate indicators of the number of correct digits in the approximate solution. Observe the effect of choosing smaller ε in the case of Stewart example. In general, calculated precision measures are on the level $\log \frac{1}{\varepsilon}$ — the same situation occurs in larger examples (c.f. [4]).

problem type	$\mathbf{L}(R\vert R)\mathbf{x} = \mathbf{b}$		$\mathbf{L}^T(R\vert R)\mathbf{x} = \mathbf{b}$	
	$\mathrm{Prec}_1(\mathbf{m}^*, \mathbf{m})$	$\mathrm{Prec}_2(\mathbf{m}^*, \mathbf{m})$	$\mathrm{Prec}_1(\boldsymbol{\pi}^*, \boldsymbol{\pi})$	$\mathrm{Prec}_2(\boldsymbol{\pi}^*, \boldsymbol{\pi})$
Courtois	4.98	4.81	4.45	4.23
modified Courtois	4.88	4.86	4.48	4.26
Stewart				
$\varepsilon = 0.01$	4.05	4.85	1.97	2.03
$\varepsilon = 0.00001$	5.95	5.49	4.20	4.79

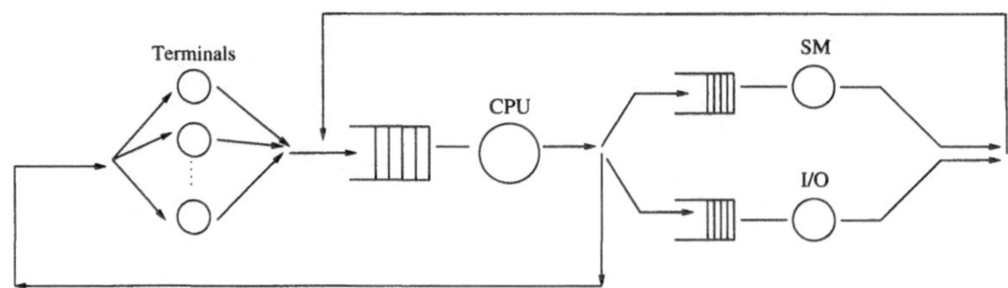

Fig. 3. Illustration for Interactive Computer System

Interactive Computer System. We consider the Markov model described in Figure 3 (for a detailed treatment see [15]). It represents a time-shared multiprogrammed, paged, virtual memory computer system, modeled as a closed queueing network. In order to perform numerical experiments we assign specific values for the parameters of the model according to [15]. The state of the system is coded by a triple $\mathbf{z} = (z_1, z_2, z_3)$ of non-negative numbers, where z_1 denotes the number of users thinking or busy at theirs terminals, z_2 and z_3 denote, respectively, the number of processes in the queue of SM and I/O. Obviously $z_1 + z_2 + z_3 \leq N$. There are at most six transition which can be made from any state.

We present outcome of the algorithm approximating mean hitting time and stationary distribution for 3, 10 and 20 users. In each case we analyze number of phases in aggregation process (denoted by p), the number of aggregates in the first step (k) and the precision measures. In the array n states for the size of the matrix and m is the number of its nonzero entries. Parameters T_{agr} and T_{gth} correspond to the time cost of aggregation algorithm and GTH procedure, respectively. Set $R = \{0, 0, 0\}$ i.e. all processes are in the CPU queue.

N	n	m	$\text{Prec}_1(\mathbf{m})$	$\text{Prec}_2(\mathbf{m})$	p	k	T_{agr}	T_{gth}
3	20	60	4.16	3.87	1	14	$0.01s$	$0.003s$
10	286	1320	4.21	4.21	1	77	$0.97s$	$25s$
20	1.771	11.011	4.97	4.97	1	252	$89s$	$> 8h$

The results for the stationary distribution are summarized in the following array.

N	n	m	$\text{Prec}_1(\boldsymbol{\pi})$	$\text{Prec}_2(\boldsymbol{\pi})$	p	k	T_{agr}	T_{gth}
3	20	60	3.4	4.49	2	4	$0.02s$	$0.003s$
10	286	1320	2.97	3.7	2	11	$0.37s$	$24s$
20	1.771	11.011	3.16	6.0	2	21	$30s$	$> 8h$

We conclude that combinatorial aggregation in the case of non-transposed system of equations (mean hitting time) is not less effective than in the case of stationary distribution. Up to now aggregation approach was used only in solving problems like stationary distribution. Combinatorial aggregation algorithms allow us to treat both type of problems in the unified manner.

5 Final Remarks

We proposed a new class of approximation algorithms based on combinatorial approach developed in [11]. We presented an algorithm approximating the mean hitting time. However after some minor modifications this algorithm is applicable to all non-transpose system, hence it can be used to calculation of other characteristics of a Markov chains, such as: $\mathbf{p} = p_{ik}(R)$ — the probability distribution (in R) in the hitting time and and $\boldsymbol{\mu} = \mu_{ij}(R)$ — the mean number of visits before absorption. We analyzed the complexity of algorithm and studied its applicability on several Markov models. Some outcomes of numerical experiments are reported. Both analytic and experimental results obtained by us classify this new method as a potentially very useful tool in practice. Our algorithm computes the asymptotic coefficients $\mathbf{b}$ and $\boldsymbol{\beta}$. It takes as an input laplacian $\mathbf{L} = (l_{ij})$

defining Markov chain; given additionally value ε it can be used to approximate mean hitting time (and other characteristics) as follows:

1. construct matrices $\boldsymbol{\Delta}$ and $\mathbf{D}$ such that: $-l_{ij} = \delta_{ij}\varepsilon^{d_{ij}}$, where $\varepsilon < \delta_{ij} \leq 1$;
2. run Algorithm 2 to compute vectors $\boldsymbol{\beta}$ and $\mathbf{b}$;
3. set $\mathbf{m}_i(R)(\varepsilon) := \beta_i\varepsilon^{b_i}$.

When $\varepsilon < \min_{ij} -(l_{ij})$, we have $d_{ij} = 0$ (for all i, j), hence $\mathbf{L} = \boldsymbol{\Delta}$ and Algorithm 2 gives the exact solution. On the other hand, larger ε's allow to profit from a specific block structure of a laplacian matrix to improve efficiency. In fact, there is a tradeoff between time/space efficiency of the algorithm and precision of approximation.

In our approach it is assumed that some preprocessing phase is needed to find an appropriate value of ε. It seems to be a challenging task to design an automatic procedure to calculate a value of ε, that would guarantee desired accuracy and time/space effectiveness of aggregation. There are several approaches aiming at defining a measure of decomposability of a transition matrix (see for example [13]). Such measures are usually closely related to parameter ε and could be probably used to find an optimal value for it. Moreover, a right value of ε can significantly improve precision of approximation as well as time and space needed for the algorithm. Another challenge is an error analysis of our algorithm. We have some preliminary results for some restricted families of Markov chains. However the obtained analytical formulas are often non-optimal and yields to significant overestimation of error level.

References

1. Courtois, P. J.: *Decomposability: Queueing and Computer System Applications*, Academic Press, New York, 1977.
2. Dayar, T. and Stewart, W.J.: On the effects of using the Grassman-Taksar-Heyman method in iterative aggregation-disaggregation, *SIAM Journal on Scientific Computing*, **vol. 17**, 1996, pp. 287-303.
3. Dayar, T. and Stewart, W.J.: Comparison of partitioning techniques for two-level iterative solvers on large, sparse Markov chains, to appear in *SIAM Journal on Scientific Computing*.
4. Gambin, A.: Combinatorial Methods in Approximation Algorithms for Markov Chains with Large State Space. *PhD thesis*, TR No. 260 Institute of Informatics, Warsaw University 1999.
5. Gambin, A. and Pokarowski, P.: A combinatorial aggregation algorithm for stationary distribution of a large Markov chain, accepted to FCT'2001.
6. W.K. Grassmann, M.I. Taksar and D.P. Heyman. Regenerative analysis and steady-state distributions for Markov chains, *Operations Research*, **vol.33**, pp 1107-1116, 1985.
7. Hassin, R. and Haviv, M.: Mean passage times and nearly uncoupled Markov chains, *SIAM Journal of Disc. Math.*,1992, **vol. 5**, pp. 386-397.
8. Iosifescu, M.: *Finite Markov Processes and Their Applications*, Wiley and Sons, 1980.

9. Kafeety, D.D., Meyer, C.D. and Stewart, W.J.: A General Framework for Iterative Aggregation/Disaggregation Methods, *Proceedings of the Fourth Copper Mountain Conference on Iterative Methods*, 1992.
10. Kemeny, J.G. and Snell, J.L.: *Finite Markov Chains.* Van Nostrand, Princeton, 1960.
11. Pokarowski, P. Directed forests and algorithms related to Markov chains, *PhD thesis* Institute of Mathematics, Polish Academy of Sciences, 1998.
12. Pokarowski, P. Directed forests with applications to algorithms related to Markov chains, *Applicationes Mathematicae*, **vol. 26**, **no. 4**, 1999, pp. 395-414.
13. Pokarowski, P.: Uncoupling measures and eigenvalues of stochastic matrices, *Journal of Applied Analysis*, **vol. 4, no. 2**, 1998, pp 259-267.
14. Shubert, B.O.: A flow-graph formula for the stationary distribution of a Markov chain, *IEEE Trans. Systems Man. Cybernet.*, 1975,**vol. 5**, pp. 565-566.
15. Stewart, W. J.: A Comparison of Numerical Techniques on Markov Modeling, *Comm. ACM* , **vol. 21**, 1978, pp. 144-152.
16. Stewart, W. J.: *Introduction to the numerical solution of Markov chains*, Princeton University Press, 1994.
17. Stewart, W. J.: Numerical methods for computing stationary distribution of finite irreducible Markov chains, *Chapter 3 of Advances in Computational Probability.* Edited by Winfried Grassmann; To be published by Kluwer Academic Publishers, 1997.

GROOME – Tool Supported Graphical Object Oriented Modelling for Computer Algebra and Scientific Computing

Victor G. Ganzha, Dmytro Chibisov[1], and Evgenii V. Vorozhtsov[2]

[1] Institute of Informatics, Technical University of Munich, Munich 80290, Arcisstr. 21, Germany; e-mail: ganzha@informatik.tu-muenchen.de
[2] Institute of Theoretical and Applied Mechanics, Russian Academy of Sciences, Novosibirsk 630090, Russia; e-mail: vorozh@itam.nsc.ru

Abstract. This paper is devoted to Object Oriented Modelling for Computer Algebra and Scientific Computing. The tool called GROOME is presented. GROOME supports object oriented problem decomposition on the basis of graphical description techniques. Such intuitive graphical description techniques are in our opinion the best medium to present the complex numerical models and methods as well as a hierarchy of them as well as relationships among them. The meaning or semantics of graphical description documents in GROOME can be completed with symbolic equations and Maple commands. Such documents can be linked together automatically and executable Maple code can be obtained.

1 Introduction

The development of software systems is a difficult and error prone task [4]. This is certainly true if systems become very large and complex. This may, however, even be true in cases where small to medium size programs have to be developed that are based on complex algorithms, data structures, or patterns of interaction between different program components.

Today software development in Scientific Computing practice is almost always done in a non-scientific manner based on pragmatics and heuristics. The software development process is structured in phases like:

- problem analysis and requirements engineering,
- specification,
- design,
- implementation and testing.

First of all we are interested to have a tool support that allows us to study the way from fluid dynamics problems and requirements analysis down to the numerical software design and program code. That would enable us to develop the correct, extensible and reusable software components that can be used for solving different fluid dynamics problems. It must be possible to link such components together with respect to a particular problem to solve. What it means is that it must be transparent to user what each component does and how it does that. For example, such strategy can be used later in so-called Problem Solving Environments (PSEs) [1,6]. The final goal of a PSE is the automation of the numerical solution of initial- and boundary-value problems for partial differential equations (PDEs), so that the PSE itself chooses a solution method for user's task. The SciNapse PSE [1] appears to be the first such PSE. This is a knowledge based program synthesis system implemented with objects, transformation rules, and a reasoning system. The system has been implemented in *Mathematica* and the template are executable Mathematica code, so they can easily be tested for correctness in *Mathematica.* The main interface to SciNapse is the "problem specification language". This language allows an initial- and boundary-value problem for a system of PDEs to be specified in terms of invariant differential operators or in a particular coordinate system. The methods for discretizing the PDEs and the boundary conditions can be specified in SciNapse by giving keywords for common method or by giving detailed mathematical replacement rules, or some combination of both.

The user who wishes to solve his specific task with the aid of SciNapse at first has to study the problem specification language of this system. In this connection it appears to be advisable to develop other, simpler projects, which would lead to publicly available systems that are easy to master.

In our report [5] the use of Graphical Description Techniques like Unified Modelling Language (UML) has been proposed to represent the complex algorithms and data structures transparent to user [8]. Furthermore the relevancy of Object Oriented Modelling Techniques in Computer Algebra System Maple to cope with the problem complexity and to obtain the correct, extensible and reusable software for Scientific Computing has been shown.

In this paper we continue this work and present a Java Tool called GROOME (GRaphical Object Oriented Modeling Environment) that supports object oriented problem decomposition and can be used to software development for solving modelling problems in Scientific Computing.

GROOME supports object oriented problem domain decomposition on the basis of graphical description techniques. Such intuitive graphical description techniques are in our opinion the best medium to present the complex numerical models and methods as well as a hierarchy of them as well as relationships among them. The meaning or semantics of graphical description documents in GROOME can be completed with symbolic equations and Maple commands.

For the graphical description of algorithms and data structures we use the class and data flow diagrams like UML diagrams, which now become more and more popular in various applications.

The GROOME user can decompose a Scientific Computing problem to solve in several subproblems by creating a hierarchy of documents. Such subproblems can be solved in a simple way, for example, with Maple. GROOME allows the solution of complete problem to be obtained by linking of subproblems solutions.

Resumed GROOME supports the user by:

– the object oriented decomposition of a Scientific Computing problem in several simpler subproblems
– the specification of solution of each original subproblem with graphical diagrams and Maple code
– the linking of the solutions together to obtain the solution of the complete problem.

Note that currently the goal of GROOME is not the computational times efficiency, but solving the problem of software complexity.

1.1 Computer Algebra versus Object Oriented Programming

The OOP paradigm has emerged in recent years and has become increasingly popular in software engineering projects whose purpose is to create large programs [9, 7]. A significant reason for this popularity is that OOP improves programming efficiency while simultaneously increasing program reliability. The key concepts in OOP are object, class, message, inheritance.

The gain in efficiency is achieved owing to the fact that one can use inheritance to produce new data types related to each other. In this way, it is possible to build a hierarchy of objects, which are obtained from one another via inheritance. Within this process, derived data types can be provided both with new data and new methods in addition to inherited features. In addition, it is possible to overwrite the already available methods [3]. The repeated use of classes and the hierarchies of classes is supported in OOP software by composition, inheritance and polymorphism. OOP implements these principles of programming best of all of the various modern programming paradigms.

At present, only two CASs, Axiom and MuPAD, completely support OOP, being designed from the very beginning as OOP systems [10]. Although one other CAS, Mathematica, is written in an object-oriented version of C, it does not provide any native support for object-oriented methods. We do note that the package `Classes.m` by Roman Maeder does implement a full-blown object-oriented extension to Mathematica [9]. This package does not make it possible to do any calculations that could not be done before, however, it makes it possible to completely rearrange the way in which they are carried out. Our package [5] makes it possible to develop the object oriented programs in CAS Maple.

1.2 Object Oriented Programming in CAS Maple

In our report [5] we have presented a package, which makes it possible to implement the object-oriented programming in Maple. In this Section, we would like to show the way in which this package functions at a relatively simple example.

At first we want to construct with the aid of Maple a base of the hat functions in the region $\theta = [0..Pi]$. We partition θ in `N=8` elements. For this purpose we place `N` nodes in θ. The intervals between these nodes are obviously the finite elements. A hat function may be assigned to each node, which takes the value 1 in the given node and vanishes in all other nodes.

A hat function is assigned to each individual element. The hat function on the element with the node `x_i=1` and the halved hat width `h_i=1` may be described and output in Maple as follows:

```
[> phi0 := piecewise(abs(x-1)<=1,1-abs(x-1),0);
[> plot(phi0, x=0..2);
```

The output is shown in Fig. 1.

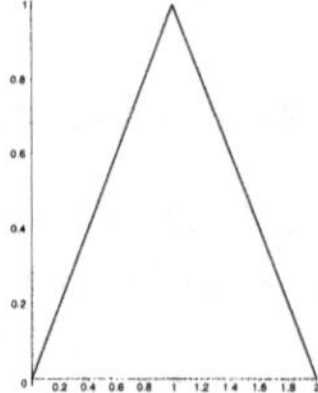

Fig. 1. A hat function

What is the way in which the grid is now stored? Before we generate the other hat functions on the grid we must specify the way in which we store the grid. The grid possesses a linear structure, hence it is natural to use a list for storing the grid. An element is described uniquely and completely in terms of two objects, the grid node and the element width. Thus, we need the following two lists:

- the list of grid nodes
- the list of the halved element widths.

Let us generate the both lists in Maple:

```
[> h0 := [Pi/8, Pi/8, Pi/8, Pi/8, Pi/8, Pi/8, Pi/8,Pi/8,Pi/8];
[> x0 := [0,Pi/8,2*Pi/8,3*Pi/8, 4*Pi/8, 5*Pi/8, 6*Pi/8,7*Pi/8,Pi];
[> N:=nops(x0);
```

In the list `h` we have stored `N` halved element widths and in the list `x0` `N` grid points. In addition, we need one more list `bs`, namely the list of the hat functions:

```
[> bs:=[seq(i,i=1..N)]:
[> for i from 1 to N do
       bs[i]:=convert(subs(x=(x-x0[i])/h0[i],phi0), piecewise,x):
   od:
```

For example, when we want to include new elements, which may be needed to refine the grid, then *all* the lists with a larger length must be again generated. It must be clear from here that when we process large data sets, the requirement of efficiency gives rise to our question and, on the other hand, the problem of memory shortage may arise. Finally, a large number of lists affect negatively their viewing, debugging, and as a result of this, the extensibility of our solution.

Therefore, it would be advisable to gather all the data as well as a code, which describe an element, in one structured data unit. This data unit obviously consists of the following parts:

- variable `xi`: grid node
- variable `hi`: element width
- variable `rbt` : boundary condition
- function `y` : ansatz function.

This data unit incorporates the properties of an abstract finite element, that is a finite element without its spatial position. It is possible for each class to generate the class instances, which correspond to a specific object. In order to define a class instance the data defined within a class receive certain values.

In our example the class must describe the element in general, that is without the input of specific values for `x_i`,`h_i` and `rbt`. For each specific value of `x_i` we will then generate an instance of this class, which in turn represents a specific hat function.

Thus, the class `FinElement` describes the following member variables:

- `xi` – the grid node
- `hi` – the halved element width
- `rbt` – the boundary condition type.

And the following methods:

- Constructor `FinElement` – Initialization of the member variable with certain values
- `y` – returns the value of the ansatz function assigned to this element.

This class incorporates the properties of various hat functions by gathering the data that are common to all the hat functions, in particular, the value in the node and the value for the hat width as well as the behavior (method `y`) depending on these data, which is also common for all the hat functions.

We at first show how the class `FinElement` can be implemented in Java and UML. After that we describe the way to implement the same in Maple.

Java. The class which describes an abstract element may be defined in Java as follows.

```
public class FinElement
{
  public float xi,hi;
  public int   rbt;

  public FinElement(float xi,float hi)
  {
    this.xi=xi;
    this.hi=hi;
  }

  public float y(float x)
  {
```

```
      if(Math.Abs(x-xi)<=hi)
         return(1-Math.Abs(x-xi)/hi);
      else
         return 0;
   }
}
```

Maple: Encapsulation of data and code in a single class. It is fortunate that Maple possesses the properties, which enable us to implement a kind of the encapsulation of data and code in a single object similarly to Java. As a matter of fact it is possible to define in a single Maple procedure the embedded functions (which are also called hidden functions). We start our package with

```
[>with(oop);
```

The class `fin_element` might be described in Maple as follows:

```
[> fin_element:=proc(x_i,h_i)
   local this;
   this:=DECLARE_CLASS();
   obj||this||xi:=x_i: obj||this||hi:=h_i:

   obj||this||phi:=proc(x)
        use oop in
    evalf(piecewise(abs(x-obj||this||xi)<=obj||this||hi,
                       1-abs(x-obj||this||xi)/obj||this||hi,0)):
        end use;
   end proc:
   ...
   RETURN(this);
   end:
```

The procedure `fin_element` obviously stores the node `x_i`, which is transferred as a parameter and to which the given basis function belongs, in the global variable `obj||this||xi` as well as the halved width `h_i` in `obj||this||hi`. In addition, this procedure defines the procedure `obj||this||phi(x)`, which computes the value of the hat function in `x` depending on the global variables `obj||this||hi` and `obj||this||xi`.

The name of the global variable for storing the own instance data is composed with the aid of the concatenation operator `||` of the local names as well as of the pointer `this`, which is uniquely assigned to each instance. The assignment of pointers is performed by our package and occurs in the procedure `DECLARE_CLASS`.

In this way we have described an abstract element, which contains both the local data and the code affecting these data. As in the Java implementation, our first Maple class `fin_element` also contains a constructor. The procedure `fin_element` itself plays the constructor role.

Generation of class instances. Our objective is now to generate an instance of `fin_element` for each basis function, in which the `m_xi` and `m_hi` take on certain numerical values.

Java. This is made in Java with the aid of the operator `new`:

```
// Generate an element with the node x=1 and the halved width h=1
FinElement elem1=new FinElement(1,1);

// Generate an element with the node x=2 and the halved width h=1
FinElement elem2=new FinElement(2,1);
 ...
```

In this way two class instances are generated. A certain memory domain is assigned to each instance, in which the instance variables and instance methods are resident. The operator `new` supplies the pointer for this memory domain. The pointers to the corresponding instances are stored in the variables `elem1` and `elem2`. In the following one can call the both class instances by using these pointers:

```
// Determine the node xi of the first element (elem1.xi==1)
float xi1=elem1.xi;

// Determine the node xi of the second element (elem2.xi==2)
float xi2=elem2.xi;
 ...
```

Maple. Our package provides similarly the procedure `new`, with the aid of which the new instances of a class can be generated:

```
[> elem1=new(fin_element,[1,1]);
```

The class name as well as the list with constructor parameters is transferred to the procedure `new`. `new` returns the pointer to the generated instance. Via this pointer one can call the methods and variables of the instance. For example, the hat function can be plotted:

```
[> plot(obj||elem||phi(x),x=0..Pi);
```

Inheritance. The gain in efficiency of OOP paradigm is achieved owing to the fact that one can use inheritance to produce new data types related to each other. In this way, it is possible to build a hierarchy of objects, which are obtained from one another via inheritance. Within this process, derived data types can be provided both with new data and new methods in addition to inherited features. In addition, it is possible to overwrite the already available methods ([3]).

Inheritance in Java. In Java the `extends` statement can be used to produce new derived classes. Consider for example the class `FinElementsList` that can be used to generate the dynamic list of finite elements:

```
public class FinElementList extends FinElement
{

  public FinElementList nextelement = null;


   public FinElementList(float xi,float hi)
  {
    super(xi,hi);
  }


  public void addElement(FinElementList element)
  {
     if(nextelement!=null)
         nextelement.addElement(element);
     else
         nextelement=element;
  }
}
```

The Java key word `super` refers to the super class. `super(x_i,h_i)` in our example refers to the constructor `FinElement(float x_i,float h_i)`. In addition to all the classes variables and methods of the class FinlElement, such as `xi`,`hi` or `public float y(float x)`, the class FinElementList provides the possibility to generate a list of finite elements:

```
// Generate the head of the list
FinElementList head_list = new FinElementList(0,1);

// Determine the node xi of the first element
float xi1=head_list.xi;

// Add new element to the head_list
head_list.addElement(new FinElementList(1,1));
```

Inheritance in Maple. Our package provides similarly to Java the statement `EXTENDS`. Consider for example the following implementation of the class `FinElementList` in Maple:

```
[> fin_element_list:=proc(x_i,h_i)
   local this,super;
```

```
this:=DECLARE_CLASS();
super:=obj||this||EXTENDS(fin_element,[x_i,h_i]);

...

RETURN(this);
end:
```

The inherited features, such as `phi(x)` or `xi`, can be used similarly to Java.

2 Object Oriented Modelling: Finite Element Method

Let the one-dimensional Poisson equation be given in the interval $[0, \pi]$:

$$-\frac{\partial^2 u}{\partial x^2} = \sin(x). \tag{1}$$

If we impose the Dirichlet boundary condition:

$$u(0) = u(\pi) = 0, \tag{2}$$

then (1) possesses the unique solution in the given interval.

To solve this equation we use the finite element method (FEM), which is being used during over 30 years as an indispensable tool for the structured modelling and simulation of many processes relevant to natural sciences and engineering. We will not dwell here on a detailed description of FEM and only briefly summarize the overall FEM procedure.The FEM itself and its implementation with the aid of OOP are presented in [3].

The advantages of the object-oriented programming in FEM code are discussed in detail in [2].

We can summarize the FEM as follows:

In the FEM a function is to be determined, which approximates a partial differential equation (PDE) with the given boundary conditions. The FEM solution technique in the simplest case may be subdivided roughly into the following steps:

- The PDE with *boundary conditions* is translated into a *variation equation* (VE).
- The spatial region is partitioned into separate *elements*, which have common vertices in *nodes*.
- The function sought for is approximated in the VE by a piecewise linear function, which is represented as a linear combination of simple piecewise linear functions with unknown coefficients.
- The VE is solved with the aid of the *Ritz-Galerkin method*. This gives rise to a linear system of equations (LSE).
- The solution of the LSE yields the coefficients for the above linear combination.

In accordance with the above we at first input equation (1) in Maple:

```
{> de:=-diff(diff(u(x),x),x)=sin(x);
```

$$-\frac{d^2}{dx^2}u(x) = \sin(x) \tag{3}$$

and approximate the solution function u by a linear combination of the ansatz functions ϕ_i:

```
[> u(x):=sum(c[i]*phi[i](x),i=1..N);
```

$$u(x) := \sum_{i=1}^{N} c_i \phi_i(x) \tag{4}$$

With regard for the boundary condition (2) we choose the ansatz functions in such a way that the following relations are valid for all j:

$$\phi_j(0) = \phi_j(Pi) = 0 \tag{5}$$

Consequently equation (3) has the following form:

$$-\sum_{i=1}^{N} c_i \frac{d^2}{dx^2}\phi_i(x) = \sin(x). \tag{6}$$

We now revert it into a variational equation:

```
[> a:=-Int(phi[j](x)*diff(diff(u(x),x),x),x=0..Pi):
[> b:=Int(phi[j](x)*sin(x)),x=0..Pi):
[> a=b;
```

$$-\int_0^\pi \phi_j(x) \sum_{i=1}^{N} c_i \frac{d^2}{dx^2}\phi_i(x)dx = \int_0^\pi \phi_j(x)\sin(x)dx \tag{7}$$

In this way we have brought equation (6) into the so-called weak form by multiplying the both sides of it by the ansatz function ϕ_j and integrating over the spatial interval.

The integration by parts of the left-hand side a of (7):

```
[> va:=intparts(lh,phi[j](x));
[> vb:=b;
```

yields

$$va := -\phi_j(\pi)\mathrm{D}(u)(\pi) + \phi_j(0)\mathrm{D}(u)(0) + \int_0^\pi \left(\frac{d}{dx}\phi_j(x)\right)\frac{d}{dx}u(x)dx \tag{8}$$

$$vb := \int_0^\pi \phi_j(x)\sin(x)dx \tag{9}$$

With regard for the condition (5)

```
[> phi[j](0):=0; phi[j](Pi):=0; va=vb;
```

the equations (8) and (9) imply the equation:

$$\int_0^\pi \left(\frac{d}{dx}\phi_j(x)\right) \frac{d}{dx}u(x)dx = \int_0^\pi \phi_j(x)\sin(x)dx. \tag{10}$$

If we again substitute instead of u the linear combination of the ansatz functions into equation (10), we finally obtain:

$$\sum_{i=1}^{N} c_i \int_0^\pi \left(\frac{d}{dx}\phi_j(x)\right) \frac{d}{dx}\phi_i(x)dx = \int_0^\pi \phi_j(x)\sin(x)dx \tag{11}$$

Or in matrix form:

$$Ac = b, \tag{12}$$

where

$$A(i,j) = \int_0^\pi \left(\frac{d}{dx}\phi_j(x)\right) \frac{d}{dx}\phi_i(x)dx \tag{13}$$

$$b(j) = \int_0^\pi \phi_j(x)f(x)dx \tag{14}$$

and c are the desired coefficients of the linear combination (4).

2.1 Object Oriented Modelling with GROOME

GROOME can be used to develop the object oriented software components for different scientific computing problems. In Fig. 2 is shown the project browser of GROOME.

Problem analysis and requirements specification First of all the particular problem to solve can be separated from the solving method. The Poisson equation with particular attributes, such as boundary conditions or dimension of equation, can be specified through a hierarchical structure.

As shown in Figure above, GROOME provides some different kinds of nodes to specify the equation with involved values, domain, constraints and values to be determined. Thereby are `lhs` und `rhs` the left and right hand side of the differential equation, respectively.

This particular problem description can be translated to Maple code by GROOME. The following class will be generated:

```
Poisson_Equation:=proc(paramlhs,paramrhs, parama, paramb,
                paramconstraint1,paramconstraint2,
                paramtodetermine)
local this:
this := DECLARE_CLASS();
```

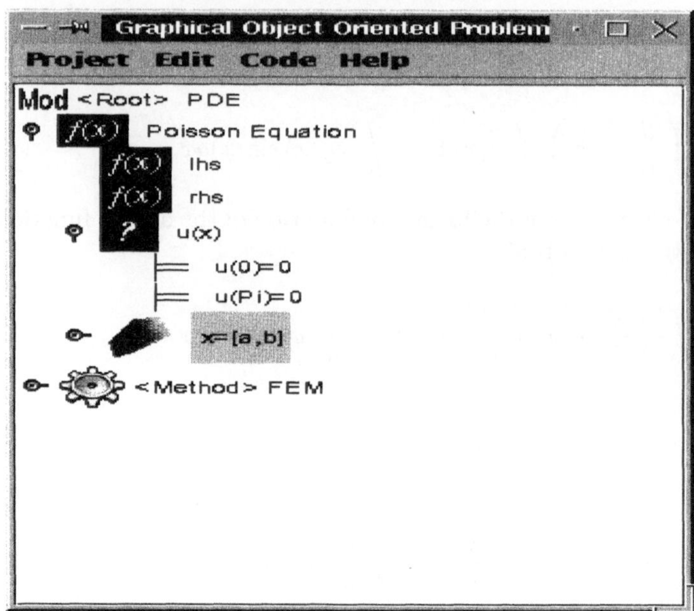

Fig. 2. GROOME: The project browser

```
# Fields declarations

obj||this||DECLARE_VAR(varlhs);
...

# Constructor

obj||this||varlhs:=varlhs;
obj||this||varlhs:=varrhs;
obj||this||vara:=parama;
obj||this||varb:=paramb;
...
RETURN(this);
end proc:
```

The instance of this class can be created with the following Maple command:

```
[> poisson_equation = new (Poisson_Equation,
           ['-diff(diff(u(x),x),x)','sin(x)',
           0,Pi,[0,0],[Pi,0],u]);
```

Solution Method The solution method, in our case, Finite Element Method can be structured hierarchically too. Using a data flow diagram (Fig. 3) the

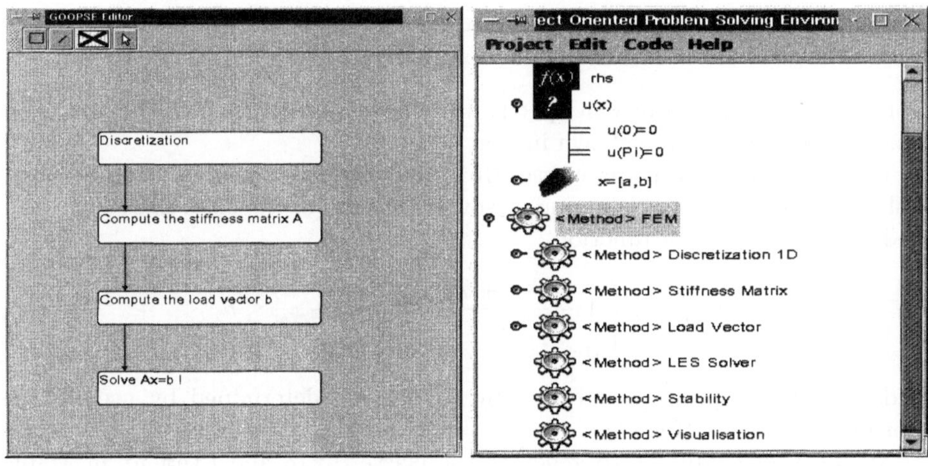

Fig. 3. Diagram to represent the data flow by using the FE method

steps to obtain the solution can be described. This diagram is associated with the node `Method FEM` in the hierarchical structure of project browser.

In this way the solver can be subdivided into several simpler methods, such

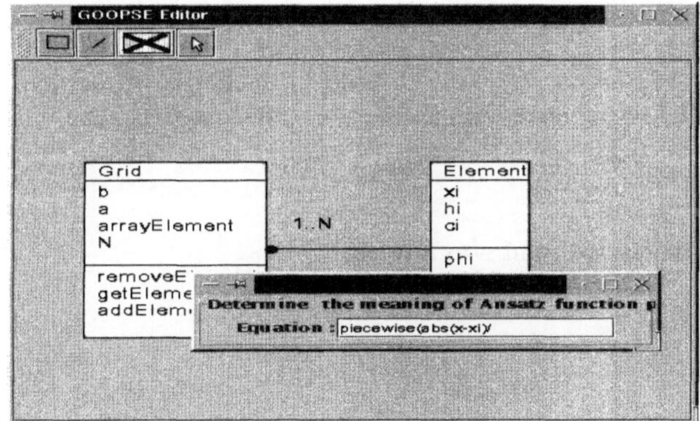

Fig. 4. Class diagram to perform the discretization

as discretization, stiffness matrix and load vector computation, solver for linear system of equations, etc.

Each step in the data flow diagram can be associated with a submethod shown in the project browser to perform this step. Consider, for example, the discretization method. To divide the 1D interval $[a..b]$ into `N` elements we need the class `Grid`, that would contain N `Element`-objects.

The individual element can be described by

– the middle point of element `x[i]`,

– the halved width of element `h[i]`,
– the assigned piecewise linear function `v[i]`.

For example, the user can specify the classes needed to describe the grid using the classes diagram shown in Fig. 4.

The relation between the classes **Grid** and **Element** means that one object of the class **Grid** may contain one or several objects of the class **Element**.

The piecewise linear function

$$\phi_i(x) := \begin{cases} 1 - \dfrac{|-x + x_i|}{h_i} & \text{if } |-x + x_i| \le h_i \\ 0 & \text{otherwise} \end{cases} \tag{15}$$

needed for the approximation of u by (4) can be determined by user in the appropriate window (Fig. 4).

All together enable us to generate the classes **Grid** and **Element** in Maple language using our OOP Package. Consider, for example, the generated class **Element**:

```
  Element:=proc(xiparam, hiparam)
  local this:

  this:=DECLARE_CLASS();

  # Class methods

  obj||this||phi:=proc(x)
   evalf(piecewise(abs(obj||this||xivar-x)<=obj||this||hivar,
                   1-abs(obj||this||xivar-x)/obj||this||hivar));
  end:


 # declaration of class methods and variables
 ...

 # Constructor

 obj||this||hivar:=hiparam;
 obj||this||xivar:=xiparam;


 RETURN(this);
 end proc:
```

At the generation of the class **Grid**, not only the user-specified variables `a` and `b` are taken into account, but also the methods **getElement**, **removeElement** and **addElement**, which are needed for the access to individual elements, are generated automatically:

```
Grid:=proc()
local this:
this:=DECLARE_CLASS();

 # Class methods

 obj||this||addElement:=proc(xiparam, hiparam)
 obj||this||arrayElementvar:=[op(obj||this||arrayElementvar),
                        new(Element,[xiparam,hiparam])];
 end:

obj||this||removeElement:=proc(i)
obj||this||arrayElementvar:=[op(obj||this||arrayElementvar[1..i-1]),
                        op(obj||this||arrayElementvar[i+1..nops
                          (obj||this||arrayElementvar)])];
 end:


obj||this||getElementCount:=proc()
   RETURN(nops(obj||this||arrayElementvar));
 end;

  # declaration of class methods and variables
 ...

 obj||this||arrayElementvar:=[];

 RETURN(this);
 end proc:
```

With the aid of the data flow diagram in Fig. 5 the grid generation on the given interval can be described.

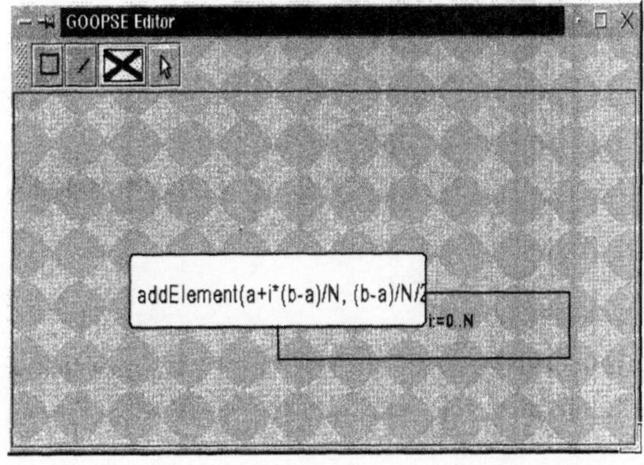

Fig. 5. Data flow diagram: Grid generation

As a result of this the class Discretization_1D will be generated:

```
Diskretization_1D:=proc(aparam,bparam,Nparam)
local this,i:
this:=DECLARE_CLASS();

 # declaration of class methods and variables

 ...

 # constructor

 obj||this||avar:=aparam;
 obj||this||bvar:=bparam;
 obj||this||Nvar:=Nparam;
 obj||this||Gridvar:=new(Grid,[]);

 for i from 0 to obj||this||Nvar do
     obj||(obj||this||Gridvar)||addElement(obj||this||avar+i*
          (obj||this||bvar-obj||this||avar)/obj||this||Nvar,
          (obj||this||bvar-obj||this||avar)/obj||this||Nvar/2]);
 od:

 RETURN(this);
 end proc:
```

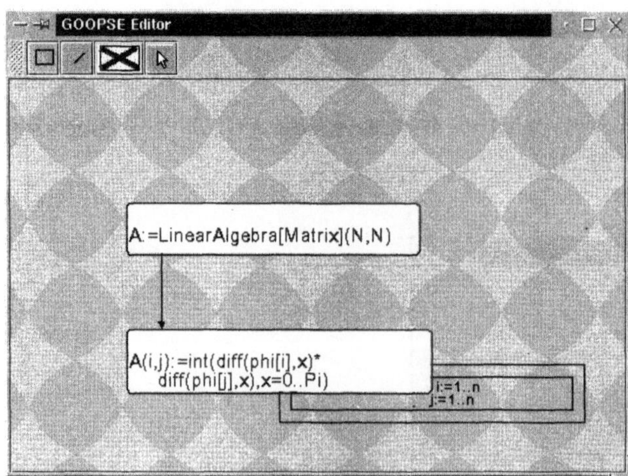

Fig. 6. Stiffness matrix computation

The methods for the computation of the stiffness matrix A and load vector b can be determined in a similar way. Consider, for example, the method *Stiffness matrix computation* in Fig. 6.

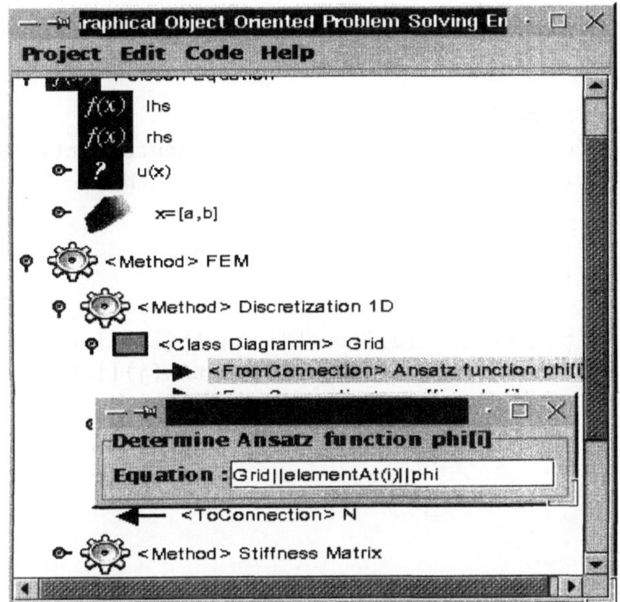

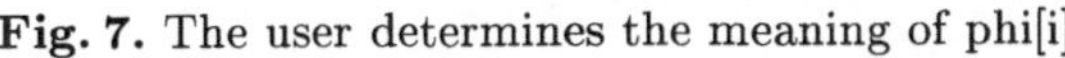

Fig. 7. The user determines the meaning of phi[i]

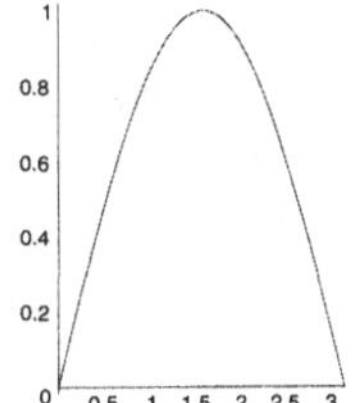

Fig. 8. The FEM approximation of the solution function by a linear combination of 31 hat functions

To link all sub-methods together GROOME needs information about the meaning of `phi[i]` and `phi[j]` in the expression to compute the stiffness matrix. To provide this information the user can create connections between different tree nodes. Such connections are presented as arrows (Fig. 7). The arrows directed to the right below a node represent variables that this node provides to use by other tree nodes. The arrows directed to the left are elements that this node needs to know. The meaning of **`phi[i]`** can be determined as shown in Fig. 7 by the provider node **`Grid`** and can be used, for example, by stiffness matrix computation.

Such connections are taken into account by code generation. Consider, for example, the class `Stifness_Matrix_Computation`. The connection `Ansatz function phi[i]` means that an object of the class `Stifness_Matrix_Computation` must be connected with an object of the class `Grid` to have the possibility to access the ansatz functions:

```
Stiffness_Matrix_Computation:=proc(paramGrid)
local this:

this := DECLARE_CLASS();

# Members declarations

...
```

```
   # Constructor

   obj||this||varGrid:=paramGrid;
   obj||this||varN:=obj||(obj||this||varGrid)||getElementCount();
   obj||this||varA:=Matrix(obj||this||varN,obj||this||varN);

for i from 1 to obj||this||varN do
for j from 1 to obj||this||varN do
 obj||this||A[i,j]:=
int(diff(obj||(obj||(obj||this||varGrid)||getElementAt(i))||phi,x)
    * diff(obj||obj||(obj||this||varGrid)||getElementAt(j)||phi,x),
          x=0..Pi);
 od:
od:

   RETURN(this);
end proc;
```

In this way the connections determine the relationships among the different classes.

The load vector `b` can be computed in similar way.

Finally the class `FEM` will be generated by GROOME:

```
FEM:=proc(Poisson_Equationparam, Nparam)
local this:

...

obj||this||Gridvar:= new (Grid,[obj||Poisson_Equationvar||avar,
                                obj||Poisson_Equationvar||bvar,
                                obj||this||Nvar]);

obj||this||Stiffness_Matrix_Computationvar:=
     new (Stiffness_Matrix_Computation, [obj||this||Gridvar]);

obj||this||Load_Vector_Computationvar:=
     new (Load_Vector_Computation, [obj||this||Gridvar]);

obj||this||LES_Solvervar:=new(LES_Solver,
    [obj||this||Stiffness_Matrix_Computationvar,
     obj||this||Load_Vector_Computationvar]);

   ...
RETURN(this);
end proc:
```

The solution of the differential equation (1) can be computed on the grid that contains 31 elements:

```
[> fem:=new(FEM,[poisson_equation,31]);
```

and be plotted (Fig. 8) by using the following command:

```
[> plot(obj||poisson_equation||u(x),x=0..Pi);
```

3 Conclusions

In this paper the Tool called GROOME has been presented. This tool can be used as a support by object oriented modelling of different scientific computing solutions. The user can decompose the problem to solve in several simpler subproblems by creating a hierarchy of documents that can contain graphical diagrams, for example, static class diagrams, data flow diagrams, or Maple language expressions. Such documents can be linked together automatically and translated into Maple code. The execution of this code enables us to validate the design of software to problem solution. The connections among different diagrams enable us to study the interaction patterns among different solution methods and develop reusable software components that can be used for the solution of different applied problems. It is, for example, possible to study the connection between FEM and equation dimensionality to develop the FEM classes in the way that they can be used in 2D und 3D cases.

References

1. Akers, R.L., Kant, E., Randall, C.J., Steinberg, S. and Young, R.L.: SciNapse: A problem-solving environment for partial differential equations. *IEEE Computational Science and Engineering* **4** (1997), No. 3, 32–42
2. Archer, G.: *Object-Oriented Finite Element Analysis.* Dissertation by University of California at Berkeley (1996)
3. Breitschuh, U., Jurisch, R.: *Die Finite-Element-Methode*, Akademie-Verlag, Berlin (1993)
4. Broy, M., Slotosch, O.: *Enriching the Software Development Process by Formal Methods*, LNCS 1641, 44-61 (http://www4.in.tum.de/proj/quest/)
5. Chibisov, D., Ganzha, V., Zenger, C.: *Objekt-Orientierte Modellierung für Wissenschaftliches Rechnen in Maple.* Preprint of the Institute of Informatics, Technical University of Munich (2001) (in press)
6. Cook, G.O., Jr.: *ALPAL: A Tool for the Development of Large-Scale Simulation Codes*, UCID-21482, Lawrence Livermore National Laboratory (1988)
7. Gray, J. W.: *Mastering Mathematica: Programming Methods and Applications*, Second Edition, San Diego, Academic Press, San Diego (1998)
8. Huber, F., Molterer, S., Rausch, A., Schötz, B., Sihling, M., Slotosch, O.: Tool supported specification and simulation of distributed systems. In: *Proc. Int. Symp. on Software Engineering for Parallel and Distributed Systems*, Krömer, B., Uchihira, N., Croll P., Russo, S. (eds.), IEEE Computer Society, Los Alamitos, California (1998) 155-164

9. Maeder, R.: *Programming in Mathematica*, Second Edition, Addison-Wesley, New York (1991)
10. Simon, B.: Symbolic math powerhouses revisited. In: *Computer Algebra Systems: A Practical Guide*, Wester, M. J. (ed.), John Wiley & Sons, Chichester, United Kingdom (1997) 1–20

Construction of Janet Bases
I. Monomial Bases

Vladimir P. Gerdt[1], Yuri A. Blinkov[2], and Denis A. Yanovich[1]

[1] Laboratory of Information Technologies
Joint Institute for Nuclear Research
141980 Dubna, Russia
gerdt@jinr.ru yan@cv.jinr.ru
[2] Department of Mathematics and Mechanics
Saratov University
410071 Saratov, Russia
BlinkovUA@info.sgu.ru

Abstract. Algorithms for computation of Janet bases for monomial ideals and implementation of these algorithms are presented. As data structures for finite monomial sets the binary trees called Janet trees are selected. An algorithm for construction of a Janet basis for the ideal generated by a finite monomial set is described. This algorithm contains as subalgorithms those to search for Janet divisor in a given tree and to insert monomials into the tree in the process of completion to involution. The algorithms presented have been implemented in C in the form of package for completion of monomial sets to Janet involutive ones. An example is given to illustrate practical efficiency of the monomial algorithms and their implementation.

1 Introduction

In papers [1–3] we invented a concept of involutive division, designed general algorithms for construction of polynomial involutive bases and studied a number of particular divisions. In paper [4] these algorithms were generalized to linear systems of partial differential equations. For Pommaret division the polynomial involutive algorithms were implemented in Reduce [1, 5]. The algorithms for construction of monomial involutive bases and the computation of the Hilbert functions and Hilbert polynomials based on them were implemented in Mathematica [6]. The latter code designed for the general involutive division and written in the Mathematica language is much slower than the straightforward implementation of Janet division in C as we demonstrated in [6]. The Mathematica implementation has been recently extended to compute involutive polynomial and linear differential bases[1].

In the given paper we present algorithms and data structures to compute monomial Janet bases with particular features of Janet division taken into account. By this reason these algorithms are more efficient then those designed for the general involutive division and applied to construction of Janet bases.

[1] The package is available at http://paul.math-inf.uni-greifswald.de/Invo/.

The paper is organized as follows. Section 2 contains the necessary definitions and notations to be used in the sequel.

In Section 3 we describe the data structure for a finite monomial set as a binary tree called Janet tree. The elements in the set are associated with the leaves of the tree. This data structure is amenable to efficient search for Janet divisor and for completion of monomial sets to Janet bases.

An algorithm of the fast search for Janet divisor in a given tree is presented in Section 4. For this algorithm we give an estimation of its complexity in comparison to that of the classical binary search algorithm.

In Section 5 we describe algorithms to insert into a given Janet tree of a monomial which has no involutive divisors among the leaf monomials in the tree and for computation of nonmultiplicative prolongations which arise from the insertion.

The main algorithm for computation of a minimal monomial Janet basis is described in Section 6. This algorithm is a specification to Janet division of the general algorithm which was put forward in [3] for the general involutive division. The given specification contains as subalgorithms all those ones described in Sections 3-4 and unlike the general algorithm there is no need in the conventional autoreduction at the initialization step.

Section 7 illustrates practical efficiency of the proposed algorithms by means of their implementation in C. One of important applications of monomial involutive bases is their convenience for computation of the Hilbert function and Hilbert polynomial [4, 7]. In this context the efficiency of monomial computations is illustrated by example from paper [8].

2 Janet Division

In this paper we use the following notations: $\mathbb{N}$ is the set of non-negative integers; $\mathbb{M} = \{x_1^{d_1} \cdots x_n^{d_n} \mid d_i \in \mathbb{N}\}$ is the set of monomials in the polynomial ring $\mathbb{R} = \mathbb{K}[x_1, \ldots, x_n]$ over zero characteristic field $\mathbb{K}$; $\deg_i(u)$ is the degree of x_i in $u \in \mathbb{M}$; $\deg(u) = \sum_{i=1}^{m} \deg_i(u)$ is the total degree of u; $\succ$ is an admissible [9–11] monomial ordering such that

$$x_1 \succ x_2 \succ \cdots \succ x_n \,. \tag{1}$$

If monomial u divides monomial v we shall write $u \mid v$.

An involutive monomial division L [1] for any finite monomial set $U \subset \mathbb{M}$ and any its element $u \in U$ assigns the partition $\{x_1, \ldots, x_n\} = M_L(u, U) \cup NM_L(u, U)$ of variables into multiplicative $x_i \in M_L(u, U)$ and nonmultiplicative $x_i \in NM_L(u, U)$. Respectively, $u \in U$ is an $L-$(involutive) divisor of $w \in \mathbb{M}$, if $u \mid w$ and monomial w/u contains only multiplicative variables for u.

A number of involutive divisions were discovered and studied to date [1–3, 7, 12]. In the present paper we consider one of the divisions which corresponds to the classical separation of variables into multiplicative and nonmultiplicative invented by Janet for the purpose of algebraic analysis of partial differential equations [13].

Definition 1. Janet division [1]. Let $U \subset \mathbb{M}$ be a finite monomial set. For each $1 \le i \le n$ divide U into groups labeled by non-negative integers $d_1, \dots, d_i$:

$$[d_1, \dots, d_i] = \{ \ u \ \in U \mid d_j = \deg_j(u), \ 1 \le j \le i \ \}. \tag{2}$$

A variable x_i is called (Janet) *multiplicative* for $u \in U$ if $i = 1$ and

$$\deg_1(u) = \max\{\deg_1(v) \mid v \in U\},$$

or if $i > 1$, $u \in [d_1, \dots, d_{i-1}]$ and $\deg_i(u) = \max\{\deg_i(v) \mid v \in [d_1, \dots, d_{i-1}]\}$. If a variable is not $J-$multiplicative for $u \in U$, it is called (Janet) *nonmultiplicative* for u. In the latter case we shall write $x_i \in NM_J(u, U)$. If $u \in U$ is a Janet divisor of monomial w we shall write $u \mid_J w$.

Definition 2. [1] A finite monomial set U is called Janet involutive if

$$(\forall u \in U) \ (\forall x \in NM_J(u, U)) \ (\exists v \in U) \ \ [\ v \mid_J (u \cdot x) \ ] \,. \tag{3}$$

To prove correctness of the main algorithm described in Section 7 we need the following lemma.

Lemma 1. *Let $U \subset \mathbb{M}$ be a finite set and the set $\tilde{U} \supseteq U$ be Janet involutive. Then $\tilde{U}$ contains any nonmultiplicative prolongation $u \cdot x$ of an element $u \in U$ ($x \in NM_J(u, U)$) such that $u \cdot x$ has no Janet divisor in U and among the conventional divisors of $u \cdot x$ there are no other nonmultiplicative prolongations of elements in U also without Janet divisors in U.*

Proof. The given property of Janet division follows from its constructivity defined in paper [1] where in the proof of Theorem 4.14 it is shown that the prolongation under consideration of the above lemma belongs to the involutive completion of the initial set.

Definition 3. [2] A Janet involutive set U is called a minimal Janet basis of the ideal $Id(U) \subset \mathbb{R}$, generated by U if for any other Janet basis V of this ideal the inclusion $U \subseteq V$ holds. This Janet basis is unique [2].

3 Data Structures

By definition of Janet division, a monomial set U is partitioned into groups (2) indexed by degrees of the first i variables. The variable x_{i+1} is multiplicative for $u \in U$ if $\deg_{i+1}(u)$ takes the maximal value among monomials in the group, and nonmultiplicative, otherwise. For the variable x_1 the whole monomial set U must be considered as a group.
Below we construct a binary tree which we shall call Janet tree. Its structure reflects the partition of elements in U into the groups (2) which are sorted in the degrees of variables within every group. We demonstrate that such a data

structure allows one to search efficiently for a Janet divisor and also to compute all the multiplicative variables for elements in U.

Before proceeding to the description of the general structure of Janet trees we explain it in terms of the following specific example.

Example 1. $U = \{x^2y, xz, y^2, yz, z^2\}$ $(x \succ y \succ z)$.

Monomial	*Variables*	
	nonmultiplicative	*multiplicative*
x^2y	—	x, y, z
xz	x	y, z
y^2	x	y, z
yz	x, y	z
z^2	x, y	z

Portray the monomial set of Example 1 in the form of Janet tree as indicated in the following picture. As shown in this picture, monomials of the given set U are

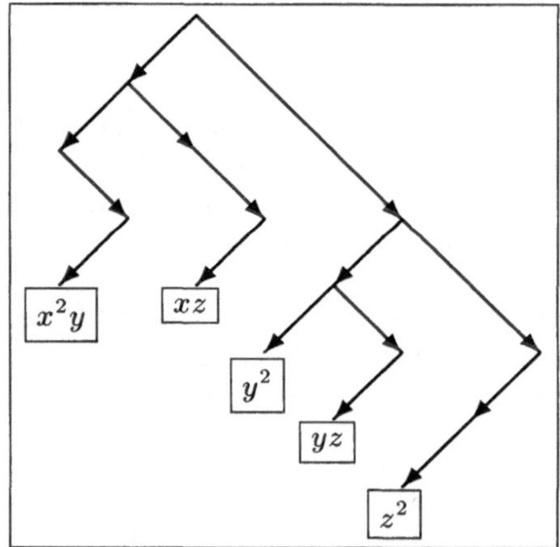

assigned to the leaves of the tree. The monomial with increased by one degree of the current variable is assigned to the left child whereas the right child points at the next variable with respect to chosen ordering.

Now consider the structure of Janet tree of the general form as a set $JT := \cup\{\nu\}$ of internal nodes and leaves which corresponds to the nonempty monomial set. To every element ν of the tree we shall assign the set of three elements $\nu = \{u, nd, nv\}$ with the following structure:

$$\begin{array}{lll} \mathrm{mon}(\nu) & = u & \text{is the monomial assigned to } \nu, \\ \mathrm{ndg}(\nu) & = nd & \text{is the pointer to the next node in degree,} \\ \mathrm{nvr}(\nu) & = nv & \text{is the pointer to the next node in variable.} \end{array}$$

In the absence of a child we shall assign the value *nil* to the corresponding pointer. Wherever it does not lead to misunderstanding we shall identify the pointers nd and nv with the nodes they point out. If the subtree $jt \subset JT$ of the Janet tree is rooted at $\nu \in JT$ we shall write $jt = \mathrm{jst}(\nu)$ and $\nu = root(jt)$.

The unit monomial is assigned to the root of a Janet tree JT which will be denoted by $\mathrm{root}(JT)$. The internal nodes and leaves of tree JT are characterized by the following states:

$$\text{Internal node: } (nv \neq nil) \vee ((nd \neq nil) \wedge (\deg(\mathrm{mon}(nd)) = deg(u) + 1))$$
$$\text{Leaf: } (nv = nil) \wedge (nd = nil)$$

4 Search for Janet Divisor

Given a Janet tree JT and a monomial w, the following algorithm provides a fast search for Janet divisor of w among the monomials assigned to leaves of JT.

Algorithm: J-divisor(JT, w)

```
Input: JT, a Janet tree; w, a monomial
Output: the leaf monomial u |_J w in JT or nil, otherwise
 1: ν := root(JT)
 2: while mon(ν) | w do
 3:    while ndg(ν) and mon(ndg(ν)) | w  do
 4:       ν := ndg(ν)
 5:    od
 6:    if nvr(ν) then
 7:       ν := nvr(ν)
 8:    elif ndg(ν) then
 9:       return nil
10:    else
11:       return mon(ν)
12:    fi
13: od
```

Prove correctness of this algorithm. By Definition 1, to be a Janet divisor of w, the monomial $u \in U$ must be the lexicographically highest either among all the monomials in U or among monomials in the group (2) with $d_1 = \deg_1(w), \dots$, $d_i = \deg_i(w)$ and some $1 \leq i < n$. As this takes place $\deg_j(w) \geq \deg_j(u) + 1$ for $j > i$.

Therefore, the correctness of algorithm **J-divisor** follows from the fact that after termination of the inner **while**-loop 3–5 there may be three alternate situations.

1. $(\mathrm{ndg}(\nu) = nil) \wedge (\mathrm{nvr}(\nu) \neq nil)$. In this case one needs to continue the search for divisor by transition to the next, by order (1), variable that is done in line 7.

2. $ndg(\nu) \neq nil$. If this holds, then there is no Janet divisor. Indeed, the sequential transition to the left in line 7 would end up with the leaf whose monomial has a higher degree in the current variable than w. Apparently, such a monomial cannot divide w both conventionally and involutively.
3. $(\mathrm{ndg}(\nu) = nil) \wedge (\mathrm{nvr}(\nu) = nil)$. This means that ν is a leaf of the tree and the monomial assigned to this leaf if just the desired Janet divisor.

The next theorem shows that complexity bound of the proposed algorithm **J-divisor** is lower than that of the classical binary search algorithm for Janet divisor in the lexicographically sorted array of monomials has been implemented earlier [6].

Theorem 1. *Let d be the maximal total degree of monomials in n variables which constitute the finite set U. Then the complexity bound of the algorithm* **J – divisor** *and the binary search algorithm is given by*

$$\begin{aligned} t_{\mathbf{J-divisor}(U)} &= O(d+n), \\ t_{\mathbf{BinarySearch}(U)} &= O(n((d+n)\log(d+n) - n\log(n) - d\log(d))). \end{aligned}$$

Proof. The complexity of the algorithm **J – divisor** is determined by the total length of walk in the related Janet tree. Since the total degree of an assigned monomial is bounded by d, the total length of walk in the tree is obviously bounded by $d+n$.

Furthermore, the structure of the Janet tree is such that when the transition to the next monomial holds in the process of the Janet divisor search this new monomial is different from the foregoing one by degree in a single variable only. This allows one to verify the Janet divisibility only with respect to that variable. As a result the computing time of the search is restricted by $t_{\mathbf{J-divisor}(G)} = O(d+n)$ instead of $O\left((d+n)\,n\right)$.

The number of distinct monomials in n variables and of the total degree $\leq d$ is $\binom{d+n}{n}$. Complexity of the binary search in a sorted list or array of k elements is $O(\log_2 k)$. Let k be a number of elements in U. Then $k \leq \binom{d+n}{n}$. To estimate the expression

$$O\left(\log_2 \binom{d+n}{n}\right) = O\left(\log\left(\frac{(d+n)!}{n!d!}\right)\right)$$

we use the Stirling formula in the form

$$\log(n!) - ((n+1/2)\log(n) - n) = O(1)\,.$$

Then, taking into account the fact that the comparison of monomials which is done in the course of the binary search algorithm has complexity $O(n)$, we obtain:

$$t_{\mathbf{BinarySearch}(U)} = O\left(n\left((d+n)\log(d+n) - n\log(n) - d\log(d)\right)\right).$$

5 Insertion of Monomials into Janet Tree

Describe now an algorithm **OneleafTree**(u, i, w) which for $w \mid u$ constructs the rooted binary tree with a single leaf. The constructed tree is rooted at monomial w, and its leaf monomial is $w \prod_{j=i}^{n} x_i^{\deg_j(u)-\deg_j(w)}$.

The internal nodes of the constructed tree receive the structure similar to the structure of Janet trees described in Section 3. In so doing the inner **while**-loop successively generates the left descendants of the current node with growing degrees of their monomials in the current variable x_i $(1 \leq i \leq n)$. Generation of the right children is done in line 13 of the outer **while**-loop 3–16. Correctness of this algorithm is obvious.

Algorithm: OneleafTree(u, i, w)

Input: u, a monomial; i, the index of a variable x_i; w, a divisor of u
Output: *tree*, the one-leaf tree rooted at w and having $w \prod_{k=i}^{n} x_k^{deg_k(u)}$ in the leaf
1: $\nu := \{w,\ nil,\ nil\}$
2: $tree := \{\nu\}$
3: **while** $u \neq w$ **and** $i \leq n$ **do**
4:   **while** $w \cdot x_i \mid u$ **do**
5:     $w := w \cdot x_i$
6:     $\mathrm{ndg}(\nu) := \{w,\ nil,\ nil\}$
7:     $\nu := \mathrm{ndg}(\nu)$
8:     $tree := tree \cup \{\nu\}$
9:   **od**
10:   **if** $u \neq w$ **then**
11:     $\mathrm{nvr}(\nu) := \{w,\ nil,\ nil\}$
12:     $\nu := \mathrm{nvr}(\nu)$
13:     $tree := tree \cup \{\nu\}$
14:     $i := i + 1$
15:   **fi**
16: **od**
17: **return** *tree*

The following algorithm **J-prolong**(jt, i, V) is recursive. It is applied to the subtree of a Janet tree such that variable x_i is nonmultiplicative for all the leaf monomials in the subtree. As part of the algorithm all nonmultiplicative prolongations of these monomials in x_i are determined and added to the set V which is used in the below described main algorithm **MonomialJanetBasis**.

It might be well to point out that the structure of the algorithm **J-prolong** is inherent in the bypassing algorithms for Janet trees. This structure shows that all the leaves of the input subtree jt are reached as part of the algorithms. Therefore, its correctness follows from the fact that at every leaf reaching the

Algorithm: J-prolong(jt, i, V)

Input: jt, a Janet subtree; i, the index of variable x_i which is nonmultiplicative for all the leaf monomials in jt; V, a monomial set
Output: the set V enlarged with nonmultiplicative prolongations in x_i of the leaf monomials in jt

```
1:  ν := root(jt)
2:  while ndg(ν) do
3:     if nvr(ν) then
4:        J-prolong(jst(ν), i, V)
5:     fi
6:     ν := ndg(ν)
7:  od
8:  if nvr(ν) then
9:     J-prolong(jst(ν), i, V)
10: else
11:    V := V ∪ {mon(ν) · x_i}
12: fi
13: return V
```

nonmultiplicative prolongation of the corresponding monomial in x_i is computed in line 11 and added to the monomial set V.

The algorithms **OneleafTree** and **J-prolong** enter as subalgorithms in the algorithm **J-insert** which inserts a monomial in a given Janet tree.

As the algorithm **J-insert** works, besides adding on the new subtrees to the Janet tree due to the insertion of monomial u, all the new nonmultiplicative prolongations induced by the insertion are also computed. These prolongations are added to the set V. It should be noted that the lack of Janet divisors of u among the leaf monomials in the tree is essentially exploited when u is inserted. This is indicated explicitly in the input data of the algorithm.

To show correctness of the algorithm **J-insert** consider two alternatives which may occur on completion in the **while**-loop 5-7:

1. $(\deg_i(\mathrm{mon}(\nu)) < \deg_i(u)) \wedge (\mathrm{ndg}(\nu) = nil)$. In this case the inserted monomial u has a higher degree in x_i than monomial in the current node of the tree. In so doing the equality $\deg_j(\mathrm{mon}(\nu)) = \deg_j(u)$ holds for all $j < i$ if $i > 1$. This is ensured by the assignment in line 13 which provides quitting the loop and transition to line 29 after execution of lines 9-12. Thereby, the insertion leads to conversion of x_i from multiplicative to nonmultiplicative for all the leaf monomials in the subtree rooted at that node. Respectively, in line 9 computation of all the new nonmultiplicative prolongations for the current branch is done. In lines 10–12 the tree is added on with the one-leaf tree having u in the leaf.
2. $\deg_i(\mathrm{mon}(\nu)) = \deg_i(u)$. If $\mathrm{ndg}(\nu) \neq nil$, then x_i is a nonmultiplicative variable for monomial u inserted in the tree, and the corresponding prolongation

Algorithm: J-insert(JT, u, V)

Input: JT, a Janet tree; u, a monomial to be inserted in J and such that *J-divisor*(JT, u) returns *nil*; V, a monomial set

Output: $JTree$, the Janet tree obtained from JT by inserting u; the set V, enlarged with nonmultiplicative prolongations of the leaf monomials in $JTree$ arising from insertion of u

```
i := 1
JTree := JT
ν := root(JTree)
while i ≤ n do
  while deg_i (mon(ν)) < deg_i(u) and ndg(ν) do
    ν := ndg(ν)
  od
  if deg_i(mon(ν)) < deg_i(u) then
    J-prolong(jst(ν), i, V)
    jtree := OneleafTree(u, i, mon(ν) · x_i)
    ndg(ν) := root(jtree)
    JTree := JTree ∪ jtree
    i := n + 1
  else
    if ndg(ν) then
      V := V ∪ {u · x_i}
    fi
    if nvr(ν) then
      i := i + 1
      ν := nvr(ν)
    else
      if i < n then
        jtree := OneleafTree(u, i + 1, mon(ν))
        ndg(ν) := root(jtree)
        JTree := JTree ∪ jtree
        i := n + 1
      fi
    fi
  fi
od
return JTree, V
```

is added to V in line 16. Furthermore, if $\mathrm{nvr}(\nu) \neq nil$ the outer **while**-loop is being continued by transition to the next variable. Otherwise one needs only add on the one-leaf tree as done in lines 23-25 and quit the outer loops afterwards that is provided by assignment in line 25.

Therefore, the correctness of algorithm **J-insert** follows from correctness of its subalgorithms **OneleafTree** and **J-prolong** has been established.

Apparently, all the considered algorithms are terminated that governs termination of the main algorithm **MonomialJanetBasis** described in the next section.

6 Main Algorithm

Given a finite monomial set U the below algorithm is a specific form of the general involutive algorithm [3] for construction of the minimal Janet basis of ideal $Id(U)$.

The initial Janet tree defined in line 1 of the algorithm has the root only. As prescribed in Section 3, the unit monomial is assigned to the root. If the input monomial set contains the unit monomial that is verified in lines 3-4, then the algorithm outputs $\bar{U} = \{1\}$. Otherwise, in the **while**-loop the set $\bar{U}$ initialized in line 6 as an empty set is augmented by the elements in the set V without Janet divisors in the current Janet tree JT.

Due to line 12, as explained in the previous section, the set V, in addition to elements in U which have not been chosen in line 8 yet, contains nonmultiplicative prolongations of elements in $\bar{U}$. These prolongations according to Definition 2 must be checked for existence of their Janet divisors in $\bar{U}$. Such a check is done in line 10, and if the nonmultiplicative prolongation under consideration has no Janet divisor it is inserted in $\bar{U}$ in line 11. Similarly, the elements of the initial monomial set U are proceeded in lines 10-11 in addition to the nonmultiplicative prolongations.

Let us now prove the correctness of the algorithm **MonomialJanetBasis**. By the condition in line 8, all those monomials in U which have no conventional divisors among elements in $\bar{U}$ are to be inserted in $\bar{U}$ since every Janet divisor is also the conventional divisor [1]. Therefore, the output set $\bar{U}$ is a basis of the monomial ideal $Id(U)$.

We must prove that this output basis satisfies the conditions (3) of Janet involutivity and is minimal. Denote by U_0 the minimal Janet basis of $Id(U)$ and by W the set $\bar{U}$ in some intermediate stage of the algorithm. Assume that $W \subset U_0$. This obviously holds after insertion of the first element in the initially empty set $\bar{U}$.

Let set V contain the nonmultiplicative prolongation $u \cdot x_i \in V$ of an element $u \in W$ such that there are no conventional divisors of this prolongation in V except the prolongation $u \cdot x_i$ itself. Then, if this prolongation is free from Janet divisor in W, by Lemma 1, the inclusion $u \cdot x_i \in U_0$ holds, and, hence, $W \cup \{u \cdot x_i\} \subseteq U_0$ in compliance with insertion done in line 12.

Algorithm: MonomialJanetBasis(U)

```
Input: U, a finite set of distinct monomials
Output: Ū, the minimal Janet basis of ideal Id(U)
 1: JT := {{1, nil, nil}}
 2: V := U
 3: if 1 ∈ V then
 4:    Ū := {1}
 5: else
 6:    Ū := ∅
 7:    while V ≠ ∅ do
 8:       choose u ∈ V without divisor in V \ {u}
 9:       V := V \ {u}
10:       if J-divisor(JT, u) = nil then
11:          Ū := Ū ∪ {u}
12:          J-insert(JT, u, V)
13:       fi
14:    od
15: fi
16: return Ū
```

Suppose now that V contains an element $u \in U$ such that u is free from the conventional divisors in $V \setminus \{u\}$ but there is such a divisor in W. If u has several such divisors, let v be the highest of them with respect to the lexicographical ordering induced by (1). Then from the lack of Janet divisors in W for u it follows that $u = v \cdot w$, where $v \in W$ and w contains variables nonmultiplicative for v. Let $x_k \in NM_J(v, W)$ be the highest (with the least subscript k) among such variables. By Definition 1 of Janet division, there is a monomial $u_1 \in W$ in the group containing v and such that $\deg_k(v) < \deg_k(u_1)$. But then the nonmultiplicative prolongation $v \cdot x_k$ must be examined in line 8 before monomial u. As this takes place, $v \cdot x_k \notin W$. Otherwise, since $v \cdot x_k \mid u$, the monomial v could not be lexicographically highest among divisors of u. On the other hand, if W would contain a Janet divisor v_1 of $v \cdot x_k$, then v_1 would be lexicographically higher than v and would satisfy $v_1 \mid u$, a contradiction. It follows that u has a Janet divisor in W.

Thus, we have shown that the initial conventional autoreduction of the input monomial set needs not be performed in the given algorithm as distinct from the general involutive algorithm described in [3].

Therefore, the correctness of the main algorithm **MonomialJanetBasis** reduces to correctness of its subalgorithms **J-divisor** and **J-insert** which has been proved. Termination of the main algorithm for arbitrary finite monomial set U follows from termination of the indicated subalgorithms and noetherity of the Janet division [1].

7 Implementation in C

The above described algorithms and data structures have been implemented in C. In the same manner as in the previous C implementation [6] the presentation of monomial $x_1^{i_1} \cdots x_n^{i_n}$ by the array $[i_1, \ldots, i_n]$ of its exponents is used. However, in contrast to the previous implementation where the classical binary search algorithm for Janet divisor was used, the given implementation which rests on the above algorithm **J-divisor** is far more efficient. This is in agreement with theoretical comparison of both search algorithms discussed in Section 4. Experimental comparison of the previous and new implementations shows that the latter is several times faster.

Besides, as well as in the previous implementation the Pommaret division [1] was also implemented. This division also admits data structures in the form of binary trees which are slightly different from Janet trees. We implemented the Pommaret division for the purpose of experimental comparison of both divisions in computation of involutive bases. Our experiment has shown that computation of monomial Pommaret bases does not reveal notable distinction in speed in comparison with Janet bases. In doing so, a minimal Janet basis coincides with a Pommaret basis whenever the latter is finite [14]. Though Janet division, unlike Pommaret division, calls for extra computation to recalculate the separation of variables into multiplicative and nonmultiplicative caused by insertion of a new monomial in the binary tree, for Janet division one often needs examine less number of prolongations than for Pommaret division. This is because the algorithm **MonomialJanetBasis** enables the fulfillment of conditions stated in Proposition 3.10 of paper [1] which imply that for any monomial in the intermediate basis its Janet nonmultiplicative variable is also Pommaret nonmultiplicative. As this takes place, the converse is not generally true.

In order to provide non-existence of conventional divisors for the choice of element in V in line 8 of the algorithm **MonomialJanetBasis** it suffices to sort V with respect to any admissible monomial ordering and choose the lowest monomial. In the given implementation the following three admissible orderings which are widely used in computation of Gröbner bases [10, 11] were built-in:

- lexicographical,
- degree lexicographical,
- degree reverse lexicographical.

In the monomial case, in contrast to the polynomial case, the choice of admissible ordering for selection of nonmultiplicative prolongations is not of crucial importance. This is because not only the output Janet basis which is uniquely defined by the input monomial set [2] but also the number of steps which are performed in the **while**-loop 7-14, are invariant under the choice of ordering [3]. The speed of computation may nevertheless depend on choice of ordering, and for Janet division the lexicographical ordering is most optimal [3]. The other built-in orderings are important in polynomial computations, for which the present implementation is an intermediate stage.

To illustrate practical efficiency of the proposed algorithms and their implementation we consider the following example, taken from paper [8] where it is used for computation of the Hilbert series implicitly representing the Hilbert function [10, 11].

Example 2. Let U_m ($m \in \mathbb{N}$) be the leading monomial set for the reduced Gröbner basis of the polynomial ideal generated by the elements of the squared $m \times m$ matrix A

$$A = \left\| \begin{matrix} x_{1,1} & x_{1,2} & \cdots & x_{1,m} \\ x_{2,1} & x_{2,2} & \cdots & x_{2,m} \\ \multicolumn{4}{c}{\dots\dots\dots\dots\dots} \\ x_{m,1} & x_{m,2} & \cdots & x_{m,m} \end{matrix} \right\| ,$$

whose m^2 elements are considered as variables. The Gröbner basis is computed for the degree reverse lexicographical ordering extending the following ordering on the variables

$$x_{1,1} \succ x_{1,2} \succ \cdots \succ x_{1,m} \succ x_{2,1} \succ \cdots \succ x_{2,m} \succ \cdots \succ x_{m,1} \succ \cdots \succ x_{m,m} \, .$$

The below table gives the quantitative characteristics of the input monomial set U_m for $m = 4, 5$, the output Janet basis $\bar{U}_m$ and the computing times on a Pentium III 700 Mhz based PC running under RedHat Linux 6.2. In the last column of the table the dimension of the monomial ideal $Id(U_m)$ is given.

m	*Number of variables*	*Size of U_m*	*Size of $\bar{U}_m$*	*Computing time (sec)*	*Dimension of ideal*
4	*16*	*161*	*1324*	*0.03*	*8*
5	*25*	*1372*	*72302*	*14.55*	*12*

The dimension of the ideals listed in the table can be determined as the degree of the Hilbert polynomial $HP(s)$, which as well as the Hilbert function $HF(s)$ are easily computed in terms of the monomial Janet basis $\bar{U}$ by the explicit formulae [4, 7]

$$HP(s) = \binom{n+s}{s} - \sum_{u \in \bar{U}} \binom{s - \deg(u) + \mu(u)}{\mu(u)} ,$$

$$HF(s) = \binom{n+s}{s} - \sum_{i=0}^{s} \sum_{u \in \bar{U}} \binom{i - \deg(u) + \mu(u) - 1}{\mu(u) - 1} .$$

Here n is the number of variables and $\mu(u)$ is the number of multiplicative variables for monomial u.

For the monomial ideals generated by U_4 and U_5 in Example 2, the Hilbert polynomials $HP_4(s)$ and $HP_5(s)$ are given by

$$HP_4(s) = \frac{1}{1440}s^8 + \frac{1}{72}s^7 + \frac{8}{45}s^6 + \frac{113}{144}s^5 + \frac{3259}{1440}s^4 + \frac{617}{144}s^3 + \frac{607}{120}s^2 + \frac{41}{12}s + 3$$

$$HP_5(s) = \frac{1}{1088640}s^{12} + \frac{1}{30240}s^{11} + \frac{121}{217728}s^{10} + \frac{209}{36288}s^9 + \frac{3223}{72576}s^8 + \frac{3173}{15120}s^7 + \frac{174367}{217728}s^6 + \frac{134699}{60480}s^5 + \frac{2480627}{544320}s^4 + \frac{152101}{22680}s^3 + \frac{19949}{3024}s^2 + \frac{1213}{315}s + 3$$

8 Conclusion

The procedures implementing slightly modified versions of algorithms described in Sections 4 and 5 form a part of our Reduce, C and C++ packages for computation of polynomial Janet bases. Instead of algorithm **MonomialJanetBasis** in the polynomial case the algorithm described in [3] is used with some extra optimizations. This improved algorithm together with its implementation and the timings for numerous benchmarks widely used by designers of the Gröbner bases software is presented in our second paper devoted to computation of polynomial Janet bases [15].

As we noticed in Sect.7 the binary tree data structure is also inherent in Pommaret division. We do not know to which extent such a structure is also inherent in other involutive divisions constructed in [1–3, 12]. But the recent results obtained in [16] outline a way to construction such a tree for a general involutive division.

This work was partially supported by the RFBR grants 00-15-96691, 01-01-00708 and by grant INTAS 99-1222.

References

1. Gerdt, V. P., Blinkov, Yu. A.: Involutive Bases of Polynomial Ideals. *Math. Comp. Simul.* **45** (1998) 519–542
2. Gerdt, V. P., Blinkov, Yu. A.: Minimal Involutive Bases. *Math. Comp. Simul.* **45** (1998) 543–560
3. Gerdt, V. P.: Involutive Division Technique: Some Generalizations and Optimizations, *Zapiski Nauchnykh Seminarov POMI (St.Petersburg)* **258** (1999) 185–206. To be published in *J. Math. Sci.*
4. Gerdt, V. P.: Completion of Linear Differential Systems to Involution. In: *Computer Algebra in Scientific Computing / CASC'99*, V.G. Ganzha, E.W. Mayr and E.V. Vorozhtsov (Eds.), Springer-Verlag, Berlin (1999) 115–137
5. Zharkov, A. Yu., Blinkov, Yu. A.: Involutive Approach to Investigating Polynomial Systems. *Math. Comp. Simul.* **42** (1996) 323–332

6. Gerdt, V. P., Kornyak, V.V., Berth, M., Czichowsky G.: Construction of Involutive Monomial Sets for Different Involutive Divisions. In: *Computer Algebra in Scientific Computing / CASC'99*, V. G. Ganzha, E.W. Mayr and E.V. Vorozhtsov (Eds.), Springer-Verlag, Berlin (1999) 147–157
7. Apel, J.: Theory of Involutive Divisions and an Application to Hilbert Function. *J. Symb. Comp.* **25** (1998) 683–704
8. Bayer, D., Stillman, M.: Computation of Hilbert Functions. *J. Symb. Comp.* **14** (1992) 31–50
9. Buchberger, B.: Gröbner Bases: an Algorithmic Method in Polynomial Ideal Theory. In: *Recent Trends in Multidimensional System Theory*, N.K. Bose (ed.), Reidel, Dordrecht (1985) 184–232
10. Becker, T., Weispfenning, V., Kredel, H.: *Gröbner Bases. A Computational Approach to Commutative Algebra.* Graduate Texts in Mathematics **141**, Springer-Verlag, New York (1993)
11. Cox, D., Little, J., O'Shea, D.: *Ideals, Varieties and Algorithms. An Introduction to Computational Algebraic Geometry and Commutative Algebra.* 2nd Edition, Springer-Verlag, New-York (1996)
12. Chen Yu-Fu, Gao Xiao-Shan.: Vector Representation of Involutive Divisions. *Mathematics-Mechanization Research Preprints* **No. 18**, Beijing (1999) 9–22
13. Janet, M.: *Leçons sur les Systèmes d'Equations aux Dérivées Partielles*, Cahiers Scientifiques, IV, Gauthier-Villars, Paris (1929)
14. Gerdt, V. P.: On the Relation Between Pommaret and Janet Bases. In: *Computer Algebra in Scientific Computing/ CASC 2000*, V.G. Ganzha, E.W. Mayr, E.V. Vorozhtsov (Eds.), Springer-Verlag, Berlin (2000) 167–181
15. Gerdt, V. P., Blinkov, Yu. A., Yanovich, D. A.: Construction of Janet Bases. II. Polynomial Bases. This volume.
16. Blinkov, Yu. A.: The Method of Separative Monomials for Involutive Divisions. *Programming and Computer Software* **3** (2001) 43–45

Construction of Janet Bases II. Polynomial Bases

Vladimir P. Gerdt[1], Yuri A. Blinkov[2], and Denis A. Yanovich[1]

[1] Laboratory of Information Technologies
Joint Institute for Nuclear Research
141980 Dubna, Russia
gerdt@jinr.ru yan@cv.jinr.ru
[2] Department of Mathematics and Mechanics
Saratov University
410071 Saratov, Russia
BlinkovUA@info.sgu.ru

Abstract. In this paper we give a detailed description of an algorithm for computation of polynomial Janet bases and present its implementation in C, C++ and Reduce. This algorithm extends to polynomial ideals the algorithm for computation of monomial Janet bases presented in the first part of our paper and improves the specialization to Janet division of the general algorithm for computation of involutive polynomial bases. The computational efficiency of our codes is compared with that of computer algebra system Singular which provides one of the best implementations of the Buchberger algorithm for computation of Gröbner bases.

1 Introduction

Involutive bases of polynomial ideals [1,2] and linear differential ideals [3] are Gröbner bases [4,5] of some special form. Though involutive bases are generally redundant as Gröbner ones, their use makes more accessible the structural information of polynomial and differential ideals and the properties of solutions of algebraic and differential systems. Janet bases came from the pioneering study of differential systems in [6] and may be cited as typical representatives of involutive bases [1]. One more typical representative of involutive bases is Pommaret basis which is the main research tool in formal theory of linear differential equations [7,8]. When a Pommaret basis is finite it is also a minimal Janet basis [9].

In this paper we present and describe in detail a new optimized version of algorithm [10] for computation of minimal Janet bases for polynomial ideals. The new algorithm extends to polynomials the monomial algorithm described in [11]. As well as for the monomial case we use Janet trees as the basic data structures for intermediate polynomial sets.

We have implemented the optimized algorithm in *C*, *C++* and *Reduce*. The timings for these three codes are compared with those for the latest release of computer algebra system *Singular* [12] (version 2-0-0) running on the same computer platform. As benchmarks we use the collection in [13] and the data base in [14] which are widely used for testing the Gröbner bases software and numerical routines for solving polynomial systems.

2 Polynomial Janet Bases

Throughout this paper we shall follow where it is possible to definitions, notations and conventions used in [11]. To describe polynomial algorithms and the related data types we need, however, some extra definitions and notations [9].

A divisor u of a monomial v will be called a *proper* divisor if $deg(u) < deg(v)$. The leading monomial of the polynomial $f \in \mathbb{R} \equiv \mathbb{K}[x_1, \dots, x_n]$ with respect to a given admissible monomial ordering $\succ$ will be denoted by $\mathrm{lm}(f)$. Similarly, the leading term of f as a product of the leading coefficient and the leading monomial will be denoted by $\mathrm{lt}(f)$. If $F \subset \mathbb{R}$ is a polynomial set, then by $\mathrm{lm}(F)$ we denote the leading monomial set for F, and by $Id(F)$ we denote the ideal in R generated by F.

Given a finite polynomial set $F \subset \mathbb{R}$ and a polynomial $f \in F$, the Janet separation of variables into multiplicative and nonmultiplicative is done in accordance with Definition 1 of Janet division in [11] with $u = \mathrm{lm}(f)$ and $U = \mathrm{lm}(F)$. We shall denote by $NM_J(f, F)$ the sets of Janet nonmultiplicative variables for f.

A finite polynomial set F is called *Janet autoreduced* if each term in every $f \in F$ has no Janet divisors among $\mathrm{lm}(F) \setminus \mathrm{lm}(f)$. A polynomial $h \in \mathbb{R}$ is said to be in the *Janet normal form modulo* F if every term in h has no $J-$ divisors in $\mathrm{lm}(F)$. In the following, the definition $NF_J(f, F)$ denotes the Janet normal form of polynomial f modulo F.

A Janet autoreduced set F is called a *Janet basis* of $Id(F)$ if

$$(\forall f \in F)\ (\forall x \in NM_J(f, F))\ \ [\ NF_J(f \cdot x, F) = 0\ ]. \tag{1}$$

A Janet basis G of ideal $Id(G)$ is called *minimal* if for any other Janet basis F of the ideal the inclusion $lm(G) \subseteq \mathrm{lm}(F)$ holds.

Janet division is a typical representative of involutive divisions [1] and a minimal Janet basis can be computed by means of general involutive algorithm described in [2, 10] when its input involutive division L is Janet division.

In its simplified form without such important optimizations as avoidance of repeated prolongations and the use of criteria to avoid unnecessary multiplicative reductions the general involutive algorithm may be written in the form of below algorithm **InvolutiveBasis**.

The correctness of this algorithm and its termination for any constructive noetherian involutive division [1] follow from those for its improved version [2, 10] and Lemma 1 in [11] which ensures correctness of the selection strategy used in line 6.

The main attributes of algorithm **InvolutiveBasis** are:

- Partition of polynomials in the intermediate basis into two disjoint sets T and Q with shifting from T to Q those polynomials whose leading terms are

Algorithm: InvolutiveBasis(F, L, $\prec$)

Input: $F \in \mathbb{R}\backslash\{0\}$, a finite polynomial set; L, an involutive division; $\prec$, an admissible monomial ordering
Output: G, a minimal involutive basis of $Id(F)$

choose $f \in F$ with the lowest $lm(f)$ w.r.t. $\prec$
$G := \{f\}$; $Q := F \setminus G$
do
  $h := 0$
  while $Q \neq \emptyset$ **and** $h = 0$ **do**
    choose $p \in Q$ with $\mathrm{lm}(p)$ without proper divisors in $\{\mathrm{lm}(q) \mid q \in Q \setminus \{p\}\}$
    $Q := Q \setminus \{p\}$
    $h := NF_L(p, G)$
  od
  if $h \neq 0$ **then**
    $G := G \cup \{h\}$
    for all $\{g \in G \mid \mathrm{lm}(g) \succ \mathrm{lm}(h)\}$ **do**
      $Q := Q \cup \{g\}$; $G := G \setminus \{g\}$
    od
    $Q := Q \cup \{NF_J(g \cdot x, G) \mid g \in G,\, x \in NM_L(g, G),\ NF_L(g \cdot x, G) \neq 0\}$
  fi
od while $Q \neq \emptyset$

higher (w.r.t. $\prec$) then the leading term of nonvanishing involutive normal form computed modulo set T for the currently examined element in Q.
- Inclusion in Q all nonmultiplicative prolongations of polynomials in T to be examined.
- Selection of a polynomial in Q to be examined such that its leading term has no proper divisors among leading terms of other polynomials in Q and then elimination of this polynomial from Q.
- Termination of the algorithm when the set Q becomes empty.

Below we present an advanced form of the above algorithm which preserves the listed attributes and is intended to compute Janet bases. Such a specialization allows one to use appropriate data structures for set T in the form of Janet trees and to design efficient algorithms relied on them. These data structures and algorithms generalize to the polynomial case those we describe in [11] for construction of Janet bases for monomial ideals.

3 Data Structures and Algorithms

Given a finite polynomial set $F \subset \mathbb{R}$ and a monomial ordering $\prec$, we shall endow with every polynomial $p \in F$ the triple structure

$$ps = \{p,\ u,\ vars\}$$

such that

$$\begin{aligned}
\mathrm{pol}(ps) &= p \in \mathbb{R} \text{ is a polynomial,}\\
\mathrm{anc}(ps) &= u \in \mathbb{M} \text{ is a monomial,}\\
\mathrm{nmp}(ps) &= vars \text{ is a (possible empty) subset of variables.}
\end{aligned}$$

In accordance with the involutivity conditions (1) in the course of construction of a Janet basis all the nonmultiplicative prolongations of intermediate polynomials must be examined with respect to their reducibility by multiplicative prolongations of other elements in the intermediate basis [1]. For this reason the involutive algorithms for construction of Janet bases must sequentially examine nonmultiplicative prolongations. Correspondingly, we shall assign to the second and third elements in the above triple the following values:

- $\mathrm{anc}(ps)$ is the lowest (w.r.t. $\prec$) monomial $lm(f)$ ($f \in F$) such that polynomial p was obtained from f by means of the nonmultiplicative prolongations. This means that $\mathrm{lm}(f) \mid \mathrm{lm}(p)$, and we shall call f an *ancestor* of p. If p has no proper *ancestor* in F than we shall assign $\mathrm{anc}(ps) = lm(p)$.
- $\mathrm{nmp}(ps)$ is the set of nonmultiplicative variables which were used for generation and examination of nonmultiplcative prolongations of p.

Consider now the structure of Janet tree JT [11] adopted for the polynomial case as a set $JT := \cup\{\nu\}$ of internal nodes and leaves which correspond to a finite nonempty polynomial set endowed with the above triple structure. Now we will assign to every node $\nu \in JT$ the set of four elements $\nu = \{u, nd, nv, ps\}$ as follows:

$$\begin{aligned}
\mathrm{mon}(\nu) &= u \quad \text{is the monomial assigned to } \nu,\\
\mathrm{ndg}(\nu) &= nd \ \text{is the pointer to the next node in degree,}\\
\mathrm{nvr}(\nu) &= nv \ \text{is the pointer to the next node in variable,}\\
\mathrm{ps}(\nu) &= ps \ \text{is the pointer to the triple } ps.
\end{aligned}$$

In addition to our notations and conventions for monomial Janet trees [11] we shall also assign the value *nil* to the pointer ps if ν is an internal node to which we shall not associate any triple.

Then the internal nodes and leaves of a polynomial Janet tree JT are to be characterized by the following states:

$$\begin{aligned}
\text{Internal node: } & ((nv \neq nil) \vee ((nd \neq nil) \wedge (\deg(\mathrm{mon}(nd)) = deg(u) + 1)) \wedge \\
& \wedge (ps = nil)\\
\text{Leaf: } & (nv = nil) \wedge (nd = nil) \wedge (ps \neq nil) \wedge (u = \mathrm{lm}(\mathrm{pol}(ps)))
\end{aligned}$$

In doing so, the following algorithm **J-divisor** for the input Janet tree JT and monomial w returns either the leaf triple of JT with the polynomial whose leading term is a $J-$divisor of w or *nil* if there are no such polynomials in JT. The structure of the algorithm is identical to that of its monomial analogue [11] except the output for the existing Janet divisor. In this case the polynomial version of the algorithm returns the corresponding triple whereas the monomial version returns the divisor itself.

Algorithm: J-divisor(JT, w)

Input: JT, a Janet tree; w, a monomial
Output: the leaf triple p in JT such that $\mathrm{lm}(pol(p)) \mid_J w$ or nil, otherwise

```
ν := root(JT)
while mon(ν) | w do
   while ndg(ν) and mon(ndg(ν)) | w do
      ν := ndg(ν)
   od
   if nvr(ν) then
      ν := nvr(ν)
   elif ndg(ν) then
      return nil
   else
      return ps(ν)
   fi
od
```

Similarly, the polynomial versions of algorithms **OneleafTree** and **J-prolong** are immediate adaptations of their monomial analogs [11] to the above described data structures. By these means the algorithm **OneleafTree** now returns a tree whose extra pointer in the leaf points out at the input triple; the algorithm **J-prolong** augments the input set Q with new triples corresponding to nonmultiplicative prolongations of the leaf polynomials with respect to x_i.

As well as in the monomial case, the algorithms **OneleafTree** and **J-prolong** enter as subalgorithms in the algorithm **J-insert** whose job is also to add new subtrees to the input Janet tree due to the insertion of a new polynomial in the leaf set and to adjust properly the set Q. If the insertion leads to converting of a nonmultiplicative variable into multiplicative then this variable must be eliminated from the relevant third element in the triple (c.f. [10]). This elimination is made in lines 14, 21 and 31.

The next algorithm **J-reduce** takes a set Q of triples and a Janet tree JT as an input and returns set Q with the polynomials in its triples whose leading terms reduced modulo the leaf polynomials in JT. This algorithm enters as a subalgorithm in the main algorithm **JanetBasis** described below. By construction in the latter algorithm, set Q contains the triples with nonmultiplicative prolongations to be examined.

The outer loop in algorithm **J-reduce** for every triple $f \in Q$ verifies in line 2 the existence of Janet divisor among the leading monomials of leaf polynomials in JT. If such a divisor exists, then the inequality $\mathrm{lm}(\mathrm{pol}(f)) \neq \mathrm{anc}(f)$ in line 4 is checked for the selected triple f. This inequality means that $\mathrm{pol}(f)$ has an ancestor in the intermediate basis. These elements of this basis are contained in the leaf triples of JT and there may be another part among the first elements of triples in Q.

Algorithm: OneleafTree(ps, i, w)

Input: ps, the triple structure for a polynomial pol(ps); i, the index of a variable x_i; w, a divisor of lm(pol(ps))

Output: $tree$, the one-leaf tree which is rooted at w and has monomial $w\prod_{k=i}^{n} x_k^{deg_k(\mathrm{lm}(\mathrm{pol}(ps)))}$ and the pointer to ps in its leaf

```
ν := {w, nil, nil, nil}
u := lm(pol(ps))
tree := {ν}
while u ≠ w and i ≤ n do
   while w · x_i | u do
      w := w · x_i
      ndg(ν) := {w, nil, nil, nil}
      ν := ndg(ν)
      tree := tree ∪ {ν}
   od
   if u ≠ w then
      nvr(ν) := {w, nil, nil, nil}
      ν := nvr(ν)
      tree := tree ∪ {ν}
      i := i + 1
   fi
od
ps(ν) := ps
return tree
```

Algorithm: J-prolong(jt, i, Q)

Input: jt, a Janet subtree; i, the index of variable x_i which is nonmultiplicative for all the leaf polynomials in jt; Q, a set of triples

Output: the set Q, enlarged with nonmultiplicative prolongations in x_i of the leaf polynomials in jt

```
ν := root(jt)
while ndg(ν) do
   if nvr(nu) then
      J-prolong(jst(ν), i, Q)
   fi
   ν := ndg(ν)
od
if nvr(ν) then
   J-prolong(jst(ν), i, Q)
else
   Q := Q ∪ {pol(ps(ν)) · x_i, anc(ps(ν)), ∅}
   nmp(ps(ν)) := nmp(ps(ν)) ∪ {x_i}
fi
return Q
```

Algorithm: J-insert(JT, ps, Q)

Input: JT, a Janet tree; ps, the structure with a polynomial pol(ps) such that **J-divisor**(JT, lm(pol(ps))) returns *nil*; Q, a set of triples

Output: $JTree$, the Janet tree obtained from JT by inserting ps; the set Q enlarged with the triples complying with nonmultiplicative prolongations of the leaf polynomials in $JTree$ arising from the insertion of ps

```
i := 1
JTree := JT
ν := root(JTree)
while i ≤ n do
  while deg_i (mon(ν)) < deg_i(lm(pol(ps))) and ndg(ν) do
    ν := ndg(ν)
  od
  if deg_i(mon(ν)) < deg_i(lm(pol(ps))) then
    J-prolong(jst(ν), i, Q)
    jtree := OneleafTree(ps, i, mon(ν) · x_i)
    ndg(ν) := root(jtree)
    JTree := JTree ∪ jtree
    for j from  i+1 to n do
      nmp(ps)) := nmp(ps) \ {x_j}
    od
    i := n + 1
  else
    if ndg(ν) then
      Q := Q ∪ {pol(ps) · x_i , anc(ps), nmp(ps) ∪ {x_i}}
    else
      nmp(ps) := nmp(ps) \ {x_i}
    fi
    if nvr(ν) then
      i := i + 1
      ν := nvr(ν)
    else
      jtree := OneleafTree(ps, i + 1, mon(ν))
      ndg(ν) := root(jtree)
      JTree := JTree ∪ jtree
      for j from  i+1 to n do
        nmp(ps)) := nmp(ps) \ {x_j}
      od
      i := n + 1
    fi
  fi
od
return JTree, Q
```

For the prolongation pol(f) the following two criteria are verified in line 5:

- **CriterionI**(f, g) is true iff $anc(f) \cdot \mathrm{anc}(g) \mid \mathrm{lm}(pol(f))$.
- **CriterionII**(f, g) is true iff $\deg(\mathrm{lcm}(\mathrm{anc}(f) \cdot \mathrm{anc}(g))) < \deg(\mathrm{lm}(pol(f))$.

These two criteria are the Buchberger criteria [4, 5] adapted to the involutive procedure. Indeed, it is easy to see that the first criterion subject to the conditions $g \neq nil$ and $\mathrm{lm}(\mathrm{pol}(f)) \neq \mathrm{anc}(f)$ is true iff $\mathrm{anc}(f)$ and $\mathrm{anc}(g)$ are coprime. Then, the related $S-$polynomial which arises in the right hand side of assignment in line 11 may be discarded. The second criterion is an involutive analogue of the Buchberger chain criterion [1, 2, 10]. One should note that the the criterion comes into action more rarely than in the course of Buchberger algorithm: many unnecessary $S-$polynomials do not manifest themselves in terms of involutive reduction of nonmultiplicative prolongations.

Algorithm: J-reduce(Q, JT)

```
Input: Q, a set of triples; JT, a Janet tree
Output: the set Q with leading terms of its polynomials reduced modulo set of the
    leaf polynomials in JT
 1: for all f ∈ Q do
 2:   g :=J-divisor(JT, lm(pol(f)))
 3:   if g ≠ nil then
 4:     if lm(pol(f)) ≠ anc(f) then
 5:       if CriterionI(f, g) or CriterionII(f, g) then
 6:         pol(f) := 0
 7:         g := nil
 8:       fi
 9:     fi
10:     h := pol(f)
11:     while g ≠ nil do
12:       h := h − pol(g) · lt(h)/ lt(pol(g))
13:       if h = 0 then
14:         g := nil
15:       else
16:         g :=J-divisor(JT, lm(h))
17:       fi
18:     od
19:     Q := Q \ {f}
20:     if h ≠ 0 then
21:       Q := Q ∪ {h, lm(h), ∅}
22:     fi
23:   fi
24: od
25: return Q
```

The **while**-loop 11–18 performs the leading term reduction of $h = \mathrm{pol}(f)$, and every reduced and nonvanishing h replaces the triple f in Q with $\{h, lm(h), \emptyset\}$ as is done in lines 19 and 21.

Correctness and termination of algorithm **J-reduce** follow immediately from those for algorithm **J-divisor** [11] and for the leading term reduction in line 12.

The following main algorithm **JanetBasis** is an advanced form of both algorithms **InvolutiveBasis** (Section 2) and that in [10] when their input division is Janet division.

Algorithm: JanetBasis(F, $\prec$)

```
Input: F ∈ ℝ \ {0}, a finite polynomial set; ≺, an admissible monomial ordering
Output: G, a minimal Janet basis of Id(F)
choose f ∈ F with the lowest lm(g) w.r.t. ≻
p := {f, lm(f), ∅}
T := {p}
Q := {{q, lm(q), ∅} | q ∈ F \ {f}}
JT := OneleafTree(p, 1, 1)
Q := J-reduce(Q, JT)
while Q ≠ ∅ do
  choose p ∈ Q with lm(pol(p)) without proper divisors in
          {lm(pol(q)) | q ∈ Q \ {p}}
  if lm(pol(p)) = 1 then
     return {1}
  else
     Q := Q \ {p}
     pol(p) := NF_J(pol(p), {pol(f) | f ∈ T})
     if lm(pol(p)) = anc(p) then
        for all {f ∈ T | lm(pol(f)) ≻ lm(pol(p))} do
           Q := Q ∪ {f}
           T := T \ {f}
        od
        JT := {{1, nil, nil, nil}}
        for all f ∈ T do
           J-insert(JT, f, Q)
        od
     fi
  fi
  J-insert(JT, p, Q)
  T := T ∪ p
  Q := J-reduce(Q, JT)
od
G := {pol(f) | f ∈ T}
return G
```

Initially, in lines 1–6, a Janet tree is constructed with the only leaf polynomial f which has the lowest leading term among the input polynomial set. All other polynomials are placed in Q and their leading monomials are reduced modulo f.

After the initialization step the main loop 7–28 runs while set Q is nonempty. Along with algorithms **InvolutiveBasis** and **MonomimalJanetBasis** [11] the selection strategy of triple $p \in Q$ is used in line 8 providing the lack of proper divisors of $\mathrm{lm}(\mathrm{pol}(p))$ among the leading terms of other polynomials in Q. If the selected polynomial is element in $\mathbb{K} \setminus \{0\}$ that is verified in line 9, then the algorithm returns $\{1\}$ as a Janet basis. Otherwise, p is eliminated from Q in line 12 and the Janet normal form of p modulo the leaf polynomials in JT is computed. Since $\mathrm{lm}(\mathrm{pol}(p))$ is Janet irreducible due to the job of algorithm **J-reduce** in lines 6 and 27, the normal form computation is in fact the tail Janet reduction.

Unlike monomial case when the input set V, which is a monomial analogue [11] of Q and may only be enlarged, in the polynomial case set Q is susceptible to more radical change. This change is made in line 13 of algorithm **InvolutiveBasis** and in line 16 of the current algorithm. Such a transfer of elements $f \in T$ to Q under the condition that $\mathrm{pol}(p)$ is not prolongation of another polynomial in $T \cup Q$, that is verified in line 14, and $\mathrm{lm}(\mathrm{pol}(f)) \succ \mathrm{lm}(\mathrm{pol}(p)$ (line 15) provides minimality of the output Janet basis [2, 10]. After the transfer, the Janet tree JT is fully reconstructed in lines 19–21 for the remaining elements in T if any. Then p is inserted in JT and T in lines 25–26, and polynomials in Q are head reduced in line 27.

For the proof of correctness and termination of algorithm **JanetBasis** we refer to the reasoning exploited in papers [2, 10] and to the above established correctness and termination of algorithm **J-reduce** and algorithms **OneleafTree** and **J-insert** [11].

It should be noted that the leading term reduction of polynomials in Q before their treatment in the **while**-loop decreases the number of transfers of elements in T to Q and related reconstructions of Janet trees. This is a notable optimization of the version of algorithm presented in [10]. Without the pre-reduction of leading terms in Q it may happen that every element from the whole group of elements selected in Q causes transfer whereas after the pre-reduction only one of them does it.

4 Implementation

We have implemented the algorithm **JanetBasis** in the symbolic mode language of Reduce 3.7 , C and C++ for the degree-reverse-lexicographical monomial ordering. For this reason we have replaced in our codes the condition of line 15 for moving an element $f \in T$ to Q with $\deg(\mathrm{lm}(\mathrm{pol}(f))) > \deg(\mathrm{lm}(\mathrm{pol}(p)))$. As well as in monomial case [11], the monomials are represented by the array of exponents and polynomials by the lists of their terms. The triple sets Q and T are also represented by lists. In the Reduce code for integer arithmetic, the built-in routines are used whereas for the codes written in C and C++ all the arithmetic operations over integers including modular arithmetic are performed by the GMP MP package version 3.1.1 [15]. In our implementations we use the following selection strategy for elements in Q in line 8 of algorithm **JanetBasis**. First, we chose a triple $p \in Q$ with the minimal $\deg(\mathrm{pol}(p))$. If there are several such triples in Q, then we chose that with the minimal number of terms in $\mathrm{pol}(p)$. If there are still several such triples we select the latest one in the list of Q.

The leftmost column in the below two tables contains a collection of examples extracted[1] from the polynomial systems data bases [13] and [14]. These data bases are widely used as benchmarks for experimentation with Gröbner bases software and with numerical routines for finding roots. In the first table we present the timings in seconds for computing Janet bases with our codes and those for computing Gröbner bases with the computer algebra system *Singular* version 2-0-0 [12]. We have chosen *Singular* for comparison with our codes since it is free and provides one of the best implementations of the Buchberger algorithm.

The timings were measured for the degree-reverse-lexicographical monomial ordering compatible with $x_1 \succ x_2 \succ \cdots \succ x_n$ on a Pentium III 700 Mhz with 546 Mb RAM computer running under Red Hat Linux release 6.2. For compiling the C code we used Intel C/C++ compiler version 5.0.1 beta whereas the C++ code was compiled with gcc compiler version egcs-2.91.66. The columns marked with subscript **m** contain timings for computation modulo prime number 31013. The other columns contain timings for computation over integers. The asterisk (*) marks the cases with running out of memory. As this takes place, the memory bound for running Reduce and C++ is 128 Mb and 546 Mb, respectively. For Singular and our C code a 1 Gb swap is also included. The second table contains the following information on the selected benchmarks:

n is the number of variables.
d is the degree of a system, that is, the maximal total degree of monomials which enter the system.
HP is the Hilbert polynomial of the ideal generated by polynomials in a system. Its degree is the dimension of the ideal. In zero-dimensional case its integer value gives the number of roots of the system.
GB is the number of polynomials in a reduced Gröbner basis.
JB is the number of polynomials in a minimal Janet basis.

Comparison of the timings and memory requirements for our codes with those for Singular shows that in the modular computations *Singular* is significantly faster than our packages with two exceptions only. As regards the computation over integers, our C and C++ codes are faster than Singular for most examples in the table. *Singular* also somewhat more often runs out of memory[2]. Such a variation in comparison of our codes and **Singular** for modular computations and for those over integers enables one to speculate that the algorithm **Janet-Basis** has a more smooth behaviour with respect to the intermediate coefficients swell than the Buchberger algorithm.

[1] We include in our tables only those examples from [13, 14] for which Singular needs on our computer more than half a second to compute Gröbner basis over integers.

[2] Our C++ code runs out of memory for redeco13 nothing but the lack of swap.

example	Reduce	Reduce$_m$	C	C$_m$	C++	C++$_m$	Singular	Singular$_m$
assur44	1 187	23	224	1	238	2	2 774	1
butcher8	15	3	4	0	4	0	10	0
chemequs	52	0	13	0	6	0	7	0
chemkin	205	29	160	3	157	5	7	2
cohn3	*	52	1 109	7	2 373	5	2 167	1
cpdm5	19	14	8	1	8	2	8	1
cyclic6	3	3	1	0	1	0	1	0
cyclic7	2 805	310	527	20	608	37	*	12
cyclic8	*	*	*	751	*	1 415	*	367
d1	235	3	27	0	30	0	38	0
des18_3	4	1	1	0	1	0	1	0
des22_24	12	2	4	0	4	0	4	0
dl	1 151	985	551	114	550	157	242	21
discret3	*	46	*	3	*	6	*	2
eco8	4	4	2	0	2	1	1	0
eco9	23	29	19	3	17	5	14	2
eco10	277	269	196	30	200	54	797	24
eco11	*	2 311	2 242	279	2 229	710	*	404
eco12	*	*	30 950	3 459	24 038	4 444	*	4 215
extcyc5	18	9	5	0	6	1	13	1
extcyc6	*	472	4 566	30	*	58	*	23
extcyc7	*	*	*	12 919	*	28 538	*	3 495
f744	62	46	28	4	29	7	8	1
f855	1 168	309	513	36	466	54	2 603	8
fabrice24	3 276	11	729	1	565	1	30 001	0
filter9	658	10	40	4	46	3	127	1
hairer2	*	296	602	57	386	71	3 986	9
hairer3	*	256	198	35	444	91	3 668	177
hcyclic7	2 566	243	499	21	512	30	21 187	11
hcyclic8	*	*	*	1 207	*	1 432	*	396
hf744	63	48	45	19	28	11	22	2
hf855	*	1 551	4 177	1 608	2 237	430	9 607	55
hietarinta1	1 432	9	20	1	23	2	28	1
i1	*	21	1 240	2	1 301	2	277	1
ilias_k_2	*	240	*	19	*	34	174 576	6
ilias_k_3	*	230	*	28	*	32	116 526	7
ilias12	*	2 337	*	203	*	327	*	52
ilias13	*	497	*	48	*	75	253 351	19
jcf26	6 261	16	1 001	1	1 135	2	3 331	0
katsura7	21	16	7	1	9	2	11	1
katsura8	212	154	102	11	100	21	114	9
katsura9	2 389	1 332	1 181	108	1 170	198	3 230	74
katsura10	*	*	18 313	1 036	21 063	1 964	76 630	651
kin1	1 609	2	164	0	159	0	228	0
kotsireas	601	25	118	2	118	3	38	1
noon7	82	85	88	17	114	39	4	3
noon8	2 736	2 331	2 584	560	4 808	1 765	39	34
noon9	*	*	*	*	*	*	665	445
pinchon1	239	195	197	55	210	91	31	3
rbpl	865	9	1 425	4	1 145	7	5 655	2
rbpl24	3 275	11	728	1	559	1	29 999	0
redcyc6	5	2	1	0	1	0	1	0
redcyc7	*	589	*	38	*	82	*	8
redcyc8	*	*	*	8 110	*	*	*	253
redeco9	10	16	8	3	7	2	2	1
redeco10	96	128	66	12	65	23	21	5
redeco11	873	1 104	606	114	601	207	207	46
redeco12	*	*	6 155	1 108	6 570	2 009	1 979	351
redeco13	*	*	74 765	13 489	*	22 864	19 843	2 986
reimer5	4	5	3	0	2	1	4	0
reimer6	145	144	91	11	89	23	*	38
reimer7	*	7 541	10 801	818	9 790	1 630	*	*
virasoro	91	66	40	3	39	7	161	3

example	n	d	HP	GB	JB
assur44	8	3	56	87	94
butcher8	8	4	$\frac{1}{2}s^3+\frac{5}{2}s^2+4s+14$	54	64
chemequs	5	3	16	20	20
chemkin	10	3	80	90	117
cohn3	4	6	$10s+163$	92	106
cpdm5	5	3	213	77	83
cyclic6	6	6	156	45	46
cyclic7	7	7	924	209	210
cyclic8	8	8	$144s+760$	372	384
d1	12	3	48	86	104
des18_3	8	3	46	39	104
des22_24	10	2	42	45	129
dl	8	6	468	323	538
discret3	8	2	128	147	164
eco8	8	3	64	59	86
eco9	9	3	128	106	172
eco10	10	3	256	203	336
eco11	11	3	512	389	662
eco12	12	3	1024	743	1329
extcyc5	6	5	350	102	102
extcyc6	7	6	936	242	243
extcyc7	8	7	6468	1371	1373
f744	12	3	$40s-30$	87	162
f855	14	4	$52s-36$	148	264
fabrice24	9	3	40	54	55
filter9	9	5	192	153	232
hairer2	13	4	$31s^2-40s+10$	128	453
hairer3	14	5	0	1	1
hcyclic7	8	7	$196s^2-546s-7056$	443	1182
hcyclic8	9	8	$\frac{176}{3}s^3-392s^2+\frac{5944}{3}s-23936$	1182	4392
hf744	13	3	$\frac{1}{60}s^5+\frac{17}{12}s^4+\frac{35}{4}s^3+\frac{487}{12}s^2-\frac{10103}{30}s+570$	204	1244
hf855	15	4	$\frac{1}{360}s^6+\frac{47}{120}s^5+\frac{395}{72}s^4+\frac{425}{24}s^3-\frac{5962}{45}s^2-\frac{151}{10}s+749$	688	4745
hietarinta1	10	3	$\frac{3}{2}s^2+\frac{23}{2}s+1$	51	52
i1	12	3	$33s^2-40s+19$	118	147
ilias_k_2	8	5	144	125	185
ilias_k_3	8	5	168	140	205
ilias12	8	5	468	231	314
ilias13	7	5	468	246	285
jcf26	34	4	40	79	80
katsura7	8	2	128	74	79
katsura8	9	2	256	143	151
katsura9	10	2	512	272	287
katsura10	11	2	1024	537	559
kin1	12	3	48	88	120
kotsireas	7	5	$108s-257$	70	78
noon7	7	3	2173	495	1157
noon8	8	3	6545	1338	3385
noon9	9	3	19665	3682	10001
pinchon1	29	16	78	108	145
rbpl	7	3	$\frac{10}{3}s^3+37s^2-\frac{427}{3}s+137$	126	309
rbpl24	9	3	40	54	55
redcyc6	7	11	$156s-588$	21	46
redcyc7	8	13	$924s-5257$	78	210
redcyc8	9	16	$92s^2+474s-6023$	193	371
redeco10	10	2	256	257	384
redeco11	11	2	512	513	768
redeco12	12	2	1024	1025	1536
redeco13	13	2	2048	2049	3072
reimer5	5	6	144	38	55
reimer6	6	7	576	95	199
reimer7	7	8	2880	227	775
virasoro	8	2	256	128	136

5 Conclusion

The results obtained in this paper give an optimized algorithm for computing the minimal Janet basis of a polynomial ideal and present the first implementation of this algorithm. Its computational efficiency is demonstrated on various examples and compared with that of the Gröbner basis module in *Singular*. The comparison shows the superiority of *Singular* over our codes for modular computations when arithmetic is suppressed. On the other hand our C/C++ codes offer some advantage over *Singular* in computation over integers.

We suggest, and our computer experiments provide support for this suggestion, that the involutive algorithms reveal a more smooth growth of the intermediate coefficients than the Buchberger algorithm. The latter admits much larger arbitrariness in selection and reduction strategies, and various strategies are used in practice to control the intermediate arithmetic. These strategies are essentially heuristic and often tailored for certain classes of examples. In algorithms **InvolutiveBasis** and **JanetBasis** a possible selection of an element in Q is much more restrictive. Furthermore, the leading term pre-reduction of polynomials in Q, as done in algorithm *JanetBases*, facilitates the choice of a triple in Q whose polynomial reduces other polynomials. Generally, the earlier such a polynomial is selected, the less is the growth of intermediate coefficients. The choice we implemented (Section 4) more commonly works in the right direction.

To increase the speed of modular computation we have to pay special attention to internal monomial representation and the related operations over them as it is done in the *Singular* team [16]. Another improvement which may be especially effective in modular computations [17] is planned to be implemented in the next version of our codes. This improvement is suitable for any constructive involutive division [1] and enables one to decrease the number of transfers from set T to set Q performed in algorithms **InvolutiveBasis** and **JanetBasis**.

Since the Pommaret division also admits the tree structure similar to the Janet tree [11] one could draw on the ideas and methods of the present paper to design and implement a special version of algorithm for construction of Pommaret bases. However, this does not seem worthy. Firstly, any finite Pommaret basis is always a minimal Janet basis [9] and, hence, can be computed by algorithm **JanetBasis**. Secondly, having computed a minimal Janet basis it is easy to check whether it is a Pommaret basis: it suffices to verify the coincidence of Janet and Pommaret multiplicative (or nonmultiplicative) variables for every element in the basis [18]. Thirdly, even in the monomial case the construction of finite Pommaret bases with the use of Pommaret trees and Pommaret division does not reveal any notable superiority in the speed of computation over algorithm **MonomialJanetBasis** [11]. Thus, in the polynomial computations it is not to be expected to obtain any gain from the use of Pommaret division instead of Janet division. As to other involutive divisions, the study started in [19] directed on the way to extension of the results of this paper to those divisions.

This work was partially supported by the RFBR grants 00-15-96691, 01-01-00708 and by grant INTAS 99-1222.

References

1. Gerdt, V. P., Blinkov, Yu. A.: Involutive bases of polynomial ideals. *Math. Comp. Simul.* **45** (1998) 519–542
2. Gerdt, V. P., Blinkov, Yu. A.: Minimal involutive bases. *Math. Comp. Simul.* **45** (1998) 543–560
3. Gerdt, V. P.: Completion of linear differential systems to involution. In: *Computer Algebra in Scientific Computing / CASC'99*, V.G. Ganzha, E.W. Mayr and E.V. Vorozhtsov (Eds.), Springer-Verlag, Berlin (1999) 115–137
4. Buchberger, B.: Gröbner bases: an algorithmic method in polynomial ideal theory. In: *Recent Trends in Multidimensional System Theory*, N.K. Bose (ed.), Reidel, Dordrecht (1985) 184–232
5. Becker, T., Weispfenning, V., Kredel, H.: *Gröbner bases. A computational approach to commutative algebra. Graduate Texts in Mathematics* **141**, Springer-Verlag, New York (1993)
6. Janet, M.: *Leçons sur les Systèmes d'Equations aux Dérivées Partielles*, Cahiers Scientifiques, IV, Gauthier-Villars, Paris (1929)
7. Pommaret, J.-F.: *Partial Differential Equations and Group Theory. New Perspectives for Applications*, Kluwer, Dordrecht (1994)
8. Seiler, W. M.: *Analysis and Application of the Formal Theory of Partial Diferential Equations.* PhD thesis, School of Physics and Materials, Lancaster University (1994)
9. Gerdt, V. P.: On the relation between Pommaret and Janet bases. In: *Computer Algebra in Scientific Computing/ CASC 2000*, V.G. Ganzha, E.W. Mayr, E.V. Vorozhtsov (Eds.), Springer-Verlag, Berlin (2000) 167–181
10. Gerdt, V. P.: Involutive division technique: Some generalizations and optimizations, *Zapiski Nauchnykh Seminarov POMI (St.Petersburg)* **258** (1999) 185–206. To be published in *J. Math. Sci.*
11. Gerdt, V. P., Blinkov, Yu. A., Yanovich, D. A.: Construction of Janet bases. I. Monomial bases. This volume.
12. Greuel, G.-M., Pfister, G., Schöenemann, H.: *Singular: A Computer Algebra System for Polynomial Computation*, Department of Mathematics, University of Keiserslautern (2001) http://www.singular.uni-kl.de/Manual/2-0-0/.
13. Bini, D., Mourrain, B.: *Polynomial Test Suite* (1996) http://www-sop.inria.fr/saga/POL.
14. Verschelde, J.: *The Database with Test Examples.* http://www.math.uic.edu/~jan/demo.html
15. *The GNU Multiple Precision Arithmetic Library.* Edition 3.1.1, 18 September 2000 http://www.gnu.org/manual/gmp/.
16. Bachmann, O., Schönemann, H.: Monomial representations for Gröbner bases computations. In: *Proceedings of ISSAC'98*, ACM Press (1998) 309–321
17. Gerdt, V. P.: On Algorithmic Optimization in Computation of Involutive Bases. In preparation.
18. Seiler, W. M.: A Combinatorial Approach to Involution and δ-regularity. *Preprint*, Universität Mannheim (2000)
19. Blinkov, Yu. A.: The Method of Separative Monomials for Involutive Divisions. *Programming and Computer Software* **3** (2001) 43–45

Low-Dimensional Quasi-Filiform Lie Algebras with Great Length

J. R. Gómez[1], A. Jiménez-Merchán[1], and J. Reyes[2]

[1] Dpto. Matemática Aplicada I, Universidad de Sevilla, Avda. Reina Mercedes s.n., 41012 Sevilla, Spain
{jrgomez, ajimenez}@us.es
[2] Dpto. de Matemáticas, Escuela Politécnica Superior, Universidad de Huelva, 21810 Huelva, Spain
reyes@uhu.es

Abstract. We consider the connected integer gradations on a type of n-dimensional nilpotent Lie algebras. We study the case where the number of non trivial subspaces is $n-1$ for those gradations, when the nilindex of the algebras is $n-2$ (quasi-filiform Lie algebras). We show how Symbolic Calculus can be useful to obtain the classification of such a family of graded algebras, which is determined for $n \leq 15$.

1 Introduction

If $\mathfrak{g}$ is a nilpotent Lie algebra of dimension n and nilindex k (index of nilpotency), it is naturally filtered by the descending central sequence of $\mathfrak{g}$, $\{C^i\mathfrak{g}\}_{0\leq i\leq k}$, ($C^0\mathfrak{g}=\mathfrak{g}$, $C^i\mathfrak{g}=[\mathfrak{g},C^{i-1}\mathfrak{g}]$). We consider the filtration given by $\{S_{i+1}\}$, where $S_{i+1}=\mathfrak{g}$, if $i\leq 0$; $S_{i+1}=C^i\mathfrak{g}$, if $1\leq i\leq k-1$, and $S_{i+1}=\{0\}$, if $i\geq k$. Associated to $\mathfrak{g}$ there exists a graded Lie algebra $\operatorname{gr}\mathfrak{g}=\oplus_{i\in\mathbb{Z}}\mathfrak{g}_i$, taking $\mathfrak{g}_i=S_i/S_{i+1}$. When $\operatorname{gr}\mathfrak{g}$ and $\mathfrak{g}$ are isomorphic, denoted by $\operatorname{gr}\mathfrak{g}=\mathfrak{g}$, it is said that $\mathfrak{g}$ is naturally graded.

The filiform case, characterized for its maximum nilindex $k=n-1$, in each dimension ($\mathfrak{g}$ is filiform if $\dim(C^i\mathfrak{g})=n-i-1$, for $1\leq i\leq n-1$), was studied by Vergne [11], who introduced the notation, proving that there are only two algebras of this type, L_n and Q_n, when the dimension of the algebra is even, and there is only one, L_n, if n is odd. The algebras L_n and Q_n are those defined on the basis $(X_0,X_1,\dots,X_{n-1})$, by

$$L_n=\{[X_0,X_i]=X_{i+1},\qquad 1\leq i\leq n-2.$$

$$Q_n=\begin{cases}[X_0,X_i]=X_{i+1}, & 1\leq i\leq n-2,\\ [X_i,X_{n-1-i}]=(-1)^{i-1}X_{n-1}, & 1\leq i\leq q-1,\, n=2q.\end{cases}$$

(The undefined brackets, except for those expressing antisymmetry, are supposed to vanish).

The naturally graded Lie algebras for the filiform class has permitted Vergne to study some problems for the considered class [11]. Furthermore, Goze and

Khakimdjanov [7] have proved how one can make use of the classification of the naturally graded filiform Lie algebras in other type of problems. This has motivated the study of this type of gradation on nilpotent Lie algebras of nilindex different to the filiform [4].

The subspaces in that natural gradation on nilpotent Lie algebras, and the existence of an appropriate homogeneous basis (necessary to obtain the classification) are a natural consequence of the central descending sequence of the considered algebras. However, it introduces a restriction by fixing the number of subspaces in the gradation by means of the nilindex.

In more recent works, there has been a fair amount of interest in graded Lie algebras which possess not just one natural gradation, but a gradation with a large number of subspaces (see [9], [2], [8]). For such a "length" of the gradation, the main interest is in algebras whose length is as large as possible. Thus, we have considered in this way the filiform Lie algebras ([5]).

On the other hand, for the quasi-filiform Lie algebras (nilindex of $\mathfrak{g}$ equal to $\dim \mathfrak{g} - 2$), we have proved that there are either only two n-dimensional algebras of maximum length, if n is odd, or only three, if n even, when $n > 12$ [6]. In addition to the Lie algebras with maximum length (it has to be n), in the quasi-filiform case, the algebras with length equal to $n-1$ could also provide useful information to know better the complete variety.

The aim of this paper is to study those graded Lie algebras, with length equal to $n-1$, providing the classification in low dimension, which has been obtained by using symbolic calculus, in particular the programming language *Mathematica* [12], to assist the computations.

This paper is structured in the following way. In Section 2, we introduce the appropriate notation, and we obtain some of the algebras in which we are interested. We discuss in Section 3 all the low-dimensional cases until dimension $n \leq 15$, and we show how the programming language *Mathematica* [12] has been used as an assistant with packages elaborated ad hoc. These packages can be readapted in order to study similar problems.

2 Length of Nilpotent Lie Algebras

We will now introduce the notation for the type of connected gradation which we are going to consider over the nilpotent Lie algebras.

2.1 Connected Gradations

We suppose, in this paper, that all Lie algebras are defined over the field of complex number $\mathbb{C}$. We also consider $\mathbb{Z}$-graded Lie algebras; that is, admitting a decomposition $\mathfrak{g} = \bigoplus_{i \in \mathbb{Z}} \mathfrak{g}_i$, where the subspaces $\mathfrak{g}_i$ satisfy $[\mathfrak{g}_i, \mathfrak{g}_j] \subset \mathfrak{g}_{i+j}$ for all $i, j \in \mathbb{Z}$. Moreover, the gradations will be *finite*, with a finite number of nonzero subspaces $\mathfrak{g}_i$.

We will say that a nilpotent Lie algebra $\mathfrak{g}$ admits a *connected* gradation $\mathfrak{g} = \mathfrak{g}_{n_1} \oplus \cdots \oplus \mathfrak{g}_{n_2}$, when each $\mathfrak{g}_i$, with $n_1 \leq i \leq n_2$, is nonzero. The number of subspaces $l(\oplus\mathfrak{g}) = n_2 - n_1 + 1$ will be called the length of the gradation. That is, the length of a gradation is given by the number of independent gradation blocks.

Definition 2.1. *The* length $l(\mathfrak{g})$ *of a Lie algebra* $\mathfrak{g}$ *is defined as*

$$l(\mathfrak{g}) = \max\{l(\oplus\mathfrak{g}) = n_2 - n_1 + 1 : \mathfrak{g} = \mathfrak{g}_{n_1} \oplus \cdots \oplus \mathfrak{g}_{n_2} \text{ is a connected gradation}\}.$$

This means that $l(\mathfrak{g})$ is the greatest number of subspaces from the connected gradations which can be obtained in $\mathfrak{g}$. Thus, every Lie algebra $\mathfrak{g}$ has at least length equal to 1, because we can consider the connected trivial gradation $\mathfrak{g} = \mathfrak{g}_0$. On the other hand, an algebra cannot have length greater than its dimension. Thus, for every nilpotent Lie algebra $\mathfrak{g}$ we have $1 \leq l(\mathfrak{g}) \leq \dim \mathfrak{g}$.

For example, the algebra L_n admits the gradation $L_n = \mathfrak{g}_1 \oplus \cdots \oplus \mathfrak{g}_n$, being the subspace $\mathfrak{g}_i = < X_{i-1} >$, with $1 \leq i \leq n$, generated by the element X_{i-1} of the basis $(X_0, X_1, \ldots, X_{n-1})$ in which the algebra L_n is defined, so it is $l(L_n) = n$. For the algebra Q_n the natural gradation has length $n-1$ and we cannot obtain another gradation of Q_n with a greater number of nonzero subspaces, so it is $l(Q_n) = n - 1$.

Some filiform Lie algebras with maximum length have demonstrated their importance and usefulness in the study of several cohomological problems on the variety of laws of nilpotent Lie algebras $\mathfrak{N}_n$ [9]. These algebras was been studied in another way (see [1], [3],[8]). Actually, those filiform Lie algebras were obtained by considering the algebras verifying $[X_i, X_j] = a_{ij}\ X_{i+j+1}$ in a certain basis.

We will now restrict our attention to the quasi-filiform case.

Definition 2.2. *An n-dimensional nilpotent Lie algebra* $\mathfrak{g}$ *is said to be* quasi-filiform *if* $C^{n-3}\mathfrak{g} \neq 0$ *and* $C^{n-2}\mathfrak{g} = 0$, *where* $C^0\mathfrak{g} = \mathfrak{g}$, $C^i\mathfrak{g} = [\mathfrak{g}, C^{i-1}\mathfrak{g}]$.

The quasi-filiform Lie algebras of maximum length which are non-split (non-trivial extension of the filiform ones) has been obtained in [6], and they are defined in a basis $(X_0, X_1, \ldots X_{n-2}, Y)$ as follows:

If $n \geq 5$ and n is odd, the algebra

$$\mathfrak{g}^1 = \begin{cases} [X_0, X_i] = X_{i+1}, & 1 \leq i \leq n-3, \\ [X_i, X_{n-2-i}] = (-1)^{i-1}Y, & 1 \leq i \leq \frac{n-3}{2}. \end{cases}$$

If $n \geq 5$, the algebra

$$\mathfrak{g}^2 = \begin{cases} [X_0, X_i] = X_{i+1},\ 1 \leq i \leq n-3, \\ [X_i, Y] = X_{i+2},\ \ 1 \leq i \leq n-4. \end{cases}$$

If $n \geq 7$, the algebra

$$\mathfrak{g}^3 = \begin{cases} [X_0, X_i] = X_{i+1},\ 1 \leq i \leq n-3, \\ [X_i, Y] = X_{i+2},\ \ 1 \leq i \leq n-4, \\ [X_1, X_i] = X_{i+3},\ 2 \leq i \leq n-5. \end{cases}$$

There are no n-dimensional quasi-filiform Lie algebras of maximum length different from the algebras described above, when $n > 12$. But, in low dimension we have to consider some additional Lie algebras to complete the classification [6].

2.2 Graded Quasi-Filiform Lie Algebras of Length $n-1$ from Extensions

One can easily check that the basis $(X_0, X_1, \ldots X_{n-2}, Y)$ is an *adapted* homogeneous basis for each quasi-filiform algebra $\mathfrak{g}^1, \mathfrak{g}^2, \mathfrak{g}^3$ of maximum length above defined, i.e., an homogeneous basis for the gradation $\mathfrak{g} = \mathfrak{g}_1 \oplus \mathfrak{g}_2 \oplus \cdots \oplus \mathfrak{g}_n$ such that

$$[X_0, X_i] = X_{i+1}, \quad 1 \leq i \leq n-3$$
$$[X_0, X_{n-2}] = [X_0, Y] = 0.$$

If $\mathfrak{g}$ is an n-dimensional quasi-filiform Lie algebra of length $n-1$, then we can always choose an adapted basis formed by homogeneous vectors. It will allow us to obtain the structure of the Lie algebras which will be considered.

Lemma 2.1. [10] *Let $\mathfrak{g} = \mathfrak{g}_1 \oplus \mathfrak{g}_2 \oplus \cdots \oplus \mathfrak{g}_{n-1}$ be an n-dimensional non-split quasi-filiform Lie algebra with length $l(\mathfrak{g}) = n-1$. Then, there exists an adapted and homogeneous basis $(X_0, X_1, \ldots, X_{n-2}, Y)$ of $\mathfrak{g}$ such that*

$$\mathfrak{g} =< X_0 > \oplus < X_1 > \oplus \cdots \oplus < X_{m-1}, Y > \oplus \cdots \oplus < X_{n-2} >$$

with $\mathfrak{g}_m =< X_{m-1}, Y >$ and $1 \leq m \leq n-1$.

To distinguish the subspace $\mathfrak{g}_i$ containing the vector Y from the adapted homogeneous basis, we will say that $\mathfrak{g} = \mathfrak{g}_1 \oplus \mathfrak{g}_2 \oplus \cdots \oplus \mathfrak{g}_{n-1}$ is an *algebra of type* $\mathfrak{g} = \mathfrak{g}_{(n,p)}$ when $Y \in \mathfrak{g}_p$ (so, $\dim \mathfrak{g}_p = 2$).

The trivial algebraic extension of the filiform Lie algebra Q_{n-1} (if n is even) is a quasi-filiform Lie algebra of length $n-1$ with $\dim(C^1\mathfrak{g}) = n-3$. In addition to the algebra $Q_{n-1} \oplus \mathbb{C}$, there are other quasi-filiform Lie algebras of length $n-1$ and $\dim(C^1\mathfrak{g}) = n-3$, which can be obtained easily from the filiform Lie algebras of maximum length. In particular, denoting by $\lfloor \alpha \rfloor$ the integral part of α, the algebras $\mathfrak{g}^i_{(n,p)}$ with $1 \leq i \leq 5$, defined below, in a basis $(X_0, X_1, \ldots, X_{n-2}, Y)$, have type $\mathfrak{g} = \mathfrak{g}_{(n,p)}$.

If $n \geq 6$, for $3 \leq p \leq n-3$,

$$\mathfrak{g}^1_{(n,p)} = \begin{cases} [X_0, X_i] = X_{i+1}, \ 1 \leq i \leq n-3, \\ [X_i, Y] = X_{i+p}, \quad 1 \leq i \leq n-p-2. \end{cases}$$

If $n \geq 6$, for $3 \leq p \leq n-3$,

$$\mathfrak{g}^2_{(n,p)} = \begin{cases} [X_0, X_i] = X_{i+1}, \ 1 \leq i \leq n-3, \\ [X_i, Y] = X_{i+p}, \quad 1 \leq i \leq n-p-2, \\ [X_1, X_i] = X_{i+2}, \ 2 \leq i \leq n-4. \end{cases}$$

If $n \geq 8$ and even, for $3 \leq p \leq n-3$ and odd,

$$\mathfrak{g}^3_{(n,p)} = \begin{cases} [X_0, X_i] = X_{i+1}, & 1 \leq i \leq n-3, \\ [X_i, Y] = X_{i+p}, & 1 \leq i \leq n-p-2, \\ [X_i, X_{n-3-i}] = (-1)^{i-1} X_{n-2}, & 1 \leq i \leq \frac{n-4}{2}. \end{cases}$$

If $n \geq 9$, for $n-5 \leq p \leq n-3$,

$$\mathfrak{g}^4_{(n,p)} = \begin{cases} [X_0, X_i] = X_{i+1}, & 1 \leq i \leq n-3, \\ \left[X_i, X_{2\lfloor\frac{n-3}{2}\rfloor-1-i}\right] = (-1)^{i-1} X_{2\lfloor\frac{n-3}{2}\rfloor}, & 1 \leq i \leq \lfloor\frac{n-5}{2}\rfloor, \\ \left[X_i, X_{2\lfloor\frac{n-3}{2}\rfloor-i}\right] = (-1)^{i-1}\left(\lfloor\frac{n-3}{2}\rfloor - i\right) X_{2\lfloor\frac{n-3}{2}\rfloor+1}, & 1 \leq i \leq \lfloor\frac{n-5}{2}\rfloor, \\ [X_i, X_{n-3-i}] = (-1)^{i\,\frac{(i-1)(n-4-i)}{2}}\, \alpha\, X_{n-2}, & 2 \leq i \leq \frac{n-4}{2}, \\ [X_i, Y] = X_{i+p}, & 1 \leq i \leq n-p-2. \end{cases}$$

where $\alpha = 0$, if n is odd, and $\alpha = 1$, if n is even.

If $n \geq 13$, for $n-5 \leq p \leq n-3$,

$$\mathfrak{g}^5_{(n,p)} = \begin{cases} [X_0, X_i] = X_{i+1}, & 1 \leq i \leq n-3, \\ [X_i, X_j] = \frac{6\,(i-1)!\,(j-1)!\,(j-i)}{(i+j)!} X_{i+j+1}, & 1 \leq i \leq \lfloor\frac{n-4}{2}\rfloor, i < j \leq n-3-i, \\ [X_i, Y] = X_{i+p}, & 1 \leq i \leq n-p-2, \end{cases}$$

These algebras can be obtained by considering extensions by derivations of the filiform Lie algebra of maximum length, which was been studied in this way by the authors of this paper in [5]. Thus, for instance the quasi-filiform Lie algebra $\mathfrak{g}^5_{(n,p)}$ can be obtained from the filiform algebra W_n, namely by the adjunction of an obvious operation.

However, as happened on the case of the quasi-filiform Lie algebras of maximum length, in low dimensions some algebras other than $\mathfrak{g}^i_{(n,p)}$, $i = 1, \ldots, 5$, could appear. Even more, we have not found yet any algebra of length $l(\mathfrak{g}) = n-1$ with $\dim(C^1\mathfrak{g}) = n-2$.

In next Section we will show how the quasi-filiform algebras of such a length $l(\mathfrak{g}) = n-1$ can be determined in a concrete dimension by using symbolic calculus, and we will study the particular cases where the algebras have dimension $n \leq 15$.

3 Low-Dimensional Quasi-Filiform Lie Algebras of Type $\mathfrak{g} = \mathfrak{g}_{(n,p)}$

In this section we will consider that all the n-dimensional algebras $\mathfrak{g}$ has length $l(\mathfrak{g}) = n-1$ and type $\mathfrak{g} = \mathfrak{g}_{(n,p)}$. First, we will describe the structure for the family, and how we could use symbolic calculus to obtain the classification of that family. Eventually, we present the classification.

3.1 Symbolic Calculus on Lie Algebras

We start obtaining some properties of the algebras which they will permit to develop the appropriate algorithms to reach the goal of the classification in concrete dimension.

Lemma 3.1. [10] *Let* $(X_0, X_1, \ldots, X_{n-2}, Y)$ *be an adapted basis of a quasi-filiform Lie algebra of type* $\mathfrak{g} = \mathfrak{g}_{(n,p)}$ *with* $p \geq 1$. *Then the structure of* $\mathfrak{g}$ *is:*

$$\mathfrak{g} = \begin{cases} [X_0, X_i] = X_{i+1}, & 1 \leq i \leq n-3, \\ [X_i, X_j] = a_{i,j}\, X_{i+j+1} + \alpha\delta_{p-2-i,j} B_i\, Y, & 1 \leq i < j \leq n-3-i, \\ [X_i, Y] = A_i\, X_{i+p}, & 1 \leq i \leq n-2-p, \end{cases}$$

where $\alpha = 0$ *if* $p \leq 4$, $\delta_{i,j} = 1$ *if* $i = j$ *and* $\delta_{i,j} = 0$ *if* $i \neq j$, *and the structure constants* $\{a_{i,j}, B_i, A_i\}$ *verify the Jacobi's relations.*

Now we can obtain the relationship among the structure constants $a_{i,j}$ for a quasi-filiform Lie algebra of type $\mathfrak{g} = \mathfrak{g}_{(n,p)}$.

Proposition 3.1. *Let* $\mathfrak{g}$ *be a quasi-filiform Lie algebra of type* $\mathfrak{g} = \mathfrak{g}_{(n,p)}$ *with* $1 \leq p \leq n-1$, *as in Lemma 3.1. Then:*

(a) $A_i = A_1$ *for* $1 \leq i \leq n-2-p$.

(b) $$\begin{cases} B_i = (-1)^{i-1} B_1, \ i \leq \left\lfloor \frac{p-3}{2} \right\rfloor, & \text{if } p \text{ is odd;} \\ B_i = 0, & \text{if } p \text{ is even.} \end{cases}$$

(c) $$a_{i,j} = \sum_{k=0}^{i-1} (-1)^k \binom{i-1}{k} a_{1,j+k}, \quad \textit{with } 2 \leq i < j \leq n-3-i.$$

(d) $$\sum_{k=0}^{i} (-1)^k \binom{i}{k} a_{1,i+1+k} = 0, \quad \textit{for } 1 \leq i \leq \left\lfloor \tfrac{n-5}{2} \right\rfloor.$$

Proof. Let $(X_0, X_1, \ldots, X_{n-2}, Y)$ an adapted basis of a quasi-filiform Lie algebra $\mathfrak{g}$ of type $\mathfrak{g} = \mathfrak{g}_{(n,p)}$. Then $\mathfrak{g}$ belongs to the family given in Lemma 3.1. The proof is a consequence of some Jacobi identities $[X,[Y,Z]]+[Y,[Z,X]]+[Z,[X,Y]] = 0$ (denoted by $J(X,Y,Z) = 0$ from now on) for appropriate elements $X, Y, Z \in \mathfrak{g}$.

For instance, the Jacobi relations $J(X_0, X_i, Y) = 0$ for $1 \leq l < i \leq n-3-p$ imply $A_i = A_1$, so (a) is verified. To prove (b) consider the Jacobi relations $J(X_0, X_i, X_{p-3-i}) = 0$ for $1 \leq i \leq \lfloor (p-5)/2 \rfloor$. Then, we have that $B_i = (-1)^{i-1} B_1$ for $\lfloor (p-3)/2 \rfloor$; if p is even, since the Jacobi relation $J(X_0, X_{\frac{p-4}{2}}, X_{\frac{p-2}{2}}) = 0$ implies $B_1 = 0$, we have all $B_i = 0$. Now, we will consider Jacobi's relations $J(X_0, X_m, X_j) = 0$ for $1 \leq m < j \leq n-3-m$ and we will prove (c) by induction. Indeed, from the Jacobi relations $J(X_0, X_m, X_j) = 0$, $1 \leq m < j-1$, we have that $a_{m+1,j} = a_{m,j} - a_{m,j+1}$, and therefore the expression is true for the particular case $m = 1$, that is $i = 2$, and every j. We suppose now for $2 \leq i \leq s$ and every j that

$$a_{s,j} = \sum_{k=0}^{s-1} (-1)^k \binom{s-1}{k} a_{1,j+k}.$$

and we will obtain $a_{s+1,j}$ for every j. In fact, the Jacobi relation $J(X_0, X_s, X_j) = 0$ implies $a_{s+1,j} = a_{s,j} - a_{s,j+1}$, and we have

$$a_{s+1,j} = \sum_{k=0}^{s-1} (-1)^k \binom{s-1}{k} a_{1,j+k} - \sum_{k=0}^{s-1} (-1)^k \binom{s-1}{k} a_{1,j+1+k}.$$

Hence

$$(-1)^0 \tbinom{s-1}{0} a_{1,j} = (-1)^0 \tbinom{s}{0} a_{1,j},$$

$$(-1)^k \left(\tbinom{s-1}{k} + \tbinom{s-1}{k-1} \right) a_{1,j+k} = (-1)^k \tbinom{s}{k} a_{1,j+k},\ 1 \leq k \leq s-1,$$

$$(-1)^s \tbinom{s-1}{s-1} a_{1,j+s} = (-1)^s \tbinom{s}{s} a_{1,j+s}.$$

Thus

$$a_{s+1,j} = \sum_{k=0}^{s} (-1)^k \binom{s}{k} a_{1,j+k}.$$

Finally, (d) is obtained in a similar way from $J(X_0, X_i, X_{i+1}) = 0$, $1 \leq i \leq \lfloor (n-5)/2 \rfloor$, by expressing

$$a_{1,i+1} + \sum_{k=1}^{i-1} (-1)^k \left(\binom{i-1}{k-1} + \binom{i-1}{k} \right) a_{1,i+1+k} + (-1)^i a_{1,2i+1} = 0,$$

in the appropriate form. □

We remark that when $A_1 \neq 0$ or $B_1 \neq 0$, we can always consider $A_1 = A = 1$ and $B_1 = B = 1$ in Proposition 3.1, by changing to the appropriate adapted homogeneous basis if it is necessary. Thus, we can summarize the result above as

Corollary 3.1. *All brackets of an n-dimensional quasi-filiform Lie algebra of length $n-1$ in an adapted homogeneous basis $(X_0, X_1, \ldots, X_{n-2}, Y)$ are determined by the brackets*

$$\begin{aligned}
[X_1, X_2] &= a_{1,2}\, X_4 + \delta_{2,p-3}\, B\, Y, \\
[X_1, X_4] &= a_{1,4}\, X_6 + \delta_{4,p-3}\, B\, Y, \\
&\ \ \vdots \qquad\qquad \vdots \\
[X_1, X_{2\lfloor \frac{n-4}{2} \rfloor}] &= a_{1,2\lfloor \frac{n-4}{2} \rfloor}\, X_{2\lfloor \frac{n-4}{2} \rfloor + 2} + \delta_{2\lfloor \frac{n-4}{2} \rfloor, p-3}\, B\, Y, \\
[X_1, Y] &= A\, X_{1+p}.
\end{aligned}$$

Proof. Indeed, $[X_0, X_i] = X_{i+1}$, $1 \leq i \leq n-3$, because of the adapted homogeneous basis chosen. Since (c) and (d) in Proposition 3.1 the coefficients $a_{i,j}$ are obtained from $a_{1,k}$, and the coefficients $a_{1,2k+1}$ are function of $a_{1,2i}$, with $1 \leq i \leq k$. Finally, from (a) the constants of structure A_i of brackets $[X_i, Y]$, $2 \leq i \leq n-p-2$, are $A_i = A_1 = A$. □

The brackets in Corollary 3.1 above together with the brackets $[X_0, X_i]$ determine the quasi-filiform Lie algebras that we are looking for.

Definition 3.1. Let $(X_0, X_1, \ldots, X_{n-2}, Y)$ be an adapted basis of a quasi-filiform Lie algebra of type $\mathfrak{g} = \mathfrak{g}_{(n,p)}$ with $p \geq 1$. Then, the brackets $[X_0, X_i] = X_{i+1}$, $1 \leq i \leq n-3$, and the brackets defined in Corollary 3.1 will be called *fundamental brackets* of the algebra. The algebra $\mathfrak{g}$ will be denoted by

$$\mathfrak{g} = FB\left([X_1, X_2], [X_1, X_4], \ldots, [X_1, X_{2\lfloor\frac{n-4}{2}\rfloor}], [X_1, Y]\right).$$

Now we finish this section with some logical functions of formal calculus which have been developed by using the software *Mathematica* [12]. These functions permit to use the computer as assistant for the study of the graded Lie algebras. The package permit among other utilities:

(1) To generate the initial family of Lie algebras to consider in every dimension.
(2) To compute the Jacobi identities for the family.
(3) To simplify the restriction given by the structure constants.
(4) To substitute the simplified parameter to reduce the family of Lie algebras.

With the simplification obtained in items above we can study the resulting family to get the classification.

The following functions show how the law of the algebras has been generated. First, we obtain the brackets with the vector X_0 of the adapted homogeneous basis.

```
For[i=1, i<=dim-3, i++, mu[X[0], X[i]] = X[i+1]  ];
mu[X[0], X[dim-2]]=0;  mu[X[0], X[dim-1]]=0;
```

Then, we assume the conditions for a graded quasi-filiform Lie algebra of type $\mathfrak{g} = \mathfrak{g}_{(n,p)}$, where the vector X_{n-1} is the vector Y of the adapted basis.

```
For[i=1, i<=dim-2, i++,
  If[i<=dim-grad-2,
    mu[X[i],X[dim-1]]=X[i+grad], mu[X[i],X[dim-1]]=0  ]  ];
```

Finally, we compute the rest of the bracket with the vectors of the basis, by using Proposition 3.1. For instance, when the vector Y is not in the derived algebra we have:

```
For[i=1, i<=dim-3, i++, mu[X[i],X[dim-2]]=0 ];

For[j=2, j<=dim-3, j++,
  If[j<=dim-4,
    If[j+3 !=grad, mu[X[1],X[j]]= a[1,j] X[j+2],
      mu[X[1],X[j]]= a[1,j] X[grad-1] ],
    mu[X[1],X[j]]= 0  ] ];

For[i=2, i<=dim-4, i++,
  For[j=i+1, j<= dim-3, j++,
    If[i+j<= dim-3,
     If[i+j+1 != grad-1,
       mu[X[i],X[j]]=
        Sum[(-1)^k Binomial[i-1,k] a[1,j+k],{k,0,i-1}] X[i+j+1],
       mu[X[i],X[j]]=
        Sum[(-1)^k Binomial[i-1,k] a[1,j+k],{k,0,i-1}] X[grad-1]],
     mu[X[i],X[j]]=0    ]  ]  ];
```

3.2 Classification's Theorem

The algebras of the List of Laws in the next theorem which are not extension of the filiform ones will be expressed as they are denoted in Definition 3.1.

Theorem 3.1. (Quasi-filiform algebras with length $n-1$) *Every n-dimensional complex quasi-filiform Lie algebra law of length $l(\mathfrak{g}) = n-1$ and type $\mathfrak{g} = \mathfrak{g}_{(n,p)}$, with $n \leq 15$, is isomorphic to a law $\mathfrak{g}_n^i$ of the following List of Laws. The laws $\mathfrak{g}_n^i$ and $\mathfrak{g}_n^j$ are not isomorphic for $i \neq j$. Two laws of the same family $\mathfrak{g}_n^i(\alpha)$ and $\mathfrak{g}_n^i(\alpha')$, with $\alpha \neq \alpha'$ are also non-isomorphic.*

List of Laws

Dimension 6.

$\mathfrak{g}_6^{1+i} = \mathfrak{g}_{(6,3)}^{1+i}$, *for* $i = 0, 1$.

$\mathfrak{g}_6^3 = FB([X_1, X_2] = Y)$.

Dimension 7.

$\mathfrak{g}_7^{1+i} = \mathfrak{g}_{(7,3+k)}^{1+j}$, *where* $i = 2k + j$ *for* $k = 0, 1$ *and* $j = 0, 1$.

$\mathfrak{g}_7^5 = FB([X_1, X_2] = Y)$.

$\mathfrak{g}_7^6 = FB([X_1, X_2] = X_4 + Y)$.

Dimension 8.

$\mathfrak{g}_8^{1+i} = \mathfrak{g}_{(8,3+k)}^{1+j}$, *where* $i = 3k + j$ *for* $k = 0, 1, 2$ *and* $j = 0, 1, 2$.

$\mathfrak{g}_8^{10} = FB([X_1, X_2] = X_4,\ [X_1, X_4] = \alpha X_6,\ [X_1, Y] = X_4)$, $\alpha \in \mathbb{C} - \{1\}$.

$\mathfrak{g}_8^{11} = FB([X_1, X_2] = X_4,\ [X_1, X_4] = \alpha X_6,\ [X_1, Y] = X_5),\ \alpha \in \mathbb{C} - \{1\}.$
$\mathfrak{g}_8^{12} = FB([X_1, X_2] = Y).$
$\mathfrak{g}_8^{13} = FB([X_1, X_2] = Y,\ [X_1, X_4] = X_6).$
$\mathfrak{g}_8^{14} = FB([X_1, X_2] = X_4,\ [X_1, X_4] = \alpha X_6,\ [X_1, Y] = X_6),\ \alpha \in \mathbb{C} - \{1\}.$
$\mathfrak{g}_8^{15} = FB([X_1, X_2] = X_4 + Y,\ [X_1, X_4] = \alpha X_6),\ \alpha \in \mathbb{C}.$
$\mathfrak{g}_8^{16} = FB([X_1, X_2] = \alpha X_4 + Y,\ [X_1, X_4] = \beta X_6,\ [X_1, Y] = X_6)$, *where* $\alpha = a\, e^{i\theta}$, $a \geq 0,\ -\pi/2 \leq \theta < \pi/2$ *and* $\beta \in \mathbb{C}.$
$\mathfrak{g}_8^{17} = FB([X_1, X_4] = Y).$
$\mathfrak{g}_8^{18} = FB([X_1, X_2] = X_4,\ [X_1, X_4] = Y).$

Dimension 9.

$\mathfrak{g}_9^{1+i} = \mathfrak{g}_{(9,3+k)}^{1+j}$, *where* $i = 2k + j$ *for* $k = 0, 1, 2, 3$ *and* $j = 0, 1.$
$\mathfrak{g}_9^{9+i} = \mathfrak{g}_{(9,4+i)}^{4}$, *for* $i = 0, 1, 2.$
$\mathfrak{g}_9^{12} = FB([X_1, X_2] = X_4,\ [X_1, X_4] = \alpha X_6,\ [X_1, Y] = X_5),\ \alpha \in \mathbb{C} - \{1\}.$
$\mathfrak{g}_9^{13} = FB([X_1, X_2] = Y).$
$\mathfrak{g}_9^{14} = FB([X_1, X_2] = Y,\ [X_1, X_4] = X_6).$
$\mathfrak{g}_9^{15} = FB([X_1, X_2] = X_4,\ [X_1, X_4] = \alpha X_6,\ [X_1, Y] = X_6),\ \alpha \in \mathbb{C} - \{1\}.$
$\mathfrak{g}_9^{16} = FB([X_1, X_2] = X_4 + Y,\ [X_1, X_4] = \alpha X_6),\ \alpha \in \mathbb{C}.$
$\mathfrak{g}_9^{17} = FB([X_1, X_2] = \alpha X_4 + Y,\ [X_1, X_4] = \beta X_6,\ [X_1, Y] = X_6)$, *where* $\alpha = a\, e^{i\theta}$, $a \geq 0,\ -\pi/2 \leq \theta < \pi/2$ *and* $\beta \in \mathbb{C}.$
$\mathfrak{g}_9^{18} = FB([X_1, X_2] = X_4,\ [X_1, X_4] = \alpha X_6,\ [X_1, Y] = X_7),\ \alpha \in \mathbb{C} - \{1\}.$
$\mathfrak{g}_9^{19} = FB([X_1, X_4] = Y).$
$\mathfrak{g}_9^{20} = FB([X_1, X_4] = X_6 + Y).$
$\mathfrak{g}_9^{21} = FB([X_1, X_2] = X_4,\ [X_1, X_4] = \alpha X_6 + Y),\ \alpha \in \mathbb{C}.$

Dimension 10.

$\mathfrak{g}_{10}^{1+i} = \mathfrak{g}_{(10,3+k)}^{1+j}$, *where* $i = 2k + j$ *for* $k = 0, 1, 2, 3, 4$ *and* $j = 0, 1.$
$\mathfrak{g}_{10}^{11} = \mathfrak{g}_{(10,3)}^{3}.$
$\mathfrak{g}_{10}^{12+i} = \mathfrak{g}_{(10,5+k)}^{3+j}$, *where* $i = 2k + j$ *for* $k = 0, 1, 2$ *and* $j = 0, 1.$
$\mathfrak{g}_{10}^{18} = FB([X_1, X_2] = X_4,\ [X_1, X_4] = -3X_6,\ [X_1, X_6] = -3X_8,\ [X_1, Y] = X_5).$
$\mathfrak{g}_{10}^{19} = FB([X_1, X_2] = Y).$
$\mathfrak{g}_{10}^{20} = FB([X_1, X_2] = Y,\ [X_1, X_4] = X_6).$
$\mathfrak{g}_{10}^{21} = FB([X_1, X_2] = Y,\ [X_1, X_6] = X_8).$
$\mathfrak{g}_{10}^{22} = FB([X_1, X_2] = X_4,\ [X_1, X_4] = \alpha X_6,\ [X_1, X_6] = \frac{5\alpha - 3}{3 - \alpha} X_8,\ [X_1, Y] = X_6),$ $\alpha \in \mathbb{C} - \{1, 3\}.$
$\mathfrak{g}_{10}^{23} = FB([X_1, X_2] = X_4 + Y, [X_1, X_4] = \alpha X_6, [X_1, X_6] = \frac{5\alpha - 3}{3 - \alpha} X_8), \alpha \in \mathbb{C} - \{3\}.$
$\mathfrak{g}_{10}^{24} = FB([X_1, X_2] = \frac{1}{2\sqrt{3}} X_4 + Y, [X_1, X_4] = \frac{\sqrt{3}}{2} X_6, [X_1, X_6] = \alpha X_8,$

$[X_1, Y] = X_6)$, $\alpha \in \mathbb{C}$.

$\mathfrak{g}_{10}^{25} = FB([X_1, X_2] = \alpha X_4 + Y,\ [X_1, X_4] = \beta X_6,\ [X_1, X_6] = \frac{-1-3\alpha^2+5\alpha\beta}{3\alpha-\beta} X_8,$ $[X_1, Y] = X_6), 3\alpha - \beta \neq 0,$ *where* $\alpha = a\, e^{i\theta},\ a \geq 0,\ -\pi/2 \leq \theta < \pi/2$ *and* $\beta \in \mathbb{C}$.

$\mathfrak{g}_{10}^{26} = FB([X_1, X_2] = X_4,\ [X_1, X_4] = \alpha X_6,\ [X_1, X_6] = \frac{5\alpha-3}{3-\alpha} X_8,\ [X_1, Y] = X_7),$ $\alpha \in \mathbb{C} - \{1, 3\}$.

$\mathfrak{g}_{10}^{27} = FB([X_1, X_4] = Y)$.

$\mathfrak{g}_{10}^{28} = FB([X_1, X_4] = Y,\ [X_1, X_6] = X_8)$.

$\mathfrak{g}_{10}^{29} = FB([X_1, X_4] = X_6 + Y)$.

$\mathfrak{g}_{10}^{30} = FB([X_1, X_2] = X_4,\ [X_1, X_4] = \alpha X_6,\ [X_1, X_6] = \frac{5\alpha-3}{3-\alpha} X_8,\ [X_1, Y] = X_8),$ $\alpha \in \mathbb{C} - \{1, 3\}$.

$\mathfrak{g}_{10}^{31} = FB([X_1, X_2] = X_4, [X_1, X_4] = \alpha X_6 + Y, [X_1, X_6] = \frac{5\alpha-3}{3-\alpha} X_8), \alpha \in \mathbb{C} - \{3\}$.

$\mathfrak{g}_{10}^{32} = FB([X_1, X_2] = \alpha X_4,\ [X_1, X_4] = \beta X_6 + Y,\ [X_1, X_6] = \frac{1-3\alpha^2+5\alpha\beta}{3\alpha-\beta} X_8,$ $[X_1, Y] = X_8), 3\alpha - \beta \neq 0,$ *where* $\alpha = a\, e^{i\theta},\ a \geq 0,\ -\pi/2 \leq \theta < \pi/2$ *and* $\beta \in \mathbb{C}$.

$\mathfrak{g}_{10}^{33} = FB([X_1, X_2] = \frac{i}{2\sqrt{3}} X_4,\ [X_1, X_4] = \frac{i\sqrt{3}}{2} X_6 + Y,\ [X_1, X_6] = \alpha X_8,$ $[X_1, Y] = X_8), \alpha \in \mathbb{C}$.

$\mathfrak{g}_{10}^{34} = FB([X_1, X_6] = Y)$.

Dimension 11.

$\mathfrak{g}_{11}^{1+i} = \mathfrak{g}_{(11,3+k)}^{1+j}$, *where* $i = 2k + j$ *for* $k = 0, 1, 2, 3, 4, 5$ *and* $j = 0, 1$.

$\mathfrak{g}_{11}^{13+i} = \mathfrak{g}_{(11,6+i)}^{4}$, *for* $i = 0, 1, 2$.

$\mathfrak{g}_{11}^{16} = FB([X_1, X_2] = X_4,\ [X_1, X_4] = -3X_6,\ [X_1, X_6] = -3X_8,\ [X_1, Y] = X_5)$.

$\mathfrak{g}_{11}^{17} = FB([X_1, X_2] = X_4,\ [X_1, X_6] = -X_8,\ [X_1, Y] = X_6)$.

$\mathfrak{g}_{11}^{18} = FB([X_1, X_2] = Y)$.

$\mathfrak{g}_{11}^{19} = FB([X_1, X_2] = Y,\ [X_1, X_4] = X_6)$.

$\mathfrak{g}_{11}^{20} = FB([X_1, X_2] = Y,\ [X_1, X_6] = X_8)$.

$\mathfrak{g}_{11}^{21} = FB([X_1, X_2] = X_4 + Y, [X_1, X_4] = \alpha X_6, [X_1, X_6] = \frac{5\alpha-3}{3-\alpha} X_8), \alpha \in \mathbb{C} - \{3\}$.

$\mathfrak{g}_{11}^{22} = FB([X_1, X_2] = \frac{2\alpha^2-1}{2\alpha} X_4 + Y,\ [X_1, X_4] = \alpha X_6,\ [X_1, X_6] = \frac{2\alpha^2+1}{2\alpha} X_8,$ $[X_1, Y] = X_6)$, $\alpha = a\, e^{i\theta}$ *with* $a > 0$ *and* $-\pi/2 \leq \theta < \pi/2$.

$\mathfrak{g}_{11}^{23} = FB([X_1, X_4] = X_6,\ [X_1, Y] = X_7)$.

$\mathfrak{g}_{11}^{24} = FB([X_1, X_2] = X_4,\ [X_1, X_4] = \alpha X_6,\ [X_1, X_6] = \frac{5\alpha-3}{3-\alpha} X_8,\ [X_1, Y] = X_7),$ $\alpha \in \mathbb{C} - \{1, 3\}$.

$\mathfrak{g}_{11}^{25} = FB([X_1, X_4] = X_6,\ [X_1, Y] = X_8)$.

$\mathfrak{g}_{11}^{26} = FB([X_1, X_4] = Y)$.

$\mathfrak{g}_{11}^{27} = FB([X_1, X_4] = X_6 + Y)$.

$\mathfrak{g}_{11}^{28} = FB([X_1, X_4] = Y,\ [X_1, X_6] = X_8)$.

$\mathfrak{g}_{11}^{29} = FB([X_1, X_2] = X_4,\ [X_1, X_4] = \alpha X_6,\ [X_1, X_6] = \frac{5\alpha-3}{3-\alpha} X_8,\ [X_1, Y] = X_8),$ $\alpha \in \mathbb{C} - \{1, 3\}.$

$\mathfrak{g}_{11}^{30} = FB([X_1, X_2] = X_4, [X_1, X_4] = \alpha X_6 + Y, [X_1, X_6] = \frac{5\alpha-3}{3-\alpha} X_8), \alpha \in \mathbb{C} - \{3\}.$

$\mathfrak{g}_{11}^{31} = FB([X_1, X_2] = \frac{i}{2\sqrt{3}} X_4,\ [X_1, X_4] = \frac{i}{2}\sqrt{3} X_6 + Y,\ [X_1, X_6] = \alpha X_8,$ $[X_1, Y] = X_8),\ \alpha \in \mathbb{C}.$

$\mathfrak{g}_{11}^{32} = FB([X_1, X_2] = \alpha X_4,\ [X_1, X_4] = \beta X_6 + Y,\ [X_1, X_6] = \frac{1-3\alpha^2+5\alpha\beta}{3\alpha-\beta} X_8,$ $[X_1, Y] = X_8),\ 3\alpha - \beta \neq 0,$ *where* $\alpha = a\, e^{i\theta},\ a \geq 0,\ -\pi/2 \leq \theta < \pi/2$ *and* $\beta \in \mathbb{C}.$

$\mathfrak{g}_{11}^{33} = FB([X_1, X_4] = X_6,\ [X_1, Y] = X_9).$

$\mathfrak{g}_{11}^{34} = FB([X_1, X_2] = X_4,\ [X_1, X_4] = \alpha X_6,\ [X_1, X_6] = \frac{5\alpha-3}{3-\alpha} X_8,\ [X_1, Y] = X_9),$ $\alpha \in \mathbb{C} - \{1, 3\}.$

$\mathfrak{g}_{11}^{35} = FB([X_1, X_6] = Y).$

$\mathfrak{g}_{11}^{36} = FB([X_1, X_6] = X_8 + Y).$

Dimension 12.

$\mathfrak{g}_{12}^{1+i} = \mathfrak{g}_{(12,3+k)}^{1+j}$, *where* $i = 2k + j$ *for* $k = 0, 1, 2, 3, 4, 5, 6$ *and* $j = 0, 1.$

$\mathfrak{g}_{12}^{15+i} = \mathfrak{g}_{(12,2i+3)}^{3}$, *for* $i = 0, 1.$

$\mathfrak{g}_{12}^{17+i} = \mathfrak{g}_{(12,7+k)}^{3+j}$, *where* $i = 2k + j$ *for* $k = 0, 1, 2$ *and* $j = 0, 1.$

$\mathfrak{g}_{12}^{23} = FB([X_1, X_2] = Y).$

$\mathfrak{g}_{12}^{24} = FB([X_1, X_2] = Y,\ [X_1, X_6] = X_8).$

$\mathfrak{g}_{12}^{25} = FB([X_1, X_2] = Y,\ [X_1, X_8] = X_{10}).$

$\mathfrak{g}_{12}^{26} = FB([X_1, X_2] = Y,\ [X_1, X_4] = X_6, [X_1, X_8] = \frac{-5}{2} X_{10}).$

$\mathfrak{g}_{12}^{27} = FB([X_1, X_2] = X_4 + Y,\ [X_1, X_4] = \alpha X_6,\ [X_1, X_6] = \frac{5\alpha-3}{3-\alpha} X_8,\ [X_1, X_8] =$ $\frac{5\alpha^3+\alpha^2-7\alpha+3}{2\alpha(2-\alpha)} X_{10}),\ \alpha \in \mathbb{C} - \{0, 2, 3\}.$

$\mathfrak{g}_{12}^{28} = FB([X_1, X_2] = \frac{2\alpha^2-1}{2\alpha} X_4 + Y,\ [X_1, X_4] = \alpha X_6,\ [X_1, X_6] = \frac{2\alpha^2+1}{2\alpha} X_8,$ $[X_1, X_8] = \frac{(1-2\alpha^2)^2}{4\alpha^3} X_{10},\ [X_1, Y] = X_6), \alpha = a\, e^{i\theta}, a > 0, -\pi/2 \leq \theta < \pi/2.$

$\mathfrak{g}_{12}^{29} = FB([X_1, X_4] = X_6,\ [X_1, X_8] = -\frac{5}{2} X_{10},\ [X_1, Y] = X_7).$

$\mathfrak{g}_{12}^{30} = FB([X_1, X_2] = X_4,\ [X_1, X_4] = \frac{3}{7} X_6,\ [X_1, X_6] = -\frac{1}{3} X_8,\ [X_1, X_8] = \frac{3}{7} X_{10},$ $[X_1, Y] = X_7).$

$\mathfrak{g}_{12}^{31} = FB([X_1, X_4] = X_6,\ [X_1, X_8] = -\frac{5}{2} X_{10},\ [X_1, Y] = X_8).$

$\mathfrak{g}_{12}^{32} = FB([X_1, X_4] = Y).$

$\mathfrak{g}_{12}^{33} = FB([X_1, X_4] = Y,\ [X_1, X_6] = X_8).$

$\mathfrak{g}_{12}^{34} = FB([X_1, X_4] = Y,\ [X_1, X_8] = X_{10}).$

$\mathfrak{g}_{12}^{35} = FB([X_1, X_4] = X_6 + Y,\ [X_1, X_8] = -\frac{5}{2} X_{10}).$

$\mathfrak{g}_{12}^{36} = FB([X_1, X_2] = X_4,\ [X_1, X_4] = \alpha X_6,\ [X_1, X_6] = \frac{5\alpha-3}{3-\alpha} X_8,\ [X_1, X_8] =$ $\frac{5\alpha^3+\alpha^2-7\alpha+3}{2\alpha(2-\alpha)} X_{10},\ [X_1, Y] = X_8),\ \alpha \in \mathbb{C} - \{0, 1, 2, 3\}.$

$\mathfrak{g}_{12}^{37} = FB([X_1,X_2] = X_4,\ [X_1,X_4] = \alpha X_6 + Y,\ [X_1,X_6] = \frac{5\alpha-3}{3-\alpha}X_8,\ [X_1,X_8] = \frac{5\alpha^3+\alpha^2-7\alpha+3}{2\alpha(2-\alpha)}X_{10}),\ \alpha \in \mathbb{C} - \{0,2,3\}$.

$\mathfrak{g}_{12}^{38} = FB([X_1,X_2] = \frac{1}{2}\sqrt{\frac{-9+16i\sqrt{6}}{33}}X_4,\ [X_1,X_4] = \frac{1}{2}\sqrt{\frac{-3-2i\sqrt{6}}{33}}X_6 + Y,$
$[X_1,X_6] = \frac{1}{2}\sqrt{\frac{(-177-8i\sqrt{6}}{33}}X_8, [X_1,X_8] = \alpha X_{10}),\ \alpha \in \mathbb{C}$.

$\mathfrak{g}_{12}^{39} = FB([X_1,X_2] = \frac{1}{2}\sqrt{\frac{-9-16i\sqrt{6}}{33}}X_4,\ [X_1,X_4] = \frac{1}{2}\sqrt{\frac{-3+2i\sqrt{6}}{33}}X_6 + Y,$
$[X_1,X_6] = \frac{1}{2}\sqrt{\frac{-177+8i\sqrt{6}}{33}}X_8, [X_1,X_8] = \alpha X_{10}), \alpha \in \mathbb{C}$.

$\mathfrak{g}_{12}^{40} = FB([X_1,X_2] = \frac{i}{2\sqrt{3}}X_4, [X_1,X_4] = \alpha X_6 + Y, [X_1,X_6] = \frac{-5i}{3\sqrt{3}}X_8, [X_1,X_8] = \frac{5i+32\sqrt{3}\alpha-60i\alpha^2}{-4\sqrt{3}+24i\alpha}X_{10},\ [X_1,Y] = X_8)$, *where* $\alpha = ae^{i\theta}$, $a \geq 0$,
$-\pi/2 \leq \theta < \pi/2$ *and* $\alpha \neq \frac{-i}{2\sqrt{3}}$.

$\mathfrak{g}_{12}^{41} = FB([X_1,X_2] = \alpha X_4,\ [X_1,X_4] = \frac{-1+3\alpha^2+3\alpha\beta}{5\alpha+\beta}X_6 + Y,\ [X_1,X_6] = \beta X_8,$
$[X_1,X_8] = \frac{5(-2+14\alpha^2+\alpha^4)+\alpha(32-\alpha^2)\beta+(47\alpha^2-6)\beta^2+21\alpha\beta^3}{(5\alpha+\beta)((7\alpha-\beta)(\alpha+\beta)-4)}X_{10}, [X_1,Y] = X_8),$
$5\alpha+\beta \neq 0$ *and* $4 - 7\alpha^2 - 6\alpha\beta + \beta^2 \neq 0$, *where* $\alpha = ae^{i\theta}$, $a \geq 0$,
$-\pi/2 \leq \theta < \pi/2$ *and* $\beta \in \mathbb{C}$.

$\mathfrak{g}_{12}^{42} = FB([X_1,X_4] = X_6,\ [X_1,X_8] = -\frac{5}{2}X_{10},\ [X_1,Y] = X_9)$.

$\mathfrak{g}_{12}^{43} = FB([X_1,X_2] = X_4,\ [X_1,X_4] = \alpha X_6,\ [X_1,X_6] = \frac{5\alpha-3}{3-\alpha}X_8,\ [X_1,X_8] = \frac{5\alpha^3+\alpha^2-7\alpha+3}{2\alpha(2-\alpha)}X_{10},\ [X_1,Y] = X_9),\ \alpha \in \mathbb{C} - \{0,1,2,3\}$.

$\mathfrak{g}_{12}^{44} = FB([X_1,X_4] = X_6,\ [X_1,X_8] = -\frac{5}{2}X_{10},\ [X_1,Y] = X_{10})$.

$\mathfrak{g}_{12}^{45} = FB([X_1,X_6] = Y)$.

$\mathfrak{g}_{12}^{46} = FB([X_1,X_6] = X_8 + Y)$.

$\mathfrak{g}_{12}^{47} = FB([X_1,X_6] = Y,\ [X_1,X_8] = X_{10})$.

$\mathfrak{g}_{12}^{48} = FB([X_1,X_2] = X_4,\ [X_1,X_4] = \alpha X_6,\ [X_1,X_6] = \frac{5\alpha-3}{3-\alpha}X_8,\ [X_1,X_8] = \frac{5\alpha^3+\alpha^2-7\alpha+3}{2\alpha(2-\alpha)}X_{10},\ [X_1,Y] = X_{10}),\ \alpha \in \mathbb{C} - \{0,1,2,3\}$.

$\mathfrak{g}_{12}^{49} = FB([X_1,X_6] = \alpha X_8 + Y, [X_1,X_8] = -\frac{1}{\alpha}X_{10}, [X_1,Y] = X_{10}),\ \alpha \in \mathbb{C} - \{0\}$.

$\mathfrak{g}_{12}^{50} = FB([X_1,X_8] = Y)$.

Dimension 13.

$\mathfrak{g}_{13}^{1+i} = \mathfrak{g}_{(13,3+k)}^{1+j}$, *where* $i = 2k + j$ *for* $k = 0,1,2,3,4,5,6,7$ *and* $j = 0,1$.

$\mathfrak{g}_{13}^{17+i} = \mathfrak{g}_{(13,8+k)}^{4+j}$, *where* $i = 2k + j$ *for* $k = 0,1,2$ *and* $j = 0,1$.

$\mathfrak{g}_{13}^{23} = FB([X_1,X_2] = Y)$.

$\mathfrak{g}_{13}^{24} = FB([X_1,X_2] = Y,\ [X_1,X_8] = X_{10})$.

$\mathfrak{g}_{13}^{25} = FB([X_1,X_2] = X_4 + Y,\ [X_1,X_4] = X_6,\ [X_1,X_6] = X_8,\ [X_1,X_8] = X_{10})$.

$\mathfrak{g}_{13}^{26} = FB([X_1,X_2] = X_4+Y, [X_1,X_4] = \frac{9}{10}X_6, [X_1,X_6] = \frac{5}{7}X_8, [X_1,X_8] = \frac{7}{12}X_{10})$.

$\mathfrak{g}_{13}^{27} = FB([X_1,X_4] = Y)$.

$\mathfrak{g}_{13}^{28} = FB([X_1,X_4] = Y,\ [X_1,X_8] = X_{10})$.

$\mathfrak{g}_{13}^{29} = FB([X_1, X_2] = X_4,\ [X_1, X_4] = X_6 + Y,\ [X_1, X_6] = X_8,\ [X_1, X_8] = X_{10})$.

$\mathfrak{g}_{13}^{30} = FB([X_1, X_2] = X_4, [X_1, X_4] = \frac{9}{10}X_6 + Y, [X_1, X_6] = \frac{5}{7}X_8, [X_1, X_8] = \frac{7}{12}X_{10})$.

$\mathfrak{g}_{13}^{31} = FB([X_1, X_2] = -\frac{2i}{\sqrt{3}}X_4,\ [X_1, X_4] = -\frac{i}{\sqrt{3}}X_6 + Y,\ [X_1, X_6] = \frac{i}{\sqrt{3}}X_8,$
$[X_1, X_8] = \frac{2i}{\sqrt{3}}X_{10},\ [X_1, Y] = X_8)$.

$\mathfrak{g}_{13}^{32} = FB([X_1, X_2] = \sqrt{\frac{-23+43i\sqrt{3}}{62}}X_4,\ [X_1, X_4] = \sqrt{\frac{2+3i\sqrt{3}}{31}}X_6 + Y,$
$[X_1, X_6] = \sqrt{\frac{2(-11-i\sqrt{3})}{31}}X_8,\ [X_1, X_8] = \sqrt{\frac{1-45i\sqrt{3}}{62}}X_{10},\ [X_1, Y] = X_8)$.

$\mathfrak{g}_{13}^{33} = FB([X_1, X_2] = \sqrt{\frac{-23-43i\sqrt{3}}{62}}X_4,\ [X_1, X_4] = \sqrt{\frac{2-3i\sqrt{3}}{31}}X_6 + Y,$
$[X_1, X_6] = \sqrt{\frac{2(-11+i\sqrt{3})}{31}}X_8,\ [X_1, X_8] = \sqrt{\frac{1+45i\sqrt{3}}{62}}X_{10}, [X_1, Y] = X_8)$.

$\mathfrak{g}_{13}^{34} = FB([X_1, X_6] = Y)$.

$\mathfrak{g}_{13}^{35} = FB([X_1, X_6] = Y,\ [X_1, X_8] = X_{10})$.

$\mathfrak{g}_{13}^{36} = FB([X_1, X_6] = i\sqrt{\frac{2}{7}}X_8 + Y,\ [X_1, X_8] = i\sqrt{\frac{7}{2}}X_{10},\ [X_1, Y] = X_{10})$.

$\mathfrak{g}_{13}^{37} = FB([X_1, X_8] = Y)$.

$\mathfrak{g}_{13}^{38} = FB([X_1, X_8] = X_{10} + Y)$.

Dimension 14.

$\mathfrak{g}_{14}^{1+i} = \mathfrak{g}_{(14,3+k)}^{1+j}$, *where* $i = 2k + j$ *for* $k = 0, 1, 2, 3, 4, 5$ *and* $j = 0, 1$.

$\mathfrak{g}_{14}^{13+i} = \mathfrak{g}_{(14,2i+3)}^{3}$, *for* $i = 0, 1, 2$.

$\mathfrak{g}_{14}^{16+i} = \mathfrak{g}_{(14,9+k)}^{1+j}$, *where* $i = 5k + j$ *for* $k = 0, 1, 2$ *and* $j = 0, 1, 2, 3, 4$.

$\mathfrak{g}_{14}^{31} = FB([X_1, X_2] = Y)$.

$\mathfrak{g}_{14}^{32} = FB([X_1, X_2] = Y,\ [X_1, X_8] = X_{10})$.

$\mathfrak{g}_{14}^{33} = FB([X_1, X_2] = Y,\ [X_1, X_{10}] = X_{12})$.

$\mathfrak{g}_{14}^{34} = FB([X_1, X_2] = X_4 + Y,\ [X_1, X_4] = X_6,\ [X_1, X_6] = X_8,\ [X_1, X_8] = X_{10},$
$[X_1, X_{10}] = X_{12})$.

$\mathfrak{g}_{14}^{35} = FB([X_1, X_2] = X_4 + Y,\ [X_1, X_4] = \frac{9}{10}X_6,\ [X_1, X_6] = \frac{5}{7}X_8,\ [X_1, X_8] =$
$\frac{7}{12}X_{10},\ [X_1, X_{10}] = \frac{27}{55}X_{12})$.

$\mathfrak{g}_{14}^{36} = FB([X_1, X_4] = Y)$.

$\mathfrak{g}_{14}^{37} = FB([X_1, X_4] = Y,\ [X_1, X_8] = X_{10})$.

$\mathfrak{g}_{14}^{38} = FB([X_1, X_4] = Y,\ [X_1, X_{10}] = X_{12})$.

$\mathfrak{g}_{14}^{39} = FB([X_1, X_2] = X_4,\ [X_1, X_4] = X_6 + Y,\ [X_1, X_6] = X_8,\ [X_1, X_8] = X_{10},$
$[X_1, X_{10}] = X_{12})$.

$\mathfrak{g}_{14}^{40} = FB([X_1, X_2] = X_4,\ [X_1, X_4] = \frac{9}{10}X_6 + Y,\ [X_1, X_6] = \frac{5}{7}X_8,\ [X_1, X_8] =$
$\frac{7}{12}X_{10},\ [X_1, X_{10}] = \frac{27}{55}X_{12})$.

$\mathfrak{g}_{14}^{41} = FB([X_1, X_2] = \sqrt{\frac{-23+43i\sqrt{3}}{62}}X_4,\ [X_1, X_4] = \sqrt{\frac{2+3i\sqrt{3}}{31}}X_6 + Y,\ [X_1, X_6] =$
$\sqrt{\frac{2(-11-i\sqrt{3})}{31}}X_8,\ [X_1, X_8] = \sqrt{\frac{1-45i\sqrt{3}}{62}}X_{10},\ [X_1, X_{10}] = -\sqrt{\frac{38-5i\sqrt{3}}{31}}X_{12},$

$[X_1, Y] = X_8)$.

$\mathfrak{g}_{14}^{42} = FB([X_1, X_2] = \sqrt{\frac{-23-43i\sqrt{3}}{62}} X_4,\ [X_1, X_4] = \sqrt{\frac{2-3i\sqrt{3}}{31}} X_6 + Y,\ [X_1, X_6] = \sqrt{\frac{2(-11+i\sqrt{3})}{31}} X_8,\ [X_1, X_8] = \sqrt{\frac{(1+45i\sqrt{3})}{62}} X_{10},\ [X_1, X_{10}] = -\sqrt{\frac{38+5i\sqrt{3}}{31}} X_{12},\ [X_1, Y] = X_8)$.

$\mathfrak{g}_{14}^{43} = FB([X_1, X_6] = Y)$.

$\mathfrak{g}_{14}^{44} = FB([X_1, X_6] = Y,\ [X_1, X_8] = X_{10})$.

$\mathfrak{g}_{14}^{45} = FB([X_1, X_6] = Y,\ [X_1, X_{10}] = X_{12})$.

$\mathfrak{g}_{14}^{46} = FB([X_1, X_6] = i\sqrt{\frac{2}{7}} X_8 + Y,\ [X_1, X_8] = i\sqrt{\frac{7}{2}} X_{10},\ [X_1, Y] = X_{10})$.

$\mathfrak{g}_{14}^{47} = FB([X_1, X_8] = Y)$.

$\mathfrak{g}_{14}^{48} = FB([X_1, X_8] = X_{10} + Y)$.

$\mathfrak{g}_{14}^{49} = FB([X_1, X_8] = Y,\ [X_1, X_{10}] = X_{12})$.

$\mathfrak{g}_{14}^{50} = FB([X_1, X_8] = \alpha X_{10} + Y, [X_1, X_{10}] = -\frac{1}{\alpha} X_{12}, [X_1, Y] = X_{12}), \alpha \in \mathbb{C} - \{0\}$.

$\mathfrak{g}_{14}^{51} = FB([X_1, X_{10}] = Y)$.

Dimension 15.

$\mathfrak{g}_{15}^{1+i} = \mathfrak{g}_{(15,3+k)}^{1+j}$, *where* $i = 2k + j$ *for* $k = 0, 1, 2, 3, 4, 5, 6, 7, 8, 9$ *and* $j = 0, 1$.

$\mathfrak{g}_{15}^{21+i} = \mathfrak{g}_{(15,10+k)}^{4+j}$, *where* $i = 2k + j$ *for* $k = 0, 1, 2$ *and* $j = 0, 1$.

$\mathfrak{g}_{15}^{27} = FB([X_1, X_2] = Y)$.

$\mathfrak{g}_{15}^{28} = FB([X_1, X_2] = Y,\ [X_1, X_{10}] = X_{12})$.

$\mathfrak{g}_{15}^{29} = FB([X_1, X_2] = X_4 + Y,\ [X_1, X_4] = X_6,\ [X_1, X_6] = X_8,\ [X_1, X_8] = X_{10},\ [X_1, X_{10}] = X_{12})$.

$\mathfrak{g}_{15}^{30} = FB([X_1, X_2] = X_4 + Y, [X_1, X_4] = \frac{9}{10} X_6, [X_1, X_6] = \frac{5}{7} X_8, [X_1, X_8] = \frac{7}{12} X_{10}, [X_1, X_{10}] = \frac{27}{55} X_{12})$.

$\mathfrak{g}_{15}^{31} = FB([X_1, X_4] = Y)$.

$\mathfrak{g}_{15}^{32} = FB([X_1, X_4] = Y,\ [X_1, X_{10}] = X_{12})$.

$\mathfrak{g}_{15}^{33} = FB([X_1, X_2] = X_4,\ [X_1, X_4] = X_6 + Y,\ [X_1, X_6] = X_8,\ [X_1, X_8] = X_{10},\ [X_1, X_{10}] = X_{12})$.

$\mathfrak{g}_{15}^{34} = FB([X_1, X_2] = X_4, [X_1, X_4] = \frac{9}{10} X_6 + Y, [X_1, X_6] = \frac{5}{7} X_8, [X_1, X_8] = \frac{7}{12} X_{10}, [X_1, X_{10}] = \frac{27}{55} X_{12})$.

$\mathfrak{g}_{15}^{35} = FB([X_1, X_6] = Y)$.

$\mathfrak{g}_{15}^{36} = FB([X_1, X_6] = Y,\ [X_1, X_{10}] = X_{12})$.

$\mathfrak{g}_{15}^{37} = FB([X_1, X_8] = Y)$.

$\mathfrak{g}_{15}^{38} = FB([X_1, X_8] = Y,\ [X_1, X_{10}] = X_{12})$.

$\mathfrak{g}_{15}^{39} = FB([X_1, X_8] = i\sqrt{\frac{1}{6}} X_{10} + Y,\ [X_1, X_{10}] = i\sqrt{6} X_{12},\ [X_1, Y] = X_{12})$.

$\mathfrak{g}_{15}^{40} = FB([X_1, X_{10}] = Y)$.

$\mathfrak{g}_{15}^{41} = FB([X_1, X_{10}] = X_{12} + Y)$.

3.3 On the Classification of the Quasi-Filiform Lie Algebra of Length $n-1$

In every dimension, we have obtained the classification in a similar way for any type of algebras that we must consider. Now, we summarize how the quasi-filiform Lie algebras with length $n-1$, where $\dim \mathfrak{g} \leq 15$, have been obtained.

• First, we fix the dimensions of the algebras $\mathfrak{g}$, $[\mathfrak{g}, \mathfrak{g}]$, the dimension of the center $\mathbf{Z}(\mathfrak{g})$ and the type of gradation.

• Second, we consider all of the possible types of gradations and by using symbolic calculus we generate the family. Then, we proceed to study each subfamily, by reducing the parameters to get an "easy" probable final expression of the algebras.

• Finally, we complete the classification by proving that the obtained algebras are non isomorphic by using change of basis if that is necessary.

For instance, in dimension 12, the algebras $\mathfrak{g}_{12}^{11}$, $\mathfrak{g}_{12}^{12}$, $\mathfrak{g}_{12}^{19}$, $\mathfrak{g}_{12}^{20}$, $\mathfrak{g}_{12}^{42}$, $\mathfrak{g}_{12}^{43}$, of the list in Theorem 3.1 has been obtained in the following way:

Suppose $\mathfrak{g}$ be an algebra verifying $\dim \mathfrak{g} = 12$, $\dim[\mathfrak{g}, \mathfrak{g}] = 9$, $\dim \mathbf{Z}(\mathfrak{g})$=1, and type $\mathfrak{g} = \mathfrak{g}_{(12,8)}$. Then, the law of the algebra is given by:

$$\mathfrak{g} = \begin{cases} [X_0, X_i] = X_{i+1}, & 1 \leq i \leq 9, \\ [X_i, X_j] = a_{i,j} X_{i+j+1}, & 1 \leq i < j \leq 9-i, \\ [X_i, Y] = X_{i+8}, & 1 \leq i \leq 2. \end{cases}$$

For this type of algebra, the program elaborated ad hoc generates such a family of algebras. Then, from the Jacobi identities that the algebras of the family must verify, we obtain the following relations betwen the structure constants.

$a_{1,2} - a_{1,3} = 0,$
$a_{1,3} - 2\,a_{1,4} + a_{1,5} = 0,$
$a_{1,4} - 3\,a_{1,5} + 3\,a_{1,6} - a_{1,7} = 0,$
$a_{1,3}\,(a_{1,5} - a_{1,6}) - a_{1,6}\,(a_{1,3} - a_{1,4}) - a_{1,2}\,(a_{1,4} - 2\,a_{1,5} + a_{1,6}) = 0,$
$a_{1,4}\,(a_{1,6} - a_{1,7}) - a_{1,7}\,(a_{1,4} - a_{1,5}) = 0,$
$a_{1,5}\,(a_{1,7} - a_{1,8}) + a_{1,2}\,(a_{1,5} - 3\,a_{1,6} + 3\,a_{1,7} - a_{1,8}) - a_{1,8}\,(a_{1,5} - a_{1,6}) = 0,$
$a_{1,3}\,(a_{1,5}-3\,a_{1,6}+3\,a_{1,7}-a_{1,8})+a_{1,8}\,(a_{1,4}-2\,a_{1,5}+a_{1,6})-a_{1,4}\,(a_{1,6}-2\,a_{1,7}+a_{1,8}) = 0.$

Thus, the general family of laws, $\mathfrak{g} = \mathfrak{g}(a_{1,2}, a_{1,3}, a_{1,4}, a_{1,5}, a_{1,6}, a_{1,7}, a_{1,8})$, is determined for the seven parameters $a_{1,j}$, verifying the above equations. Now, we use *Mathematica* [12] as a computer assistant to simplify the set of equations. Indeed, if $a_{1,2} = 0$, by reducing the parameters we have some particular solutions for the algebra $\mathfrak{g}$. Then, it has to be $\mathfrak{g} = \mathfrak{g}_{12}^{11}$, $\mathfrak{g} = \mathfrak{g}_{12}^{19}$, $\mathfrak{g} = \mathfrak{g}_{12}^{20}$ or $\mathfrak{g} = \mathfrak{g}_{12}^{42}$. If $a_{1,2} \neq 0$, the algebra is $\mathfrak{g} = \mathfrak{g}_{12}^{12}$, or we obtain a two-parameter family of algebras,

$\mathfrak{g} = \mathfrak{g}(a_{1,5}, a_{1,6})$. Then, in the bracket

$$[X_1, X_2] = \frac{5a_{1,5} - 5a_{1,6} + \sqrt{25a_{1,5}^2 - 46a_{1,5}a_{1,6} + 25{a_{1,6}}^2}}{2} X_4,$$

we can suppose the coefficient $a_{1,2}$ of X_4 to be 1 (by changing the basis if that is necessary), and we can write

$$\mathfrak{g} = [X_1, X_4] = \alpha X_6,\ [X_1, X_6] = \frac{5\alpha - 3}{3 - \alpha} X_8,\ [X_1, X_8] = \frac{5\alpha^3 + \alpha^2 - 7\alpha + 3}{2\alpha(2 - \alpha)} X_{10},$$

where $\alpha \in \mathbb{C} - \{0, 2, 3\}$. In this case, if $\alpha = 1$, it is $\mathfrak{g} = \mathfrak{g}_{12}^{12}$, and for the other possible values of α the algebra is $\mathfrak{g} = \mathfrak{g}_{12}^{43}$. Now, to prove that every pair of algebras $\mathfrak{g}_{12}^{43}(\alpha_1)$, $\mathfrak{g}_{12}^{43}(\alpha_2)$, are non isomorphic if and only if $\alpha_1 \neq \alpha_2$, we consider the general change of basis $X_0' = \sum_{i=0}^{10} A_i X_i + A_{11} Y$, $X_1' = \sum_{i=0}^{10} B_i X_i + B_{11} Y$, $Y' = \sum_{i=0}^{10} C_i X_i + C_{11} Y$. From the brackets $[X_0', X_i'] = X_{i+1}'$, for $1 \leq i \leq 9$, we obtain the vectors $X_2', X_3', \ldots, X_{10}'$. Since it is also $[X_1', X_2'] = X_4'$ and $[X_1', X_9'] = 0$, we obtain $B_0 = 0$ and $B_1 = A_0^2$. Finally, from the fundamental bracket $[X_1', X_4']$ we obtain $\alpha_1 = \alpha_2$.

3.4 Conclusion

We have seen that Symbolic Calculus can be useful in order to obtain the classification of a family of Lie algebras in low dimensions. In those dimensions usually there exist algebras which do not appear on the general case; and so the computations needed to obtain a classification use to be very complicated. In fact, this approach is also useful to conjecture the existence of certain *patterns* on a family of algebras. And these patterns could lead to obtain the classification of that family in any arbitrary dimension. Some of the results introduced in Section 2 were obtained by using this approach, which can be used to study similar problems. For instance, the n-dimensional quasi-filiform Lie algebras of length $l(\mathfrak{g}) = n-1$, with $n > 15$, seem to match a pattern with some algebras in every dimension. We can see, for example, that the algebras $\mathfrak{g}_{15}^i$, for $i = 27, 31, 35, 37, 40$, are some particular cases of the family $\mathfrak{g}(n,p)$ with $5 \leq p \leq n-1$ and p is odd, defined by:

$$\mathfrak{g}(n,p) = \begin{cases} [X_0, X_i] = X_{i+1}, & 1 \leq i \leq n-3, \\ [X_i, X_{p-2-i}] = (-1)^{i-1} Y, & 1 \leq i \leq \frac{p-3}{2}. \end{cases}$$

Thus, the next step will be to obtain the classification of the quasi-filiform Lie algebras in each dimension with length equal to $n-1$, where $n > 15$.

References

1. F. Bratzlavsky. *Sur les algèbres Admettant un Tore de d'Automoorphismes Donné.* Journal of Algebra 30 (1974) 305–316.

2. J. M. Cabezas, J. R. Gómez, Jiménez-Merchán. Family of p-filiform Lie algebras in *Algebra and Operator Theory.* Kluwer Academics Publishers (1998) 93–102.
3. R. Carles Sur certaines classes d'algèbres de Lie rigides in *Mathematische Annalen.* Springer-Verlag (1985) 477–488.
4. J. R. Gómez, A. Jiménez-Merchán. The graded algebras of a class of Lie algebras in *Mathematics with Vision.* Computational Mechanics Publications (1995) 151–158.
5. J. R. Gómez, A. Jiménez-Merchán, J. Reyes. *Filiform Lie Algebras of maximun Length.* Extracta Mathematicae (to appear).
6. J. R. Gómez, A. Jiménez-Merchán, J. Reyes. *Quasi-Filiform Lie Algebras of Maximum Length.* Linear Algebra and its Applications (to appear).
7. M. Goze, Y. Khakimdjanov. *Sur les algèbres de Lie nilpotentes admettant un tore de dérivations.* Manuscripta Mathematica 84 (1994) 115–224.
8. M. Goze, Y. Khakimdjanov *Nilpotent Lie algebras.* Kluwer A. P. (1996).
9. Y. Khakimdjanov. *Variétés des lois d'algèbres de Lie nilpotentes.* Geometria Dedicata 40, **3** (1991) 269–295.
10. J. Reyes *Álgebras de Lie Casifiliforme Graduadas de Longitud Maximal.* Ph.D. Thesis. Sevilla (1998).
11. M. Vergne. *Cohomologie des algèbres de Lie nilpotentes. Application à l'étude de la variété des algèbres de Lie nilpotentes.* Bulletin Société Mathematique de la France 98 (1970) 81–116.
12. S. Wolfram *The Mathematica Book, 4th. ed.* Cambrige University Press (1999).

Algebraic Methods for Sectioning Parametric Surfaces

Jesus Espinola*, Laureano Gonzalez–Vega* and Ioana Necula*

Departamento de Matemáticas, Estadística y Computación,
Universidad de Cantabria, Santander (Spain)
espinolj@unican.es, gvega@matesco.unican.es and ioana@matesco.unican.es

Abstract. This paper is devoted to show how the already widely used Scientific Computing Systems (in our case `Maple`) integrating symbolic and numeric capabilities can be used to develop Problem Solving Environments very useful to solve problems into a CAD/CAM framework before they are integrated into a concrete CAD/CAM system. It is shown and motivated how algebraic techniques and Scientific Computing Systems can be very useful in Computer Aided Geometric Design by presenting how generic implicitation can be used to solve problems arising when computing the planar section of a parametric surface and to the computation (topologically exact) of a revolution surface sectioning.

1 Introduction

In this paper it is shown how widely used Scientific Computing Systems (in our case `Maple`) integrating symbolic and numeric facilities can be used to develop Problem Solving Environments which are very useful for solving problems into a CAD/CAM framework. The utility of CAD/CAM systems as a way of increasing the efficiency of simulation and design processes in the productive area is nowadays unanswerable. Advantages as production time reduction, final product improvement and cost reduction are frequentely called as the greatest benefits produced by introducing CAD/CAM systems in an industrial environment.

This paper gathers three research lines:

- Studying and improving the graphic tools provided by the actual Scientific Computing Systems (`Mathematica`, `Matlab`, `Maple` and `Axiom`) together with simulation module development in these systems, for geometric modelling and visualization.
- Adapting, developing and integrating the algebraic equation system solution manipulation techniques developed in the framework of FRISCO project (ESPRIT/LTR 21024: European Union) to solve efficiently problems concerning parametric curve and surface manipulation.

* Partially supported by the FEDER Project 1FD97-0409 and by DGESIC PB 98-0713-C02-02 (Ministerio de Educación y Cultura)

- Solving a set of concrete problems (unsolved satisfactorily at this moment) for the CAD/CAM environment CSIS used by CANDEMAT (company dedicated to make dies for cars). This research line is approached by including the developed techniques inside the simulation methods mentioned above and/or including techniques or software before mentioned.

The Computer Aided Geometric Design systems traditionally have been developed using programming languages with good characteristics for scientific computing (`C`, `C++`, `Fortran`, `Basic`, `Pascal`,...). While the hardware technology is progressing rapidly, the software evolution goes on more slowly, although continuously too. This kind of software has been reaching a large diffusion and importance in the last few years. Together with their basic general statements, these systems offer additional modules which include more specific applications. That is the case of `Mathematica` packages and `Matlab` toolboxes.

In `Mathematica`, there is a package called "Graphics Spline", having limited capacities for generating basic entities like Bezier curves and cubic splines. The strong graphical capacities on parametrics, operators, etc. allow these primitives to be applied for interesting but limited models. In `Matlab` there is a basic function for generating cubic splines and (since 1990) a toolbox called "Spline" which contains B-spline functions for curves and surfaces, following the classical work of Carl de Boor ([2], [3]).

Moreover in the last few years the international scientific community has looking for efficient (i.e. rapid) and as precise as possible (i.e. guaranteed validity for the obtained solutions) algorithms/methods for solving algebraic or polynomial equations and for manipulating the sets of its solutions. The exact meaning of the phrase above can be observed more clearly in the following example with a Computer Aided Geometric Design flavour.

Example 1.
We consider the parametric equation of the bicubic surface S

$$\begin{aligned}
x &= 3t(t-1)^2 + (s-1)^3 + 3s \\
y &= 3s(s-1)^2 + t^3 + 3t \\
z &= -3s(s^2-5s+5)t^3 - 3(s^3+6s^2-9s+1)t^2 + \\
&\qquad +t(6s^3+9s^2-18s+3) - 3s(s-1)
\end{aligned}$$

for parameter values between 0 and 1 and displayed in Figure 1.

A first problem which can arise in this situation is how to determine the intersection points of the surface S with the straight line $x = u, y = u, z = u$ (if such a solution exists). That means solving the following equation system:

$$\begin{aligned}
u &= 3t(t-1)^2 + (s-1)^3 + 3s \\
u &= 3s(s-1)^2 + t^3 + 3t \\
u &= -3s(s^2-5s+5)t^3 - 3(s^3+6s^2-9s+1)t^2 \\
&\qquad +t(6s^3+9s^2-18s+3) - 3s(s-1)
\end{aligned}$$

In this particular case we get only one intersection point, having coordinates

$$(0.5561, 0.5561, 0.5561)$$

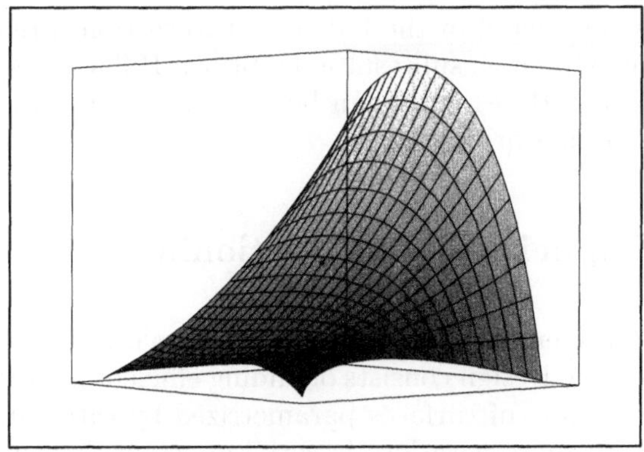

Fig. 1. Bicubic surface presented in Example 1

reached at $s = 0.2748$ and $t = 0.0408$. Another way of solving this problem consists of determining the implicit equation of $\mathcal{S}$. If this equation has been computed, the previous problem reduces to solving the equation $H(u,u,u) = 0$. Analogously, every intersection of $\mathcal{S}$ with any curve can be dealt by solving a single variable equation. In this particular case, the implicit equation of $\mathcal{S}$ has the following structure:

$$H(x,y,z) = z^9 + \sum_{i=1}^{9} r_i(x,y) z^{9-i}$$

$$\begin{aligned}
\mathbf{r}_1(x,y) &= -\frac{233469x}{2048} + \frac{188595y}{2048} - \frac{112832595}{262144} - \frac{81x^2}{64} + \frac{135xy}{32} - \frac{81y^2}{64} \\
\mathbf{r}_2(x,y) &= -\frac{20972672709381x}{536870912} + \frac{17975329363179y}{536870912} - \frac{729y^4}{8192} - \frac{729x^4}{8192} + \frac{1215x^3y}{2048} - \\
&\quad \frac{4779x^2y^2}{4096} + \frac{1215xy^3}{2048} - \frac{4105971x^3}{65536} + \frac{3129597y^3}{65536} + \frac{14456151x^2y}{65536} - \\
&\quad \frac{13181049xy^2}{65536} - \frac{54187594407x^2}{16777216} + \frac{48101467761xy}{8388608} - \frac{38812918311y^2}{16777216} - \\
&\quad \frac{22656991982391171}{137438953472} - \frac{1}{2}\left(\frac{233469x}{2048} - \frac{188595y}{2048} + \frac{112832595}{262144} + \frac{81x^2}{64} - \right. \\
&\quad \left. \frac{135xy}{32} + \frac{81y^2}{64}\right)\left(-\frac{233469x}{2048} + \frac{188595y}{2048} - \frac{112832595}{262144} - \frac{81x^2}{64} + \right. \\
&\quad \left. \frac{135xy}{32} - \frac{81y^2}{64}\right) \\
\mathbf{r}_3(x,y) &= \ldots\ldots\ldots\ldots\ldots\ldots\ldots\ldots\ldots\ldots\ldots\ldots\ldots\ldots\ldots\ldots\ldots\ldots\ldots\ldots
\end{aligned}$$

and by substituting $x = u, y = u, z = u$ we obtain one equation of degree 18 in u, very easy to solve:

$$5159780352u^{18} - 609499054080u^{17} + \ldots + 3707912273492242256259566313 = 0$$

The problem which occurs with this philosophy of work resides in the fact that, although there are methods for computing the implicit equation $H(x,y,z)$, these methods are very inefficient and can hardly be integrated in the CAD/CAM packages, where the solutions for any user request must be computed in real time.

The problems considered in the following two sections are examples of how algebraic techniques (such as subresultants, Sturm-Habicht sequences, symmetric functions, Gröbner Bases, etc.) can be used for the resolution of Computer Aided Geometric Design problems.

2 Generic Implicitation for Sectioning

One of the main problems in the manipulation of parametric surfaces in Computer Aided Geometric Design consists of finding efficient algorithms for computing the implicit equation of surfaces parametrized by rational functions. Being in possession of both representations is very important due to the fact that, for instance, for tracing the considered surface the parametric representation is the most convenient, meanwhile for deciding efficiently the position of a point with respect to the considered surface the implicit representation is desired.

There are many algebraic algorithms solving this problem but most of them depend strongly on the fact that the parametrization coefficients are known exactly. To overcome this drawback we propose the use of symbolic methods to compute in a generic way the implicit equation of different classes of parametric surfaces and then the computation of the implicit equation of the consireded surface via a well–controlled evaluation.

In case that the implicit object is too complicated to deal with, a degree-change approach is applied allowing also the possibility of replacing a complicated rational description by an integral one, i.e. without denominators.

The advantages and disadvantages of this approach have been considered through the study of a concrete set of problems and examples coming from CANDEMAT, a company devoted to car dies manufacturing together with the impact that this new approach has produced over some useful techniques in CAD/CAM such as sectioning and offsetting.

2.1 Implicitation of B–Spline Surfaces

Let $\mathcal{S}$ be an elementary bicubic B–spline surface defined by:

$$\begin{pmatrix} x \\ y \\ z \end{pmatrix} = (n_0(u)\; n_1(u)\; n_2(u)\; n_3(u)) \cdot \begin{pmatrix} \mathbf{P}_{00} & \mathbf{P}_{01} & \mathbf{P}_{02} & \mathbf{P}_{03} \\ \mathbf{P}_{10} & \mathbf{P}_{11} & \mathbf{P}_{12} & \mathbf{P}_{13} \\ \mathbf{P}_{20} & \mathbf{P}_{21} & \mathbf{P}_{22} & \mathbf{P}_{23} \\ \mathbf{P}_{30} & \mathbf{P}_{31} & \mathbf{P}_{32} & \mathbf{P}_{33} \end{pmatrix} \cdot \begin{pmatrix} n_0(v) \\ n_1(v) \\ n_2(v) \\ n_3(v) \end{pmatrix}$$

with $0 \leq u, v \leq 1$, $\mathbf{P}_{ij} \in \mathbb{R}^3$, and

$$n_0(s) = \frac{(1-s)^3}{6} \qquad n_1(s) = \frac{3s^3 - 6s^2 + 4}{6}$$

$$n_2(s) = \frac{-3s^3 + 3s^2 + 3s + 1}{6} \qquad n_3(s) = \frac{s^3}{6}$$

The problem consists in computing a polynomial $H(x, y, z)$ with the smallest possible degree such that

$$\mathcal{S} \subseteq \{(\alpha, \beta, \gamma) \in \mathbb{R}^3 : H(\alpha, \beta, \gamma) = 0\}.$$

The implicitation problem represents a classical question in Algebraic Geometry (Elimination Theory), which has been approached by different methods (see for example [4]):

- through multivariate resultants and/or moving surfaces,
- through Gröbner Basis or characteristic sets,
- through deformations and/or "ad-hoc" techniques .

If available, the implicit equation is very useful for point-surface positioning, surface-surface intersections, surface-curve intersections, sectioning, sculptured solid manipulation, surface trimming, etc (see [9]).

Two main difficulties were encountered when trying to use inside a software system for CAD/CAM the usual elimination techniques to deal with the implicitation problem:

- it is usually a very costly algebraic operation
- the coefficients of the polynomials involved in the parametrization are usually floating-point real numbers.

The second difficulty was overcome by taking into account that, in general, a concrete object to model (and then to construct) is made by several hundreds (or thousands) of small patches, all of them sharing the same algebraic structure. For such an object a database is constructed containing the implicit equation of every kind of patch appearing in its definition. This database must be pruned "through evaluation" in order to avoid specialization problems. Moreover the database generated for a specific object is kept into a bigger and general database for a further use.

In the particular case of the objects provided by the CANDEMAT company in IGES format (see [11]), the implicitation process was accomplished by using Sylvester resultants, Groebner Basis computations and ad-hoc techniques for specific cases. In meny cases the whole computation for a concrete family of parametric surfaces took several hours.

For instance, the implicit equation of the parametric patch

$$x = x_{00}\frac{t_2 - t}{t_2 - t_1} + x_{01}\frac{t - t_1}{t_2 - t_1}$$
$$y = y_{00}\frac{s_2 - s}{s_2 - s_1} + y_{11}\frac{s - s_1}{s_2 - s_1}$$
$$z = \left[z_{00}\frac{s_2 - s}{s_2 - s_1} + z_{10}\frac{s - s_1}{s_2 - s_1}\right]\frac{t_2 - t}{t_2 - t_1} + \left[z_{10}\frac{s_2 - s}{s_2 - s_1} + z_{11}\frac{s - s_1}{s_2 - s_1}\right]\frac{t - t_1}{t_2 - t_1}$$

appears in the database as:

$$(z_{00} - 2z_{10} + z_{11})xy + +(y_{00}z_{10} + y_{11}z_{10} - y_{00}z_{11} - y_{11}z_{00})x$$
$$+(x_{00}z_{10} + x_{01}z_{10} - x_{00}z_{11} - x_{01}z_{00})y + (x_{00}y_{11} + x_{01}y_{00} - x_{00}y_{00} - x_{01}y_{11})z$$
$$+x_{01}y_{11}z_{00} + x_{00}y_{00}z_{11} - x_{00}y_{11}z_{10} - x_{01}y_{00}z_{10}.$$

A solution to the first problem before mentioned was provided by the solution given to the second one: since the implicitation procedure is a preprocessing step, no computing time (other than the one used for navigating into the database, evaluating the generic implicit equation and verifying its goodness) is spent when the CAD/CAM user is working.

Some algebraic structures are not easy to generically implicitize: for example

$$x = \frac{f_1(s,t)}{q(s,t)}, \quad y = \frac{f_2(s,t)}{q(s,t)}, \quad z = \frac{f_3(s,t)}{q(s,t)}$$

with:

$$\begin{aligned} q(s,t) = &\ s^3(A_3t^2 + B_3t + C_3) + s^2(A_2t^2 + B_2t + C_2) + \\ &\qquad + s(A_1t^2 + B_1t + C_1) + (A_0t^2 + B_0t + C_0) \\ f_i(s,t) = &\ s^3(\alpha_{3i}t^2 + \beta_{3i}t + \gamma_{3i}) + s^2(\alpha_{2i}t^2 + \beta_{2i}t + \gamma_{2i}) + \\ &\qquad + s(\alpha_{1i}t^2 + \beta_{1i}t + \gamma_{1i}) + (\alpha_{0i}t^2 + \beta_{0i}t + \gamma_{0i}) \end{aligned}$$

These rather complicated structures have been treated by approximating the surface with a polynomially parametrized surface (if it is presented in a rational way) or, if its degree is too high, with a surface defined by a parametrization of smaller degree. The approximation procedure (see [1], [12], [13]) involves the resolution of a non–linear system of equations. We recall that this kind of approach of the approximate B–spline conversion problem is not unique, for instance in [10] the problem is solved by imposing osculating conditions and minimizing the error.

2.2 A Concrete Example from the CANDEMAT Company

The considered surface is defined by a set of 9 patches given by the parametrizations:

$$x = p_1(s,t), \quad y = p_2(s,t), \quad z = p_3(s,t)$$

defined on the domain:

$$s_0 = s_1 < s_2 < s_3 < s_4 = s_5, \quad t_0 = t_1 < t_2 < t_3 < t_4 = t_5$$

where

$$\begin{aligned} p_1 = &(x_{00}f_0 + x_{10}f_1 + x_{20}f_2 + x_{30}f_3)g_0 + (x_{01}f_0 + x_{11}f_1 + x_{21}f_2 + x_{31}f_3)g_1 \\ &+ (x_{02}f_0 + x_{12}f_1 + x_{22}f_2 + x_{32}f_3)g_2 + (x_{03}f_0 + x_{13}f_1 + x_{23}f_2 + x_{33}f_3)g_3 \\ p_2 = &(y_{00}f_0 + y_{10}f_1 + y_{20}f_2 + y_{30}f_3)g_0 + (y_{01}f_0 + y_{11}f_1 + y_{21}f_2 + y_{31}f_3)g_1 \\ &+ (y_{02}f_0 + y_{12}f_1 + y_{22}f_2 + y_{32}f_3)g_2 + (y_{03}f_0 + y_{13}f_1 + y_{23}f_2 + y_{33}f_3)g_3 \\ p_3 = &(z_{00}f_0 + z_{10}f_1 + z_{20}f_2 + z_{30}f_3)g_0 + (z_{01}f_0 + z_{11}f_1 + z_{21}f_2 + z_{31}f_3)g_1 \\ &+ (z_{02}f_0 + z_{12}f_1 + z_{22}f_2 + z_{32}f_3)g_2 + (z_{03}f_0 + z_{13}f_1 + z_{23}f_2 + z_{33}f_3)g_3 \end{aligned}$$

and

$$\left[\begin{array}{lcl} f_0 & = & \frac{s_2-s}{s_2-s_0} \\ f_1 & = & \frac{s-s_0}{s_2-s_0} \\ f_2 & = & 0.0 \\ f_3 & = & 0.0 \end{array}\right\} \text{ if } s_0 \le s < s_2; \quad \left[\begin{array}{lcl} f_0 & = & 0.0 \\ f_1 & = & \frac{s_3-s}{s_3-s_2} \\ f_2 & = & \frac{s-s_2}{s_3-s_2} \\ f_3 & = & 0.0 \end{array}\right\} \text{ if } s_2 \le s < s_3$$

$$\left[\begin{array}{lcl} f_0 & = & 0.0 \\ f_1 & = & 0.0 \\ f_2 & = & \frac{s_4-s}{s_4-s_3} \\ f_3 & = & \frac{s-s_3}{s_4-s_3} \end{array}\right\} \text{ if } s_3 \le s < s_4; \quad \left[\begin{array}{lcl} g_0 & = & \frac{t_2-t}{t_2-t_0} \\ g_1 & = & \frac{t-t_0}{t_2-t_0} \\ g_2 & = & 0.0 \\ g_3 & = & 0.0 \end{array}\right\} \text{ if } t_0 \le t < t_2$$

$$\left[\begin{array}{lcl} g_0 & = & 0.0 \\ g_1 & = & \frac{t_3-t}{t_3-t_2} \\ g_2 & = & \frac{t-t_2}{t_3-t_2} \\ g_3 & = & 0.0 \end{array}\right\} \text{ if } t_2 \le t < t_3; \quad \left[\begin{array}{lcl} g_0 & = & 0.0 \\ g_1 & = & 0.0 \\ g_2 & = & \frac{t_4-t}{t_4-t_3} \\ g_3 & = & \frac{t-t_3}{t_4-t_3} \end{array}\right\} \text{ if } t_3 \le t < t_4$$

The concrete values of the paramaters $x_{ij}, y_{ij}, z_{ij}, s_k, t_l$ appearing in the previous equations are the following ones:

$$\begin{array}{ll} s_0 = s_1 = 0.0 & s_2 = 0.333333333333333 \\ s_3 = 0.666666666666666 & s_4 = s_5 = 1.0 \\ t_0 = t_1 = -1.14816198642716 & t_2 = 0.333333333333333 \\ t_3 = 0.666666666666666 & t_4 = t_5 = 2.14816198642715 \end{array}$$

$$\begin{array}{lll}
x_{00} = -402.396210422345 & y_{00} = 125.604939074148 & z_{00} = 76.0994513388919 \\
x_{10} = -261.17588628063 & y_{10} = 120.007923434398 & z_{10} = 80.2658802206019 \\
x_{20} = -119.951398608196 & y_{20} = 114.410862351266 & z_{20} = 84.4324639780064 \\
x_{30} = 21.2789144514636 & y_{30} = 108.813733891507 & z_{30} = 88.5992634123272 \\
x_{01} = -402.301520929449 & y_{01} = 57.6898233810474 & z_{01} = 57.9054501358602 \\
x_{11} = -261.108042469145 & y_{11} = 53.9585848013379 & z_{11} = 62.570735352082 \\
x_{21} = -119.914564008841 & y_{21} = 50.2273462216285 & z_{21} = 67.2360205683038 \\
x_{31} = 21.2789144514635 & y_{31} = 46.496107641919 & z_{31} = 71.9013057845257 \\
x_{02} = -402.274168598392 & y_{02} = 38.0716305763488 & z_{02} = 52.6498686670318 \\
x_{12} = -261.089807581774 & y_{12} = 36.206011286494 & z_{12} = 57.8146792486379 \\
x_{22} = -119.905446565155 & y_{22} = 34.3403919966393 & z_{22} = 62.979489830244 \\
x_{32} = 21.2789144514634 & y_{32} = 32.4747727067845 & z_{32} = 68.1443004118501 \\
x_{03} = -402.179479105497 & y_{03} = -29.8434851167516 & z_{03} = 34.4558674640001 \\
x_{13} = -261.021963770289 & y_{13} = -29.8433273465661 & z_{13} = 40.119534380118 \\
x_{23} = -119.8686119658 & y_{23} = -29.8431241329984 & z_{23} = 45.7830464205415 \\
x_{33} = 21.2789144514633 & y_{33} = -29.8428535428032 & z_{33} = 51.4463427840486
\end{array}$$

The generic implicit equation of this surface is

$$c_{10}^2 c_{22}^2 + ((c_{11}^2 - 2c_{10}c_{12})c_{20} - c_{10}c_{11}c_{21})c_{22} + c_{10}c_{12}c_{21}^2 - c_{11}c_{12}c_{20}c_{21} + c_{12}^2 c_{20}^2$$

where

- c_{10}:

```
((s2-s0)*t0*t0*x01+(-s2+s0)*t0*t2*x00+((s2-s0)*t0*t2+(-s2+s0)*t0*t0)*x)*z11
  +((-s2+s0)*t0*t2*x01+(s2-s0)*t2*t2*x00+((-s2+s0)*t2*t2+(s2-s0)*t0*t2)*x)*z10
  +((-s2+s0)*t0*t0*x11+(s2-s0)*t0*t2*x10+((-s2+s0)*t0*t2+(s2-s0)*t0*t0)*x)*z01
  +((s2-s0)*t0*t2*x11+(-s2+s0)*t2*t2*x10+((s2-s0)*t2*t2+(-s2+s0)*t0*t2)*x)*z00
  +(((-s2+s0)*t0*t2+(s2-s0)*t0*t0)*x11+((s2-s0)*t2*t2+(-s2+s0)*t0*t2)*x10
  +((s2-s0)*t0*t2+(-s2+s0)*t0*t0)*x01+((-s2+s0)*t2*t2+(s2-s0)*t0*t2)*x00)*z
```

- c_{11}:

```
(((-2*s2)+2*s0)*t0*x01+((s2-s0)*t2+(s2-s0)*t0)*x00+((-s2+s0)*t2+(s2-s0)*t0)*x)*z11
  +(((s2-s0)*t2+(s2-s0)*t0)*x01+((-2*s2)+2*s0)*t2*x00+((s2-s0)*t2+(-s2+s0)*t0)*x)*z10
  +((2*s2+(-2*s0))*t0*x11+((-s2+s0)*t2+(-s2+s0)*t0)*x10+((s2-s0)*t2+(s0-s2)*t0)*x)*z01
  +(((s0-s2)*t2+(s0-s2)*t0)*x11+2*(s2-s0)*t2*x10+((-s2+s0)*t2+(s2-s0)*t0)*x)*z00
  +(((s2-s0)*t2+(-s2+s0)*t0)*x11+((-s2+s0)*t2+(s2-s0)*t0)*x10
  +((-s2+s0)*t2+(s2-s0)*t0)*x01+((s2-s0)*t2+(-s2+s0)*t0)*x00)*z
```

- c_{12}:

```
((s2-s0)*x01+(-s2+s0)*x00)*z11+((-s2+s0)*x01+(s2-s0)*x00)*z10
  +((-s2+s0)*x11+(s2-s0)*x10)*z01+((s2-s0)*x11+(-s2+s0)*x10)*z00
```

- c_{20}:

```
((s2-s0)*t0*t0*y01+(-s2+s0)*t0*t2*y00+((s2-s0)*t0*t2+(s0-s2)*t0*t0)*y)*z11
  +((-s2+s0)*t0*t2*y01+(s2-s0)*t2*t2*y00+((-s2+s0)*t2*t2+(s2-s0)*t0*t2)*y)*z10
  +((-s2+s0)*t0*t0*y11+(s2-s0)*t0*t2*y10+((-s2+s0)*t0*t2+(s2-s0)*t0*t0)*y)*z01
  +((s2-s0)*t0*t2*y11+(-s2+s0)*t2*t2*y10+((s2-s0)*t2*t2+(-s2+s0)*t0*t2)*y)*z00
  +(((-s2+s0)*t0*t2+(s2-s0)*t0*t0)*y11+((s2-s0)*t2*t2+(-s2+s0)*t0*t2)*y10
  +((s2-s0)*t0*t2+(-s2+s0)*t0*t0)*y01+((-s2+s0)*t2*t2+(s2-s0)*t0*t2)*y00)*z
```

- c_{21}:

```
(((-2*s2)+2*s0)*t0*y01+((s2-s0)*t2+(s2-s0)*t0)*y00+((-s2+s0)*t2
   +(s2-s0)*t0)*y)*z11+(((s2-s0)*t2+(s2-s0)*t0)*y01+((-2*s2)+2*s0)*t2*y00
   +((s2-s0)*t2+(-s2+s0)*t0)*y)*z10+((2*s2+(-2*s0))*t0*y11+((-s2+s0)*t2
   +(s0-s2)*t0)*y10+((s2-s0)*t2+(s0-s2)*t0)*y)*z01+(((s0-s2)*t2+(s0-s2)*t0)*y11
   +(2*s2+(-2*s0))*t2*y10+((-s2+s0)*t2+(s2-s0)*t0)*y)*z00+(((s2-s0)*t2
   +(-s2+s0)*t0)*y11+((-s2+s0)*t2+(s2-s0)*t0)*y10+((-s2+s0)*t2+(s2-s0)*t0)*y01
   +((s2-s0)*t2+(-s2+s0)*t0)*y00)*z
```

- c_{22}:

```
((s2-s0)*y01+(-s2+s0)*y00)*z11+((-s2+s0)*y01+(s2-s0)*y00)*z10+((-s2+s0)*y11
   +(s2-s0)*y10)*z01+((s2-s0)*y11+(-s2+s0)*y10)*z00
```

Considering the concrete values of the parameters $x_{ij}, y_{ij}, z_{ij}, s_k, t_l$, we obtain the concrete implicit equation:

$$\begin{aligned} H(x,y,z) = {} & -12.86405481303769z^3 + [6.8938261970858576y+ \\ & 1.0322819534013938x + 4441.0953941817452]z^2 - [0.92359758113477841y^2+ \\ & (0.27659911577510787x + 1976.073857399886)y + 0.020708984196767578x^2+ \\ & 295.89741941049658x + 387924.22771793336]z + [(3.4188084068007236 \cdot 10^{-18}x+ \\ & 210.63109093399621)y^2 + (-6.1121850507742521 \cdot 10^{-19}x^2+ \\ & 63.079824695413464x + 92111.888130195322)y + 4.7227883903081773x^2+ \\ & 13792.839722405979x + 10070450.541378483] \end{aligned}$$

The "error" can be computed, in this case, either symbolically:

$$\begin{aligned} H(p_1,p_2,p_3) = {} & (-1.7970904234100502 \cdot 10^{-15}s^3- \\ & 3.5539035642606386 \cdot 10^{-14}s^2 + 1.3866796036202881 \cdot 10^{-12}s- \\ & 8.3765572022996195 \cdot 10^{-12})t^3 + (5.3954545200343888 \cdot 10^{-14}s^3- \\ & 2.7284841053187847 \cdot 10^{-12}s^2 + 1.1641532182693481 \cdot 10^{-10})t^2+ \\ & (-1.1527137854777534 \cdot 10^{-12}s^3 + 2.9103830456733704 \cdot 10^{-11}s^2- \\ & 1.1641532182693481 \cdot 10^{-10}s - 2.7939677238464355 \cdot 10^{-9})t- \\ & 1.7043986506752499 \cdot 10^{-11}s^3 + 4.6566128730773926 \cdot 10^{-9}s+ \\ & 3.7252902984619141 \cdot 10^{-9} \end{aligned}$$

or numerically:

$$\frac{\sum_{i=1}^{N} H(x_i,y_i,z_i)}{N} = 5.5134296417236327 \cdot 10^{-9}$$

where $N = 100$, $x_i = p_1(s_i,t_i), y_i = p_2(s_i,t_i), z_i = p_3(s_i,t_i)$ and the values of the parameters s_i and t_i are uniformly generated in $[0,1]$.

2.3 Sectioning an Implicit B–Spline Surface

Consider now all the patches (implicitly represented) defining the B–spline surface to be sectioned by the plane $x = k$. For each patch, and with the equation $x(u,v) = k$, we compute the intersection of this curve (into the $u - v$ domain) with the boundary of the definition domain (i.e. starting with $u = 0$, then $u = 1$, $v = 0$ and $v = 1$, usually two points are determined at most). By evaluating these points in the parametrization we obtain the extremes of the section on the B–spline surface. With each point computed before, and by using the implicit equation, every component of the section is discretized (always inside the plane $x = k$). The points computed before are interpolated by using a cubic spline curve representing the section of the considered patch. The previous steps are repeated for every patch of the considered surface.

Next figure shows how looks like the sectioning, by using the generic implicitation, of a concrete object in the CAD/CAM environment CSIS of the company CANDEMAT.

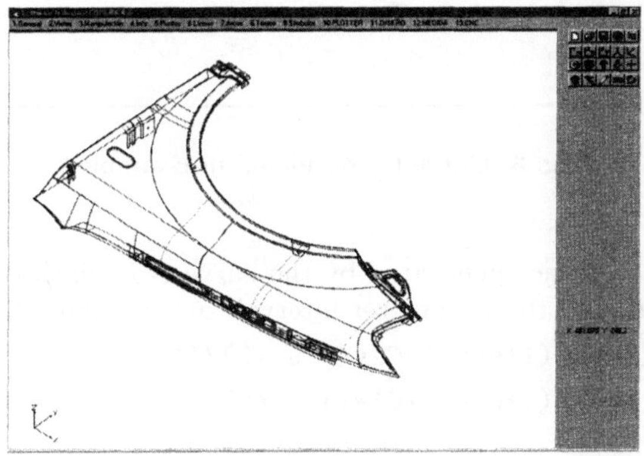

Fig. 2. Real Surface Sectioning

3 Revolution Surface Sectioning

The problem considered in this section is the `Maple` computation, topologically exact, of the section for a revolution surface by using the generic implicitation methodology before mentioned. For that it is used a series of algebraic techniques in order to determine the exact shape of the considered sectioning. The details of the algorithm can be found in [7] and [6].

Example 2.
The curve in the plane $y = 0$ defined by the parametrization $x = C_1(t)$, $z = C_3(t)$ will be rotated with respect to the OZ axis. In Figure 3 it is displayed the 3–dimensional curve C defined by the parametrization $(C_1(t), 0, C_2(t))$.

```
> plots[setoptions3d](scaling=CONSTRAINED,axes=FRAMED);
> plots[setoptions](scaling=CONSTRAINED,axes=FRAMED);
> C1:=(2*t-1)/(1+t**2); C2:=0;
> C3:=(1-t+t**2)/(2+t+t**2);
> plot3d([C1,C2,C3],t=-15..15,s=-1..1,orientation=[60,75],
> grid=[175,2]);
```

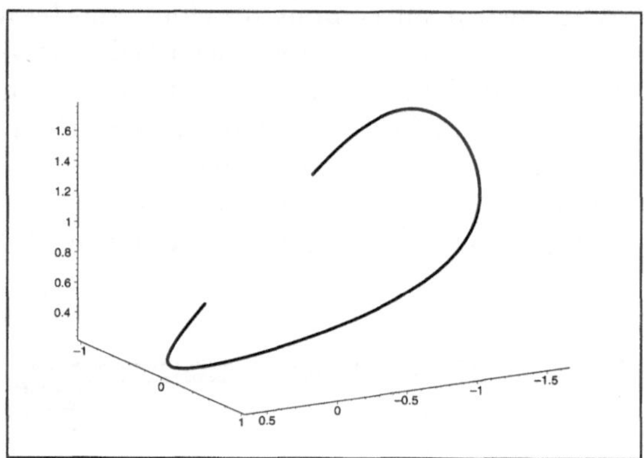

Fig. 3. Curve C considered in Example 2

The revolution surface generated by the curve C is displayed as follows (see figure 4). The limits of the parameter intervals are ± 3 and ± 6 respectively.

```
> T1:=C1*(2*s)/(1+s**2)-C2*(1-s**2)/(1+s**2):
> T2:=C2*(2*s)/(1+s**2)+C1*(1-s**2)/(1+s**2):
> T3:=C3:
> P1:=(a,b)->plot3d([T1,T2,T3],s=a..b,t=-6..6,grid=[50,50],
> axes=BOXED,scaling=UNCONSTRAINED,projection=1,
> tickmarks=[0,0,0],orientation=[59,75]):
> P1(-3,3);
```

For studying the sections of this revolution surface, we firstly compute its implicit equation by using the generic implicitation method, which is based on manipulating efficiently mutivariate symmetric functions (see for example [8]). We have implemented this method en `Maple` and named it `ImplicitSuperfRevol`

```
> toro:=ImplicitSuperfRevol(C1,C2,C3,x,y,z,t);
```

$$\begin{aligned} toro := {} & 9 + 8\,z\,x^2\,y^2 + 262\,z^2 - 308\,z^3 - 84\,z + 8\,z^4\,x^2\,y^2 + 121\,z^4 + \\ & 16\,z^2\,x^2\,y^2 + 20\,z\,x^2 + 20\,z\,y^2 - 3\,x^2 - 3\,y^2 + x^4 + y^4 + 16\,z^3\,x^2\,y^2 + \\ & 19\,z^4\,x^2 + 19\,z^4\,y^2 + 2\,x^2\,y^2 + 4\,z^4\,x^4 + 4\,z^4\,y^4 + 8\,z^3\,x^4 + \\ & 8\,z^3\,y^4 - 72\,z^3\,x^2 - 72\,z^3\,y^2 + 8\,z^2\,x^4 + 8\,z^2\,y^4 - \\ & 28\,z^2\,x^2 - 28\,z^2\,y^2 + 4\,z\,x^4 + 4\,z\,y^4 \end{aligned}$$

Next the section of this surface with the plane $y = 1/2$ is computed.

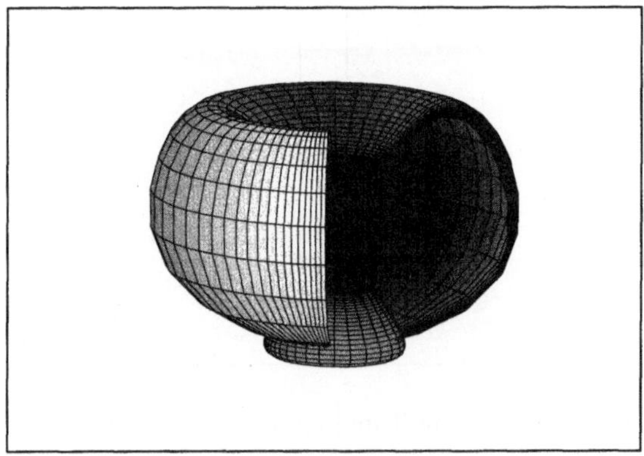

Fig. 4. Revolution Surface Generated by C

```
> Section:=subs(y=1/2,toro);
```

$$Section := -\frac{315}{4}z + 21z^4x^2 + \frac{511}{2}z^2 - \frac{651}{2}z^3 + 126z^4 + x^4 - \frac{5}{2}x^2 + 4z^4x^4 \\ -68z^3x^2 + 8z^3x^4 - 24z^2x^2 + 8z^2x^4 + 4zx^4 + 22zx^2 + \frac{133}{16}$$

Initially, the topology of this sectioning is determined (it is important to mention that it has two isolated points: coming from the complex part of the surface after implicititation), as displayed in Figure 5(a). In this step we are interested only in the critical and regular points of the curve, which are the graph points. The behaviour of the curve between these points is completely determined (taking into account the branch counting in each interval) so we can connect them by edges. The algorithm we have used to determine the topological structure of the curve is different from the classical algorithms because by a simple change of coordinates we can work with a curve which fulfils the following condition: for every root of the discriminant, we have at most a critical point of the curve (see [7] and [6]). After determining the topology of the curve, we are able to draw it exactly using for example the Newton method, without losing isolated or small components (Figure 5(b)). The `Maple` implementations of the topology computation of the curve and of the curve drawing algorithm are contained in the files `prGrafos.txt` and `prDibujos.txt` respectively.

```
> read "prGrafos.txt";
> principal(Section,x,z,20,'black');

> read "prDibujos.txt";
> principal(Section,x,z,20,500,10,'black');

> P2:=y->plot3d([s,y,t],s=-4..4,t=-2..2,style=PATCHNOGRID):
```

In this way and in an exact way, the section of our revolution surface we are interested about has been computed (Figure 6).

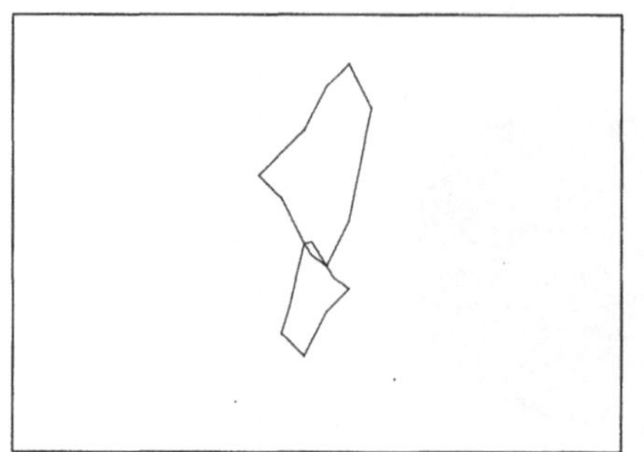

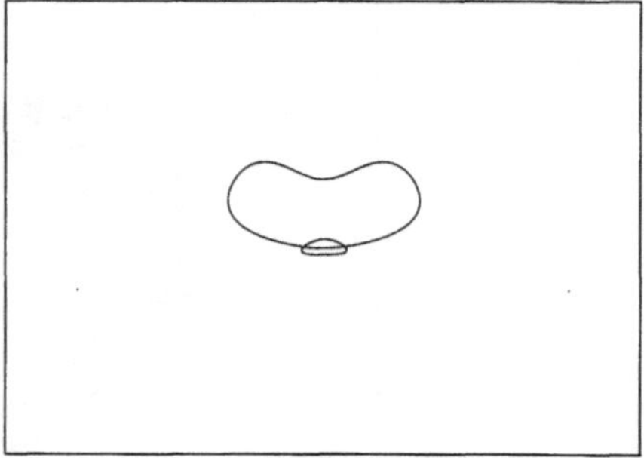

Fig. 5. (a) Sectioning Topology and (b) Sectioning Curve

```
> plots[display](P1(-4,4),P2(1),grid=[50,50],axes=BOXED,
> scaling=UNCONSTRAINED,projection=1,tickmarks=[0,0,0],
> orientation=[13,130]);
```

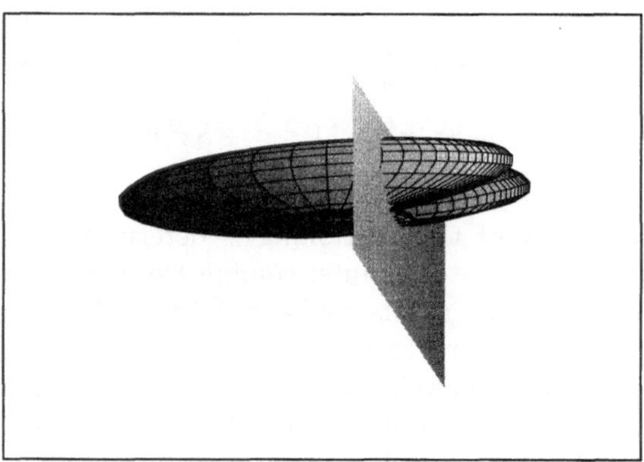

Fig. 6. Revolution Surface Sectioning

References

1. Bardis L., Patrikalakis N.: Approximate conversion of rational B–spline patches. Computer Aided Geometric Design **6** (1989) 189–204
2. de Boor C.: On calculating with B–splines. Journal of Approximation Theory **6** (1972) 50–62
3. de Boor C.: A Practical Guide to Splines. Applied Mathematical Sciences **27** (1978) Springer–Verlag
4. Cox D., Little J., O'Shea D.: Ideals, Varieties and Algorithms. Undergraduate Texts in Mathematics (1993) Springer–Verlag
5. Golub G., van Loan C.: Matrix Computations. Johns Hopkins Studies in the Mathematical Sciences (1996) Johns Hopkins University Press

6. Gonzalez–Vega L., El Kahoui M.: An Improved Upper Complexity Bound for the Topology Computation of a Real Algebraic Plane Curve. Journal of Complexity **12** (1996) 527–544
7. Gonzalez–Vega L., Necula I.: Computing the Topology of Implicit Algebraic Plane Curves. Preprint (2001)
8. Gonzalez–Vega L.: Implicitization of Parametric Curves and Surfaces by using Multidimensional Newton formulae. Journal of Symbolic Computation **23** (1997) 137–151
9. Hoffmann C. M.: Geometric and Solid Modeling: An Introduction. Morgan Kaufmann Publishers (1989)
10. Hoschek J., Schneider F.J.: Aproximate Conversion for Integral and Rational Bezier and B–spline surfaces. Geometry Processing for Design and Manufacturing (1992) 45–86, SIAM
11. The Initial Graphics Exchange Especification (IGES). Version 5.2 (1993) ANSI Y14.26M
12. Patrikalakis N.: Approximate conversion of rational splines, Computer Aided Geometric Design **6** (1989) 155–165
13. Piegl L., Tiller W.: The NURBS book. Monographs in Visual Communications (1997) Springer–Verlag
14. VDA Working Group 'CAD/CAM': VDA Surface, Data Interface (VDAFS) Version 2.0 (1987)

6. Gonzalez-Vega L., El Kahoui M.: An Improved Upper Complexity Bound for the Topology Computation of a Real Algebraic Plane Curve. Journal of Complexity 12 (1996) [illegible]
7. Gonzalez-Vega L., Necula I.: Computing the Topology of Implicit Algebraic Plane Curves. Preprint (2001).
8. Gonzalez-Vega L.: Implicitization of Parametric Curves and Surfaces by Using Multidimensional Newton Formulae. Journal of Symbolic Computation 23 (1997) 137–151.
9. Hoffmann C. M.: Geometric and Solid Modeling: An Introduction. Morgan Kaufmann Publishers (1989).
10. [illegible]
11. [illegible]
12. [illegible]
13. [illegible]
14. [illegible]

The Methods of Computer Algebra and the Arnold–Moser Theorem

E.A. Grebenikov

CS RAS, Moscow, greben@ext1.ccas.ru
University of Podlasie, Poland, imif@wsrp.siedlce.pl

Abstract. It is well known that the problem of stability of the particular solutions of Hamilton systems in Lyapunov sense cannot be solved within the framework of the classical stability theory. For Hamilton systems with two degrees of freedom the above problem is investigated within the framework of the KAM theory on the basis of the well-known theorem of Arnold–Moser about stability in the so-called "elliptic case". We formulate and investigate the problem of the stability of equilibrium state for dynamic models termed restricted problems of many ($n > 3$) bodies. All necessary analytical transformations, switching of linearization of the differential equations, and the normalization of hamiltonians after Birkhoff are executed with the help of the System of Symbolical Calculations (SSC) "Mathematica".

1 Introduction

The existence of two new dynamic models of space dynamics, termed the restricted Newton problems of many ($n > 3$) bodies, has been proved in [1]. The motion of passively gravitational point (its mass is equal to zero) in the gravity field of many bodies having equal masses, forming a correct polygon and rotating around the common barycenter, was studied within the framework of Lagrange-Wintner model. In the second model, which was non-inertial, the motion of zero mass was investigated in the gravity field of many bodies, among which one of them has the mass m_0, and the other have the mass $m \neq 0$ and form a correct polygon rotating around the geometrical center, in which there is a mass m_0.

The existence of the particular solutions of the problem of many bodies forming correct polygons follows from the well-known results of A.Wintner stated in his monograph [2] as well as from the research of V. Elmabsout [3] and E. Grebenikov [4]. Note that at $n = 3$ we have the famous restricted problem of three bodies, to which a large number of the publications has been devoted. The restricted problem with $n > 3$ bodies is typically hamiltonian, and hence, differential equations governing the dynamics within the framework of the above models, can be written in canonical form.

It follows from this, in particular, that the problem of stability of their particular solutions can be solved only within the framework of the theory [5,6,7] with preliminary normalization of Hamiltonians in Birkhoff's sense [8]. The abbrevitation "KAM theory" means the association of well-known results under the theory of the conditional-periodic solutions of hamilton systems given on multi-dimensional toruses and advanced by A. N. Kolmogorov, V.I. Arnold and J. Moser in 50–70es years of the XXth century.

The present paper is devoted to stability investigation of equilibrium states of hamilton systems [9] describing the dynamics of a zero mass in a symmetrical field of attraction formed by n masses which are placed in the vertices of a regular polygon turning around the center of gravity of n masses determined by E. Grebenikov and B. Elmabsout.

2 Determination of Equilibrium States

We now describe a method for finding the equilibrium (equilibrium solutions) of a system of differential equations of the restricted problem of $n > 3$ bodies, which have been written down in the uniformly rotating coordinate system P_0xyz, where P_0 is one of gravitational material points with mass m_0. The specified system of equations is as follows [1]:

$$\begin{cases} \frac{d^2x}{dt^2} - 2\omega_n \frac{dy}{dt} = \frac{\partial R}{\partial x}, \\ \frac{d^2y}{dt^2} + 2\omega_n \frac{dy}{dt} = \frac{\partial R}{\partial y}, \\ \frac{d^2z}{dt^2} = \frac{\partial R}{\partial y}, \end{cases} \tag{1}$$

$$R(x,y,z) = \frac{\omega_n^2}{2}(x^2+y^2) + \frac{m_0}{\Delta} + m\sum_{k=1}^{n-1}\left[\frac{1}{\Delta_k} - \frac{xx_k + yy_k + zz_k}{r_k^3}\right], \tag{2}$$

$$\begin{cases} \Delta_k^2 = (x-x_k)^2 + (y-y_k)^2 + (z-z_k)^2, \quad \Delta^2 = x^2+y^2+z^2, \\ r_k^2 = x_k^2 + y_k^2 + z_k^2, \quad k = 1,\dots,n-1, \\ x_k = a_0 \cos\frac{2\pi(k-1)}{n-1}, \quad y_k = a_0 \sin\frac{2\pi(k-1)}{n-1}, \quad z_k = 0, \quad k = 1,\dots,n-1, \end{cases} \tag{3}$$

$$\omega_n = \sqrt{\frac{1}{a_0^3}\left[m_0 + \frac{m}{4}\sum_{k=2}^{n-1}\left(\sin\frac{\pi(k-1)}{n-1}\right)^{-1}\right]}, \tag{4}$$

where x, y, z are the coordinates of passive gravitational point P with mass equal to zero, ω_n is the angular velocity of coordinate system P_0xyz with respect to initial non-rotating coordinate system with its origin at point P_0.

The quantity ω_n is obviously also the angular velocity of rotation of regular polygon $P_1P_2\dots P_{n-1}$, in whose vertices there are the masses $m_1 = m_2 = \dots = m_{n-1} = m \neq 0$, around the central body P_0. It was proved in [4] that such a dynamic model does not contradict the equations of the Newton mechanics, and a configuration of bodies $P_1,\dots,P_{n-1}$ having the form of a regular polygon is homographical solution of the differential equations of the problem of n bodies (1) in the Wintner's sense [2].

The determination of equilibrium state of system (1) reduces eventually [10] to the solution of the following nonlinear system of functional equations:

$$\frac{\partial R}{\partial x} = \frac{\partial R}{\partial y} = \frac{\partial R}{\partial z} = 0, \tag{5}$$

or

$$\begin{cases} \omega_n^2 x - \frac{m_0 x}{\Delta^3} + m \sum\limits_{k=1}^{n-1} \left[\frac{x_k - x}{\Delta_k^3} - \frac{1}{a_0^2} \cos \frac{2\pi(k-1)}{n-1} \right] = 0, \\ \omega_n^2 y - \frac{m_0 y}{\Delta^3} + m \sum\limits_{k=1}^{n-1} \left[\frac{y_k - y}{\Delta_k^3} - \frac{1}{a_0^2} \sin \frac{2\pi(k-1)}{n-1} \right] = 0, \\ -\frac{m_0 z}{\Delta^3} + m \sum\limits_{k=1}^{n-1} \frac{z}{\Delta_k^3} = 0. \end{cases} \tag{6}$$

In equations (6), the quantities x, y, and z are unknown. The last equation of system (6) at $z = 0$ is satisfied always. It means that all equilibrium states of system (1) are in the P_0xy plane. It was shown in [10] that for any n, system (6) (at $z = 0$) is equivalent to the following system:

$$\begin{cases} \omega_n^2 x - \frac{m_0 x}{\Delta^3} + m \sum\limits_{k=1}^{n-1} \frac{x_k - x}{\Delta_k^3} = 0, \\ \omega_n^2 y - \frac{m_0 y}{\Delta^3} + m \sum\limits_{k=1}^{n-1} \frac{y_k - y}{\Delta_k^3} = 0. \end{cases} \tag{7}$$

The solvability conditions for the system of nonlinear algebraic equations (7) represent also necessary and sufficient conditions for existence of equilibrium states of the system of differential equations (1). Unfortunately, the analytical methods for finding the solutions of this system are unknown, therefore, its solutions can indeed be found only with the help of SSC.

Using the SSC "Mathematica" [16], we have calculated the coordinates of equilibrium states in the restricted problems of 4, 5, 6, and 7 bodies presented in Figs. 1–4.

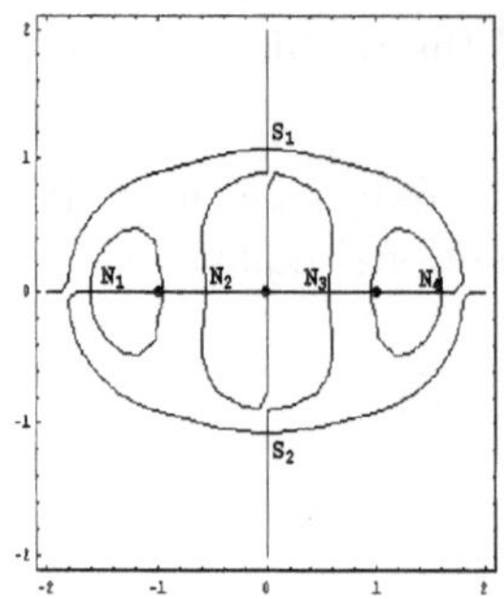

Fig. 1

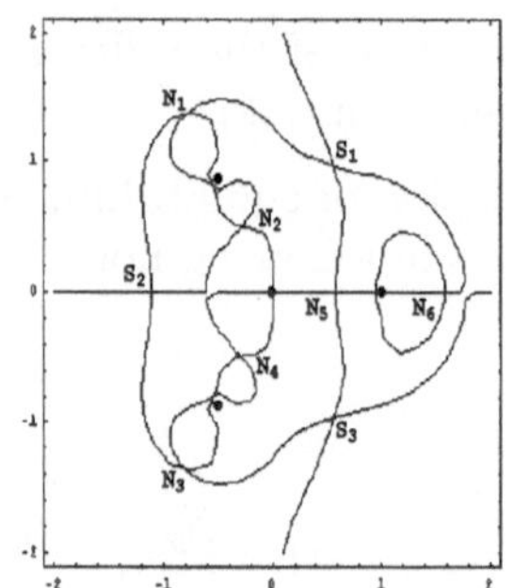

Fig. 2

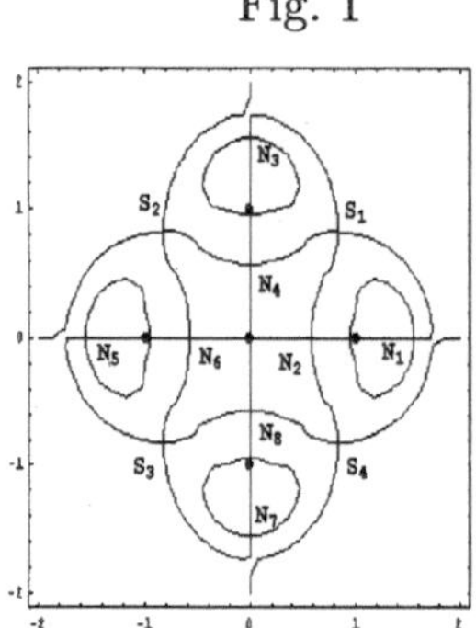

Fig. 3

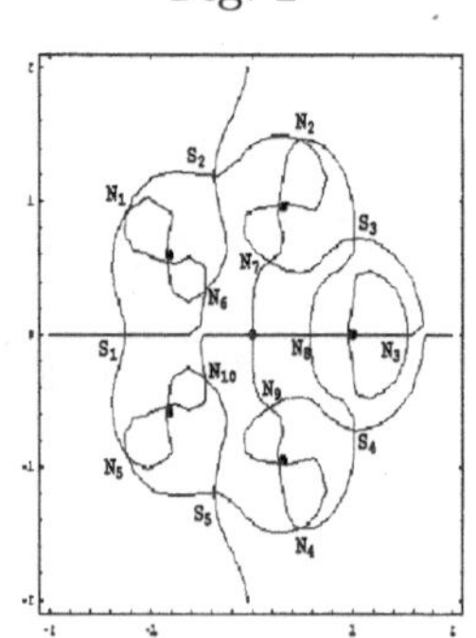

Fig. 4

In these figures, the equilibrium states are represented by points of intersection of curves $\phi_1(x,y)=0$ and $\phi_2(x,y)=0$ (where ϕ_1 and ϕ_2 are the left-hand sides of equations (7)).

It was proved in [11–14] that the stationary points N_i are unstable for all values of m, and the points S_i are stable in the first approximation for $0 \le m < m^*$ given in the table:

$n+1$	$0 \le m < m^*$
4	$0 \le m < 0.085\ldots$
5	$0 \le m < 0.023\ldots$
6	$0 \le m < 0.0094\ldots$
7	$0 \le m < 0.0047\ldots$

As is seen from figures 1–4, in the restricted problem of 4 bodies there exist two stable equilibrium states as the first approximation (points S_1 and S_2 in Fig. 1), in the restricted problem of 5 bodies there are three steady equilibrium states as the first approximation (points S_1, S_2 and S_3 in Fig. 2), in the restricted problem of 6 bodies there are four such equilibrium states (points S_1–S_4 in Fig. 3), and in the restricted problem of 7 bodies there are five such equilibrium states (points S_1–S_5 in Fig. 4). The coordinates of points S_i depend on the value of parameter m from the specified table and have been calculated by Newton's method. The valuess of these coordinates are given in [11,12,13].

3 Construction of Birkhoff Transformations

The application of the Arnold-Moser theorem [7,15] probably only after its realization in the vicinity of the equilibrium state is stable in the first approximation of normalization Hamiltonians after Birkhoff [8] to within the fourth degree of local coordinates. The algebraic operations necessary for it were executed in SSC "Mathematica" [16].

At an example of the restricted problem of 7 bodies it will be shown that in an interval of the table $0 \leq m < m^* = 0.0047$ (except for two so-called resonant points), the equilibrium states S_i are stable after Lyapunov.

The Hamiltonian of the restricted problem of 7 and 6 bodies (n = 6) in the Cartesian coordinate system P_0xyz is as follows [14]:

$$
\begin{aligned}
H(6) = &-(x^2+y^2)^{-1/2} - m\Bigg(((x-1)^2+y^2)^{-1/2} + \\
&+ \left(\left(\frac{1}{4}(1+\sqrt{5})+x\right)^2 + \left(-\frac{1}{2}\sqrt{\frac{1}{2}(5-\sqrt{5}}+y\right)^2\right)^{-1/2} \\
&+ \left(\left(\frac{1}{4}(1+\sqrt{5})+x\right)^2 + \left(\frac{1}{2}\sqrt{\frac{1}{2}(5-\sqrt{5}}+y\right)^2\right)^{-1/2} \\
&+ \left(\left(\frac{1}{4}(1-\sqrt{5})+x\right)^2 + \left(-\frac{1}{2}\sqrt{\frac{1}{2}(5+\sqrt{5}}+y\right)^2\right)^{-1/2} \\
&+ \left(\left(\frac{1}{4}(1-\sqrt{5})+x\right)^2 + \left(\frac{1}{2}\sqrt{\frac{1}{2}(5+\sqrt{5}}+y\right)^2\right)^{-1/2}\Bigg) \\
&+ \omega_5(yp_x - xp_y) + \frac{1}{2}(p_x^2+p_y^2).
\end{aligned}
\tag{8}
$$

At first we shall translate the coordinate origin from point P_0 to a point, at which the stability is investigated, S_i, with coordinates x^*, y^* by using the formulas

$$
\begin{aligned}
X &= x - x^*, \\
Y &= y - y^*, \\
P_X &= p_x - p_x^*, \\
P_Y &= p_y - p_y^*,
\end{aligned}
\tag{9}
$$

and we now introduce a new initial variable (X, Y, P_X, P_Y). Then the Hamiltonian $H(6)$ of the restricted problem of 7 bodies is given by the formula:

$$\begin{aligned}
H(6) = &-((X+x^*)^2+(Y+y^*)^2)^{-1/2} - m\Bigg(((X+x^*-1)^2+(Y+y^*)^2)^{-1/2} \\
&+\left(\left(\frac{1}{4}(1+\sqrt{5})+X+x^*\right)^2+\left(-\frac{1}{2}\sqrt{\frac{1}{2}(5-\sqrt{5}}+Y+y^*\right)^2\right)^{-1/2} \\
&+\left(\left(\frac{1}{4}(1+\sqrt{5})+X+x^*\right)^2+\left(\frac{1}{2}\sqrt{\frac{1}{2}(5-\sqrt{5}}+Y+y^*\right)^2\right)^{-1/2} \\
&+\left(\left(\frac{1}{4}(1-\sqrt{5})+X+x^*\right)^2+\left(-\frac{1}{2}\sqrt{\frac{1}{2}(5+\sqrt{5}}+Y+y^*\right)^2\right)^{-1/2} \\
&+\left(\left(\frac{1}{4}(1-\sqrt{5})+X+x^*\right)^2+\left(\frac{1}{2}\sqrt{\frac{1}{2}(5+\sqrt{5}}+Y+y^*\right)^2\right)^{-1/2}\Bigg) \\
&+\omega_5((Y+y^*)(P_X+p_x^*)-(X+x^*)(P_Y+p_y^*)) \\
&+\frac{1}{2}((P_X+p_x^*)^2+(P_Y+p_y^*)^2).
\end{aligned} \tag{10}$$

The hamilton differential equations of the restricted problem of 7 bodies are as follows:

$$\begin{aligned}
\tfrac{dP_X}{dt} &= -\tfrac{\partial H}{\partial X}, & \tfrac{dX}{dt} &= \tfrac{\partial H}{\partial P_X}, \\
\tfrac{dP_Y}{dt} &= -\tfrac{\partial H}{\partial Y}, & \tfrac{dY}{dt} &= \tfrac{\partial H}{\partial P_Y}.
\end{aligned} \tag{11}$$

And they obviously have the particular solutions

$$X = Y = P_X = P_Y. \tag{12}$$

The Hamiltonian (10) is not a definitely positive function of variables (X, Y, P_X, P_Y) [9], therefore, further transformations are necessary to study the Lyapunov stability. The performance of normalization of the equations after Birkhoff depends on the coordinates of a specific equilibrium state. Consider further steady point S_3 as the first approximation (see Fig. 4) with coordinates

$$x^* = 0.809317, \quad y^* = 0.588003, \tag{13}$$

calculated at $m = 0.001$.

In a sufficiently small neighbourhood of point S_3 we will present the Hamiltonian (10) with the help of a converging sedate series

$$H = H_2(X, Y, P_X, P_Y) + H_3(X, Y) + H_4(X, Y) + \ldots, \tag{14}$$

where H_k is a homogeneous form of kth degree, and

$$H_2 = -0.48197X^2 - 0.0220063Y^2 + 0.5(P_X^2 + P_Y^2) - 1.41562XY + \omega_5(YP_X - XP_Y), \tag{15}$$

$$H_3 = 0.113608X^2 - 0.365825Y^2 + 1.98185X^2Y + 0.876412XY^2, \tag{16}$$

$$H_4 = 0.206391X^4 + 0.395727Y^4 - 1.87471X^3Y + 0.709283XY^3 - 2.95444X^2Y^2. \tag{17}$$

Expression (15) shows that the square-law form $H_2(X, Y, P_X, P_Y)$ contains a summand, which is the first obstacle on the way of research of Lyapunov's stability. Therefore, it is at first feasible to find such a non-singular canonical transformation $(X, Y, P_X, P_Y) \mapsto (q_1, q_2, p_1, p_2)$

$$\begin{bmatrix} X \\ Y \\ P_X \\ P_Y \end{bmatrix} = A \cdot \begin{bmatrix} q_1 \\ q_2 \\ p_1 \\ p_2 \end{bmatrix} \tag{18}$$

where the matrix A is determined in such a way that the new transformed Hamiltonian K: $(H(X, Y, P_X, P_Y) \mapsto K(q_1, q_2, p_1, p_2))$ had the form

$$K(q_1, q_2, p_1, p_2) = K_2(q_1, q_2, p_1, p_2) + K_3(q_1, q_2, p_1, p_2) + K_4(q_1, q_2, p_1, p_2) + \dots \tag{19}$$

and its square-law form K_2 would not contain the expressions q_1p_2, q_2p_2, q_2p_1, q_1p_1, i.e. expressions being the products of pulses and of Lagrange coordinates.

The finding of transformation (18) is equivalent to definition of a matrix A of dimension 4×4 with 16 unknown elements. The solution of a system of homogeneous linear algebraic equations of 16th order has appeared practically possible only using SSC.

The realization of canonical transformation (18) gives for the forms K_2, K_3, and K_4 the following expressions:

$$K_2 = 0.485052p_1^2 - 0.115866p_2^2 + 0.485052q_1^2 - 0.115866q_2^2, \tag{20}$$

$$\begin{aligned} K_3 = {} & -2.52816p_1^3 - 2.51376p_2^3 + 1.19723q_1^3 + 0.339633q_2^3 + \\ & +2.35999q_1^2q_2 + 1.55067q_1q_2^2 + p_1^2(-12.0813p_2 - 0.600941q_1 - \\ & -0.39486q_2) + p_2^2(-14.7484p_1 + 16.6356q_1 + 10.9308q_2) + \\ & +p_1(8.12181p_2q_1 + 5.5534q_1^2 + 5.33659p_2q_2 + 7.29794q_1q_2 + \\ & +2.39762q_2^2) + p^2(11.6697q_1^2 + 15.3356q_1q_2 + 5.03827q_2^2), \end{aligned} \tag{21}$$

$$
\begin{aligned}
K_4 = {} & -0.599766p_1^4 - 77.3193p_2^4 + 1.9228q_1^4 + 0.358408q_2^4 + \\
& p_1^3(-16.9111p_2 - 14.8001q_1 - 9.72468q_2) + p_2^3(-139.675p_1 - \\
& -107.938q_1 - 70.9224q_2) + p_1^2(-82.2071p_2^2 - 97.3317p_2q_1 - \\
& -16.6075q_1^2 - 63.9536p_2q_2 - 21.8245q_1q_2 - 7.1701q_2^2) + \\
& +p_2^2(-188.407p_1q_1 - 20.6904q_1^2 - 123.796p_1q_2 - 27.1901q_1q_2 - \\
& -8.93287q_2^2) + p_1(-44.955p_2q_1^2 + 2.0817q_1^3 + 4.10347q_1^2q_2 - \\
& -59.077p_2q_1q_2 + 19.4088p_2q_2^2 + 2.69626q_1q_2^2 + 0.590543q_2^3) + \\
& +p_2(10.059q_1^3 + 19.8283q_1^2q_2 + 13.0286q_1q_2^2 + 2.85356q_2^3) + \\
& +5.05365q_1^3q_2 + 4.9809q_1^2q_2^2 + 2.18186q_1q_2^3.
\end{aligned}
\tag{22}
$$

It is the first important step in the research of Lyapunov stability. The canonical variables (q_1, q_2, p_1, p_2) are not variable such as "action - angle", as K_2 depends not only on pulses p_1, p_2, but also on coordinates q_1, q_2. Therefore, it is further necessary to proceed from canonical variable (q_1, q_2, p_1, p_2) to new a canonical variable $(\theta_1, \theta_2, \tau_1, \tau_2)$ by using the Birkhoff formulas [8]:

$$
\begin{cases}
q_1 = \sqrt{2\tau_1}\sin\theta_1, \\
q_2 = \sqrt{2\tau_2}\sin\theta_2, \\
p_1 = \sqrt{2\tau_1}\cos\theta_1, \\
p_2 = \sqrt{2\tau_2}\cos\theta_2.
\end{cases}
\tag{23}
$$

Transformation (23) "liquidates" in a square-law part of a new Hamiltonian $F(K(q_1, q_2, p_1, p_2) \mapsto F(\theta_1, \theta_2, \tau_1, \tau_2))$ the expressions with coordinates θ_1, θ_2. If we present the new Hamiltonian F in the form

$$
F(\theta_1, \theta_2, \tau_1, \tau_2) = F_2(\tau_1, \tau_2) + F_3(\theta_1, \theta_2, \tau_1, \tau 2) + F_4(\theta_1, \theta_2, \tau_1, \tau 2) + \ldots, \tag{24}
$$

then its square-law form F_2 will not depend on phase angles θ_1, θ_2, and depends only on new pulses τ_1, τ_2. The forms F_2, F_3, F_4 are expressed by the following equalities:

$$
F_2 = 0.970103\tau_1 - 0.231732\tau_2, \tag{25}
$$

$$
\begin{aligned}
F_3 = {} & -1.4362\tau_1^{3/2}\cos\theta_1 - 17.4666\sqrt{\tau_1}\tau_2\cos\theta_1 - 5.71453\tau_1^{3/2}\cos 3\theta_1 - \\
& -1.28015\sqrt{\tau_1}\tau_2\cos(\theta_1 - 2\theta_2) - 11.6341\tau_1\sqrt{\tau_2}\cos(2\theta_1 - \theta_2) - \\
& -0.582121\tau_1\sqrt{\tau_2}\cos\theta_2 - 1.7699\tau_2^{3/2}\cos\theta_2 - 5.34009\tau_2^{3/2}\cos 3\theta_2 - \\
& -21.9549\tau_1\sqrt{\tau_2}\cos(2\theta_1 + \theta_2) - 22.968\sqrt{\tau_1}\tau_2\cos(\theta_1 + 2\theta_2) + \\
& +2.11478\tau_1^{3/2}\sin\theta_1 + 25.7193\sqrt{\tau_1}\tau_2\sin\theta_1 - 1.2715\tau_1^{3/2}\sin 2\theta_1 + \\
& +6.89313\sqrt{\tau_1}\tau_2\sin(\theta_1 - 2\theta_2) + 7.69096\tau_1\sqrt{\tau_2}\sin(2\theta_1 - \theta_2) + \\
& +2.77911\tau_1\sqrt{\tau_2}\sin\theta_2 + 8.44968\tau_2^{3/2}\sin\theta_2 + 7.48905\tau_2^{3/2}\sin 3\theta_2 + \\
& +3.79502\tau_1\sqrt{\tau_2}\sin(2\theta_1 + \theta_2) + 14.4402\sqrt{\tau_1}\tau_2\sin(\theta_1 + 2\theta_2),
\end{aligned}
\tag{26}
$$

$$
\begin{aligned}
F_4 = &-6.31918\tau_1^2 - 105.087\tau_1\tau_2 - 119.908\tau_2^2 - 5.04513\tau_1^2\cos 2\theta_1 - \\
&-73.6677\tau_1\tau_2\cos 2\theta_1 + 8.96525\tau_1^2\cos 4\theta_1 - 74.8189\sqrt{\tau_1}\tau_2^{3/2}\cos(\theta_1 - 3\theta_2) - \\
&-54.2213\tau_1\tau_2\cos(2\theta_1 - 2\theta_2) - 51.1759\tau_1^{3/2}\sqrt{\tau_2}\cos(\theta_1 - \theta_2) - \\
&-229.539\sqrt{\tau_1}\tau_2^{3/2}\cos(\theta_1 - \theta_2) + 0.582859\tau_1^{3/2}\sqrt{\tau_2}\cos(3\theta_1 - \theta_2) - \\
&-100.708\tau_1\tau_2\cos 2\theta_2 - 155.355\tau_2^2\cos 2\theta_2 - 34.014\tau_2^2\cos 4\theta_2 - \\
&-44.5123\tau_1^{3/2}\sqrt{\tau_2}\cos(\theta_1 + \theta_2) - 208.894\sqrt{\tau_1}\tau_2^{3/2}\cos(\theta_1 + \theta_2) + \\
&+27.461\tau_1^{3/2}\sqrt{\tau_2}\cos(3\theta_1 + \theta_2) + 4.85566\tau_1\tau_2\cos(2\theta_1 + 2\theta_2) - \\
&-45.4469\sqrt{\tau_1}\tau_2^{3/2}\cos(\theta_1 + 3\theta_2) - 12.7184\tau_1^2\sin 2\theta_1 - \\
&-185.71\tau_1\tau_2\sin 2\theta_1 - 8.4409\tau_1^2\sin 4\tau_1 + 1.71031\sqrt{\tau_1}\tau_2^{3/2}\sin(\theta_1 - 3\theta_2) - \\
&-53.6605\tau_1\tau_2\sin(2\theta_1 - 2\theta_2) - 21.0421\tau_1^{3/2}\sqrt{\tau_2}\sin(\theta_1 - \theta_2) - \\
&-94.3798\sqrt{\tau_1}\tau_2^{3/2}\sin(\theta_1 - \theta_2) - 46.7812\tau_1^{3/2}\sqrt{\tau_2}\sin(3\theta_1 - \theta_2) - \\
&-44.1253\tau_1\tau_2\sin 2\theta_2) - 68.0689\tau_2^2\sin 2\theta_2 - 36.888\tau_2^2\sin 4\theta_2 - \\
&-46.1126\tau_1^{3/2}\sqrt{\tau_2}\sin(\theta_1 + \theta_2) - 216.404\sqrt{\tau_1}\tau_2^{3/2}\sin(\theta_1 + \theta_2) - \\
&-60.6094\tau_1^{3/2}\sqrt{\tau_2}\sin(3\theta_1 + \theta_2) - 137.442\tau_1\tau_2\sin(2\theta_1 + 2\theta_2) - \\
&-122.676\sqrt{\tau_1}\tau_2^{3/2}\sin(\theta_1 + 3\theta_2).
\end{aligned} \tag{27}
$$

The canonical variables $(\theta_1, \theta_2, \tau_1, \tau_2)$ are the variables such as "action – angle" and the hamilton equations in the vicinity of the equilibrium state S_3 are expressed by the following equality:

$$
\left\{
\begin{aligned}
\frac{d\tau_1}{dt} &= -\frac{\partial F_3}{\partial \theta_1} - \frac{\partial F_4}{\partial \theta_1} + \cdots, \\
\frac{d\tau_2}{dt} &= -\frac{\partial F_3}{\partial \theta_2} - \frac{\partial F_4}{\partial \theta_2} + \cdots,
\end{aligned}
\right.
\qquad
\left\{
\begin{aligned}
\frac{d\theta_1}{dt} &= \frac{\partial F_2}{\partial \tau_1} + \frac{\partial F_3}{\partial \tau_1} + \frac{\partial F_4}{\partial \tau_1} + \cdots, \\
\frac{d\theta_2}{dt} &= \frac{\partial F_2}{\partial \tau_2} + \frac{\partial F_3}{\partial \tau_2} + \frac{\partial F_4}{\partial \tau_2} + \cdots,
\end{aligned}
\right.
\tag{28}
$$

where the functions F_2, F_3, F_4 are expressed by equalities (25)–(27).

Let us now present the formulation of the Arnold–Moser theorem [6].

Theorem.

Let the Hamiltonian

$$
\begin{aligned}
&W(\psi_1, \psi_2, T_1, T_2) = W_2(T_1, T_2) + W_4(T_1, T_2) + \ldots, \\
&W_2(T_1, T_2) = \sigma_1 T_1, \qquad W_4 = c_{20}T_1^2 + c_{11}T_1T_2 + c_{02}T_2^2
\end{aligned}
\tag{29}
$$

be such that:

1. *The eigenvalues of the linear system*

$$
\begin{aligned}
\frac{dT_1}{dt} &= -\frac{\partial W_2}{\partial \psi_1} = 0, &\quad \frac{d\psi_1}{dt} &= \frac{\partial W_2}{\partial T_1}, \\
\frac{dT_2}{dt} &= -\frac{\partial W_2}{\partial \psi_2} = 0, &\quad \frac{d\psi_2}{dt} &= \frac{\partial W_2}{\partial T_2},
\end{aligned}
\tag{30}
$$

are the numbers $\pm i\sigma_1, \quad \pm i\sigma_2$;

2. $\quad n_1\sigma_1 + n_2\sigma_2 \neq 0, \quad$ *for* $\quad 0 < |n_1| + |n_2| \leq 4;$ (31)
3. $\quad c_{20}\sigma_1^2 + c_{11}\sigma_1\sigma_2 + c_{02}\sigma_2^2 \neq 0$ (32)

Then the equilibrium state

$$T_1 = T_2 = \psi_1 = \psi_2 = 0 \tag{33}$$

of the hamilton system with Hamiltonian (29) is stable in the sense of Lyapunov.

It may be seen from the formulation of the Arnold–Moser theorem that the 1 and 2 forms of decomposition of Hamiltonian W involving the variables T_1 and T_2 (which correspond to the 2nd and 4th forms of its decomposition involving the variables (q_1, q_2, p_1, p_2)) do not contain the phase angles ψ_1 and ψ_2, and the form order 3/2 rather (T_1, T_2) is in general absent. Therefore, the hamilton equations (28) for the present do not satisfy the conditions of the above theorem.

It follows from here that it is necessary to construct one more canonical transformation $(\theta_1, \theta_2, \tau_1, \tau_2) \mapsto (\psi_1, \psi_2, T_1, T_2)$, which reduces to zero the form order 3/2, i.e. it will transform $F_3(\theta_1, \theta_2, \tau_1, \tau_2)$ to $W_3(\psi_1, \psi_2, T_1, T_2) = 0$, and the form of order 2 $F_4(\theta_1, \theta_2, \tau_1, \tau_2)$ will transform to $W_4(T_1, T_2)$.

Thus, the canonical transformation $(\theta_1, \theta_2, \tau_1, \tau_2) \mapsto (\psi_1, \psi_2, T_1, T_2)$ can keep phase angles ψ_1, ψ_2 only in the forms of decomposition of higher order (W_5, W_6 and so on) of the Hamiltonian $W(\psi_1, \psi_2, T_1, T_2)$.

We will search for the last canonical transformation (with necessary accuracy) as

$$\begin{aligned}
\theta_1 &= \psi_1 + V_{13}(T_1, T_2, \psi_1, \psi_2) + V_{14}(T_1, T_2, \psi_1, \psi_2),\\
\theta_2 &= \psi_2 + V_{23}(T_1, T_2, \psi_1, \psi_2) + V_{24}(T_1, T_2, \psi_1, \psi_2),\\
\tau_1 &= T_1 + U_{13}(T_1, T_2, \psi_1, \psi_2) + U_{14}(T_1, T_2, \psi_1, \psi_2),\\
\tau_2 &= T_2 + U_{23}(T_1, T_2, \psi_1, \psi_2) + U_{24}(T_1, T_2, \psi_1, \psi_2),
\end{aligned} \tag{33}$$

where $U_{13}, U_{23}, U_{14}, U_{24}, V_{13}, V_{23}, V_{14}, V_{24}$ are determined from some linear partial differential equations. For their solution we have used the method of asymptotic multi-frequency integration of systems of the differential equations developed in [17]. For example, the equation for unknown function U_{13} is as follows:

$$\frac{\partial U_{13}}{\partial \psi_1}\sigma_1 - \frac{\partial U_{13}}{\partial \psi_2}\sigma_2 = A_{13}(\psi_1, \psi_2, T_1, t_2), \tag{34}$$

where A_{13} is expressed as:

$$\begin{aligned}
A_{13} = &-2.115478T_1^{3/2}\cos\psi_1 - 25.7193\sqrt{T_1}T_2\cos\psi_1 + 3.8145T_1^{3/2}\cos 3\psi_1 -\\
&-6.89313\sqrt{T_1}T_2\cos(\psi_1 - 3\psi_2) - 15.3819T_1\sqrt{T_2}\cos(2\psi_1 - \psi_2) -\\
&-7.59004T_1\sqrt{T_2}\cos(2\psi_1 + \psi_2) - 14.4402\sqrt{T_1}T_2\cos(\psi_1 + 2\psi_2) -\\
&-1.4362T_1^{3/2}\sin\psi_1 - 17.4666\sqrt{T_1}T_2\sin\psi_1 - 17.1436T_1^{3/2}\sin 3\psi_1 -\\
&-1.28015\sqrt{T_1}T_2\sin(\psi_1 - 2\psi_2) - 23.2682T_1\sqrt{T_2}\sin(2\psi_1 - \psi_2) -\\
&-43.9099T_1\sqrt{T_2}\sin(2\psi_1 + \psi_2) - 22.968\sqrt{T_1}T_2\sin(\psi_1 + 2\psi_2).
\end{aligned} \tag{35}$$

Now it is necessary to choose from all solutions of (35) (their infinite set) a solution, which provides the form (29) of a new Hamiltonian:

$$W_2 = \sigma_1 T_1 - \sigma_2 T_2, \qquad W_4 = c_{20}T_1^2 + c_{11}T_1T_2 + c_{02}T_2^2,$$

where

$$\sigma_1 = 0.970103, \quad \sigma_2 = 0.231732,$$
$$c_{20} = -152.527, \quad c_{11} = -33654.64, \quad c_{02} = -331.57.$$

Such a solution is as follows:

$$\begin{aligned} U_{13} = & -2.11478T_1^{3/2}\frac{\sin\psi_1}{\sigma_1} - 25.7193\sqrt{T_1}T_2\frac{\sin\psi_1}{\sigma_1} + 3.8145T_1^{3/2}\frac{\sin 3\psi_1}{3\sigma_1} \\ & -6.89313\sqrt{T_1}T_2\frac{\sin(\psi_1 - 2\psi_2)}{\sigma_1 + 2\sigma_2} - 15.3819T_1\sqrt{T_2}\frac{\sin(2\psi_1 - \psi_2)}{2\sigma_1 + \sigma_2} \\ & -7.59004T_1\sqrt{T_2}\frac{\sin(2\psi_1 + \psi_2)}{2\sigma_1 - \sigma_2} - 14.4402\sqrt{T_1}T_2\frac{\sin(\psi_1 + 2\psi_2)}{\sigma_1 - 2\sigma_2} \\ & +1.4362T_1^{3/2}\frac{\cos\psi_1}{\sigma_1} + 17.4666\sqrt{T_1}T_2\frac{\cos\psi_1}{\sigma_1} + 17.1436T_1^{3/2}\frac{\cos 3\psi_1}{3\sigma_1} \\ & +1.28015\sqrt{T_1}T_2\frac{\cos(\psi_1 - 2\psi_2)}{\sigma_1 + 2\sigma_2} + 23.2682T_1\sqrt{T_2}\frac{\cos(2\psi_1 - \psi_2)}{2\sigma_1 + \sigma_2} \\ & +43.9099T_1\sqrt{T_2}\frac{\cos(2\psi_1 + \psi_2)}{2\sigma_1 - \sigma_2} + 22.968\sqrt{T_1}T_2\frac{\cos(\psi_1 + 2\psi_2)}{\sigma_1 - 2\sigma_2}. \end{aligned} \tag{36}$$

Thus, the decomposition of Hamiltonian of the restricted problem of seven bodies in the vicinity of equilibrium state S_3 with coordinates

$$x^* = 0.809317, \quad y^* = 0.588003,$$

represented at the end by means of canonical variables $(\psi_1, \psi_2, T_1, T_2)$, satisfies the conditions of the Arnold–Moser theorem (30)–(32), and, hence, the equilibrium state S_3 is stable in Lyapunov's sense.

In the interval $0 \le m \le 0.0047$ there are two "resonant" values of parameter m [18] $(m_1 \approx 0.0019, m2 \approx 0.0031)$, for which the question of Lyapunov's stability of point S_3 remains open. It can be proven in a similar way that the other equilibrium states S_i represented in Figs. 1–4 are stable in Lyapunov's sense.

In summary we would like to note that with increase of number n, the amount of symbolic calculations necessary to perform the operations of linearization and normalization of Hamiltonians in the vicinity of equilibrium states increases by a large factor.

References

1. Grebenicov, E.: Two new dynamical models in celestial mechanics. Rom. Astron. J. **8** (1998) 13–19
2. Wintner, A.: The Analytical Foundations of Celestial Mechanics, Princeton, Princeton Univ. Press (1941)
3. Elmabsout, B.: Sur l'existence de certaines configurations d'equillibre relatif dans le probleme des n corps. Celestial Mechanics and Dynamical Astronomy **4** (1988) 131–151
4. Grebenicov, E.: Existence of the exact symmetric solutions in flat Newton problem of many bodies. Mathematical Modeling **10**, No. 8 (1998) 74–80 (in Russian)
5. Arnold, V.: Mathematical Methods of the Classical Mechanics, Nauka, Moscow (1974) (in Russian)
6. Markeev, A.: The Points of Fluctuation in the Celestial Mechanics and Astrodynamics, Nauka, Moscow (1978) (in Russian)

7. Moser, Ju.: Lecture about Hamilton Systems, Mir, Moscow (1973)
8. Birkhoff, D.: Dynamic Systems, GITTL, Leningrad (1941) (in Russian)
9. Lyapunov, A.: General Problem about Stability of Motion, Proc. of AS USSR, vol. 1, Moscow (1954) (in Russian)
10. Grebenicov, E., Gadomski, L.: The existence conditions for equilibrium in a bounded circular many - body problem, In: Proc. of the Steklov Inst. of Math. **223** (1998) 159–162 (in Russian)
11. Kozak, D., Oniszk, E.: Equilibrium points in the restricted four-body problem: sufficcient conditiones for linear stability. Rom. Astron. J. **8** (1998) No. 2
12. Kozak, D., Oniszk, E.: Application of computer system MATHEMATICS for determination of coordinates of equilibrium state in gravitational model of four bodies of Grebenicov-Elmabsout. Nonlinear Fluctuations **2** (1999), No. 1
13. Jakubiak, M.: Sufficient conditions of linear stability of equilibrium states in gravitational Newton model of six bodies. Nonlinear Fluctuations **2** (1999), No. 1
14. Grebenicov, E., Kozak-Skoworodkin, D., Jakubiak, M.: The algebraic problems of the normalization in Hamiltonian theory. In: Proc. Second Internat. Workshop on "MATHEMATICA" System in Teaching and Research (2000) 73–90
15. Arnold, B.: About stability of equilibrium states in hamilton system of ordinary differential equations in a general elliptic case. Moscow, Ed. of a Far Eastern Branch of the USSR Acad. Sci. **137**, No. 2 (1961) 255–257
16. Wolfram, S.: The Mathematica - Book, Cambridge, Cambridge University Press (1996)
17. Grebenicov, E., Mitropolski, Ju., Rjabov, Ju.: Introduction in Resonant Analytical Dynamics, Janus-K, Moscow (1999) (in Russian)
18. Kozak, D., Jakubiak, M.: Problem of resonances frequencies existence in restricted Newton problem of six and seven bodies. "Nonlinear Analysis and Homographical Dynamics", No. 1, Ed. CC of RAS, Moscow (1999) 20–40

Symbolic Algorithms of Algebraic Perturbation Theory: Hydrogen Atom in the Field of Distant Charge

Alexander Gusev[1], Valentin Samoilov[1], Vitaly Rostovtsev[2], and Sergue Vinitsky[3]

[1] Scientific Center for Applied Research, Joint Institute for Nuclear Research, Dubna, Moscow Region 141980, Russia
[2] Laboratory of Information Technologies, Joint Institute for Nuclear Research, Dubna, Moscow Region 141980, Russia
rost@jinr.ru
[3] Bogoliubov Laboratory of Theoretical Physics, Joint Institute for Nuclear Research, Dubna, Moscow Region 141980, Russia
vinitsky@thsun1.jinr.ru

Abstract. We present symbolic algorithms implemented in REDUCE 3.6 for evaluation of eigenvalues and eigenfunctions of a hydrogen atom in the field of a distant point charge. These solutions are presented as perturbation series by a small parameter. This parameter is the inverse value of the separation between the distant point charge and the hydrogen like atom nucleus. Algebraic perturbation theory schemes are built up using the irreducible representations of the dynamical symmetry algebra $so(4,2)$. The representations are connected by the tilting transformations with wave functions of the discrete spectrum of the hydrogen atom in an arbitrary bound state characterized by a set of parabolic or spherical quantum numbers. Such a construction is based on a reduction of the unperturbed Hamiltonian and polynomial perturbation operators via generators of the algebra. The efficiency of the proposed perturbation scheme and symbolic algorithm is demonstrated by calculating the coefficients of the high order perturbation series via the parabolic quantum numbers.

1 Introduction

The problem of a hydrogen atom in the field of a distant point charge and its applications in atomic and molecular physics have been under investigation for more than 60 years [1]-[9]. However, a simple representation of solutions of the problem as the perturbation series in a symbolic form [10], which should be suitable for its further incorporating in some computer numerical codes applied in the real atomic and molecular calculations [9], has not been constructed. So, in papers [4], [5] it has been pointed out that conventional perturbation schemes [1]-[3] did not yield a true closed analytical representation of the solutions characterized by parabolic and spherical quantum numbers. A true but rather a complicated algorithm and representation of the solutions classified by spheroidal quantum numbers have been given in [6].

The aim of this talk is to describe the universal symbolic computer algebra algorithms and simple conventional algebraic perturbation schemes, which yield a closed analytical representation of the solutions labelled by parabolic or spherical quantum numbers in an arbitrary high order perturbation series by a small parameter. This parameter is the inverse value of the separation between the distant point charge and the hydrogen like atom nucleus. We use the irreducible representations of algebra $so(4,2)$ which are related by the tilting transformation with eigenfunctions of the discrete spectrum of a hydrogen atom characterized by parabolic or spherical quantum numbers according to [11],[12],[13]. This circumstance allows us to construct an algebraic perturbation scheme via substitution of the polynomial perturbations for an appropriate combination of generators of the algebra. To solve a linear eigenvalue problem without troubles, we use an appropriate choice of the normalized factor without fractional powers of parabolic or spherical quantum numbers involved in the evaluation process and to renormalize only final eigenvectors. To simplify our presentation, we implement the scheme for calculating the solutions of the above perturbation problem in the parabolic representation only taking into account that it is connected by the known orthogonal transformation with the spherical one [14]. This representation provides a closed form of the zero-order wave functions in a diagonal representation of both parabolic and spherical quantum numbers for the perturbation operator under consideration. To evaluate perturbation corrections to an eigenvalue in a more efficient way, we describe an appropriate minimal set of unknown coefficients in the expansions of eigenvector corrections over the irreducible representation, as has been proposed in [15].

Sections 2 and 3 give a description of the hydrogen atom wave functions of a discrete spectrum in parabolic and spherical coordinates with the help of the irreducible representations of the algebra $so(4,2)$ for the free and perturbation cases. Section 4 deals with the new efficient recursive algorithm in the framework of an algebraic version of the conventional perturbation theory. In Section 5, we display the program **Pointfield** and present the analytical expressions for the energy spectrum up to the seventh order. Correspondingly, the corrections of the seventh order of eigenvalues and the fourth order of eigenfunctions have been evaluated. The proposed algorithm was implemented with the help of REDUCE 3.6[1].

2 Parabolic and Spherical Bases of Hydrogen Atom

The Lie algebra $so(4,2)$ comprises 15 generators

$$L_{\alpha\beta} = -L_{\beta\alpha}, \quad [L_{\alpha\beta}, L_{\alpha\gamma}] = \imath g_{\alpha\alpha} L_{\beta\gamma}, \quad \alpha, \beta, \gamma = 1, ..., 6, \tag{1}$$

[1] The authors are trying to implement the same procedure in Maple V

where $g_{\alpha\alpha} = (1,1,1,1,-1,-1)$. In the configuration space the generators $L_{\alpha\beta}$ are defined by the relations [15]

$$
\begin{gathered}
L_{ij} = x_i p_j - x_j p_i = \varepsilon_{ijk} L_k, \quad i,j,k = 1,2,3, \\
L_{i4} = \frac{1}{2}(x_i \mathbf{p}^2 + 2\imath p_i - 2\mathbf{xp}p_i - x_i) = A_i, \quad L_{46} = \frac{1}{2}(r\mathbf{p}^2 - r), \\
L_{i5} = \frac{1}{2}(x_i \mathbf{p}^2 + 2\imath p_i - 2\mathbf{xp}p_i + x_i), \quad L_{56} = \frac{1}{2}(r\mathbf{p}^2 + r), \\
L_{45} = -\imath(1 + \imath\mathbf{xp}), \qquad L_{i6} = -rp_i,
\end{gathered} \tag{2}
$$

where

$$
p_k = -\imath\frac{\partial}{\partial x_k}, \quad r = \sqrt{x_1^2 + x_2^2 + x_3^2}.
$$

Note that the coordinates may be expressed in terms of the generators

$$
x_i = L_{i5} - L_{i4}, \quad r = L_{56} - L_{46}. \tag{3}
$$

The above operators act in the Hilbert space of functions with the inner scalar product

$$
\langle f|g\rangle = \int f^*(\mathbf{x}) r^{-1} g(\mathbf{x}) d\mathbf{x} \tag{4}
$$

with respect to which they are Hermitian. The basis vectors in the representation space with given n will be classified over eigenvalues of the operators L_{56}, L_{34}, L_{12}, if we use a parabolic basis $|n_1, n_2, m\rangle$

$$
\begin{gathered}
L_{56}|n_1 n_2 m\rangle = n|n_1 n_2 m\rangle, \quad L_{34}|n_1 n_2 m\rangle = (-n_1 + n_2)|n_1 n_2 m\rangle, \\
L_{12}|n_1 n_2 m\rangle = m|n_1 n_2 m\rangle.
\end{gathered}
$$

In this case, as the basis that realizes the infinite-dimensional irreducible representation of the algebra $so(4,2)$, we take eigenfunctions of these operators, which in the parabolic coordinates $\{y_1, y_2, y_3\}$

$$
y_1 = r + x_3, \quad y_2 = r - x_3, \quad y_3 = arctg(x_2/x_1), \quad d^3y = \frac{1}{4}(y_1 + y_2) dy_1 dy_2 dy_3
$$

have the form

$$
\Phi^{(0)}_{n_1 n_2 m}(\mathbf{y}) = C_{n_1 n_2 |m|} \phi_{n_1|m|}(y_1) \phi_{n_2|m|}(y_2) \frac{e^{\imath m y_3}}{\sqrt{2\pi}} \begin{cases} -1, & m > 0 \\ 1, & m \le 0 \end{cases}, \tag{5}
$$

$$
\begin{gathered}
C_{n_1 n_2 |m|} = 2^{\frac{1}{2}} C_{n_1|m|} C_{n_2|m|}, \quad C_{n_j|m|} = \left[\frac{n_j!}{(n_j + |m|)!}\right]^{\frac{1}{2}}, \\
\phi_{n_j|m|}(y_j) = [(n_1 + |m|)!]^{-1} y_j^{\frac{|m|}{2}} e^{-\frac{y_j}{2}} L^{|m|}_{n_j+|m|}(y_j), \quad j = 1,2.
\end{gathered}
$$

Here $L^{|m|}_{n_j+|m|}$ are the Laguerre polynomials [10], and the basis functions $\Phi^{(0)}_{n_1 n_2 m}(\mathbf{y})$ are orthonormalised according to (4) with $d\mathbf{x} = d^3y$.

Another possibility is to classify the basis vectors in the representation space with given n by eigenvalues of the operators $L_{56}, \mathbf{L}^2, L_{12}$, where $\mathbf{L}^2 = L_{12}^2 + L_{13}^2 + L_{23}^2$, if we will use a spherical basis $|n,l,m\rangle$

$$L_{56}|nlm\rangle = n|nlm\rangle, \quad \mathbf{L}^2|nlm\rangle = l(l+1)|nlm\rangle, \quad L_{12}|nlm\rangle = m|nlm\rangle.$$

In this case, as the basis that realizes the infinite-dimensional irreducible representation of the algebra $so(4,2)$, we take eigenfunctions of these operators, which in the spherical coordinates $\{y_1,\ y_2\ y_3\}$

$$y_1 = r \quad y_2 = \cos\theta = x_3/r, \quad y_3 = arctg(x_2/x_1), \quad d^3y = y_1^2 dy_1 dy_2 dy_3,$$

have the form

$$\Phi^{(0)}_{nlm}(\mathbf{y}) = C_{nl|m|}\phi(y_1)\phi(y_2)\frac{e^{\imath m y_3}}{\sqrt{2\pi}}\begin{cases} -1, & m > 0 \\ 1, & m \le 0 \end{cases}, \tag{6}$$

$$C_{nl|m|} = C_{nl}C_{l|m|}$$

$$C_{nl} = 2^{\frac{3}{2}}\left[\frac{1}{2}\frac{(n-l-1)!}{(n+l)!}\right]^{\frac{1}{2}}, C_{l|m|} = \left[\frac{(2l+1)(l-|m|)!}{2(l+|m|)!}\right]^{\frac{1}{2}},$$

$$\phi(y_1) = [(n+l)!]^{-1}(2r)^l e^{-r} L^{2l+1}_{n+l}(2r), \quad \phi(y_2) = P_{l|m|}(\cos\theta).$$

Here L^{2l+1}_{n+l} are the Laguerre polynomials and $P_l^{|m|}$ are the Legendre polynomials [10], and the basis functions $\Phi^{(0)}_{nlm}(\mathbf{y})$ are orthonormalized according to (4) with $d\mathbf{x} = d^3y$.

As a consequence of the $so(4)$ accidental symmetry the above basis functions in the layer with a fixed value $n = n_1 + n_2 + |m| + 1$ are connected by the orthogonal transformation [11]

$$\Phi^{(0)}_{n_1 n_2 m}(\mathbf{y}) = \sum_{l=|m|}^{n-1} A^{n_1 n_2}_{nlm}\Phi^{(0)}_{nlm}(\mathbf{y}), \qquad \Phi^{(0)}_{nlm}(\mathbf{y}) = \sum_{n_1,n_2=0}^{n-|m|} A^{n_1 n_2}_{nlm}\Phi^{(0)}_{n_1 n_2 m}(\mathbf{y}),$$

Here the matrix elements of the orthogonal transformation A read as [14]

$$A^{n_1 n_2}_{nlm} = C^{n_1 n_2}_{nlm}\ {}_3F_2\left[\begin{matrix}|m|+l+1, -l+|m|+1, -n_2; 1\\ |m|+1, -n+|m|+1,\end{matrix}\right], \tag{7}$$

where $C^{n_1 n_2}_{nlm}$ is a normalization factor

$$C^{n_1 n_2}_{nlm} = (-1)^{l-|m|}\frac{(N-|m|-1)!}{|m|!}\left[\frac{(2l+1)(l+|m|)!(n_1+|m|)!(n_2+|m|)!}{(n+l)!(l-|m|)!(n-l-1)!(n_1)!(n_2)!}\right]^{\frac{1}{2}},$$

and ${}_3F_2$ is the hypergeometric function.

It is known that the discrete spectrum of a hydrogen atom with charge Z_a (atomic units are used), $E^{(0)} = -\frac{Z_a^2}{2n^2}$ is numbered by eigenvalues n of the operator L_{56}. The corresponding wave functions are normalized by a standard condition

$$\langle \overline{i_1, i_2, m} \,|\overline{i'_1, i'_2, m'}\rangle = \int \overline{\Phi}^{*(0)}_{i_1 i_2 m}(\mathbf{y})\overline{\Phi}^{(0)}_{i'_1 i'_2 m'}(\mathbf{y})d^3y = \delta_{i_1 i'_1}\delta_{i_2 i'_2}\delta_{mm'},$$

where $\langle \mathbf{y}|\overline{i_1, i_2, m}\,\rangle \equiv \overline{\Phi}^{(0)}_{n_1,n_2,m}(\mathbf{y})$ for the parabolic basis or $\langle \boldsymbol{y}|\overline{i_1, i_2, m}\,\rangle \equiv \overline{\Phi}^{(0)}_{n,l,m}(\boldsymbol{y})$ for the spherical basis. The wave functions $\langle \mathbf{y}|\overline{i_1, i_2, m}\,\rangle$ of the hydrogen atom are related to the eigenfunctions $\langle \mathbf{y}|i_1, i_2, m\,\rangle$ (Eq. (5) or Eq. (6)) of the operator L_{56} by the tilting transformation

$$\begin{aligned} \langle \mathbf{y}|\overline{i_1, i_2, m}\,\rangle &= \frac{(-2E^{(0)})^{1/4}}{n^{1/2}}U^{-1}\langle \mathbf{y}|i_1, i_2, m\rangle \\ &= \frac{(-2E^{(0)})^{3/4}}{n^{1/2}}\langle \sqrt{(-2E^{(0)})}\mathbf{y}|i_1, i_2, m\rangle. \end{aligned} \tag{8}$$

Here U is the unitary operator for $E^{(0)} < 0$

$$U = \exp(\imath\theta^{(0)}L_{45}), \quad \operatorname{th}\theta^{(0)} = \frac{1/2 + E^{(0)}}{1/2 - E^{(0)}} \tag{8a}$$

which provides the passage to new variables [15]

$$U^{-1}x_k U = \sqrt{-2E^{(0)}}x_k. \tag{9}$$

If we take into account (8), (8a) and scalar product (4), the standard normalization condition for $|\overline{\Phi}^{(0)}\rangle$ assumes the form

$$\begin{aligned} \langle \overline{i_1\,, i_2,\; m}|\overline{i'_1,\; i'_2,\; m'}\rangle &= \int \overline{\Phi}^{(0)*}_{i_1 i_2 m}(\mathbf{y})\overline{\Phi}^{(0)}_{i'_1 i'_2 m}(\mathbf{y})d^3y \\ = \frac{1}{n}\int \Phi^{(0)*}_{i_1 i_2 m}(\sqrt{-2E^{(0)}}\mathbf{y})&\Phi^{(0)}_{i'_1 i'_2 m'}(\sqrt{-2E^{(0)}}\mathbf{y})(\sqrt{-2E^{(0)}})^3 d^3y \\ &= \frac{1}{n}\int \Phi^{(0)*}_{i_1 i_2 m}(\mathbf{y})r\Phi^{(0)}_{i'_1 i'_2 m'}(\mathbf{y})r^{-1}d^3y \\ = \frac{1}{n}\langle i_1, i_2, m|L_{56} &- L_{46}|i'_1, i'_2, m'\rangle = \delta_{i_1 i'_1}\delta_{i_2 i'_2}\delta_{mm'}. \end{aligned} \tag{10}$$

Here, we have used the relation

$$\langle i_1, i_2, m|L_{56} - L_{46}|i'_1, i'_2, m'\rangle = n\delta_{i_1 i'_1}\delta_{i_2 i'_2}\delta_{mm'}. \tag{10a}$$

So, using the representation of the algebra $so(4,2)$, we can readily find the wave functions of both parabolic and spherical representations normalized by the standard condition and the energy of the discrete spectrum of a hydrogen-like atom.

3 Formulation of the Problem

The Shroedinger equation of a hydrogen-like atom with the charge Z_a in nonuniform electric field of the distant point charge Z_b in atomic units takes the form

$$\left(\frac{1}{2}\mathbf{p}^2 - \frac{Z_a}{r} + \frac{Z_b(Z_a-1)}{R} - \sum_{k=1} \frac{Z_b r^{k-1}}{R^k} P_{k-1}(\frac{x_3}{r}) - E\right)|\overline{\Phi}\rangle = 0, \qquad (11)$$

where $P_{k-1}(y_2)$ are Legendre polynomials. Under condition $R \gg 1$, the wave functions $|\overline{\Phi}\rangle$ and the energy $E = E(R)$ are sought for in the form

$$|\overline{\Phi}\rangle = \sum_{k=0} R^{-k}|\overline{\Phi}^{(k)}\rangle, \quad E = \sum_{k=0} R^{-k} E^{(k)}. \qquad (12)$$

Let us now multiply equation (11) by r and rewrite it using definitions (12)

$$\left(\frac{1}{2}r\mathbf{p}^2 - Z_a - \sum_{k=1} R^{-k} V^{(k)}(x_3, r) - rE^{(0)}\right)|\overline{\Phi}\rangle = 0, \qquad (13)$$

where

$$V^{(1)}(x_3, r) = r(E^{(1)} - Z_b(Z_a - 1)),$$

$$V^{(k)}(x_3, r) = r(E^{(k)} - Z_b r^{k-1} P_{k-1}(x_3/r)), \quad k \geq 2. \qquad (13a)$$

Then we can write (13) in notations (2) and (3). Making use of the tilting transformation (8), we go over from the hydrogen states $|\overline{\Phi}\rangle$ to the basis states $|\Phi\rangle$, and taking into account (8a), (9), we arrive at the following equation:

$$\left[L_{56} - \frac{Z_a}{\sqrt{-2E^{(0)}}} - \frac{1}{\sqrt{-2E^{(0)}}} \sum_{k=1} R^{-k} V^{(k)}\left(\frac{L_{35} - L_{34}}{\sqrt{-2E^{(0)}}}, \frac{L_{56} - L_{46}}{\sqrt{-2E^{(0)}}}\right)\right]|\Phi\rangle = 0. \quad (14)$$

The normalization condition for basis states $|\overline{\Phi}\rangle$ in this equation in terms of scalar product (4) for states $|\Phi\rangle$ is analogous to the condition of (10)

$$\langle\overline{\Phi}|\overline{\Phi}\rangle = \frac{\sqrt{-2E^{(0)}}}{n}\langle\Phi|U(L_{56} - L_{46})U^{-1}|\Phi\rangle$$

$$= \frac{\sqrt{-2E^{(0)}}}{n}\langle\Phi|\frac{L_{56} - L_{46}}{\sqrt{-2E^{(0)}}}|\Phi\rangle = \frac{1}{n}\langle\Phi|L_{56} - L_{46}|\Phi\rangle = 1. \qquad (10b)$$

Hence, it follows that in the course of the passage from the wave functions $|\Phi\rangle$ to the basis states $|\overline{\Phi}\rangle$, the first one should be normalized by the condition

$$\langle\Phi|L_{56} - L_{46}|\Phi\rangle = n, \qquad (15)$$

which becomes an identity for states $|\Phi^{(0)}\rangle$.

4 The Scheme and Algorithm of Perturbation Theory

We look for a solution to equation (14) in the form of the perturbation series

$$|\Phi\rangle = \sum_{k=0}^{k_{max}} R^{-k}|\Phi^{(k)}\rangle. \tag{16}$$

The unknown coefficients $|\Phi^{(k)}\rangle$ satisfy the system of inhomogeneous differential equations

$$L(n)|\Phi^{(0)}\rangle = (L_{56} - n)|\Phi^{(0)}\rangle = 0 \equiv f^{(0)}, \tag{17}$$

$$L(n)|\Phi^{(1)}\rangle = \frac{L_{56} - L_{46}}{(-2E^{(0)})}[E^{(1)} - Z_b(Z_a - 1)]|\Phi^{(0)}\rangle \equiv f^{(1)}, \tag{17a}$$

$$L(n)|\Phi^{(k)}\rangle = \frac{L_{56} - L_{46}}{(-2E^{(0)})}\sum_{p=0}^{k-1} V^{(k-p)}(\frac{L_{35} - L_{34}}{\sqrt{-2E^{(0)}}}, \frac{L_{56} - L_{46}}{\sqrt{-2E^{(0)}}})|\Phi^{(p)}\rangle \equiv f^{(k)}. \tag{17b}$$

As the basis that realizes the infinite-dimensional irreducible representation of the algebra $so(4,2)$ in x-space, we take the eigenfunctions $\langle \boldsymbol{y}|s,t\rangle$ of commuting operators L_{56}, A_3 and L_3, which differ from the above introduced basis functions $\langle \boldsymbol{y}|n_1 + s, n_2 + t, m\rangle$ only by the normalization factor and coincide with them at $s = t = 0$

$$\langle \boldsymbol{y}|s,t\rangle = \frac{C_{n_1 n_2 |m|}}{C_{n_1+s, n_2+t, |m|}}\langle \boldsymbol{y}|n_1 + s, n_2 + t, m\rangle. \tag{18}$$

The operators $L_{56}, L_{34}, L_{46}, L_{35}$ and L_{12} on the functions $\langle \boldsymbol{y}|s,t\rangle$ are defined by the relations without fractional powers of parabolic quantum numbers

$$L_{56}|s,t\rangle = (n_1 + n_2 + |m| + 1 + s + t)|s,t\rangle = (n + s + t)|s,t\rangle, \tag{19}$$

$$L_{34}|s,t\rangle = (-(n_1 + s) + (n_2 + t))|s,t\rangle, \quad L_{12}|s,t\rangle = |m||s,t\rangle, \tag{19a}$$

$$\begin{aligned} L_{45}|s,t\rangle = \tfrac{1}{2}(\quad & (n_1 + s + |m|)|s-1,t\rangle + (n_1 + s + 1)|s+1,t\rangle \\ & +(n_2 + t + |m|)|s,t-1\rangle + (n_2 + t + 1)|s,t+1\rangle), \end{aligned} \tag{19c}$$

$$\begin{aligned} L_{35}|s,t\rangle = \tfrac{1}{2} \quad & (-(n_1 + s + |m|)|s-1,t\rangle - (n_1 + s + 1)|s+1t\rangle \\ & +(n_2 + t + |m|)|s,t-1\rangle + (n_2 + t + 1)|s,t+1\rangle). \end{aligned} \tag{19d}$$

We assume that the azimuthal quantum number m will be positive $m = |m|$. Applying relations (19)–(19d), we expand the right-hand side $f^{(k)}$ and solutions $|\Phi^{(k)}\rangle$ of (17) over basis states $|s,t\rangle$ (18):

$$f^{(k)} = \sum_{s,t} f_{st}^{(k)}|s,t\rangle, \quad |\Phi^{(k)}\rangle = \sum_{s,t} b_{st}^{(k)}|s,t\rangle. \tag{20}$$

Substituting (20) into (17)–(17b) and taking into account the relation

$$L(n)|s,t\rangle = (s+t)|s,t\rangle$$

and orthogonality condition (4) of basis (18) for each $k \geq 1$, we obtain the system of linear algebraic equations for unknown coefficients $b_{st}^{(k)}$ and perturbation corrections $E^{(k)}$

$$(s+t)b_{st}^{(k)} - f_{st}^{(k)} = 0, \quad \min(|s|,|t|,|s+t|) \leq k. \tag{21}$$

It enables us to find the coefficients $b_{st}^{(k)}$ using known coefficients $f_{st}^{(k)}$ from the above definitions. In the second order $k = 2$, we obtain the energy correction $E^{(2)} = 3ndZ_b/2Z_a$ at $s = t = 0$ and eight coefficients $b_{st}^{(2)}$ at $s + t \neq 0$:

$$\begin{aligned}
b_{0,-1}^{(2)} &= -\frac{n^3 Z_b}{8Z_a^3}(d-m-n+1)(d+2n-2)),\\
b_{0,1}^{(2)} &= \frac{n^3 Z_b}{8Z_a^3}(d+m-n-1)(d+2n+2)),\\
b_{0,2}^{(2)} &= \frac{n^3 Z_b}{32Z_a^3}(d+m-n-1)(d+m-n-3)),\\
b_{0,-2}^{(2)} &= -\frac{n^3 Z_b}{32Z_a^3}(d-m-n+3)(d-m-n+1)),\\
b_{1,-1}^{(1)} &= \frac{n}{8Z_a}(d-m+n+1)(d-m-n+1)),\\
b_{1,-1}^{(2)} &= -\frac{n^2}{8Z_a^2}(d-m+n+1)(d-m-n+1)(d+1)),\\
b_{2,-2}^{(2)} &= \frac{n^2}{128Z_a^2}(d-m+n+3)(d-m+n+1)(d-m-n+3)(d-m-n+1)).
\end{aligned}$$

The remaining coefficients are obtained up to the sign by interchanging n_1 by n_2 ($d = n_2 - n_1 \to -d$), $b_{st}^{(k)}(d) = (-1)^{k+1} b_{ts}^{(k)}(-d)$, $b_{s,-s}^{(k)}(d) = (-1)^k b_{-s,s}^{(k)}(-d)$. Indeed, at each k the functions $f_{st}^{(k)}$ depend on the unknown correction $E^{(k)}$ and known coefficients $E^{(p)}$ and $b_{st}^{(p)}$ for $p = 0, 1, \cdots, k-1$, which are evaluated from previous $k-1$ equations. The coefficients $b_{st}^{(p)}$ also depend on the corrections $E^{(q)}$ and $b_{st}^{(q)}$ with $q = 0, 1, \cdots, p-1$, which are evaluated from previous equations (21) by recurrence. Thus, in each order ($k \geq 1$) we calculate step-by-step the needed corrections $E^{(k)}, b_{s,-s}^{(k-2)}, b_{s,t}^{(k)}$ by solving the following algebraic equations:

$$f_{00}^{(k)}\left(E^{(k)}, E^{(p)}, b_{s't'}^{(p)}, 0 \leq p \leq k-1\right) = 0 \to E^{(k)}, \tag{22}$$

$$f_{s-s}^{(k)}\left(E^{(k)}, E^{(p)}, b_{s',t'}^{(p)}, 1 \leq p \leq k-1\right) = 0 \to b_{s-s}^{(k-2)}, \tag{22a}$$

$$b_{s,t}^{(k)} = (s+t)^{-1} f_{st}^{(k)}\left(E^{(k)}, E^{(p)}, b_{s',t'}^{(p)}, 0 \leq p \leq k-1\right). \tag{22b}$$

The initial conditions for the recurrence procedure are given by

$$E^{(0)} = -\frac{Z_a^2}{2n^2}, \quad b_{0,0}^{(0)} = 1, \quad b_{s,t}^{(0)} = 0 \quad \text{for} \quad s,t \neq 0, \quad b_{0,0}^{(k)} = 0. \tag{23}$$

To obtain the normalized wave function Φ up to the kth order, we must redefine the coefficient $b_{0,0}^{(k)}$ by the following relation:

$$b_{00}^{(k)} = -\frac{Z_a}{2n^2}\sum_{p=0}^{k}\sum_{s',t'}\sum_{s,t} b_{st}^{(k-p)}\langle s,t|\frac{L_{56}-L_{46}}{\sqrt{-2E^{(0)}}}|s',t'\rangle b_{s't'}^{(p)}. \quad (24)$$

In particular, for coefficient $b_{0,0}^{(1)} = 0$ and $b_{0,0}^{(2)}$ is defined by

$$b_{0,0}^{(2)} = \frac{n^2}{64Z_a^2}(-d^4+(2d^2-m^2)(m^2+n^2-3)+m^2(3n^2-1)-(n^2-1)^2)+\frac{3dn^3Z_b}{4Z_a^3}. \quad (25)$$

The passage from the kth order approximate solution Φ in the form (16), (20) and (24) to the approximate solution $\overline{\Phi}$ of equation (13) normalized by condition (10b) with the same accuracy is made by means of tilting transformation (8), (8a). To have the approximate solution in the spherical representation (6), the additional orthogonal transformation (7) should be applied. So, we finish the description of the symbolic algorithm which yields the evaluation of the perturbation series of both parabolic and spherical representations.

5 Program Pointfield and Results

In this Section we show the main program **Pointfield** and auxiliary program **Generatev** realized in REDUCE 3.6. A test run output file for the energy corrections **ee(i)**= $E^{(i)}$ of the hydrogen-like atoms with charge Z_a in the field of a distant charge Z_b, atomic units is also used here.

PROGRAM POINTFIELD

```
% PROGRAM POINTFIELD
kmax:=7; off list,echo, nat;on nero;
operator x3,r,f,b,ee,ket,v;
in pointfieldcorr;
for all k let f(k)=n/za*(v(k,x3,r)*ket(0,0)
 +for p1:=1:k-2 sum v(k-p1,x3,r)*
   (for s:=-p1:p1 sum
     for tt:=-min(p1,p1+s):min(p1,p1-s)sum b(p1,s,tt)*ket(s,tt)));

let x3=x3(), r=r();
for all x,y let x3()*ket(x,y)=x3(x,y), r()*ket(x,y)=r(x,y);
for all x,y let r(x,y)=((n+x+y)*ket(x,y)
 -1/2*((x+n1+m)*ket(x-1,y)+(x+n1+1)*ket(x+1,y)
 +(y+n2+m)*ket(x,y-1)+(y+n2+1)*ket(x,y+1)))*n/za;
for all x,y let x3(x,y)=
(1/2*(-(x+n1+m)*ket(x-1,y)-(x+n1+1)*ket(x+1,y)
 +(y+n2+m)*ket(x,y-1)+(y+n2+1)*ket(x,y+1))-(y-x-d)*ket(x,y))*n/za;
```

```
procedure fk(k);
   begin scalar u,u1,u2;
     u:=f(k);u1:=den u; u:=num u;
     u2:=coeffn(u,ket(0,0),1); u2:=-u2/coeffn(u2,ee(k),1)+ee(k);
     ee(k):=u2;

     for i:=1:k-2 do
      for each i2 in {-1,1} do
      <<i1:=i*i2;
        u2:=coeffn(u,ket(i1,-i1),1)/u1;
        write b(k-2,i1,-i1):=
          sub(solve(u2,b(k-2,i1,-i1)),b(k-2,i1,-i1));
        u:=sub(ket(i1,-i1)=0,u);
     >>;
     b(k,0,0):=0;
     if k>kmax-2 then goto 1;
     for i1:=1/2 step 1/2 until (k/2) do
       for j1:=-k+i1:k-i1 do for each i2 in {-1,1} do
        <<ii:=i1*i2+j1;jj:=i1*i2-j1;
          u2:=coeffn(u,ket(ii,jj),1)/u1;
          write b(k,ii,jj):=u2/(ii+jj);
          u:=sub(ket(ii,jj)=0,u);
        >>;
1:end;
let n1=(n+d-m-1)/2, n2=(n-d-m-1)/2;
out pointfieldout;
for i:=1:kmax do fk(i);
for i:=1:kmax do write "e(",i,"):=",ee(i);
showtime;
shut pointfieldout;
bye;
```

Comments to the program
The file "pointfieldcorr" contains the substitution rules for operator v(k,x3,r) for k=1:kmax. The operator v(k,x3,r) realizes relations (13a) for function $V^{(k)}(x_3, r)$. The file "pointfieldcorr" should be generated by a preliminary run of the auxiliary program GENERATEV:

```
% PROGRAM GENERATEV
 kmax:=7;
 off echo,div,nat;operator v,y,x;
 let rr^(kmax+1)=0;
 for k:=1:kmax do v(k):=y*(
   coeffn((-za*zb*rr+zb*rr*(1+(for i:=1:kmax sum  (for j:=1:i product
    ((1/2-j)*(2*x*rr+y^2*rr^2)/j))) )),rr,k)+ee(k));

 out  poinfieldcorr;
```

```
  for k:=1:kmax do  write "for all x,y let v(",k,",x,y)=",v(k);
 shut poinfieldcorr;
bye;
```

The variables **f(k)**=$f^{(k)}$ are the right-hand sides in equations (17)–(17b). The procedure fk(k) solves equations (17)–(17b) with substitution (20), and in the k-th order, after calculation of the **f(k)**, finds sequentially the **ee(k)**= $E^{(k)}$, **b(k-1,s,-s)**= $b^{(k-1)}_{s,-s}$ and **b(k,s,tt)**= $b^{(k)}_{st}$ in accordance with eqs. (22).

Note that $b^{(k)}_{00} = 0$ so if $\min(|s|,|t|,|s+t|) > k$, then $b^{(k)}_{st} = f^{(k)}_{st} = 0$. If we wish to apply the above algorithm for the evaluation of only the energy spectrum corrections up to $k_{max} = 2k+1$, we can restrict the range of summation for s,t in equations (20), (21),(22a) and (22b) for each kth order to the region of $0 \le min(|s|,|t|,|s+t|) \le \min(k, k_{max}-k)$ (see Figs. 1,2). It means that in this case, we only use a minimal set of coefficients $b^{(k)}_{st}$ needed for the calculation of $E^{(k_{max})}$. It gives significant savings in computer resources.

s \ t	-6	-5	-4	-3	-2	-1	0	1	2	3	4	5	6
6							6						
5						6	5	6					
4					6	5	4	5	6				
3					5	4	3	4	5				
2			6	5	5	3	2	3	5	5	6		
1		6	5	4	3	3	2	3	3	4	5	6	
0	6	5	4	3	2	2	0	2	2	3	4	5	6
-1		6	5	4	3	3	2	3	3	4	5	6	
-2			6	5	5	3	2	3	5	5	6		
-3					5	4	3	4	5				
-4					6	5	4	5	6				
-5						6	5	6					
-6							6						

Fig. 1 In a row and column we point out the values of s and t characterized a set of basis vectors $|s,t\rangle$ making up an image domain of operator $V^{(k)}$ (13a), i.e. homogeneous polynomial of two variables of degree k, as a result of the action of $V^{(k)}$ on the ground basis vector $|0,0\rangle$; the numbers in the table are shown for a fixed value k for which the coefficient in front of basis vector $|s,t\rangle$ with given values s and t is not equal to zero.

s,t	-5	-4	-3	-2	-1	0	1	2	3	4	5
5	5		5	5	5	5					
4		4	5	4	4	4	5				
3	5	5	3	4	3	3	4	5			
2	5	4	4	2	3	2	3	4	5		
1	5	4	3	3	1	2	3	3	4	5	
0	5	4	3	2	2	0	2	2	3	4	5
-1		5	4	3	3	2	1	3	3	4	5
-2			5	4	3	2	3	2	4	4	5
-3				5	4	3	3	4	3	5	5

```
-4              5  4  4  4  5  4
-5                 5  5  5  5     5
```

Fig. 2 In a row and column we point out the values of s and t characterized a set of the coefficients $b_{s,t}^{(k)}$ and defined really a domain of summation in Eqs. (20); the numbers in the table are shown a fixed value k for which the coefficient $b_{s,t}^{(k)}$ with given values s, t is not equal zero.

Other variables are auxiliary. We consider a class of the polynomial perturbations in the x-representation. It takes the form of the multiply operators, i.e. $[x_k, r] = 0 => [L_{56} - L_{46}, L_{35} - L_{34}] = 0$. In this case, we do not need to use the declaration **noncom** x_k, r, while we use such a declaration when we calculate the inner scalar product in calculation of the element $b_{0,0}^{(k)}$ of (24). In the case of a nonlocal perturbation operator, for an example, depending on L_{45}, etc., it is necessary to apply the declaration **noncom** and the algebraic relations (1).

TEST RUN OUTPUT

```
e(1):=zb*(za - 1)$
e(2):=(3*d*n*zb)/(2*za)$
e(3):=(n**2*zb*( - 6*d**2+ n**2 - 1))/(2*za**2)$

e(4):=(n**3*zb*(109*d**3*za + 3*d**2*n*zb - 9*d*m**2*za - 39*d*n**2*za
+ 59*d*za  + 9*m**2*n*zb - 17*n**3*zb - 19*n*zb))/(16*za**4)$

e(5):=(3*n**4*zb*( - 355*d**4*za - 28*d**3*n*zb + 78*d**2*m**2*za
+ 198*d**2*n**2*za - 410*d**2*za - 84*d*m**2*n*zb + 148*d*n**3*zb
+ 252*d*n*zb - 3*m**4*za  + 6*m**2*n**2*za + 6*m**2*za - 11*n**4*za
+ 46*n**2*za - 35*za))/(64*za**5)$

e(6):=(n**5*zb*(2727*d**5*za**2 + 414*d**4*n*za*zb - 1056*d**3*m**2*za**2
- 2076*d**3*n**2*za**2 - 6*d**3*n**2*zb**2 + 5544*d**3*za**2
+ 1152*d**2*m**2*n*za*zb - 2088*d**2*n**3*za*zb - 4872*d**2*n*za*zb
+ 93*d*m**4*za**2 - 78*d*m**2*n**2*za**2 + 66*d*m**2*n**2*zb**2
- 450*d*m**2*za**2 + 273*d*n**4*za**2 + 138*d*n**4*zb**2
- 1470*d*n**2*za**2 + 234*d*n**2*zb**2 + 1533*d*za**2
- 30*m**4*n*za*zb - 84*m**2*n**3*za*zb + 324*m**2*n*za*zb
+ 178*n**5*za*zb + 84*n**3*za*zb - 742*n*za*zb))/(64*za**7)$

e(7):=(n**6*zb*( - 28861*d**6*za**2 - 7074*d**5*n*za*zb
+ 16485*d**4*m**2*za**2 + 27885*d**4*n**2*za**2 + 240*d**4*n**2*zb**2
- 92305*d**4*za**2 - 17460*d**3*m**2*n*za*zb + 34580*d**3*n**3*za*zb
+ 103140*d**3*n*za*zb - 2451*d**2*m**4*za**2
+ 270*d**2*m**2*n**2*za**2 - 2880*d**2*m**2*n**2*zb**2
+ 17610*d**2*m**2*za**2 - 6051*d**2*n**4*za**2 - 5664*d**2*n**4*zb**2
+ 40410*d**2*n**2*za**2 - 11760*d**2*n**2*zb**2 - 54487*d**2*za**2
+ 1494*d*m**4*n*za*zb + 2940*d*m**2*n**3*za*zb - 19380*d*m**2*n*za*zb
- 7122*d*n**5*za*zb - 5900*d*n**3*za*zb + 51102*d*n*za*zb
+ 43*m**6*za**2 - 3*m**4*n**2*za**2 - 240*m**4*n**2*zb**2
- 497*m**4*za**2 - 123*m**2*n**4*za**2 + 768*m**2*n**4*zb**2
```

```
+ 570*m**2*n**2*za**2  + 3600*m**2*n**2*zb**2  + 865*m**2*za**2
+ 163*n**6*za**2 - 400*n**6*zb**2 - 1697*n**4*za**2 - 6896*n**4*zb**2
+ 4825*n**2*za**2 - 6336*n**2*zb**2 - 3291*za**2))/(256*za**8)$

Time: 2647370 ms  plus GC time: 10000770 ms
```

This test was calculated on computer PC-2 350MHz 64MB memory. In the above formulas, we use the following notations: **n**$\equiv n = n_1 + n_2 + |m| + 1$, **m**$\equiv |m|$, **d**$\equiv n_1 - n_2$, **za**$\equiv Z_a$, **zb**$\equiv Z_b$. Note that our results coincide up to the sixth order with [6].

6 Conclusion

We have demonstrated the efficiency of the proposed recursive symbolic algorithm in the framework of an algebraic version of the conventional perturbation theory without assumption on separation of independent variables in both parabolic and spherical representations, which are needed for applications [7],[9]. A development of this algorithm in the case of hydrogen atom in the other nonuniform fields like a two-dimensional hydrogen atom in the electric and magnetic fields [16], for which independent variables are separated in the elliptic coordinates, will be presented elsewhere. It is interesting to compare such an approach with the quantization procedures [17] in the framework of the normal form method [18].

Acknowledgements

We are grateful to Professor V.P. Gerdt for his support of this work by the joint project in the development of computer algebra algorithms for integrable systems in external fields (in the framework of the themes N° 04-6-0996-93/2001 LIT and N° 01-03-1028-99/2003 BLTP of JINR). The authors (AG, VR and SV) appreciate support of Russian Foundation for Basic Research (grants No. 00-01-00617, No. 00-02-16337, and No. 00-02-81023-Bel2000_a.)

References

1. Coulson, C.A.: The Van der Waals force between a proton and a hydrogen atom. Proc. R. Soc. Edinburgh **A61** (1941) 20–26
2. Krogdahl, M.: The interaction of a proton and a hydrogen atom in its excited states. Astrophys. J. **100** (1944) 311–332
3. Coulson, C. A., Gilliam, C. M.: The Van der Waals force between a proton and a Hydrogen atom. II. Excited States. Proc. R. Soc. **A62** (1947) 360–368
4. Coulson, C. A., Robinson, P. D.: Wave functions for the hydrogen atom in spheroidal coordinates. I: The derivation and properties of the functions. Proc. Phys. Soc. London **A71** (1958) 815–827
5. Robinson, P. D.: Wave functions for the hydrogen atom in spheroidal coordinates. II: Interaction with a Point Charge with a Dipole. Proc. Phys. Soc. London **A71** (1958) 828–842

6. Power, J. D.: Fixed nuclei of two center problem in quantum mechanics. Phyl. Trans. Soc. London **A274** No 1246 (1973) 663–702.
7. Faifman, M. P., Ponomarev, L. I., Vinitsky, S. I.: Asymptotic form of effective potential of the Coulomb three-body problem in the adiabatic representation. J. Phys. B: Atom. Molec. Phys. **9** (1976) 2255–2268
8. Kadomtsev, M. B., Vinitsky, S. I.: Perturbation theory within the O(4,2) group for hydrogen atom in the field of distant charge. J. Phys. A: Math. Gen. **18** (1985) L689–L695
9. Abrashkevich, A.G., Puzynin, I.V., Vinitsky, S. I.: ASYMPT: a program for calculating asymptotics of hyperspherical potential curves and adiabatic potentials. Computer Physics Communications **125** (2000) 259–281
10. Courant, R., Hilbert, D.: Methods of Mathematical Physics. Interscience publishers, New York, London (1953)
11. Engerfield, M. J.: Group Theory and the Coulomb Problem. Monash university, Victoria (1972)
12. Malkin, I.A., Man'ko, V. I.: Dynamical symmetry and coherent states of quantum systems. Nauka, Moscow (1979) (in Russian)
13. Adams, B.G., Cizek, J., Paldus, J.: Lie algebraic methods and their applications to simple quantum systems. In: Per-Olov Lowdin (Ed.): Advances in Quantum Chemistry. Academic Press, New York **18** (1988) 1–85
14. Tarter, C.B.: Coefficients connected the Stark and fieldfree functions for Hydrogen. J. Math. Phys. **11** (1970) 3192–3195
15. Gusev, A. A., Samoilov, V. N., Rostovtsev, V. A., Vinitsky, S. I.: Symbolic Algorithm of Algebraic Perturbation Theory of a Hydrogen Atom: the Stark effect. In: Computer Algebra in Scientific Computing, V.G. Ganzha, E.W. Mayr,E.V. Vorozhtsov (eds.), Springer-Verlag, Berlin (2000) 219–231
16. Rakovic, M. J., Uzer, T., Farrelly, D.: Classical and quantum mechanics of an integrable limit of the hydrogen atom in combined circularly polarized microwave and magnetic fields. Phys. Rev. **A57** (1998) 2814–2831
17. Gusev, A. A., Chekanov, N. A., Baumann, G., Rostovtsev, V. A., Vinitsky, S. I.: On quantization of the planar hydrogen atom in uniform magnetic fields. In: Symmetries and Integrable Systems: Proceedings of Seminar, A. N. Sissakian (ed.), JINR D2-99-310 Dubna (1999) 51–62
18. Uwano, Y., Chekanov, N. A., Rostovtsev, V. A., Vinitsky, S. I.: On normalization of a class of polynimial Hamiltonians: From ordinary and inverse points of view. In: Computer Algebra in Scientific Computing, V.G. Ganzha, E.W. Mayr, E.V. Vorozhtsov (eds.), Springer-Verlag, Berlin (1999) 441–461

Perturbation versus Differentiation Indices*

Marcus Hausdorf and Werner M. Seiler

Lehrstuhl für Mathematik I, Universität Mannheim, 68131 Mannheim, Germany,
Email: {hausdorf,werner.seiler}@math.uni-mannheim.de,
WWW: http://www.math.uni-mannheim.de/~wms

Abstract. We discuss the relation between perturbation and differentiation indices for overdetermined systems of differential equations based on the formal theory. We show how the Cartan normal form of an involutive system can be used to extend the notion of an underlying equation from differential algebraic equations to partial differential equations. This allows us to generalise results of Campbell and Gear on the relation between perturbation and differentiation indices to systems of partial differential equations.

1 Introduction

This article is a sequel to [17] and further elaborates and rigorously proves some statements given there. We are especially concerned with the relation of two different kinds of index concepts for overdetermined systems of differential equations. *Differentiation indices* essentially count the number of differentiations required until one obtains a system with certain properties. This makes them comparatively easy to compute and thus they are much used. On the other hand, one may wonder why they should give any indication of difficulties appearing in the numerical integration of such systems. This becomes much more evident with *perturbation indices* based on estimates of the difference between solutions of the exact and a perturbed system. But such estimates are often rather hard to establish. For this reason it is of interest to relate differentiation and perturbation indices.

For the case of overdetermined systems of ordinary differential equations, often called *differential algebraic equations* [2,6], this problem has eventually been solved by Campbell and Gear [4]. The main purpose of this article is to extend their work to non-normal systems of partial differential equations. For this we first review their results in a more geometric manner that clarifies the main ingredients needed; in particular the important role of underlying equations is emphasised. Then we show how these ingredients can be defined for partial differential equations using the so-called Cartan normal form of an involutive system. This allows us, at least for linear systems, to extend their proof straightforwardly.

* This work has been supported by Deutsche Forschungsgemeinschaft, Landesgraduiertenförderung Baden-Württemberg and INTAS grant 99-1222.

2 Geometry of Differential Algebraic Equations

Most approaches to differential algebraic equations are purely computational but some geometric ones [12, 13] have been developed, too. Although these are typically a bit more restricted in their applicability (often they implicitly make some constant rank assumptions), they have the advantage to exhibit more clearly the basic ideas. For this reason we will mainly work geometrically in this article. We will furthermore assume in this section that we are dealing with autonomous systems; this is only for simplifying the presentation and not a real restriction.

Given a normal system of ordinary differential equations $\boldsymbol{y}' = \boldsymbol{Y}(\boldsymbol{y})$, its geometric counterpart is a vector field $\boldsymbol{Y}$ living on some manifold M (often simply $\mathbb{R}^m$). A differential algebraic equation is a more complicated object. Instead of a single vector field we are now dealing with a whole *pencil* of vector fields on the ambient manifold M and in addition we are given a submanifold $N_0 \subset M$ (see Fig. 1).

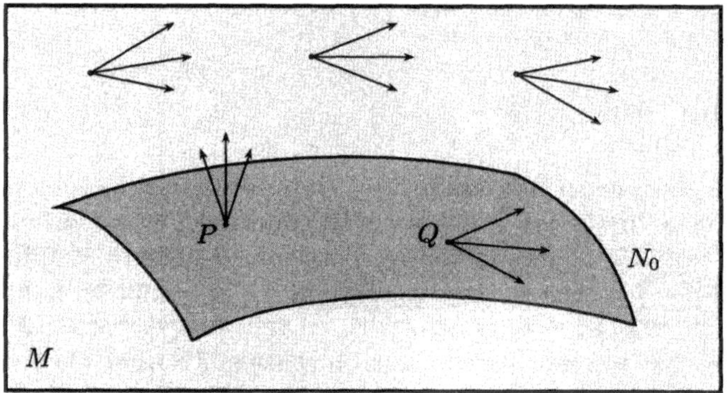

Fig. 1. Vector field pencil and submanifold

Example 1. The pencil and the submanifold are easily read off, if the differential algebraic equation is given in semi-explicit form. As a trivial example we consider the linear equation $y_2' = y_1 - 2$, $y_3' = -y_2$ and $y_2 = 3$. Here $M = \mathbb{R}^3$ and the submanifold N_0 is the plane described by the constraint equation $y_2 = 3$. We have a pencil of vector fields, as we do not know the y_1-component. The fields contained in the pencil can be written in parametric form as $\boldsymbol{Y}_a(\boldsymbol{y}) = (a(\boldsymbol{y}), y_1 - 2, -y_2)^t$ where a is an arbitrary function. ◁

Given a differential algebraic equation it is not at all clear that it actually possesses solutions and if yes where they live. For this reason one must first *complete* the system to a *formally integrable* or *involutive* one. One may also consider this operation as a consistency check. In our geometric framework it amounts to imposing *tangency conditions*.

Any solution of the differential algebraic equation must satisfy two conditions: it must define an integral curve for a vector field in the pencil and this curve must lie in the submanifold N_0. But these two conditions can be fulfilled simultaneously only, if at least some vector fields in the pencil are tangent to the submanifold N_0.

As one can see in Fig. 1, we must distinguish two cases. If all vector fields in the pencil are transversal to N_0 at some point $P \in N_0$, then no integral curve through P can lie in N_0 and we must eliminate this point. These eliminations lead to a smaller submanifold $N_1 \subset N_0$. From a computational point of view, this means that there are some "hidden" *algebraic* constraints. The second case arises, if the pencil contains at a point $Q \in N_0$ vector fields transversal to N_0. Then we must remove these and shrink the pencil to the tangential fields. In concrete computations this means that differentiating the constraints yields new *differential* equations.

After having imposed the first tangency condition, we deal in general with a smaller submanifold $N_1 \subseteq N_0$ and a shrunk vector field pencil. In case that N_1 is a proper submanifold of N_0, we must again analyse the tangency of the pencil, but this time to the *smaller* submanifold N_1. This might lead to a yet smaller submanifold N_2 and a further shrinking of the pencil.

After a finite number of steps we reach a final submanifold[1] N_k and a vector field pencil where all contained fields are tangent to N_k. This is called an *involutive* system. In most applications, the final pencil consists only of a single vector field as shown in Fig. 2. Otherwise we are dealing with an underdetermined system. The final submanifold consists of all points which may be used as initial values for the given differential algebraic equation.

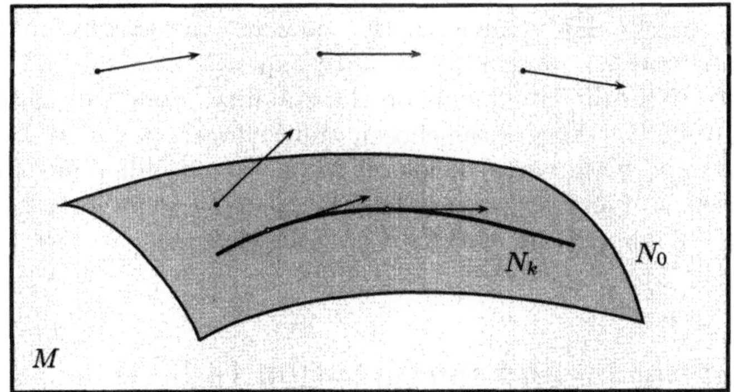

Fig. 2. Completed system

[1] If the system is not consistent, i.e. if it does not possess any solutions, then we obtain at some stage an empty submanifold.

If the final pencil contains only a single vector field, it is often called an *underlying* field for the differential algebraic equation. One should note that such a field is uniquely defined only *on* the final submanifold N_k. If we add to an underlying vector field any field that vanishes on N_k, we obtain again an underlying field: for the differential algebraic equation it is only of relevance what happens on N_k.

Example 2. In our simple example above, the completion needs two steps. In the first step we require that the vector fields $\boldsymbol{Y}_a$ are tangent to the plane N_0 or equivalently that the y_2-component vanishes on N_0. This cannot be achieved by shrinking the pencil, i. e. by choosing a special form of the function $a(\boldsymbol{y})$, but only by restricting to the straight line $N_1 \subset N_0$ described by the two equations $y_2 = 3$ and $y_1 = 2$.

In the second step we check whether all vector fields $\boldsymbol{Y}_a$ in the pencil are tangential to N_1. Obviously, this requires that the y_1-component must vanish on N_1. This time this can be achieved by shrinking the pencil, namely by setting $a(\boldsymbol{y}) = 0$. As $N_2 = N_1$, the completion process stops here. The final submanifold is the straight line N_1 and the pencil has shrunk to a single underlying vector field, e. g. $\boldsymbol{Y} = (0, 0, -3)^t$. More generally, any vector field on $\mathbb{R}^3$ that takes this form restricted to N_1 is an underlying field. We could also choose $\boldsymbol{Y} = (y_1 - 2, y_2 - 3, y_1 - y_2 - 2)^t$. ◁

Note an obvious but very important property of the final constraint manifold N_k: it is an *invariant manifold* for the flow of any underlying vector field $\boldsymbol{Y}$: if $\phi(\boldsymbol{y}) = 0$ belongs to the equations describing N_k, its derivative along flow lines, $\dot{\phi} = \boldsymbol{Y}\phi$, vanishes on N_k. Thus ϕ is a *weak invariant*[2].

A simple approach to the numerical integration of a differential algebraic equation is to take an underlying vector field and to compute its integral curve through a point on N_k. Because of the weak invariance, the integral curve will completely lie on N_k. Unfortunately, this does not necessarily hold for its numerical approximations; in general, we must expect a *drift* off N_k.

How severe this drift is depends on the stability properties of the invariant manifold N_k under the flow of the chosen underlying vector field. Here the non-uniqueness of underlying vector fields off N_k is crucial; different extensions will possess different stability properties. For the special case of constrained Hamiltonian systems a comparison of different underlying fields can be found in [16, 18].

3 Differentiation and Perturbation Indices

As already indicated in [17], the basic idea behind *differentiation indices* is to count the number of tangency conditions that have to be imposed until a certain state is reached. Looking at our geometric description in the last section, we see that there are two natural choices.

[2] It is not a first integral, as generally $\boldsymbol{Y}\phi$ does not vanish everywhere on M.

Definition 1. *The* involution index ν_i *is the minimal number of tangency conditions that must be imposed until an involutive system is reached, i. e. the final submanifold is* N_{ν_i}.

The involution index is defined for any differential algebraic equation, as any system can be completed in a finite number of steps. It has been introduced by several authors under different names. It is for example contained in the geometric works [12, 13] and it is equivalent to the strangeness index of Kunkel and Mehrmann [8, 9].

Definition 2. *The* determinacy index ν_d *is the minimal number of tangency conditions that must be imposed until the pencil has shrunk to a single vector field, i. e. until an underlying vector field has been found.*

The determinacy index is *not* always defined: for underdetermined systems no underlying vector field exists; this property explains the name. It is equivalent to "the" differentiation index, the most used index concept for differential algebraic equations. It is usually defined via so-called derivative arrays, but these represent just a computational approach to the geometric tangency conditions we impose.

In our example in the last section we have $\nu_d = \nu_i = 2$. In general, we have obviously $\nu_d \leq \nu_i$. An existence theory can only be built upon the involution index, as only after the full completion we know whether the given differential algebraic equation is consistent. So one may wonder about the importance of the determinacy index. We will answer this question in the next section.

The notion of a *perturbation index* was introduced by Hairer, Lubich and Roche [5]. It is a purely analytic concept based on estimating the effect of perturbing a given differential algebraic equation and measures up to which order derivatives of the perturbation enter the estimate.

Definition 3. *Let* $\boldsymbol{y}(x)$ *be a solution of the implicit differential algebraic equation*[3] $\boldsymbol{F}(x, \boldsymbol{y}, \boldsymbol{y}') = 0$ *and* $\hat{\boldsymbol{y}}(x)$ *a solution of the perturbed equation* $\boldsymbol{F}(x, \hat{\boldsymbol{y}}, \hat{\boldsymbol{y}}') = \boldsymbol{\delta}(x)$ *where* $\boldsymbol{\delta}(x)$ *is a sufficiently often differentiable function. Assuming that both solutions are defined on the interval* $[0, X]$, *we say that the equation* $\boldsymbol{F} = 0$ *has* perturbation index ν_p *along the solution* $\boldsymbol{y}(x)$, *if* ν_p *is the smallest integer such that the following estimate holds for all* $x \in [0, X]$ *provided the right hand side is sufficiently small:*

$$|\boldsymbol{y}(x) - \hat{\boldsymbol{y}}(x)| \leq C \left(|\boldsymbol{y}(0) - \hat{\boldsymbol{y}}(0)| + \|\boldsymbol{\delta}\|_{\nu_p - 1} \right) . \tag{1}$$

Here the constant C *may depend only on* $\boldsymbol{F}$ *and the length* X *of the interval.* $|\cdot|$ *denotes some vector norm on* $\mathbb{R}^m$ *and* $\|\cdot\|_k$ *is a kind of Sobolev norm* $\|\boldsymbol{\delta}\|_k = \sum_{i=0}^{k} \|\boldsymbol{\delta}^{(i)}\|$ *based on a norm* $\|\cdot\|$ *on* $C([0, X], \mathbb{R}^t)$ *which is usually either the maximum norm* $\|\cdot\|_{L^\infty}$ *or the uniform norm* $\|\cdot\|_{L^1}$.

[3] We do not require here that there are as many equations as unknowns; the values of $\boldsymbol{y}$ are m-dimensional vectors, those of $\boldsymbol{F}$ are t-dimensional.

Whereas it is not so clear why the differentiation indices should indicate the degree of difficulty of the numerical integration of a differential algebraic equation, this is rather obvious for the perturbation index. If we take for $\hat{\boldsymbol{y}}$ an approximate solution of $\boldsymbol{F} = 0$, we may interpret the perturbation $\boldsymbol{\delta}$ as the residual obtained by entering it into the equation. (1) tells us that for an equation with $\nu_p > 1$ it does not suffice to keep this residual as small as possible, as also some of its derivatives enter the estimate. Strictly speaking, this implies that the initial value problem for such an equation is ill-posed.

In general there is no relation between ν_d, ν_i and ν_p, as demonstrated with some examples by Campbell and Gear [4]. This is not so surprising, as the perturbation index is based on a perturbed equation and the perturbation affects for non-linear equations the completion.

Example 3. Consider the following simple quasi-linear system [4]:

$$y_2 y_2' + y_1 = 0\,, \quad y_2 = 0\,. \tag{2}$$

Obviously this system is equivalent to $y_1 = y_2 = 0$. Thus its indices are $\nu_d = \nu_i = 1$, as after one step we obtain an underlying equation $y_1' = y_2' = 0$.

For the perturbed system

$$y_2 y_2' + y_1 = \delta_1\,, \quad y_2 = \delta_2 \tag{3}$$

the situation is different. The first tangency condition yields $y_2' = \delta_2'$ and then the first equation of (3) is equivalent to $y_1 = \delta_1 - \delta_2\delta_2'$. Imposing a second tangency condition leads to $y_1' = \delta_1' - (\delta_2\delta_2')'$ and thus an underlying equation. The indices of the perturbed system are $\tilde{\nu}_d = \tilde{\nu}_i = 2$.

As shown in [4] this example can be extended. Using an m-dimensional vector of unknowns $\boldsymbol{y}$, one can construct systems such that for the unperturbed system still $\nu_d = \nu_i = 1$ but for the perturbed one $\tilde{\nu}_d = \tilde{\nu}_i = m$. Thus the difference between the unperturbed and the perturbed indices may become arbitrarily large. ◁

Definition 4. *Let $\boldsymbol{F}(x, \boldsymbol{y}, \boldsymbol{y}') = 0$ be a differential algebraic equation. Its* perturbed determinacy index $\tilde{\nu}_d$ *and its* perturbed involution index $\tilde{\nu}_i$*, respectively, are defined as the indices of the perturbed differential algebraic equation $\boldsymbol{F}(x, \boldsymbol{y}, \boldsymbol{y}') = \boldsymbol{\delta}(x)$.*

This definition implicitly assumes that the completion process of the perturbed equation is independent of the precise form of $\boldsymbol{\delta}$. In most applications this is probably the case. Otherwise, the perturbed indices are computed with the generic branch of the completion, i. e. under the assumption that the perturbation does *not* satisfy any special conditions.

4 Estimating the Perturbation Index

We have seen in Example 3 that the indices ν_d, ν_i and ν_p, respectively, are in general not related. It is now a natural thought to search for a relation between

$\tilde{\nu}_d$, $\tilde{\nu}_i$ and ν_p, as all these indices are based on the perturbed system. But before we recall some auxiliary results.

Lemma 1 (Gronwall). *Let $I = [a, b] \subseteq \mathbb{R}$ and let $\phi : I \to \mathbb{R}$ be a continuous function. Suppose that on I*

$$\phi(x) \leq A + B \int_a^x \phi(\xi) d\xi \tag{4}$$

for some constants $A, B \geq 0$. Then for all $x \in I$

$$\phi(x) \leq A e^{B(x-a)} . \tag{5}$$

This is a classical result that can be found with proof in almost any textbook on ordinary differential equations (see e. g. [22]). Based on it we can compute ν_p for normal systems of ordinary differential equations.

Proposition 1. *Let $\boldsymbol{F}(x, \boldsymbol{y}, \boldsymbol{y}') = 0$ be an implicit ordinary differential equation such that the Jacobian $\partial \boldsymbol{F} / \partial \boldsymbol{y}'$ has a bounded inverse on $[0, X]$. Furthermore let $\boldsymbol{F}$ be Lipschitz continuous in $\boldsymbol{y}$ with constant L_0 and let $\hat{\boldsymbol{y}}(x)$ be a solution of the perturbed equation $\boldsymbol{F}(x, \hat{\boldsymbol{y}}, \hat{\boldsymbol{y}}') = \boldsymbol{\delta}(x)$. Then for all $x \in [0, X]$*

$$|\boldsymbol{y}(x) - \hat{\boldsymbol{y}}(x)| \leq C \left(|\boldsymbol{y}(0) - \hat{\boldsymbol{y}}(0)| + \int_0^x |\boldsymbol{\delta}(\xi)| \, d\xi \right) \tag{6}$$

where the constant C depends only on $\boldsymbol{F}$ and the length X of the interval.

Proof. We solve the original and the perturbed equation for the derivatives $\boldsymbol{y}'$ and $\hat{\boldsymbol{y}}'$, respectively, which is always possible under the made assumptions. This yields the equations $\hat{\boldsymbol{y}}' = \boldsymbol{f}(x, \hat{\boldsymbol{y}}, \boldsymbol{\delta})$ and $\boldsymbol{y}' = \boldsymbol{f}(x, \boldsymbol{y}, 0)$. The function $\boldsymbol{f}$ satisfies a Lipschitz condition in both $\hat{\boldsymbol{y}}$ and $\boldsymbol{\delta}$ with a constant L_1 depending on L_0 and the bound of $(\partial \boldsymbol{F} / \partial \boldsymbol{y}')^{-1}$. Now we integrate both equations from 0 to x, subtract them and take the norm:

$$|\boldsymbol{y}(x) - \hat{\boldsymbol{y}}(x)| \leq |\boldsymbol{y}(0) - \hat{\boldsymbol{y}}(0)| + \left| \int_0^x [\boldsymbol{f}(\xi, \boldsymbol{y}(\xi), 0) - \boldsymbol{f}(\xi, \hat{\boldsymbol{y}}(\xi), \boldsymbol{\delta}(\xi))] d\xi \right| . \tag{7}$$

Interchanging the integration and the norm, we can exploit the Lipschitz continuity of $\boldsymbol{f}$ to obtain

$$|\boldsymbol{y}(x) - \hat{\boldsymbol{y}}(x)| \leq |\boldsymbol{y}(0) - \hat{\boldsymbol{y}}(0)| + L_1 \int_0^x \Big(|\boldsymbol{y}(\xi) - \hat{\boldsymbol{y}}(\xi)| + |\boldsymbol{\delta}(\xi)| \Big) d\xi . \tag{8}$$

Obviously, $\int_0^x |\boldsymbol{\delta}(\xi)| \, d\xi$ is a monotonically increasing function of x on the interval $[0, X]$ taking its maximal value at $x = X$. Applying Gronwall's Lemma with $\phi(x) = |\boldsymbol{y}(x) - \hat{\boldsymbol{y}}(x)|$ and $A = |\boldsymbol{y}(0) - \hat{\boldsymbol{y}}(0)| + L_1 \int_0^X |\boldsymbol{\delta}(\xi)| \, d\xi$, $B = L_1$ yields immediately the claimed estimate with $C = \max(1, L_1) \exp(L_1 X)$. □

Thus a fully implicit system has generally the perturbation index $\nu_p = 1$, as the right hand side of (6) is the L^1 norm of $\boldsymbol{\delta}$. But note that in order to derive (8) we had to interchange in (7) the integration and the norm. If we start with an *explicit* system, so that the perturbed form is $\hat{\boldsymbol{y}}' = \boldsymbol{f}(x, \hat{\boldsymbol{y}}) + \boldsymbol{\delta}$, such an interchange is not necessary and we obtain on the right hand side of (6) a term of the form $\left|\int_0^x \boldsymbol{\delta}(\xi)d\xi\right|$. Interpreting $\|\boldsymbol{\delta}\|_{-1}$ as the (maximum) norm of $\int_0^x \boldsymbol{\delta}(\xi)d\xi$, we see that such a system has in fact the perturbation index $\nu_p = 0$.

Theorem 1. *Let $\boldsymbol{F}(x, \boldsymbol{y}, \boldsymbol{y}') = 0$ be a differential algebraic equation. Then*

$$\tilde{\nu}_d \leq \nu_p \leq \tilde{\nu}_d + 1\,. \tag{9}$$

Proof. The essential step of the proof is to show that the completion of the perturbed form of the given differential algebraic equation leads to an underlying equation of the form

$$\boldsymbol{F}^c(x, \hat{\boldsymbol{y}}, \hat{\boldsymbol{y}}', \boldsymbol{\delta}, \boldsymbol{\delta}', \dots, \boldsymbol{\delta}^{(\tilde{\nu}_d)}) = 0 \tag{10}$$

with $\partial \boldsymbol{F}^c / \partial \boldsymbol{\delta}^{(\tilde{\nu}_d)} \neq 0$. Then we can apply the proposition above and get an estimate of the form (1) with $\nu_p - 1 = \tilde{\nu}_d$. The lower bound is obtained by taking into account the remark above: if the integration can be left within the norm, it kills one derivative. ν_p cannot be less than $\tilde{\nu}_d$, as the solution of (10) depends at least on $\boldsymbol{\delta}^{(\tilde{\nu}_d - 1)}$, and by definition $\tilde{\nu}_d$ is the smallest number of differentiations leading to an underlying equation.

Before the completion we transform the perturbed differential algebraic equation into semi-explicit form; this is a purely algebraic process which does not introduce any derivatives. As outlined in Section 2, the completion process takes then the following simple form: the algebraic equations are differentiated and it is checked whether this yields any new equations; the completion stops as soon as no new *algebraic* equations arise.

In the first step the algebraic part of the differential algebraic equation is differentiated. If this leads to any new equations, they contain a first order derivative of a perturbation. If some of the new equations are algebraic, they are differentiated again. Thus any new equations arising out of them contains a second order derivative of a perturbation and so on. The determinacy index $\tilde{\nu}_d$ counts the last completion step where new *differential* equations appear, i. e. equations that are part of the underlying equation $\boldsymbol{F}^c = 0$. Thus these equations depend effectively on derivatives of order $\tilde{\nu}_d$ of the perturbations and $\boldsymbol{F}^c$ is of the form claimed above. □

Example 4. We consider again the system of Example 3. From its closed form solution one can easily see that $\nu_p = 2$. Thus $\nu_p = \tilde{\nu}_d$ in agreement with our theorem. We are at the lower bound, as the underlying equation is of the *explicit* form $y_1' = \delta_1' - (\delta_2\delta_2')'$, $y_2' = \delta_2'$ and we can thus perform the integration within the norm. ◁

5 Involutive Systems of Partial Differential Equations

The formal theory of differential equations provides an intrinsic approach to very general systems of partial differential equations. In particular, it is able to handle effectively under- and overdetermined systems based on the notion of *involution*. This is now a considerably more complicated concept as in the case of ordinary differential equations comprising a mixture of geometric and algebraic ideas. For lack of space it is not possible to discuss here the formal theory in detail. We must refer to the literature [11, 15]; some basic notions can also be found in [17]. A simple introduction into a combinatorial approach to involution is given in [3]; a deeper analysis is contained in [19].

Although we will describe everything in local coordinates, one should note that the formal theory is completely intrinsic and can be presented without any coordinates in a purely geometric manner. However, this requires the introduction of a certain machinery like jet bundles which we want to avoid here. For simplicity, we furthermore assume that we are given a first order system for m unknown functions u_α in n independent variables x_i without any algebraic equations.

For the analysis of differential algebraic equations it is very convenient to have them in semi-explicit form. Its analogue for systems of partial differential equations is the following triangular form where the numbers $\beta^{(i)}$ satisfy $0 \leq \beta^{(1)} \leq \cdots \leq \beta^{(n)} \leq m$.

$$\begin{aligned}
\frac{\partial u_\alpha}{\partial x_n} &= \phi_{n\alpha}\left(\boldsymbol{x}, \boldsymbol{u}, \frac{\partial \boldsymbol{u}}{\partial x_j}, \frac{\partial u_\delta}{\partial x_n}\right), &&\begin{cases} 1 \leq \alpha \leq \beta^{(n)}, \\ 1 \leq j < n, \\ \beta^{(n)} < \delta \leq m, \end{cases} \\
\frac{\partial u_\alpha}{\partial x_{n-1}} &= \phi_{n-1,\alpha}\left(\boldsymbol{x}, \boldsymbol{u}, \frac{\partial \boldsymbol{u}}{\partial x_j}, \frac{\partial u_\delta}{\partial x_{n-1}}\right), &&\begin{cases} 1 \leq \alpha \leq \beta^{(n-1)}, \\ 1 \leq j < n-1, \\ \beta^{(n-1)} < \delta \leq m, \end{cases} \\
&\vdots \\
\frac{\partial u_\alpha}{\partial x_1} &= \phi_{1\alpha}\left(\boldsymbol{x}, \boldsymbol{u}, \frac{\partial u_\delta}{\partial x_1}\right), &&\begin{cases} 1 \leq \alpha \leq \beta^{(1)}, \\ \beta^{(1)} < \delta \leq m. \end{cases}
\end{aligned} \tag{11}$$

Here as many equations as possible have been solved for an x_n-derivative. Of the remaining ones as many as possible have been solved for an x_{n-1}-derivative and so on. The classical case of a system in Cauchy–Kovalevskaya form arises, if $\beta^{(n)} = m$ and all other $\beta^{(i)}$ vanish.

We say that those equations in system (11) which are solved for an x_k-derivative are of *class* k and assign them the *multiplicative variables* $x_1, \ldots, x_k$. The system is *involutive*, if the following property holds: any equation obtained by differentiating one of the equations in (11) with respect to a non-multiplicative variable can be written as a linear combination of the equations in (11) plus those equations obtained by differentiating them with respect to their multiplicative variables only. For an involutive system, (11) is also called its *Cartan normal form*.

Example 5. Consider the following linear system with constant coefficients in three independent variables:

$$\boldsymbol{u}_{x_3} = A_1\boldsymbol{u}_{x_1} + A_2\boldsymbol{u}_{x_2}\,, \tag{12}$$

$$0 = B_1\boldsymbol{u}_{x_1} + B_2\boldsymbol{u}_{x_2}\,. \tag{13}$$

Here $A_1, A_2 \in \mathbb{R}^{m\times m}$ and $B_1, B_2 \in \mathbb{R}^{r\times m}$ for some $r \leq m$. The system is already in triangular form (though not explicitly solved for the x_2-derivatives) provided that the rank of B_2 is r what we assume. The equations in (12) are all of class 3 and thus all independent variables are multiplicative for them. The equations in (13) are only of class 2 and hence x_3 is non-multiplicative for them. There are no equations of class 1.

In order to check whether or not the system is involutive, we must analyse what happens, if we differentiate the equations in (13) with respect to x_3. This yields the equations $B_1\boldsymbol{u}_{x_1x_3} + B_2\boldsymbol{u}_{x_2x_3} = 0$. The second derivatives can be eliminated by subtracting x_1- and x_2-differentiations of the equations in (12) and we obtain $B_1A_1\boldsymbol{u}_{x_1x_1} + (B_1A_2 + B_2A_1)\boldsymbol{u}_{x_1x_2} + B_2A_2\boldsymbol{u}_{x_2x_2} = 0$. The system is involutive, if and only if these equations can be written as linear combinations of x_1- and x_2-differentiations of the equations in (13), i.e. if and only if two matrices $M_1, M_2 \in \mathbb{R}^{r\times m}$ exist such that $B_1A_1 = M_1B_1$, $B_2A_2 = M_2B_2$ and $M_1B_2 + M_2B_1 = B_1A_2 + B_2A_1$. If this is the case, we can write

$$\begin{aligned}\frac{\partial}{\partial x_3}(B_1\boldsymbol{u}_{x_1} + B_2\boldsymbol{u}_{x_2}) = {}& \left[B_1\frac{\partial}{\partial x_1} + B_2\frac{\partial}{\partial x_2}\right](\boldsymbol{u}_{x_3} - A_1\boldsymbol{u}_{x_1} + A_2\boldsymbol{u}_{x_2}) - \\ & \left[M_1\frac{\partial}{\partial x_1} + M_2\frac{\partial}{\partial x_2}\right](B_1\boldsymbol{u}_{x_1} + B_2\boldsymbol{u}_{x_2})\,.\end{aligned} \tag{14}$$

Thus we expressed the non-multiplicative differentiation on the left hand side as a linear combination of multiplicative ones. ◁

The *Cartan-Kähler Theorem* states that the (non-characteristic) initial value problem for an analytic involutive system with analytic initial data possesses a unique analytic solution. Its proof is based on the Cartan normal form and an iterative application of the classical Cauchy–Kovalevskaya Theorem to the subsystems of the normal form.

The *Cartan–Kuranishi Theorem* asserts that (under some regularity assumptions) any system of partial differential equations can be completed to an equivalent involutive one. Its proof is constructive. The arising completion method is *not* a straightforward generalisation of the simple procedure for differential algebraic equations outlined in Section 2. An efficient algorithmic realisation of it was derived in [7] combining geometric and algebraic ideas. It has been implemented in the computer algebra system *MuPAD* [1].

6 Indices for Partial Differential Equations

It is straightforward to extend the involution index to partial differential equations based on the Cartan-Kuranishi Theorem. Essentially, it requires only a reformulation of the definition above such that it can be applied to both ordinary and partial differential equations. It is explicitly shown in [17] that the following definition is for ordinary differential equations equivalent to the one given above.

Definition 5. *The* involution index ν_i *of a system of differential equations is the minimal number of steps required by the Cartan–Kuranishi completion to an involutive system.*

But we saw above that for estimating the perturbation index the determinacy index $\tilde{\nu}_d$ and not the involution index $\tilde{\nu}_i$ is decisive. Thus the question arises how this index can be generalised. [17] gave only a purely formal answer to this. In our discussion above it has clearly emerged that the fundamental concept behind the determinacy index is the underlying equation. We will demonstrate how this notion can be extended to partial differential equations using the Cartan normal form of an involutive system.

Our claim is that in the Cartan normal form the subsystem consisting of all equations of class n may be considered as an underlying equation. We will restrict in the sequel to systems which are not underdetermined, i. e. where $\beta^{(n)} = m$. This is only for simplifying the presentation and not because of any fundamental problems with underdetermined systems. Under this assumption, the subsystem of class n is normal, i. e. in Cauchy–Kovalevskaya form. An important property of an underlying equation of a differential algebraic equation is that the constraint manifold is invariant under its flow. We show now that our subsystem satisfies a similar property; however, for a rigorous proof we restrict to linear systems with analytic coefficients.

Proposition 2. *Let (11) be an involutive system and let the functions* $\phi_{k\alpha}$ *be linear in* $\boldsymbol{u}$*,* $\partial\boldsymbol{u}/\partial\boldsymbol{x}$ *and analytic in* $\boldsymbol{x}$*. If the hyperplane* $x_n = 0$ *is non-characteristic and* $\boldsymbol{U}(\boldsymbol{x})$ *is a smooth solution of the subsystem of class n which satisfies for* $x_n = 0$ *all other equations in the system, then it satisfies these equations for all values of* x_n*.*

Proof. We must analyse what happens upon entering the solution $\boldsymbol{U}(\boldsymbol{x})$ into the equations of lower class. So we introduce for $1 \leq k < n$ and $1 \leq \alpha \leq \beta^{(k)}$ the residuals

$$\Delta_{k\alpha}(\boldsymbol{x}) = \frac{\partial U_\alpha}{\partial x_k}(\boldsymbol{x}) - \phi_{k\alpha}\left(\boldsymbol{x}, \boldsymbol{U}(\boldsymbol{x}), \frac{\partial \boldsymbol{U}}{\partial \boldsymbol{x}}(\boldsymbol{x})\right). \tag{15}$$

Obviously, they are smooth functions. The proposition claims that if the functions $\Delta_{k\alpha}$ vanish for $x_n = 0$, they vanish for all values of x_n. We will prove it by deriving a system of partial differential equations for the residuals.

By assumption, (11) is a linear involutive system with analytic coefficients. As x_n is a non-multiplicative variable for all equations of class less than n, there must exist analytic functions $A_{i\beta j}(\boldsymbol{x})$ and $B_{i\beta}(\boldsymbol{x})$ such that[4]

$$\begin{aligned}\frac{\partial}{\partial x_n}\left(\frac{\partial u_\alpha}{\partial x_k} - \phi_{k\alpha}(\boldsymbol{x},\boldsymbol{u},\partial\boldsymbol{u}/\partial\boldsymbol{x})\right) = \\ \sum_{i=1}^{n}\sum_{\beta=1}^{\beta^{(i)}}\sum_{j=1}^{i} A_{i\beta j}(\boldsymbol{x})\frac{\partial}{\partial x_j}\left(\frac{\partial u_\beta}{\partial x_i} - \phi_{i\beta}(\boldsymbol{x},\boldsymbol{u},\partial\boldsymbol{u}/\partial\boldsymbol{x})\right) + \\ \sum_{i=1}^{n}\sum_{\beta=1}^{\beta^{(i)}} B_{i\beta}(\boldsymbol{x})\left(\frac{\partial u_\beta}{\partial x_i} - \phi_{i\beta}(\boldsymbol{x},\boldsymbol{u},\partial\boldsymbol{u}/\partial\boldsymbol{x})\right).\end{aligned} \tag{16}$$

Entering our solution $\boldsymbol{U}(\boldsymbol{x})$ into these relations yields the following *linear* homogeneous system with analytic coefficients for the residuals $\Delta_{k\alpha}$

$$\frac{\partial \Delta_{k\alpha}}{\partial x_n} = \sum_{i=1}^{n-1}\sum_{\beta=1}^{\beta^{(i)}}\left\{\sum_{j=1}^{i} A_{i\beta j}(\boldsymbol{x})\frac{\partial \Delta_{i\beta}}{\partial x_j} + B_{i\beta}(\boldsymbol{x})\Delta_{i\beta}\right\}. \tag{17}$$

Here the summands for $i = n$ disappeared, as the function $\boldsymbol{U}(\boldsymbol{x})$ is a solution of the subsystem of class n. The system (17) is in Cauchy–Kovalevskaya form and we are interested in its smooth solutions for the initial conditions $\Delta_{k\alpha}(x_n = 0) = 0$, as by assumption $\boldsymbol{U}$ satisfies for $x_n = 0$ all equations of lower class. Because of the homogeneity, $\Delta_{k\alpha}(\boldsymbol{x}) = 0$ is obviously a solution. As (17) is analytic, the Holmgren Theorem tells us that this is the only smooth solution of the initial value problem. Thus the residuals vanish for all values of x_n and $\boldsymbol{U}(\boldsymbol{x})$ is a smooth solution of the full system (11). □

For simplicity, we have formulated this proposition only for smooth solutions. In fact, the Holmgren Theorem gives uniqueness for $\mathcal{C}^1$ solutions (there are also extensions to Sobolev spaces). So the result holds for much larger classes of solutions including weak ones. Unfortunately, the restriction to analytic coefficients is not so easy to remove.[5] The Calderon Theorem provides uniqueness also for linear systems with smooth coefficients; however, the system must satisfy some technical conditions [21, 23]. In any case, the above used technique of proof reduces the problem even for *non-linear* systems always to the question of the uniqueness of the zero solution of a *linear* homogeneous system.

Definition 6. *The* determinacy index ν_d *of a system of differential equations is the number of steps required in the Cartan–Kuranishi completion until a system is reached with* $\beta^{(n)} = m$.

[4] For general systems, the functions $\boldsymbol{A}$ and $\boldsymbol{B}$ could also depend on $\boldsymbol{u}$ and $\partial\boldsymbol{u}/\partial\boldsymbol{x}$. Because of the assumed linearity of (11) it suffices to consider here functions depending only on $\boldsymbol{x}$.

[5] This is also the main obstacle for extending Proposition 2 to non-linear systems. Only for analytic solutions of analytic systems the extension is straightforward using the Cauchy–Kovalevskaya Theorem.

As in the case of differential algebraic equations, the determinacy index is only defined for systems which are not underdetermined, as otherwise the condition $\beta^{(n)} = m$ will never be satisfied.[6] The idea behind this definition is exactly the same as in Definition 2: as soon as the condition is met, we know an underlying equation. Under additional hypotheses one can now derive results similar to Theorem 1. We give here only a simple example.

Theorem 2. *Let (11) be as in Proposition 2 and let again the hyperplane $x_n = 0$ be non-characteristic for it. Assume furthermore that the right hand side of the subsystem of class n can be written in the form $A\boldsymbol{u}$ where A is the generator of a strongly continuous semigroup acting on an appropriately chosen Banach space. Then $\tilde{\nu}_d \leq \nu_p \leq \tilde{\nu}_d + 1$.*

Proof. We slightly change our notation: we write t for x_n and $\dot{\boldsymbol{u}}$ for the partial derivatives $\partial \boldsymbol{u}/\partial x_n$. Then we may rewrite the subsystem of class n as $\dot{\boldsymbol{u}} = A\boldsymbol{u} + \boldsymbol{\delta}$ where $\boldsymbol{u}$ and $\boldsymbol{\delta}$ are considered as functions of t taking values in the chosen Banach space. It is an elementary result in the theory of strongly continuous semigroups [10, 14] that the solution $\hat{\boldsymbol{u}}(t)$ of this equation is

$$\hat{\boldsymbol{u}}(t) = S(t)\hat{\boldsymbol{u}}(0) + \int_0^t S(t-\tau)\boldsymbol{\delta}(\tau)d\tau \tag{18}$$

where $S(t)$ is the semigroup generated by A and $\hat{\boldsymbol{u}}(0)$ an arbitrary element of the domain of A.

Now we can easily prove an analogous estimate to Proposition 1. We compare the solution $\hat{\boldsymbol{u}}(t)$ given by (18) with a solution $\boldsymbol{u}(t)$ for $\boldsymbol{\delta}(t) = 0$. If $\|\cdot\|$ denotes the norm of the used Banach space, we get

$$\|\boldsymbol{u}(t) - \hat{\boldsymbol{u}}(t)\| \leq C\left(\|\boldsymbol{u}(0) - \hat{\boldsymbol{u}}(0)\| + \int_0^t \|\boldsymbol{\delta}(\tau)\|\, d\tau\right) \tag{19}$$

for some constant C, as the norms of the operators $S(t)$ of a strongly continuous semigroup are bounded on any finite interval.

As the original system has the perturbed determinacy index $\tilde{\nu}_d$, the perturbations $\boldsymbol{\delta}$ in (18) depend on derivatives of this order of the perturbations of the original system. Thus we have $\nu_p \leq \tilde{\nu}_d + 1$. By exactly the same reasoning as in the finite-dimensional case it can happen, if $S(t)$ possess a particularly simple structure, that the integration can be performed within the norm and thus kills one differentiation. Hence $\nu_p \geq \tilde{\nu}_d$. □

[6] One can easily verify that if $\beta^{(n)} < m$ in an involutive system, then some of the components of $\boldsymbol{u}$ can be chosen completely arbitrary, and thus this condition represents a simple criterion for an underdetermined system of partial differential equations [15]. The classical "counting rule" comparing the number of equations with the number of unknowns can be misleading, as the Yang-Mills equations show: although they comprise as many equations as unknowns, they are underdetermined due to the gauge symmetry.

7 Conclusions

We have shown that the formal theory of differential equations provides a convenient framework for analysing very general systems. In particular, as it is a theory for partial differential equations it allows us to present classical results on differential algebraic equations in a manner that can be straightforwardly extended to multivariate systems.

Of course, it is much harder to derive estimates of solutions of partial differential equations than of solutions of ordinary differential equations. Thus one cannot expect to obtain results in the same generality as Theorem 1. However, using the formal theory we have at least reduced the problem to analysing a normal system instead of an overdetermined one. A huge amount of results is available on such systems. We applied here only some elementary semigroup theory, as it permits us to treat the given partial differential equation as an ordinary differential equation on an appropriately chosen Banach space and thus to straightforwardly extend Theorem 1.

The key for extending our approach to larger classes of systems is Proposition 2. Its proof depends on the uniqueness of the zero solution of (17). Thus the crucial point is to understand the properties of this system which may be considered as a differential form of syzygies. An example of how it inherits properties of the original system can be found in [20]. There systems are studied where the underlying equation is (strictly) hyperbolic and the constraints form an elliptic system. Then one can show that (17) is again a (strictly) hyperbolic system.

Most approaches to indices for general systems of partial differential equations are based on first reducing the system to a differential algebraic equation by semi-discretisations, integral transforms or something similar and then determining an index of this differential algebraic equation. In [20] we have shown for linear systems with constant coefficients how the thus obtained indices are related to indices directly determined from the original system of partial differential equations.

References

1. J. Belanger, M. Hausdorf and W.M. Seiler. A *MuPAD* library for differential equations. This proceedings.
2. K.E. Brenan, S.L. Campbell, and L.R. Petzold. *Numerical Solution of Initial-Value Problems in Differential-Algebraic Equations.* Classics in Applied Mathematics 14. SIAM, Philadelphia, 1996.
3. J. Calmet, M. Hausdorf, and W.M. Seiler. A constructive introduction to involution. In *Proc. Int. Symp. Applications of Computer Algebra – ISACA '2000.* World Scientific, Singapore, to appear.
4. S.L. Campbell and C.W. Gear. The index of general nonlinear DAEs. *Numer. Math.*, 72:173–196, 1995.
5. E. Hairer, C. Lubich, and M. Roche. *The Numerical Solution of Differential-Algebraic Equations by Runge-Kutta Methods.* Lecture Notes in Mathematics 1409. Springer-Verlag, Berlin, 1989.

6. E. Hairer and G. Wanner. *Solving Ordinary Differential Equations II.* Springer Series in Computational Mathematics 14. Springer-Verlag, Berlin, 1996.
7. M. Hausdorf and W.M. Seiler. An efficient algebraic algorithm for the geometric completion to involution. Preprint Universität Mannheim, 2000.
8. P. Kunkel and V. Mehrmann. Canonical forms for linear differential-algebraic equations with variable coefficients. *J. Comp. Appl. Math.*, 56:225–251, 1994.
9. P. Kunkel and V. Mehrmann. A new look at pencils of matrix valued functions. *Lin. Alg. Appl.*, 212/213:215–248, 1994.
10. A. Pazy. *Semigroups of Linear Operators and Applications to Partial Differential Equations.* Applied Mathematical Sciences 44. Springer, New York, 1983.
11. J.F. Pommaret. *Systems of Partial Differential Equations and Lie Pseudogroups.* Gordon & Breach, London, 1978.
12. P.J. Rabier and W.C. Rheinboldt. A geometric treatment of implicit differential algebraic equations. *J. Diff. Eq.*, 109:110–146, 1994.
13. S. Reich. On an existence and uniqueness theory for nonlinear differential-algebraic equations. *Circ. Sys. Sig. Proc.*, 10:343–359, 1991.
14. M. Renardy and R.C. Rogers. *An Introduction to Partial Differential Equations.* Texts in Applied Mathematics 13. Springer, New York, 1993.
15. W.M. Seiler. *Analysis and Application of the Formal Theory of Partial Differential Equations.* PhD thesis, School of Physics and Materials, Lancaster University, 1994.
16. W.M. Seiler. Momentum versus position projections for constrained Hamiltonian systems. *Num. Algo.*, 19:223–234, 1998.
17. W.M. Seiler. Indices and solvability for general systems of differential equations. In V.G. Ghanza, E.W. Mayr, and E.V. Vorozhtsov, editors, *Computer Algebra in Scientific Computing — CASC '99*, pages 365–385. Springer, Berlin, 1999.
18. W.M. Seiler. Numerical integration of constrained Hamiltonian systems using Dirac brackets. *Math. Comp.*, 68:661–681, 1999.
19. W.M. Seiler. A combinatorial approach to involution and δ-regularity. Preprint Universität Mannheim, 2000.
20. W.M. Seiler. Completion to involution and semi-discretisations. *Appl. Num. Math.*, to appear, 2001.
21. M.E. Taylor. *Pseudodifferential Operators.* Princeton Mathematical Series 34. Princeton University Press, Princeton, 1981.
22. F. Verhulst. *Nonlinear Differential Equations and Dynamical Systems.* Springer, Berlin, 2nd edition, 1996.
23. C. Zuily. *Uniqueness and Non-Uniqueness in the Cauchy Problem.* Progress in Mathematics 33. Birkhäuser, Boston, 1983.

Employment of the Gröbner Bases in Analysis of Systems Having Algebraic First Integrals

Valentin Irtegov and Tatyana Titorenko

Institute of Systems Dynamics and Control Theory SB RAS,
134, Lermontov str., Irkutsk, 664033, Russia,
irteg@icc.ru

Abstract. The paper discusses some problems of analysis of the phase space of mechanical systems having algebraic first integrals with the aid of the Gröbner bases. An attempt has been made to perform the analytical computations needed in this case by the system of computer algebra available.

Qualitative investigations of the phase space of conservative nonlinear mechanical systems require as a rule, in each specific case, specialization of the approach and cumbersome computations already for the problems of not large dimensions.

If the consideration is restricted to the systems of differential equations with polynomial first integrals, then the capabilities of employment of standard algebraic aids combined with those of computer algebra are substantially increased.

In the present paper, the authors have made an attempt to demonstrate the specifics of usage of the Gröbner bases in the analysis of phase space for conservative mechanical systems on the basis of Routh - Lyapunov's technique [1]. Special attention has been paid to the problems of finding the stationary sets of the first integrals algebra of a problem and further investigation of obtained invariant (due to Routh-Lyapunov's theorem) manifolds in the aspect of revealing their stability and bifurcation.

It is well known that a sufficiently large number of integrals available allows us to obtain even some larger information on the structure of the phase space of the system of differential equations. For example, this is almost obvious for sufficiently integrable systems.

If the system of differential equations

$$\dot{x}_i = X_i(x_1, \ldots, x_n) \quad (i = 1, \ldots, n)$$

assumes the first integrals $V_1(x) = c_1, \ldots, V_m(x) = c_m$ then the integral $K = (V_1(x) - c_1)^2 + \ldots + (V_m(x) - c_m)^2$ can be put in correspondence to the family of sets, wherein it acquires a stationary value. The equations, which define the latter family, will satisfy the necessary conditions of extremum K:

$$\frac{\partial K}{\partial x_i} = 2(V_1(x) - c_1)\frac{\partial V_1}{\partial x_i} + \ldots + 2(V_m(x) - c_m)\frac{\partial V_m}{\partial x_i} = 0 \quad (i = 1, \ldots, n).$$

As obvious from above conditions, one of the sought collections of equations describing properties of the stationary sets K writes:

$$V_1(x_1,\dots,x_n)=c_1,\dots,V_m(x_1,\dots,x_n)=c_m.$$

These equations have the constants of the first integrals in the capacity of the parameters and define an invariant manifold of steady motions (IMSM) of the original system of differential equations for each fixed collection of these constants.

Obviously, each of such IMSMs (invariant manifolds of steady motions) is stable with respect to m variables: $c_1,\dots,c_m$. If one declares the IMSMs obtained above as IMSMs of zero level then it is possible to speak also of first-level IMSMs giving a stationary value to the problem's first integrals or to their bundle, for example, to the linear bundle of basic first integrals:

$$K_1=\sum_{i=1}^{m}\lambda_i V_i(x),\quad \text{where } \lambda_i=\text{const}.$$

To the end of demonstrating the technique and results of finding and analysis of the first-level IMSMs consider the following problem, i.e. Kovalevskaya's case of motion of a rigid body with a fixed point. In this case, the differential equations of motion

$$2\dot p=qr,\ 2\dot q=-rp+x_0\gamma_3,\ \dot r=-x_0\gamma_3,\ \dot\gamma_1=\gamma_2 r-\gamma_3 q,$$
$$\dot\gamma_2=\gamma_3 p-\gamma_1 r,\ \dot\gamma_3=\gamma_1 q-\gamma_2 p,$$

assume the four first integrals:

$$2H=2p^2+2q^2+r^2+2x_0\gamma_1=2h,\ V_1=2p\gamma_1+2q\gamma_2+r\gamma_3=m,$$
$$V_2=(p^2-q^2-x_0\gamma_1)^2+(2pq-x_0\gamma_2)^2=k^2,\ V_3=\gamma_1^2+\gamma_2^2+\gamma_3^2=1. \quad (1)$$

Here: p,q,r are projections of angular velocity of the body onto its main inertia axes; $\gamma_1,\gamma_2,\gamma_3$ are directional cosines of the vertical in the above axes; x_0 is the normalized coordinate of the body's mass center.

Consider the problem of obtaining the IMSMs of the above system, which give a stationary value to the complete linear bundle of the first integrals (1). In accordance with somewhat extended Routh-Lyapunov's theorem [2], such IMSMs will be defined by solutions of the system of necessary conditions of the extremum of the integrals from the family:

$$K_1=\frac{1}{2}\lambda_0(2p^2+2q^2+r^2+2x_0\gamma_1)-\lambda_1(2p\gamma_1+2q\gamma_2+r\gamma_3)$$
$$-\frac{1}{2}\lambda_2[(p^2-q^2-x_0\gamma_1)^2+(2pq-x_0\gamma_2)^2]+\frac{1}{2}\lambda_3(\gamma_1^2+\gamma_2^2+\gamma_3^2).$$

These nonlinear algebraic relations write:

$$\frac{\partial K_1}{\partial p}=2[(\lambda_2 x_0 p-\lambda_1)\gamma_1+\lambda_0 p+\lambda_2 x_0 q\gamma_2-\lambda_2 p(p^2+q^2)]=0,$$

$$\begin{aligned}
\frac{\partial K_1}{\partial q} &= 2[\lambda_2 x_0 p - \lambda_1)\gamma_2 + \lambda_0 q - \lambda_2 x_0 q \gamma_1 - \lambda_2 q(p^2+q^2)] = 0, \\
\frac{\partial K_1}{\partial \gamma_1} &= -(\lambda_3 + \lambda_2 x_0^2)\gamma_1 + \lambda_0 x_0 - 2\lambda_1 p + \lambda_2 x_0 (p^2 - q^2)] = 0, \\
\frac{\partial K_1}{\partial \gamma_2} &= -(\lambda_3 + \lambda_2 x_0^2)\gamma_2 - 2q(\lambda_1 - \lambda_2 x_0 p) = 0, \\
\frac{\partial K_1}{\partial r} &= \lambda_0 r - \lambda_1 \gamma_3 = 0, \quad \frac{\partial K_1}{\partial \gamma_3} = -\lambda_1 r - \lambda_3 \gamma_3 = 0.
\end{aligned} \tag{2}$$

The Jacobian of the latter system (for $\lambda_0 = 1$) can be written as follows:

$$\begin{aligned}
\det \Delta = -4(\lambda_3 + \lambda_1^2)\{&[2\lambda_1^2 + \lambda_3 + x_0^2\lambda_2 - 4\lambda_1\lambda_2 x_0 p - 2\lambda_3\lambda_2(p^2+q^2)]^2 \\
&-(\lambda_3 + \lambda_2 x_0^2)^2 \lambda_2^2 k^2\}.
\end{aligned}$$

The following two families of ordinary solutions (of permanent rotations of a body about its vertical central inertia axis) are well known for the initial system of differential equations:

$$q = r = \gamma_2 = \gamma_3 = 0, \ p = \text{const}, \ \gamma_1^2 = 1. \tag{3}$$

Substitution of the first one ($\gamma_1 = 1$) of these solutions into (2) gives the following restrictions imposed on the parameters λ_i ($i = 1, 2, 3$) from K_1:

$$x_0 - \lambda_3 - 2\lambda_1 p - \lambda_2 x_0 (x_0 - p^2) = 0, \quad -\lambda_1 + p + \lambda_2 p\,(x_0 - p^2) = 0. \tag{4}$$

Availability of freedom of selection of the Lagrange multipliers in the problem of finding the extremum for solutions (3) means that the latter give stationary values to several different first integrals from the family K_1 (to be exact, to an infinite number of such integrals). Such solutions of equations (2) will be called special.

In the authors' paper [3], a method of finding the solutions of the system of form (2), for which some ordinary "special" family of solutions of form (3) is known, has been proposed. To be specific, the preliminary finding of all the bifurcation values of the parameters λ_i on the elements of family (3) and, next, finding (with the aid of (2)) the IMSMs, which correspond to these bifurcation values of parameters and adjunct to the family of specific motions, was implied.

To the end of realization of the plan appointed it is necessary to investigate solutions of system (4), to which the conditions providing the zero Jacobian ($det\Delta = 0$) have been added. Consider the following extended system (4):

$$\begin{aligned}
&x_0 - \lambda_3 - 2\lambda_1 p - \lambda_2 x_0 (x_0 - p^2) = 0, \ -\lambda_1 + p + \lambda_2 p(x_0 - p^2) = 0, \\
&\lambda_3 + \lambda_1^2 = 0,
\end{aligned} \tag{5}$$

where the latter added equation is the one of possible conditions of obtaining the zero Jacobian of system (2). Solutions of this system will define for us the bifurcation values of the parameters.

Let us construct the Gröbner bases for the system (5) under different permutations of the variables $(\lambda_1, \lambda_2, \lambda_3, p)$ in order to choose later on a "most suitable" of them for finding solutions (from now on for obtaining the Gröbner bases the lexicographic ordering is used). There are 4!=24 such bases. The number of equations in them varies from 3 to 13. The computation time for one basis is within 0,05 – 0,1 sec. (from now on for the purpose of computations Pentium III with a 500 MHz Celeron processor, 64 MB memory and the computer algebra systems Mathematica 4.0, Maple 6 were applied). We have obtained 4 bases with the number of equations equal to 3. All of them are practically equivalent up to the accuracy of permutations of the equations.

Consider one of these bases, the one, which corresponds to the order of the variables $(\lambda_3, \lambda_1, p, \lambda_2)$ (the rest ones have the order, respectively, $(\lambda_3, \lambda_1, \lambda_2, p)$, $(\lambda_1, \lambda_3, \lambda_2, p)$, $(\lambda_1, \lambda_3, p, \lambda_2)$):

$$(\lambda_2 p^2 - 1)(1 + \lambda_2(p^2 - x_0)(p^2 - x_0) = 0, \quad \lambda_1 + p(-1 + \lambda_2(p^2 - x_0)) = 0,$$
$$\lambda_3 - 2\lambda_2 p^4 - x_0(1 - \lambda_2 x_0) + p^2(2 + \lambda_2 x_0) = 0.$$

The latter system can easily be solved with the (or without any) aid of a computer because the first equation has been factorized.

So, there are the following three possibilities:

1) $p^2 = x_0$ (obviously, here it is necessary to put $x_0 > 0$). For these values of p the second equation yields $\lambda_1 = p$, and the third one – $\lambda_3 = -x_0 = -\lambda_1^2$, λ_2 remains arbitrary.

2) $p^2 = 1/\lambda_2$ $(\lambda_2 > 0)$. The second equation leads to the relationship $\lambda_1^2 - x_0^2\lambda_2 = 0$ for critical values of parameters, and the third equation – to $\lambda_3 + x_0^2\lambda_2 = 0$.

3) $p^2 = x_0 - 1/\lambda_2$. The second equation yields $2p = \lambda_1$, and the third one leads to the relationship $\lambda_3 = 4(1/\lambda_2 - x_0)$. Having excluded p from both the first and second equations, we have: $\lambda_1^2 = 4(x_0 - 1/\lambda_2)$.

Note, in the third case, it is possible that $p = 0, \lambda_1 = \lambda_3 = 0, \lambda_2 = 1/x_0$. The second case remains reasonable for $p^2 = x_0, \lambda_1^2 = x_0 = -\lambda_3, \lambda_2 = 1/x_0$. The third case has only the asymptotics to the first one: $p^2 \to x_0, \lambda_2 \to \infty$.

For the purpose of revealing IMSMs adjunct to the elements of family (3) under the bifurcation values of parameters obtained λ_i, it is necessary to turn back to the system of equations (2), while adding (to it) the above condition of making the zero Jacobian ($det\Delta = 0$) - $\lambda_3 + \lambda_1^2 = 0$. Since in system (2) there are 2 latter equations for r, γ_3, which can be considered independent of the rest ones, let us avoid taking them in consideration in the process of constructing the bases. For the system formed like that let us construct the Gröbner bases. For various permutations of the variables, there are 6!=720 of such bases. The number of equations in them varies from 12 to 49. The total time of constructing them is $\simeq 14$ min. Consider one of these systems composed of 12 equations and start substituting the bifurcation values of parameters obtained above into it. After substituting the values of parameters from the first case, we have the following family of IMSMs $p = \lambda_1\gamma_1, q = \gamma_2 = 0, r = \lambda_1\gamma_3$, which branches off the permanent rotation with $p = \sqrt{x_0}$.

It can be readily seen that for $\lambda_0 = \lambda_1 = \lambda_3 = 0$ another IMSM can be obtained, which branches off the permanent rotation (3) with the angular velocity $p = \sqrt{x_0}$. This is the Delonay IMSM defined by the equations:

$$p^2 - q^2 - x_0\gamma_1 = 0, \quad 2pq - x_0\gamma_2 = 0.$$

It is possible to proceed differently: firstly, to substitute the bifurcation values of the parameters into the first four equations of eq. (2), and, next, to construct the Gröbner bases for them. This is how we intend to proceed in the remaining two cases. In the second case, $(\lambda_1^2 - x_0^2\lambda_2 = 0,\ \lambda_3 + x_0^2\lambda_2 = 0)$, and after the elimination of $\lambda_2,\ \lambda_3$, equations (2) will be reduced to:

$$\begin{aligned}
&\frac{2}{x_0^2}[x_0^2(p - \lambda_1\gamma_1) - \lambda_1(p + pq^2 - \gamma_1 x_0 p - \gamma_2 x_0 q)] = 0,\\
&\frac{-2}{x_0^2}[\lambda_1\gamma_2 x_0(x_0 - \lambda_1 p) - q(x_0^2 - \lambda_1^2(p^2 + q^2 + \gamma_1 x_0))] = 0,\\
&\frac{1}{x_0}[\lambda_1^2(p^2 - q^2) - 2\lambda_1 p x_0 + x_0^2] = 0,\ \frac{2}{x_0^2}\lambda_1 q(\lambda_1 p - x_0) = 0,\\
&r - \lambda_1\gamma_3 = 0, \quad \lambda_1(\lambda_1\gamma_3 - r) = 0.
\end{aligned} \tag{6}$$

One of the "minimum" bases for the first four equations (6) (there are 120 equations in the total) writes:

$$\begin{aligned}
&q^2 = 0, \quad q(x_0 - \lambda_1 p) = 0, \quad (x_0 - \lambda_1 p)^2 = 0, \gamma_2(\lambda_1 p - x_0) - \lambda_1\gamma_1 q = 0,\\
&x_0^2\gamma_2 q + p(x_0 - \lambda_1 p)(\gamma_1 x_0 + 2p^2) = 0.
\end{aligned}$$

From the latter system we have the following family of one parameter IMSMs in the capacity of the solution:

$$q = 0, \quad p = \frac{x_0}{\lambda_1}, \quad r = \lambda_1\gamma_3. \tag{7}$$

The equation for $r,\ \gamma_3$ has been added from the system (6) for obvious reasons. The variables, which remained free, are interconnected with the following differential equations

$$\dot{r} = -x_0\gamma_1, \ \dot{\gamma_1} = r\gamma_2, \ \dot{\gamma_2} = r(\frac{x_0}{\lambda_1^2} - \gamma_1)$$

on the obtained family of IMSMs. As obvious from the equations defining IMSMs (7), such IMSMs branch off each permanent rotation (3) under the respective value of the parameter λ_1.

Now consider the third case. Eliminate the parameters λ_2 and λ_3 with the aid of the relations $\lambda_1^2 + \lambda_3 = 0,\ 1/\lambda_2 = x_0 - \lambda_1^2/4$ from eqs. (2). For the system obtained (like in case 2) let us construct the Gröbner bases. Here, the time needed for constructing separate bases reached about 20 to 30 mins. Among the bases obtained (8 bases in this case) there were the bases composed of 7 equations. Consideration of the latter allows one to indicate the following family of IMSMs:

$$\begin{aligned}
&(x_0 - \frac{1}{4}\lambda_1^2)\lambda_1 - (2x_0 p - \lambda_1(p^2+q^2)) = 0,\\
&(x_0 - \frac{1}{2}\lambda_1^2)^2 \gamma_1 = (x_0 - 2\lambda_1 p)(x_0 - \frac{1}{4}\lambda_1^2) + x_0(p^2 - q^2),\\
&(x_0 - \frac{1}{2}\lambda_1^2)^2 \gamma_2 = 2q[x_0 p - \lambda_1(x_0 - \frac{1}{4}\lambda_1^2)],\quad r = \lambda_1\gamma_3
\end{aligned} \tag{8}$$

which adjunct to all the elements of family (3). Meanwhile, the Gröbner bases of the initial system (2) for $\lambda_1^2 + \lambda_3 = 0$ allow one to separate the two-parameter family of IMSMs:

$$\begin{aligned}
&\lambda_1 - 2\lambda_1 x_0 p + \lambda_1\lambda_2(p^2+q^2) = 0,\\
&(\lambda_1^2 - x_0^2\lambda_2)\gamma_1 + x_0 - 2\lambda_1 p + \lambda_2 x_0(p^2 - q^2) = 0,\\
&(\lambda_1^2 - x_0^2\lambda_2)\gamma_2 - 2q(\lambda_1 - \lambda_2 x_0 p) = 0,\quad r = \lambda_1\gamma_3.
\end{aligned} \tag{9}$$

The above one-parameter family of IMSMs (8) is a subfamily of the two-parameter family of IMSMs (9) when there exist links between the parameters $1/\lambda_2 = x_0 - \lambda_1^2/4$. The elements of family (9) already do not adjunct to the family of permanent rotations (3).

After all, consider a limit situation when $p = 0$. Proceeding likewise above, it is possible to show that there are the two types of IMSMs branching off the equilibrium position $p = q = r = \gamma_2 = \gamma_3 = 0,\ \gamma_1 = 1$:

1) the family of pendular oscillations of a body about a fixed (in the body) axis y. The equations defining the family are as follows:

$$\lambda_1 = \lambda_3 = 0,\ p = r = \gamma_2 = 0,\ 1 - \lambda_2(q^2 + x_0\gamma_1) = 0 \tag{10}$$

having the vector field on this family:

$$2\dot{q} = x_0\gamma_1,\ \dot{\gamma_1} = -q\gamma_3,\ \dot{\gamma_3} = q\gamma_1;$$

2) the second IMSM, which branches off the equilibrium position, is the manifold of pendular oscillations of the body about a fixed axis z bound up with this body. The corresponding equations for the IMSM indicated are:

$$p = q = \gamma_3 = 0. \tag{11}$$

The vector field on this IMSM is defined by the equations:

$$\dot{r} = x_0\gamma_2,\ \dot{\gamma_2} = -r\gamma_1,\ \dot{\gamma_1} = q\gamma_2.$$

It can readily be verified that the family of IMSMs (9) for $p \to 0,\ \lambda_1 \to 0$ adjuncts to family (10), and its subfamily (8) – also to the equilibrium position. Family (7) adjuncts to the IMSM of pendular oscillations (11) under $\lambda_0 \to 0,\ \lambda_1 \to 0$.

Note in conclusion that the first integral reaching a stationary value on some of separated IMSMs can have an extremum for them. This allows one to obtain sufficient conditions of stability of above IMSMs. Thus, having introduced the new variables

$$y_1 = p^2 - q^2 - x_0\gamma_1,\ y_2 = 2pq - x_0\gamma_2,$$

it is possible to write the equations defining the family (10) as follows:

$$p = r = y_2 = 0,\ y_1 = \frac{1}{\lambda_2}.$$

The first integral, which corresponds to this family of IMSMs in the neighbourhood of above IMSMs, in terms of new variables writes:

$$\Delta K_1 = 2p^2 + \frac{1}{2}r^2 - \frac{1}{\lambda_2}(\tilde{y}_1^2 + y_2^2).$$

Here by $\tilde{y}_1$ we denote the deviation y_1 from its nondisturbed value. From the representation of the quadratic form ΔK_1 it immediately follows that for $\lambda_2 < 0$ the form will be positive definite, and hence, the elements of the corresponding subfamily of IMSMs ($\lambda_2 < 0$) are stable.

For the elements of the family of IMSMs (9) it is also possible to similarly obtain sufficient conditions of stability. Let us introduce the deviation from above IMSMs:

$$\begin{aligned}
\varphi_1 &= \lambda_1 - 2\lambda_1 x_0 p + \lambda_1\lambda_2(p^2 + q^2),\\
\varphi_2 &= (\lambda_1^2 - x_0^2\lambda_2)\gamma_1 + x_0 - 2\lambda_1 p + \lambda_2 x_0(p^2 - q^2),\\
\varphi_3 &= (\lambda_1^2 - x_0^2\lambda_2)\gamma_2 - 2q(\lambda_1 - \lambda_2 x_0 p),\ \varphi_4 = r - \lambda_1\gamma_3.
\end{aligned}$$

In terms of these variables, the first integral, which reaches a stationary value on the IMSM (9), writes:

$$\Delta K = \frac{1}{2\lambda_2(x_0^2\lambda_2 - \lambda_1^2)}\varphi_1^2 - \frac{1}{2}(x_0^2\lambda_2 - \lambda_1^2)(\varphi_2^2 + \varphi_3^2) + \frac{1}{2}\varphi_4^2.$$

The quadratic form obtained is positive definite under the conditions:

$$x_0^2\lambda_2 - \lambda_1^2 < 0,\quad \lambda_2 < 0.$$

Consequently, the IMSMs included in the family (9), for which the values of parameters satisfy the inequalities (13), are stable.

Note that stable IMSMs from the family (9) adjunct to the stable IMSMs (10). As far as the rest of separated IMSMs, their first integrals do not provide signdefinite forms in the neighbourhood of the IMSMs.

As obvious from above considerations, usage of the aids of computer algebra has given the possibility to execute the computational part of the problem of separating and analysis of properties of stationary solutions in the Kovalevskaya's problem almost completely by computer. This gives the hope that the technique proposed can be efficiently used also in solving problems with more "complex" first integrals as well as the problems of larger dimension.

References

1. Lyapunov, A.M.: On permanant screw motion of a rigid body in liquids. Collected Works, Moscow, USSR Acad. Sci. Publ., **1** (1954) 276–319 (in Russian)
2. Irtegov, V.D.: Invariant Manifolds of Steady Motions and Their Stability. Novosibirsk, Nauka Publ. (1985) (in Russian)
3. Irtegov, V.D., Titorenko, T.N.: On modelling and investigation of some problems with the aid of computer algebra. Programming **1** (1997) 68 – 74 (in Russian)

"Coalgebra" Structures on 1–Homological Models for Commutative Differential Graded Algebras

Jiménez M.J. and Real P. *

Dpto. de Matemática Aplicada I, Universidad de Sevilla,
Avda. Reina Mercedes, 41012 Sevilla, Spain,
{majiro, real}@cica.es

Abstract. In [3] "small" 1–homological model H of a commutative differential graded algebra is described. Homological Perturbation Theory (HPT) [7–9] provides an explicit description of an A_∞–coalgebra structure $(\Delta_1, \Delta_2, \Delta_3, \ldots)$ of H. In this paper, we are mainly interested in the determination of the map $\Delta_2 : H \to H \otimes H$ as a first step in the study of this structure. Developing the techniques given in [20] (inversion theory), we get an important improvement in the computation of Δ_2 with regard to the first formula given by HPT. In the case of purely quadratic algebras, we sketch a procedure for giving the complete Hopf algebra structure of its 1–homology.

1 Introduction

In recent years, the relevance of Homological Algebra in the field of Theoretical Physics becomes more and more apparent. New emergent areas, such as Cohomological Physics [23] and Secondary Calculus [25], make use of notions from Homological Algebra to clearly describe a series of interesting physical problems. In particular, the role of A_∞-structures [22, 19] in mathematical physics has enormously increased at the beginning of the nineties [6, 17, 24]. Example of this was M. Kontsevich's talk [15] at the International Congress of Mathematicians in 1994, in which he gave a conjectural interpretation of mirror symmetry as the "shadow" of an equivalence between two triangulated categories associated with A_∞-categories. His conjecture was proved in the case of elliptic curves by A. Polishchuk and E. Zaslow [18].

An A_∞–algebra $(A, m_1, m_2, m_3, \ldots)$ (see, for example, [14]) is a graded module $A = \oplus_{i=0}^{n} A_i$ endowed with graded maps $m_i : A^{\otimes n} \to A$, $n \geq 1$ of degree $n - 2$ satisfying for $n \geq 1$:

$$\sum \pm m_u(1^{\otimes r} \otimes m_s \otimes 1^{\otimes t}) = 0,$$

* Authors are partially supported by the PAICYT research project FQM-296 from Junta de Andalucia and the DGES–SEUID research project PB98–1621–C02–02 from Education and Science Ministry (Spain).

where the sum runs over all decompositions $n = r + s + t$ and we put $u = r + 1 + t$. Therefore, an A_∞-algebra is a differential (m_1) graded algebra with multiplication m_2, strongly homotopy associative (this lack of associativity is measured by a series of morphisms $(m_3, m_4, \ldots)$). In an analogous way, it is possible to define an A_∞-coalgebra $(C, \Delta_1, \Delta_2, \Delta_3, \ldots)$.

We are interested in the computation of the A_∞–algebra and A_∞–coalgebra structures of the 1–homology of a commutative differential graded algebra (briefly called CDGA) working with coefficients in a commutative ring Λ with $1 \neq 0$ (usually, the ground ring will be $\mathbf{Z}$). Let us recall that the 1-homology of a CDGA A is the homology of the reduced bar construction $\bar{B}(A)$ of A. The complex $\bar{B}(A)$ is a Hopf algebra, that is, it has both algebra and coalgebra structures such that they are compatible in some sense. Consequently, its homology carries A_∞–algebra and A_∞–coalgebra structures (unique, up to isomorphism [13]), both transferred from respective structures on $\bar{B}(A)$.

The problem of computing the structure of the A_∞-algebra of a "small" 1-homological model HBA (that means that there exists a homotopy equivalence between $\bar{B}(A)$ and HBA, such that HBA has less algebra generators than $\bar{B}(A)$), using homological perturbation tools [10, 20] is attacked in [3]. There, it was realized that this A_∞-structure reduces to a simple algebra structure $(HBA, m_1, m_2, 0, 0, \ldots)$, where m_1 represents the differential of the complex and m_2 is the associative product. The determination of m_2 is immediate and the attention is focused on the consecution of an "economical" formulation of m_1.

Here, we are interested in the dual, but extremely complicated, problem of calculating the A_∞-coalgebra structure of HBA. In this structure, $(HBA, \Delta_1, \Delta_2, \Delta_3, \ldots)$, Δ_1 coincides with the differential m_1, and the first step in solving this question consists of getting an "efficient" description of $\Delta_2 : HBA \to HBA \otimes HBA$. More concretely, in the present paper we obtain an important improvement in the computation of Δ_2 with regard to the initial formula provided by Homological Perturbation Theory [10] using the techniques which are comprised under the name of inversion theory [20, 3]. In some particular cases, such as purely quadratic algebras, the model HBA represents the actual 1-homology of A (that is, $m_1 = \Delta_1 = 0$). In that context, we sketch a "reasonable" algorithm for giving the complete Hopf algebra structure of HBA.

Finally, let us emphasize that our approach could be useful in providing new insights on the difficult problem of defining the A_∞-Hopf algebra structure [21].

2 Preliminaries

The algebraic setting and notation we need in this paper is conveniently described in [3]. In order to put into context the problem we deal with, most relevant notions of our framework are reviewed.

Let Λ be a commutative ring with non zero unit which is considered as ground ring. $(A, d_A, *_A, \xi_A, \eta_A)$ denotes a commutative differential graded algebra, endowed with an augmentation ξ_A and a cougmentation η_A. We will respect Koszul conventions.

Examples of CDGAs are the monogenic algebras: exterior algebra $E(x, 2n+1)$, polynomial algebra $P(y, 2n)$ and divided power algebra $\Gamma(y, 2n)$; all of them with trivial differential.

The *reduced bar construction* [16] associated to a CDGA A is defined as the differential graded bimodule $\bar{B}(A)$:

$$\bar{B}(A) = \Lambda \oplus \text{Ker } \xi_A \oplus (\text{Ker } \xi_A \otimes \text{Ker } \xi_A) \oplus \cdots \oplus (\text{Ker } \xi_A \otimes \cdots \otimes \text{Ker } \xi_A) \oplus \cdots .$$

An element from $\bar{B}(A)$ is denoted by $\bar{a} = [a_1|\cdots|a_n]$. There is a *tensor graduation* given by $|\bar{a}|_t = \sum_{i=1}^{n} |a_i|$; as well as a *simplicial graduation* ($|\bar{a}|_s = |[a_1|\cdots|a_n]|_s = n$). The total differential is given by the sum of the *tensor* one, which depends on the differential of A, and the *simplicial* differential, which acts by using the product of A.

When the algebra A is commutative, it is possible to define a multiplicative structure upon $\bar{B}(A)$ (via an operator called *shuffle product*), so that the reduced bar construction also becomes a CDGA.

Given two non–negative integers p and q , a (p, q)*–shuffle* is defined as a permutation π of the set $\{0, \ldots, p+q-1\}$, such that $\pi(i) < \pi(j)$ when $0 \leq i < j \leq p-1$ or $p \leq i < j \leq p+q-1$.

Let us observe that there are $\binom{p+q}{p}$ different (p, q)–shuffles.

So, given a CDGA A, the *shuffle product* $\star : \bar{B}(A) \otimes \bar{B}(A) \longrightarrow \bar{B}(A)$, is defined, up to sign, by:

$$[a_1|\cdots|a_p] \star [b_1|\cdots|b_q] = \sum_{\pi \in \{(p,q)\text{–shuffles}\}} \pm[c_{\pi(0)}|\cdots|c_{\pi(p+q-1)}];$$

where $(c_0, \ldots, c_{p-1}, c_p, \ldots, c_{p+q-1}) = (a_1, \ldots, a_p, b_1, \ldots b_q)$.

On the other hand, a coproduct can be define on $\bar{B}(A)$ which provides it a coalgebra structure,

$$\Delta([a_1|\cdots|a_n]) = \sum_{i=0}^{n} [a_1|\cdots|a_i] \otimes [a_{i+1}|\cdots|a_n] \tag{1}$$

Both structures of algebra and coalgebra are compatible in the sense that $\bar{B}(A)$ is a Hopf algebra, that is, $\Delta\star = (\star \otimes \star)(1 \otimes T \otimes 1)(\Delta \otimes \Delta)$, where $T : \bar{B}(A) \otimes \bar{B}(A) \to \bar{B}(A) \otimes \bar{B}(A)$ is the morphism that interchanges the factors.

A *contraction* $C : \{N, M, f, g, \phi\}$ [4, 11], also denoted by $(f, g, \phi) : N \overset{C}{\Rightarrow} M$, from a differential graded module (N, d_N) to a differential graded module (M, d_M) consists in a homotopy equivalence determined by three morphisms f, g and ϕ; $f : N_* \to M_*$ (projection) and $g : M_* \to N_*$ (inclusion) being two differential graded module morphisms and $\phi : N_* \to N_{*+1}$ a homotopy operator. Moreover, these data are required to satisfy the following rules:

$$\text{(c1)}\, fg = 1_M, \;\; \text{(c2)}\, \phi d_N + d_N \phi + gf = 1_N, \;\; \text{(c3)}\, f\phi = 0, \;\; \text{(c4)}\, \phi g = 0, \;\; \text{(c5)}\, \phi\phi = 0 .$$

Therefore, the homology groups of M and N coincide.

The Basic Perturbation Lemma (BPL) [7] states that given a contraction $C : \{N, M, f, g, \phi\}$ of chain complexes and a perturbation δ of d_N (that is, $(d_N + \delta)^2 = 0$), then there exists a new contraction $C_\delta = (f_\delta, g_\delta, \phi_\delta)$ from $(N, d_N + \delta)$ to $(M, d_M + d_\delta)$, verifying that

$$f_\delta = f(1 - \delta \Sigma_c^\delta \phi)\,, \quad g_\delta = \Sigma_c^\delta g\,, \quad \phi_\delta = \Sigma_c^\delta \phi\,, \quad d_\delta = f\delta \Sigma_c^\delta g\,, \tag{2}$$

where $\Sigma_c^\delta = \sum_{i \geq 0} (-1)^i \, (\phi\delta)^i = 1 - \phi\delta + \phi\delta\phi\delta - \cdots + (-1)^i (\phi\delta)^i + \cdots$.

It is commonly known that every CDGA A "factors", up to homotopy equivalence, into a *twisted tensor product* (or TTP) of exterior and polynomial algebras $\tilde{\otimes}^\rho_{i \in I} A_i$, that is, a tensor product of these algebras whose differential structure is enriched with a differential–derivation, ρ (see, for example, [16]). We will always assume that any CDGA A considered in this paper, is factored as a twisted tensor product $\tilde{\otimes}^\rho_{i \in I} A_i$ (being I a finite set of indexes, $I = \{1, 2, \ldots, n\}$) of exterior and polynomial algebras, A_i, of generators x_i, such that $|x_i| \leq |x_{i+1}|$. Notice that, consequently, an order is fixed on the factors.

Given two CDGAs A and A', a *semi-full algebra contraction* $(f, g, \phi) : A \Rightarrow A'$ [20, 2] consists of

- an inclusion, g, which is a morphism of DGAs (i.e., a multiplicative morphism);
- a *quasi-algebra projection* f, that is, $f *_A (\phi \otimes \phi) = 0$, $f *_A (\phi \otimes g) = 0$, $f *_A (g \otimes \phi) = 0$.
- and a *quasi-algebra homotopy* ϕ, that is, $\phi *_A (\phi \otimes \phi) = 0$, $\phi *_A (\phi \otimes g) = 0$, $\phi *_A (g \otimes \phi) = 0$.

The class of all semi-full algebra contractions is closed under composition, tensor product of contractions and perturbation.

Theorem 1. *[20]*
Let $C : \{N, M, f, g, \phi\}$ be a semi-full algebra contraction and $\delta : N \to N$ be a perturbation-derivation of d_N. Then, the perturbed contraction C_δ, is a new semi-full algebra contraction.

To obtain a 1-*homological model* for a CDGA A consists in establishing **a "chain" of *semi-full algebra contractions*** starting at the reduced bar construction $\bar{B}(A)$ and ending up at a CDGA HBA that is free and of finite type as graded module. An algorithm for computing 1-homological models for CDGAs was given in [1]. Now we recall the main steps on this algorithm which are essential in our work.

Three *almost-full* algebra contractions (that is, semi-full algebra contractions endowed with multiplicative projections) are used for this purpose:

- The contraction defined in [5] from $\bar{B}(A \otimes A')$ to $\bar{B}(A) \otimes \bar{B}(A')$, where A and A' are two CDGAs;

$$C_{\bar{B}\otimes} : \{\bar{B}(A \otimes A'), \bar{B}(A) \otimes \bar{B}(A'), f_{\bar{B}\otimes}, g_{\bar{B}\otimes}, \phi_{\bar{B}\otimes}\}\,;$$

- $f_{\bar{B}\otimes}([a_1 \otimes a'_1|\cdots|a_n \otimes a'_n])$

$$= \sum_{i=0}^{n} \xi_A(a_{i+1} *_A \cdots a_n)\xi_{A'}(a'_1 *_{A'} \cdots a'_i)[a_1|\cdots|a_i] \otimes [a'_{i+1}|\cdots|a'_n] . \quad (3)$$

- $g_{\bar{B}\otimes}([a_1|\cdots|a_n] \otimes [a'_1|\cdots|a'_m]) = [a_1|\cdots|a_n] \star [a'_1|\cdots|a'_m]$

  with $[a_1|\cdots|a_n] \in \bar{B}(A)\,, \quad [a'_1|\cdots|a'_m] \in \bar{B}(A')\,;$

  $[a_1|\cdots|a_n] = [a_1 \otimes \theta'|\cdots|a_n \otimes \theta']\,, \quad [a'_1|\cdots|a'_n] = [\theta \otimes a'_1|\cdots|\theta \otimes a'_n]\,,$

  where θ and θ' are units on A and A', respectively.
- up to sign, $\phi_{\bar{B}\otimes}([a_1 \otimes a'_1|\cdots|a_n \otimes a'_n])$

$$\begin{aligned} = \sum_{0\le p\le n-q-1\le n-1} & \pm\xi_A(a_{n-q+1} *_A \cdots a_n)[a_1 \otimes a'_1|\cdots|a_{\bar{n}-1} \otimes a'_{\bar{n}-1} \\ & |(a'_{\bar{n}} *_{A'} \cdots *_{A'} a'_{n-q})|c_0|\cdots|c_{p+q}]\,, \end{aligned} \quad (4)$$

  where $\bar{n} = n - p - q$; $(c_{\pi(0)}, \ldots, c_{\pi(p+q)}) = (a_{\bar{n}}, \ldots, a_{n-q}, a'_{n-q+1}, \ldots a'_n)$; π is a $(p+1, q)$–shuffle and $0 \le p \le n-q-1 \le n-1$.

Given a tensor product $\otimes_{i\in I} A_i$ of CDGAs, a contraction from $\bar{B}(\otimes_{i\in I} A_i)$ to $\otimes_{i\in I}\bar{B}(A_i)$ is easily determined, by applying $C_{\bar{B}\otimes}$ several times in a suitable way. This new contraction is also denoted by $C_{\bar{B}\otimes}$.

– The isomorphism of DGAs described in [5]

$$C_{\bar{B}E} : \{\bar{B}(E(u, 2n+1)), \Gamma(\underline{u}, 2n+2), f_{\bar{B}E}, g_{\bar{B}E}, 0\}\,,$$

where

$$f_{\bar{B}E}([u|\overset{m \text{ times}}{\cdots}|u]) = \underline{u}^{(m)};\ \ g_{\bar{B}E}(\underline{u}^{(m)}) = [u|\overset{m \text{ times}}{\cdots}|u]\,.$$

– The contraction also stated in [5]

$$C_{\bar{B}P} : \{\bar{B}(P(v, 2n)), E(\underline{v}, 2n+1), f_{\bar{B}P}, g_{\bar{B}P}, \phi_{\bar{B}P}\}\,,$$

where

$$f_{\bar{B}P}([v^r]) = \begin{cases} 0 \text{ if } r \ne 1 \\ \underline{v} \text{ if } r = 1 \end{cases};\ \ f_{\bar{B}P}([v^{r_1}|\cdots|v^{r_m}]) = 0$$

$$g_{\bar{B}P}(\underline{v}) = [v];\ \ \phi_{\bar{B}P}([v^{r_1}|\cdots|v^{r_m}]) = [v|v^{r_1-1}|\cdots|v^{r_m}]\,.$$

If $\tilde{\otimes}^{\rho}_{i\in I} A_i$ is a twisted tensor product of exterior and polynomial algebras, the perturbation ρ produces a perturbation–derivation δ on the tensor differential of $\bar{B}(\otimes_{i\in I} A_i)$. Thanks to the contractions above, it is possible to establish, by composition and tensor product of contractions in a recurring way, a new semi–full algebra contraction

$C^n : (f^n, g^n, \phi^n)$ from $\bar{B}(\otimes_{i=1}^n A_i)$ to $\otimes_{i=1}^n HBA_i$,

$$\bar{B}((\otimes_{i=1}^{n-1} A_i) \otimes A_n) \Rightarrow (\otimes_{i=1}^{n-1}\bar{B}(A_i)) \otimes \bar{B}(A_n) \Rightarrow (\otimes_{i=1}^{n-1} HBA_i) \otimes HBA_n\,, \quad (5)$$

where

$$f^n = (f^{n-1} \otimes f_{\bar{B}A_n}) f_{\bar{B}\otimes}$$
$$g^n = g_{\bar{B}\otimes}(g^{n-1} \otimes g_{\bar{B}A_n})$$
$$\phi^n = \phi_{\bar{B}\otimes} + g_{\bar{B}\otimes}(\phi^{n-1} \otimes g_{\bar{B}A_n} f_{\bar{B}A_n} + 1 \otimes \phi_{\bar{B}A_n}) f_{\bar{B}\otimes} \,,$$

HBA_i are exterior or divided power algebras and $(f^{n-1}, g^{n-1}, \phi^{n-1}) : \bar{B}(\otimes_{i=1}^{n-1} A_i) \overset{C_{n-1}}{\Longrightarrow} \otimes_{i=1}^{n-1} HBA_i$. A 1–homological model, HBA, of $A = \tilde{\otimes}_{i\in I}^{\rho} A_i$ is then obtained by perturbing this contraction.

$$C_\delta^n : (f_\delta^n, g_\delta^n, \phi_\delta^n) \;:\; \bar{B}(\tilde{\otimes}_{i\in I}^{\rho} A_i) \Rightarrow (\otimes_{i=1}^{n} HBA_i, d_\delta) \,, \tag{6}$$

The differential d_δ as well as the morphims which compose the new contraction, are determined by the Basic Perturbation Lemma.

3 Inversion Theory

This section is devoted to the theory that initially appeared in [20] and which was later used in [3] for the simplification in the computation of d_δ on the 1–homological model (6). We further develop inversion theory, though the proof of almost all the results we state here, are only briefly sketched. Complete proofs will be widely showed in [12]. These techniques will allows to prove that the projection f_δ in (6) is multiplicative. This fact has important repercussion on the computation of the A_∞–structure of the 1–homological model.

We consider a commutative differential graded algebra under the conditions already described, $\tilde{\otimes}_{i\in I}^{\rho} A_i$, with A_i exterior or polynomial algebras of generators x_i. In order to shorten notation, we will consider $A = \otimes_{i=1}^{n-1} A_i$ and $A' = A_n$. The following definition complements the one given in [3] for inversions. Notice that $\bar{a} \in \bar{B}(A \otimes A')$ will have *inversions* caused by the "first" algebra, A; inversions caused by the tensor product; and inversions caused by the "last" algebra, A'.

Definition 1. Let $A \otimes A'$ be a CDGA under conditions described above with $n = 2$ ($A = A_1$ and $A' = A_2$) and let us consider a homogeneous element $[a_1 \otimes a_1' | a_2 \otimes a_2' | \cdots | a_n \otimes a_n']$ from $\bar{B}(A \otimes A')$. We say that a component $a_i \otimes a_i'$ from that element is responsible for an *inversion*, if some of these cases takes place:

- $a_i = \theta$ and there exists an index $j > i$ with $a_j \neq \theta$. Then a_i' is responsible for a $\otimes$*–inversion.*
- Whenever A is a polynomial algebra, $a_i \neq \theta$ and there exists an index $j > i$ such that $a_j \neq \theta$. Then a_i is responsible for a $p1$*–inversion.*
- Whenever A' is a polynomial algebra, $a_i' \neq \theta'$ and $a_{i-1} = \theta$, $a_i = \theta, \ldots a_n = \theta$. Then a_i' is responsible for a p*–inversion.*

In the case that $n > 2$, inversions of $A = \otimes_{i=1}^{n-1} A_i$ are those of the first factor of $\lambda(\bar{a})$ in $\bar{B}(\otimes_{i=1}^{n-1} A_i)$, with (up to sign)

$$\lambda : \bar{B}(A \otimes A') \to \bar{B}(A) \times \bar{B}(A') : \lambda[a_1 \otimes a_1' | \cdots | a_n \otimes a_n'] = [a_1 | \cdots | a_n] \times [a_1' | \cdots | a_n'] \,.$$

So, an element has k inversions if there exist k components responsible for an inversion. We say an element from $\bar{B}(A \otimes A')$ has k inversions, if it is the sum of homogeneous elements with, at least, k inversions each one of them.

Let us consider the contraction $C_{\bar{B}\otimes}$,

$$(f_{\bar{B}\otimes}, g_{\bar{B}\otimes}, \phi_{\bar{B}\otimes}) : \bar{B}(A \otimes A') \Rightarrow \bar{B}(A) \otimes \bar{B}(A')$$

We analyze the behavior of the component morphisms with respect to the different types of inversions. For this purpose, we won't take into account signs in the referred formulas.

- **Lemma 1.** *$f_{\bar{B}\otimes}$ preserves the* p1*–inversions and* p*–inversions and is null if the element had one $\otimes$–inversion.*
  Recalling the formulation for this morphism (3), we realize that only the elements with the structure $[a_1|\cdots|a_k|a'_{k+1}|\cdots|a'_n]$, with $a_i \in A$ and $a'_i \in A'$, don't belong to the kernel of this morphism.
- **Lemma 2.** *$g_{\bar{B}\otimes}$ preserves the number of inversions of the factors (canonically included in $\bar{B}(A \otimes A')$).*
  Shuffle product doesn't change either the number of components from each algebra nor the relative position in between them, so the number of inversions remains the same or increases.
- **Lemma 3.** *$\phi_{\bar{B}\otimes}$ produces elements with, at least one more inversion than the original one.*
  After evaluating $\phi_{\bar{B}\otimes}$ over an element $[a_1 \otimes a'_1|a_2 \otimes a'_2|\cdots|a_n \otimes a'_n]$, one obtains a sum of elements, that can be sketched as follows:
  $$\pm\xi_A(a_{n-q+1} *_A \cdots *_A a_n)[a_1 \otimes a'_1|\cdots|a_{n-p-q-1} \otimes a'_{n-p-q-1}|(a'_{\bar{n}} *_{A'} \cdots *_{A'} a'_{n-q})|$$
  $$[a_{n-p-q}|\cdots|a_{n-q}] \star [a'_{n-q+1}|\cdots|a'_n]] \,. \tag{7}$$

  We can check that the number of non degenerate components from the first algebra stay the same at each term, so the number of $p1$–inversions doesn't come affected.
  In the case that A' is a polynomial algebra, $\phi_{\bar{B}\otimes}$ leaves each p-inversion the same or changes it into a $\otimes$-inversion. If the original element has k p-inversions, each term of the resultant sum will have $k - i$ p-inversions and, at least, i $\otimes$-inversions, where $0 \leq i \leq k$. So the k initial inversions are preserved.
  Concerning $\otimes$–inversions, the component $(a'_{\bar{n}} *_{A'} \cdots *_{A'} a'_{n-q})$ is always responsible for a new inversion of this kind, as well as those of the shuffle product indicated in (7).

Now we consider the already described contraction (5):

$$(f, g, \phi) : \bar{B}(A \otimes A') \Rightarrow \bar{B}(A) \otimes \bar{B}(A') \Rightarrow HBA \otimes HBA'$$

where

$$f = f^n = (f^{n-1} \otimes f_{\bar{B}A'})f_{\bar{B}\otimes},$$
$$g = g^n = g_{\bar{B}\otimes}(g^{n-1} \otimes g_{\bar{B}A'})$$
$$\phi = \phi^n = \phi_{\bar{B}\otimes} + g_{\bar{B}\otimes}(\phi^{n-1} \otimes g_{\bar{B}A'}f_{\bar{B}A'} + 1_{\bar{B}A} \otimes \phi_{\bar{B}A'})f_{\bar{B}\otimes}$$

Now, we will study the behavior of f and ϕ with respect to inversions.

- **Lemma 4.** *The evaluation of f over an element from $\bar{B}(A \otimes A')$ is null whenever such an element has, at least, one inversion.*
  This is clear since the formula for f is $f = (f_{\bar{B}A} \otimes f_{\bar{B}A'})f_{\bar{B}\otimes}$, so the lemma follows from lemma 1 and the fact that $f_{\bar{B}P}([v^{r_1}|\cdots|v^{r_k}]) = 0$ whenever $k > 1$.
- **Lemma 5.** *ϕ increases the number of inversions, at least, by one.*
  The case $n = 2$ can be proved taking into account that both $\phi_{\bar{B}P}$ and $\phi_{\bar{B}\otimes}$ (lemma 3) satisfy this condition (recall that $\phi_{\bar{B}E} = 0$) and that shuffle product preserves inversions (lemma 2); case $n > 2$ is proved, then, by induction.

Now we assume there is a perturbation ρ for the tensor product of $A \otimes A'$, which induces a perturbation–derivation, δ, on $\bar{B}(A \otimes A')$. We analyze also, the behavior of the latter with respect to inversions, coming to the following conclusion.

Lemma 6. *The evaluation of δ over a homogeneous element from $\bar{B}(A \otimes A')$ with k inversions, produces an element with at least $k - 1$ inversions.*

Let us notice that each component of a homogeneous element from $\bar{B}(A \otimes A')$ is responsible for, at most, one inversion. Since the action of δ is reduced to the application of ρ to each component $a_i \otimes a'_i$ of the element $[a_1 \otimes a'_1|\cdots|a_n \otimes a'_n]$, then, only one inversion can be killed, if any.

4 An A_∞–Coalgebra Structure

We recall the definition of A_∞–coalgebra given in [19].

An A_∞–coalgebra is a graded Λ-module C endowed with a locally finite family of morphisms $\Delta_i : C \longrightarrow C^{\otimes i}$, $i \geq 1$, such that the degree of Δ_i is $i - 2$ and

$$\sum_{k=1}^{n}\sum_{\lambda=0}^{n-k}(-1)^{k+\lambda+\lambda k}(1^{\otimes(n-\lambda-k)} \otimes \Delta_k \otimes 1^{\otimes\lambda})\Delta_{n+k+1} = 0.$$

Recall that, given a morphism of DG–modules $h : M \to N$, the notation $h^{\otimes i}$ is used for the morphism $h\otimes \overset{\text{i times}}{\cdots} \otimes h$.

Let us observe that in the case $n = 3$, the following expression is obtained:

$$(1\otimes\Delta_2)\Delta_2-(\Delta_2\otimes1)\Delta_2 = \Delta_3\Delta_1+(1^{\otimes2}\otimes\Delta_1)\Delta_3+(1\otimes\Delta_1\otimes1)\Delta_3+(\Delta_1\otimes1^{\otimes2})\Delta_3$$

That is, the morphism Δ_3 *measures* the coassociativity of Δ_2.

In the case of the reduced bar construction of a CDGA, the structure of coalgebra can be trivially considered as an A_∞–coalgebra with the differential $d = \Delta_1$, the coproduct $\Delta = \Delta_2$ and $\Delta_i = 0$, $i \geq 3$.

We are interested in the transference of this coalgebra structure from the reduced bar construction, $\bar{B}(A)$, of a CDGA, A, to the 1–homological model, HBA, described in section 2. Using HPT (see [10]), it is clear how to get an A_∞–coalgebra structure $(\Delta_1, \Delta_2, \Delta_3, \ldots)$ on the 1–homological model, where $\Delta_1 = d_\delta$ and whose formulation for Δ_2 and Δ_3 is:

$$\Delta_2 = (f_\delta \otimes f_\delta)\, \Delta\, g_\delta \tag{8}$$

$$\Delta_3 = f_\delta^{\otimes 3}\, (-\Delta \otimes 1 + 1 \otimes \Delta)\, (\phi_\delta \otimes g_\delta f_\delta + 1 \otimes \phi_\delta)\, \Delta\, g_\delta \tag{9}$$

We focus on the computation of Δ_2. At first, it would be necessary to evaluate this morphism over all the module generators of HBA. Nevertheless, the following proposition implies the compatibility of Δ_2 with the product. This makes possible to do this calculation only for the algebra generators. This proposition also allows to construct a **test of coassociativity** for Δ_2.

Proposition 1. *Let us consider the almost–full algebra contraction, $C : \{\bar{B}(A), HBA, f, g, \phi\}$ and $\delta : \bar{B}(A) \to \bar{B}(A)$ a data perturbation for C. Then, the perturbed contraction*

$$C_\delta : \{(\bar{B}(A), d_{\bar{B}} + \delta),\ (HBA, d_\delta),\ f_\delta,\ g_\delta,\ \phi_\delta\}$$

is also an almost–full algebra contraction.

Proof. For proving that f_δ is a morphism of DGAs, let us consider the contraction

$$C_\delta \otimes C_\delta : \{\, \bar{B}(A) \otimes \bar{B}(A),\ HBA \otimes HBA,\ f_\delta \otimes f_\delta,\ g_\delta \otimes g_\delta,\ \phi_\delta^{[\otimes 2]} = \phi_\delta \otimes g_\delta f_\delta + 1 \otimes \phi_\delta \,\}\,.$$

Taking into account condition $(c2)$ from the definition of contraction,

$$\phi_\delta^{[\otimes 2]} \delta^{[2]} + \delta^{[2]} \phi_\delta^{[\otimes 2]} = 1^{\otimes 2} - g_\delta^{\otimes 2} f_\delta^{\otimes 2}\,,$$

where $\delta^{[2]} = 1 \otimes \delta + \delta \otimes 1$ is the differential–perturbation on $A \otimes A$.

By composing to the left with the shuffle product $\star$ and then with the morphism f_δ, we obtain the expression:

$$f_\delta \star \phi_\delta^{[\otimes 2]} \delta^{[2]} + f_\delta \star \delta^{[2]} \phi_\delta^{[\otimes 2]} = f_\delta \star - f_\delta \star g_\delta^{\otimes 2} f_\delta^{\otimes 2} \tag{10}$$

Taking into account that g_δ is a morphism of DGAs and condition $(c1)$, the second member on the right hand side satisfies

$$f_\delta \star g_\delta^{\otimes 2} f_\delta^{\otimes 2} = f_\delta\, g_\delta \bullet f_\delta^{\otimes 2} = \bullet f_\delta^{\otimes 2}\,,$$

where $\bullet : HBA \otimes HBA \to HBA$ is the product of the CDGA HBA.

So, if we prove that the first member of equality (10) is null, f_δ will be a morphism of DGAs. But the first term in the sum can be written

$$f_\delta \star (\phi_\delta \otimes g_\delta f_\delta + 1 \otimes \phi_\delta)\delta^{[2]} = f_\delta \star (\phi_\delta \otimes g_\delta f_\delta)\delta^{[2]} + f_\delta \star (1 \otimes \phi_\delta)\delta^{[2]}$$

where the first term on the right hand side is null because of the fact that f_δ is a quasi–algebra projection, in particular, that $f_\delta \star (\phi_\delta \otimes g_\delta) = 0$. On the other hand, the evaluation of $f_\delta \star (1 \otimes \phi_\delta)\delta^{[2]}$ is always null since ϕ, and hence ϕ_δ, produces elements with one inversion (lemma 5) that is preserved by the shuffle product (lemma 2).

Finally, the second summand in the first member of (10) can be expressed by:

$$f_\delta \star \delta^{[2]} \phi_\delta^{[\otimes 2]} = f_\delta \delta \star \phi_\delta^{[\otimes 2]} = d_\delta f_\delta \star \phi_\delta^{[\otimes 2]}$$

(since $\delta^{[2]}$ is a derivation and f_δ is a morphism of DG-modules). And now we reason in the same way as before to conclude that the evaluation of $f_\delta \star \phi_\delta^{[\otimes 2]}$ is null.

Taking into account this proposition, we can state the following one:

Proposition 2. *Let HBA be the 1–homological model of a CDGA A and $\Delta_2 = (f_\delta \otimes f_\delta)\, \Delta\, g_\delta$. The following property is satisfied:*

$$\Delta_2 \bullet = (\bullet \otimes \bullet)\,(1 \otimes T \otimes 1)\,(\Delta_2 \otimes \Delta_2)\,.$$

Proof. For the proof, it is only necessary to notice that g_δ and f_δ are both morphisms of DGAs and the fact that $\bar{B}(A)$ is a Hopf algebra. So the following chain of equalities can be established:

$$\begin{aligned}
\Delta_2 \bullet = (f_\delta \otimes f_\delta)\, \Delta\, g_\delta \bullet &= (f_\delta \otimes f_\delta)\, \Delta \,\star\, g_\delta^{\otimes 2} \\
&= (f_\delta \otimes f_\delta)\,(\star \otimes \star)\,(1 \otimes T \otimes 1)\,(\Delta \otimes \Delta)\, g_\delta^{\otimes 2} \\
&= (\bullet \otimes \bullet)\,(1 \otimes T \otimes 1)\, f_\delta^{\otimes 4}\, \Delta^{\otimes 2}\, g_\delta^{\otimes 2} \\
&= (\bullet \otimes \bullet)\,(1 \otimes T \otimes 1)\,((f_\delta \otimes f_\delta)\, \Delta\, g_\delta) \otimes ((f_\delta \otimes f_\delta)\, \Delta\, g_\delta) \\
&= (\bullet \otimes \bullet)\,(1 \otimes T \otimes 1)\,(\Delta_2 \otimes \Delta_2)
\end{aligned}$$

From now on, we will denote $(\bullet \otimes \bullet)\,(1 \otimes T \otimes 1)$ by $\bullet_\otimes$.

In particular, this relationship means that, for determining the morphism Δ_2, we only need to evaluate it over the algebra generators of HBA (a finite number of elements!).

On the other hand, strict coassociativity of Δ_2 would guarantee the coalgebra structure on HBA. Proposition 1 allows to claim the following one:

Proposition 3. *If the morphism Δ_2 on the 1–homological model HBA of a CDGA A is coassociative for the algebra generators, then so it is for the rest of elements of HBA.*

Proof. It is sufficient to prove

$$(\Delta_2 \otimes 1)\Delta_2 \bullet = (1 \otimes \Delta_2)\Delta_2 \bullet . \tag{11}$$

Starting from the first term and applying proposition 2,

$$\begin{aligned}
(\Delta_2 \otimes 1)\, \Delta_2 \bullet &= (\Delta_2 \otimes 1)\,(\bullet \otimes \bullet)\,(1 \otimes T \otimes 1)\,(\Delta_2 \otimes \Delta_2) \\
&= ((\bullet \otimes \bullet)\,(1 \otimes T \otimes 1)\,(\Delta_2 \otimes \Delta_2) \otimes \bullet)(1 \otimes T \otimes 1)\,(\Delta_2 \otimes \Delta_2) \\
&= ((\bullet \otimes \bullet)\,(1 \otimes T \otimes 1)\ \otimes \bullet)((\Delta_2 \otimes \Delta_2) \otimes 1)(1 \otimes T \otimes 1)\,(\Delta_2 \otimes \Delta_2) \\
&= (\bullet_{\otimes} \otimes \bullet)\,(1^{\otimes 2} \otimes (1 \otimes T)(T \otimes 1) \otimes 1)\,((\Delta_2 \otimes 1)^{\otimes 2})\,\Delta_2^{\otimes 2} \\
&= (\bullet_{\otimes} \otimes \bullet)\,(1^{\otimes 2} \otimes (1 \otimes T)(T \otimes 1) \otimes 1)\,((\Delta_2 \otimes 1)\,\Delta_2)^{\otimes 2} \\
&= (\bullet_{\otimes} \otimes \bullet)\,(1^{\otimes 2} \otimes (1 \otimes T)(T \otimes 1) \otimes 1)\,((1 \otimes \Delta_2)\,\Delta_2)^{\otimes 2} .
\end{aligned}$$

An analogous treatment can be done for the right hand side of (11) in order to get the same result.

Now, we attack the problem of the complexity of the associated algorithm to the explicit formula of Δ_2 and we considerably reduce the amount of elementary operations (in comparison with the algorithm derived from the initial formulation (8)) that are necessary to compute this morphism over an algebra generator. The responsible for the high complexity in the evaluation of Δ_2 is the homotopy operator, ϕ, due, essentially, to the shuffles that are involved in the formulas of $\phi_{B\otimes}$ and $g_{B\otimes}$. We intend to reduce this complexity by eliminating unnecessary terms, and, for this aim, we use the inversion theory given in section 3.

We recall the formula for Δ_2:

$$\Delta_2 = ((f - f\delta\phi + f\delta\phi\delta\phi - \cdots) \otimes (f - f\delta\phi + f\delta\phi\delta\phi - \cdots))\, \Delta\, (g - \phi\delta g + \phi\delta\phi\delta g - \cdots)$$

We can observe that f is the last morphism applied. If, at the last stage, after applying Δ, the element y obtained by ϕ, has more than one inversion, then $\delta(y)$ will have at least one inversion. This way, each time we go on applying $\delta \circ \phi$, we obtain an element with at least one inversion (lemmas 5 and 6), and, therefore, the final evaluation by f is null (lemma 4). This means that, for the application of $f_\delta \otimes f_\delta$, we only have to consider summands of ϕ which produce elements with, at most, one inversion.

In consequence, we can establish the following theorem by which we considerably reduce the complexity in the computation of Δ_2.

Theorem 2. *When Δ_2 is applied to an algebra generator from the 1–homological model for a CDGA $A \otimes A'$, the formula for ϕ that is involved in the definition of $f_\delta \otimes f_\delta$, can be reduced to the following one:*

$$\phi = \bar{\phi}_{\bar{B}\otimes} + \bar{g}_{\bar{B}\otimes}(\phi_{\bar{B}A} \otimes g_{\bar{B}A'} f_{\bar{B}A'} + 1 \otimes \phi_{\bar{B}A'}) f_{\bar{B}\otimes},$$

where

– $\bar{\phi}_{\bar{B}\otimes}$ *is the one given in [3] for the simplification in the calculation of* d_δ.
– $\bar{g}_{\bar{B}\otimes}([a_1|\cdots|a_n]\otimes[a'_1|\cdots|a'_m]) = [a_1|\cdots|a_n|a'_1|\cdots|a'_m]$.

When Δ_2 *is applied to a product of two generators,* $z = x \bullet y$*, then,* $\Delta_2(z) = (\bullet\otimes\bullet)\,(1\otimes T\otimes 1)\,(\Delta_2(x)\otimes\Delta_2(y))$.

Recall that, the number of terms in the formula for $\bar{\phi}_{\bar{B}\otimes}$ is

$$\sum_{q=0}^{n-1}\sum_{p=0}^{n-q-1} 1 = \frac{n^2+n}{2},$$

in contrast to the original number of terms:

$$\sum_{q=0}^{n-1}\sum_{p=0}^{n-q-1}\binom{p+q+1}{q} \;=\; 2^{n+1}-n-2\,.$$

Besides, the formula for $g_{\bar{B}\otimes}$ is reduced now to 1 term in the sum instead of $\binom{m+n}{n}$ (the reduction for this morphism in [3] was of n terms).

Even though the reduction in complexity is considerable, the algorithm for computing Δ_2 is still an extremely expensive procedure, due, mainly, to the morphism ϕ in g_δ. This fact forced us to look for cases of CDGAs in which this calculation becomes more reasonable.

5 The Case of Purely Quadratic CDGAs

We consider here an important subset of CDGAs for which $\Delta_1 = 0$. Therefore, the 1–homological model for these algebras coincides with its 1–homology, so if we study the A_∞–coalgebra structure on the model, we are actually considering such a structure on the 1–homology of a CDGA.

This is the case of the commutative differential algebras whose differential–perturbation, ρ, does'nt have any *linear* summand, that is, if $x, x_{i_1}, \ldots x_{i_k}$ are algebra generators, then $\rho(x) = \sum_i \lambda_i\, x_{i_1}^{r_{i_1}}\otimes\cdots\otimes x_{i_k}^{r_{i_k}}$, with $k\geq 2$. We denote this set of algebras by CDGA_0, for which we can state the following proposition.

Proposition 4. *If* A *is a* $CDGA_0$ *under the conditions described in section 2, the formula for* Δ_2 *in the* 1*–homological model,* HBA*, is*

$$\Delta_2 \;=\; (f\otimes f)\,\Delta\, g_\delta\,.$$

Proof. We recall that $f_\delta = \sum(-1)^i f\,(\delta\phi)^i$. Notice that the morphism $(f\,\delta)$ is applied at each summand, except for the first one. Such a composition is null since δ produces elements with, at least, one non-linear component and, hence, $f_{\bar{B}\otimes}$ becomes zero.

Definition 2. Let (A,ρ) be a CDGA, we say A is *quadratic*, if the action of the differential–perturbation on every algebra generator x is

$$\rho(x) = \sum_i \lambda_i \, x_{i_1}^{r_{i_1}} \otimes x_{i_2}^{r_{i_2}} ,$$

where x_{i_1} , x_{i_2} are also algebra generators and $\lambda_i \in \Lambda$.

We say that A is *purely quadratic* if $r_{i_1} = 1$, $r_{i_2} = 1$ for all the indexes i.

Theorem 3. *Let A be a purely quadratic CDGA, the A_∞–coalgebra structure of its 1–homology, HBA, reduces to that of coalgebra.*

Notice that, in that case, HBA is a Hopf algebra.

Proof. It is sufficient to show that $\phi_\delta^{[\otimes 2]} \, \Delta \, g_\delta \; = \; 0$, what implies that $\Delta_3 = 0 = \Delta_4 = \cdots$ and, in particular, since Δ_3 is null, Δ_2 is coassociative and, hence, a real coproduct on HBA.

We say a homogeneous element from $\bar{B}(A)$ is *simple* if all the components of such an element are linear. We extent this concept in the natural way to any element from $\bar{B}(A)$.

We show that we always obtain simple elements by applying g_δ to the algebra generators. As a consequence, this property will be true for all the elements g_δ is applied to, since this morphism is multiplicative (theorem 1).

$$g_\delta(\underline{x}) = \sum_{i \geq 0} (-1)^i \; (\phi\delta)^i \; g\,(\underline{x})$$

We will prove, by induction on i, that $(\phi\delta)^i g(\underline{x}) \; = \; (\phi\delta)^i[x]$ are simple.

- $\phi\delta g(\underline{x}) \; = \; \phi\delta[x] \; = \; \sum_i \lambda_i \, \phi[x_{i_1} \otimes x_{i_2}] \; = \; \sum_i \lambda_i \, [x_{i_2}|x_{i_1}]$, which are simple.
- Let us assume that $(\phi\delta)^{i-1}[x]$ is simple. Then, each term of $\delta(\phi\delta)^{i-1}[x]$ is 1–*simple*, that is, has a unique quadratic component, $x_{k_1} \otimes x_{k_2}$,

  $$[x_1|\cdots|x_{k-1}|x_{k_1} \otimes x_{k_2}|x_{k+1}|\cdots|x_m] \, . \tag{12}$$

  Recall that A is factored as $A = \tilde{\otimes}_{i \in I}^{\rho} A_i$, with $I = \{1, 2, \ldots, n\}$ and, therefore, the homotopy operator ϕ can be written as follows,

  $$\phi \; = \; \phi^n \; = \; \phi_{\bar{B}\otimes} \; + \; g_{\bar{B}\otimes}(\phi^{n-1} \otimes g_{\bar{B}A_n} f_{\bar{B}A_n} + 1 \otimes \phi_{\bar{B}A_n}) f_{\bar{B}\otimes}$$

  Let us show, by induction on n, that ϕ turns that quadratic component into two linear components.
  - Case $n = 2$, A_1 and A_2 are a polynomial or an exterior algebra each one of them.

    It is clear that $f_{\bar{B}\otimes}$ is null when applied to elements like (12). Concerning $\phi_{\bar{B}\otimes}$, if we pay attention to the formula (4),we can check that the resultant element is is a sum of homogeneous elements with the structure

    $$[x_1|\cdots|x_{k-j}|x_{k_2}|\,[x_{k-j+1}|\cdots|x_{k_1}|\cdots x_{k+i}] \, \star \, [x_{k+i+1}|\cdots|x_m]] \, ,$$

    which are simple.

- Case $n > 2$,
  1. If $x_{k_2} \in A_n$, $\phi_{\bar{B}\otimes}$ acts in the same way as before, as well as $f_{\bar{B}\otimes}$, which is null.
  2. If x_{k_2} is the algebra generator of A_j with $j < n$, $\phi_{\bar{B}\otimes}$ will be null. On the other hand, $f_{\bar{B}\otimes}$ will only be no null if the components corresponding to A_n are just the last, for example, i components, $x_{m-i+1}, \ldots, x_m$. In this case,

$$f_{\bar{B}\otimes}([x_1|\cdots|x_{k-1}|x_{k_1} \otimes x_{k_2}|\cdots|x_{m-i}|\cdots|x_m])$$
$$= [x_1|\cdots|x_{k_1} \otimes x_{k_2}|\cdots|x_{m-i}] \otimes [x_{m-i+1}|\cdots|x_m] \quad (13)$$

Now, it is easy to verify that

$$(1 \otimes \phi_{\bar{B}A_n})([x_1|\cdots|x_{k_1} \otimes x_{k_2}|\cdots|x_{m-i}] \otimes [x_{m-i+1}|\cdots|x_m]) = 0\,,$$

since, on one hand, $\phi_{\bar{B}P}$ is null when applied to simple elements, and on the other hand, $\phi_{\bar{B}E} = 0$.
As for $g_{\bar{B}\otimes}(\phi^{n-1} \otimes g_{\bar{B}A_n} f_{\bar{B}A_n})$, its application to the element (13) gives place to simple elements, since if $g_{\bar{B}A_n} f_{\bar{B}A_n}[x_{m-i+1}|\cdots|x_m]$ is not null, it is clear that is simple; on the other hand, $\phi^{n-1}([x_1|\cdots|x_{k_1} \otimes x_{k_2}|\cdots|x_{m-i}]$ is simple by induction hypotheses. Finally, by shuffle product we obtain simple elements.

After g_δ, the morphism Δ, factors these simple elements as tensor product of pairs of simple factors.

Finally, we must show that ϕ is null when applied to simple elements and so will be $\phi_\delta^{[\otimes 2]}$. But, considering, again, the formula of ϕ, we realize that it is easy to prove it by induction. The key is that both $\phi_{\bar{B}\otimes}$ and $\phi_{\bar{B}A_n}$ are null when applied to simple elements.

Therefore, it is possible to derive an algorithm for computing the complete Hopf algebra structure of the 1–homology of a purely quadratic algebra. We intend to implement this algorithm in the near future starting from the program used in [3].

References

1. V. Álvarez, A. Armario, P. Real y B. Silva. *Homological Perturbation Theory and computability of Hochschild and cyclic homologies of CDGAs.* The international Conference on Secondary Calculus and Cohomological Physics. EMIS Electronic Proceedings. http://www.emis.de/proceedings/SCCP97
2. A. Armario, P. Real and B. Silva. *On p–minimal homological models of TTPs of elementary complexes at a prime.* Contemporary Mathematics, **vol. 227** (1999), 303-314.
3. CHATA group. *Computing "small 1–homological models for CDGAs.* CASC-2000 Proceedings. Springer-Verlag (2000), 87-100.

4. S. Eilenberg and S. Mac Lane. *On the groups $H(\pi, n)$, I.* Annals of Math., **v. 58** (1953), 55-106 .
5. S. Eilenberg and S. Mac Lane. *On the groups $H(\pi, n)$, II.* Annals of Math., **v. 60** (1954), 49-139.
6. E. Getzler and J.D.S. Jones. *A_∞-algebras and the cyclic bar complex.* Illinois J. Math., **34** (1990) 256-283.
7. V. K. A. M. Gugenheim. *On the chain complex of a fibration.* Illinois J. Math. **3** (1972) 398-414.
8. V. K. A. M. Gugenheim and L. Lambe. *Perturbation theory in Differential Homological Algebra, I.* Illinois J. Math **33** (1989) 56-582.
9. V. K. A. M. Gugenheim, L. Lambe and J. Stasheff. *Perturbation theory in Differential Homological Algebra, II.* Illinois J. Math. **35** n. 3 (1991) 357-373.
10. V. K. A. M. Gugenheim and J. Stasheff. *On Perturbations and A_∞–structures.* Bull. Soc. Math. Belg., vol. 38 (1986) 237-246.
11. J. Huebschmann and T. Kadeishvili. *Small models for chain algebras.* Math. Zeit. **v. 207**(1991), pp. 245-280.
12. M.J. Jiménez and P. Real. *Inversion theory for Commutative Differential Graded Algebras.* In preparation.
13. T.V. Kadeishvili. *The algebraic structure in the homology of an $A(\infty)$-algebra.* (Russian), Soobshch. Akad. Nauk. Gruzin. SSR **108** (1982), 249-252.
14. B. Keller. *Introduction to A_∞ Algebras and Modules.* Homology, Homotopy and Applications. **vol 3**, No 1 (2001), pp 1-35.
15. M. Kontsevich. *Homological Algebra of Mirror Symmetry.* Proceedings of the International Congress of Mathematicians, Zürich, Switzerland 1994, vol. 1, Birkhäuser Basel, 1995, 120-139.
16. S. Mac Lane. *Homology.* Classics in Mathematics. Springer-Verlag, Berlin (1995). Reprint of the 1975 edition.
17. J. McCleary (Ed.), Higher homotopy structures in Topology and Mathematical Physics. Contemp. Math., **227**, A.M.S. Providence, RI, 1999.
18. A. Polishchuk, E. Zaslow. *Categorical Mirror Symmetry: the Elliptic Curve.* http://xxx.lanl.gov/abs/math/9801119.
19. A. Prouté. *Algèbres différentielles fortement homotopiquement associatives.* PhD thesis, Université Paris VII, (1984)
20. P. Real. *Homological Perturbation Theory and associativity.* Homology, Homotopy and its Applications, vol. 2,No5, (2000) pp. 51-88.
21. S. Saneblidze and R. Umble. *The category of A_∞-Hopf algebras.* In preparation.
22. J.D. Stasheff. *Homotopy associativity of H–spaces I, II.* Trans. Amer. Math. Soc. **108** (1963), 275-313.
23. J. D. Stasheff. *The (secret?) homological algebra of the Batalin-Vilkovisky approach.* Contemp. Math. **219** (1998), pp 195-210
24. J.D. Stasheff. *Differential graded Lie algebras, quasi-Hopf algebras, and higher homotopy algebras.* Quantum Groups, Proc. workshops, Euler Inst. Math. Ins., Leningrad 1990, Lecture Notes in Mathematics **1510**, Springer 1992, 120-137
25. A. Vinogradov. *Introduction to Secondary Calculus.* Contemp. Math. **219** (1998), pp 241-272.

Conservative Finite Difference Schemes for Cosymmetric Systems

Bülent Karasözen[1] and Vyacheslav G. Tsybulin[2]

[1] Mathematics Department, Middle East Technical University, Turkey
bulent@math.edu.tr
[2] Department of Mathematics and Mechanics, Rostov State University, Russia
tsybulin@math.rsu.ru

Abstract. We consider the application of computer algebra for the derivation of the formula for the preservation of the cosymmetry property through discretization of partial differential equations. The finite difference approximations of differential operators for both regular and staggered grids are derived and applied to the planar filtration-convection problem.

1 Introduction

In recent years, there is an intensive research on so called geometric integrators, which are based on the preservation of essential qualitative features of the continuous problem (ordinary or partial differential equation). On the other hand there is permanent interest in the construction of finite-difference or finite-element schemes approximating conservative properties of important partial differential equations [8], [9],[11]. Many differential equations have symmetries which are fundamental to the physics of the underlying systems. Symmetry preserving numerical schemes are stable in long term computations and present the qualitative features of the continuous system better then the standard methods which are based only on accuracy and error control mechanism (see for example [2]).

The advanced computer algebra systems (CAS) like Maple, Mathematica, Macsyma provide powerful tools for the derivation of approximation schemes and for studying the stability properties of numerical methods [3], [7]. The derivation of finite-dimensional analogs of underlying continuous systems and subsequent code generation are also current application areas of CAS.

We are interested in the conservation of the cosymmetry property for a system of partial differential equations (the planar filtration-convection problem) through finite-difference approximation. Following [12], for a dynamical system $\dot{y} = F(y)$ in a Banach space H we call $L(y)$ a cosymmetry when $(F(y), L(y))_H = 0$ for all $y \in H$, where $(\cdot, \cdot)_H$ denotes some inner product. The cosymmetry may be the reason for the onset of continuous families of stationary regimes with varying spectrum; it differs from the symmetry, for which all regimes are identical [13]. It was shown in [6] for the planar filtration-convection problem that an inappropriate approximation leads to the destruction of the cosymmetry and consequently to degeneration of the family of equilibria to isolated stationary regimes. Nevertheless it was shown that the Arakawa approximation [1] preserves the cosymmetry in the planar filtration-convection problem.

We stress that Arakawa's work [1] was a pioneering one for finite-difference approximations of ideal incompressible fluid equations with conservative quantities, like kinetic energy, enstrophy and mean value of vortices. Since then a number of extensions were derived (see [8], [9], [10] and references therein).

In this work we present an application of computer algebra for the derivation the finite-difference operators preserving the cosymmetry property in the planar filtration-convection problem. We study regular and staggered grids. The computation of the family of stationary regimes is described and various numerical results are presented.

2 Filtration-Convection Problem

We start with the exposition of the planar filtration-convection problem for incompressible fluid saturated with a porous medium in a rectangular container which is uniformly heated below. The dimensionless equations of filtration convection problem [13] in the rectangle domain $\mathcal{D} = [0,a] \times [0,b]$ may be written as:

$$\frac{\partial \theta}{\partial t} = \Delta\theta + \lambda\frac{\partial \psi}{\partial x} - J(\psi,\theta) := F \tag{1}$$

$$0 = \Delta\psi - \frac{\partial \theta}{\partial x} := G, \quad \Delta = \frac{\partial^2}{\partial x^2} + \frac{\partial^2}{\partial y^2}, \tag{2}$$

$$\theta = 0, \qquad \psi = 0 \qquad \text{on} \quad \partial\mathcal{D}. \tag{3}$$

where $\psi(x,y,t)$ and $\theta(x,y,t)$ denote perturbations of the stream function and temperature; λ is the Rayleigh number, Δ is the Laplacian and $J(\theta,\psi)$ denotes the Jacobian operator over (x,y):

$$J(\theta,\psi) = \frac{\partial \theta}{\partial x}\frac{\partial \psi}{\partial y} - \frac{\partial \theta}{\partial y}\frac{\partial \psi}{\partial x}. \tag{4}$$

The initial condition is only defined for the temperature

$$\theta(x,y,0) = \theta_0(x,y), \tag{5}$$

where θ_0 denotes the initial temperature distribution. For a given θ_0, the stream function ψ can be obtained from (2) and (5) as the solution of the Dirichlet problem via Green's operator $\psi = G\theta_x$.

The linear cosymmetry of the right hand parts of (1)-(3) can be given with the pair $(\psi, -\theta)$:

$$\int_{\mathcal{D}} (F\psi - G\theta)dxdy = 0. \tag{6}$$

Indeed, multiply (1) by ψ, (2) by $-\theta$, sum them and integrate over the domain $\mathcal{D}$. Then using Green's formula, integration by parts and taking into account the boundary conditions (3) we obtain that result. Particularly, (6) holds because the Jacobian obeys the following equality

$$\int_{\mathcal{D}} J(\psi,\theta)\psi dxdy = 0. \tag{7}$$

We mention that the finite-difference approximation of the Jacobian must preserve the finite-dimensional analog of (7) and it is directly connected with cosymmetry conservation. It is desirable also to require that the Jacobian approximation nullifies the discrete analog of the integral

$$\int_{\mathcal{D}} J(\psi,\theta)\theta dxdy = 0, \tag{8}$$

and the skew-symmetry of the Jacobian $J(f,g) = -\ J(g,f)$ must hold.

In the following sections we investigate various finite difference approximations for the system (1)-(5) from the point of view of cosymmetry preservation and correct computation of the family of stationary regimes.

3 Regular Grids

The filtration convection equations (1)-(5) are discretized in space using a conservative finite-difference scheme. We start with uniform regular grids for both spatial coordinates:

$$\{(x_n,y_m)|\ x_n = nh_1,\ n=-N \div N,\ y_m = mh_2,\ m=-M \div M\}$$

here $h_1 = a/N,\ h_2 = b/M$ are the uniform mesh step sizes in directions x and y respectively. The temperature θ and the stream function ψ are computed at the nodes (x_n, y_m) and are denoted by θ_{nm} and ψ_{nm}. The discretized boundary conditions are given by

$$\begin{aligned}
\theta_{n,-M} = \theta_{nM} = \psi_{n,-M} = \psi_{nM} = 0, \quad n &= -N+1,\dots,N-1,\\
\theta_{-N,m} = \theta_{Nm} = \psi_{-N,m} = \psi_{Nm} = 0, \quad m &= -M,\dots,M.
\end{aligned}$$

The Laplacian Δ and the first derivatives are replaced by second order centered finite differences:

$$\Delta_h f_{nm} = \frac{f_{n+1,m} - 2f_{n,m} + f_{n-1,m}}{h_1^2} + \frac{f_{n,m+1} - 2f_{n,m} + f_{n,m-1}}{h_2^2}, \tag{9}$$

$$D_1 f_{nm} = \frac{f_{n+1,m} - f_{n-1,m}}{2h_1}, \quad D_2 f_{nm} = \frac{f_{n,m+1} - f_{n,m-1}}{2h_2}, \tag{10}$$

where f stands for θ and ψ.

Using (9) and (10) and including the boundary conditions we obtain the following system of ordinary differential equations as the semi-discretized form of (1)-(2) in space

$$\dot{\theta}_{nm} = F_{nm} \equiv (\Delta_h\theta + \lambda D_1\psi + J)_{nm}, \tag{11}$$

$$0 = G_{nm} \equiv (\Delta_h\psi - D_1\theta)_{nm}. \tag{12}$$

Here J_{nm} denotes the approximation of the Jacobian which will be given in the following sections.

The finite-dimensional analog of (6) is given by

$$\sum_{n=-N+1}^{N-1} \sum_{m=-M+1}^{M-1} (F\psi - G\theta)_{nm} = 0. \tag{13}$$

Using the difference version of Green's formula and summation by parts (the discrete version of integration by parts) we deduce that the quadratic terms in the summation (13) are nullified. So, the equation (11)-(12) will have the same cosymmetry as (1)-(2) when the following equality holds

$$\sum_{n=-N+1}^{N-1} \sum_{m=-M+1}^{M-1} J_{nm}(\theta, \psi)\ \psi_{nm} = 0. \tag{14}$$

We seek also a such Jacobian approximation that the antisymmetry property holds and the nullification of convective terms in the difference version of energy equation (gyroscopic effect) takes place

$$\sum_{n=-N+1}^{N-1} \sum_{m=-M+1}^{M-1} J_{nm}(\theta, \psi)\ \theta_{nm} = 0. \tag{15}$$

It was shown in [6] that the violation of (15) may leads to some discrepancies in the computation of the family of equilibria.

3.1 Jacobian Approximation

We describe now the derivation of the Jacobian approximation on a regular grid by the direct method of free (indefinite) parameters using Maple. The analytical approach started with the famous Arakawa scheme [1] which was widely used in computational hydrodynamics for two-dimensional Euler and Navier-Stokes equations. It provides two quadratic conservation laws corresponding to energy and enstrophy. This approach was further generalized to arbitrary grids by Salmon and Talley [10]. Systematic formulation for any space dimensions and number of integrals was proposed in [8].

We use a nine-node stencil to obtain a Jacobian approximation of second order accuracy which preserve (7), (8) and satisfy the antisymmetry property $J_{nm}(\theta, \psi) = -\, J_{nm}(\psi, \theta)$. For a given node (x_n, y_m) the Jacobian can be written as

$$J_{nm}(\theta, \psi) = \sum_{k,l,j,i=-1}^{1} A_{klji} \theta_{n+k,m+j} \psi_{n+l,m+i}. \tag{16}$$

Firstly, we find Taylor series for $J_{nm}(\theta, \psi) - J(\theta, \psi)|_{(x_n, y_m)}$ and formulate the system for unknown parameters A_{klji} resulting from the nullification of the coefficients at the terms up to third order. Then we add the equations obtained by preservation of discrete versions of the equalities (7) and (8). After substitution (16) to (14) and (15) we may rewrite them as

$$\sum_{n,m} P_{nm}(\theta)\psi_{n,m} = 0, \quad P_{nm}(\theta) = \sum_{k,l,j,i=-1}^{1} B_{klji}^{nm}\theta_{n+k,m+j}\theta_{n+l,m+i}, \tag{17}$$

$$\sum_{n,m} Q_{nm}(\psi)\theta_{n,m} = 0, \quad Q_{nm}(\psi) = \sum_{k,l,j,i=-1}^{1} C_{klji}^{nm}\psi_{n+k,m+j}\psi_{n+l,m+i}, \tag{18}$$

We sum for all internal nodes and coefficients B_{klji}^{nm} (C_{klji}^{nm}) which do not depend on node number. The equality (17) must hold for any θ_{nm}, therefore $B_{klji}^{nm} = 0$. Analogically we conclude that $C_{klji}^{nm} = 0$ from (18). Obviously some coefficients A_{klji} are zero. For example, in (17) we have that $B_{-1,-1,1,0}^{nm} = A_{-1,-1,1,0}$ and thus $A_{-1,-1,1,0} = 0$. Thus, the terms $\theta_{n-1,m-1}\psi_{n+1,m}$, $\theta_{n-1,m-1}\psi_{n+1,m+1}$ and the similar can be omitted in (16). So, in Maple 6 we form a generic expression for the Jacobian approximation on a 9-node rectangular stencil

```
> eds:=0:AA:=NULL: for k from -1 to 1 do for j from -1 to 1 do
> for kk from max(-1,k-1) to min(1,k+1) do
> for jj from max(-1,j-1) to min(1,j+1) do
> aa:=A.(kk+1+3*(k+1))||(jj+1+3*(j+1));
> eds:=eds+aa*v(x+k*h,y+j*g)*u(x+kk*h,y+jj*g);
> AA:=AA,aa; od: od: od: od:
```

We compute the Taylor series and deduce the system for unknowns A_{klji}

```
> sds:=mtaylor(D[1](u)(0,0)*D[2](v)(0,0)-D[2](u)(0,0)*D[1](v)(0,0)
> -subs(x=0,y=0,eds),[h,g],4):
> cho:=seq(combinat[choose]([1,1,1,2,2,2],m)[],m=0..3);
> kk:=0: su:=NULL: sv:=NULL: for s in cho do kk:=kk+1;
> su:=su,D[s[]](u)(0,0)=u||kk; sv:=sv,D[s[]](v)(0,0)=v||kk; od:
> uu:=map(rhs,[su]); vv:=map(rhs,[sv]); SDS:=subs(su,sv,sds):
> eq:={}:for ut in uu do for vt in vv do t:=diff(diff(SDS,ut),vt);
> if t<>0 then if has(t,AA) then eq:=eq union t; fi; fi; od: od;
```

Then we apply standard command solve and specify Jacobian expression

```
> ss:=solve(eq,AA): par:={}:
> for s in ss do
> if rhs(s)=lhs(s) then par:=par union rhs(s); fi; od:
> JJ:=unapply(subs(ss,eds),x,y):
> sv:=seq(seq(v(k*h,j*h)=v||k||j,j=-1..1),k=-1..1):
> su:=seq(seq(u(k*h,j*h)=u||k||j,j=-1..1),k=-1..1):
```

After that we use the equalities (14) and (15)

```
> s1:=0: s2:=0:
> for k from -1 to 1 do for j from -1 to 1 do x:=k*h; y:=j*g;
> s1:=s1+diff(subs(u(0,0)=z,JJ(x,y)),z)*v(x,y);
> s2:=s2+diff(subs(v(0,0)=z,JJ(x,y)),z)*u(x,y); od: od;
```

Then the coefficients in P_{nm} (17) and Q_{nm} (18) are obtained and the following systems of additional equations will be formed

```
> he1:=map(factor,linalg[hessian](subs(sv,s1),vv)): EQ1:=NULL:
> he2:=map(factor,linalg[hessian](subs(su,s2),uu)): E21:=NULL:
> for k from 1 to nops(vv) do for j from k to nops(vv) do
> EQ1:=EQ1,factor(he1[k,j]); EQ2:=EQ2,factor(he2[k,j]); od: od:
```

We note that here we work with a stencil consisting of equal number of nodes for both θ and ψ, in contrast to the staggered grids where different number of the nodes for the variables θ and ψ are used. Finally we solve the full system and obtain one-parameter family of Jacobian approximations:

```
> so:=solve({EQ1,EQ2},par); pa:={}:
> for s in so do
> if rhs(s)=lhs(s) then pa:=pa union rhs(s); fi; od:
> ja:=subs(op(pa)=-A/g/h/6,subs(so,subs(ss,eds)));
```

Now we show that this approximation follows from the general scheme proposed in [10]. This approximation was derived using the finite-element method which is obtained as a linear combination of two finite-element schemes, one bilinear in rectangles, the other linear in triangles [5]. We consider the node (x_n, y_m) and organize two sets of the same type triangles as shown in Fig. 1. For the each set a stencil for the node (x_n, y_m) consists of 7 nodes. Considering right set of triangles (see Fig. 1) we derive the Jacobian approximation by the following formula [5].

$$J_{nm}^{<1>} = \frac{1}{6h_1h_2}\sum_{\alpha=1}^{6}\theta_{nm}^{(\alpha)}(\psi_{nm}^{(\alpha+1)} - \psi_{nm}^{(\alpha-1)}), \tag{19}$$

with

$$\begin{aligned} &f_{nm}^{(1)} \equiv f_{nm}^{(7)} = f_{n+1,m}, \quad f_{nm}^{(2)} = f_{n,m+1}, \quad f_{nm}^{(3)} = f_{n-1,m+1}, \\ &f_{nm}^{(6)} \equiv f_{nm}^{(0)} = f_{n+1,m-1}, f_{nm}^{(4)} = f_{n-1,m}, \quad f_{nm}^{(5)} = f_{n,m-1}. \end{aligned}$$

For the left set of triangles in Fig. 1 we obtain

$$J_{nm}^{<2>} = \frac{1}{6h_1h_2}\sum_{\alpha=1}^{6}\theta_{nm}^{(\alpha)}(\psi_{nm}^{(\alpha+1)} - \psi_{nm}^{(\alpha-1)}), \tag{20}$$

$$\begin{aligned} &f_{nm}^{(1)} \equiv f_{nm}^{(7)} = f_{n+1,m}, \quad f_{nm}^{(2)} = f_{n+1,m+1}, \quad f_{nm}^{(3)} = f_{n,m+1}, \\ &f_{nm}^{(6)} \equiv f_{nm}^{(0)} = f_{n,m-1}, \quad f_{nm}^{(4)} = f_{n-1,m}, \quad f_{nm}^{(5)} = f_{n-1,m-1}. \end{aligned}$$

Combining formulas (19) and (20) we finally obtain the Jacobian approximation

$$J_{nm} = \beta J_{nm}^{<1>} + (1-\beta)J_{nm}^{<2>}. \tag{21}$$

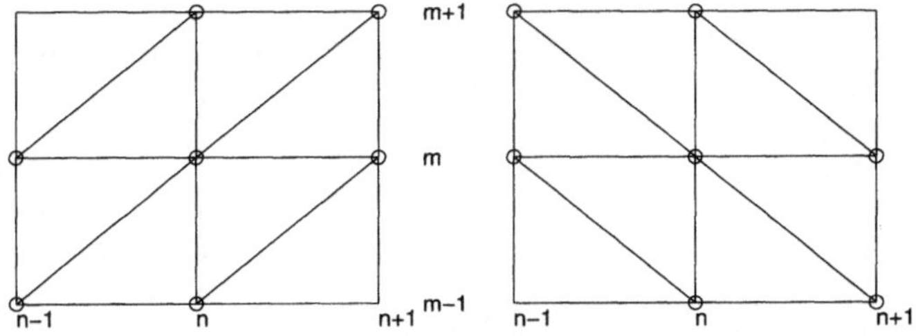

Fig. 1. The sets of triangles and stencils

The planar filtration-convection problem (1)-(5) is also invariant with respect to two discrete symmetries

$$R_x : \{x, y, \theta, \psi\} \mapsto \{-x, y, \theta, -\psi\}, \tag{22}$$

$$R_y : \{x, y, \theta, \psi\} \mapsto \{x, -y, -\theta, -\psi\}. \tag{23}$$

As one can see the linear terms of (11)-(12) are invariant with respect to the finite-difference analogs of transformation (22)-(23) given by

$$R_x^h : \{\theta_{nm}, \psi_{nm}\} \mapsto \{\theta_{-n,m}, -\psi_{-n,m}\}, \tag{24}$$

$$R_y^h : \{\theta_{nm}, \psi_{nm}\} \mapsto \{-\theta_{n,-m}, -\psi_{n,-m}\}. \tag{25}$$

The substitution of the Jacobian approximation (21) in (11)-(12) and the application of the transformation R_x^h lead to

$$\dot{\theta}_{-n,m} = \left(\Delta_h\theta + \lambda D_1\psi + (1-\beta)J^{<1>} + \beta J^{<2>}\right)_{-n,m}. \tag{26}$$

However for the node (x_{-n}, y_m) we find the coefficient β staying at $J^{<1>}_{-n,m}$ and $(1-\beta)$ at $J^{<2>}_{-n,m}$. So, the invariance of (11)-(12) under the symmetry transformation (24) will take place only for $\beta = 1/2$. This value corresponds to the Arakawa scheme [1]. Thus, in the case of a uniform regular grid the difference operators (9)-(10) and (21) for $\beta = 1/2$ guarantee a second order approximation of the system (1)-(2), to satisfy the difference equalities (14), (15) and the discrete analogs of the symmetry transformations (22)-(23).

4 Staggered Grids

In hydrodynamics, staggered grids are used very often for the discretization of partial differential equations. We give an extension of the technique which was developed for cosymmetry preservation on the regular grid. We consider two kinds of staggered grids which are sketched in Fig. 2. The node for each unknown is denoted by the corresponding letter and lines mark the boundary. We arrange the mesh in such a manner as to locate nodes for the temperature θ

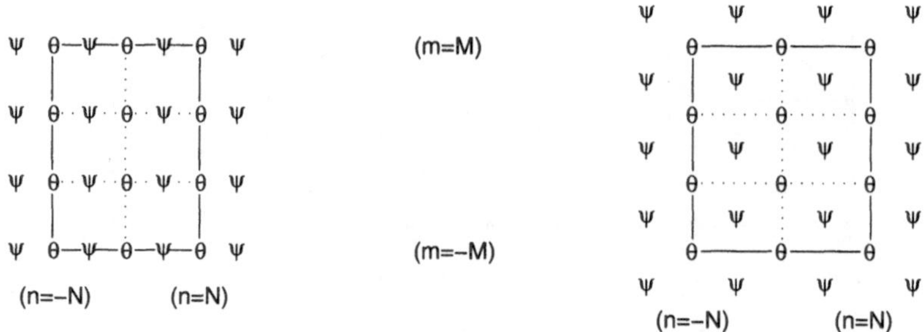

Fig. 2. Staggered grid organization

on the boundary, and fictitious nodes for ψ have to be introduced to take into account the Dirichlet boundary condition .

We start by presenting the case when the staggered grid follows from a regular grid by shifting on half step in the x direction. So, we introduce the auxiliary mesh on x:

$$x_{n-1/2} = \frac{(x_n + x_{n-1})}{2}, \quad x_{\pm N \pm 1/2} = x_{\pm N} \pm \frac{h_1}{2}$$

The temperature θ is defined at the nodes (x_n, y_m) and the stream function ψ is defined at staggered grid nodes $(x_{n-1/2}, y_m)$. The stream function ψ at fictitious nodes is computed from the Dirichlet boundary conditions

$$\psi_{\pm N \pm 1/2, m} = -\psi_{\pm N \mp 1/2, m}.$$

Analogically to the case of regular grid, we derive the finite-difference formulas preserving a cosymmetry and discrete symmetries for the underlying system (1)-(2). Using an approach of indefinite coefficients and the computer algebra system Maple we deduce the appropriate approximations. We omit the details of derivation and mention only that the realization of the Jacobian approximation follows from (21) for $\beta = 1/2$ if one substitutes

$$\psi_{n,m} \to \frac{1}{2}(\psi_{n-1/2,m} + \psi_{n+1/2,m}).$$

For a second order approximation of the Jacobian we need a stencil containing 9 nodes for the temperature θ and 6 nodes for the stream function ψ. But for the preservation of the cosymmetry property (7), and (8) and the discrete symmetries (24), (25) we have to enlarge a stencil with 6 additional nodes for ψ. In addition, the first derivative ψ_x in (1) must be approximated by the 4-node stencil

$$d_1(\psi)_{nm} = \frac{\psi_{n+3/2,m} + \psi_{n+1/2,m} - \psi_{n-1/2,m} - \psi_{n-3/2,m}}{4h_1}, \tag{27}$$

but for the derivative θ_x in (2) it is enough to apply the usual two-point stencil

$$\widetilde{D}_1(\theta)_{n-1/2,m} = \frac{\theta_{n,m} - \theta_{n-1,m}}{h_1}. \tag{28}$$

We obtain then the following system of ordinary differential equations

$$\dot{\theta}_{nm} = F_{nm} \equiv (\Delta_h\theta + \lambda d_1\psi + J)_{nm}, \quad (29)$$

$$0 = G_{n-1/2,m} \equiv \left(\Delta_h\psi - \widetilde{D}_1\theta\right)_{n-1/2,m}, \quad (30)$$

and the finite-dimensional analog of cosymmetry becomes

$$\sum_{m=-M+1}^{M-1}\left(\sum_{n=-N+1}^{N-1} F_{nm}\frac{\psi_{n-1/2,m}+\psi_{n+1/2,m}}{2} - \sum_{n=-N}^{N-1} G_{n+1/2,m}\frac{\theta_{nm}+\theta_{n+1,m}}{2}\right) = 0.$$

The discretization for the grid presented on the right in Fig. 2 can be performed in a similar way. Here we use fictitious nodes for ψ at whole boundary and introduce difference operators for the first derivatives ψ_x and θ_x defined respectively on 8-node stencil and 4-node stencil. This case corresponds to the Jacobian approximation with stencil consisting of 9 nodes for the temperature θ and 16 nodes for the stream function ψ.

5 Numerical Results

For all values of the Rayleigh number λ the system (1)-(5) has the trivial equilibrium $\theta = \psi = 0$. When the parameter λ passes $\lambda_{11} = 4\pi^2(a^{-2} + b^{-2})$ a one-parameter family of stationary solutions emerges in the form of a closed curve in the phase space of the system. It was shown in [13] that the spectrum varies along this family and therefore this family can not be an orbit of the action of any symmetry group.

To compute the family of equilibria for the finite-difference approximation of the original partial differential equation, we apply the technique based on the cosymmetric version of the implicit function theorem [14]. Different realizations for ordinary and partial differential equations were developed in [4], [6] and we briefly describe this technique.

When λ is slightly larger than λ_{11}, then all points of the family are stable [13]. Starting from the vicinity of unstable zero equilibrium we integrate the ode system (18) by the classical Runge-Kutta method up to a point Θ_0 close to a stable equilibrium on a family. Then the algorithm to family computation may be formulated as the following:

1. Correct the point Θ_0 using the modified Newton method.
2. Determine the kernel of the linearization matrix at the point Θ_0.
3. Predict the next point on the family by the Adams-Bashford method.
4. Repeat the steps 1-3 until a closed curve is obtained.

The procedure of Newton iteration contains a number of stages. If the guess point is rather far from the family we apply the usual Newton iterations. Then Several iterations can be used without repeated computation of Jacobi matrix. Finally, we correct the solution of the reduced system, because close to the family the Jacobian becomes degenerate.

Based on this approach a program for the computation of continuous families of stationary regimes was developed in [6]. The program was written initially in MATLAB and then transferred with improvements to separate code on C by A. Kurdyumov.

Our results show that rather coarse grids allow the prediction of the threshold values describing the onset of the instability on the family and study the phenomena of families collision and aggregations. We present some figures and comments to justify that only an appropriate approximation of the Jacobian and other differential operators provide us with the finite-dimensional system keeping cosymmetry and continuous families of equilibria (stationary regimes).

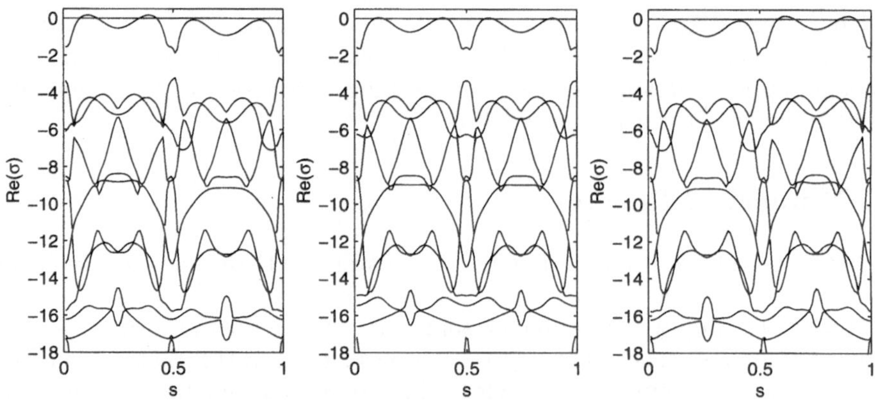

Fig. 3. Real part of spectrum for the equilibria for the families computed at $\lambda = 26.0$ for $\beta = 0$ (left), $\beta = 0.5$ (center) and $\beta = 1$ (right), s is the coordinate along the family of equilibria; $a = 2$, $b = 5$, mesh 12×12

When the Jacobian approximation does not preserve the discrete symmetry (22) then the following phenomena were observed. For the container with aspect ratio $b/a = 2.5$ both results based on spectral computation [4] and finite-difference discretization [6] predict the onset of instability on the family simultaneously in four points. In contrast, the computation with (21) for $\beta \neq 1/2$ predicts instability only at two points. The computed families are asymmetrical and the magnitude of asymmetry depends on the β. This effect is illustrated in Fig. 3 where the spectra against coordinate s (continual number of equilibrium on the family) are given for three values of β. We show a number of eigenvalues with largest real part are plotted for Rayleigh number $\lambda = 26$ characterized by the formation of unstable arcs on the family. It is clearly seen that the approximation for values apart from $\beta = 1/2$ gives an asymmetrical picture for the spectra of equilibria. The magnitude of this effect is not significant but it is easy to predict this effect in FEM-approximation when the system of triangles with some orientation is often used.

In Fig. 3 we display the evolution of the family of equilibria at increasing values of Rayleigh parameter λ. The computed families are presented in the Nusselt coordinates used in [4]:

$$Nu_h = \frac{h_2}{2h_1} \sum_{m=-M+1}^{M-1} (\theta_{1,m} - \theta_{-1,m}) \approx \int_{-b}^{b} \frac{\partial \theta}{\partial x} |_{x=0} \, dy,$$

$$Nu_v = \frac{h_1}{2h_2} \sum_{n=-N+1}^{N-1} (4\theta_{n,-M+1} - \theta_{n,-M+2}) \approx \int_{-a}^{a} \frac{\partial \theta}{\partial y} |_{y=-b} \, dx.$$

Here Nu_h corresponds to the combined heat flux through central vertical section of rectangle container. The value Nu_v presents a heat flux calculated on the bottom of container.

Each family is given by a continuum of stationary regimes and each regime may be realized with appropriate choice of the initial state. A number of computed families are presented in Fig. 4 (left part). It is clearly seen how stability changes on the family: arcs of stable equilibria (solid lines) interfere with those consisting of unstable stationary regimes (dotted lines). When the parameter λ increases we observe the onset of the limit cycle ($\lambda \approx 26$), its growth and its subsequent transformation to the chaotic regime. In the right part of Fig. 4 we display two families of equilibria (dotted lines) and a limit cycle (solid line) computed at $\lambda = 28$. In Fig. 5 the stream functions for several stationary states from the family computed for $\lambda = 28$ (marked by numbers in Fig. 4) are shown (top picture). The spectra distributions that correspond to each regime are given at the bottom Fig. 5.

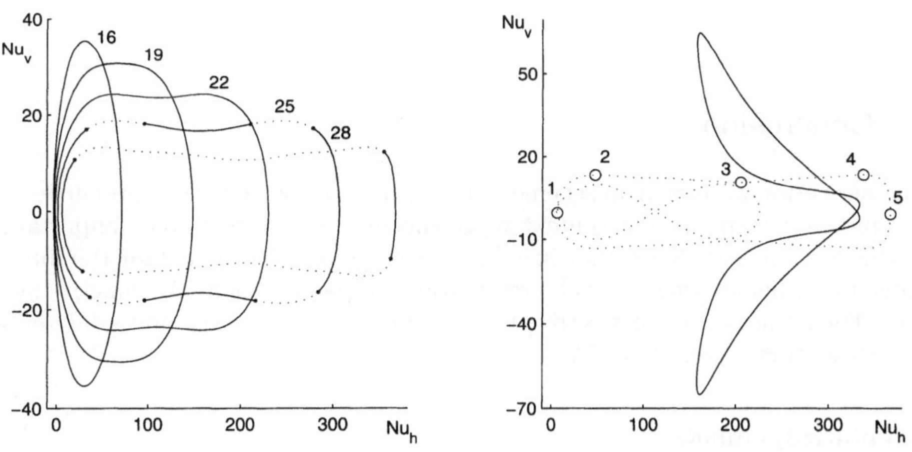

Fig. 4. Continuous families of equilibria for different parameter λ values (number near family), (left); two families of equilibria (dotted) and stable limit cycle (solid), $\lambda = 28$, $a = 2$, $b = 5$, mesh 20×10

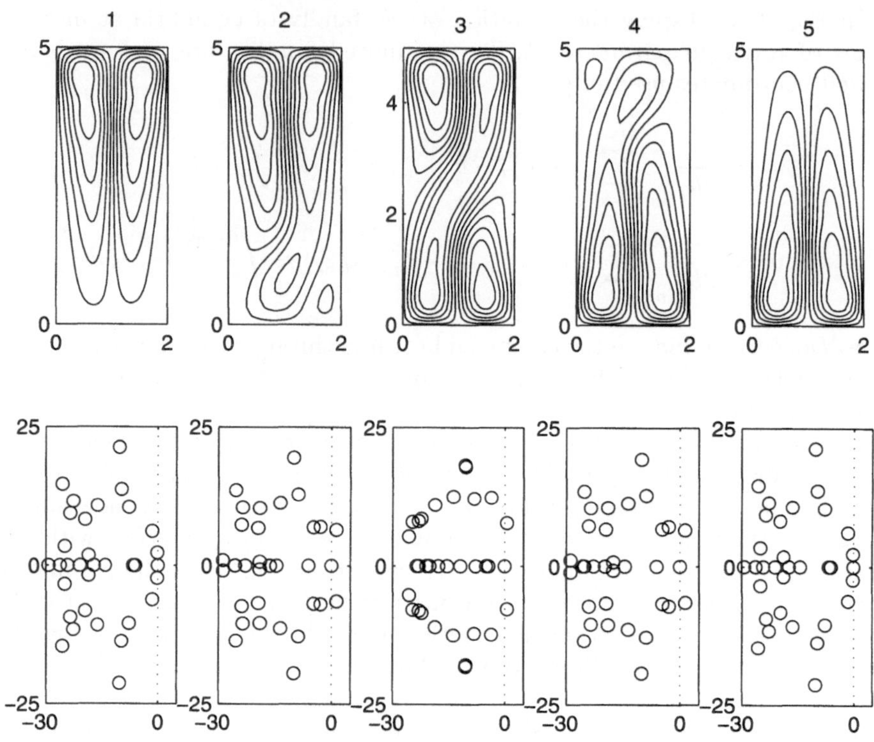

Fig. 5. Streamlines (top) and spectrum (bottom) for the equilibria from family at $\lambda = 28$, mesh 20×10, $a = 2$, $b = 5$

6 Conclusion

The derivation of numerical schemes which preserve the geometric properties of continuum systems in the context of geometric integration is very important. In this work it was shown that just CAS provide an efficient tool for the finite-difference schemes with several conservation properties for a cosymmetric system. The numerical results verify the predicted behavior of the finite-difference approximations derived by CAS.

Acknowledgements

The authors acknowledge the support of NATO-PC Advanced Fellowship Programme of TÜBITAK (Turkish Scientific Research Council). V. Tsybulin was partially supported by the Program "Russian Universities - Fundamental Studies" (# 4087) and Russian Foundation for Basic Research (# 00-15-96188, #

99-01-01023). We are thankful to the referees for their comments and their careful reading of original version of the paper and A. Kurdyumov for the help in the programming on C and computations.

References

1. A. Arakawa, Computational design for long-term numerical integration of the Equations of Fluid Motion: two-dimensional incompressible flow. Part I, *J. Comput. Phys.* **1** (1966) 119-143.
2. C. J. Budd and M. D. Piggott, The geometric integration of scale-invariant ordinary and partial differential equations, *J. Comput. Appl. Math.* **128** (2001) 399-422
3. V. G. Ganzha and E. V. Vorozhtsov. *Numerical Solutions for Partial Differential Equations. Problem Solving Using Mathematica.* CRC Press, Boca Raton, New York, London, 1996.
4. V. G. Govorukhin, Numerical simulation of the loss of stability for secondary steady regimes in the Darcy plane-convection problem, *Doklady Akademii Nauk* **363** (1998) 806-808.
5. D. C. Jespersen, Arakawa's Method is a Finite-Element Method, *J. Comput. Phys.* **16** (1974), 383–390.
6. B. Karasözen, V. G. Tsybulin, Finite-Difference Approximations and Cosymmetry Conservation in Filtration Convection Problem. *Physics Letters* **A 262** (1999) 321–329.
7. R. Liska, B. Wendroff, Where Numerics Can Benefit from Computer Algebra in Finite Difference Modelling of Fluid Flows, In: *Computer Algebra in Scientific Computing*, Ganzha V. G. Mayr E. W. and E. V. Vorozhtsov, Springer-Verlag, Berlin, 1999, 269-286.
8. R. I. McLachlan, Spatial discretization of partial differential equations with integrals, *preprint* (1998) (see http://www.massey.ac.nz/ RMcLachl/)
9. Y. Morinishi, T. S. Lund, O. V. Vasilyev and P. Moin, Fully conservative higher order finite difference schemes for incompressible flow. *J. Comput. Phys.* **143** (1998), no. 1, 90–124.
10. R. Salmon and L. D. Talley, Generalization of Arakawa's Jacobian, *J. Computat. Phys.* **83** (1989) 247–259.
11. M. I. Shashkov, Conservative finite-difference methods on general grids CRC Press, 1996.
12. V. I. Yudovich, Cosymmetry, degeneracy of the solutions of operator equations, and the onset of filtrational convection, *Matem. Zametki* **49**, (1991) 142–160.
13. V. I. Yudovich, Secondary cycle of equilibria in a system with cosymmetry, its creation by bifurcation and impossibility of symmetric treatment of it, *Chaos* **5** (1995) 402–441.
14. V. I. Yudovich, An implicit function theorem for cosymmetric equations, (Russian), *Mat. Zametki* **60** (1996), 313–317. (Translation in *Math. Notes* **60** (1996), 235–238.)

A Mathematica Solver for Two-Point Singularly-Perturbed Boundary Value Problems.

Raya Khanin

Department of Applied Mathematics and Theoretical Physics,
University of Cambridge
R.Khanin@damtp.cam.ac.uk
www.damtp.cam.ac.uk/user/bioguest/rk259.html

Abstract. This paper introduces a *Mathematica* solver which constructs approximate symbolic solutions of second-order singularly-perturbed boundary value problems. The employed symbolic technique is based on the method of matched asymptotic expansion and the Nipp polyhedron algorithm for finding appropriate scalings in local approximating systems. The main command *BVPSolve* uses the standard *Mathematica* syntax of the *Solve*-type commands. The package finds approximate solutions of second-order linear, quasilinear and semilinear singularly perturbed boundary value problems. Several illustrative examples are shown. The package can be used in research and teaching. It can also be used for initialising numerical solvers for boundary-value problems.

1 Introduction

Singularly perturbed boundary value problems occur in various areas of science and engineering: fluid dynamics, elasticity, quantum mechanics, chemical reactions to name just a few. Such problems exhibit multiscale character wherein the solution changes very rapidly in a thin layer(s) and varies relatively slowly while away from such layer(s). Much of the computational effort for dealing with singularly perturbed (or stiff) problems is focused on their numerical treatment as the presence of boundary and interior layers makes it hard to obtain accurate numerical solution. For example, the MATLAB.6 solver for two-point boundary value problems for ordinary differential equations, *bvp4c*, requires initialisation of the mesh structure by the user who is supposed either to guess it or to make calculations on paper (see tutorial on ftp.mathworks.com/pub/doc/papers/bvp/).

Substantial theoretical background exists for finding asymptotic symbolic solutions of singular perturbation problems including boundary value problems (see, for example, [2], [4], [9]). Due to their huge area of applications and an algorithmic structure common to all perturbation methods, examples of singular perturbation problems have long been solved with the aid of computer algebra systems. Examples of computer algebra implementation of the singular perturbation solution procedure (in MACSYMA) have first been developed by Rand and Armbruster [10]. Kauffman (www.ifm.ethz.ch/~kaufmann/news.html) has developed several *Mathematica* tools to generate educational examples of perturbation expansions, including singular perturbation problems. The functions in his *Perturbation* package include finding boundary layer, simple matching, and constructing composite expansions. Still, however, modern computer algebra systems do not have automatic facilities to find approximate symbolic solutions of singularly perturbed initial and boundary value problems.

The novelty of the present paper is that it introduces an automatic solver for two-point singularly perturbed boundary value problems. The solver effectively finds approximate symbolic solutions of linear problems up to a specified order. It also constructs perturbation equations to quasilinear boundary value problems and makes an attempt to solve them. The package also deals with semilinear boundary value problems by using symbolic-numerical approach.

2 Theoretical Background

A constructive algorithm for finding analytical solutions of linear and quasilinear second-order boundary value problems is well developed ([2], [4], [9]). The main technique for dealing with such problems is based on the method *of matched asymptotic expansions* which finds outer and inner solutions and matches them in the overlap region. To find the appropriate scalings for independent and dependent variables in the local expansions the Nipp's polyhedron algorithm can be employed ([5], [8]). This algorithm reduces the problem of finding the scalings to the linear programming problem.

Two main classes of linear boundary value problems are distinguished as being non-self-adjoint and self-adjoint problems. In addition, they are classified according to the existence of turning points or shocks in the interval. Below, an analytical treatment of non-self-adjoint problems is considered in detail. Self-adjoint problems are solved in a very similar fashion. In addition, the approximate solutions of quasilinear (and semilinear) problems can be constructed based on the known solutions of linear problems.

2.1 Non Self-Adjoint Problems

Consider non-self adjoint problems of the type

$$\epsilon y''(x) + p(x)y'(x) + q(x)y(x) = f(x), \; y(a) = A(\epsilon), \; y(b) = B(\epsilon) \; . \qquad (1)$$

Depending upon the sign of $p(x)$ on the interval, there is a boundary layer of rapid changes in the solution at either left end-point ($x = a$) if $p(x) > 0$ for $\forall x \in [a, b]$ or right end-point ($x = b$) if $p(x) < 0$ for $\forall x \in [a, b]$. For definiteness consider the function $p(x)$ to be positively defined. The asymptotic solution of this problem exists, is unique and consists of at least two separate parts [9]

$$y(x; \epsilon) = Y(x; \epsilon) + \zeta(\tau; \epsilon) \; , \qquad (2)$$

where the outer solution $Y(x; \epsilon)$ and the inner solution $\xi(\tau; \epsilon)$ have power series expansions

$$Y(x; \epsilon) = \sum_{j=0}^{\infty} Y_j(x)\epsilon^j, \; \xi(\tau; \epsilon) = \sum_{j=0}^{\infty} \xi_j(\tau)\delta_j(\epsilon) \qquad (3)$$

with an asymptotic sequence $\delta_j(\epsilon)$ and the stretched time scale

$$\tau = (x - x^*)/\epsilon^{\alpha} \; . \tag{4}$$

Here x^* is the singular point. The commonly used asymptotic sequence is $\delta_0 = 1$, $\delta_1(\epsilon) = \epsilon$, $\delta_2(\epsilon) = \epsilon^2 \ldots$ (the power series of ϵ) and $\tau = (x - x^*)/\epsilon$, i.e. $\alpha = 1$. Generally speaking, however, the scaling is sought based on either qualitative argument or using the Nipp polyhedron algorithm described elsewhere ([8], [5]). The asymptotic solution is obtained by substituting the expansion series (3) in the solution. Two types of situation have to be distinguished here: no turning points and one (or more) turning points on the interval $[a, b]$. A turning point is the point when the coefficient at the first derivative changes its sign, i.e. a zero of the function $p(x)$ or . Let us consider the two cases separately.

No Turning Points If $p(x) > 0$ does not have any zeros on the interval $[a, b]$, the outer problem is given by

$$\epsilon Y''(x) + p(x)Y'(x) + q(x)Y(x) = f(x) \; , Y(b) = B(\epsilon) \; . \tag{5}$$

Substituting the outer expansion (3) into equation (5) and equating coefficients term-wise yields the following outer perturbation equations. At the zero-th order of ϵ

$$p(x)Y_0'(x) + q(x)Y_0(x) = f(x), \quad Y(1) = B_0 \tag{6}$$

and at the higher-orders of ϵ, $\forall j > 0$

$$p(x)Y_j'(x) + q(x)Y_j(x) = -Y_j''(x), \quad Y(1) = B_j \; . \tag{7}$$

Here B_j are the terms in the series of expansion of $B(\epsilon)$ in powers of ϵ.

The inner equations are obtained by introducing the stretched variable τ (4) with $x^* = a$, $\alpha = 1$ and representing functions $p(\tau)$ and $q(\tau)$ by the Taylor series in the vicinity of the left end-point (for $p(x) > 0$)

$$\xi''(\tau) + \xi'(\tau) \sum_{m=0}^{n} \frac{1}{m!} \epsilon^m p^m(\tilde{a})(\tau - \tilde{a})^m + \xi(\tau)\epsilon \sum_{m=0} \frac{1}{m!} q^m(\tilde{a})(\tau - \tilde{a})^m = 0 \; . \tag{8}$$

Here $\tilde{a} = a/\epsilon^{\alpha}$. The zero-th order of ϵ (the reduced equation) is given by

$$\xi_0''(\tau) + p(\tilde{a})\xi_0'(\tau) = 0 \tag{9}$$

and at the higher orders of ϵ the solutions satisfy

$$\xi_j''(\tau) + p(\tilde{a})\xi_j'(\tau) = -\beta_{j-1}(\tau). \tag{10}$$

Here the terms β_{j-1} are known successively as they depend on the terms found on the previous steps.

There are two principal methods to determine the boundary conditions for the inner problem. One way is to use the fact that the sum of outer and inner solution constitutes the solution of the whole problem and therefore satisfies the boundary condition at the left

$$\xi_0(\tilde{a}) = A - Y_0(a) . \tag{11}$$

Employing the assumption that inner solutions are valid in thin layer only, condition that they decay to zero away from the layer should be imposed:

$$\xi_j(\infty) = 0 . \tag{12}$$

This method of imposing boundary condition is easier to use. It is not, however, very general and works only for the linear type of problems with no turning points. Another way of imposing conditions in the inner problem is to satisfy the boundary conditions together with the matching of the inner and outer solutions. Such conditions can be used in various types of problems but they require a subtle procedure of matching which usually needs the introduction of intermediate variables. The inner solutions should satisfy the end-point boundary condition

$$\xi_0(\tilde{a}) = A_0, \quad \xi_j(\tilde{a}) = A_j(\epsilon) . \tag{13}$$

In addition, the inner solutions should match with the outer solutions in the overlap region (the matching condition)

$$\xi_0(\infty) = Y_0(a) , \sum_{j=0}^{k} \xi_j(\infty)\delta_j(\epsilon) = \sum_{j=0}^{k} \epsilon^j Y_j(a) . \tag{14}$$

The usual procedure of matching is based on an assumption that the two local approximations have an overlap domain of validity where they are equal. To construct representation of the inner and outer solutions in the overlap domain, the intermediate variable η is introduced in the vicinity of the singular point x^*:

$$\eta = \frac{x - x^*}{\epsilon^\beta} , 0 < \beta < \alpha . \tag{15}$$

The interval for β comes from the requirement that the scaling for η lies between the outer and the inner scalings. Variable in the outer expansion is changed from x to η, and variable in the inner expansion is changed from τ to η. The terms of the respective order must be equal. This is how the unknown constants and β itself are found. It is important that the matching does not depend on the specific choice of η.

If $p(x) < 0$ on $[a, b]$, the outer solutions should satisfy the left end-point boundary condition, i.e. $Y(a) = A(\epsilon)$. The stretched variable, τ, in the inner solution has the form $\tau = (b - x)/\epsilon^\alpha$. The rest of the solution procedure is exactly the same as for $p(x) > 0$.

With One Turning Point If the coefficient at the first derivative changes its sign on the interval of interest, i.e. $p(x)$ has a zero at the interval $[a, b]$, the system (1) has much different behaviour from the one discussed above. In this case, there two outer solutions in the interval with a transition or shock layer connecting them at the turning point. Suppose $x^* \in [a, b]$ is a zero of function $p(x)$. Then, the two limiting outer solutions of the left and to the right of the turning point are given by

$$\begin{aligned} p(x)Y_{L0}'(x) + q(x)Y_{L0} &= f(x), \quad Y_{L0} = A, \quad a < x < x^* \\ p(x)Y_{R0}'(x) + q(x)Y_{R0} &= f(x), \quad Y_{R0} = B, \quad x^* < x < b \,. \end{aligned} \tag{16}$$

Provided $Y_{L0}(x^*) \neq Y_{R0}(x^*)$, a transition solution is needed. Its leading terms are given by the reduced equation

$$\xi_0''(\tau) + p(x^*)\xi_0'(\tau) = 0, \tau = (x - x^*)/\epsilon^\alpha \,. \tag{17}$$

In this case, no boundary conditions imposed on the inner solution. The unknown constants in it are found by the matching of the inner solution with both outer solutions. At the leading order we have:

$$\xi_0(-\infty) = Y_{L0}(x^*) = \lim_{x \to x^*} Y_{L0}(x), \quad \xi_0(\infty) = Y_{R0}(x^*) = \lim_{x \to x^*} Y_{R0}(x) \,.$$

2.2 With Many Turning Points

If the coefficient $p(x)$ has more than one zeros in the interval $[a, b]$, the solution has several turning points on the interval. In this case, the algorithm for finding an approximate solution is similar to the one with a single turning point. Suppose equation (1) has J turning points on the interval $[a, b]$: $x_1^*, x_2^* \dots x_J^*$. Then, the full approximate solution of the problem consists of $J + 1$ outer solutions $(Y^1, \dots Y^{J+1})$ and J inner solutions $(\xi^1, \dots \xi^J)$

$$y(x; \epsilon) = \sum_{j=1}^{J+1} Y_j + \sum_{j=1}^{J} \xi_j \,. \tag{18}$$

Expansion terms in the outer solution Y_j satisfy equations of the type (6, 7) while expansion terms in the inner solution ξ_j satisfies equations of the type (9, 10). The boundary condition for the first outer limiting solution Y_0^1 is given by $Y_0^1(a) = A$ and for the last one Y_0^{J+1} is given by $Y_0^1(B) = B$. Other conditions on outer and inner solutions are imposed from the matching:

$$\begin{aligned} \xi_0^1(-\infty) &= Y_0^1(x_1^*), \quad \xi_0^1(\infty) = Y_0^2(x_1^*), \\ \xi_0^2(-\infty) &= Y_0^2(x_2^*), \quad \dots \\ \xi_0^j(-\infty) &= Y_0^j(x_j^*), \quad \xi_0^{j+1}(\infty) = Y_0^{j+1}(x_{j+1}^*), \\ \dots & \qquad \xi_0^J(\infty) = Y_0^{J+1}(x_J^*) \,. \end{aligned} \tag{19}$$

This is a system of algebraic equations with respect to the unknown constants C_i and C_o which appear in the general solutions of the outer and inner perturbation equations.

2.3 Self-Adjoint Problems

Self-adjoint problems are presented by equations of the type

$$\epsilon y''(x) + q(x)y(x) = f(x), \quad y(a) = A, \; y(b) = B \; . \tag{20}$$

They have boundary layers at both ends of the interval. They are similar to non-self adjoint problems with a turning point at one of the end points. An algorithm for solving such equations is exactly the same as for non self-adjoint problems.

2.4 Quasilinear Problems

For quasilinear problems described by

$$\epsilon y''(x) + p(y,x)y'(x) + f(y,x) = 0, \; y(a) = A, \; y(b) = B \tag{21}$$

the approximate solutions are also sought as the sum of the outer solution and the inner layer correction (2). The limiting outer solution satisfies the reduced equation

$$p(Y_0, x)Y_0'(x) + f(Y_0, x) = 0, \quad Y_0(b) = B_0 \tag{22}$$

and the succeeding Y_j terms satisfy linear variational equations of the form

$$\begin{aligned} p(Y_0,x)Y_j'(x) + p_y(Y_0,x)Y_j(x)Y_0'(x) + f_x(Y_0,x)X_j &= h_{j-1}(x) \; , \\ Y_j(b) = B_j(\epsilon) \; , \end{aligned} \tag{23}$$

where terms in the right-hand side $h_{j-1}(x)$ are determined by the preceding approximations. If the solution $Y_0(x)$of the reduced problem (22) is found and $p(Y_0(x), x) \neq 0$ throughout the interval (a, b) the successive terms $Y_j(x)$ can be found by integrating linear differential equations (24). To satisfy the boundary layer stability assumption the following condition should hold true:

$$p(Y_0(x), x) > 0, \quad \forall x \in [a, b] \; . \tag{24}$$

The inner solution is sought in the usual way by introducing the stretched variable $\tau = (t - x^*)/\epsilon^\alpha$. The leading term of the inner solution satisfies the nonlinear initial value problem

$$\begin{aligned} &\xi_0''(\tau) \; + p(Y_0(a) + \xi_0, a)\xi_0'(\tau) = 0, \xi_0(a) = y(a) - Y_0(a) \; , \\ &\xi_0(\infty) = 0 \; . \end{aligned} \tag{25}$$

The higher order terms of ϵ of the inner equation satisfy the linear variational problems

$$\begin{aligned} &\xi_j''(\tau) \; + p(Y_0(a) + \xi_0, a) + p_y(Y_0(a) + \xi_0, a)\xi_j\xi_0' = k_{j-1}(\tau), \\ &\xi_j(a) = -Y_j(a), \quad \xi_j(\infty) = 0 \; . \end{aligned} \tag{26}$$

Note, that the reduced problem (22) is generally a nonlinear equation and its closed form solution cannot necessarily be found. In this case, the solver returns a list of (non-linear) perturbation equations which do not contain parameter ϵ. The treatment of such non-stiff equations by numerical methods is computationally easier than the numerical integration of the original stiff problem (21).

Turning Points If $p(Y_0, x)$ has zeros at the interval $[a, b]$, then the full equation (21) has turning points. The algorithm of solving quasilinear equation with turning point is similar to the one for solving non self-adjoint problems. Here the outer equations are given by (22, 24) and the inner equations are given by (26, 27). Analogous methods can be used when y is a vector and the matrix $p(y, x)$ is positive definite [3]. At present, it is not, however, included in the solver.

3 Package BVPSolver

Based on the theoretical results summarised above, author has developed a *Mathematica* package *BVPSolver*. The package constructs approximate symbolic solutions of the second-order linear boundary value problems. Linear non-self-adjoint and self-adjoint problems are solved, including those with one or several turning points. Quasilinear problems are also attempted to be solved. The package also includes routines to solve semilinear second order problems (see Example 4).

DSolve command is used to solve the reduced outer or inner equation. The solutions obtained from solving inner and outer perturbation equations contain unknown constants C_i which are to be found using the matching procedure (or in simple cases by employing the decaying assumption (11, 12)). It is performed using *Matching* function which has been developed as part of *BPSolver* package. This function introduces intermediate variables (15), substitutes them into the adjacent outer and inner expansions and attempts to find unknown constants by equating the two. The composite solution of the problem is constructed by adding all local approximate solutions and subtracting their common parts.

The main command of the solver *BVPSolve* has syntax of the built-in *Mathematica* function *DSolve* and enhances its functionality by allowing to find approximate solutions of the two-point boundary-value problems. Four examples are now worked out to illustrate several features of the solution procedure and the usage of the package.

3.1 Example 1

Consider a simple non-self adjoint problem without turning points

$$example1 = \{\epsilon * y''[x] \ + 2 * y'[x] \ + 2 * y[x] == 0, \ y[0] == 0, \ y[1] == 1\} \quad (27)$$

which can be solved exactly using *DSolve* command.

```
exactsol = y[x] /. DSolve[example1, y[x], x][[1]] // Simplify
```

$$\frac{e^{-\frac{(-1+x)\left(1+\sqrt{1-2\epsilon}\right)}{\epsilon}}\left(-1+e^{\frac{2x\sqrt{1-2\epsilon}}{\epsilon}}\right)}{-1+e^{\frac{2\sqrt{1-2\epsilon}}{\epsilon}}} \quad (28)$$

Function *BVPSolve* finds asymptotic solution of equation (27) up to the second-order of small parameter $\epsilon \ll 1$. The default order of approximation is 1.

```
apprsol = BVPSolve[example1, y, x, EpsOrder->2, PrintAll->True]
this is non self-adjoint problem
no turning points
Solving outer equation of order 0
```

$$\{2\,Y[0][x] + 2\,Y[0]'[x] == 0, Y[0][1] == 1\}$$

$$\{Y[0][x] \to (e^{1-x})\}$$

```
Solving inner equation of order 0
```

$$yy[0]'[s] + yy[0]''[s] == 0, Y[0][0] + yy[0][0] == 0$$

$$\{yy[0][s] \to (-\,e^{1-2\,s})\}$$

```
Solving outer equation of order 1
```

$$\{e^{1-x} + 2\,Y[1][x] + 2\,Y[1]'[x] == 0, Y[1][1] == 0\}$$

$$\{Y[1][x] \to \left(-\frac{1}{2}\,e^{-x}\,(-e + e^{x})\right)\}$$

```
Solving inner equation of order 1
```

$$yy[0][s] + 2yy[1]'[s] + yy[1]''[s] == 0, Y[1][0] + yy[1][0] == 0$$

$$\{yy[1][s] \to \left(-\frac{1}{2}\,e^{1-2\,s}\,(1 + 2\,s)\right)\}$$

```
Solving outer equation of order 2
```

$$\{e^{1-x} - \frac{1}{2}\,e^{-x}\,(-e + e^{x}) + 2\,Y[2][x] + 2\,Y[2]'[x] == 0, Y[2][1] == 0\}$$

$$\{Y[2][x] \to \left(\frac{1}{8}\,e^{-x}\,(5\,e - 6\,e\,x + e\,x^2)\right)\}$$

```
Solving inner equation of order 2
```

$$\{2yy[1][s] + 2\,yy[2]'[s] + yy[2]''[s] == 0, Y[2][0] + yy[2][0] == 0\}$$

$$\{yy[2][s] \to \left(-\frac{1}{8}\,e^{1-2\,s}\,(5 + 8\,s + 4\,s^2)\right)\}$$

$$e^{1-x} - e^{1-\frac{2x}{\epsilon}} - \frac{1}{2}\,e^{-x}\,(-e + e\,x)\,\epsilon - \frac{1}{2}\,e^{1-\frac{2x}{\epsilon}}\left(1 + \frac{2\,x}{\epsilon}\right)\epsilon + \frac{1}{8}\,e^{-x}\,(5\,e - 6\,e\,x + e\,x^2)\,\epsilon^2 - \frac{1}{8}\,e^{1-\frac{2x}{\epsilon}}\left(5 + \frac{4\,x^2}{\epsilon^2} + \frac{8\,x}{\epsilon}\right)\epsilon^2 \quad (29)$$

Figure 1 shows comparison between exact (28) and approximate (29) solutions of equation (27). It is seen that for this simple problem the approximate solution is indistinguishable from the exact one already for $\epsilon = 0.2$.

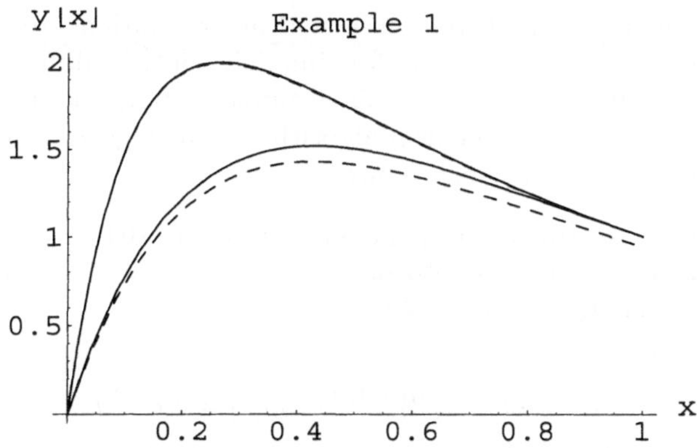

Fig. 1. Comparison between exact (28) and approximate solutions (29) of equation (29) for two values of ϵ. Lower curve corresponds to $\epsilon = 0.4$ (solid line is for exact solution and dashing line is for approximate one); upper curve corresponds to $\epsilon = 0.2$ (two solutions are indistinguishable)

3.2 Example 2

Here is another example of non-self adjoint problem ($\alpha > 0$) ([4]):

$$example2 = \{\epsilon y''(x) + (1+\alpha x)y'(x) + y(x) == 0, \;\; y(0) = 0, \;\; y(1) = 1\} \,. \quad (30)$$

The approximate solution (with up to the first order of ϵ) is given by

```
apprsol = BVPSolver[example2,  y[x], x]
this is non self-adjoint problem
no turning points
```

$$-e^{-\frac{x}{\epsilon}}\,(1+\alpha) + \frac{1+\alpha}{1+x\,\alpha} + \frac{\left(\alpha+\alpha^2 - \frac{\alpha}{1+\alpha} - \frac{2\,x\,\alpha^2}{1+\alpha} - \frac{x^2\,\alpha^3}{1+\alpha}\right)\epsilon}{(1+x\,\alpha)^3} + \frac{e^{-\frac{x}{\epsilon}}\,\alpha\left(-2\,\alpha\,(2+\alpha) + \frac{x^2\,(1+\alpha)^2}{\epsilon^2}\right)\epsilon}{2\,(1+\alpha)}$$

3.3 Example 3

Consider now an example of the linear boundary value problem with a turning point. The following equation has a turning point at $x = 1/2$ on the interval $[0,1]$ ([2]):

$$example3 = \{\epsilon y''(x) + (x - 1/2)y'(x) - y(x) = 0, \;\; y(0) = 2, \;\; y(0) = 3\} \quad (31)$$

has a turning point at $x = 1/2$ on the interval $[0,1]$ (corner layer). The outer solutions of the left and right of the turning point are named Yl and Yr respectively. It is automatically done in the *BVPSolve* function. In the case of J

singular points in the interval, the outer and inner solutions are named Y_1, Y_2, ..., Y_{J+1} and yy_1, yy_2, ..., yy_J. Here for simplicity only leading terms of the approximation are found. Note that in this example scalings for the inner problem are found using the Nipp polyhedron algorithm. This algorithm is implemented using the *LinearProgramming* procedure.

```
apprsol = BVPSolver[example3, y, x, EpsOrder->0]
this is non self-adjoint problem
there is one turning point: 1/2
inner scalings:
```

$$s == (x - 1/2)/\epsilon^{1/2}, \quad yy[0] == \epsilon^{1/2} y$$

$$\{yy[0][s] \to \left(\frac{1}{2}e^{-\frac{s^2}{2}}\left(-2\,CI[0,2]+2e^{\frac{s^2}{2}}CI[0,1]s-e^{\frac{s^2}{2}}\sqrt{2\pi}CI[0,2]\,Erf\,[\frac{s}{\sqrt{2}}]\,s\right)\right)\},$$

$$\{Yl[0][x] \to (-2\,(-1+2\,x)),\quad Yr[0][x] \to (3\,(-1+2\,x))\} \tag{32}$$

```
Matching to the zero-th order...
```

$$\{CI[0,1] \to 1, CI[0,2] \to -5\,\sqrt{\frac{2}{\pi}}\}$$

```
Common part to the left
```

$$-4\,\sqrt{\epsilon}\eta$$

```
Common part to the right
```

$$6\,\sqrt{\epsilon}\eta$$

$$\{yy[0][s] \to 5\,e^{-\frac{s^2}{2}}\,\sqrt{\frac{2}{\pi}} + s + 5\,s\,Erf\,[\frac{s}{\sqrt{2}}]\} \tag{33}$$

$$-\frac{1}{2} + x + 5\,e^{-\frac{\left(-\frac{1}{2}+x\right)^2}{2\,\epsilon}}\,\sqrt{\frac{2}{\pi}}\,\sqrt{\epsilon} + \left(-\frac{5}{2} + 5\,x\right)\,Erf\,[\frac{-\frac{1}{2}+x}{\sqrt{2}\,\sqrt{\epsilon}}] \tag{34}$$

Figure 2 shows an approximate solution (34) of the corner layer problem (31). The transition solution provides a smooth transition between the two linear outer solutions.

3.4 Example 4

3.5 Semilinear Problems

This example demonstrates construction of the approximate solution of the semilinear problem using symbolic-numerical approach. An algorithm employs the Newton's quasilinearisation technique which is routinely used in numerical schemes ([1]) and easy-to-find approximate solutions of the self-adjoint problems ([7]). Consider the semilinear boundary value problem:

$$\epsilon y''(x) = F(y, x), \quad y(a) = A, \; y(b) = B \; . \tag{35}$$

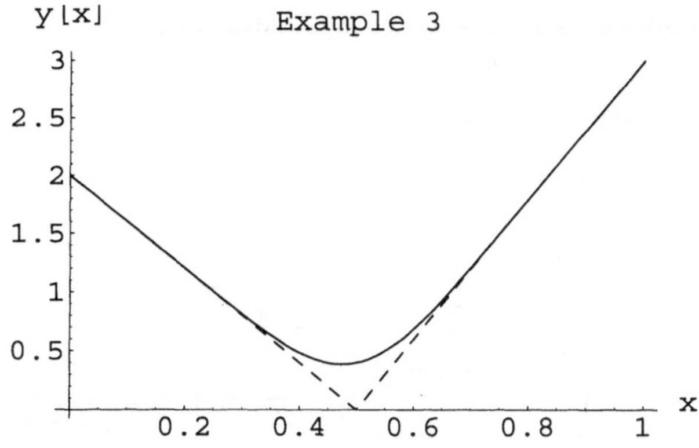

Fig. 2. The approximate solution (34) (solid line) and outer solutions (32) Yl for $0 \le x \le 1/2$ and Yr for $1/2 \le x \le 1$ (dashed line) for the corner layer example (31). In these calculations $\epsilon = 0.01$

Suppose $F(y,x)$ is a smooth function with a positively defined derivative

$$F_y(y,x) \ge b > 0, \ \forall x \in [a,b] \ . \tag{36}$$

Construct the sequence of functions $\{y_m\}_0^\infty$ such that the initial guess $y_0(x) = Y_0(x)$ is the smooth solution of the reduced algebraic problem: $F(Y_0, x) = 0$. For every already computed function y_m, the function y_{m+1} is the solution of the following self-adjoint problem

$$\epsilon y''_{m+1} + b_m(x) y_{m+1} = F_m(x) \ , \tag{37}$$

where

$$b_m(x) = F_y(x, y_m), \ \ F_m(x) = F_y(x, y_m) - F(x, y^m) \tag{38}$$

Following the method of proof for the Newton's quasilinearisation method (see [1]), it can be shown that the sequence $\{y_m\}_0^\infty$ converges to the solution of the original problem $y(x)$. In addition, from the condition (36) it follows that

$$b_m(x) \ge F_y(x, y_m) \ge b > 0, \ \ \forall x \in [a,b] \ . \tag{39}$$

Therefore, equations (37) for y_m are self-adjoint problems whose approximate solutions are readily found. The number of iterations required to obtain accurate solutions is essentially independent from small parameter ϵ. For semilinear problems (35) with negatively defined derivative $F(x,y)$ the scheme for finding the solution is essentially the same.

Consider the following semilinear problem

$$\mathit{example4} = \{\epsilon^2 y''(x) = -y^2(x) + x + 1, \ \ y(0) = -2, \ \ y(1) = -2\} \tag{40}$$

BVPSolve constructs the sequence of approximations to the problem $\{y_m\}_0^\infty$. Here only leading order terms of self-adjoint equations (37) are taken into account. By default, three approximations for constructing $\{y_m\}_0^\infty$ are constructed (option *SemiLinSteps*).

```
apprsol = BVPSolver[example4, x, EpsOrder->0]
this is semilinear problem
```

The first approximation has the form

```
Subscript[y, 1] /. apprsol
```

$$e^{-\frac{2^{3/4}\,(1-x)}{\epsilon}}\left(-2+\sqrt{2}+\frac{e^{-\frac{\sqrt{2}}{\epsilon}}}{1-e^{\frac{-\sqrt{2}-2^{3/4}}{\epsilon}}}\right)+$$
$$e^{-\frac{\sqrt{2}\,x}{\epsilon}}\left(-1-\frac{(-2+\sqrt{2})\,e^{-\frac{2^{3/4}}{\epsilon}}}{1-e^{\frac{-\sqrt{2}-2^{3/4}}{\epsilon}}}\right)+\frac{-2-2\,x}{2\,\sqrt{1+x}}$$

Three approximation to the semilinear problem (40) are shown on Figure 3. It is seen that the second and third approximation are very close. The lengthy formulas for the second and third approximations are not given here.

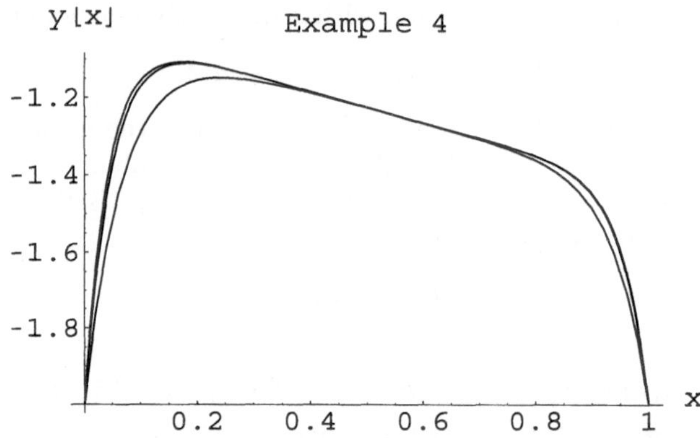

Fig. 3. Three approximations to semilinear problem (40) calculated by the algorithm described above. The lower curve corresponds to the first approximation, the two close curves correspond to the second and the third approximation respectively. In these calculations $\epsilon = 0.1$

4 Summary and Discussion

This paper presents a *Mathematica* package *BVPSolver* for construction of approximate solutions of linear two-point boundary value problems including those with turning points. Appropriate scalings for local approximations are found using the polyhedron algorithm. For quasilinear problems the solver generates perturbation equations and attempts to find their symbolic solutions. Approximate solutions of semilinear problems are constructed using symbolic-numerical approach.

The capabilities of the *BVPSolver* package for dealing with the second-order singularly-perturbed boundary value problems are mainly limited by the capabilities of the *DSolve* command to find closed-form solution of second or first order differential equations. The main difficulty arises while attempting to solve non-linear reduced equation in the procedure of solving quasilinear problems. For instance, *DSolve* does not always compute all solutions for non-linear problems. Other problems are encountered while attempting to solve linear equations.

The matching procedure sometimes gets tricky and there are plenty of counterexamples wherein the straightforward matching does not work ([6]). Occasionally, it happens that an expansion produces a term of an order that the other does not have. The process of having to insert scales into an expansion because of what is happening in another layer is called *switchbacking*. Author is looking at the ways of automating steps in the switchbacking procedure. There are refined matching procedures for special cases wherein, for instance, the location of the turning point is not known. Inclusion of such refined matching methods into the solver is in the pipeline. At present, however, if the matching is not successful, the solver returns solutions which include unknown constants.

Some other restrictions of the solver are associated with the standard *Limit* function which often cannot find limits to some well-known expressions. Several rules for finding limits in the cases commonly encountered in the solutions of linear differential equations (exponents, for example) have been included in *BVPSolver.*

Author works on an extension of *BVPSolver* package to apply it to systems of linear two-point boundary value problems. An algorithm for constructing approximate solutions is somewhat similar to the scalar case. Author expects to make the version of the *BVPSolver* package available to *MathSource* users by the time of the conference (September, 2001).

References

1. E.P.Doolan, Miller, J., and Schilders, W. *Uniform numerical methods for problems with initial and boundary layers.* Boole Press.
2. Holmes, M. *Introduction to perturbation methods.* Springer, New York, 1995.
3. Howes, F., and O'Malley, R. Singular perturbations of semilinear second order systems. *Lecture Notes in Mathematics 827* (1980), 130–150.
4. Kevorkian, J., and Cole, J. *Perturbation Methods in Applied Mathematics.* Springer-Verlag, New York, 1996.
5. Khanin, R. On Nipp polyhedron algorithm for solving singular perturbation problems. *Mathematics and Computers in Simulation.* in press
6. Lagerstrom, P., and Casten, R. Basic concepts underlying singular perturbation techniques. *SIAM Rev. 14*, 1 (1972), 63–120.
7. Natesan, S., and Ramanujam, N. Improvement of numerical solution of self-adjoint perturbation problems by incorporation of asymptotic approximations. *Appl. Math. and Comp. 98* (1999), 119–137.
8. Nipp, K. An algorithmic approach for solving singularly perturbed initial value problems. *Dynamics Reported 1.* Eds:Kirchgraber, U. and Walther H.O., John Wiley & Sons, Chichester, 1988.

9. O'Malley, R. *Singular perturbation methods for ordinary differential equations.* Springer-Verlag, New York, 1991.
10. Rand, R. *Perturbation methods, bifurcation theory and computer algebra.* Springer-Verlag, New York, 1987.

A New Algorithm for Computing Cohomologies of Lie Superalgebras

Vladimir V. Kornyak

Laboratory of Information Technologies
Joint Institute for Nuclear Research
141980 Dubna, Russia
kornyak@jinr.ru

Abstract. In this paper two versions of a new algorithm for computing cohomologies of Lie (super)algebras are described. The algorithm is based on splitting the full cochain complex into smaller subcomplexes. This approach makes the computation essentially more efficient since in most cases the dimensions of full cochain spaces included in the complexes are very high and these high-dimensional spaces can be divided into much smaller subspaces. Examples illustrating the work of the algorithm are given.

1 Introduction

In papers [1–4] we presented an algorithm for computation of Lie (super)algebra cohomologies. These papers contain also a description of its C implementation and some results obtained with the help of codes designed. This algorithm computes cohomology of Lie (super)algebra L over module X in a straightforward way, i. e., for cochain complex (see definitions, e. g., in [5])

$$0 \to C^0 \xrightarrow{d^0} \cdots \xrightarrow{d^{k-2}} C^{k-1} \xrightarrow{d^{k-1}} C^k \xrightarrow{d^k} C^{k+1} \xrightarrow{d^{k+1}} \cdots, \tag{1}$$

the algorithm constructs the full set of basis monomials forming the space C^k, generates subsequently all basis monomials in the space C^{k+1}, computes the differentials corresponding to these monomials to obtain the set of linear equations determining the space of cocycles

$$Z^k = \mathrm{Ker}\ d^k = \{C^k \mid dC^k = 0\}, \tag{2}$$

computes the space of coboundaries

$$B^k = \mathrm{Im}\ d^{k-1} = \{C^k \mid C^k = dC^{k-1}\}. \tag{3}$$

Finally, the algorithm constructs the basis elements of quotient space

$$H^k(L;X) = Z^k/B^k. \tag{4}$$

This last step is based on the Gauss elimination procedure.

The main difficulty in computing cohomology comes from very high dimensions of the spaces C^k: for n-dimensional ordinary Lie algebra and p-dimensional module $\dim C^k = p\binom{n}{k}$, and for $(n|m)$-dimensional Lie superalgebra $\dim C^k = p\binom{n}{k} + p\sum_{i=1}^{k}\binom{n}{k-i}\binom{m+i-1}{i}$.

In many cases it is possible to extract some easier to handle subcomplexes of the full cochain complex (1). The partition of cochain complex for a graded algebra and module into homogeneous components is a typical example. In many papers (see, e. g., [6–8]) more special subcomplexes were used successfully to obtain new results in the theory of cohomology of Lie (super)algebras.

The main idea of our new algorithms is to extract the minimal possible (in the sense explained below) subcomplexes from complex (1) and to carry out computations within these subcomplexes.

2 Sketch of Algorithms

In this section we describe shortly two algorithms: **ComputeCohomology** and **SearchCohomology.** The algorithm **ComputeCohomology** is applicable to the cases when it is possible to handle the full space of cochains C^k, i. e., $\dim C^k$ should be moderate enough to keep the basis of C^k in the memory of a computer. This algorithm allows to obtain the full set of basis elements of cohomology for a given cochain complex.

On the other hand, the algorithm **SearchCohomology** allows to obtain partial information about cohomology in high-dimensional complexes. In some cases this algorithm can be applied even to infinite-dimensional complexes.

The input data for both algorithms should include:

1. The Lie (super)algebra L over the module X. L and X should be presented as sets of basis elements, their commutator and action tables. Our C implementation of the algorithms is able to construct these basis elements and tables for most important algebras and modules.
2. Integer non-negative number k, which is the degree (or dimension) of cohomology.
3. Integer number g giving the grade of cohomology. Most important algebras and modules are graded (if not, one can always prescribe zero grade to all elements of algebra and module) and this grading induces grading in the cohomology.

For the algorithm **SearchCohomology** a positive number n should additionally be input. It restricts the number of attempts to find non-trivial cohomological classes.

The output for both algorithms is a set BH_g^k of basis elements of cohomology. This set is full for the algorithm **ComputeCohomology** and partial for the algorithm **SearchCohomology**.

Let $\{e_i\}$ and $\{a_\alpha\}$ be the sets of basis elements of Lie (super)algebra L and module X, correspondingly. The following super skew-symmetric monomials

$$C(e_{i_1},\ldots,e_{i_k};a_\alpha) \equiv C(e_{i_1})\wedge\cdots\wedge C(e_{i_k})\otimes a_\alpha \equiv e'_{i_1}\wedge\cdots\wedge e'_{i_k}\otimes a_\alpha$$

form the basis of the cochain space C^k. Here $i_1 \leq \cdots \leq i_k$ and e'_i is the dual to e_i element. We use the notation m_g^k for such monomials in both algorithms.

Algorithm: ComputeCohomology

Input: L, Lie (super) algebra; X, module; k, cohomology degree; g, grade

Output: BH_g^k, set of basis cohomological classes

Local: M_g^k, full set of k-cochain monomials (basis of C_g^k);
s, current subcomplex: $C_{g,s}^{k-1} \xrightarrow{d_{g,s}^{k-1}} C_{g,s}^k \xrightarrow{d_{g,s}^k} C_{g,s}^{k+1}$;
$m_g^k \in M_g^k$, starting monomial for constructing subcomplex s;
$M_{g,s}^k$, set of k-cochain monomials involved in subcomplex s;
$BH_{g,s}^k$, set of basis cohomological classes in subcomplex s

```
BH_g^k := ∅
M_g^k := GenerateMonomials(L, X, k, g)
while M_g^k ≠ ∅ do
   m_g^k := ChooseMonomial(M_g^k)
   {s, M_{g,s}^k} := ConstructSubcomplex(m_g^k)
   M_g^k := M_g^k \ M_{g,s}^k
   BH_{g,s}^k := ComputeCohomologyInSubcomplex(s)
   if BH_{g,s}^k ≠ ∅ then
      BH_g^k := BH_g^k ∪ BH_{g,s}^k
   fi
od
return BH_g^k
```

Both algorithms construct the following local (working) objects:

1. The subcomplex s constructed by the subalgorithm **ConstructSubcomplex.**
2. The k-cochain monomial m_g^k. This monomial is input for the subalgorithm **ConstructSubcomplex.** The algorithm **ComputeCohomology** takes the monomials m_g^k from the set M_g^k by the subalgorithm **ChooseMonomial,** whereas the algorithm **SearchCohomology** generates m_g^k by the subalgorithm **NewMonomial**. This last subalgorithm generates random monomial m_g^k and checks its absence in the set of already used monomials M_g^k.
3. The set $M_{g,s}^k$ of monomials involved in the current subcomplex s. This set is subtracted from the set M_g^k in the algorithm **ComputeCohomology** (and is one of constituent parts of s in the algorithm **SearchCohomology**).
4. The set $BH_{g,s}^k$ of basis elements of cohomology in the subcomplex s.

For the work of algorithm **ComputeCohomology** the full set M_g^k of basis monomials for the space of k-cochains is generated by the subalgorithm **GenerateMonomials** at the start of computation.

Algorithm: SearchCohomology

Input: L, Lie (super) algebra; X, module; k, cohomology degree; g, grade; $n > 0$ – number of generations of subcomplexes
Output: BH_g^k, set of basis cohomological classes
Local: s, current subcomplex: $C_{g,s}^{k-1} \xrightarrow{d_{g,s}^{k-1}} C_{g,s}^{k} \xrightarrow{d_{g,s}^{k}} C_{g,s}^{k+1}$;
m_g^k, starting monomial for constructing subcomplex s;
$M_{g,s}^k$, set of k-cochain monomials involved in subcomplex s;
$BH_{g,s}^k$, set of basis cohomological classes in subcomplex s

$BH_g^k := \emptyset$
do
  $m_g^k := NewMonomial(M_g^k)$
  $\{s,\, M_{g,s}^k\} := ConstructSubcomplex(m_g^k)$
  $M_g^k := M_g^k \cup M_{g,s}^k$
  $BH_{g,s}^k := ComputeCohomologyInSubcomplex(s)$
  if $BH_{g,s}^k \neq \emptyset$ **and** $BH_{g,s}^k \not\subseteq BH_g^k$ **then**
    $BH_g^k := BH_g^k \cup BH_{g,s}^k$
  fi
  $n := n - 1$
od while $n \neq 0$
return BH_g^k

The subalgorithms **ConstructSubcomplex** and **ComputeCohomologyInSubcomplex** are central parts of both algorithms. The subalgorithm **ComputeCohomologyInSubcomplex** computes the set of basis elements of cohomology in a given subcomplex in a standard way as is described in Introduction.

The most important part of our approach is the subalgorithm **ConstructSubcomplex.** Starting with arbitrary monomial m_g^k this subalgorithm constructs the minimal subcomplex s of the form

$$C_{g,s}^{k-1} \xrightarrow{d_{g,s}^{k-1}} C_{g,s}^{k} \xrightarrow{d_{g,s}^{k}} C_{g,s}^{k+1} \tag{5}$$

involving the monomial m_g^k. The subalgorithm **ConstructSubcomplex** is based on the formula for differential d^k for Lie (super)algebras (see, e. g., [3]). This formula is a set of relations connecting the k- and $(k+1)$-monomials. The relations containing the monomial m_g^k contain also other monomials and we can add into consideration the relations for these monomials. The process terminates if any given basis element e_i of algebra appears only in a finite number of the right hand sides of commutators and similar property (with respect to action of algebra on module) holds for basis elements of module a_α. Note that the subalgorithm **ConstructSubcomplex** can construct finite-dimensional subcomplexes in some cases when grading does not allow finite-dimensional subcomplexes, e. g., when computing cohomology in adjoint module for infinite-dimensional graded Lie (super)algebras. Note also that the subcomplexes constructed by the subalgorithm **ConstructSubcomplex** are not invariant, i. e., they depend on the choice of bases for algebra and module, and they not always can be extended to full sequence (1) in such a way that part (5) remains unchanged, but for our purposes this is not important.

3 Examples: Computation of $\mathbf{H^3(Poincare(1,3))}$ and $\mathbf{H^3(SH(0|4))}$

Both examples of computations in this section are carried out in zero grade complexes and we shall omit subscripts $g = 0$ everywhere below. As the first example we consider the computation of third cohomology in the trivial module for 10-dimensional Poincaré algebra. The basis of this algebra contains the following generators of: translations P_0, P_1, P_2, P_3; rotations R_1, R_2, R_3; Lorentz transformations L_1, L_2, L_3. These basis elements satisfy the relations:

$$[P_\mu, P_\nu] = 0, \qquad \mu, \nu = 0, \dots, 3;$$

$$[P_0, L_i] = P_i, \quad [P_i, L_j] = \delta_{ij} P_0, \quad [P_i, R_j] = \epsilon_{ijk} P_k,$$

$$[R_i, R_j] = \epsilon_{ijk} R_k, \quad [R_i, L_j] = \epsilon_{ijk} L_k, \quad [L_i, L_j] = -\epsilon_{ijk} R_k, \qquad i, j, k = 1, \dots, 3.$$

Here ϵ_{ijk} is the permutation symbol with $\epsilon_{123} = 1$.

We consider the part

$$C^2 \xrightarrow{d^2} C^3 \xrightarrow{d^3} C^4 \tag{6}$$

of cochain complex (1). The spaces of cochains have the following dimensions:

$$\dim C^2 = 45, \qquad \dim C^3 = 120, \qquad \dim C^4 = 210.$$

The system of equations for 3-cocycles $d^3(C^3) = 0$ contains initially 187 equations for 120 variables. The Gauss elimination process shows that the rank of this system is equal to 83, i. e., the 3-cocycles form 37-dimensional subspace in 120-dimensional space C^3. The expressions $d^2(C^2)$ determine parametrically 35-dimensional subspace of coboundaries in the subspace of 3-cocycles, i. e., we get finally the cohomology as a 2-dimensional quotient space of these subspaces.

The algorithm **ComputeCohomology** divides complex (6) into 28 subcomplexes as shown in Table 1. As can be seen from Table 1, the nontrivial cohomological classes arise only in subcomplexes 17 and 18. Dimensions of subspaces C^3_{17} and C^3_{18} are equal to 4. Subcomplexes 17 and 18 are generated starting with the monomials $C(L_1, L_2, L_3)$ and $C(L_1, L_2, R_3)$ correspondingly.
The set of equations for cocycles $d^3(C^3_{17}) = 0$ is empty and expressions for coboundaries $d^2(C^2_{17})$ take the form

$$\begin{aligned}
C(L_1, L_2, L_3) &= -C(L_1, R_1) - C(L_2, R_2) - C(L_3, R_3), \\
C(L_1, R_2, R_3) &= C(L_1, R_1) - C(L_2, R_2) - C(L_3, R_3), \\
C(L_2, R_1, R_3) &= C(L_1, R_1) - C(L_2, R_2) + C(L_3, R_3), \\
C(L_3, R_1, R_2) &= -C(L_1, R_1) - C(L_2, R_2) + C(L_3, R_3),
\end{aligned}$$

Table 1. Subcomplex structure for complex (6)

s	Starting monomial	$\dim C^3_s$	$\dim Z^3_s$	$\dim B^3_s$	$\dim H^3_s$	BH^3_s
1	$C(P_0, P_1, P_2)$	1	0	0	0	
2	$C(P_0, P_1, P_3)$	1	0	0	0	
3	$C(P_0, P_2, P_3)$	1	0	0	0	
4	$C(P_1, P_2, P_3)$	1	0	0	0	
5	$C(L_1, L_2, R_1)$	2	1	1	0	
6	$C(L_1, L_2, R_2)$	2	1	1	0	
7	$C(L_1, L_3, R_1)$	2	1	1	0	
8	$C(L_1, R_1, R_2)$	2	1	1	0	
9	$C(L_1, R_1, R_3)$	2	1	1	0	
10	$C(L_2, R_1, R_2)$	2	1	1	0	
11	$C(P_0, P_1, L_2)$	4	1	1	0	
12	$C(P_0, P_1, L_3)$	4	1	1	0	
13	$C(P_0, P_1, R_2)$	4	1	1	0	
14	$C(P_0, P_1, R_3)$	4	1	1	0	
15	$C(P_0, P_2, L_3)$	4	1	1	0	
16	$C(P_0, P_2, R_3)$	4	1	1	0	
17	$C(L_1, L_2, L_3)$	4	4	3	1	h^3_{17}
18	$C(L_1, L_2, R_3)$	4	1	0	1	h^3_{18}
19	$C(P_0, L_1, R_2)$	6	2	2	0	
20	$C(P_0, L_1, R_3)$	6	2	2	0	
21	$C(P_0, L_2, R_3)$	6	2	2	0	
22	$C(P_1, L_2, R_3)$	6	2	2	0	
23	$C(P_0, P_1, L_1)$	6	0	0	0	
24	$C(P_0, P_1, R_1)$	6	0	0	0	
25	$C(P_0, L_1, L_2)$	9	3	3	0	
26	$C(P_0, L_1, L_3)$	9	3	3	0	
27	$C(P_0, L_1, R_1)$	9	3	3	0	
28	$C(P_0, L_2, L_3)$	9	3	3	0	

where 2-cochains should be treated as arbitrary parameters. These relations determine 3-dimensional subspace and basis element h^3_{17} of cohomology for subcomplex 17 can be expressed in the form

$$h^3_{17} = C(L_1, L_2, L_3) - C(L_1, R_2, R_3) + C(L_2, R_1, R_3) - C(L_3, R_1, R_2).$$

For subcomplex 18 equations $d^3(C^3_{18}) = 0$ determining cocycles take the form

$$\begin{aligned} C(L_1, L_2, R_1, R_2) &= -C(L_1, L_2, R_3) - C(L_1, L_3, R_2) + C(L_2, L_3, R_1) \\ &\quad + C(R_1, R_2, R_3) = 0, \\ C(L_1, L_3, R_1, R_3) &= C(L_1, L_2, R_3) + C(L_1, L_3, R_2) + C(L_2, L_3, R_1) \\ &\quad + C(R_1, R_2, R_3) = 0, \end{aligned}$$

$$C(L_2,L_3,R_2,R_3) = C(L_1,L_2,R_3) - C(L_1,L_3,R_2) - C(L_2,L_3,R_1) \\ +C(R_1,R_2,R_3) = 0.$$

This system of equations determines 1-dimensional subspace. Expressions for coboundaries $d^2(C^2_{18})$ determine 0-dimensional subspace taking the form

$$C(L_1,L_2,R_3) = 0,$$
$$C(L_1,L_3,R_2) = 0,$$
$$C(L_2,L_3,R_1) = 0,$$
$$C(R_1,R_2,R_3) = 0,$$

and basis element h^3_{18} of cohomology can be chosen as

$$h^3_{18} = C(R_1,R_2,R_3)$$

or as linear combination

$$h^3_{18} = \alpha C(R_1,R_2,R_3) + \beta C(L_1,L_2,R_3) + \gamma C(L_1,L_3,R_2) + \delta C(L_2,L_3,R_1)$$

with $\alpha - \beta + \gamma - \delta \neq 0$.

Let us consider also an example of computation that is typical of Lie superalgebras. Computation of the third cohomology in the trivial module and in zero grading for special Hamiltonian algebra SH(0|4) is a task of approximately the same complexity as in the above example. This (6|8)-dimensional superalgebra has the following even $E_1 = \theta_1\theta_2$, $E_2 = \theta_1\theta_3$, $E_3 = \theta_2\theta_3$, $E_4 = \theta_1\theta_4$, $E_5 = \theta_2\theta_4$, $E_6 = \theta_3\theta_4$ and odd $O_1 = \theta_1$, $O_2 = \theta_2$, $O_3 = \theta_3$, $O_4 = \theta_4$, $O_5 = \theta_1\theta_2\theta_3$, $O_6 = \theta_1\theta_2\theta_4$, $O_7 = \theta_1\theta_3\theta_4$, $O_8 = \theta_2\theta_3\theta_4$ basis elements. Here θ_1, θ_2, θ_3, θ_4 are the Grassmann variables and Lie product coincides with the Poisson bracket $\{\cdot,\cdot\}_{Pb}$. For this particular case

$$\{f,g\}_{Pb} = -(-1)^{p(f)}\Sigma^4_{i=1}\partial f/\partial\theta_i\partial g/\partial\theta_i,$$

where f and g are functions of $\theta_1, \ldots \theta_4$ and $p(f)$ is parity of function f. In the case of Lie superalgebra the cochain complex can be split into two subcomplexes, even and odd. In our example only even subspace of C^3 contains nontrivial cocycle. The spaces of even subcomplex have the following dimensions:

$$\dim C^2 = 31, \qquad \dim C^3 = 116, \qquad \dim C^4 = 355.$$

The initial number of determining equations for 3-cocycles is equal to 351. This system of equations determines a 25-dimensional space of cocycles. The expressions for coboundaries describe a 24-dimensional subspace, hence the cohomology is 1-dimensional.

The algorithm **ComputeCohomology** generates eight subcomplexes. Six of them have the following characteristics:

$$\dim C^3_s = 14, \qquad \dim Z^3_s = \dim B^3_s = 3, \qquad \dim H^3_s = 0.$$

There are also two subcomplexes with $\dim C_s^3 = 16$, one of them has $\dim Z_s^3 = \dim B_s^3 = \dim H_s^3 = 0$. And finally, the remaining subcomplex (with 30 equations for 3-cocycles) contains 1-dimensional cohomology: $\dim Z_s^3 = 7, \dim B_s^3 = 6, \dim H_s^3 = 1$. We do not present here the basis element of this cohomology explicitly because its expression is rather long.

4 Conclusion

As one can see from the two above, chosen at random, examples, cochain complexes in many cases can effectively be divided into smaller subcomplexes. One can show that generally the efficiency of such division grows with increase in cochain degree k, i. e., the dimensions of subcomplexes grow slower than the dimension of full complex. We are implementing the algorithms **Compute-Cohomology** and **SearchCohomology** as different regimes of one C program. The elaboration of the program is now close to be finished and we hope to make some progress in computation of Lie superalgebra cohomologies with the help of this program.

Acknowledgements

This work was partially supported by the grants RFBR 01-01-00708 and INTAS 99-1222.

References

1. Kornyak, V.V.: A program for Computing Cohomologies of Lie Superalgebras of Vector Fields. *Zapiski nauchnyh seminarov POMI. St.Petersburg.* **258** (1999) 148–160
2. Kornyak, V.V.: Cohomology of Lie Superalgebras of Hamiltonian Vector Fields: Computer Analysis. In: *Computer Algebra in Scientific Computing / CASC'99*, V.G. Ganzha, E.W. Mayr and E.V. Vorozhtsov (Eds.), Springer-Verlag, Berlin, (1999) 241–249; math.SC/9906046
3. Kornyak, V.V.: Computation of Cohomology of Lie Superalgebras of Vector Fields. *Int. J. of Mod. Phys. C.* **11** (2000) 397–414; arXiv: math.SC/0002210
4. Kornyak, V.V.: Computation of Cohomology of Lie Superalgebras: Algorithm and Implementation. *Russian Journal for Computer Science ("Programmirovanie")* **3** (2001) 46–50 (in Russian)
5. Fuks, D.B.: *Cohomology of Infinite Dimensional Lie Algebras.* Consultants Bureau, New York (1987)
6. Gel'fand, I.M., Kalinin, D.I., Fuks, D.B.: On Cohomology of Lie Algebra of Hamiltonian Formal Vector Fields. *Funkts. Anal. Prilozhen.* **6** (1972) 25–29 (in Russian)
7. Perchik, J.: Cohomology of Hamiltonian and related formal vector fields Lie algebras. *Topology* **15**, 4 (1976) 395–404
8. Guillemin, V.M., Shnider, S.D.: Some stable results on the cohomology of classical infinite dimensional Lie algebras. *Trans. Amer. Math. Soc.* **179** (1973) 275–280

Parallel Computing with Mathematica

Roman Maeder

MathConsult Dr. R. Mäder, 8832 Wollerau, Switzerland
`http://www.mathconsult.ch`

Abstract. The MathLink communication protocol can be used to control several Mathematica kernel processes from within Mathematica. This feature allows the implementation of an environment for parallel programming, the *Parallel Computing Toolkit*, using processes with distributed memory. The toolkit is written completely in Mathematica in a machine-independent way, allowing its use in heterogeneous networks without common file systems, as well as on multi-processor machines. All library and application code is distributed through MathLink.
Mathematica's high-level language allows for the easy and natural expression of parallel programming paradigms and for the parallelization of other than purely numeric computations. Examples presented include virtual shared memory, fault-tolerance, and automatic parallelization of functional programming constructs.

Solution of Systems of Linear Diophantine Equations

Gennadi I. Malaschonok

Tambov State University,
33 Internatsionalnaya, 392622 Tambov, Russia
malaschonok@math-iu.tambov.su

Abstract. Two new methods to solve linear systems of Diophantine equations are proposed – modular (CRT) and p-adic (Hensel). Each of them allows to obtain solutions of a system with the size $n \times m$ with the complexity $O(n^\beta m)$. For quasi-square systems, the p-adic method allows to obtain solution with the complexity $O(n^3)$, and the modular method with complexity $O(n^{\beta+1})$. Both estimates have the accuracy up to the logarithmic multipliers, β being the power in the estimation of matrix multiplication time.

1 Introduction

The usual method to solve systems of linear Diophantine equations is based on the computation of Smith form of the coefficient matrix [6], [5]. In the recent papers [11], [10] the *probabilistic* methods for finding the solutions of Diophantine systems were presented. The main lack of these methods is the probabilistic character of the solution and the absence of a capability to prove the incompatibility of system. These methods do not allow to distinguish a compatible system, in which the solution is not obtained in the process of solving, from an incompatible system.

The main result of the article is as follows.

We propose the deterministic p-adic and modular methods which are independent of the Smith form computation. Both methods reduce the problem to the solution of systems of linear equations in the residue ring.

The p-adic method is based on the linear p-adic Hensel lifting. This is a way to obtain the basic set of system solutions in the plane of all solutions. Then we compute all integer points on this plane.

The second method is based on the Chinese remainder theorem. The result of its application is the diagonal form of a non-singular submatrix of maximal rank. Then we solve the system in modulus which equals the determinant of this submatrix.

The complexity of the solution is estimated for systems in the ring of integers $\mathbb{Z}$ and in the ring of polynomials $F[x]$ over a field F.

For essentially rectangular systems, when $m > 2n$, both methods have the same order of complexity $O(n^{\beta+1}m)$ within a logarithmic factor. For quasi-square systems the complexity of a p-adic method is $O(n^3)$ and the complexity of the modular method is $O(n^{\beta+1})$. Here β is the power in the estimation of complexity of matrix multiplication. The value of β known at present is less than 2.356 [3].

The next Section has an auxiliary character, it is devoted to the problem of solving systems of linear equations in a residue ring. Here is presented a method with complexity $O(n^\beta)$ ring operations. For a system of order n the asymptotic evaluation of this method is less than $6.54M(n)$, where $M(n)$ is the asymptotic evaluation of Strassen multiplication of matrices of order n.

2 QuasiEuclidean Rings

Let $\mathbf{R}$ be a commutative ring with identity, and let $\alpha \in \mathbf{R}^2$ be a non-zero pair. An unimodular matrix $E_\alpha \in \mathbf{R}^{2\times 2}$ is called the *Diophantine matrix* of α if

$$E_\alpha \alpha = \begin{pmatrix} g \\ 0 \end{pmatrix}, \; g \neq 0, \; \det E_\alpha = 1.$$

The *unit matrix* is called the Diophantine matrix of a *zero* pair.

We must select a special class of rings, in which an algorithm for computing the Diophantine matrix exists.

Definition 1 . *A ring* $\mathbf{R}$ *is called quasiEuclidean, if an algorithm for computing the Diophantine matrix for any* $\alpha \in \mathbf{R}^2$ *exists.*

If $\alpha = \begin{pmatrix} p \\ q \end{pmatrix}$, then $\begin{pmatrix} p \\ q \end{pmatrix} = E_\alpha^{-1} \begin{pmatrix} g \\ 0 \end{pmatrix}$. So the ideal (p, q) is a principal ideal generated by g. Therefore, each of the quasiEuclidean rings is a ring of principal ideals, where the algorithm for computation of ideal generator exists.

On the contrary, let $\mathbf{R}$ be the ring of principal ideals, in which an algorithm for calculation of the generator g of an ideal (p, q), $p, q \in \mathbf{R}$ exists, denote it GI algorithm:

$$\{g, a, b\} = GI(p, q), \; (g) = (p, q), \; ap + bq = g.$$

For a non-zero pair $\alpha = \begin{pmatrix} p \\ q \end{pmatrix}$, a Diophantine matrix is $E_\alpha = \begin{pmatrix} a & b \\ -q/g & p/g \end{pmatrix}$.

Therefore, *a commutative ring is quasiEuclidean if and only if it is a ring of principal ideals, where an algorithm for computing of ideal generator exists.*

Note that the numbers a, b and g are not uniquely defined. Therefore, the Diophantine matrix is not uniquely defined. The Diophantine matrix can be determined uniquely if an order is introduced into $\mathbf{R}$.

If $\mathbf{R}$ is the Euclidean ring, then the extended Euclidean algorithm may be taken as GI algorithm.

Proposal 1 . *QuasiEuclidivity of a ring is inherited by its quotient-ring.*

A quotient-ring of the ring of principal ideal is a ring of principal ideal. So to prove this proposal we need to show how to construct the Diophantine matrices in quotient-ring.

Let $\mathbf{R}'$ be a quotient-ring of a quasiEuclidean ring $\mathbf{R}$, and $\phi : \mathbf{R} \to \mathbf{R}'$ be a canonical homomorphism. Let us introduce an order relation in R, and define a function

$$\psi : \mathbf{R}' \to \mathbf{R},$$

selecting a representative of a coset for $p' \in \mathbf{R}'$ in a pre-image $\phi^{-1}(p')$ by a standard way: $\psi(p') = min(\phi^{-1}(p'))$. Let

$$\psi(p') = p, \quad \psi(q') = q, \quad \alpha = \begin{pmatrix} p \\ q \end{pmatrix},$$

and let E_α be the Diophantine matrix for α and $E_\alpha \alpha = \begin{pmatrix} g \\ 0 \end{pmatrix}$.

For a non-zero pair $\alpha' = \begin{pmatrix} p' \\ q' \end{pmatrix} \in \mathbf{R}'^2$ we define a Diophantine matrix $E_{\alpha'}$ as follows: $E_{\alpha'} = \phi(E_\alpha)$. Such definition is correct since

$$E_{\alpha'}\alpha' = \begin{pmatrix} g' \\ 0 \end{pmatrix}, \quad \det E_{\alpha'} = 1, \; g' \neq 0.$$

Let us show that $g' = \phi(g) \neq 0$. Due to the invertibility of Diophantine matrices, p and q are divided by g. Therefore, if $\phi(g) = 0$, then $\phi(p) = \phi(q) = 0$. But by the condition, at least one of the numbers $p' = \phi(p)$, $q' = \phi(q)$ is not zero.

Therefore, a set of quasiEuclidean rings contains all Euclidean rings and their quotient-rings.

Let now $\mathbf{R}$ be a quasiEuclidean ring, $A \in \mathbf{R}^{n \times m}$. A matrix A is called *upper triangular*, if all elements under the main diagonal are zeros: $(i, j) = 0$ for all $n \geq j > i \geq 1$. (We say nothing about other matrix elements.)

We say that the matrix $\mathcal{A}$ is *decomposable*, if it is a product of an unimodular matrix $\tilde{\mathcal{A}}$ and an upper triangular matrix $\bar{\mathcal{A}}$: $\mathcal{A} = \tilde{\mathcal{A}}\bar{\mathcal{A}}$.

Proposal 2 . *In a quasiEuclidean ring any matrix is decomposable.*

Proof. We compute such decomposition for a matrix $A = (a_{ij})$ of the size $n \times m$. Let $\alpha_{ij} = \begin{pmatrix} a_{jj} \\ a_{ij} \end{pmatrix}$, $i > j$, $E_{\alpha_{i,j}}$ be a Diophantine matrix for α_{ij}. Unimodular matrix $E^n_{\alpha_{i,j}}$ is a matrix that is obtained from the unit matrix I_n of the order n after replacement of its four elements (jj), (ji), (ij), (ii) by the four elements (18), (19), (21), (22) of matrix $E_{\alpha_{i,j}}$, correspondingly.

Let $E_j = E_{\alpha_{n,j}} \cdots E_{\alpha_{j+1,j}}$,

$$U = E_{n-1} \cdots E_1.$$

Then U is a unimodular matrix and UA is an upper triangular matrix. Such computation leads to the elimination of non-zero elements under the diagonal element in the first column, then in the second column, and so on up to the $(n-1)$th column.

Denote by C_D an asymptotic estimate of the complexity of a Diophantine matrix calculation, and let $M(n)$ be an asymptotic estimate of the complexity of multiplying matrices of the order n, $M(n) = \alpha n^\beta$. The best value of β known today is less than 2.356 [3].

Lemma 1. *Let* $\mathbf{R}$ *be a quasiEuclidean ring* $A, C \in \mathbf{R}^{n\times n}$, *and let* A *be an upper square matrix. Then the matrix* $\begin{pmatrix} A \\ C \end{pmatrix}$ *may be decomposed as* $\begin{pmatrix} A \\ C \end{pmatrix} = Q \begin{pmatrix} \bar{A} \\ 0 \end{pmatrix}$ *with an asymptotic estimate of complexity which is no more superior than*

$$C_s(n) = n^2 C_D + \alpha\lambda_\beta n^\beta, \tag{1}$$

here $\lambda_\beta = 7(2^{\beta-2} - 1)^{-1}$.

Proof. The decomposition of the 2×1 matrix needs the multiplication by its Diophantine matrix. Without loss of generality we suppose that $n = 2^p$ and decompose the matrices A and C into four square blocks:

$$A = \begin{pmatrix} a & b \\ 0 & d \end{pmatrix}, \quad C = \begin{pmatrix} e & f \\ g & h \end{pmatrix}.$$

Let us have four decompositions.

$$Q_1 \begin{pmatrix} a \\ e \end{pmatrix} = \begin{pmatrix} a_1 \\ 0 \end{pmatrix}, \text{ denote } \begin{pmatrix} b_1 \\ f_1 \end{pmatrix} = Q_1 \begin{pmatrix} b \\ f \end{pmatrix}.$$

$$Q_2 \begin{pmatrix} a_1 \\ g \end{pmatrix} = \begin{pmatrix} a_2 \\ 0 \end{pmatrix}, \text{ denote } \begin{pmatrix} b_2 \\ h_1 \end{pmatrix} = Q_2 \begin{pmatrix} b_1 \\ h \end{pmatrix}.$$

$$Q_3 \begin{pmatrix} d \\ f_1 \end{pmatrix} = \begin{pmatrix} d_1 \\ 0 \end{pmatrix}.$$

$$Q_4 \begin{pmatrix} d_1 \\ h_1 \end{pmatrix} = \begin{pmatrix} d_2 \\ 0 \end{pmatrix}.$$

The matrices

$$Q_i = \begin{pmatrix} \alpha_i & \beta_i \\ \gamma_i & \delta_i \end{pmatrix}, \quad i = 1, \dots, 4,$$

are unimodular, a, a_1, a_2, d, d_1, d_2 are upper triangular matrices. Then we obtain:

$$\bar{A} = \begin{pmatrix} a_2 & b_2 \\ 0 & d_2 \end{pmatrix} \quad \text{and } Q = \begin{pmatrix} \alpha_1\alpha_2 & 0 & \beta_1\alpha_2 & \beta_2 \\ \alpha_4\beta_3\gamma_1 + \beta_4\gamma_2\alpha_1 & \alpha_4\alpha_3 & \alpha_4\beta_3\delta_1 + \beta_4\gamma_2\beta_1 & \beta_4\delta_2 \\ \delta_3\gamma_1 & \gamma_3 & \delta_3\delta_1 & 0 \\ \gamma_4\beta_3\gamma_1 + \delta_4\gamma_2\alpha_1 & \gamma_4\alpha_3 & \gamma_4\beta_3\delta_1 + \delta_4\gamma_2\beta_1 & \delta_4\delta_2 \end{pmatrix}.$$

We obtain the matrix Q as a product

$$\begin{pmatrix} 1 & 0 & 0 & 0 \\ 0 & \alpha_4 & 0 & \beta_4 \\ 0 & 0 & 1 & 0 \\ 0 & \gamma_4 & 0 & \delta_4 \end{pmatrix} \begin{pmatrix} 1 & 0 & 0 & 0 \\ 0 & \alpha_3 & \beta_2 & 0 \\ 0 & \gamma_2 & \delta_2 & 0 \\ 0 & 0 & 0 & 1 \end{pmatrix} \begin{pmatrix} \alpha_2 & 0 & 0 & \beta_2 \\ 0 & 1 & 0 & 0 \\ 0 & 0 & 1 & 0 \\ \gamma_3 & 0 & 0 & \delta_3 \end{pmatrix} \begin{pmatrix} \alpha_1 & 0 & \beta_1 & 0 \\ 0 & 1 & 0 & 0 \\ \gamma_1 & 0 & \delta_1 & 0 \\ 0 & 0 & 0 & 1 \end{pmatrix}.$$

Therefore, these calculations need 28 multiplications of blocks of the order $n/2$ and four decompositions of blocks of size $n \times n/2$.

We reduce the calculations step by step to the blocks of size 2×1. Finally we obtain the following asymptotic estimate of decomposition:

$$C_s(n) = n^2 C_D + 7\sum_{i=1}^{p} 4^i M(\frac{n}{2^i}) < n^2 C_D + \alpha\lambda_\beta n^\beta.$$

Here the number $\lambda_\beta = 7(2^{\beta-2} - 1)^{-1}$ lies between $\lambda_3 = 7$ and $\lambda_{\log 6} = 14$.

Corollary 1 . *Let* $\mathbf{R}$ *be a quasiEuclidean ring, and let* A *be an upper square matrix,* $A \in \mathbf{R}^{m\times m}$, $C \in \mathbf{R}^{(n-m)\times m}$. *Then the matrix* $\begin{pmatrix} A \\ C \end{pmatrix}$ *may be decomposed with an asymptotic estimate of complexity which is no more superior than*

$$C_s^1(n,m) = (n-m)mC_D + \alpha m^{\beta-1}(\lambda_\beta(n-m) + 4n - 8m),$$

for $n \geq 2m$ *and*

$$C_s^2(n,m) = (n-m)mC_D + \alpha(n-m)^{\beta-2}(\lambda_\beta m(n-m) + 10mn - 4n^2 - 4m^2),$$

for $m \leq n \leq 2m$.

Proof.

a. Let $n \geq 2m$ and m be a divisor of n. Let C_i, $i = 1, \ldots, (n-m)/m$, be an mth order block of the matrix C. We count the matrix C blocks from top to bottom. And then we decompose sequentially the matrices:

$$\begin{pmatrix} A_{i+1} \\ 0 \end{pmatrix} = Q_i \begin{pmatrix} A_i \\ C_i \end{pmatrix}, \quad i = 1, \ldots, (n-2m)/m,$$

where $A_1 = A$.

We need four multiplications of matrices of size $m \times m$ at each step.

So the total number of operations is

$$C_s^1(n,m) = C_s(m)\frac{(n-m)}{m} + 4M(m)\frac{(n-2m)}{m}.$$

b. Let $m \leq n \leq 2m$ and $n - m$ be a divisor of m. Let us decompose the matrices A and C into blocks of size $(n-m) \times (n-m)$. The matrix C contains one block row. We denote $C = (C_1, \bar{C}_1)$, here C_1 is a left block of the matrix C. The matrix A contains $k = m/(n-m)$ block rows and columns.

At first we decompose the matrix that is formed by the upper left diagonal block A_1 of the matrix A and the left block C_1 of the matrix C:

$$\begin{pmatrix} A_1 \\ C_1 \end{pmatrix} = Q_1 \begin{pmatrix} \bar{A}_1 \\ 0 \end{pmatrix},$$

where Q_1 is a unimodular matrix.

We compute the product

$$\begin{pmatrix} A'_1 \\ C'_2 \end{pmatrix} = Q_1 \begin{pmatrix} \bar{A}_1 \\ \bar{C}_1 \end{pmatrix},$$

here $\bar{A}_1$ is a part of the first block row of the matrix A situated on the left from the diagonal block A_1. Denote $C'_2=(C_2,\bar{C}_2)$, here C_2 is a left square block. We need to do $4(k-1)$ block multiplications.

Then we sequentially decompose the matrices which are formed by the ith diagonal block A_i of the matrix A and the left block C_i of the matrix C'_i: $\begin{pmatrix} A_i \\ C_i \end{pmatrix} = Q_i \begin{pmatrix} \bar{A}_i \\ 0 \end{pmatrix}$, and compute the product

$$\begin{pmatrix} A'_i \\ C_{i+1} \end{pmatrix} = Q_i \begin{pmatrix} \bar{A}_i \\ \bar{C}_i \end{pmatrix}, \quad i = 2, \ldots, k.$$

Here Q_i is a unimodular matrix, $\bar{A}_i$ is part of the i-th block row of the matrix A situated on the left from the diagonal block A_i, $C'_i=(C_i,\bar{C}_i)$.

Substitute the blocks (1,1), (1,2), (2,1), (2,2) of the matrix Q_i instead of the blocks (i,i), (i,k+1), (k+1,i), (k+1,k+1) of the unit matrix I_n respectively. As a result we obtain the matrix Q'_i. To obtain the unimodular matrix Q it is required to multiply the unimodular matrices Q'_i:

$$Q = Q'_k \cdots Q'_2 Q'_1.$$

One such matrix multiplication needs four block multiplications and the ith step needs $4(k-i)+4$ block multiplications, $i = 2, \ldots, k$.

So the total number of operations is

$$\begin{aligned}
&(\sum_{i=1}^{k} 4(k-i+1) - 4)M(n-m) + kC_s(n-m) \\
&+ kC_s(n-m) + 2(k^2+k-2)M(n-m) \\
&= m(n-m)C_D + M(n-m)(\lambda_\beta m/(n-m) \\
&+ 2(5mn - 2n^2 - 2m^2)/(n-m)^2) = C_s^2(n,m).
\end{aligned}$$

Remark. *It is easy to show that we can change the upper triangular matrix A to a lower triangular matrix in the condition of Lemma 1. We can also change the order of multipliers in the decomposition formula, the evaluations in the lemma and in the corollary will be the same.*

Theorem 1. *In a quasiEuclidean ring any quadratic matrix A of the order n may be decomposed with an asymptotic estimation of complexity which is no more superior than*

$$C_t(n) = \mu_\beta M(n) + \frac{n^2-n}{2} C_D, \tag{2}$$

where

$$\mu_\beta = \frac{9 \cdot 2^{\beta-2} + 19}{(2^\beta - 2)(2^{\beta-2} - 1)}.$$

Proof. The decomposition of a second order matrix needs the multiplication of its first column by the Diophantine matrix.

We suppose that $n = 2^p$ and decompose the matrix $\mathcal{A}$ into four square blocks:

$$\mathcal{A} = \begin{pmatrix} A & B \\ C & D \end{pmatrix}.$$

1. Let the block A be decomposable:

$$A = \tilde{A}^{-1}\bar{A}.$$

Denote $A_1 = \tilde{A}A$, $B_1 = \tilde{A}B$.

2. Let the block $\begin{pmatrix} A_1 \\ C \end{pmatrix}$ be decomposable:

$$\begin{pmatrix} A_1 \\ C \end{pmatrix} = Q^{-1}\begin{pmatrix} A_2 \\ 0 \end{pmatrix}, \tag{3}$$

where Q is a unimodular matrix, A_2 is an upper triangular matrix. Denote

$$\begin{pmatrix} B_2 \\ D_1 \end{pmatrix} = Q\begin{pmatrix} B_1 \\ D \end{pmatrix}.$$

3. Let the block D_1 be decomposable: $D_1 = \tilde{D}_1^{-1}\bar{D}_1$. Denote $D_2 = \bar{D}_1$.

Then the matrix $\mathcal{A}$ is decomposable:

$$\mathcal{A} = \tilde{\mathcal{A}}^{-1}\bar{\mathcal{A}}, \quad \tilde{\mathcal{A}} = \operatorname{diag}(I, \tilde{D}_1) \cdot Q \cdot \operatorname{diag}(\tilde{A}, I), \quad \bar{\mathcal{A}} = \begin{pmatrix} A_2 & B_2 \\ 0 & D_2 \end{pmatrix}.$$

Such a decomposition requires 9 block multiplications, two block decompositions and one decomposition of the rectangular $n \times n/2$ block.

We denote by $C_t(n)$ the complexity of decomposition of a matrix of the order n. Then we obtain according to (1)

$$C_t(n) = 2C_t(n/2) + 9M(n/2) + C_s(n).$$

According to lemma 1 we obtain:

$$C_t(n) = 2C_t(n/2) + (n/2)^2 C_D + \alpha(9 + \lambda_\beta)(n/2)^\beta.$$

We shall decompose the matrices A and D into four square blocks, and continue such procedure up to the second order blocks.

The asymptotic estimate of complexity is

$$\sum_{i=1}^{p} 2^{i-1}\left(\left(\frac{n}{2^i}\right)^2 C_D + \alpha(9 + \lambda_\beta)(n/2^i)^\beta\right) < \frac{n^2 - n}{2} C_D + \alpha\mu_\beta n^\beta = C_t(n),$$

where the number $\mu_\beta = \frac{1}{2^\beta - 2}\left(9 + \frac{7}{2^{\beta-2} - 1}\right)$ lies between $\mu_3 = 8/3$ and $\mu_{\log 6} = 23/4$.

Corollary 2 *In a quasiEuclidean ring any rectangular matrix $\mathcal{A}$ of size $n \times m$ is decomposable. It is possible to execute such a decomposition with an asymptotic estimation of complexity which is not higher than*

$C_t^0(n,m) = C_D(n^2-n)/2 + \alpha n^\beta(\mu_\beta + (m-n)/n)$, *for* $n < m$,
$C_t^1(n,m) = C_D(nm - (m^2+m)/2) + \alpha\mu_\beta m^\beta + \alpha(n-m)^{\beta-1}(\lambda_\beta m(n-m) + 10mn - 4n^2 - 4m^2)$, *for* $m \le n \le 2m$ *and*
$C_t^2(n,m) = C_D(nm-(m^2+m)/2) + \alpha m^{\beta-1}(\mu_\beta m + \lambda_\beta(n-m) + 4n - 8m)$, *for* $n \ge 2m$.

Proof. The proof follows from the estimates for $C_t(n)$ and $C_s(n)$ in (1) and (2).

For $n < m$ it is enough to decompose the left square block and then to multiply the left block from $m-n$ columns of the matrix $\mathcal{A}$ by $\tilde{A}$.

For $n \ge m$ it is sufficient to decompose the upper square block and to use the corollary of Lemma 1.

Theorem 2. *In a quasiEuclidean ring any system of linear equations $\mathcal{A}y = c$ with the upper triangular coefficient matrix $\mathcal{A} \in \mathbf{R}^{n\times n}$ may be solved with an asymptotic estimate of complexity*

$$C_n = C_D(n^2-n)/2 + \alpha\phi_\beta n^\beta, \quad \phi_\beta = \frac{1}{2^\beta - 2}(5 + \frac{7}{2^{\beta-2}-1}). \tag{4}$$

Proof. Solving the system $\mathcal{A}y = c$ is reduced to solving two systems of twice less size and one matrix decomposition.

Let

$$\begin{pmatrix} A & B \\ 0 & D \end{pmatrix}\begin{pmatrix} y_1 \\ y_0 \end{pmatrix} = \begin{pmatrix} f \\ h \end{pmatrix}$$

be the block form of the system $\mathcal{A}y = c$. All the matrix $\mathcal{A}$ blocks are square and A, D are the upper triangular blocks.

(1). Let $y_0 = Vs_0 + v$ be a solution of the system $Dy_0 = h$, $V \in \mathbf{R}^{n/2\times n/2}$, $v, s_0 \in \mathbf{R}^{n/2}$, s_0 is a vector of parameters. Substituting it into the system $Ay_1 + By_0 = f$ we obtain $Ay_1 + Gs_0 = g$, here $G = BV$, $g = f - Bv$.

(2). Decompose the matrix (A, G): $(A,G)P = (\bar{A}, 0)$. P is a unimodular matrix, $\bar{A}$ is an upper triangular matrix. The complexity of such decomposition is $C_s(n/2)$ according to Lemma 1 and the remark.

(3). Let $z_1 = Wt_1 + w$ be a solution of the system $\bar{A}z_1 = g$, $W \in \mathbf{R}^{n/2\times n/2}$, $w, t_1 \in \mathbf{R}^{n/2}$, t_1 is a vector of parameters.

Then

$$\begin{pmatrix} z_1 \\ z_0 \end{pmatrix} = \text{diag}(W, I)\begin{pmatrix} t_1 \\ t_0 \end{pmatrix} + \begin{pmatrix} w \\ 0 \end{pmatrix}$$

is the solution of the system

$$(\bar{A}, 0)\begin{pmatrix} z_1 \\ z_0 \end{pmatrix} = g.$$

Let $P = \begin{pmatrix} P_a & P_b \\ P_c & P_d \end{pmatrix}$ be the block form of the matrix P. Since $\begin{pmatrix} y_1 \\ s_0 \end{pmatrix} = P \begin{pmatrix} z_1 \\ z_0 \end{pmatrix}$ and $y_0 = Vs_0 + v$, the solution of the initial system $\mathcal{A}y = c$ is obtained in the form

$$y = \begin{pmatrix} P_a W & P_b \\ VP_c W & VP_d \end{pmatrix} \begin{pmatrix} t_1 \\ t_0 \end{pmatrix} + \begin{pmatrix} P_a w \\ VP_c w + v \end{pmatrix}.$$

We need five block multiplications in total. Denote by C_n the complexity of solving a system with square coefficient matrix of the order n. Then

$$C_n = 2C_{n/2} + 5M(n/2) + C_s(n/2) = 2C_{n/2} + \alpha(5 + \lambda_\beta)n^{n/2} + (n/2)^2 C_D.$$

Hence we obtain the asymptotic estimate

$$C_n = C_D(n^2 - n)/2 + \alpha\phi_\beta n^\beta, \quad \text{where } \phi_\beta = \frac{1}{2^\beta - 2}\left(5 + \frac{7}{2^{\beta-2} - 1}\right).$$

Theorem 3. *In a quasiEuclidean ring any system of linear equations* $\mathcal{A}x = c$, $\mathcal{A} \in \mathbf{R}^{n\times m}$ *may be solved with an asymptotic estimate of complexity* C_{nm}:

$$C_{nm} = (nm - l)C_D + \alpha\psi_\beta l^\beta,$$

$\psi_\beta = \mu_\beta + \phi_\beta + \lambda_\beta \frac{n-m}{m} + \frac{4n}{m} - 8$ *for* $2m \le n$,
$\psi_\beta = \mu_\beta + \phi_\beta + \lambda_\beta \frac{m-n}{n} + \frac{6m}{n} - 9$ *for* $2n \le m$;

$$C_{nm} = (nm - l)C_D + \alpha\psi_\beta l^\beta + \alpha|n - m|^{\beta-2}(\lambda_\beta l|m - n| + 10nm - 4m^2 - 4n^2),$$

$\psi_\beta = \mu_\beta + \phi_\beta$ *for* $m \le n \le 2m$,
$\psi_\beta = \mu_\beta + \phi_\beta + \frac{2m}{n} - 1$ *for* $n < m \le 2n$,
here $l = \min(m, n)$.

Proof. The decomposition $(\mathcal{A}, c) = P(\mathcal{A}', c')$ with the upper triangular matrix $(\mathcal{A}', c')$ may be obtained with the complexity $C_t^*(n, m)$ according to Corollary 2.

Therefore, it is sufficient to solve the system

$$\mathcal{A}'x = c' \tag{5}$$

with upper triangular coefficient matrix.

a. Let $m > n$, $\mathcal{A}' = (A, B)$, A be a square block with an upper triangular form.

According to Corollary 1 and the remark we can obtain a unimodular matrix Q such that $(A, B)Q = (\bar{A}, 0)$, with the complexity $C_s^*(n, m)$. Here $\bar{A}$ is an upper triangular matrix.

The solutions of (5) and of the system $\bar{A}y = c'$ are connected according to the equation: $x = Q\begin{pmatrix} y \\ t_0 \end{pmatrix}$, where t_0 is a vector of parameters.

Let $Ut_1 + u$ be a solution of the system $\bar{A}y = c'$, $U \in \mathbf{R}^{n\times n}$, $t_1, u, y \in \mathbf{R}^n$, t_1 be a vector of parameters, $t_1 \in \mathbf{R}^n$.

Then the solution of initial system will be $x = Q\,\mathrm{diag}(U,I)t + Q\begin{pmatrix} u \\ 0 \end{pmatrix}$, where $t \in \mathbf{R}^m$ is a vector of parameters. The computation of matrix product $Q\,\mathrm{diag}(U,I)$ needs $\alpha n^{\beta-1}m$ operations. According to Theorem 2 the solution of the system $\bar{A}y = c'$ may be obtained with the complexity C_n.

We obtain in total the evaluation of complexity of the system $\mathcal{A}x = c$ solving

$$C_{nm} = C_t^0(n,m) + C_s^i(m,n) + C_n + M(n)(m/n),$$

$i = 1,2$, where $i = 1$ in the case $m \geq 2n$ and $i = 2$ in the case $2n \geq m > n$.

b. Let $n \geq m$, $(\mathcal{A}', c') = \begin{pmatrix} A & c_A \\ 0 & c_B \end{pmatrix}$, A be a square block with upper triangular form. The system is incompatible if the vector c_B is non-zero.

According to Theorem 2 we can obtain the solution of the system $Ax = c_A$ with complexity C_m.

So the total complexity is

$$C_{nm} = C_t^i(n,m) + C_m \quad i = 1,2, \tag{6}$$

where $i = 1$ in the case $2m \geq n \geq m$ and $i = 2$ in the case $n \geq 2m$.

Therefore, the theorem assertion follows.

In particular, for the system with square coefficient matrix we obtain:

$$C_{nn} = n^2 C_D + \alpha\psi_\beta n^\beta, \quad \psi_\beta = \frac{7 \cdot 2^{\beta-1}}{(2^\beta - 2)(2^{\beta-1} - 2)}.$$

We have $\psi_3 < 4.67$, $\psi_{log7} < 6.54$, $\psi_{log6} = 10.5$.

3 Hensel Lift

Let $\mathbf{R}$ be a commutative domain with identity, $\mathbf{K}$ be a quotient field of $\mathbf{R}$, $A \in \mathbf{R}^{n\times m}$, $c \in \mathbf{R}^n$. Let

$$Ax = c \tag{7}$$

be a system of linear equations in $\mathbf{R}$.

Suppose that $n < m$, $\mathrm{rank}A = n$, $k = m - n + 1$ is a number of the linearly independent solutions of (7). Let the matrix $A = (A_0, A_1)$ have a square block A_0, $\det A_0 \neq 0$, and assume that the last row of the matrix $A_0^{-1}(A_1, c)$ does not contain zeroes. The system in general may be reduced to such a case (it may be done, for example, as in [10]).

Let $\mathbf{e}_j$ be the jth column of the unit matrix $I_k = (\mathbf{e}_1, \ldots, \mathbf{e}_k)$ of the order k. We seek the solution of the system (7) in the form $\begin{pmatrix} x_j \\ \xi_j \mathbf{e}_j \end{pmatrix}$, $x_j \in \mathbf{K}^{n-1}$, $\xi_j \in \mathbf{K}\setminus 0$, $(j = 1, \ldots, k)$. Then we obtain a determined system of the order n.

It is known that such system may be solved by p-adic lift [4]. If R is a ring of integers, the solution may be obtained with

$$\mathcal{D}_Z = O(n^3(\log n + \log ||A|| + \log p)^2)$$

bits operations. If $\mathbf{R}$ is a ring of polynomials over a field F, then solving the system needs

$$\mathcal{D}_{F[x]} = O(n^3(\deg A + \deg p)^2)$$

operations in the field F [11], [4].

To solve k such systems we need

$$k\mathcal{D}_Z, \quad \text{and} \quad k\mathcal{D}_{F[x]} \tag{8}$$

operations correspondingly.

As $\xi_j \neq 0$, $j = 1, \ldots, k$, the solutions of these k systems are linearly independent. Reduce the components of each solution to their common denominator (it is a divisor of some minor of the matrix A), and write the solutions in the following way: $x_j = \begin{pmatrix} \mathbf{x}_j \\ \alpha_j \mathbf{e}_j \end{pmatrix} \chi_j^{-1}$, $j = 1, \ldots, k$, $\mathbf{x}_j \in \mathbf{R}^{n-1}$, $\alpha_j, \chi_j \in \mathbf{R} \backslash 0$.

The **probabilistic** approach to finding Diophantine solutions consists in the following.

Denote $\begin{pmatrix} \mathbf{x}_j \\ \alpha_j \mathbf{e}_j \end{pmatrix}$ by $\mathbf{w}_j$. It is easy to show that if $\{\mathbf{w}_j {\chi_j}^{-1} | i = 1, 2, \ldots, k\}$ is a basis set of solutions of the system (7), then

$$\left\{ \frac{\sum_{j=1}^k \mathbf{w}_j \eta_j}{\sum_{j=1}^k \chi_j \eta_j} \middle| \ \eta_j \in \mathbf{R}, \sum_{j=1}^k \chi_j \eta_j \neq 0 \right\}$$

is the set of all solutions of this system [10].

So the system has Diophantine solutions if the ideal generated by the denominators of a basis set of solutions is unit. In this case we can find such $\eta_j \in \mathbf{R}$ that $\sum_{j=1}^k \chi_j \eta_j = 1$ and $\sum_{j=1}^k \mathbf{w}_j \eta_j$ is a Diophantine solution. Moreover if $x_i = \mathbf{w}_i \chi_i^{-1}$, $i = 1, 2, \ldots, h$ is a basis set of solutions and $\chi_1 = 1$, then the set $\mathbf{w}_1$, $\mathbf{w}_i - \mathbf{w}_1(\chi_i - 1)$, $i = 2, 3, \ldots, h$, is an integer basis [10].

If the ideal generated by the denominators of a basis set of solutions is not unity then we can rearrange columns in the matrix A and repeat the whole computation process. It might be necessary to repeat such procedure several times.

An expected number of such computations for the ring $\mathbb{Z}$ is $O(k^{-1}(\log n + \log\log \|(A, c)\|))$ [10], [11].

The **deterministic** approach to finding Diophantine solutions consists in the following.

The number $\mathrm{LCM}(\alpha_1, \ldots, \alpha_k)$ is a divisor of the minor of a matrix (A, c) constructed from the first $n-1$ columns and the last column, according to Cramer's rule. Denote $\delta = \mathrm{LCM}(\alpha_1, \ldots, \alpha_k)$.

Multiply the numerator and the denominator of the solution x_j by $\delta\alpha_j^{-1}$ and denote $\mathbf{x}_j \delta \alpha_j^{-1} = \mathbf{v}_j$, $\chi_j \delta \alpha_j^{-1} = \delta_j$ for each j. We obtain the basic set of the solutions of the system (7) in the form $x_j = \begin{pmatrix} \mathbf{v}_j \\ \delta \mathbf{e}_j \end{pmatrix} \delta_j^{-1}$.

Denote the vector $(\delta_1, \ldots, \delta_k)$ by $\mathbf{u}$, $\mathbf{u} \in \mathbf{R}^k$, and the matrix $(\mathbf{v}_1, \ldots, \mathbf{v}_k)$ by V, $V \in \mathbf{R}^{(n-1)\times k}$.

Theorem 4. *The system (7) and*

$$\begin{cases} Vy = 0 \bmod \delta \\ \mathbf{u}y = \delta \end{cases} \tag{9}$$

are consistent simultaneously in the ring $\mathbf{R}$.

If Y *is the set of the solutions of (9), then*

$X = \left\{ \begin{pmatrix} y \\ Vy\delta^{-1} \end{pmatrix} \mid y \in Y \right\}$ *is the set of the solutions of (7).*

Proof. The set of the solutions of the linear system (7) is a plane of the dimension $m-n$ in the space $\mathbf{K}^m$. It is defined by $m-n+1$ linearly independent solutions. Therefore, for any solution $z \in \mathbf{K}^m$ of (7) the numbers $\sigma_i \in \mathbf{K}$, $i = 1, \ldots, k$, exist, such that

$$\sum_{i=1}^{k} x_i\sigma_i = z \quad \text{and} \quad \sum_{i=1}^{k} \sigma_i = 1.$$

Denote $\sigma_i\delta_i^{-1} = \tau_i \in \mathbf{K}$ and $z = (y, w)$, $y \in \mathbf{R}^k$, $w \in \mathbf{R}^{m-k}$, $t = (\tau_1, \ldots, \tau_k)$.

Then

$$\sum_{i=1}^{k} \delta_i\tau_i x_i = z \quad \text{and} \quad \mathbf{u}t = 1,$$

and these equations may be written as the system:

$$\begin{pmatrix} \delta I_k & -I_k & 0 \\ V & 0 & -I_{n-1} \\ \mathbf{u} & 0 & 0 \end{pmatrix} \begin{pmatrix} t \\ y \\ w \end{pmatrix} = \begin{pmatrix} 0 \\ 0 \\ 1 \end{pmatrix}$$

Multiplying each part of this system by the matrix

$$\begin{pmatrix} I_k & 0 & 0 \\ -V & \delta I_{n-1} & 0 \\ -\mathbf{u} & 0 & \delta \end{pmatrix},$$

invertible in $\mathbf{K}$, we obtain the system

$$\begin{pmatrix} \delta I_k & -I_k & 0 \\ 0 & V & -\delta I_{n-1} \\ 0 & \mathbf{u} & 0 \end{pmatrix} \begin{pmatrix} t \\ y \\ w \end{pmatrix} = \begin{pmatrix} 0 \\ 0 \\ \delta \end{pmatrix}. \tag{10}$$

Let the system (7) have the Diophantine solution $\mathbf{z} = (\mathbf{y}, \mathbf{w}) \in \mathbf{R}^m$. Then (10) is consistent. Therefore, the system (9) is also consistent.

On the contrary, if (9) is consistent and $\mathbf{y} \in \mathbf{R}^k$ is its solution, then $(\delta^{-1}\mathbf{y}, \mathbf{y}, \delta^{-1}V\mathbf{y})$ is the solution of (10). Therefore, $\mathbf{z} = (\mathbf{y}, \delta^{-1}V\mathbf{y})$ is the Diophantine solution of (7).

Now we estimate the complexity of the solution of (9) in the ring of integers and in the ring of polynomials. The matrix V has the size $(n-1) \times k$, and the module δ is a divisor of a minor of the order n. Therefore, for $\mathbf{R} = \mathbb{Z}$

$$\log \delta \leq n(\log ||A|| + 0.5 \log n), \tag{11}$$

where $||A||$ is an absolute value of the greatest element of the matrix A. For $\mathbf{R} = F[z]$

$$\deg \delta \leq n \deg ||A||, \tag{12}$$

where $\deg ||A||$ is the greatest degree of entries of matrix A.

Let us use the standard algorithm for multiplying integers and polynomials. Then the multiplication of N bit numbers requires N^2 bit operations, the multiplication of two polynomials of degree N requires N^2 multiplications in the field F.

By Theorem 3 we obtain the following asymptotic estimates of operations for $\mathbb{Z}$ and $F[x]$, respectively:

$$\eta_\beta knM(n)(\log ||A|| + 0.5 \log n)^2, \quad \eta_\beta knM(n) \deg^2 ||A||, \tag{13}$$

for $k \geq n$, and for $k < n$ we obtain:

$$\delta k^{\beta-1}(\log ||A|| + (\log n)/2)^2(n^3(4+\lambda_\beta) + n^2 k(\nu_\beta + \mu_\beta - \lambda_\beta - 4)), \tag{14}$$

$$\delta k^{\beta-1} \deg^2 ||A||(n^3(4+\lambda_\beta) + n^2 k(\nu_\beta + \mu_\beta - \lambda_\beta - 4)). \tag{15}$$

These estimates in general do not depend on the complexity (8) of k solutions of (7).

However, for quasi-square systems it is not so, the contribution of the value of (8) is essential. In this case for $\mathbb{Z}$ and $F[x]$ we have the following estimates:

$$O(n^3(\log n + \log ||A|| + \log p)^2), \quad O(n^3(\deg ||A|| + \deg p)^2). \tag{16}$$

Fast methods of multiplying integers and polynomials correspondingly improve these estimations. For example, the Karasuba method demands $O(N^{\log 3})$ operations and the FFT method demands $O(N \log N \log\log N)$ operations.

4 Modular Method

Let us solve (7) assuming that $A = (A_0, A_1)$, block A_0 is a square block, $\det A_0 = \delta \neq 0$. A recursive method [9] allows to transform (7) to the diagonal form:

$$(\delta I_n, A'_1)x = c' \tag{17}$$

with $O(n^{\beta-1}m)$ operations in the ring $\mathbf{R}$. The methods described in [7] and [8] can do this transformation with $O(n^2 m)$ operations.

All coefficients of (17) are minors of the order n of matrix (A, c). Therefore, the estimates (11) and (12) are suitable for the ring of integers and ring of polynomials, respectively.

The calculation needs

$$O(mn^{\beta+1}(\log n + \log ||A||)^2) \quad \text{and} \quad O(mn^{\beta+1} \deg^2 ||A||),$$

operations in the rings $\mathbb{Z}$ and $F[x]$, respectively.

The Chinese remainder theorem allows to reduce the amount of operations to

$$O(mn^{\beta}(\log n + \log ||A||)^2) \text{ and } O(mn^{\beta} \deg^2 \ |A\ |). \quad (18)$$

The solution of (17) is reduced to solving the system

$$A'_1 y = c' \bmod \delta. \quad (19)$$

in the quotient-rings $\mathbf{R}/\delta\mathbf{R}$.

Let Y be a set of the solutions of (19), then

$$X = \left\{ \begin{pmatrix} (c' - A'_1 y)\delta^{-1} \\ y \end{pmatrix} \middle| \ y \in Y \right\}$$

is the set of the solutions of (17).

Let us estimate the complexity of the solution of (19). The matrix A' has the size $n \times (m - n)$, and the module δ is a minor of the order n.

By Theorem 3 we obtain the following estimates for $\mathbb{Z}$ and $F[x]$, respectively:

$$O(n^{\beta+1}(m-n)(\log ||A|| + \log n)^2), \quad O(n^{\beta+1}(m-n) \deg^2 ||A||). \quad (20)$$

for $m - n \geq n$, and for $m - n \leq n$

$$O(n^2 m(m-n)^{\beta-1}(\log ||A|| + \log n)^2), \quad O(n^2 m(m-n)^{\beta-1} \deg^2 ||A||). \quad (21)$$

In general (20) and (21) are the estimates for the modular method. However, for quasi-square systems the value $m - n$ can be neglected and the estimate (18) will give the basic contribution. In this case we obtain the following estimates for the number of operations:

$$O(n^{\beta+1}(\log n + \log ||A||)^2) \quad \text{and} \quad O(n^{\beta+1} \deg^2 ||A||) \quad (22)$$

for $\mathbb{Z}$ and $F[x]$, respectively.

5 Summary

The estimates (13)–(16) and (20)–(22) obtained in the last two Sections enable us to formulate the following theorem about the complexity (number of bit operations) of solving a linear Diophantine system of size $n \times m$.

Theorem 5. *For essentially rectangular systems, when $m > 2n$, both methods have the identical complexity $O(mn^{\beta+1})$. For quasi-square systems the complexity of the p-adic method is $O(n^3)$, and the complexity of the modular method is $O(n^{\beta+1})$.*

All estimates in the Theorem are given within logarithmic factors. In the case of the polynomial ring $F[x]$, an analogous theorem is valid for the evaluation of the number of operations in the field F.

The comparison of the p-adic and the modular methods shows that for quasi-square systems the p-adic method has the advantage, it is $O(n^{\beta-2})$ times as fast.

The other advantage of the p-adic method for systems of the arbitrary size is the possibility to use the probability approach. The general solution of system can be obtained with a high probability with $O((m-n+1)n^3)$ operations [10]. An expected amount of attempts will be $O((\log n + \log\log ||A||)/(m-n+1))$.

These estimates can still be improved, if the fast algorithms of multiplication would be used.

References

1. Abbott, J., Bronstein, M., and Mulders, T.: Fast deterministic computation of determinants of dense matrices. In: Proc. ISSAC'99. ACM Press, Vancouver (1999) 197–204
2. Bareiss, E. H.: Computational solutions of matrix problems over an integral domain. J. Inst. Maths Applics **10** (1972) 68–104
3. Coppersmith, D. and Winograd, S.: Matrix multiplication via arithmetic progressions. J. Symbolic Computation **9** (1990) 251–280
4. Dixon, J.: Exact solution of linear equations using p-adic expansions. Numer. Math. **40** (1982) 137-141
5. Gregory, R. T., Krishnamurthy E. V.: Methods and Applications of Error-Free Computation, Springer-Verlag, Berlin (1984)
6. Hurt, M. F., Waid, C. A.: A generalized inverse which give all the integral solutions to a system of linear equations. SIAM J. Appl. Math. **10**, (1970) 547–550
7. Malaschonok, G. I.: Solution of a system of linear equations in an integral domain. USSR Journal of Computational Mathematics and Mathematical Physics **23** (1983) 1497–1500
8. Malaschonok, G. I.: Algorithms for the solution of systems of linear equations in commutative rings. In: Mora, T. and Traverso, C. (eds.): Effective Methods in Algebraic Geometry. Progress in Mathematics 94, Birkhauser (1991) 289–298
9. Malaschonok, G. I.: Recursive Method for the Solution of Systems of Linear Equations. In: Sydow, A. (ed.): Computational Mathematics (Proceedings of the 15th IMACS World Congress, Vol. I, Berlin, August 1997). Wissenschaft & Technik Verlag, Berlin (1997) 475–480
10. Malaschonok, G. I.: Effective Matrix Methods in Commutative Domains, In: Krob, D., Mikhalev, A. A., Mikhalev, A. V. (eds.): Formal Power Series and Algebraic Combinatorics. Springer (2000) 506–517
11. Mulders, T., Storjohann, A.: Diophantine Linear System Solving. In: Proc. ISSAC'99. ACM Press, Vancouver (1999) 181–188

SYMOPT: Symbolic Parametric Mathematical Programming

Isolde Mazzucco

Universität Passau, FMI
D-94030 Passau, Germany
mazzucco@fmi.uni-passau.de

Abstract. We present the REDUCE-package SYMOPT that is devoted to symbolic parametric mathematical programming i.e. the solution of parametric optimization problems. This paper formulates problem types that are solvable with SYMOPT, explains the solving methods, demonstrates various types of usage and the handling of solving processes depending on the users options.

1 Introduction

SYMOPT - the abbreviation for SYMbolic OPTimization - is a REDUCE -package for symbolic parametric mathematical programming. It based on the techniques of the REDLOG-package that extends REDUCE to a computer logic system [1]. Parametric mathematical programming describes the optimization of parametric objective functions subject to parametric constraints. Therefore SYMOPT provides symbolic algorithms for the minimization of parametric linear, hyperbolic or quadratic objective functions w.r.t a not necessarily convex boolean combination of parametric linear constraints. Moreover it contains algorithms for testing the feasibility of a not necessarily convex boolean combination of parametric linear constraints. Thus, this package is applicable to a wide range of feasibility and mainly of optimization problems. This paper gives an overview about the various problem types that are solvable with SYMOPT. It explains the basic solving method, its extensions and speeding up measures and you get a general idea in the types of usage and the handling of solving processes depending on users options.
All SYMOPT algorithms use a variable elimination method as basic solving method. This method is derived from a quantifier elimination method for linear problems and has single exponential complexity in the number of variables, polynomial in all other data [11]. We extend this method by taking advantage of the special problem types (linear, hyperbolic, quadratic). The efficiency of the variable elimination method depends strongly on the order in which variables are eliminated. Various measures and strategies are implemented that support different elimination orders. Part of the package are default parametric mathematical programming algorithms as well as user dependent specific programming. Users

can influence solving processes in several ways. They commit the computation structure, select strategies to gain efficiency and apply optional data to manipulate results.
Because of all these features SYMOPT promises an effective and efficient computation with few efforts.
Our implementations are tested on benchmark examples from the netlib and finally some of our experiments illustrate various applications.
Conclusions finish the presentation and give an outlook for future developments.

Informations about SYMOPT are documented as written manual, as INFO-file and as HTML-file.

2 Parametric Mathematical Programming

In contract to other well-known optimization solvers SYMOPT is not restricted to conjunctive constraints. Instead, input constraints are boolean combinations of parametric linear equations and inequalities. Thus, input constraints don't have to be convex and may involve additive parameters.
Input objective functions are linear, hyperbolic or quadratic terms also involving additive parameters.
SYMOPT algorithms are implemented in order to minimize an objective function w.r.t input constraints.
The obligatory data are of the following form.

Input Constraints:
Constraints are input as a first order formula that will be called a constraint formula - built with the n-ary infix operators *and* and *or* in an arbitrary nesting. Each atomic constraint is of the following form:
$f_i(x, u) = a_i * x + b_i(u) \geq 0 \quad i = 1, \ldots, m.$

Input Objective Function:
Objective functions are input as a parametric linear, hyperbolic or quadratic term $q(x, u)$ as follows:

$q(x, u) = c * x + d(u)$ ... linear case

$q(x, u) = (c_1 * x + d_1(u))/(c_2 * x + d_2(u))$ ... hyperbolic case

$q(x, u) = \sum c_{i,j}(u) * x_i * x_j + \sum c_i(u) * x_i + c_0(u)$ ... quadratic case.

where:
$x_1, \ldots, x_n$ are variables,
$u_1, \ldots, u_r$ are parameters,
$a_1, \ldots, a_n$ are n-tupels of rational numbers,
$b_i(u)$ are polynomials in u with rational coefficient for all i.
c, c_1, c_2 are n-tupels of rational numbers,
d, d_1, d_2, c_0 are polynomials in u with rational coefficients.
$c_{i,j}, c_i$ for $i, j = 1, \ldots, n$ are polynomials in u with a rational coefficient.

Input Parameters:
In solving parametric optimization problems parameters are also obligatory input data. They have to be distinguished from variables; otherwise all algorithms would handle them as variables and eliminate them. Parameters are declared by a list.

Optional Input Data:
Optional input data and control parameters like switches support users in solving their problems.

At present input and output of SYMOPT algorithms are performed in the algebraic mode of REDUCE. Input data are internally transformed into the symbolic mode of REDUCE and after computation the result is transformed back into the algebraic mode. A detailed description of the procedures, control parameters and input/output format is documented in[8].

3 Types of Computations

The basic solving method in solving parametric decision and optimization problems is a variable elimination method developed and presented by Weispfenning [12]. He proved that the variable elimination method has single exponential complexity in the number of variables, polynomial in all other data also in solving parametric linear, hyperbolic and quadratic optimization problems subject to parametric linear constraints [11]. The idea of this method is to eliminate variables using a finite selection of parametric test points.
The efficiency of the variable elimination method is strongly influenced by the number of test points in each elimination step and by the elimination order of the variables. To speed up all solving algorithms we implemented measures that varify the test point selection and the elimination order.
In the following the solving method, its extensions and improvements are briefly explained.

Basic Solving Method
The elimination method successively eliminates variables from a constraint formula. For this purpose a subset of constraints is step by step selected and then each constraint is transformed into an equation. As result of a solution of this linear equation a variable is replaced by the linear term in the remaining variables and parameters (the test point). This term is called an elimination term of a variable x. We refer to an elimination set as a subset of all elimination terms of a variable. Thus, an elimination set consists of all candidates for upper and lower bounds for a variable. A process that yield the elimination of one variable is called elimination step. A variable x is eliminated, if its occurences in the constraint formula is replaced by all candidates for its bounds [11].
The elimination method successively constructs an elimination tree and each elimination term is symbolized by a node of the tree. Therefore the out-degree of the tree depends on the number of elimination terms (number of test points).

A solving process stops, if the end of all branches of the elimination tree is reached [11].

Measures in order to speed up the computation:

Transformation
A standard transformation in solving linear and quadratic optimization problems is: Introducing a new objective function and extending the input constraints by a new upper bound.
$newq(z) \ := \ z$
$newF \ := \ F \ and \ z - q \ >= \ 0.$
In computing hyperbolic problems this modification is not allowed, because it would destroy the special type of a linear fractional objective function.

Upper Bounds on the objective function
If the end of a branch of the elimination tree is feasible, i.e. a feasible point of the constraints is reached, the value *val* of the objective function q is computed by substituting the point coordinates. The most recently computed value of the objective function is introduced as new upper bound for q into all remaining constraint formulas in order to force algorithms to compute only values smaller or equal than the current value.
The following upper bounds are conjunctively introduced:

in the linear/quadratic case:
$val \ - \ q \ >= \ 0,$

in the hyperbolic case:
$(de(q) \ >= \ 0 \ and \ val * de(q) - num(q) \ >= \ 0) \ or$
$(de(q) \ <= \ 0 \ and \ val * de(q) - num(q) \ <= \ 0)$

where:
val ... the computed value of q,
$de(q)$ ... the denominator of the hyperbolic q,
$num(q)$ ... the numerator of the hyperbolic q [5],[6].

Measures in order to construct elimination sets containing a small number of terms:

Passive List
Constructing and applying a passive list is a very important treatment to improve efficiency. A passive list stores step by step elimination terms. At the beginning of each elimination step terms, contained in the passive list, are either compared with all terms obtained from the constraints or with terms obtained from an already constructed elimination set. Terms that occur in both compared lists are discarded [12]. This procedure avoids to reach a point (feasible or not) more than once. The comparison with all terms of a constraint formula or with terms stored in an elimination set is also a variant that reduce elimination sets with different effects [5].
The construction of a passive list yields an administrative overhead that is com-

pensated by its efficiency effect. Nevertheless it is optional to compute problems also without using a passive list.

Linear Optimization
In minimizing linear optimization problems we restrict elimination sets to upper **or** lower bounds for a variable. The elimination set consists only of terms obtained from one-sided bounds. If there aren't any upper (lower) bounds w.r.t a variable x, then we test the unboundedness of q by substituting $+(-)\infty$ for x [12]. If the constraint formula is evaluated to $TRUE$ then q is unbounded from below and the algorithms stops. Otherwise the elimination process is continued.

Hyperbolic Optimization
In minimizing hyperbolic objective functions it is optional to consider partial derivatives. Considering the partial derivative of a variable x w.r.t q, its numerator is introduced both as upper and as lower bound into the constraint formula. If the extended constraint formula is evaluated to $FALSE$, then branches that are constructed using candidates for upper (lower) bounds to eliminate x, are not relevant to reach the optimal value [6]. These branches are discarded.
Applying partial derivatives means, testing the boundedness of the hyperbolic objective function as early as possible during the elimination process.
This measure involves also an administrative overhead, but sometimes the efficiency effect is stronger.

Quadratic Optimization
In solving quadratic optimization problems the apex of the objective function symbolized as $dq/dx = 0$ w.r.t a variable x must be considered as additional elimination equation. But elimination terms arising from these equations are only important, if the corresponding coefficient of x^2 in the objective function is positive. Only in this case the apex is a potential minimal value of q w.r.t the constraints. If the coefficient of x^2 in q is negative, then the apex is a local maximum of q and we only have to test the unboundedness from below.
To reduce the size of elimination sets we consider two additional strategies.
Upper (lower) bounds of a variable x are not relevant for the optimal value of q, if the apex is smaller (greater) or equal than **one** upper (lower) bound. The elimination set becomes smaller by discarding the upper (lower) bounds. These bounds are unnecessary as candidate for an optimal point.
Moreover, if the "local optimal branch" is identified and its end is reached, then all next elimination steps are omitted. A branch is called locally optimal, when it is constructed by replacing each variable with the term obtained from $dq/dx = 0$. We identify this branch as early as possible during the elimination process.

Measures in order to varify elimination orders:

Selection Strategies
Selection strategies search for favourable elimination orders. We distinguish three types of constraint selection and four types of variable selections by weights. One constraint selection strategy for example is to use the Gauss elimination, when-

ever variables occur in equations of the constraints. A special variable selection strategy by weights for example is picking all constraints containing a variable x and determining for each variable y, with $y \neq x$, the number of occurences in such constraints and sum over all these y. The variable x with the smallest weight will be chosen for the elimination.
Some strategies are combinable and some depend on problem types. Their influence is enormously. A precised description is documented in [7].

Duality
Sometimes it is more efficient to consider the dual problem of a parametric linear optimization problem with conjunctive constraints. This interesting effect depends on the dimension of a problem, especially on the number of variables that must be eliminated.

Primal and dual problems are described as follows:
If $min\ (max)\ \{c^t * x \mid A * x\ >= (<=)\ b\ and\ x\ >=\ 0\}$ is the **primal** problem, then its **dual** problem is $max\ (min)\ \{b^t * y \mid A^t * y\ <= (>=)\ c\ and\ y\ >=\ 0\}$,

where:
A ... (nxm)-coefficient matrix,
c ... n-vector,
b ... m-vector,
$x_1, \ldots, x_n$ are variables,
$y_1, \ldots, y_m$ are variables. [9]

The connection between primal and dual problems allows to solve the dual problem in order to optimize a primal problem:
1.If x^o is a feasible point for a primal problem and y^o for its dual, then the following is true: $c^t * x^o\ >= (<=)\ b^t * y$.
2. If a primal problem has an optimal solution, then there is an optimal solution to the dual problem.
3. If x^* is optimal for the primal problem and y^* is optimal for the dual, then is true: $c^t * x^*\ =\ b^t * y^*$ [2].

SYMOPT contains an algorithm that transforms a linear parametric optimization problem into its dual and computes the optimal value considering the dual constraints and dual objective function. This implementation works as follows:

- The primal (min-) problem is transformed into its dual (max-) problem.
- To fulfill the input data structure of SYMOPT, we convert the dual again into a minimization problem by multiplying the dual objective function with -1.
- The optimal value q^* of the converted dual problem is computed using the variable elimination method - $(-q^*)$ is the optimal value of the primal problem.
- The corresponding point coordinates w.r.t the primal constraints are optionally computed using the SYMOPT decision algorithm (s.section 4).

The following assumptions justify the implementation of the dual algorithm:
The solving process of dual problems usually runs faster than of primal problems,

if the number of variables after the primal/dual-transformation becomes smaller. Moreover, the elimination of dual variables yields a more efficient elimination tree.

4 Types of Usage

SYMOPT offers default and user dependent parametric mathematical programming. Furthermore users can control solving processes. We summerize and formulate the various types of usage subdivided in default procedures, users options and users control.

4.1 Default Procedures

1. **Minimizing parametric linear, hyperbolic and quadratic** objective functions w.r.t a **not necessarily boolean combination** of **parametric** linear constraints.
2. **Solving the dual problem** of a parametric linear optimization problem with a conjunctive constraint formula.
3. **Deciding feasibility** of a boolean combination of parametric linear constraints - combinable with the computation of a feasible point.
4. **Analysing** conjunctive input constraints to get information about the structure of the problem (i.e. number of variables, number of constraints, number of NONZERO elements of the coefficient matrix, density of the matrix, etc).

After the computation of solvable problems formatted results are output. The format differs in solving parameter free or parametric problems. In the case of parameter free optimization problems we get a **tupel** consisting of the optimal value and the corresponding point coordinates. In solving parametric optimization problems the result contains a list of **triplets**. Each triplet consists of conditions on parameters together with a parametric value for the objective function and corresponding parametric point coordinates. For more information see [8].

4.2 Users Option

Sometimes users are interested in the behaviour of the optimum after slightly modifying the input data. To avoid a new formulation of the optimization problem, SYMOPT provides parametric optimization. A modification is realized using additive parameters; multiplicative parameters are not permitted. We consider changes in the constraints as well as changes in the objective functions. Moreover SYMOPT algorithms allow to input optional constraints for parameters. Constraints for parameters are linear inequalities and equations.
In the following we specify the options concerning the use of parameters.

1. **Time-Dependent Optimization**:
   Using only one parameter (time-parameter) and study the effect on optimal

solutions is called time-dependent optimization. In this case we get information about the behaviour of an objective function w.r.t the constraints depending on the time axis. The result triplets are chronologically ordered and the conditions on the time-parameter are pairwise disjointed intervals. The parametric time-dependent term in each triplet represents the optimal value of q in an interval [4].

2. **Parametric Optimization**:
   It is possible to modify input data using more than one parameter. The output list consists usually of a long list of triplets. Users shorten an output list by specifiying bounds for parameters.
3. **Bounds For Parameters**:
   All SYMOPT algorithms consider bounds for parameters that are input as list of inequalities. All computed conditions on parameters - stored in the output triplets - involve these bounds.
4. **Sensitivity Region**:
   A special effect is obtained, if parameters are associated with fixed values. Associated values are introduced as list of equations. This suggested point for the parameters is substituted into the parametric value of the objective function, whenever a feasible parametric point is reached, and compared with a previous computed value. Only the currently best value is stored. The result remains parametric and yields a feasible point in a specified parameter region (sensitivity region) including the suggested point. Moreover, for parameter values at the suggested point the result is optimal.

4.3 Users Control

Users control can influence solving processes by activating switches and by entering optional data. The following options are possible:

1. **Fixing the search structure**:
   Users decide, whether elimination trees are constructed with depth first search, or a mixture between depth and breadth first search. Default is the more storage-efficient depth first search, but there are problems that run faster using also breadth first search.
2. **Selecting strategies**:
   Users activate strategies for constraint and variable selection by weights to solve their problems in the most efficient way.
3. **Extending the constraint formula**:
   If an upper bound for the optimal value is already known, it can be input as optional data. The input constraint formula is conjunctively extended by this bound and an algorithm forced to compute only values smaller or equal than this bound.
4. **Restricting the computation**:
   The number of potential feasible points that an algorithm computes is restrictable . Users can control this number, if they are satisfied with a value for the objective function that doesn't need to be optimal.

5. **Calling statistical information**
Optionally users get information about the realtime and the elimination steps, subdivided into selection strategies, that are necessary to reach the optimal value. Furthermore the number of infeasible and feasible points is count in order to illustrate how many values are (unnecessarily) computed.

5 A Computational Example

We demonstrate various types of usage of SYMOPT using a benchmark linear example from the netlib (ftp://ftp.zib.de/pub/mp-testdata/lp/netlib-lp/) called AFIRO, some of its variants and its quadratic version QAFIRO provided by the Brunel optimization group (ftp://ftp.sztaki.hu/pub/oplab/QPDATA). All examples contain the same conjunctive input constraints, exceptionally a time parameter, and are computed on a SUN ULTRASPARC 1, 140mH, 16MB memory.

Structure of the Problems

Number of all variables: 32
Number of all parameters: 0 (1)
Number of equations: 8
Number of inequalities: 51
$\longrightarrow$
Number of variables after the first Gauss elimination: 24
$\longrightarrow$
Coefficient matrix w.r.t the inequalities: 51 x 24 = 1224 entries
NONZERO: 81
DENSITY of the coefficient matrix: 6 %

Objective Function

linear:
$1/25 * (-10 * vx02 - 8 * vx14 - 15 * vx23 - 12 * vx36 + 250 * vx39)$

quadratic:
$1/25 * (125*vx01 + 25*vx01*vx02 + 25*vx01*vx03 + 125*vx02 + 25*vx02* vx03 - 10*vx02 + 125*vx03 - 8*vx14 - 15*vx23 - 12*vx36 + 250*vx39)$

5.1 Default Solving - AFIRO

```
DEPTH FIRST SEARCH
→ ELIMINATION STEPS
Number of steps: 151
C1-strategy (Gauss): 12
C2-strategy (counting boundaries): 139
C3-strategy (scalarproduct): 0
V1-strategy (counting variables) 0
→ RESULTS
infeasible results: 62
feasible results: 3
best result: 3.
```

{−406659/875,
$\{vx01 = 80, vx02 = 51/2, vx03 = 109/2, vx04 = 424/5, vx07 = 0, vx08 = 0, vx09 = 0, vx10 = 0, vx11 = 0, vx12 = 0, vx13 = 0, vx14 = 255/14, vx15 = 0, vx16 = 2703/140, vx22 = 500, vx23 = 11898/25, vx24 = 602/25, vx25 = 0, vx26 = 215, vx28 = 0, vx29 = 0, vx30 = 0, vx31 = 0, vx32 = 0, vx33 = 0, vx34 = 0, vx35 = 0, vx36 = 11898/35, vx37 = 13438/35, vx38 = 0, vx39 = 0\}\}$

Realtime: 1s

Interpretation:
The solving process constructs an elimination tree using depth first search and needs totally 151 elimination steps to solve the problem. Two constraint selection strategies are combined - Gauss elimination is used 12 times, an additional constraint selection strategy is used 139 times. The elimination tree ends 62 times in an infeasible and 3 times in a feasible point. The third feasible point delivers the optimal value for the objective function.
SYMOPT only needs **1s** to reach the optimal value. The Simplex algorithm implemented in the linear algebra package of REDUCE needs **100s**.

5.2 Solving the Dual - DAFIRO

DEPTH FIRST SEARCH
⟶ ELIMINATION STEPS
Number of steps: 73
C1-strategy (Gauss): 2
C2-strategy (counting boundaries): 71
C3-strategy (scalarproduct): 0
V1-strategy (counting variables) 0
⟶ RESULTS
infeasible results: 27
feasible results: 2
best result: 2.

{−406659.875}

Realtime: 1s

Interpretation:
The efficiency in solving AFIRO using the standard algorithm is satisfactory, but the dual is solved faster.
Although the number of variables doesn't become smaller after the primal/dual transformation (the dual problem consists of 35 variables), the elimination order of the dual variables yields a more efficient branching out of the elimination tree (recognizable by the total number of elimination steps).

5.3 Time-Dependent Optimization - TDAFIRO

Considering a time parameter the output list is chronologically ordered and the parametric output formulas describe pairwise disjointed intervals.

DEPTH FIRST SEARCH
⟶ ELIMINATION STEPS
Number of steps: 187
C1-strategy (Gauss): 12
C2-strategy (counting boundaries): 175
C3-strategy (scalarproduct): 0
V1-strategy (counting variables) 0

$\longrightarrow$ RESULTS
infeasible results: 270
feasible results: 5

$\{\{49 * t + 1020 \leq 0,$
$(7641631 * t - 1736542500)/3710000,$
$\{\{vx01 = (-57 * t + 9500)/106, vx02 =$
$(-22723 * t + 1861500)/53000, vx03 = (-109 * t + 54500)/1000, vx04 = (-57 * t + 9500)/100, vx06 =$
$(-22723 * t + 1861500)/74200, vx07 = 0, vx08 = 0, vx09 = 0, vx10 = 0, vx11 = 0, vx12 = 0, vx13 =$
$0, vx14 = (-22723 * t + 1861500)/74200, vx15 = 0, vx16 = (-22723 * t + 1861500)/70000, vx22 =$
$-t + 500, vx23 = (-88843 * t + 50140500)/106000, vx24 = (-17157 * t + 2859500)/106000, vx25 =$
$0, vx26 = (-43 * t + 21500)/100, vx28 = 0, vx29 = 0, vx30 = 0, vx31 = 0, vx32 = 0, vx33 = 0, vx34 =$
$0, vx35 = 0, vx36 = (-88843 * t + 50140500)/148400, vx37 = (-237243 * t + 56670100)148400, vx38 =$
$0, vx39 = 0\}\}\},$
$\{12971209885133 * t - 266526390810640 \leq 0 \; and \; 49 * t + 1020 \geq 0 ,$
$(3467191 * t - 77375280)/2500000 ,$
$\{\{vx01 = -t + 80, vx02 =$
$(-967191 * t + 77375280)/1000000, vx03 = (-32809 * t + 2624720)/1000000, vx04 = (-53 * t +$
$4240)/50, vx06 = 0, vx07 = 0, vx08 = 0, vx09 = 0, vx10 = 0, vx11 = 0, vx12 = 0, vx13 =$
$0, vx14 = 0, vx15 = 0, vx16 = 0, vx22 = (-301 * t + 24080)/1000, vx23 = 0, vx24 = (-301 * t +$
$24080)/1000, vx25 = 0, vx26 = (-12943 * t + 1035440)/100000, vx28 = 0, vx29 = 0, vx30 = 0, vx31 =$
$0, vx32 = 0, vx33 = 0, vx34 = 0, vx35 = 0, vx36 = 0, vx37 = -t + 44, vx38 = 0, vx39 = 0\}\}\},$
$\{158410787 * t - 5472862960 \geq 0 \; and \; 1119791 * t - 84089680 \leq 0 ,$
$(24190825203 * t - 585741616240)/36148315000 ,$
$\{\{vx01 = -t + 80, vx02 =$
Umbruch Meixner $(-1119791 * t + 84089680)/1032809, vx03 = (86982 * t - 1464960)/1032809, vx04 =$
$(-53 * t + 4240)/50, vx06 = (-5598955 * t + 420448400)/7229663, vx07 = 0, vx08 = 0, vx09 =$
$0, vx10 = 0, vx11 = 0, vx12 = 0, vx13 = 0, vx14 = (-5598955 * t + 420448400)/7229663, vx15 =$
$0, vx16 = (-59348923 * t + 4456753040)/72296630, vx22 = (798000 * t - 13440000)/1032809, vx23 =$
$(1108875509 * t - 38310040720)/1032809000, vx24 = (-301 * t + 24080)/1000, vx25 = 0, vx26 =$
$(343140 * t - 5779200)/1032809, vx28 = 0, vx29 = 0, vx30 = 0, vx31 = 0, vx32 = 0, vx33 =$
$0, vx34 = 0, vx35 = 0, vx36 = (158410787 * t - 5472862960)/206561800 vx37 = (-48151013 * t +$
$3615856240)/206561800, vx38 = 0, vx39 = 0\}\}\}\}.$

Realtime: 3 s

Interpretation:
Five feasible points are reached that means the result consists of five triplets containing not necessarily disjointed intervals (s. statistical information). After the chronologically sorting of the intervals, considering the behaviour of the objective function, there are only 3 pairwise disjointed intervals. The first both border on each other, the third is separated. The parametric term in each triplet symbolizes the optimal value of q w.r.t the interval.

5.4 Restricting the Computation - QAFIRO

To get a feasible solution of the quadratic problem the number of potential feasible points that the algorithm should comput is restricted by seven.

DEPTH FIRST SEARCH
$\longrightarrow$ ELIMINATION STEPS
Number of steps: 369
C1-strategy (Gauss): 66
C2-strategy (counting boundaries): 303
C3-strategy (scalarproduct): 0
V1-strategy (counting variables) 0
$\longrightarrow$ RESULTS
infeasible results: 320
feasible results: 7
best result: 7.

{-92610384617619/58216900000000,
{$vx01$ = 2901573/7630000, $vx02$ = 0, $vx03$ = 2901573/7630000, $vx04$ = 153783369/381500000, $vx06$ = 80, $vx07$ = 0, $vx08$ = 0, $vx09$ = 0, $vx10$ = 0, $vx11$ = 0, $vx12$ = 0, $vx13$ = 0, $vx14$ = 0, $vx15$ = 80, $vx16$ = 424/5, $vx22$ = 2901573/831670, $vx23$ = 2806375291443/831670000000, $vx24$ = 124767639/1090000000, $vx25$ = 0, $vx26$ = 124767639/83167000,
$vx28$ = 25999988251443/1164338000000, $vx29$ = 0, $vx30$ = 0, $vx31$ = 0,
$vx32$ =, $vx33$ = 0, $vx34$ = 0, $vx35$ = 0, $vx36$ = 2806375291443/1164338000000, $vx37$ = 602/25, $vx38$ = 1117999494812049/116433800000000, $vx39$ = 0}}

Realtime: 7s

Interpretation:
The process immediately stops after reaching 7 feasible points. The user can decide, whether this computed value for the objective function is small enough. The greatest integer less than the 7.th computed value is **-1**, so its difference to the optimal solution is **-0,59**.

6 Conclusions

The development of SYMOPT hasn't finished yet. But now the package has reached a stage of development that justifies publication.
We are continually looking for additional measures to alter elimination orders of variables and therefore to compute an optimal value more directly. Especially in solving nonlinear optimization problems the efficiency should be enhanced. For this purpose we are going to implement an algorithm for dual quadratic optimization problems [10].
Furthermore we will extend SYMOPT w.r.t more general types of objective functions - for example piecewise linear functions.
In order to make SYMOPT applicable for interested people not familiar with REDUCE, we are going to implement interfaces to a modeling language. This modeling language helps to develop and formulate decision and optimization problems [3].
We also plan a SYMOPT home-page that makes the package and some examples publically available, so that users can test algorithms and apply them for their own problems.

References

[1] Andreas Dolzmann and Thomas Sturm. Redlog—computer algebra meets computer logic. Technical Report MIP-9603, FMI, Universität Passau, D-94030 Passau, Germany, February 1996.
[2] Wolfgang Domschke. *Einführung in die Operations Research.* Springer-Verlag, Berlin, Heidelberg, 1990.
[3] Robert Fourer, David Gay, and Brian Kernighan. *A Modeling Language For Mathematical Programming.* The Scientific Press, San Franzisco, 1993.
[4] R.G. Jeroslow. Linear programs dependent on a single parameter. *Discrete Mathematics*, 6:119–140, 1973.
[5] Isolde Mazzucco. Linear optimization by variable elimination. Preliminary Report, November 1998.

[6] Isolde Mazzucco. Hyperbolic optimization by variable elimination. Preliminary Report, March 1999.

[7] Isolde Mazzucco. Optimization by variable elimination based on redlog. *Proceedings of RWCA'00*, pages 157–166, March 2000.

[8] Isolde Mazzucco. *User Manual - SYMOPT*. FMI, Universität Passau, Passau, 2001.

[9] Daniel Solow. *Linear Programming*. North-Holland, New York, Amsterdam, Oxford, 1984.

[10] Manfred Walk. *Theory of Duality in Mathematical Programming*. Springer Verlag, Wien, New York, 1989.

[11] Volker Weispfenning. Parametric linear and quadratic optimization by elimination. Technical Report MIP-9404, FMI, Universität Passau, D-94030 Passau, Germany, April 1994.

[12] Volker Weispfenning. Simulation and optimization by quantifier elimination. *Journal of Symbolic Computation*, 24(2):189–208, August 1997. Special issue on applications of quantifier elimination.

Representing Graph Properties by Polynomial Ideals

Michal Mnuk*

Institut für Informatik
Technische Universität München
D-80290 Munich, Germany
mnuk@in.tum.de

Abstract. Computer algebra provides means to obtain a diversity of results in many areas of science. In this paper we explore the possibility of representing properties of graphs by polynomial ideals. We show that several properties (emptiness, colorability) admit such representation, but we also work out limitations of this approach.

1 Introduction

Many areas of science benefit from the ability to describe their objects in various ways using formalisms from fields which are seemingly far away from each other. In most cases, this does not introduce a mere redundancy, but frequently opens new views and many times provides for unforseen insights.

Motivated by the work of Yu. Matiyasevich ([8]), we are going to describe an algebraic method to study properties of graphs. This method is developed to provide means to solve the decision problem whether a given graph G has some property P. Especially, we do *not* intend to design an all-embracing general solution but rather adapt some concepts from algebra to graph theory where these naturally fit.

Given a decision problem on graphs, we will transform each instance of it into an instance of the polynomial ideal membership problem by capturing the essence of the property P in a polynomial ideal and encoding the graph G as a homogeneous polynomial. Then, ideally, the graph G satisfies the property P if and only if the polynomial corresponding to G is contained in the ideal corresponding to P.

What are the supposed merits of such approach? First, numerous methods for solving the polynomial ideal membership problem are available, some of them have efficient implementation in software. Second, polynomial ideals are in most cases not "flat" but rather "structured" objects. This structure can possibly be transformed back into the graph theory where it could provide new insights. And, finally, even though graph properties and polynomial ideals are used to select (basically) equal subsets of the set of all graphs, these two concepts stress

* This work was partially supported by the *Euler International Mathematical Institute* at St. Petersburg, Russia.

different aspects. Hence, it is not unreasonable to assume that there will be classes of graphs which can be described and/or studied using polynomial ideals more easily than using logical formulas.

The approach we will be using here is as follows: Let G be a graph with n vertices. We assign each vertex of G a variable. The graph G will then be associated with a polynomial $f_G \in k[x_1, \dots, x_n]$, and we construct an ideal $\mathfrak{I}$ in $k[x_1, \dots, x_n]$ (depending on the property under consideration P, the number n, and usually on some other parameters) such that an assertion of the type

$$P(G) \Longleftrightarrow f_G \in \mathfrak{I} \tag{1}$$

is valid. However, we will see that this situation is not reachable under all circumstances. We will even prove that in some interesting cases the above relationship is impossible.

The next chapter gives a short introduction to polynomial ideals. After some general considerations regarding the properties which are suitable to be described by polynomial ideals we will start with representing the property of a graph to be nonempty. After showing that the colorability with a fixed number of colors is a property which can be described by a relationship as in (1), we will finally consider the property of a graph to contain a complete subgraph. This property is tightly connected with the colorability and the expectation is that they will behave similarly. However, it turns out that the containment of complete graphs cannot be expressed by means of polynomial ideals.

In Section 3 the notion of a Gröbner basis is extensively used. Readers not familiar with Gröbner bases may consult some of various text books such as [4].

2 Basic Notions and Concepts

A *graph* $G = (V, E)$ is a pair of finite sets V (vertices) and E (edges) with $E \subseteq \{\{u, v\} |\ u \neq v,\ u, v \in V\}$, i.e. all considered graphs are undirected with no loops or multiple edges. For the sake of simplicity we assume that the vertices of G are named $\{1, \dots, n\}$ where $n = |V|$ unless explicitly stated otherwise. We write $[n]$ to denote the set $\{1, \dots, n\}$.

The following definition associates every graph with a polynomial. It is originally motivated by the wish to express whether the vertices of a graph can be colored in such a way that no two adjacent vertices get the same color. However, same objects will prove useful also for deciding other properties, like the emptiness or whether a graph contains a complete subgraph.

Definition 1. *Let $G = (V, E)$ be a graph on n vertices $\{1, \dots, n\}$. With each vertex $i \in V$ we associate the variable x_i. Let $\preceq$ be a connex partial order on the set of variables $\{x_1, \dots, x_n\}$ (i.e. a reflexive, transitive, and antisymmetric order with $x_i \preceq x_j$ or $x_j \preceq x_i$ for any i, j). The* graph polynomial *f_G of G is given by*

$$f_G(x_1, \dots, x_r) = \prod_{\substack{\{i,j\} \in E \\ x_i \succeq x_j}} (x_i - x_j).$$

Note that the polynomial f_G is a homogeneous polynomial of degree $|E|$.

Remark 2. There are several ways to associate a graph with a polynomial. All of them aim at encoding graph properties. The approaches differ in the way which parts of the polynomial are used to encode the desired information. One approach is to use values of the polynomial at certain distinguished points – examples are the *Penrose* and the *Tutte* polynomial (cf. [10], [1]). Another way is to use polynomials as generating functions and to use coefficients – examples for this kind of encoding is the classical *characteristic* polynomial, the *matching* polynomial, and others (cf. [9], [5], [2]).

The graph polynomial defined above carries the information in its coefficients. This makes it possible to vary the ground field according to particular needs. Section 3.2 provides an example where the freedom of choosing the ground field makes solving a problem easier.

Throughout the paper we will denote by k an arbitrary field, and by $k[x_1, \dots, x_n]$ the ring of polynomials over k. Moreover, we assume that the variables are ordered $x_1 \preceq x_2 \preceq \dots \preceq x_n$. At this moment it may not be clear why an ordering on variables is needed – except for eliminating redundant factors from the graph polynomial. The reason becomes apparent when Gröbner bases come into play later in the paper.

As we already mentioned, the properties will be encoded in *polynomial ideals*. The structure of an ideal makes it possible to circumvent the problem which arises from the assignment of variables to vertices. This step virtually introduces labels – something we want to avoid. On the other hand, ideals constitute a framework where efficient computations can be performed and whose structure is to a large extent compatible with problems we intend to solve.

Definition 3. *Let R be a commutative ring. An* ideal $\mathfrak{a}$ *in R is an additive subgroup of R such that $\mathfrak{a}R \subseteq \mathfrak{a}$.*

Notation. *Given a (finite) set of elements $A = \{a_1, \dots, a_n\} \subseteq R$ we denote by $(a_1, \dots, a_n)$ the smallest ideal of R containing A, i.e. the set of all finite sums $\sum a_i r_i$, $r_i \in R$.*

After having defined basic ingredients we are going to clarify the precise meaning of the relationship (1). There are some questions to answer:

- In order to be able to define a graph polynomial we introduced an assignment of variables to vertices of G. Since this assignment is arbitrary, it should not have any influence on the result, and we require that the defined ideals do not depend on it.
- Since we are working with polynomials with a fixed number of variables, a property P usually cannot be captured in a single ideal. It will be described by defining an infinite sequence of ideals – one for each number of vertices of G.

Definition 4. *Let P be a predicate (property) defined on the set of all unlabeled graphs. We say that the sequence of ideals $(\mathfrak{I}_n)_{n\in\mathbb{N}}$* represents *$P$ if for all $n \in \mathbb{N}$ and for all unlabeled graphs G with n vertices the following holds*

$$P(G) \iff f_{\pi(G)} \in \mathfrak{I}_n \text{ for all permutations } \pi \in S_n,$$

where $\pi(G)$ corresponds to G where each vertex i has been renamed to $\pi(i)$, and S_n denotes the permutation group on n elements.

Whenever it is clear (or irrelevant) what the parameter n is and how the sequence $\mathfrak{I}_n$ is constructed, we will omit the reference to n saying that "the ideal $\mathfrak{I}$" describes the property P.

Before we proceed, let us see what implications arise from the decision to choose ideals as the underlying structure. Suppose, the ideal $\mathfrak{I} \subseteq k[x_1, \ldots, x_n]$ represents a property P. This means that the graph G satisfies P if and only if the graph polynomial f_G lies in $\mathfrak{I}$. Now, let $g \in k[x_1, \ldots, x_n]$ be a polynomial such that the product gf_G turns out to be a graph polynomial (i.e. is a square-free product of differences of variables) of some graph G' (which contains G). Since $\mathfrak{I}$ is an ideal, the product gf_G belongs to $\mathfrak{I}$ and thus G' has to satisfy P too. Since this holds for arbitrary polynomials g, all properties P to be represented by ideals must be *monotone*, i.e. they satisfy the following condition:

> Let G and G' be two graphs such that G is a subgraph of G'. If P holds for G then it holds also for G'.

Fortunately, quite some properties, which we are interested in, are monotone. Some of them will be studied in the next section.

3 Special Graph Properties

In this section we consider three properties of graphs for which we would like to obtain an algebraic description in the sense of Definition 4:

(1) the graph contains at least one edge;
(2) the vertices of a graph are not properly colorable by a fixed number of colors;
(3) the graph contains a complete subgraph.

Next, we will construct ideals representing the first two properties, and we will prove that the third one does not admit such representation.

3.1 Non-Emptiness

We start with a simple property of a graph to contain at least one edge. For any n we want to find an ideal $\mathfrak{E}_n$ such that the following holds: Let G be a graph with n vertices and f_G the graph polynomial corresponding to some assignment of variables $\{x_1, \ldots, x_n\}$ to vertices of G. Then G is not empty if and only if $f_G \in \mathfrak{E}_n$. Note that if G contains at least one edge, the polynomial f_G has degree at least 1 and vice versa. Hence, we are looking for an ideal containing all polynomials of degree at least 1.

Obviously, $(x_1, \dots, x_n) \subseteq k[x_1, \dots, x_n]$ has the required property. If a graph contains at least one edge, its graph polynomial is a homogeneous polynomial of degree at least one and hence lying in $(x_1, \dots, x_n)$. On the other hand, if a graph has no edges its graph polynomial is 1 which is not contained in the ideal $(x_1, \dots, x_n)$. Otherwise we could write 1 as a linear combination of $\{x_1, \dots, x_n\}$ with polynomial coefficients. Setting all variables to zero would yield a contradiction.

We note that this ideal – and, in general, all ideals which represent some property – is not unique as it can be replaced for example by $(x_n - x_1, x_n - x_2, \dots, x_n - x_{n-1})$. This ideal represents the non-emptiness of a graph as well.

3.2 Coloring of Graphs

In 1974 Yu. Matiyasevich ([8]) discovered a way to describe proper vertex coloring of graphs in terms of properties of coefficients of graph polynomials (see [8]). In this section we use these ideas to construct an ideal which represents the property of a graph to be *not* colorable with a fixed number of colors.

Let $G = (V, E)$ be a graph with n vertices and let $1 \leq r \leq n$ be an integer. A *proper coloring* of vertices of G with colors $\{c_1, \dots, c_r\}$ is an assignment $\varphi : V \to \{c_1, \dots, c_r\}$ of colors to vertices of G such that no two vertices connected with an edge are assigned the same color, i.e. $(\forall i \neq j)(\{v_i, v_j\} \in E \Rightarrow \varphi(v_i) \neq \varphi(v_j))$. Let us denote by $\bar{C}_r^n(G)$ the property that a graph G with n vertices has *no* proper coloring with at most r colors. Since a graph which can be properly colored with a single color has no edges, we assume from now on that $r \geq 2$.

Theorem 5. *Let*

$$U_r^n := (x_1^r - 1, \dots, x_n^r - 1) \subseteq k[x_1, \dots, x_n].$$

The ideal U_r^n represents the property of a graph to be not properly colorable by at most r colors.

Proof. Let G be a graph with n vertices such that

$$\bar{C}_r^n(G) \text{ holds,} \tag{2}$$

i.e. there is no proper coloring of G with at most r colors. We will show that (2) is true if and only if its graph polynomial lies in U_r^n.

Let F denote some field. If the colors c_i's are elements from F, the property (2) means that for all subsets $\Gamma \subseteq F$ with at most r elements and for all assignments $\psi : \{x_1, \dots, x_n\} \to \Gamma$ of colors to variables we obtain

$$f_G(\psi(x_1), \dots, \psi(x_n)) = 0, \tag{3}$$

if f_G is considered as a polynomial over F. Due to the special form of f_G and due to the fact that we are working in a field it is easy to see that the above assertion is equivalent to the fact that (3) is true for *some* set Γ with r pairwise different elements and for all assignments $\psi : \{x_1, \dots, x_n\} \to \Gamma$. Especially we can choose Γ to contain only nonzero elements of F. Thus (2) is equivalent to

$$f_G(x_1, \dots, x_n) \equiv 0 \text{ on } \Gamma^n. \tag{4}$$

Now we are going to specify precisely how the colors $c_1, \dots, c_r$ in Γ are selected.

Let q be a natural number with $q \equiv 1 \bmod r$ which is a power of some prime p. (If p is chosen so that it does not divide r, the cyclic subgroup of the multiplicative group $\mathbb{Z}_r^*$ generated by p is obviously finite. It is easy to see that we may set q to be $p^{\mathrm{ord}(p)}$, where $\mathrm{ord}(p) > 0$ is the *order* of p in $\mathbb{Z}_r^*$.) Then there is a finite field $\mathrm{GF}(q)$ with q elements (for a summary of properties of finite fields see e.g. [7]). Now, the key idea is to select the colors $\{c_1, \dots, c_r\}$ to be certain distinguished elements of $\mathrm{GF}(q)$.

We recall that every finite field $\mathrm{GF}(q)$ has a primitive element, i.e. an $\alpha \in \mathrm{GF}(q)$ such that any nonzero element $a \in \mathrm{GF}(q)$ can be written as $a = \alpha^i$ for some integer i. Moreover, such elements satisfy the equation

$$X^{q-1} = 1. \tag{5}$$

Let m be such that $q = rm + 1$. We set

$$c_i := \alpha^{(i-1)m}, \text{ for } 1 \le i \le r.$$

Since α is primitive, the set

$$\Gamma := \{1, \alpha^m, \alpha^{2m}, \dots, \alpha^{(r-1)m}\} \tag{6}$$

consists of r pairwise different elements of $\mathrm{GF}(q)$.

Now, if the graph polynomial f_G of G is fully reduced modulo U_r^n (i.e. every occurrence of x_i^r in f_G is replaced by 1), every variable has degree at most $r-1$. The reduction process corresponds to subtraction of elements of U_r^n from f_G. Hence, f_G can be written as

$$f_G = \tilde{f}_G + u, \tag{7}$$

for some $u \in U_r^n$. We will prove that $\tilde{f}_G$ must be zero, i.e. all coefficients of $\tilde{f}_G$ vanish.

We assumed that G has no proper coloring with (at most) r colors. As we saw above, this is equivalent to the fact that f_G vanishes in all points of Γ^n. As $u \in U_r^n$, it is easily seen that the same is true for the polynomial u. Hence, the equation (7) shows that also $\tilde{f}_G$ vanishes on Γ^n. However, the following lemma (see e.g.[3]) shows that this is impossible unless it is identically zero.

Lemma 6. *Let $f(x_1, \dots, x_n) \in k[x_1, \dots, x_n]$ be a polynomial where the degree $\deg_{x_i}(f)$ of the variable x_i is bounded by t_i and in which every variable x_i occurs in some term with a nonzero coefficient. Let $\{S_i\}_{i=1}^n$ be a set of subsets of k such that $|S_i| > t_i$ for each $1 \le i \le n$. Then there is a collection $(s_1, \dots, s_n) \in S_1 \times S_2 \times \dots \times S_n$ with $f(s_1, \dots, s_n) \ne 0$.*

Proof. If $n = 1$, the assertion follows from the well known fact that a univariate nonzero polynomial of degree t has at most t roots. The general case is inferred by induction. □

Now, let $\{x_{i_1}, \ldots, x_{i_s}\}$ be the set of variables which effectively occur in $\tilde{f}_G$. Since the degree of any variable in $\tilde{f}_G$ is at most $r - 1$ and this polynomial vanishes on Γ^n, the previous lemma restricted to $\{x_{i_1}, \ldots, x_{i_s}\}$ yields that $\tilde{f}_G$ must vanish identically. Then (7) finally implies

$$f_G \in (x_1^r - 1, \ldots, x_n^r - 1). \tag{8}$$

On the other hand, if (8) holds then there is obviously no proper coloring of G with at most r colors. □

To conclude, we proved that for a graph $G = (V, E)$ on n vertices and for any assignment $\psi : V \to \{x_1, \ldots, x_n\}$ of variables to vertices the graph G has no proper coloring with at most r colors if and only if the corresponding graph polynomial f_G lies in the ideal U_r^n:

$$\bar{C}_r^n(G) \Longleftrightarrow f_G \in U_r^n. \tag{9}$$

3.3 Representation of Subgraphs

After we have affirmatively answered the question about representing non-emptiness and colorability of graphs by ideals, we will consider another interesting property. Let G and H be two graphs with n and r vertices, resp. We are looking for an ideal which would represent the property of H being a "subgraph" of G (as an unlabeled structure). To be more precise, we require G to contain an isomorphic copy of H. Let us elaborate this point.

Both G and H are given by sets of vertices and edges: $G = (V, E)$, $H = (W, F)$. Then, by definition, H is a *subgraph* of G if $W \subseteq V$ and $F \subseteq E$. However, here the names of vertices play an important role. We want to represent a similar property without referring to these names.

Definition 7. *Let G and H be as above. The graph G* contains an isomorphic copy *of H if there is an injective map $\rho : W \to V$ such that $\rho(H)$ is a subgraph of G (by $\rho(H)$ we denote the graph H after renaming each vertex $i \in W$ to $\rho(i) \in V$). This property will be denoted by $H \lesssim G$.*

Remark 8. To keep the notation simple we will use the term "subgraph" in the more general sense of "contains an isomorphic copy".

How should an ideal $\mathfrak{J}$ corresponding to graphs containing H as a subgraph look like? If each vertex $i \in W$ is assigned a variable x_i we obtain a graph polynomial $f_H \in k[x_1, \ldots, x_r]$. The polynomial f_H has to lie in $\mathfrak{J}$. In Definition 4 we stipulated that the construction is to be independent of a particular naming of vertices. Hence, $\mathfrak{J}$ has to contain $f_{\pi(H)}$ for all permutations $\pi \in S_r$ on $[r]$. Moreover, according to Definition 7, the ideal also should not depend on the names of vertices. This yields a motivation for the following definition.

Definition 9. *For two graphs H and H' we write $H \cong H'$ if H' arises from H by renaming its vertices or, in other words, if H contains an isomorphic copy of H' and vice versa.*

Let H be a graph with r vertices and $n \geq r$ an integer. The graph ideal of H in $k[x_1, \ldots, x_n]$ *denoted by $\mathfrak{I}_H^n$ is given by*

$$\mathfrak{I}_H^n := (\{f_{H'} \,|\, H' = (W, F),\, W \subseteq \{1, \ldots, n\},\, H' \cong H\}) \subseteq k[x_1, \ldots, x_n].$$

The hope is that the ideal $\mathfrak{I}_H^n$ will contain exactly graph polynomials of those graphs which contain H.

The non-emptiness problem we considered above is a special instance of the subgraph problem. So let us apply the definition to the case where H is a single edge.

Example 10. Let G be a graph with n vertices and let P_1 denote a graph consisting of a single edge. Then the graph ideal of P_1 in $k[x_1, \ldots, x_n]$ is

$$\mathfrak{I}_{P_1}^n = (\{x_i - x_j \,|\, 1 \leq i, j \leq n\}).$$

Let $f_i := x_n - x_i$, for $1 \leq i \leq n-1$. Then $x_i - x_j = f_j - f_i$ and hence

$$\mathfrak{I}_{P_1}^n \subseteq (x_n - x_1, x_n - x_2, \ldots, x_n - x_{n-1}).$$

The other inclusion is obvious and we obtain the equality of both ideals.

Thus the graph ideal $\mathfrak{I}_{P_1}^n$ represents correctly the property of a graph to contain at least one edge.

Graphs Containing a Complete Subgraph. An important special instance of the subgraph problem is the question whether a given graph contains a complete subgraph, i.e. a graph with maximum possible number of edges. We will show that in general the graph ideal of a complete graph does not fully satisfy our requirements specified in the Definition 4.

Let K_r denote the complete graph on r vertices. The specialization of Definition 9 for this case yields the graph ideal $\mathfrak{K}_r^n := \mathfrak{I}_{K_r}^n$ generated by the set

$$\mathcal{B}_{K_r}^n := \{f_H \,|\, H = (W, F),\, W \subseteq \{1, \ldots, n\},\, H \cong K_r\}. \tag{10}$$

In other words, the ideal $\mathfrak{K}_r^n$ is generated by graph polynomials of all complete graphs on r vertices which are named by a subset of $[n]$ with r elements.

It is a remarkable fact that $\mathcal{B}_{K_r}^n$ is a Gröbner basis of $\mathfrak{K}_r^n$ with respect to the degree lexicographical ordering. This implies that, first, $\mathcal{B}_{K_r}^n$ is a concise description of the ideal, and, second, due to the strength of Gröbner bases many properties of $\mathfrak{K}_r^n$ may be inferred by inspecting merely $\mathcal{B}_{K_r}^n$.

Theorem 11. *The set $\mathcal{B}_{K_r}^n$ is a Gröbner basis of $\mathfrak{K}_r^n$ with respect to the degree lexicographic ordering.*

Proof. Let $\preceq$ denote the degree lexicographical ordering. A Gröbner basis of an ideal $\mathfrak{I}$ is a set of polynomials such that the leading terms of these polynomials generate a monomial ideal which is equal to $\mathrm{lt}(\mathfrak{I})$ (w.r.t. some term ordering), where the *leading term* of a polynomial is its largest term (w.r.t. $\preceq$) and

$$\mathrm{lt}(\mathfrak{I}) := \{\mathrm{lt}(f) | \, f \in \mathfrak{I}\}.$$

From this it is clear that any set of polynomials containing a Gröbner basis is again a Gröbner basis. We will show that the set

$$\mathcal{G}^n_{K_r} := \{f_H | \, H = (\{1\} \cup W', F),\, W' \subseteq \{2, \dots, n\},\, H \cong K_r\} \subseteq \mathcal{B}^n_{K_r}, \tag{11}$$

i.e. the set of graph polynomials of all labelings of K_r containing the vertex 1 is a reduced Gröbner basis of $\mathfrak{K}^n_r$ with respect to the degree lexicographic ordering $\preceq$.

Without loss of generality we assume $x_1 \preceq x_2 \preceq \dots \preceq x_n$. In the sequel we will index the elements of $\mathcal{B}_{K_r}$ by natural numbers instead of graphs. We also assume that all notions which require an ordering refer to the degree lexicographical order. Let $\mathcal{G}^n_{K_r} = \{g_1, \dots, g_t\}$. We will show

$$\mathrm{lt}(\mathfrak{K}^n_r) = (\mathrm{lt}(g_1), \dots, \mathrm{lt}(g_t)). \tag{12}$$

Since $\mathcal{G}^n_{K_r} \subseteq \mathcal{B}^n_{K_r}$ the inclusion $(\mathrm{lt}(g_1), \dots, \mathrm{lt}(g_t)) \subseteq \mathrm{lt}(\mathfrak{K}^n_r)$ follows immediately. To show the equality, let $f = \sum_{i=1}^{s} p_i f_i \in \mathfrak{K}^n_r$, $p_i \in k[x_1, \dots, x_n]$, and $f_i \in \mathcal{B}^n_{K_r}$. We will prove that there exists an i such that $\mathrm{lt}(g_i) | \, \mathrm{lt}(f)$. Then $\mathrm{lt}(\mathfrak{K}^n_r) \subseteq (\mathrm{lt}(g_1), \dots, \mathrm{lt}(g_t))$ and the equality (12) follows. Since all polynomials are defined over a field, for the sake of simplicity we will work with leading *monomials* (product of powers of variables without the coefficient) instead of leading *terms*.

How do the leading monomials of polynomials from $\mathcal{B}^n_{K_r}$ look like? First, since a graph polynomial is a product of $\binom{r}{2}$ linear factors, it is homogeneous. If a polynomial f involves variables $x_{i_r}, \dots, x_{i_1}$, where $i_r > i_{r-1} > \dots > i_1$, its leading monomial is

$$x_{i_r}^{r-1} x_{i_{r-1}}^{r-2} \cdot \ldots \cdot x_{i_3}^2 x_{i_2}$$

(note that $i_2 > 1$). Let $\mathrm{lm}^{(i)}(f)$ denotes the i-th largest monomial of f. Then, e.g.,

$$\mathrm{lm}^{(2)}(f) = x_{i_r}^{r-1} x_{i_{r-1}}^{r-2} \cdot \ldots \cdot x_{i_3}^2 x_{i_1}.$$

If we consider the $r-1$ exponents of variables in each term of f as a vector in $\mathbb{N}^{r-1}$, the set E of such vectors is a subset of $\mathbb{N}^{r-1}$ consisting of all vectors $(e_{k_1}, \dots, e_{k_{r-1}})$ with

$$\sum_{l=1}^{r-1} e_{k_l} = \binom{r}{2} \quad \text{and} \quad e_{k_l} \le r-1, \text{ for all } 1 \le l \le r-1.$$

The terms of f are sorted in the decreasing lexicographical order of their exponent vectors.

To see that for $f = \sum_{i=1}^{s} p_i f_i \in \mathfrak{K}_r^n$ the leading monomial $\mathrm{lm}(f)$ is divisible by some $\mathrm{lm}(g_i)$, we distinguish two cases depending on whether $\mathrm{lm}(f)$ stems from the leading monomial of a single term $p_i f_i$ or whether it arouse by canceling leading monomials of some terms.

Case 1. ($\mathrm{lm}(f) = \mathrm{lm}(p_i f_i)$ for some i) In this case $\mathrm{lm}(f_i) | \mathrm{lm}(f)$ and since $\mathrm{lm}(f_i)$ is equal to some $\mathrm{lm}(g_j)$ the assertion follows immediately.

Case 2. ($\mathrm{lm}(f)$ is different from all $\mathrm{lm}(p_i f_i)$) Then there are indices $i < j$ such that $\mathrm{lm}(p_i f_i) = \mathrm{lm}(p_j f_j)$ (there may be more than two indices with this property). Let $\{i_1, \dots, i_r\}$ and $\{j_1, \dots, j_r\}$ be the variables involved in $\mathrm{lm}(p_i f_i)$ and $\mathrm{lm}(p_j f_j)$, resp. We assume $i_r > i_{r-1} > \dots > i_1$ and $j_r > j_{r-1} > \dots > j_1$. Then we have for some monomials Y and Y'

$$\begin{aligned} \mathrm{lm}(p_i f_i) &= Y x_{i_r}^{r-1} x_{i_{r-1}}^{r-2} \cdot \dots \cdot x_{i_3}^2 x_{i_2} \\ \mathrm{lm}(p_j f_j) &= Y' x_{j_r}^{r-1} x_{j_{r-1}}^{r-2} \cdot \dots \cdot x_{j_3}^2 x_{j_2}. \end{aligned}$$

Let $l \geq 1$ be the least index such that $i_l \neq j_l$ and $i_k = j_k$, for $1 \leq k < l$. Without loss of generality we assume $i_l > j_l$, i.e. $x_{i_l} \succeq x_{j_l}$. The leading monomials of $p_i f_i$ and $p_j f_j$ have the form

$$\mathrm{lm}(p_i f_i) = \mathrm{lm}(p_j f_j) = U x_{i_l}^{l-1} x_{j_l}^{l-1} \cdot \prod_{k=1}^{l-1} x_{i_k}^{k-1},$$

where U is divisible by

$$\prod_{t \in \{\{i_{l+1}, \dots, i_r\} \cup \{j_{l+1}, \dots, j_r\}} x_t^{\epsilon_t}, \tag{13}$$

with $\epsilon_t = k - 1$ if $t = i_k$ or $t = j_k$.

Let l be as above and let

$$\sigma := |\{(e_1, \dots, e_{l-1}) \subseteq \mathbb{N}^{l-1} \,|\, e_i \leq r - 1, \sum_{s=1}^{l-1} e_s = \binom{l}{2}\}|.$$

It is not hard to see that the σ largest monomials of $p_i f_i$ and $p_j f_j$ will be identical and thus will not appear in f. Moreover,

$$\begin{aligned} \mathrm{lm}^{(\sigma+1)}(p_i f_i) &= U x_{i_l}^{l-2} x_{j_l}^{l-1} x_{i_{l-1}} \cdot \prod_{k=1}^{l-1} x_{i_k}^{k-1} \\ \mathrm{lm}^{(\sigma+1)}(p_j f_j) &= U x_{i_l}^{l-1} x_{j_l}^{l-2} x_{j_{l-1}} \cdot \prod_{k=1}^{l-1} x_{i_k}^{k-1} = U x_{i_l}^{l-1} x_{j_l}^{l-2} x_{i_{l-1}} \cdot \prod_{k=1}^{l-1} x_{i_k}^{k-1}. \end{aligned}$$

Thus $\mathrm{lm}^{(\sigma+1)}(p_j f_j)$ becomes the leading monomial of f. Using (13) we see that $\mathrm{lm}(f)$ is divisible by

$$\mathrm{lm}(f_i) = \prod_{k=1}^{r} x_{i_k}^{k-1}$$

and hence by $\mathrm{lm}(g_i)$ for some $g_i \in \mathcal{G}_{K_r}$.

□

The Case K_2 and K_3. In the special case $r = 2$, the basis $\mathcal{B}^n_{K_2}$ of the ideal $\mathfrak{I}^n_{K_2}$ consists of graph polynomials of all edges $\{\{i,j\} \mid i,j \in [n], i \neq j\}$. This is the ideal $\mathfrak{I}^n_{P_1}$ considered in the Example 10. Hence, the ideal $\mathfrak{I}^n_{K_2}$ correctly represents the property of a graph on n vertices to contain an edge.

Now, in general, does $\mathfrak{K}^n_r$ represent the property of a graph with n vertices to contain an isomorphic copy of a complete graph K_r? We will show that the answer is – unfortunately – *No*. The counterexample yields already the ideal $\mathfrak{K}^n_3$, generated by all graph polynomials of triangles.

First, it is clear that if a graph contains a triangle then its graph polynomial (regardless of naming of vertices) is in $\mathfrak{K}^n_3$. However, the reverse direction is not true.

Theorem 12. *Let C_m denote the circle with $m \geq 3$ vertices, i.e. a graph with edges $\{\{i,i+1\} \mid 1 \leq i \leq m\} \cup \{\{1,m\}\}$, and $\pi \in S_m$ a permutation on the set $[m]$. The ideal $\mathfrak{K}^n_3$ contains the graph polynomial of all $\pi(C_m)$ for any odd $m \leq n$.*

Proof. The assertion will be proved by induction on m.

When $m = 3$ the statement is trivial since the graph polynomials of all $\pi(C_3)$ are equal to $f_{C_3} \in \mathcal{B}^n_{K_3}$.

Let $5 \leq m \leq n$ be an odd positive integer and $\pi \in S_m$ a permutation. Then

$$f_{C_m} = \underbrace{(x_m - x_{m-1})(x_m - x_1)(x_{m-1} - x_{m-2}) \cdot \ldots \cdot (x_5 - x_4)}_{=:\sigma} \cdot (x_4 - x_3)(x_3 - x_2)(x_2 - x_1).$$

We obtain

$$\begin{aligned} & f_{C_m} + \sigma f_{C_3} \\ & = \sigma(x_4 - x_3)(x_3 - x_2)(x_2 - x_1) + \sigma(x_3 - x_2)(x_3 - x_1)(x_2 - x_1) \\ & = \sigma(x_4 - x_1)(x_3 - x_2)(x_2 - x_1) \\ & = f_{\pi'(C_{m-2})}(x_3 - x_2)(x_2 - x_1), \end{aligned}$$

for some $\pi' \in S_{m-2}$. Thus $f_{C_m} + \sigma f_{C_3}$ corresponds to a graph polynomial which is a multiple of $f_{\pi'(C_{m-2})}$ for some $\pi' \in S_{m-2}$. Since both f_{C_3} and $f_{\pi'(C_{m-2})}$ are in $\mathfrak{K}^n_3$ the assertion follows for the identity permutation $\pi = \mathrm{id}$. The proof for arbitrary π is done by a simple modification of the above argument. □

This is a bad news as it destroys any hope to success by "massaging" the ideal $\mathfrak{K}_3^n$. We immediately obtain:

Proposition 13. *There exists no ideal representing the property of graphs to contain a complete subgraph K_3.*

Proof. If such ideal existed it would contain graph polynomials of all triangles (with vertex names taken from $[n]$ for some n) and hence the whole $\mathfrak{K}_3^n$. Consequently, graph polynomials of circles of odd length would be contained in it as well. □

Despite of this negative statement, Theorem 12 provides for a somewhat surprising conclusion.

Corollary 14. *The ideal $\mathfrak{K}_3^n$ contains graph polynomials of all graphs on n vertices which are* not *bipartite, i.e. which cannot be properly colored with two colors.*

Proof. Let G be a graph with n vertices which is not bipartite. To keep the proof simple, we borrow a theorem from graph theory saying that every non-bipartite graph contains a circle with an odd number of edges (see e.g. [6]). Then the graph polynomial f_G is a multiple of $f_{\pi(C_m)}$ for some odd m and some $\pi \in S_m$. Thus $f_G \in \mathfrak{K}_3^n$. □

On the other hand, $\mathfrak{K}_3^n$ is generated by polynomials of the form $(x_i - x_j)(x_i - x_k)(x_j - x_k)$, for some $1 \leq k < j < i \leq n$. Taking the normal form with respect to the ideal U_2^n (see Section 3.2) – which is zero – we immediately see that every such polynomial is in U_2^n and hence

$$\mathfrak{K}_3^n \subseteq U_2^n. \tag{14}$$

Thus if G is a graph with n vertices such that $f_G \in \mathfrak{K}_3^n$ then $f_G \in U_2^n$. In Section 3.2 we proved that in this case G cannot be properly colored with two colors – it is not bipartite.

What we got is the representation of not bipartite graphs! The same property which was considered in Section 3.2.

Theorem 15. *The ideal $\mathfrak{K}_3^n$ represents the property of a graph to be not bipartite.*

To summarize this subsection, we saw that the ideal $\mathfrak{K}_2^n$ represents the property of a graph to contain K_2 (an edge). However, it proved to be impossible to extend this statement to the general case due to the fact that $\mathfrak{K}_3^n$ contains circles of odd length. Finally, we were able to prove that $\mathfrak{K}_3^n$ represents the same property as U_2^n – the property of a graph to be not bipartite.

The General Case. When graphs containing complete subgraphs K_r, for $r \geq 4$, are considered, the ideal $\mathfrak{K}_r^n$ contains more and more polynomials corresponding to "exceptional" graphs, i.e. those graphs whose graph polynomial lies in the ideal but which do not contain K_r. The nature of these exceptional graphs is diverse and does not allow a simple description as in the case $r = 3$.

Computations showed that there are e.g. 72 graphs with 6 nodes not containing K_4 but whose polynomials lie in $\mathfrak{K}_4^6$. Similarly, there are 252 graphs with 7 nodes not containing K_5 whose polynomials lie in $\mathfrak{K}_5^7$. These graphs do not fall into any single graph theoretic category. However, we saw that there is a connection between them and the corresponding complete graphs – the defining ideal. This matter requires further investigation to see whether it may be exploited to obtain something useful.

4 Conclusion

In this paper we were exploring the possibility of representing graph properties by polynomial ideals. We showed that some properties (non-emptiness, coloring) are suitable to be treated by this approach, but in some interesting cases a representation (in the sense of Definition 4) is impossible – the ideals proved to be too coarse to capture the desired property exactly. Nevertheless, they may serve as an "approximate" representation.

The proposed approach has many positive aspects, but also some limitations. On the other hand, the fact that it cannot mimic graph theory exactly is not necessarily a failure. It merely leads to new classes of graphs characterized by common properties which cannot be efficiently described in the language of graph theory.

Acknowledgements

The author would like to express his thanks to Yuri Matiyasevich for his motivation, helpfulness, and for providing an excellent environment while the the author was visiting the *Euler International Mathematical Institute* at St. Petersburg, Russia.

References

1. Martin Aigner. The Penrose polynomial of a plane graph. *Mathematische Annalen*, 307:173–189, 1997.
2. N. Alon and M. Tarsi. Colorings and orientation of graphs. *Combinatorica*, 12(2):125–134, 1992.
3. Noga Alon. Combinatorial Nullstellensatz. *Combinatorics, Probability and Computing*, 8:7–29, 1999.
4. Thomas Becker, Volker Weispfenning, and Heinz Kredel. *Gröbner bases. A computational approach to commutative algebra*, volume 141 of *Graduate Texts in Mathematics*. Springer-Verlag, New York-Berlin-Heidelberg-London-Paris-Tokyo-Hong Kong-Barcelona-Budapest, 1993.
5. Robert A. Beezer and E.J. Farell. The matching polynomial of a regular graph. *Discrete Mathematics*, 137:7–18, 1995.

6. Frank Harary. *Graph Theory*. Addison-Wesley series in mathematics. Addison-Wesley, Reading, Mass., 1972.
7. Serge Lang. *Algebra*. Addison Wesley, third edition, 1997.
8. Yuri Matiyasevich. A criterion for colorability of vertices of a graph formulated in terms of edge orientation. In *Discrete Analysis*, volume 26, pages 65–71. Mathematical Institute, Academy of Sciences of USSR, 1974. in Russian.
9. I. Sciriha and S. Fiorini. On the characteristic polynomial of homeomorphic images of a graph. *Discrete Mathematics*, 174:293–308, 1997.
10. Dominic Welsh. The Tutte polynomial. *Random Structures and Algorithms*, 15(3–4):210–228, 1999.

Parametric G^1–Blending of Several Surfaces*

Sonia Pérez–Díaz and Rafael Sendra

Dpto de Matemáticas
Universidad de Alcalá
E-28871 Madrid, Spain
{sonia.perez,rafael.sendra}@uah.es

Abstract. In this paper we present a symbolic algorithm for blending parametrically with G^1–continuity a collection of rational surfaces with rational clipping intersections. The method provides a family of parametrizations that depends on a set of parameters, which number can be controlled in advance, and that can be used further in the modeling process. For this purpose, we introduce the notion of curves being in good position, that can always be assumed w.l.o.g., and we prove the main property of a set of rational curves in good position, that guarantees that the method always provides a correct solution.

1 Introduction

One of the main problems in Computer Aided Geometric Design is modeling objects (see [9]). Usually, one models the object as a collection of surfaces. However, in many cases, one wants this collection to form a composite object whose surface is smooth. This question leads to the blending problem. In fact, a blending surface is a surface that provides a smooth transition between distinct geometric features of an object.

Roughly speaking, if $V_1, \dots, V_n$ (surfaces to be blended), and $U_1, \dots, U_n$ (clipping surfaces) are given, the blending problem consists in finding an algebraic surface V such that $C_i = U_i \cap V_i \subset V$, and V meets each V_i at C_i with "certain" smooth conditions (G^k–continuity, see [2]). In this paper we only deal with G^1–continuity (i.e. tangent plane continuity along C_i) but the method presented can be extended to G^k–continuity. Thus, the problem of blending smoothly a collection of surfaces may be decomposed into two separate subproblems. The first one focuses on finding appropriate clipping surfaces U_i (see [8], [14]), and the second assumes that the clipping surfaces are given, and deals with the question of determining the final blending surface V (see [7], [14]). In this paper we investigate on the second problem, that we will call in the sequel the blending problem.

In addition, the blending problem can be approached from two different points of view, namely, implicitly (see [8], [15]) or parametrically (see [3],[13]). In the first case, one wants to compute an implicit expression of the solution while

* Authors partially supported by DGES PB98-0713-C02-01 and DGES HU1999-0029.

in the second parametric solutions are required. Furthermore, a second consideration, depending on whether either symbolic or numerical techniques are used, can be made (see [1], [5], [9] for numerical techniques, and [9], [13] for symbolic techniques). In this paper, we are interesting in symbolic algorithms to solve the problem. For the implicit blending, it holds that the family of solutions are in the intersection of some polynomial ideals generated by the implicit equations of V_i, and powers of the equations of U_i, and therefore elimination techniques as Gröbner basis can be applied (see [14]). On the hand, if the question is approached from the parametric point of view, one may ask whether there exists rational blending solutions of the problem, and if so, derive parametrizations of them. For this purpose, one may impose that V_i are rational, that parametrizations of them are known, and that the clipping surfaces U_i are taken such that the clipping curves C_i are also rational. Some contributions for the parametric case have been done ([4], [11]), mainly on the numerical aspects, but most of the authors have addressed the implicit version of the problem ([8], [15]).

In this paper we investigate the parametric version of the problem, and we present a symbolic parametric algorithm. A parametric method, based on Hermite functions, can be found in [3]. In [6] a similar approach, to the one presented here, is given. In [6] the interpolation is done using parametrizations of the primary surfaces instead of using parametrizations of the clipping curves, and only at most two modeling parameter are free. Furthermore, the interpolating coefficients in [6] are in general rational functions. However, in our method, the number of modeling parameters can be freely chosen, and the interpolating coefficients may be forced to be polynomials if syzygies are used.

We prove that one can always take the clipping surfaces, and parametrizations of the clipping curves, in such a way that an interpolation of them, along certain sweeping curves, provides parametric blending solutions. As a consequence of these results, we present a symbolic algorithm that generates directly, from the clipping parametrizations, parametrizations of families of blending surfaces that depend on additional parameters. Furthermore, one may control in advance the number of parameters to be involved in the final solutions to afterwards utilize them to model properly the final object. The solution that we present is given for the case of G^1–continuity, but the method can be extend to the G^k–continuity. For space limitations we do not consider this here.

The structure of the paper is as follows. The second section is introductory and states formally the parametric blending problem. In the third section, we introduce the notion of curves being in good position (see Definition 2). We prove that one can always assume w.l.o.g. that the clipping surfaces are given in good position, and we state the main property of a set of rational curves being in good position (see Theorems 1 and 2, and Corollary 2), that basically ensures that certain linearly independent properties for parametrizations of them hold. These properties are fundamental to derive deterministically the blending solution (see Theorems 4 and 5, and Corollary 3). In Section 5 we show how all these results can be used to derive a symbolic algorithm for the problem, and examples are given. In the last section, we present a practical implementation of the methods and real computing times are given.

2 The Parametric Blending Problem

In order to state properly the parametric blending problem we consider a field $\mathbb{K}$ of characteristic zero, and we denote by $\overline{\mathbb{K}}$ its algebraic closure; in practical applications $\mathbb{K}$ is the field of rational numbers or the field of real numbers.

First of all, we introduce the notion of G^1-continuity. Essentially the idea is to require surfaces meeting tangentially, i.e. surfaces intersecting along a given curve with tangent plane continuity. The concept of G^k–continuity can be stated similarly. For further details we refer to [2]. More precisely, the notion of G^1–continuity can be defined as follows (see [2], [9], [10], [15]).

Definition 1. *Let V_1, V_2 be surfaces over $\overline{\mathbb{K}}$, and $C \subset V_1 \cap V_2$ an irreducible curve such that V_1, V_2 are smooth at all but finitely many points on C. Then, we say that V_1 meets V_2 along C with G^1-continuity if the tangent planes to V_1, and V_2 agree along C.*

In this situation, the parametric G^1-continuity blending problem can be stated as follows:

- Given: n **rational** surfaces $V_1, \ldots, V_n$ over $\overline{\mathbb{K}}$ defined implicitly by the polynomials $g_1, \ldots, g_n \in \mathbb{K}[x_1, x_2, x_3]$, respectively, and n auxiliary surfaces $U_1, \ldots, U_n$ over $\overline{\mathbb{K}}$ such that the corresponding intersection space curves $C_i = V_i \cap U_i$ are **parametrizable** over $\mathbb{K}$ and such that V_i are smooth at all but finitely many points on C_i.
- Compute: a rational parametrization $\mathcal{P}$ of a surface V, such that V and V_i meets with G^1-continuity along C_i for $i \in \{1, \ldots, n\}$.

We will refer to V_i as the *primary surfaces*, to the auxiliary surfaces U_i as the *clipping surfaces*, and to the intersection rational curve C_i as the *clipping curves*.

In this paper we deal with the G^1–continuity blending problem. The method presented here can be also extended to the G^k–continuity blending problem, but for space limitation reasons we focus only on the G^1–continuity.

3 Parametrizations in Good Position

In order to develop our method we need to assume some conditions on the clipping curves that we call *"good position"*. We emphasize the fact that this requirement can easily be always achieved without loss of generality. In this section, we introduce the new notion of being in *"good position"*, and we establish the main property of curves being in good position.

Definition 2. *We say that the rational space curves $\{C_1, \ldots, C_n\}$, $n \geq 2$, are in good position, if there exist at least two curves that are not lines through the*

origin and that are not on a plane passing through the origin.
In the following, whenever we refer to a set $\{C_1, \ldots, C_n\}$, $n \geq 2$, *of curves in good position, we assume that* C_1, C_2 *are the curves satisfying the conditions of the definition.*

Remark. Note that using a linear change of coordinates, one can always assume w.l.o.g that any set of rational space curves is in good position.

In order to prove the main property of curves in good position, that essentially states certain linearly independence of their parametrizations when they are rational, we start with some technical Lemmas. In these lemmas we use the expression "for almost all non-constant rational functions in $\mathbb{K}(t)$". Note that since $\mathbb{K}(t)$ is a field, it means "for all non-constant rational functions up to finitely many exceptions". In addition we denote by $\mathbb{K}(t)^*$ the set of non-zero rational functions of $\mathbb{K}(t)$, and by $\mathbb{K}(t) \setminus \mathbb{K}$ the set of non-constant rational functions.

Lemma 1. *Let* $\mathcal{P}(t) = (p_1(t)/q_1(t), p_2(t)/q_2(t)) \in \mathbb{K}(t)^2$ *be a rational parametrization in reduced form of a plane curve, and let* $G_i(t,h) = p_i(t)q_i(h) - p_i(h)q_i(t), i = 1, 2$. *Then, there exists at the most* $deg_h(gcd(G_1(t,h), G_2(t,h)))$ *rational functions* $R(t) \in \mathbb{K}(t)$, *such that* $\mathcal{P}(R(t)) = \mathcal{P}(t)$.

Proof. Follows from [12]. □

Lemma 2. *Let* $\mathcal{Q}(t) = (q_1(t), q_2(t), q_3(t)) \in \mathbb{K}(t)^3$ *be a rational parametrization of an space curve* C. *The following statements are equivalents:*

1. *There exists* $R_1(t) \in \mathbb{K}(t) \setminus \mathbb{K}$ *such that for almost all* $R_2(t) \in \mathbb{K}(t)^*$ *there exists* $\lambda_{R_2}(t) \in \mathbb{K}(t)$ *satisfying* $\mathcal{Q}(R_1(t)) = \lambda_{R_2}(t)\mathcal{Q}(R_2(t))$.
2. *There exists* $R_1(t) \in \mathbb{K}(t) \setminus \mathbb{K}$ *such that for infinitely many linear rational functions* $R_2(t) \in \mathbb{K}(t)$ *there exists* $\lambda_{R_2}(t) \in \mathbb{K}(t)$ *satisfying* $\mathcal{Q}(R_1(t)) = \lambda_{R_2}(t)\mathcal{Q}(R_2(t))$.
3. C *is a line through the origin.*
4. *There exists* $\nu(t) \in \mathbb{K}(t)$ *such that* $\mathcal{Q}(t) = \nu(t)\frac{\partial \mathcal{Q}}{\partial t}(t)$.

Proof. (1) implies (2) is trivial. Let us proof that (2) implies (3). We first observe that, since $\mathcal{Q}$ is a parametrization, at least one component of $\mathcal{Q}$ is not zero. We assume w.l.o.g that $q_1 \neq 0$. By the hypothesis it holds that for infinitely many invertible rational functions R_2, one gets that $f(R_1(t)) = f(R_2(t)), g(R_1(t)) = g(R_2(t))$, where $f = q_2/q_1$, and $g = q_3/q_1$. Thus, since R_2 is invertible, one has that $f \circ R_1 \circ R_2^{-1} = f$, and $g \circ R_1 \circ R_2^{-1} = g$. Now, we prove that the cardinality of the set $\Omega = \{R_1 \circ R_2^{-1} \mid R_2 \text{ is linear}\}$, is not finite. Indeed: If $Card(\Omega) < \infty$, then there exists a subset Ω_0 of $\mathbb{K}(t)$ consisting in linear rational functions, that is not finite, and such that for all $\overline{R}_1, \overline{R}_2 \in \Omega_0$ it holds that $R_1 \circ \overline{R}_1^{-1} = R_1 \circ \overline{R}_2^{-1}$. Hence, for all $\overline{R}_1, \overline{R}_2 \in \Omega_0$ it holds that $R_1 \circ (\overline{R}_1^{-1} \circ \overline{R}_2) = R_1$. So, there exists infinitely many linear rational functions $\overline{R}$ such that $R_1 \circ \overline{R} = R_1$. In this situation, since R_1 is non-constant we consider the rational parametrization $\mathcal{P}(t) = (R_1, \lambda)$, where $\lambda \in \mathbb{K}$. Then, one has that for infinitely rational functions $\overline{R} \in \mathbb{K}(t)$, $\mathcal{P}(\overline{R}(t)) = \mathcal{P}(t)$, which is impossible because of Lemma 1. Therefore,

$Card(\Omega) = \infty$.
In this situation we consider $\overline{\mathcal{P}}(t) = (f(t), g(t))$. Then, taking into account how f, and g has been defined, one has that for all $R(t) \in \Omega$ it holds that $\overline{\mathcal{P}}(R(t)) = \overline{\mathcal{P}}(t)$. Hence applying Lemma 1, one deduces that f and g are constant. But this implies that C is the line $\{y = fx,\ z = gx\}$, that passes through the origin.
Now we prove that (3) implies (4). Since C is a line through the origin, then C can be parametrized as $\overline{\mathcal{Q}}(t) = (tc_1, tc_2, tc_3)$, where $c_1, c_2, c_3 \in \mathbb{K}$. Thus, by Lüroth's Theorem, there exists $\phi \in \mathbb{K}(t)$ such that $\mathcal{Q}(t) = \overline{\mathcal{Q}}(\phi(t))$, and then $\mathcal{Q}(t) = \nu(t)\frac{\partial \mathcal{Q}}{\partial t}(t)$, where $\nu(t) = \frac{\phi(t)}{\frac{\partial \phi}{\partial t}(t)}$.
Finally, we prove that (4) implies (1). Let $\nu(t) \in \mathbb{K}(t)$ be such that $\mathcal{Q}(t) = \nu(t)\frac{\partial \mathcal{Q}}{\partial t}(t)$, and let us assume w.l.o.g that $q_1 \neq 0$. Then, $\mathcal{Q}$, and $\frac{\partial \mathcal{Q}}{\partial t}$ can be written as $\mathcal{Q} = (q_1, q_1 f, q_1 g)$, $\frac{\partial \mathcal{Q}}{\partial t} = (\frac{\partial q_1}{\partial t}, \frac{\partial q_1}{\partial t} f, \frac{\partial q_1}{\partial t} g)$, where $f = \frac{\partial q_2}{\partial t} / \frac{\partial q_1}{\partial t}$, $g = \frac{\partial q_3}{\partial t} / \frac{\partial q_1}{\partial t}$, $\nu = q_1 / \frac{\partial q_1}{\partial t}$. Furthermore, $\frac{\partial \mathcal{Q}}{\partial t} = \left(\frac{\partial q_1}{\partial t}, \frac{\partial q_1}{\partial t} f + \frac{\partial f}{\partial t} q_1, \frac{\partial q_1}{\partial t} g + \frac{\partial g}{\partial t} q_1\right)$, and then $\frac{\partial f}{\partial t} q_1 = 0$, $\frac{\partial g}{\partial t} q_1 = 0$. Therefore $f, g \in \mathbb{K}$. Now, fix $R_1(t) \in \mathbb{K}(t) \setminus \mathbb{K}$. Then, for every rational function $R_2(t) \in \mathbb{K}(t)^*$, it holds that $\mathcal{Q}(R_1(t)) = \lambda_{R_2}(t)\mathcal{Q}(R_2(t))$, where $\lambda_{R_2} = \frac{q_1(R_1)}{q_1(R_2)}$, and therefore statement (1) holds. □

From Lemmas 1 and 2 we deduce the following fundamental theorems on linearly independence.

Theorem 1. *Let C be a rational space curve over $\overline{\mathbb{K}}$ that is not a line passing through the origin, and let $\mathcal{P}_1$ be a rational parametrization of C. Then for every rational parametrization $\mathcal{P}_2(t) \in \mathbb{K}(t)^3$, and for all, but finitely many, linear rational functions $R(t) \in \mathbb{K}(t)$ it holds that $\{\mathcal{P}_1(R), \mathcal{P}_2\}$ are linearly independent as vectors in $\mathbb{K}(t)^3$.*

Proof. Let $\mathcal{P}_2(t) \in \mathbb{K}(t)^3$ be a rational parametrization, and let us assume that infinitely many linear rational functions $R(t) \in \mathbb{K}(t)$, one has that $\{\mathcal{P}_1(R), \mathcal{P}_2\}$ are linearly dependent. Thus for infinitely many linear $R(t) \in \mathbb{K}(t)$ there exists $\nu_R(t) \in \mathbb{K}(t)$ such that $\mathcal{P}_1(R(t)) = \nu_R(t)\mathcal{P}_2(t)$. In particular, let $R_1(t) \in \mathbb{K}(t)$ be linear such that the equality above holds. Then, for infinitely many linear $R_2(t) \in \mathbb{K}(t)$ there exists $\nu_{R_2}(t) \in \mathbb{K}(t)$ such that $\mathcal{P}_1(R_2(t)) = \nu_{R_2}(t)\mathcal{P}_2(t)$. But this implies that, for infinitely many linear $R_2(t) \in \mathbb{K}(t)$ there exists $\lambda_{R_2}(t) \in \mathbb{K}(t)$, namely $\lambda_{R_2}(t) = \frac{\nu_{R_1}(t)}{\nu_{R_2}(t)}$, such that $\mathcal{P}_1(R_1(t)) = \lambda_{R_2}(t)\mathcal{P}_1(R_2(t))$, which is impossible by Lemma 2. □

Theorem 2. *Let C_1, C_2 be rational space curves over $\overline{\mathbb{K}}$ in good position. Let $\mathcal{P}_1, \mathcal{P}_2$ be rational parametrizations of C_1, C_2, respectively. Then for every rational parametrization $\mathcal{P}_3(t) \in \mathbb{K}(t)^3$ and for all, but finitely many, linear rational functions $R(t), L(t) \in \mathbb{K}(t)$ it holds that $\{\mathcal{P}_1(R), \mathcal{P}_2(L), \mathcal{P}_3\}$ are linearly independent as vectors in $\mathbb{K}(t)^3$.*

Proof. We first observe that since C_2 is not a line through the origin, and taking into account Theorem 1, one has that for all, but finitely many, linear $L(t) \in \mathbb{K}(t)$, it holds that $\{\mathcal{P}_2(L), \mathcal{P}_3\}$ are linearly independent.
Let us assume that there exists infinitely many linear $R(t), L(t) \in \mathbb{K}(t)$, such that $\{\mathcal{P}_1(R), \mathcal{P}_2(L), \mathcal{P}_3\}$ are linearly dependent in $\mathbb{K}(t)^3$. Thus, taking into

account that for all but finitely many, linear $L(t) \in \mathbb{K}(t)$, $\{\mathcal{P}_2(L), \mathcal{P}_3\}$ are linearly independent, one has that there exist $\lambda_{L,R}(t)$, $\nu_{L,R}(t) \in \mathbb{K}(t)$ such that $\mathcal{P}_1(R(t)) = \lambda_{L,R}(t)\mathcal{P}_2(L(t)) + \nu_{L,R}(t)\mathcal{P}_3(t)$. Hence,

$$D_{L,R}(t) = \begin{vmatrix} p_{1,1}(R(t)) & p_{2,1}(L(t)) & p_{3,1}(t) \\ p_{1,2}(R(t)) & p_{2,2}(L(t)) & p_{3,2}(t) \\ p_{1,3}(R(t)) & p_{2,3}(L(t)) & p_{3,3}(t) \end{vmatrix} = 0 \qquad (I)$$

where $\mathcal{P}_i(t) = (p_{i,1}(t), p_{i,2}(t), p_{i,3}(t))$. In this situation, we consider the minors of $D_{L,1} = p_{2,2}(L)p_{3,3} - p_{2,3}(L)p_{3,2}$, $D_{L,2} = p_{2,1}(L)p_{3,3} - p_{2,3}(L)p_{3,1}$, $D_{L,3} = p_{2,1}(L)p_{3,2} - p_{2,2}(L)p_{3,1}$. Note that since $\{\mathcal{P}_2(L), \mathcal{P}_3\}$ are linearly independent, at least one of these minors does not vanish. Let us assume w.l.o.g that $D_{L,3}(t) \neq 0$. Then, from (I) one deduces that

$$p_{1,3}(R(t)) = -\frac{D_{L,1}(t)}{D_{L,3}(t)}p_{1,1}(R(t)) + \frac{D_{L,2}(t)}{D_{L,3}(t)}p_{1,2}(R(t)), \qquad (II).$$

In the following we distinguish three cases:
Case 1: If the rational functions $\frac{D_{L,1}(t)}{D_{L,3}(t)}, \frac{D_{L,2}(t)}{D_{L,3}(t)}$ are not both constant. Then, we consider the space curve $\mathcal{M}(t) = (f(t), g(t), -1)$, where $g(t) = D_{L,2}(t)/D_{L,3}(t)$, $f(t) = -D_{L,1}(t)/D_{L,3}(t)$. Note that (II) implies that $\mathcal{P}_1(R)$ is orthogonal to $\mathcal{M}(t)$. Furthermore, since $\mathcal{P}_1(R)$ is not the zero vector, one deduces that there exists a non-zero rational function $a_{R,L}(t) \in \mathbb{K}(t)$ such that $\mathcal{P}_1(R(t)) = a_{R,L}(t)(\frac{\partial f(t)}{\partial t}, \frac{\partial g(t)}{\partial t}, 0)$.
Now, we fix a linear $R_1(t) \in \mathbb{K}(t)$ such that the equality above holds, and we take $R_2(t) \in \mathbb{K}(t)$ being any other linear satisfying the equality. This implies that there exists $\lambda_{R_2,L}(t) \in \mathbb{K}(t)$, namely $\lambda_{R_2,L}(t) = \frac{a_{R_1,L}(t)}{a_{R_2,L}(t)}$, such that $\mathcal{P}_1(R_1(t)) = \lambda_{R_2,L}(t)\mathcal{P}_1(R_2(t))$. But then, from Lemma 2, it follows that C_1 is a line passing through the origin which is impossible.
Case 2: If $D_{L,1}(t) = 0$, $D_{L,2}(t) = 0$, we have that $p_{2,2}(L)p_{3,3} = p_{3,2}p_{2,3}(L)$, and $p_{2,1}(L)p_{3,3} = p_{3,1}p_{2,3}(L)$. In this situation we distinguish two subcases.
Case 2.1: If $p_{3,3}(t) = 0$, we prove that $p_{2,3}(L) \neq 0$. Indeed: If $p_{2,3}(L) = 0$, then $p_{3,3} = p_{2,3} = 0$, and thus since $\mathcal{P}_1(R) = \lambda_{L,R}\mathcal{P}_2(L) + \nu_{L,R}\mathcal{P}_3$, one would have that $p_{1,3} = 0$ which is impossible because C_1, C_2 would be in the plane $z = 0$. Thus $p_{3,1} = p_{3,2} = 0$ which is impossible because $\mathcal{P}_3$ is a parametrization.
Case 2.2: If $p_{3,3}(t) \neq 0$, we prove that $p_{3,2}(t) \neq 0$. Indeed: If $p_{3,2} = 0$, then $p_{2,2} = 0$, and thus since $\mathcal{P}_1(R) = \lambda_{L,R}\mathcal{P}_2(L) + \nu_{L,R}\mathcal{P}_3$, one would have that $p_{1,2} = 0$ which is impossible because C_1, C_2 would be in the plane $y = 0$. Similarly one also obtains that $p_{3,1} \neq 0$. Hence, $p_{2,1}(L)/p_{3,1} = p_{2,3}(L)/p_{3,3}$, $p_{2,2}(L)/p_{3,2} = p_{2,3}(L)/p_{3,3}$, and then $D_{L,3}(t) = 0$, which is impossible.
Case 3: Finally, let us assume that $\frac{D_{L,1}(t)}{D_{L,3}(t)} = c_1 \in \mathbb{K}$, $\frac{D_{L,2}(t)}{D_{L,3}(t)} = c_2 \in \mathbb{K}$, and c_1, c_2 are not both zero. Then, equality (II) is written as $p_{1,3}(R) = -c_1p_{1,1}(R) + c_2p_{1,2}(R)$. In this situation we distinguish two subcases.
Case 3.1: If $p_{1,3}(R) = 0$, then $c_2p_{1,2}(R) - c_1p_{1,1}(R) = 0$. Thus $\mathcal{P}_1$ parametrizes the line $\{z = 0,\ xc_1 = yc_2\}$ that passes through the origin, which is a contradiction.

Case 3.2: If $p_{1,3}(R) \neq 0$, then C_1 is on the plane $z = -xc_1 + yc_2$. Moreover, since $c_1 = D_{L,1}/D_{L,3}$, and $c_2 = D_{L,2}/D_{L,3}$ one has that $p_{2,3}(L) = -c_1p_{2,1}(L) + c_2p_{2,2}(L)$. Thus, C_2 is on the plane $z = -xc_1 + yc_2$. Hence, C_1, C_2 are in a plane that passes through the origin which is a contradiction.
Therefore, for all but finitely many linear $L(t)$, $R(t) \in \mathbb{K}(t)$, $\{\mathcal{P}_1(R), \mathcal{P}_2(L), \mathcal{P}_3\}$ are linearly independent. □

We now state a proposition derived from Theorem 2.

Proposition 1. *Let $\{C_1, C_2\}$ be rational space curves over $\overline{\mathbb{K}}$ in good position. Let $\mathcal{P}_1, \mathcal{P}_2$ be rational parametrizations of C_1, C_2, respectively. Then for all, but finitely many, linear rational functions $R(t), L(t) \in \mathbb{K}(t)$, it holds that $\{\mathcal{P}_1(R), \mathcal{P}_2(L), \frac{\partial \mathcal{P}_2}{\partial t}(L)\}$ are linearly independent as vectors in $\mathbb{K}(t)^3$.*

Proof. First we observe that since C_2 is not a line through the origin, and taking into account Lemma 2, one has that for all, but finitely many, linear $L(t) \in \mathbb{K}(t)$, $\{\mathcal{P}_2(L), \frac{\partial \mathcal{P}_2}{\partial t}(L)\}$ are linearly independent. Hence, reasoning similarly as in the proof of Theorem 2, the statement holds. □

Remark. Note that similarly one may deduce that for all, but finitely many, linear rational functions $R(t), L(t) \in \mathbb{K}(t)$, $\{\mathcal{P}_1(R), \mathcal{P}_2(L), \frac{\partial \mathcal{P}_1}{\partial t}(R)\}$ are linearly independent as vectors in $\mathbb{K}(t)^3$.

Corollary 1. *Let $\{C_1, \ldots, C_n\}$, $n \geq 2$, be rational space curves in good position, and for $i = 1, 2, \ldots, n$ let $\mathcal{P}_i(t) \in \mathbb{K}(t)^3$ be a rational parametrization of C_i, respectively. Then it holds that*

1. *If $n = 2$ then for all, but finitely many, linear rational functions $R(t), L(t) \in \mathbb{K}(t)$, $\{\mathcal{P}_1(R), \mathcal{P}_2(L), \frac{\partial \mathcal{P}_1}{\partial t}(R)\}$ and $\{\mathcal{P}_1(R), \mathcal{P}_2(L), \frac{\partial \mathcal{P}_2}{\partial t}(L)\}$ are both linearly independent as vectors in $\mathbb{K}(t)^3$.*
2. *If $n > 2$ then for all, but finitely many, linear rational functions $R(t), L(t) \in \mathbb{K}(t)$, $\{\mathcal{P}_1(R), \mathcal{P}_2(L), \mathcal{P}_3\}$ are linearly independent as vectors in $\mathbb{K}(t)^3$.*

Proof. It follows from Theorem 2, and Proposition 1. □

Corollary 2. (Main Property of Curves in Good Position)
Let $\{C_1, \ldots, C_n\}$, $n \geq 2$, be rational space curves in good position. Then for all proper rational parametrization $\mathcal{Q}_i$ of C_i, $i = 1, \ldots, n$, but finitely many, it holds that

1. *If $n = 2$ then $\{\mathcal{Q}_1, \mathcal{Q}_2, \frac{\partial \mathcal{Q}_1}{\partial t}\}$ and $\{\mathcal{Q}_1, \mathcal{Q}_2, \frac{\partial \mathcal{Q}_2}{\partial t}\}$ are both linearly independent as vectors in $\mathbb{K}(t)^3$.*
2. *If $n > 2$ then $\{\mathcal{Q}_1, \mathcal{Q}_2, \mathcal{Q}_3\}$ are linearly independent as vectors in $\mathbb{K}(t)^3$.*

Proof. For the case $n = 2$, let us prove that for all proper rational parametrizations $\mathcal{Q}_i$ of C_i, but finitely many exceptions, $\{\mathcal{Q}_1, \mathcal{Q}_2, \frac{\partial \mathcal{Q}_1}{\partial t}\}$ are linearly independent. For this purpose, let $\mathcal{L}$ be the set of linear rational functions of $\mathbb{K}(t)$, and let Σ be the set of all proper rational parametrizations of a given rational curve. Then, by Lüroth's Theorem it holds that there exists a bijection between Σ and $\mathcal{L}$. Indeed: If $\mathcal{P} \in \Sigma$ is fixed, then the map

$$\begin{aligned} \varphi_{\mathcal{P}} : \Sigma &\longrightarrow \mathcal{L} \\ \mathcal{Q} &\longmapsto \mathcal{P}^{-1} \circ \mathcal{Q}. \end{aligned}$$

is one to one, and $\varphi_{\mathcal{P}}^{-1}(L) = \mathcal{P} \circ L$.
Now, let us fix rational proper parametrizations $\mathcal{P}_i$ of C_i. Then, by Corollary 1, statement (1), it follows that for all, but finitely many, linear $R(t)$, $L(t) \in \mathbb{K}(t)$ then $\{\mathcal{P}_1(R), \mathcal{P}_2(L), \frac{\partial \mathcal{P}_1}{\partial t}(R)\}$ is a linear independent set. Thus, since $\mathcal{P}_1(R) = \varphi_{\mathcal{P}_1}^{-1}(R)$, and $\mathcal{P}_2(L) = \varphi_{\mathcal{P}_2}^{-1}(L)$, one deduces that for all proper rational parametrizations $\mathcal{Q}_i$ of C_i, but finitely many exceptions, there exists $R_1(t) \in \mathbb{K}(t) \setminus \mathbb{K}$ such that $\{\mathcal{Q}_1, \mathcal{Q}_2, 1/\frac{\partial R_1}{\partial t}\frac{\partial \mathcal{Q}_1}{\partial t}\}$, are linearly independent. Thus, $\{\mathcal{Q}_1, \mathcal{Q}_2, \frac{\partial \mathcal{Q}_1}{\partial t}\}$ are linearly independent.
For the case $n > 2$, the reasoning is similar. □

4 Parametric G¹–Blending Method

In this section, we present a method for blending several surfaces with G^1-continuity. The basic idea of the method is to interpolate the clipping parametrizations along a certain sweeping curve.

More precisely, let V_i, U_i, C_i as in the statement of the problem described in Section 2, and let also assume that $\{C_1, \ldots, C_n\}$ are rational space curves in good position. Furthermore, let

$$\mathcal{Q}_i(t) = (q_{i,1}(t), q_{i,2}(t), q_{i,3}(t)) \in \mathbb{K}(t)^3, \quad i = 1, \ldots, n$$

be rational proper parametrizations of C_i, such that if $n = 2$, $\{\mathcal{Q}_1, \mathcal{Q}_2, \frac{\partial \mathcal{Q}_j}{\partial t}\}$ are linearly independent for $j = 1, 2$, and if $n > 2$, $\{\mathcal{Q}_1, \mathcal{Q}_2, \mathcal{Q}_3\}$ are linearly independent as vectors in $\mathbb{K}(t)^3$. Note that this can always be assumed w.l.o.g. because of Corollary 2.

We start our analysis with a technical lemma that essentially ensures the existence of certain interpolating polynomials that will be used to generate the sweep curve in the blending method.

Lemma 3. *Let $\mathbb{L}$ be a field extension of $\mathbb{K}$, let $S = (s_0, \ldots, s_{n-1}) \subset \mathbb{L}^n$ where $s_i \neq s_j$ if $i \neq j$, and let $\Lambda = \{\lambda_1, \ldots, \lambda_n\}$ be a set of parameters. Then, for every $\ell \geq 2n - 1$ there exists polynomials $B_0^{\Lambda}(h) \in \mathbb{L}[\Lambda][h]$, and $B_1(h), \ldots, B_{\ell-2n+1}(h) \in \mathbb{L}[h]$, such that any polynomial A in $\mathbb{L}[\Lambda][h]$ satisfying that*

(1.) $deg_h(A(h)) \leq \ell$.
(2.) $A(s_0) = 0, \ A(s_i) = 1, \ i = 1, \ldots, n-1$.
(3.) $\frac{\partial A}{\partial h}(s_{i-1}) = \lambda_i, \ i = 1, \ldots, n$.

can be written as $A(h) = B_0^{\Lambda}(h) + \sigma_1 B_1(h) + \cdots + \sigma_{\ell-2n+1} B_{\ell-2n+1}(h)$, where $\sigma_i \in \overline{\mathbb{K}}$.

We will denote the generic polynomial satisfying (1), (2), (3) as $A^{\Lambda,\Sigma}(h) \in \mathbb{L}[\Lambda][\Sigma][h]$ where $\Sigma = \{\sigma_1, \ldots, \sigma_{\ell-2n+1}\}$, and we will refer to Λ and Σ as the blending parameters and modeling parameters, respectively.

Remark. Note that if $\ell = 2n - 1$ then $A^{\Lambda,\Sigma}(h)$ is unique, and only depends on the blending parameters.

Lemma 4. *Let $\mathbb{L}$ be a field extension of $\mathbb{K}$, and let $S = (s_0, \ldots, s_{n-1}) \subset \mathbb{L}^n$ where, $s_i \neq s_j$ if $i \neq j$. Then, for every $\omega \geq n-1$ there exists polynomials $B_0(h), B_1(h), \ldots, B_{\omega-n+1}(h) \in \mathbb{L}[h]$, such that any polynomial A in $\mathbb{L}[h]$ satisfying that*

(1.) $deg_h(A(h)) \leq \omega$.
(2.) $A(s_0) = 0,\ \ A(s_i) = 1,\ \ i = 1, \ldots, n-1$.

can be written as $A(h) = B_0(h) + \sigma_1 B_1(h) + \cdots + \sigma_{\omega-n+1} B_{\omega-n+1}(h)$, where $\sigma_i \in \overline{\mathbb{K}}$. We will denote the generic polynomial satisfying (1), (2) as $A^{\Sigma}(h) \in \mathbb{L}[\Sigma][h]$ where $\Sigma = \{\sigma_1, \ldots, \sigma_{\omega-n+1}\}$.

Remark. Note that the polynomial $A^{\Sigma}(h)$ only depends on the modeling parameters, and that if $\omega = n-1$ then $A^{\Sigma}(h)$ is unique, and belongs to $\mathbb{L}[h]$.

The next theorem gives sufficient conditions for a surface parametrization to meet V_i, $i = 1, \ldots, n$, with G^1-continuity along C_i.

Theorem 3. *Let $\mathcal{P}(t,h)$ be a rational parametrization of a surface V over $\overline{\mathbb{K}}$. If there exists $m_i, n_i \in \overline{\mathbb{K}}(t)$ such that $\mathcal{P}(m_i(t), n_i(t)) = \mathcal{Q}_i(t)$, for $i = 1, \ldots, n$, and $\{\frac{\partial \mathcal{P}}{\partial h}(m_i(t), n_i(t)), \frac{\partial \mathcal{P}}{\partial t}(m_i(t), n_i(t))\}$ is a linear independent set orthogonal to $\nabla g_i(\mathcal{Q}_i)$, then V meets V_i, for $i = 1, \ldots, n$, with G^1-continuity along C_i.*

Proof. Let F be the implicit equation of V. We first observe that since for $i = 1, \ldots, n$, $\mathcal{P}(m_i, n_i) = \mathcal{Q}_i$, one has that $F(\mathcal{Q}_i) = F(\mathcal{P}(m_i, n_i)) = 0$. Moreover, one deduces that $\nabla F(\mathcal{Q}_i) \neq 0$, and that $\{\frac{\partial \mathcal{P}}{\partial h}(m_i, n_i), \frac{\partial \mathcal{P}}{\partial t}(m_i, n_i)\}$ is a set orthogonal to $\nabla F(\mathcal{Q}_i)$ because $\{\frac{\partial \mathcal{P}}{\partial h}(m_i, n_i), \frac{\partial \mathcal{P}}{\partial t}(m_i, n_i)\}$ is a linear independent set. Hence, taking into account that $\{\frac{\partial \mathcal{P}}{\partial h}(m_i(t), n_i(t)), \frac{\partial \mathcal{P}}{\partial t}(m_i(t), n_i(t))\}$ is a linear independent set orthogonal to $\nabla g_i(\mathcal{Q}_i)$, and that $\nabla g_i(\mathcal{Q}_i) \neq 0$, one deduces that there exists $N_i \in \mathbb{K}^*(t)$, such that $\nabla F(\mathcal{Q}_i) = N_i(t) \nabla g_i(\mathcal{Q}_i)$. Therefore, for $i = 1, \ldots, n$, F and g_i have common tangent planes along C_i. □

Now we proceed to construct a blending parametrization from the input clipping space curve parametrizations, that depends on two families of parameters, namely, the blending parameters and the modeling parameters, that are generated applying Lemma 3 and Lemma 4. The modeling parameters can be afterwards used to impose additional conditions for modeling the final blending. We apply several times Lemma 3 and Lemma 4 to generate interpolating polynomials. We want the parameters in each interpolating polynomial to be different. For this purpose, we take n different elements $s_0, s_1, \ldots, s_{n-1} \in \mathbb{K}$, and we take three different families $\Lambda_k = \{\lambda_{k,1}, \ldots, \lambda_{k,n}\}$, $k = 1, 2, 3$, of different parameters. Moreover, we take integers $\ell_k \geq 2n-1$, and $\omega_k \geq n-1$. These integers control the number of free modeling parameters. Thus, depending on the final proposes one may enlarge ℓ_k, and ω_k. We apply Lemma 3, and Lemma 4 as follows:

1. If $n > 2$
   - For each Λ_k we consider the tuple $S_k = (s_{k-1}, s_1, \ldots, s_{k-2}, s_k, \ldots, s_{n-1})$, $k =$ 1, 2, 3, and we apply Lemma 3, to obtain a generic polynomial $A_k^{\Lambda_k, \Sigma_k}(h) \in \mathbb{K}[\Sigma_k][\Lambda_k][h]$ of degree at most ℓ_k. Note that $A_k^{\Lambda_k, \Sigma_k}(h)$ depends on n blending parameters, and on $\ell_k + 1 - 2n$ modeling parameters.

– We consider the tuple $S_k = (s_{k-1}, s_1, \dots, s_{k-2}, s_k, \dots, s_{n-1}),\ k = 4, \dots, n,$ and we apply Lemma 4 to obtain a generic polynomial $A_k^{\Sigma_k}(h) \in \mathbb{K}[\Sigma_k][h]$ of degree at most ω_k. Note that $A_k^{\Sigma_k}(h)$ depends only on $\omega_k + 1 - n$ modeling parameters. In this situation, the pattern blending parametrization is

$$\mathcal{P}_{\Lambda,\Sigma}(t,h) = A_1^{\Lambda_1,\Sigma_1}(h)\mathcal{Q}_1(t) + A_2^{\Lambda_2,\Sigma_2}(h)\mathcal{Q}_2(t) + A_3^{\Lambda_3,\Sigma_3}(h)\mathcal{Q}_3(t) +$$

$$+ A_4^{\Sigma_4}(h)\mathcal{Q}_4(t) + \cdots + A_n^{\Sigma_n}(h)\mathcal{Q}_n(t)$$

where $\Lambda = (\Lambda_1, \Lambda_2, \Lambda_3)$, and $\Sigma = (\Sigma_1, \dots, \Sigma_n)$.

2. If $n = 2$, then the polynomials $A_1^{\Lambda_1,\Sigma_1}, A_2^{\Lambda_2,\Sigma_2}$ are generated as above, and the pattern blending parametrization is

$$\mathcal{P}_{\Lambda,\Sigma}(t,h) = A_1^{\Lambda_1,\Sigma_1}(h)\mathcal{Q}_1(t) + A_2^{\Lambda_2,\Sigma_2}(h)\mathcal{Q}_2(t),$$

where $\Lambda = (\Lambda_1, \Lambda_2)$, and $\Sigma = (\Sigma_1, \Sigma_2)$.

Note that if one does not use modeling parameters, then, the pattern blending parametrization is of the form $\mathcal{P}_\Lambda(t,h) = \sum_{i=1}^n A_i^{\Lambda_i}(h)\mathcal{Q}_i(t)$.

In addition, let $V_{\Lambda,\Sigma}$ denote the surface generated by $\mathcal{P}_{\Lambda,\Sigma}(t,h)$.

From the construction we have done it is clear that the parametrization $\mathcal{P}_{\Lambda,\Sigma}(t,h)$ satisfies the following properties

Lemma 5. For $i = 1, \dots, n$, it holds that

(1) $\mathcal{P}_{\Lambda,\Sigma}(t, s_{i-1}) = \mathcal{Q}_i(t)$
(2) $\frac{\partial \mathcal{P}_{\Lambda,\Sigma}}{\partial t}(t, s_{i-1}) = \frac{\partial \mathcal{Q}_i}{\partial t}(t)$.
(3) If $n > 2$, *then* $\frac{\partial \mathcal{P}_{\Lambda,\Sigma}}{\partial h}(t, s_{i-1}) = \sum_{k=1}^3 \lambda_{k,i}\mathcal{Q}_k(t) + \sum_{k=4}^n \frac{\partial A_k^{\Sigma_k}}{\partial h}(s_{i-1})\mathcal{Q}_k(t)$
(4) If $n = 2$, *then* $\frac{\partial \mathcal{P}_{\Lambda,\Sigma}}{\partial h}(t, s_{i-1}) = \sum_{k=1}^2 \lambda_{k,i}\mathcal{Q}_k(t)$.

In order to analyze the existence of parametrizations $\mathcal{P}_{\Lambda^0,\Sigma}(t,h)$ that meet each V_i with G^1-continuity along C_i, we distinguish two cases depending on whether $n = 2$ or $n > 2$.

Theorem 4. (Blending Solution for n = 2)
Let $n = 2$. Then, for every tuple of values $\Lambda^0 = (\Lambda_1^0, \Lambda_2^0)$ of the blending parameters Λ satisfying that

$$\lambda_{1,i}\mathcal{Q}_1\nabla g_i(\mathcal{Q}_i) + \lambda_{2,i}\mathcal{Q}_2\nabla g_i(\mathcal{Q}_i) = 0, \quad i = 1, 2,$$

and such that for each $i = 1, 2$, $\lambda_{1,i}$, $\lambda_{2,i}$ are not simultaneously zero, the surface $V_{\Lambda^0,\Sigma}$ meets each V_1, V_2 with G^1-continuity along C_1, C_2, respectively. Furthermore, the number of modeling parameters Σ involved in Λ^0 is $\rho_0 \le \sum_{i=1}^2 \ell_i - 6$.

Proof. We apply Theorem 3 to $\mathcal{P}_{\Lambda,\Sigma}(t,h)$ in order to find sufficient conditions on the parameters Λ, such that $\mathcal{P}_{\Lambda,\Sigma}(t,h)$ is a blending parametrization. First of all, we observe that from Lemma 5 statement (1), one has that

$\mathcal{P}_{\Lambda,\Sigma}(t, s_{i-1}) = \mathcal{Q}_i(t)$. Therefore, taking $m_i = t$, and $n_i = s_{i-1}$ in Theorem 3, one deduces that if $\{\frac{\partial \mathcal{P}_{\Lambda,\Sigma}}{\partial t}(t, s_{i-1}), \frac{\partial \mathcal{P}_{\Lambda,\Sigma}}{\partial h}(t, s_{i-1})\}$ is a linear independent set orthogonal to $\nabla g_i(\mathcal{Q}_i)$, then $\mathcal{P}_{\Lambda,\Sigma}(t, h)$ is a blending parametrization. Furthermore, by Lemma 5 statements (2) and (4), one has that $\frac{\partial \mathcal{P}_{\Lambda,\Sigma}}{\partial t}(t, s_{i-1}) = \frac{\partial \mathcal{Q}_i}{\partial t}(t)$, and $\frac{\partial \mathcal{P}_{\Lambda,\Sigma}}{\partial h}(t, s_{i-1}) = \sum_{k=1}^{2} \lambda_{k,i}\mathcal{Q}_k(t)$. Note that $\lambda_{1,i}$, $\lambda_{2,i}$ can not be simultaneously zero since otherwise $\mathcal{P}_{\Lambda,\Sigma}(t, h)$ would be the zero vector, and hence $\mathcal{P}_{\Lambda,\Sigma}(t, h)$ would not be a blending solution.
Now we observe that $\{\mathcal{Q}_1, \mathcal{Q}_2\}$ has been taken such that $\{\mathcal{Q}_1, \mathcal{Q}_2, \frac{\partial \mathcal{Q}_i}{\partial t}\}$ are linearly independent. Therefore, the only condition that remains to check is that $\frac{\partial \mathcal{P}_{\Lambda,\Sigma}}{\partial h}(t, s_{i-1})$ is orthogonal to $\nabla g_i(\mathcal{Q}_i)$, and this is clearly equivalent to

$$\lambda_{1,i}\mathcal{Q}_1\nabla g_i(\mathcal{Q}_i) + \lambda_{2,i}\mathcal{Q}_2\nabla g_i(\mathcal{Q}_i) = 0, \quad i = 1, 2. \qquad \square$$

Theorem 5. (Blending Solution for n > 2)
Let $n > 2$. Then, for every vector $M_i(t) \in \mathbb{K}(t)^3$ orthogonal to $\nabla g_i(\mathcal{Q}_i)$ and non-proportional to $\frac{\partial \mathcal{Q}_i}{\partial t}(t)$, let Λ^0 be the solution over $\mathbb{K}(\Sigma)(t)$ of the linear systems

$$\begin{pmatrix} q_{1,1} & q_{2,1} & q_{3,1} \\ q_{1,2} & q_{2,2} & q_{3,2} \\ q_{1,3} & q_{2,3} & q_{3,3} \end{pmatrix} \cdot \begin{pmatrix} \lambda_{1,i} \\ \lambda_{2,i} \\ \lambda_{3,i} \end{pmatrix} = M_i - \sum_{k=4}^{n} \frac{\partial A_k^{\Sigma_k}}{\partial h}(s_{i-1})\mathcal{Q}_k.$$

Then, $V_{\Lambda^0,\Sigma}$ meets each V_i with G^1-continuity along C_i. Furthermore the number of modeling parameters Σ involved in Λ^0 is $\rho_0 \leq \sum_{i=1}^{3} \ell_i + \sum_{j=4}^{n} \omega_j - n^2 - 2n$.

Proof. First of all, we observe that from Lemma 5 statement (1), one has that $\mathcal{P}_{\Lambda,\Sigma}(t, s_{i-1}) = \mathcal{Q}_i(t)$. Therefore, taking $m_i = t$, and $n_i = s_{i-1}$ in Theorem 3, one deduces that if $\{\frac{\partial \mathcal{P}_{\Lambda,\Sigma}}{\partial t}(t, s_{i-1}), \frac{\partial \mathcal{P}_{\Lambda,\Sigma}}{\partial h}(t, s_{i-1})\}$ is a linear independent set orthogonal to $\nabla g_i(\mathcal{Q}_i)$, then $\mathcal{P}_{\Lambda,\Sigma}(t, h)$ is a blending parametrization. Furthermore, by Lemma 5 statement (2) one has that $\frac{\partial \mathcal{P}_{\Lambda,\Sigma}}{\partial t}(t, s_{i-1}) = \frac{\partial \mathcal{Q}_i}{\partial t}(t)$. Thus, $\frac{\partial \mathcal{P}_{\Lambda,\Sigma}}{\partial t}(t, s_{i-1})$ is orthogonal to $\nabla g_i(\mathcal{Q}_i)$. Therefore, it only remains to achieve that $\frac{\partial \mathcal{P}_{\Lambda,\Sigma}}{\partial h}(t, s_{i-1})$ is non-proportional to $\frac{\partial \mathcal{Q}_i}{\partial t}(t)$ and orthogonal to $\nabla g_i(\mathcal{Q}_i)$. For this purpose, since M_i is non-proportional to $\frac{\partial \mathcal{Q}_i}{\partial t}$ and orthogonal to $\nabla g_i(\mathcal{Q}_i)$ one gets a blending solution, forcing $\frac{\partial \mathcal{P}_{\Lambda,\Sigma}}{\partial h}(t, s_{i-1})$ to be proportional to M_i.
Let us prove that for each $i = 1, \ldots, n$, the equation $\frac{\partial \mathcal{P}_{\Lambda,\Sigma}}{\partial h}(t, s_{i-1}) = \alpha M_i$ $(\alpha \neq 0)$ is solvable. Indeed: By Lemma 5 statement (3), equality above is writing in matrix form as $N\,\bar{X} = M_i - \sum_{k=4}^{n} \frac{\partial A_k^{\Sigma_k}}{\partial h}(s_{i-1})\mathcal{Q}_k$, $i = 1, \ldots, n$, where N is the matrix of the system. Now we observe that $\{\mathcal{Q}_1, \mathcal{Q}_2\}$ has been taken such that $\{\mathcal{Q}_1, \mathcal{Q}_2, \mathcal{Q}_3\}$ are linearly independent. Therefore, we obtain that rank$(N) = 3$, and thus the system is solvable. $\square$

Corollary 3. (Blending Property of $\mathbf{V}_{\Lambda,\Sigma}$.)
There always exists blending parameters values Λ^0 of Λ such that $V_{\Lambda^0,\Sigma}$ meets each V_i with G^1-continuity along C_i.

Proof. If $n = 2$ the result is trivial from Theorem 4. If $n > 2$, since the clipping surfaces g_i are smooth at all but a finite number of points on C_i one can always take a vector $M_i(t)$ orthogonal to $\nabla g_i(\mathcal{Q}_i)$ and non-proportional to $\frac{\partial \mathcal{Q}_i}{\partial t}(t)$. □

Remark. The interpolating polynomials generated by Lemma 3 can be also be taken such that the linear constraints in Theorem 5 simplify to $N\bar{X} = M_i$. More precisely, we can also apply Lemma 3, and Lemma 4 as follows: For $k = 1,2,3$, $A_k^{\Lambda_k,\Sigma_k}(h) \in \mathbb{K}[\Sigma_k][\Lambda_k][h]$ are generated as above. For $k = 4,\ldots,n$, let us consider $\Lambda_k = \{0,\ldots,0\}$, and $S_k = (s_{k-1}, s_1,\ldots,s_{k-2}, s_k,\ldots,s_{n-1})$. We apply Lemma 3 to obtain a polynomial $A_k^{\Sigma_k}(h) \in \mathbb{K}[\Sigma_k][h]$ of degree ℓ_k, that depends only on $\ell_k + 1 - 2n$ modeling parameters.
In this situation, one has that Λ^0 is the solution over $\mathbb{K}(\Sigma)(t)$ of the linear systems $N\bar{X} = M_i$ (N is the matrix of the system of Theorem 5), and the number of modeling parameters involved is $\rho_0 \leq \sum_{k=1}^{n} \ell_k + n(1-2n)$.

5 Algorithm and Examples

The above results can be applied to derive an algorithm that solves the parametric G^1-continuity blending problem. We outline the algorithm, and we illustrate it by two examples. The input of the algorithm is as it is described in the statement of the problem (see Section 4). Additionally, we introduced a possitive integer ρ, that controls the number of modeling parameters. The output of the algorithm is a blending parametrization solution $\mathcal{P}_\Sigma$ depending at most on ρ modeling parameters.

<u>Algorithm</u> BLENDING($g_1,\ldots,g_n$, $\mathcal{Q}_1,\ldots\mathcal{Q}_n$, ρ)

(1) **Check** whether $\{C_1,\ldots,C_n\}$ are in good position. If not, do a change of coordinates.
(2) **Check** whether $\mathcal{Q}_i(t) \in \mathbb{K}(t)^3$ satisfies that if $n = 2$, $\{\mathcal{Q}_1, \mathcal{Q}_2, \frac{\partial \mathcal{Q}_j}{\partial t}\}$ are linearly independent for $j = 1,2$, and if $n > 2$, $\{\mathcal{Q}_1, \mathcal{Q}_2, \mathcal{Q}_3\}$ are linearly independent. If not, do a reparametrization of $\mathcal{Q}_1$, and $\mathcal{Q}_2$.
(3) **Take** integers ℓ_k, ω_k such that $\rho \leq \sum_{i=1}^{2} \ell_i - 6$ if $n = 2$, and $\rho \leq \sum_{i=1}^{3} \ell_i + \sum_{j=4}^{n} \omega_j - n^2 - 2n$ if $n > 2$.
(4) **Compute** the polynomials $A_k^{\Lambda_k,\Sigma_k}(h)$ of degree at most ℓ_k for $k = 1,2,3$, and $A_k^{\Sigma_k}(h)$ of degree at most ω_k for $k = 4,\ldots,n$.
(5) **Take** the parametrization $\mathcal{P}_{\Lambda,\Sigma}(t,h) = \sum_{j=1}^{3} A_j^{\Lambda_j,\Sigma_j}(h)\mathcal{Q}_j(t) +$
$+ \sum_{j=4}^{n} A_j^{\Sigma_j}(h)\mathcal{Q}_j(t)$ where $\Lambda = (\Lambda_1, \Lambda_2, \Lambda_3)$, and $\Sigma = (\Sigma_1,\ldots,\Sigma_n)$.
(6) If $n = 2$: **Compute** the vectors $\nabla g_i(\mathcal{Q}_i)$, and $\frac{\partial \mathcal{P}_{\Lambda,\Sigma}}{\partial h}(t, s_{i-1})$ for $i = 1,2$; **Compute** Λ^0 solution of the systems $\frac{\partial \mathcal{P}_{\Lambda,\Sigma}}{\partial h} . \nabla g_i(\mathcal{Q}_i) = 0$, for $i = 1,2$ (See Theorem 4.).
(7) If $n > 2$: **Compute** vectors $M_i(t) \in \mathbb{K}(t)^3$ orthogonal to $\nabla g_i(\mathcal{Q}_i)$ and non-proportional to $\frac{\partial \mathcal{Q}_i}{\partial t}$ for $i = 1,\ldots,n$; **Compute** Λ^0 solution of the systems $N\bar{X} = M_i - \sum_{k=4}^{n} \frac{\partial A_k^{\Sigma_k}}{\partial h}(s_{i-1})\mathcal{Q}_k$, for $i = 1,\ldots,n$, where N is as in Theorem 5 and $\bar{X} = (\lambda_{1,i}, \lambda_{2,i}, \lambda_{3,i})^T$.

(8) **Substitute** $\lambda_{j,i}$ into the parametrization $\mathcal{P}_{\Lambda,\Sigma}$, and **return** $\mathcal{P}_{\Lambda,\Sigma}$.

Example 1. Let V_1, V_2 be surfaces defined by $g_1 = x^2 + (z-1)^2 + y^2 - 1$, $g_2 = x^2 + y^2 - 4$, and let U_1, U_2 be the auxiliary surfaces defined by $h_1 = z - 1$, $h_2 = z$. Let us also consider the proper parametrizations of the clipping curves C_i,

$$\mathcal{Q}_1(t) = \left(\frac{t^2-1}{t^2+1}, 2\frac{t}{t^2+1}, 1\right), \quad \mathcal{Q}_2(t) = \left(2\frac{t^2-1}{t^2+1}, 4\frac{t}{t^2+1}, 0\right).$$

We apply algorithm BLENDING to V_1, V_2, with 4-modeling parameters.

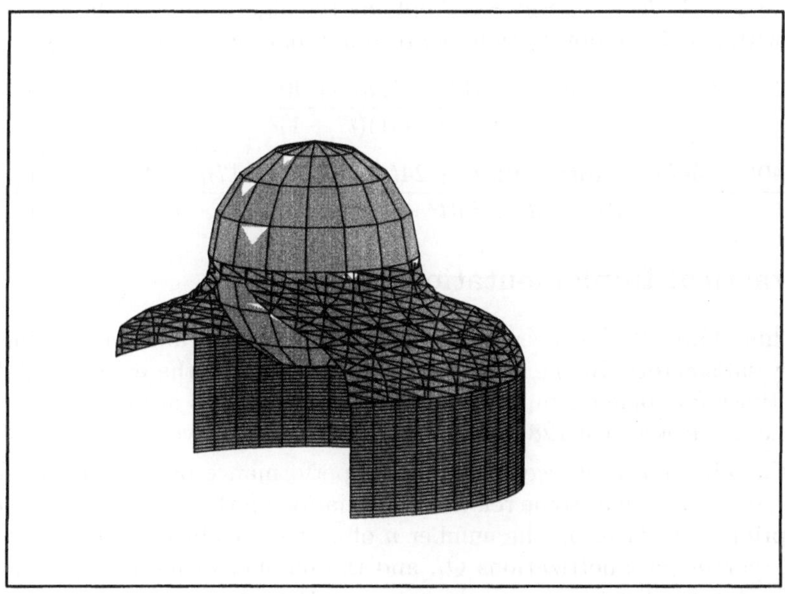

Fig. 1. Sphere, Cylinder and Blending Surface

First we note that $\{C_1, C_2\}$ are in good position, and that $\{\mathcal{Q}_1, \mathcal{Q}_2, \frac{\partial \mathcal{Q}_j}{\partial t}\}$ are linearly independent (steps 1, 2). In step 3, we take $\ell_k = 5$. In step 4, we compute the polynomials $A_k^{\Lambda_k,\Sigma_k}(h)$:

$A_1^{\Lambda_1,\Sigma_1}(h) = (1 + \lambda_{1,1}h + (-3 - 2\lambda_{1,1} - \lambda_{1,2})h^2 + (2 + \lambda_{1,1} + \lambda_{1,2})h^3) + (-2h^3 + h^4 + h^2)\sigma_{1,1} + (2h^2 + h^5 - 3h^3)\sigma_{1,2}$,

$A_2^{\Lambda_2,\Sigma_2}(h) = (\lambda_{2,1}h + (3 - 2\lambda_{1,1} - \lambda_{2,2})h^2 - (2 - \lambda_{2,1} - \lambda_{2,2})h^3) + (2h^2 + h^5 - 3h^3)\sigma_{2,2} + (-2h^3 + h^4 + h^2)\sigma_{2,1}$

where $\Sigma_i = \{\sigma_{i,1}, \sigma_{i,2}\}$,$\Lambda_i = \{\lambda_{i,1}, \lambda_{i,2}\}$. In step 5, we consider the pattern blending parametrization $\mathcal{P}_{\Lambda,\Sigma}(t,h) = \sum_{i=1}^{2} A_i^{\Lambda_i,\Sigma_i}(h)\mathcal{Q}_i(t)$, where $\Lambda = (\Lambda_1, \Lambda_2)$, and

$\Sigma = (\Sigma_1, \Sigma_2)$. Solving the system (step 6), $\lambda_{1,i}\mathcal{Q}_1\nabla g_i(\mathcal{Q}_i) + \lambda_{2,i}\mathcal{Q}_2\nabla g_i(\mathcal{Q}_i) = 0$, $i = 1, 2$, we obtain $\lambda_{1,1} = -2\lambda_{2,1}$, $\lambda_{1,2} = -2\lambda_{2,2}$.
Substituting in $\mathcal{P}_{\Lambda,\Sigma}$ (step 8), we deduce that the blending parametrization $\mathcal{P}_{\Lambda^0,\Sigma}$ is

$$\left(-\frac{(t^2-1)(-1-3h^2+2h^3)}{t^2+1}, -2\frac{t(-1-3h^2+2h^3)}{t^2+1}, 1-2h+3h^2-2h^3\right).$$

In Fig. 1 we plot together the sphere V_1, part of the cylinder V_2, and part of the blending surface $\mathcal{P}_{\Lambda^0,\Sigma}$. □

Example 2. Let V_1, V_2 be surfaces defined implicitly by the polynomials $g_1 = 2z^4 - 3z^2y + y^2 - 2y^3 + y^4 + x$, $g_2 = z^4 + 8z^2y + 2z^2y^2 + 16y^2 + 8y^3 + y^4 - 16xz^2 - 16xy^2$, and let U_1, U_2 be the auxiliary surfaces defined implicitly by $h_1 = z^2 + y^2 - 1$, $h_2 = z^2 + y^2 - 4$. Applying the algorithm to blend the surfaces V_1, V_2 with $\rho = 4$, we obtain that the blending parametrization $\mathcal{P}_{\Lambda^0,\Sigma}$ is

$$(\frac{-8t^8+17ht^8+24t^7-40ht^7+16t^6-12ht^6+40t^5-56ht^5+326ht^4-144t^4}{4(t^4+2t^2+1)(t^2+1)^2}+$$

$$+\frac{40t^3-56ht^3+16t^2-12ht^2+24t-40t-8+17h}{4(t^4+2t^2+1)(t^2+1)^2}, \frac{2t}{t^2+1}, \frac{t^2-1}{t^2+1}). \square$$

6 Practical Implementation

Algorithm BLENDING has been implemented in Maple V., and the running times are very satisfactory. In the following we briefly outline the experimental computing times for some examples. Actual computing times are measures on a PC PENTIUM III PROCESSOR 128 MB of RAM, and times are given in seconds of CPU.

In the following table we illustrate the performance of our implementation showing times as well as some relevant information on the examples as the degree of the primary surfaces g_i, the number n of surfaces to blend, the degree of the clipping curves parametrizations $\mathcal{Q}_i$, and the number of modeling parameters. Because of space limitations we do not provide here the precise input of each example.

n	2	2	2	2	2	2	3	3	3	4	6
ρ	1	2	3	5	10	2	3	3	7	5	3
D_p	[4,4]	[7,10]	[2,2]	[4,7]	[2,2]	[6,6]	[2,2,2]	[10,10,10]	[4,4,2]	[2,2,2,2]	[2,2,2,2,2,2]
D_c	[4,4]	[4,4]	[1,2]	[8,4]	[2,2]	[6,6]	[2,2,2]	[3,3,2]	[8,4,2]	[2,2,2,2]	[2,2,2,2,2,2]
t	1.466	1.515	.042	4.645	.075	3.732	13.653	.981	41.19	4.269	18.550

n=number of surfaces to blend
ρ=number of free modeling parameters
D_P= a list with the degrees of the primary surfaces g_i
D_C= a list with the degrees of the clipping curve parametrizations $\mathcal{Q}_i(t)$
t=time in seconds.

References

1. Bajaj, C., Ihm, I., and Warren, J., *Higher Order Interpolation and Least Squares Approximation Using Implicit Algebraic Surfaces.* ACM Transactions on Graphics, vol.12, N.4 , 1993, pp. 327-347.
2. DeRose, A.D. *Geometric Continuity: A Parametrization Independent Measure of Continuity for Computer Aided Geometric Design.* PhD thesis, Computer Science, Univ. of California, Berkeley, 1985.
3. Filip, D.J., *Blending Parametric Surfaces.* ACM Transactions on Graphics, vol.8, N.3 , 1989, pp. 164-173.
4. Hartmann, E. *Blending an Implicit with a Parametric Surface.* CAGD, vol. 12 ,1995, pp. 825-835.
5. Hartmann, E. *Numerical Implicitation for Intersection and G^n-Continuous Blending of Surfaces.* Computer Aided Geometric Design, vol. 15, 1998, pp. 377-397.
6. Hartmann, E. *Parametric G^n Blending of Curves and Surfaces.* The Visual Computer, vol. 17, 2001, pp. 1-13.
7. Hoffmann, C., and Hopcroft, J., *Quadratic Blending Surfaces.* Computer Aided Geometric Design, vol. 18 , 1986, pp. 301-307.
8. Hoffmann, C., and Hopcroft, J., *The Potential Method for Blending Surfaces and Corners.* In Geometric Modeling, G. Farin, Ed.SIAM, Philadelphia 1987.
9. Hoschek, J., and Lasser, D. *Fundamentals of Computer Aided Geometric Desing.* A K Peters Wellesley, Massachusetts. 1993.
10. Liang, Y., Wang, G., and Zheng, J., *GC^n continuity conditions for adjacent rational parametric surfaces.* Computer Aided Geometric Design, vol. 12 ,1995, pp. 111-129.
11. Pottman, H., and Wallner, J., *Rational Blending Surfaces Between Quadrics.* Computer Aided Geometric Design, vol. 14 , 1997, pp. 407-419.
12. Sendra, J.R., and Winkler, F., *Computation of the Degree of a Rational Map Between Curves.* Proc. ISSAC-2001 (To appear).
13. Vida, J., Martin, R.R., and Varady, T., *A Survey of Blending Methods that use Parametric Surfaces.* Computer-Aided Design, vol. 26, 1994, pp. 341-365.
14. Warren, J. *On Algebraic Surfaces Meeting with Geometric Continuity.* PhD thesis, Cornell University, 1986.
15. Warren, J. *Blending Algebraic Surfaces.* ACM Transactions on Graphics, vol 8, N.4 ,1989, pp. 263-278.

A Method of Logic Deduction and Verification in KBS Using Positive Integers*

E. Roanes-Lozano[1], E. Roanes-Macías[1], L.M. Laita[2]

[1] Universidad Complutense de Madrid, Dept. Algebra,
Edificio "La Almudena", c/ Rector Royo Villanova s/n, 28040-Madrid, Spain
{eroanes,roanes}@eucmos.sim.ucm.es
[2] Universidad Politécnica de Madrid, Dept. Artificial Intelligence,
Campus de Montegancedo, Boadilla del Monte, 28660-Madrid, Spain
laita@fi.upm.es

Abstract. Classic propositional Boolean algebra is modelized in this paper as a subset of $\mathbb{N}$ (the divisors of a certain product of prime numbers) with operations *gcd* and *lcm*. The isomorphism is constructed in a way that recalls Gödel numbers. This approach can be used to study logic deduction and to check the consistency of Rule-Based Knowledge Based Systems. An implementation in the Computer Algebra system Maple, that uses intensively exact arithmetic, is included as an appendix. Although the growth of the integers involved makes this implementation interesting only if the number of propositional variables is not greater than 8, we think its simplicity makes it very interesting to illustrate KBS behaviour.

Keywords. Knowledge Based Systems. Logic Deduction. Verification. Boolean Algebras. Boolean Logic.

1 Introduction

The straightforward way to deal with propositional logic is based on directly handling truth tables [10]. A very exciting approach relates Logic and Algebra and usually makes use of Gröbner Bases (GB) to deal with effective calculi.

The latter was introduced in [5, 6] for bivalued logics. It was extended to multivalued logics in [1, 2] and reorganized using residue class rings and interpretations borrowed from Algebraic Geometry (using algebraic varieties) in [11]. Considering a residue class ring has the advantage of providing not only effective methods but also an algebraic model. An obvious extension is to address Rule-Based Knowledge Based Systems (KBS) consistency [9]. Following this approach an interpretation of different "degrees" of inconsistency is given in [12]. A disadvantage of this method is that its theoretical background is not simple.

* Partially supported by projects TIC2000-1368-C03-03 and TIC2000-1368-C03-01 (Ministry of Science and Technology, Spain).

But other methods, such as Petri nets, are not simple either. A state of the art in KBS verification can be found in [7].

A surprisingly simple model, based on working (only) with positive integers, the operations *lcm* and *gcd* and the ordering "to be a divisor" ($|$), will be introduced in this paper. It can check KBS consistency and can be extended to the multivalued case. Unfortunately, because of the growth of the integers involved, this implementation is fast only if the number of propositional variables is not greater than 8.

Sections 2, 4 and 6 are introductory notes to lattices and Boolean algebras, Boolean logic and KBS, respectively. Proofs are omitted for the sake of brevity, but details can be found in any introductory textbook. These sections can be skipped by the aware reader.

2 Introductory Notes About Lattices and Boolean Algebras

Details can be found, for example, in [3, 4, 8, 13].

2.1 Lattices and Boolean Algebras

Definition 1. *If the operations of* $(\mathcal{A}, \sqcup, \sqcap)$ *are associative, commutative and cancelative w.r.t to the other one,* $(\mathcal{A}, \sqcup, \sqcap)$ *is said to be a "lattice".*

Definition 2. *When both operations of a lattice are distributive w.r.t. the other one and there is a complement for each element, the lattice is said to be a "Boolean algebra".*

Example 1. Well known examples of lattices are

- $\mathbb{N}$ w.r.t. *lcm* and *gcd*
- linear varieties of the vectorial plane w.r.t. sum ($+$) and intersection ($\cap$)
- convex sets in the affine plane w.r.t. convex union (minimum convex containing the union) and intersection ($\cap$)
- switching circuits w.r.t. parallel connection and series connection
- subsets of a given set w.r.t. union ($\cup$) and intersection ($\cap$)
- logic propositions w.r.t. conjunction ($\wedge$) and disjunction ($\vee$).

The three last examples are Boolean algebras (the first three are not). The complements are: the switch (or circuit) that works the opposite way, the complement of a set, the logic negation ($\neg$), respectively.

In the first example, if, instead of $\mathbb{N}$, the set of divisors of a number, l, is considered as universe, then the distributive lattice becomes a Boolean algebra (the complement of a number, n, is $\frac{l}{n}n$).

A partial ordering (such that for every two elements there is a supreme and an infimum) can be defined from the two operations of the lattice. Reciprocally, given a partial ordering (such that for every two elements there is a supreme and an infimum), two operations with these properties can be defined. The following definition illustrate how to define an ordering from the operations.

Proposition 1. *Given a lattice* $(\mathcal{A}, \sqcup, \sqcap)$, *a partial ordering,* $\sqsubset$, *can be defined as follows:* $\forall a, b \in \mathcal{A},\ a \sqsubset b \Leftrightarrow a \sqcap b = a$.

Example 2. For instance, for classic Boolean logic, $\wedge$ and $\vee$, conditional ($\rightarrow$) is obtained. For the subsets of a given set, $\cup$ and $\cap$, $\subseteq$ is obtained. For $\mathbb{N}$, *lcm* and *gcd*, "to be a divisor" ($|$) is obtained.

Definition 3. *The "atoms" of a lattice are the minimal elements in the ordering. The "co-atoms" are the maximal elements in the ordering.*

Example 3. Let us consider the classic propositional Boolean logic, being the propositional variables $\{q, r\}$. There are

- $\binom{4}{0} = 1$ infimum (contradiction, $\underline{c}$),
- $\binom{4}{1} = 4$ atoms,
- $\binom{4}{2} = 6$ "half-way elements" (among them the 4 propositional variables),
- $\binom{4}{3} = 4$ co-atoms, and
- $\binom{4}{4} = 1$ supremum (tautology, $\underline{t}$).

The ordering $\rightarrow$ is the reflexive-transitive closure of the relation represented in Figure 1.

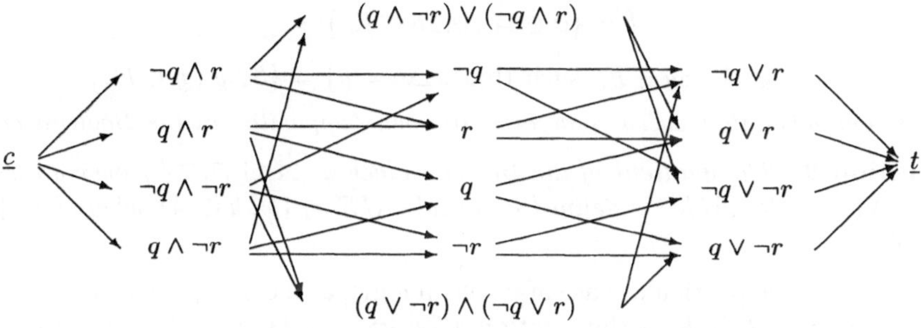

Fig. 1. Propositional logic ordered by $\rightarrow$

Remark 1. There is a big difference between the usual generation of a classic propositional Boolean logic $\mathcal{C}$ and that of the Boolean algebra $(Divisors(t), gcd, lcm)$.

The atoms of $Divisors(t)$ (minimal elements for the ordering) are given or easily obtained (prime numbers). Meanwhile, a propositional Boolean logic $\mathcal{C}$ is usually given by its propositional variables (if the lattice is ordered, propositional variables are precisely halfway between the infimum and the supremum). Atoms are not completely trivial (if the propositional variables are $\{v_1, v_2, ..., v_m\}$, they are exactly the propositions that can be expressed as $\circ v_1 \wedge \circ v_2 \wedge ... \wedge \circ v_m$, where $\circ$ is either $\neg$ or no symbol at all).

2.2 Ideals and Quotients

Proposition 2. *Given a Boolean algebra* $(\mathcal{A}, \sqcup, \sqcap, \sqsubset, ')$ *(where* $'$ *represents the complement), if operation* $\triangle$ *is defined as follows:* $P \triangle Q = (P \sqcap Q') \sqcup (Q \sqcap P')$ *, then* $(\mathcal{C}, \triangle, \sqcap)$ *is a ring (called a "Boolean ring"). Every element is its own opposite with respect to* $\triangle$.

Remark 2. In the particular case of Boolean logic, $q \leftrightarrow r$ and $q \triangle r$ are defined as $(\neg q \vee r) \wedge (q \vee \neg r)$ and $(q \wedge \neg r) \vee (\neg q \wedge r)$, respectively. Therefore, as a consequence of the De Morgan laws, $\neg(q \leftrightarrow r)$ and $q \triangle r$ are equivalent. Moreover, $q \leftrightarrow r$, $\neg(q \triangle r)$, $\neg q \triangle r$ and $q \triangle \neg r$ are also equivalent.

Definition 4. *Given a Boolean algebra* $(\mathcal{A}, \sqcup, \sqcap, \sqsubset)$*, the (principal) "ideal" generated by* $a \in \mathcal{A}$ *is* $E_a = \{x \in \mathcal{A} : x \sqsubset a\}$.

Remark 3. The ideal defined in Definition 4 corresponds to the ideal generated by this element in the corresponding Boolean ring (i.e., minimal subring to which this element belongs and with the property that the product of any element in the subring by any element in the ring remains in the subring). This justifies the following definition.

Definition 5. *The quotient of the Boolean algebra* $(\mathcal{A}, \sqcup, \sqcap, \sqsubset, ')$ *over its ideal* E_a*, denoted* $\mathcal{A}/E_a$*, is the quotient of the corresponding Boolean ring,* $(\mathcal{A}, \triangle, \sqcap)$*, over this ideal. Therefore,* $\mathcal{A}/E_a = \{\overline{b} : b \in \mathcal{A}\}$ *, where*

$$\overline{b} = \{x \in \mathcal{A} : x \triangle b \in E_a\} =$$

$$= \{x \in \mathcal{A} : \exists y \in E_a \text{ such that } x \triangle b = y\} = \{b \triangle y : y \in E_a\}$$

(let us remember that every element is its own "opposite" in the Boolean ring).

Definition 6. *The quotient of the Boolean algebra* $(\mathcal{A}, \sqcup, \sqcap, \sqsubset, ')$ *over an equivalence relation* $\mathcal{R}$*,* $\mathcal{A}/\mathcal{R}$*, is defined as usual:* $\mathcal{A}/\mathcal{R} = \{\overline{b} : b \in \mathcal{A}\}$ *where* $\overline{b} = \{x \in \mathcal{A} : x\mathcal{R}b\}$.

Remark 4. In the particular case of Boolean logic, as a consequence of the second part of Remark 2, if $\mathcal{R}$ is the relation $x\mathcal{R}y \Leftrightarrow \neg(x \leftrightarrow y) \in E_a$, then $\mathcal{C}/\mathcal{R} = \mathcal{C}/E_a$.

3 The Integer Model for Classic Boolean Logic

3.1 Defining the Model for Classic Propositional Boolean Logic

A certain subset of $\mathbb{N}$ and a tricky mapping are used. Let $(\mathcal{C}, \vee, \wedge, \rightarrow)$ be a classic propositional Boolean logic given by its propositional variables $\{v_1, v_2, ..., v_m\}$.

Let us denote by p_i the p-th prime number ($p_1 = 2, p_2 = 3, p_3 = 5...$). Let us assign

$$c = 1 \quad ; \quad t = \prod_{j=1}^{2^m} p_j$$

and let us consider the subset of $\mathbb{N}$ of the divisors of t, $S = \{n \in \mathbb{N} : n | t\}$ (that has precisely $2^{(2^m)}$ elements and 2^m atoms, as $\mathcal{C}$ does).

3.2 The Isomorphism

Definition 7. *$\psi : \mathcal{C} \longrightarrow S$ is the mapping defined as follows. For any propositional variable v_i*

$$\psi(v_i) = \prod_{j=0}^{2^m-1} p_{j+1}^{(floor(\frac{j}{2^{i-1}}))\ mod\ 2}$$

and for any formulae A, B

$$\psi(A \vee B) = lcm(\psi(A), \psi(B)) \quad ; \quad \psi(\neg A) = \frac{t}{\psi(A)} \ .$$

Remark 5. The ideas in the background of this mapping are:

- As each propositional variable is a product of exactly half of the 2^m primes, they are precisely halfway between the infimum and the supremum.
- We consider the primes ordered, and the mapping makes v_1 the product of the primes occupying even (i.e., $\dot{2}$) positions, v_2 the product of the primes occupying positions of the form $\dot{4}+3$ or $\dot{4}$, v_3 the product of the primes occupying positions of the form $\dot{8}+5$, $\dot{8}+6$, $\dot{8}+7$ or $\dot{8}$... (despite the fact that this looks initially strange, is the same way as valuations are usually ordered in truth tables).
- The image of each propositional variable has exactly one common prime factor with each other propositional variable.
- The prime decompositions of $\psi(v_i)$ and $\psi(\neg v_i)$ have no prime in common, so their *gcd* is 1.

Proposition 3. *As a consequence of the previous definition:*

$$\psi(A \wedge B) = gcd(\psi(A), \psi(B))$$
$$\psi(\underline{t}) = t \quad ; \quad \psi(\underline{c}) = c = 1$$
$$(A \rightarrow B) \Leftrightarrow (\psi(A) | \psi(B))$$
$$(A \leftrightarrow B) \Leftrightarrow (\psi(A) = \psi(B)) \ .$$

Proof.- They respectively follow from:

- the De Morgan laws in $\mathcal{C}$,
- $\underline{t} \leftrightarrow v_1 \vee \neg v_1 \quad ; \quad \underline{c} \leftrightarrow v_1 \wedge \neg v_1$,
- the definition of the ordering from the operations (see Proposition 1),
- the previous equality and $(a = b) \Leftrightarrow (a|b)$ and $(b|a)$.

Corollary 1. *ψ is a Boolean algebra homomorphism.*

Theorem 1. *$\psi : (S, lcm, gcd, |) \longrightarrow (\mathcal{C}, \vee, \wedge, \rightarrow)$ is a Boolean algebra isomorphism.*

Example 4. Using isomorphism ψ, the model for the logic of Example 3 would be $(\{n \in \mathbb{N} : n|210\}, gcd, lcm, |)$. The relation $|$ is the reflexive-transitive closure of of the relation represented in Figure 2 (corresponding elements have been located in the same place as in Figure 1).

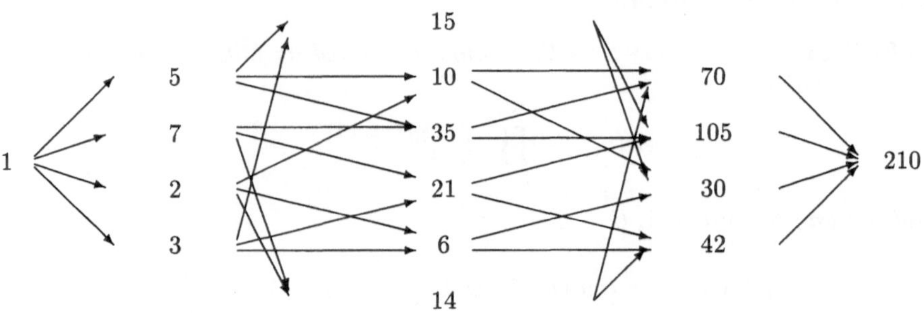

Fig. 2. $\{n : n|210\}$ ordered by |

Remark 6. Observe that, unlike what happens in polynomial ideals, the null ideal is the ideal generated by 1: $E_1 = \{1\}$ (because 1 is the infimum of the Boolean algebra, not the supremum). Similarly, E_t is the whole $\mathcal{S}$.

Remark 7. Mapping ψ also admits a simple interpretation in terms of Disjunctive Normal Forms (DNF). If a formula, A, is written in DNF as a disjunction of atoms, then $\psi(A)$ is the *lcm* of the prime numbers that are the images in ψ of these atoms. We have stated above that A is written as a disjunction of atoms because the DNF of a formula is not unique, and we are not interested here in the minimal DNF (see Example 5).

Example 5. Let us consider the notation of Examples 3 and 4. If A is formula $(q \vee \neg q) \wedge \neg r$, then both $\neg r$ and $(q \wedge \neg r) \vee (\neg q \wedge \neg r)$ are DNF of A. Then: $\psi(A) = gcd(lcm(21, 21'), 35')) = gcd(lcm(21, 10), 6)$; for its minimal DNF (the one we are not interested in) we would have: $\psi(\neg q) = 6$ and for its DNF as a disjunction of atoms we would have: $\psi((q \wedge \neg r) \vee (\neg q \wedge \neg r)) = lcm(3, 2)$.

4 Introductory Notes about Boolean Logic

Definition 8. *A* literal *is a variable preceded or not by the symbol* $\neg$.

Definition 9. *A formula* A_0 *is a tautology, denoted* $\models A_0$*, iff* A_0 *can only take the truth-value true.*

Corollary 2. $\models A_0$ *iff* $A_0 \leftrightarrow \underline{t}$ *holds.*

Definition 10. *A formula* A_0 *is a* tautological consequence *of the formulae* A_1, $A_2, ..., A_s$*, denoted* $\{A_1, A_2, ..., A_s\} \models A_0$*, iff whenever all* $A_1, A_2, ..., A_s$ *are true, then* A_0 *is true.*

Proposition 4. $\{A_1, A_2, ..., A_s\} \models A_0$ iff $A_1 \wedge A_2 \wedge ... \wedge A_s \rightarrow A_0$ *is a tautology (the logic is supposed to be classic Boolean).*

Definition 11. *In this context,* $\{A_1, A_2,...,A_s\}$ *is called a* contradictory domain *iff* $\{A_1, A_2,...,A_s\} \models A$, *where* A *is any formula of the language in which* A_1, A_2, ..., A_s *are expressed.*

The name *contradictory domain* comes from the fact that, if all formulae follow from $\{A_1, A_2, ..., A_s\}$, contradictory formulae, in particular, follow (for example a formula that can only take the truth value false, such as $P \wedge \neg P$).

5 Tautologies, Tautological Consequences and Contradictory Domains in the Integer Model

Proposition 5. $\models A_0$ iff $\psi(A_0) = t$.

Proof.- As ψ is an isomorphism, it follows from Corollary 2.

Proposition 6. $\{A_1, A_2, ..., A_s\} \models A_0$ iff $\psi((A_1 \wedge A_2 \wedge ... \wedge A_s) \to A_0) = t$.

Proof.- As ψ is an isomorphism, it follows from Proposition 4.

Proposition 7. *The three following assertions are equivalent:*

i) $\{A_1, A_2,...,A_s\}$ *is a contradictory domain*
ii) $\psi(A_1 \wedge A_2 \wedge ... \wedge A_s) = c \quad (c = 1)$
iii) $\psi(\neg A_1 \vee \neg A_2 \vee ... \vee \neg A_s) = t$ [1].

Proof.- *i)⇔ii):* $\{A_1, A_2,...,A_s\} \models A$, for any formula $A \Leftrightarrow$

$$\Leftrightarrow (A_1 \wedge A_2 \wedge ... \wedge A_s) \to A, \text{ for any formula } A \Leftrightarrow$$
$$\Leftrightarrow \psi(A_1 \wedge A_2 \wedge ... \wedge A_s) | \psi(A), \text{ for any formula } A \Leftrightarrow$$
$$\Leftrightarrow \psi(A_1 \wedge A_2 \wedge ... \wedge A_s) | n, \text{ for any number } n \in S \Leftrightarrow$$
$$\Leftrightarrow \psi(A_1 \wedge A_2 \wedge ... \wedge A_s) = 1 \ .$$

i)⇔iii): $\{A_1, A_2,...,A_s\} \models A$, for any formula $A \Leftrightarrow$

$$\Leftrightarrow (A_1 \wedge A_2 \wedge ... \wedge A_s) \to A, \text{ for any formula } A \Leftrightarrow$$
$$\Leftrightarrow (A_1 \wedge A_2 \wedge ... \wedge A_s) \to \underline{c} \Leftrightarrow$$
$$\Leftrightarrow (A_1 \wedge A_2 \wedge ... \wedge A_s) \leftrightarrow \underline{c} \Leftrightarrow$$
$$\Leftrightarrow (\neg A_1 \vee \neg A_2 \vee ... \vee \neg A_s) \leftrightarrow \underline{t} \Leftrightarrow$$
$$\Leftrightarrow \psi(\neg A_1 \vee \neg A_2 \vee ... \vee \neg A_s) = t \ .$$

6 Introductory Notes About KBS

Definition 12. *A rule-based* Knowledge Based System *or* Expert System *(denoted* KBS*) is a set of "rules", "facts" and "integrity constraints". A KBS* rule *is an implication between a conjunction of propositional variables* v_i *and a disjunction of propositional variables, such as*

$$\circ v_{i_1} \wedge \circ v_{i_2} \wedge ... \circ v_{i_s} \to \circ v_{j_1} \vee \circ v_{j_2} \vee ... \vee \circ v_{j_{s'}} \ .$$

Under a Boolean logic, the symbols "∘" can be replaced by the symbol ¬ or no symbol at all.

[1] Let us remember Remark 6; the ideal generated by t is the whole ring, so this result is not contradictory with the method presented in [9, 11].

Definition 13. *A* potential fact *in a set of rules is any literal, which stands in the antecedent of some rule(s) in the set but not in the consequent of any rule in the set. Any potential facts that are stated (in each case) will be called* facts.

Definition 14. *The experts sometimes add that, x and y, for example, cannot occur simultaneously. This is termed an* integrity constraint *(IC). Then the negation of the formula concerned $\neg(X \wedge Y)$ (denoted NIC) is added to the KBS as new information.*

Definition 15. *A rule is* fired *if all the literals in the antecedent are facts (firing corresponds to the extraction of tautological consequences). If a literal is in the consequent of one rule and is a part of the antecedent of another rule and the first rule can be fired, then we shall say that this literal is a* derived fact. *A rule is also* fired *if all the literals in the antecedent are facts or derived facts (firing corresponds to the extraction of tautological consequences).*

Example 6. Let us take a very simple example of KBS (based on Boolean logic). For the sake of simplicity, the consequents consist of only one element, despite the fact that they are generally disjunctions of elements.

Rule 1. $v_1 \wedge v_2 \to v_3$

Rule 2. $v_2 \wedge \neg v_4 \to v_5$

Rule 3. $v_5 \to \neg v_1$

Letters like v_1, or letters preceded by $\neg$, such as $\neg v_4$, are examples of literals. The potential facts of the set of rules {Rule 1, Rule 2, Rule 3} are v_1, v_2 and $\neg v_4$.

If v_2 and $\neg v_4$ are the facts, then by firing Rule 2, v_5 is obtained. v_5 appears in the antecedent of Rule 3, so the literal v_5 is a derived fact and Rule 2 can be fired. Then $\neg v_1$ is also a derived fact.

Example 7. Let us consider the KBS of Example 6 (rules 1, 2 and 3) and facts v_1, v_2 and $\neg v_4$. The firing of the rules would lead to $\neg v_1$, that is a logical contradiction with fact v_1. Therefore, this set of facts and rules is a contradictory domain.

Example 8. Let us consider the KBS formed by rules 1 and 2 of Example 6,

Rule 3b. $v_5 \to v_1$

and facts v_1, v_2 and $\neg v_4$. This set of facts and rules is not a contradictory domain.

Definition 16. *A set of facts of the KBS is said to be consistent iff it is not a contradictory domain.*

Definition 17. *A KBS is said to be inconsistent iff a consistent set of facts exists, such that the union of this set with the set of rules and integrity constraints of the KBS is a contradictory domain. Consequently, a KBS is consistent iff for each maximal consistent set of facts, its union with the set of rules and integrity constraints of the KBS is not a contradictory domain.*

7 The Integer Model for KBS

The union of a consistent set of facts with the set of rules and integrity constraints of the KBS is usually considered (because, if the set of facts is already inconsistent, any set of propositions containing it will be a contradictory domain).

7.1 Consistency in a KBS with the Integer Model

The consistency of a KBS can be studied using Proposition 7. Let us see an example.

Example 9. The value of $\psi(v_1 \wedge v_2 \wedge \neg v_4 \wedge R1 \wedge R2 \wedge R3)$ of Example 7 is 1. The value of $\psi(v_1 \wedge v_2 \wedge \neg v_4 \wedge R1 \wedge R2 \wedge R3b)$ of Example 8 is not 1 (89 is obtained if the coding presented in this paper is followed exactly).

7.2 Logic Deduction in a KBS with the Integer Model

Let us suppose that $\{H_1, H_2, ..., R_1, R_2, ..., I_1, I_2, ...\}$ are the given facts, the rules and the NICs of the KBS. Let us suppose that it is not a contradictory domain (otherwise every formulae follows). As a consequence of Proposition 6:

Corollary 3. *A formula A_0 follows from a given set of facts, the rules and the NICs of the KBS, $\{H_1, H_2, ..., R_1, R_2, ..., I_1, I_2, ...\} \models A_0$, iff $\psi(H_1 \wedge H_2 \wedge ... \wedge R_1 \wedge R_2 \wedge ... \wedge I_1 \wedge I_2 \wedge ... \to A_0) = t$.*

Example 10. In Example 8, the value of $\psi(v_1 \wedge v_2 \wedge \neg v_4 \wedge R1 \wedge R2 \wedge R3b \to v5)$ is t (i.e., 525896479052627740771371797072411912900610967452630 is obtained if the coding presented in this paper is followed exactly). Therefore $v5$ is a derived fact.

8 Integer Model for KBS as a Quotient Boolean Algebra

Let $(\mathcal{C}, \vee, \wedge, \to)$ be a classic propositional Boolean logic given by its propositional variables $\{v_1, v_2, ..., v_m\}$. A KBS is constructed upon this logic. Let us suppose that a set of facts, the rules and the NICs of the KBS are stated as true (they are supposed to be consistent).

As shown in previous papers, the KBS can be interpreted as a polynomial quotient ring (over a certain ideal) [9, 11]. The reason is that, when moving from the Boolean logic $\mathcal{C}$ to the KBS, as some facts and the rules and NICs have been stated as true, new equivalences arise. Instead of stating what is added as true, we'll determine what is added as false: the negation of the given facts, the rules and the NICs. If we denote by J the smallest ideal containing the negation of the given facts, rules and NICs stated as true, then the KBS is isomorphic to $\mathcal{C}/J$.

Observe that all ideals of Boolean algebra are principal (i.e., a single element can generate them). J is generated by the disjunction of the negation of the given facts, the rules and the NICs.

8.1 Intuitive Approach

Example 11. Let us interpret intuitively a trivial example. Initially the Boolean logic of Example 3 is considered (see Figure 1). If, for example, rules $q \to r$ and $r \to q$ are added (as "true"), then we have $q \leftrightarrow r$, and therefore

- $\underline{c} \leftrightarrow \neg q \wedge r \leftrightarrow q \wedge \neg r \leftrightarrow (q \wedge \neg r) \vee (\neg q \wedge r)$
- $q \leftrightarrow q \wedge r \leftrightarrow q \vee r \leftrightarrow r$
- $\neg q \leftrightarrow \neg q \wedge \neg r \leftrightarrow \neg q \vee \neg r \leftrightarrow \neg r$
- $\underline{t} \leftrightarrow \neg q \vee r \leftrightarrow q \vee \neg r \leftrightarrow (q \vee \neg r) \wedge (\neg q \vee r)$

so, really only 4 different elements (classes of the quotient Boolean algebra) are left. The initial logic of Figure 1 collapses into that of Figure 3.

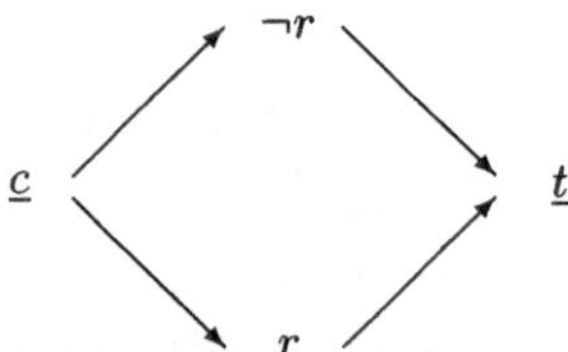

Fig. 3. Boolean logic associated to the KBS

From an intuitive point of view, the translation into the integer model is to add the equality $21 = 35$ (as $\to$ in $\mathcal{C}$ corresponds to $|$ in S, $\leftrightarrow$ in $\mathcal{C}$ corresponds to $=$ in S), i.e. $3 = 5$. Then we would have

- $1 = 3 = 5 = 15$ (because: $15 = lcm(3,5) = lcm(3,3) = 3 = 5 = gcd(5,5) = gcd(3,5) = 1$)
- $7 = 21 = 35 = 105$ (multiplying the previous equality by 7)
- $2 = 6 = 10 = 30$
- $14 = 42 = 70 = 210$

and the corresponding same 4 classes are obtained. Observe that the operations in the model, gcd and lcm, are all "multiplicative" and no simplifications related to addition are allowed.

8.2 Formal Approach

Corollary 4. *Let $a \in (S, lcm, gcd, |)$. The ideal generated by $a \in S$ is $E_a = \{x \in S : x|a\}$ (see §2.2).*

Let us suppose that $\{H_1, H_2, ..., R_1, R_2, ..., I_1, I_2, ...\}$ are the given facts, the rules and the NICs of the KBS.

Theorem 2. *The integer model for the KBS constructed by stating in $\mathcal{C}$ formulae in $\{H_1, H_2, ..., R_1, R_2, ..., I_1, I_2, ...\}$ as true, i.e., of*

$$C/E_{(\neg H_1 \vee \neg H_2 \vee \ldots \vee \neg R_1 \vee \neg R_2 \vee \ldots \vee \neg I_1 \vee \neg I_2 \vee \ldots)}$$

is

$$S/E_{\psi(\neg H_1 \vee \neg H_2 \vee \ldots \vee \neg R_1 \vee \neg R_2 \vee \ldots \vee \neg I_1 \vee \neg I_2 \vee \ldots)} \cdot$$

Proof.- It is a consequence of ψ being an isomorphism.

Example 12. Let us return to the previous example. From a formal point of view the translation into the integer model of stating $q \leftrightarrow r$ in C is to move into the quotient ring S/J, where J is the ideal of $(S, lcm, gcd, |)$ generated by $\psi(\neg(q \leftrightarrow r)) = 15$, i.e. $J = \{1, 3, 5, 15\}$. Therefore the elements of S/J are

- $\overline{\psi(\underline{c})} = \overline{1} = \{1, 3, 5, 15\}$
- $\overline{\psi(q)} = \overline{21} = \{21, 7, 35, 105\}$
- $\overline{\psi(\neg q)} = \overline{10} = \{10, 2, 6, 30\}$
- $\psi(\underline{t}) = \overline{210} = \{210, 42, 70, 14\}$

as we obtained in Example 11.

9 Generalization

It is possible to extend the integer model from classic propositional Boolean to multivalued propositional modal logics.

Let us consider for instance Kleene's three-valued logic [14]. Let $\{v_1, v_2, ..., v_m\}$ be the propositional variables. As in §3.1, let us denote by p_i the p-th prime number and let us consider the set of divisors of

$$\prod_{j=1}^{2^m} p_j^3$$

(so now we deal with products of primary numbers instead of products of prime numbers).

In this case, for any propositional variable v_i,

$$\psi(v_i) = \prod_{j=0}^{2^m - 1} p_{j+1}^{2 \cdot ((floor(\frac{j}{2^i - 1}))\ mod\ 2)}$$

The same definition for the images of $\vee$, $\wedge$ is given. The best definition we have found to introduce modal connectives "necessary" and "possible" is to raise exponents 2 to 3 and to lower exponents 2 to 1, respectively (somehow the exponents give the certainty) and to use a different definition of negation, tautology and contradiction.

Unfortunately the approach looses all its charm, as it becomes really cumbersome.

Acknowledgements

We would like to thank the referees for their most valuable comments.

10 Conclusion

We have shown here the possibility of translating logic deduction in KBS and inconsistency of KBS in terms of manipulations of integer numbers that provides an intuitive model of the behaviour of KBS.

Appendix: Maple V.5 Implementation

Let us revisit Examples 6, 7 and 8 with the Maple implementation. Observe that input lines are preceded by symbol ">" meanwhile outputs are centered. In Maple, if the input line finishes with a ":" symbol instead of a ";" symbol, the output is not shown.

First the session is restarted, the propositional variables are defined and the assignings to contradiction and tautology are given.

```
> restart;
> LV:=[seq(v.i , i=1..5)]:
> c:=1:
> t:=product(ithprime(j),j=1..2^M):
```

Then the variables can be coded by executing the following procedure.

```
> code_variables:=proc()
>     global M;
>     local i,j;
>     M:=nops(LV);
>     for i to M do
>        print(op(i,LV));
>        assign(op(i,LV) , product('ithprime(j+1)^(
>                                 floor(j/(2^(i-1))) mod 2)',
>                                 'j'=0..2^M-1));
>        print(op(i,LV));
>       od;
>   NULL;
>  end:
```

For example:

```
> code_variables():
```

v1

617903232816918931114O8549

v2

2081157658701269904O5069855

v3

1770771844810020174087496817

v4

7433774111188011802117628344 1

v5

16137160443118396219519316344231

Then the translations of connectives are given. Observe that, in Maple, functions `‘&...‘` are infix, meanwhile usual functions are prefix.

```
> neg:=p->t/p:
> ‘&or‘:=proc(p,q)
>     lcm(p,q);
>   end:
> ‘&and‘:=proc(p,q)
>     gcd(p,q);
>   end:
> ‘&imp‘:=proc(p,q)
>     neg(p) &or q;
>   end:
> ‘&iff‘:=proc(p,q)
>     (p &imp q) &and (q &imp p);
>   end:
```

To illustrate it, a very simple example is shown afterwards.

```
> v1;  v2;  v1 &and v2;
                         6179032328169189311140 8549
                        20811576587012699040506985 5
                                23101100172899
> ifactor(v1);  ifactor(v2);  ifactor(v1 &and v2);
(3)(7)(13)(19)(29)(37)(43)(53)(61)(71)(79)(89)(101)(107)(113)(131)
(5)(7)(17)(19)(31)(37)(47)(53)(67)(71)(83)(89)(103)(107)(127)(131)
   (7)    (19)    (37)    (53)    (71)    (89)    (107)    (131)
```

Next two Boolean procedures that check if two given formulae are equivalent and if a formula is a tautology, respectively, are given.

```
> are_equiv:=proc(p,q)
>     evalb(p=q);
>   end:
> is_taut:=proc(p)
>     evalb(p=t);
>   end:
```

Examples:

```
> are_equiv( neg(neg(v5)) , v5 );
                                   true
> is_taut( ((v1 &and v2) &or (v1 &and v3)) &iff
>            (v1 &and (v2 &or v3)) );
                                   true
```

Let us now consider the KBS in Example 6:

```
> R1:=(v1 &and v2) &imp v3:
> R2:=(v2 &and neg(v4)) &imp v5:
> R3:=v5 &imp neg(v1):
```

To test consistency we shall define procedure `is_contrad`, that uses an auxiliary procedure that returns the conjunction of all the elements in a list.

```
> and_all:=proc(L)
>    if nops(L)>1 then and_all( [op(1,L) &and op(2,L) ,
>                            op(3..nops(L),L)] )
>                 else op(1,L)
>     fi;
>  end:
> is_contrad:=proc(L)
>    if and_all(L)=1 then 'is a contradictory domain'
>                    else 'is not a contradictory domain' fi;
>  end:
```

Now examples 6 and 7 can be revisited.

```
> H1:=v1;
                    H1 := 617903232816918931114O8549
> H2:=v2:
> H3:=neg(v4):
> is_contrad([H1,H2,H3,R1,R2,R3]);
                              is a contradictory domain
> R3b:=v5 &imp v1:
> is_contrad([H1,H2,H3,R1,R2,R3b]);
                            is not a contradictory domain
```

In this last case we shall check logic deduction. We shall first define the procedure:

```
> follows:=proc(L::list,consec)
>    evalb(and_all(L) &imp consec = t);
>  end:
```

and then:

```
> follows([H1,H2,H3,R1,R2,R3b] , v5);
                                   true
> follows([H1,H2,H3,R1,R2,R3b] , v4);
                                  false
```

References

1. Alonso, J.A., Briales, E.: Lógicas Polivalentes y Bases de Gröbner In: Martin, C. (ed.): Actas del V Congreso de Lenguajes Naturales y Lenguajes Formales, Barcelona (1989) 307-315.

2. Chazarain, J., Riscos, A., Alonso, J.A., Briales, E.: Multi-valued Logic and Gröbner Bases with Applications to Modal Logic. J. Symb. Comp. **11**(1991) 181-194.
3. Halmos, P.R.: Lectures on Boolean Algebras. Springer-Verlag, Berlin Heidelberg New York (1974).
4. Hermes, H.: La teoría de retículos y su aplicación a la lógica matemática. CSIC, Madrid (1963).
5. Hsiang, J.: Refutational Theorem Proving using Term-Rewriting Systems. Artif. Intell. **25** (1985) 255-300.
6. Kapur, D., Narendran, P.: An Equational Approach to Theorem Proving in First-Order Predicate Calculus. 84CRD296 General Electric Corp. R&D Report, Schenectady, NY (1984). Also in: Procs. 9th IJCAI, vol. 2 (1985) 1146-1153.
7. Laita, L.M., de Ledesma, L.: Knowledge-Based Systems Verification. In: Kent, A., Williams, J.G. (eds.): Encyclopedia of Computer Science and Technology. Marcel Dekker, New York Basel Hong Kong (1997) 253-280.
8. Mendelson, E.: Bolean Algebra and Switching Circuits. McGraw-Hill, New York (1970).
9. Roanes-Lozano, E., Laita, L.M., Roanes-Macías, E.: Maple V in A.I.: The Boolean Algebra Associated to a KBS. Comp. Alg. Nederland Nieuwsbrief **14** (1995) 65-70.
10. Roanes-Lozano, E.: Introducing Propositional Multi-Valued Logics with the Help of a CAS. In: Gilbert, R.P., Kajiwara, J., Xu, Y.S. (eds.): Recent Developments in Complex Analysis and Computer Algebra. Procs. ISAAC 1997. Kluwer Academic Publishers, Dordrecht Boston London (1999) 277-290.
11. Roanes-Lozano, E., Laita, L.M., Roanes-Macías, E.: A Polynomial Model for Multivalued Logics with a Touch of Algebraic Geometry and Computer Algebra. Math. Comp. Simul. **45/1-2** (1998) 83-99.
12. Roanes-Lozano, E., Roanes-Macías, E., Laita, L.M.: Geometric Interpretation of Strong Inconsistency in Knowledge Based Systems. In: Ganzha, V.G., Mayr, E.W., Vorozhtsov, E.V. (eds.): Computer Algebra in Scientific Computing. Procs. CASC'99. Springer-Verlag, Berlin Heidelberg (1999) 349-363.
13. Roanes-Macías, E.: Retículos en la Matemática Elemental. Bol. Soc. Puig Adam **4** (1984-5) 41-48.
14. Turner, R.: Logics for Artificial Intelligence. Ellis Horwood, Chichester (1984).

Progressive Long Waves on a Slope (A New Solution to the Euler Equation?)

Alexander Shermenev

Wave Research Center, Russian Academy of Science,
Moscow, 117942, Russia,
sher@orc.ru

Abstract. Computer algebra system is used for deriving and studying high-order Boussinesq-type equations for long periodic nonlinear waves climbing a sloping beach. Potential and surface elevation for wave motion are expanded in Fourier series up to the fourth harmonic inclusively. Coefficients of these series are explicitly presented as polynomials in Bessel functions. One may speculate that the obtained expressions are the first terms of the expanded exact solution to the Euler equation for surface waves.

1 Introduction

We study the motion of periodic long nonlinear waves propagating on a sloping beach. Such propagation is governed by the so-called Boussinesq-type equations whose special feature is the possibility to reduce the dimension of problem by expanding the velocity potential in power series in the vertical coordinate. This expansion was used by Lagrange [1], developed by Boussinesq [2], and expressed in its modern form in the note of Friedrichs [3].

In 1966, Mei & Mehaute [5] extended these equations to an uneven bottom in one dimension using the potential at bottom as basic variable. Similar equations based on the depth-averaged velocity and on the velocity at the still water level were derived later (see a survey in the paper of Madsen & Schäffer [4]).

In this paper, we follow Mei & Mehaute and write down Boussinesq-type equations for the bottom potential.

There are two small parameters associated with Boussinesq-type equations: the ratio of amplitude to depth, ε; and the ratio of depth to wavelength, μ. The classical Boussinesq equations include terms of orders $O(\varepsilon)$, $O(\mu^2)$.

In papers [7] and [8], we have begun deriving and studying the higher order analog of Mei & Mehaute version of Boussinesq equation. Here we use the computer algebra system for obtaining the equations and their solutions in the next orders. The equations of Mei & Mehaute include ε and μ^2 terms. Our equations include ε^3, $\varepsilon\mu^2$, and μ^4 terms.

We consider regular periodic wave motion over the even slope sx excluding the deep water region where the shallow water assumptions are violated, and the neighborhood of $x = 0$ (shoreline) where a singularity is possible. The potential at the bottom is expanded in Fourier series

$$\begin{aligned} f(x,t) = {} & C^0(x) + S^1(x)\sin(\omega t) + C^1(x)\cos(\omega t) \\ & + S^2(x)\sin(2\omega t) + C^2(x)\cos(2\omega t) \\ & + S^3(x)\sin(3\omega t) + C^3(x)\cos(3\omega t) \\ & + S^4(x)\sin(4\omega t) + C^4(x)\cos(4\omega t)\ . \end{aligned} \tag{1}$$

The main result of this work consists in the explicit expressions (24)–(60) for functions $S^m(x)$ and $C^m(x)$ up to orders $(\varepsilon^3, \varepsilon\mu^2, \mu^4)$ which are homogenous polynomials in Bessel functions $Z_0(2\omega\sqrt{\frac{x}{s}})$ and $Z_1(2\omega\sqrt{\frac{x}{s}})$ whose coefficients are polynomials in $x^{-\frac{1}{2}}$ and $x^{\frac{1}{2}}$. A similar expansion for the surface elevation is also given. These expressions give periodic solutions to the Boussinesq equations within the same accuracy as the equations are derived. Therefore, the result can be interpreted as periodic solution of Euler equations (3)–(6) calculated up to orders ε^3, $\varepsilon\mu^2$, and μ^4. (The first term of this expansion $J_0(2\omega\sqrt{\frac{x}{s}})\sin(\omega t)$ is used, for example, in the book [6]of Mei).

There is some analogy with situation of 1847 [9] when Stokes presented the first three terms of the expanded solution describing a periodic wave motion over horizontal bottom of a finite depth. At that time, it was clear how to calculate the next terms, but convergence was proved only in 1926 [10]. The case of sloping bottom is essentially more complicated because each found coefficient was calculated as a solution of some overdetermined system of linear algebraic equations. In the considered cases, their compatibility looks as a piece of luck, and I do not see reasons why the luck will continue. Nevertheless I take the liberty to make an "experimental" conjecture that the derived terms are only the lowest terms of a new exact solution to the Euler equations (3)–(6) over a sloping beach. The explicit formulas could hardly be obtained without use of computer algebra.

The special case of standing waves was considered in note [7]. For the standing waves Carrier and Greenspan [11] have given an exact solution to Airy's approximate equations which are fully nonlinear but nondispersive (terms of order ε are retained).

For deriving and transforming the equations and for calculating the series satisfying these equations, the computer algebra system "MATHEMATICA" was used. The author is grateful to A.V. Marchenko for help in organization of these calculations.

2 Basic Equations

Non-dimensional coordinates are introduced as follows:

$$x = \frac{x'}{l'_0},\ z = \frac{z'}{h'_0},\ t = \frac{g^{\frac{1}{2}} h_0'^{\frac{1}{2}}}{l'_0} t',\ \eta = \frac{\eta'}{a'_0},$$

$$\varphi \qquad = \frac{h'_0}{a'_0 l'_0 g^{\frac{1}{2}} h'^{\frac{1}{2}}_0}\varphi', \; h = \frac{h'}{h'_0} \tag{2}$$

where the prime denotes physical variables, and a'_0, l'_0, and h'_0 denote the reference wave amplitude, depth, and wavelength, respectively; g is the acceleration of gravity; x is the horizontal coordinate; z is the vertical coordinate; and t is the time. The scaled governing equation and the boundary conditions for the irrotational wave problem read

$$\mu^2\varphi_{xx} + \varphi_{zz} = 0, \quad -h(x) < z < \varepsilon\eta(x,t) \tag{3}$$

$$\eta_t + \varepsilon\varphi_x\eta_x - \mu^{-2}\varphi_z \quad = 0, \quad z = \varepsilon\eta(x,t) \quad , \tag{4}$$

$$\varphi_t + \frac{1}{2}\varepsilon\left(\varphi_x^2 + \mu^{-2}\varphi_z^2\right) + \eta \quad = 0, \quad z = \varepsilon\eta(x,t) \quad , \tag{5}$$

$$\varphi_z \quad = -\mu^2 h_x\varphi_x, \quad z = -h(x), \tag{6}$$

where ε and μ are the measures of nonlinearity and frequency dispersion defined by

$$\varepsilon = a'_0/h'_0, \qquad \mu = h'_0/l'_0. \tag{7}$$

We expand the potential $\varphi(x,z,t)$ in powers of vertical coordinate

$$\varphi(x,z,t) = \sum_{m=0}^{\infty}(z+h(x))^m F_m(x,t) \tag{8}$$

and assume that the function defining the bottom $z = -h(x)$ is linear:

$$h(x) \equiv sx \tag{9}$$

Substituting (8) and (9) into (3) and equating to zero the coefficients of each power of $z + h(x)$, we have:

$$F_{m+2} = -\mu^2\frac{2s(m+1)F_{m+1,x} + F_{m,xx}}{(m+2)(m+1)(1+s^2\mu^2)}\,. \tag{10}$$

The boundary condition at the bottom (6) gives

$$F_1 = -\mu^2\frac{2sF_{0,x}}{1+s^2\mu^2}\,. \tag{11}$$

Denoting $f(x,t) \equiv F_0(x,t)$, and expanding all expressions in powers of μ, we obtain the first terms of φ

$$\begin{aligned}\varphi = f + \left(-s + s^3\mu^2 - s^5\mu^4\right)\mu^2(z+h)f_x + \\ \left(-\tfrac{1}{2} + \tfrac{3}{2}s^2\mu^2 - \tfrac{5}{2}s^4\mu^4\right)\mu^2(z+h)^2 f_{xx} + \\ \left(\tfrac{1}{2}s - \tfrac{5}{3}s^3\mu^2\right)\mu^4(z+h)^3 f_{xxx} +\end{aligned}$$

$$\left(\tfrac{1}{24} - \tfrac{5}{12}s^2\mu^2\right)\mu^4(z+h)^4 f_{xxxx} +$$
$$\left(-\tfrac{1}{24}s\right)\mu^6(z+h)^5 f_{xxxxx} +$$
$$\left(-\tfrac{1}{720}\right)\mu^6(z+h)^6 f_{xxxxxx}\,. \tag{12}$$

Expression (12) satisfies (3) and (6). Substitution of (12) into (4) and (5) gives the Boussinesq-type equations for potential at bottom $f(x,t)$ and for surface elevation $\eta(x,t)$:

$$\begin{gathered}
\eta_t + sf_x + sxf_{xx} \\
+\left(-s^3 f_x - 3s^3 x f_{xx} - \tfrac{3}{2}s^3x^2 f_{xxx} - \tfrac{1}{6}s^3x^3 f_{xxxx}\right)\mu^2 + \\
\begin{pmatrix} s^5 f_x + 5s^5 x f_{xx} + 5s^5x^2 f_{xxx} + \tfrac{5}{3}s^5x^3 f_{xxxx} \\ +\tfrac{5}{24}s^4x^4 f_{xxxxx} + \tfrac{1}{120}s^5x^5 f_{xxxxxx} \end{pmatrix}\mu^4 + \\
\left(f_{xx}\eta + f_x\eta_x\right)\varepsilon + \\
\begin{pmatrix} \left(-3s^2 f_{xx} - 3s^2 x f_{xxx} - \tfrac{1}{2}s^2x^2 f_{xxxx}\right)\eta \\ +\left(-s^2 f_x - 2s^2 x f_{xx} - \tfrac{1}{2}s^2x^2 f_{xxx}\right)\eta_x \end{pmatrix}\varepsilon\mu^2 \\
= 0
\end{gathered} \tag{13}$$

$$\begin{gathered}
\eta + f_t + \left(-s^2 x f_{xt} - \tfrac{1}{2}s^2x^2 f_{xxt}\right)\mu^2 + \\
\left(s^4 x f_{xt} + \tfrac{3}{2}s^4x^2 f_{xxt} + \tfrac{1}{2}s^4x^3 f_{xxxt} + \tfrac{1}{24}s^4x^4 f_{xxxxt}\right)\mu^4 \\
\left(\tfrac{1}{2}f_x^2\right)\varepsilon \\
+\begin{pmatrix} -\tfrac{1}{2}s^2 f_x^2 - s\eta f_{xt} - s^2 x f_x f_{xx} + \tfrac{1}{2}s^2x^2 f_{xx}^2 \\ -sx\eta f_{xxt} - \tfrac{1}{2}s^2x^2 f_x f_{xxx} \end{pmatrix}\varepsilon\mu^2 \\
= 0\,.
\end{gathered} \tag{14}$$

To express the surface elevation $\eta(x,t)$ in terms of $f(x,t)$ and its derivatives, we expand it in powers of μ: $\eta = \eta_0 + \eta_2\mu^2 + \eta_4\mu^4 + O\left(\mu^6\right)$. (Expansion of η_i in powers of ε is not important for this purpose.) After substitution of this expansion into (14), the following formulas are derived:

$$\begin{aligned}
\eta_0 &= -f_t - \tfrac{1}{2}f_x^2\varepsilon \\
\eta_2 &= s^2 x f_{xt} + \tfrac{1}{2}s^2x^2 f_{xxt} + \\
&\begin{pmatrix} \tfrac{1}{2}s^2 f_x^2 - sf_t f_{xt} + s^2 x f_x f_{xx} - \tfrac{1}{2}s^2x^2 f_{xx}^2 \\ -sxf_t f_{xxt} + \tfrac{1}{2}s^2x^2 f_x f_{xxx} \end{pmatrix}\varepsilon \\
\eta_4 &= -s^4 x f_{xt} - \tfrac{3}{2}s^4x^2 f_{xxt} - \tfrac{1}{2}s^4x^3 f_{xxxt} - \tfrac{1}{24}s^4x^4 f_{xxxxt}\,.
\end{aligned} \tag{15}$$

Substituting (15) into (13), we have the single equation for the function f :

$$\begin{gathered}
-f_{tt} + sf_x + sxf_{xx} \\
+ \begin{pmatrix} -s^3 f_x + s^2 x f_{xtt} - 3s^3 x f_{xx} \\ +\tfrac{1}{2}s^2x^2 f_{xxtt} - \tfrac{3}{2}s^3x^2 f_{xxx} - \tfrac{1}{6}s^3x^3 f_{xxxx} \end{pmatrix}\mu^2
\end{gathered}$$

$$
\begin{aligned}
+ & \quad (-f_{xx}f_t - 2f_x f_{xt})\,\varepsilon \\
+ & \begin{pmatrix} s^5 f_x - s^4 x f_{xtt} + 5s^5 x f_{xx} - \frac{3}{2}s^4x^2 f_{xxtt} \\ +5s^5x^2 f_{xxx} - \frac{1}{2}s^4x^3 f_{xxxtt} + \frac{5}{3}s^5x^3 f_{xxxx} \\ -\frac{1}{24}s^4x^4 f_{xxxxtt} + \frac{5}{24}s^5x^4 f_{xxxxx} \\ +\frac{1}{120}s^5x^5 f_{xxxxxx} \end{pmatrix} \mu^4 \\
+ & \begin{pmatrix} -s f_{tt} f_{xt} + 3s^2 f_x f_{xt} - s f_t f_{xtt} + 3s^2 f_t f_{xx} \\ +4s^2 x f_{xx} f_{xt} - s x f_{tt} f_{xxt} + 3s^2 x f_x f_{xxt} \\ -\frac{1}{2}s^2x^2 f_{xx} f_{xxt} - s x f_t f_{xxtt} + 3s^2 x f_t f_{xxx} \\ +s^2x^2 f_{xt} f_{xxx} + s^2x^2 f_x f_{xxxt} + \frac{1}{2}s^2x^2 f_{xxxx} f_t \end{pmatrix} \varepsilon\mu^2 \\
+ & \quad \left(\tfrac{3}{2} f_x^2 f_{xx}\right)\varepsilon^2 = 0\,.
\end{aligned}
\tag{16}
$$

3 Periodic Problem

We suppose that the solution is periodic in time and can be expanded in Fourier series in an area excluding the deep water region and a neighborhood of $x = 0$ (shore line).

$$
\begin{aligned}
f(x,t) & = u_{10}(x)\varepsilon + u_{30}(x)\varepsilon^3 \\
& + \left(S_{00}^1(x) + S_{20}^1(x)\varepsilon^2 + S_{02}^1(x)\mu^2 + S_{04}^1(x)\mu^4\right)\sin(\omega t) \\
& + \left(C_{00}^1(x) + C_{20}^1(x)\varepsilon^2 + C_{02}^1(x)\mu^2 + C_{04}^1(x)\mu^4\right)\cos(\omega t) \\
& + \left(S_{10}^2(x)\varepsilon + S_{12}^2(x)\varepsilon\mu^2 + S_{30}^2(x)\varepsilon^3\right)\sin(2\omega t) \\
& + \left(C_{10}^2(x)\varepsilon + C_{12}^2(x)\varepsilon\mu^2 + C_{30}^2(x)\varepsilon^3\right)\cos(2\omega t) \\
& + S_{20}^3(x)\varepsilon^2\sin(3\omega t) + C_{20}^3(x)\varepsilon^2\cos(3\omega t) \\
& + S_{30}^4(x)\varepsilon^3\sin(4\omega t) + C_{30}^4(x)\varepsilon^3\cos(4\omega t)\,.
\end{aligned}
\tag{17}
$$

(Forms of coefficients affecting $\sin(m\omega t)$ and $\cos(m\omega t)$ are determined by recurrence calculations when solving (16)).

We denote by $S = S(x)$ and $C = C(x)$ two solutions of equation

$$\omega^2 Z + sZ_x + sxZ_{xx} = 0, \tag{18}$$

and by S' and C' their derivatives. The functions $S(x)$, $C(x)$, $S'(x)$, and $C'(x)$ can be expressed in terms of Bessel functions as follows:

$$S(x) = a_{11}J_0\left(2\omega\sqrt{x/s}\right) + a_{12}Y_0\left(2\omega\sqrt{x/s}\right), \tag{19}$$

$$S'(x) = \omega s^{-\frac{1}{2}}x^{-\frac{1}{2}}\begin{pmatrix} -a_{11}J_1\left(2\omega\sqrt{x/s}\right) \\ -a_{12}Y_1\left(2\omega\sqrt{x/s}\right) \end{pmatrix} \tag{20}$$

$$C(x) = a_{21}J_0\left(2\omega\sqrt{x/s}\right) + a_{22}Y_0\left(2\omega\sqrt{x/s}\right), \tag{21}$$

$$C'(x) = \omega s^{-\frac{1}{2}} x^{-\frac{1}{2}} \begin{pmatrix} -a_{21} J_1 \left(2\omega\sqrt{x/s}\right) \\ -a_{22} Y_1 \left(2\omega\sqrt{x/s}\right) \end{pmatrix} \tag{22}$$

Denote by a the determinant $a_{11}a_{22} - a_{12}a_{21}$. The following expression for Wronskian $W(S(x), C(x))$ can easily be proved:

$$SC' - CS' = \frac{a}{\pi x}. \tag{23}$$

The major finding of this paper is the expressions for $S^i_{\alpha\beta}$ and $C^i_{\alpha\beta}$:

Zero Harmonic

$$u_{10} = -\frac{a\omega}{2\pi s} \cdot \frac{1}{x} \tag{24}$$

$$u_{30} = -\frac{\omega^2}{4\,s^2\,x^3}\left(C^2 + S^2\right) - \frac{3}{8\,s\,x^2}\left(C'^2 + S'^2\right). \tag{25}$$

First Harmonic

$$S^1_{00} = S \tag{26}$$

$$\begin{aligned} S^1_{20} = &-\frac{a\omega}{2sx^2}C - \frac{\omega^4}{8s^3x}S^3 - \frac{\omega^2}{8s^2x}S^2S' - \frac{\omega^2}{8s^2}SS'^2 \\ &-\frac{1}{8s}S'^3 + \left(-\frac{\omega^4}{8s^3x}C^2 - \frac{3\omega^2}{8s^2}C'^2\right)S \\ &+\left(-\frac{\omega^2}{8s^2x}C^2 + \frac{\omega^2}{4s^2}CC' + \frac{1}{8s}C'^2\right)S' \end{aligned} \tag{27}$$

$$S^1_{02} = -\frac{2s\omega^2x}{9}S + \left(\frac{7s^2x}{9} + \frac{s\omega^2x^2}{9}\right)S' \tag{28}$$

$$\begin{aligned} S^1_{04} = &\left(-\frac{479s^3\omega^2x}{1350} - \frac{29s^2\omega^4x^2}{225} - \frac{s\omega^6x^3}{162}\right)S \\ &+\left(-\frac{479s^4x}{1350} - \frac{136s^3\omega^2x^2}{675} - \frac{7s^2\omega^4x^3}{450}\right)S' \end{aligned} \tag{29}$$

$$C^1_{00} = C \tag{30}$$

$$\begin{aligned} C^1_{20} = &\frac{a\omega}{2sx^2}S - \frac{\omega^4}{8s^3x}C^3 - \frac{\omega^2}{8s^2x}C^2C' - \frac{\omega^2}{8s^2}CC'^2 \\ &-\frac{1}{8s}C'^3 + \left(-\frac{\omega^4}{8s^3x}S^2 - \frac{3\omega^2}{8s^2}S'^2\right)C \\ &+\left(-\frac{\omega^2}{8s^2x}S^2 + \frac{\omega^2}{4s^2}SS' - \frac{1}{8s}S'^2\right)C' \end{aligned} \tag{31}$$

$$C^1_{02} = -\frac{2s\omega^2x}{9}C + \left(\frac{7s^2x}{9} + \frac{s\omega^2x^2}{9}\right)C' \tag{32}$$

$$\begin{aligned} C^1_{04} = &\left(-\frac{479s^3\omega^2x}{1350} - \frac{29s^2\omega^4x^2}{225} - \frac{s\omega^6x^3}{162}\right)C \\ &+\left(-\frac{479s^4x}{1350} - \frac{136s^3\omega^2x^2}{675} - \frac{7s^2\omega^4x^3}{450}\right)C'. \end{aligned} \tag{33}$$

Second Harmonic

$$S^2_{10} = -\tfrac{\omega}{2s}SS' + \tfrac{\omega}{2s}CC' \tag{34}$$

$$\begin{aligned} S^2_{12} &= \left(\tfrac{4\omega^3}{3} + \tfrac{7\omega^5 x}{18s}\right) S^2 + \tfrac{7\omega^3 x}{3} SS' \\ &+ \left(-\tfrac{2s\omega x}{9} - \tfrac{7\omega^3 x^2}{18}\right) S'^2 + \left(-\tfrac{4\omega^3}{3} - \tfrac{7\omega^5 x}{18s}\right) C^2 \\ &- \tfrac{7\omega^3 x}{3} CC' + \left(\tfrac{2s\omega x}{9} + \tfrac{7\omega^3 x^2}{18}\right) C'^2 \end{aligned} \tag{35}$$

$$\begin{aligned} S^2_{30} = &\ \tfrac{\omega^5}{12\, s^4\, x^2}\left(S^4 - C^4\right) + \tfrac{\omega}{8\, s^2}\left(C'^4 - S'^4\right) \\ &+ \left(\tfrac{\omega^5}{6\, s^4\, x} - \tfrac{\omega^3}{12\, s^3\, x^2}\right)\left(C^3\, C' - S^3\, S'\right) \\ &+ \left(\tfrac{\alpha\, \omega^2}{s^2\, x} - \tfrac{3\, \alpha}{4\, s\, x^2}\right) C'\, S' \\ &+ \left(\tfrac{\alpha\, \omega^2}{2\, s^2\, x^3} - \tfrac{\alpha\, \omega^4}{s^3\, x^2}\right) CS \\ &- \tfrac{3\, \alpha\, \omega^2}{2\, s^2\, x^2}\left(C\, S' + C'\, S\right) \\ &+ \left(\tfrac{\omega^3}{6\, s^3} - \tfrac{\omega}{8\, s^2\, x}\right)\left(C\, C'^3 - SS'^3\right) \end{aligned} \tag{36}$$

$$C^2_{10} = -\tfrac{\omega}{2s}SC' - \tfrac{\omega}{2s}S'C \tag{37}$$

$$\begin{aligned} C^2_{12} &= \left(\tfrac{8\omega^3}{3} + \tfrac{7\omega^5 x}{9s}\right) SC + \tfrac{7\omega^3 x}{3} SC' \\ &+ \tfrac{7\omega^3 x}{3} S'C + \left(-\tfrac{4s\omega x}{9} - \tfrac{7\omega^3 x^2}{9}\right) S'C' \end{aligned} \tag{38}$$

$$\begin{aligned} C^2_{30} = &\ \tfrac{\omega^5}{6\, s^4\, x^2}\left(C^3\, S + C\, S^3\right) - \tfrac{\omega}{4\, s^2}\left(C'^3\, S' + C'\, S'^3\right) \\ &+ \left(\tfrac{\omega^3}{12\, s^3\, x^2} - \tfrac{95\, \omega^5}{288\, s^4\, x}\right)\left(S^3\, C' + C^3\, S'\right) \\ &+ \left(\tfrac{\omega^3}{12\, s^3\, x^2} - \tfrac{\omega^5}{288\, s^4\, x}\right)\left(C^2\, S\, C' + C\, S^2\, S'\right) \\ &+ \left(-\tfrac{73\, \omega^3}{144\, s^3\, x} + \tfrac{\omega^5}{432\, s^4}\right)\left(C^2\, C'\, S' + S^2\, C'\, S'\right) \\ &+ \left(\tfrac{73\, \omega^3}{144\, s^3\, x} - \tfrac{\omega^5}{432\, s^4}\right)\left(C\, S\, C'^2 + C\, S\, S'^2\right) \\ &+ \left(\tfrac{3\, \omega}{32\, s^2\, x} - \tfrac{\omega^3}{144\, s^3}\right)\left(C\, C'^2\, S' + S\, C'\, S'^2\right) \\ &+ \left(\tfrac{5\, \omega}{32\, s^2\, x} - \tfrac{47\, \omega^3}{144\, s^3}\right)\left(S\, C'^3 + C\, S'^3\right) \\ &+ \alpha\left(\tfrac{\omega^2}{6\, s^2\, x^3} - \tfrac{\omega^4}{144\, s^3\, x^2}\right)\left(C^2 - S^2\right) \\ &+ \alpha\left(-\tfrac{35\, \omega^2}{72\, s^2\, x^2} - \tfrac{\omega^4}{216\, s^3\, x}\right)\left(C\, C' - S\, S'\right) \\ &+ \alpha\left(\tfrac{-3}{16\, s\, x^2} + \tfrac{\omega^2}{72\, s^2\, x}\right)\left(C'^2 - S'^2\right). \end{aligned} \tag{39}$$

Third Harmonic

$$\begin{aligned} S^3_{20} = &-\tfrac{\omega^4}{8s^3x}S^3 - \tfrac{\omega^2}{8s^2x}S^2S' + \tfrac{3\omega^2}{8s^2}SS'^2 + \tfrac{1}{24s}S'^3 \\ &+\left(\tfrac{3\omega^4}{8s^3x}C^2 + \tfrac{\omega^2}{4s^2x}CC' - \tfrac{3\omega^2}{8s^2}C'^2\right)S \\ &+\left(\tfrac{\omega^2}{8s^2x}C^2 - \tfrac{3\omega^2}{4s^2}CC' - \tfrac{1}{8s}C'^2\right)S' \end{aligned} \tag{40}$$

$$\begin{aligned} C^3_{20} = &\tfrac{\omega^4}{8s^3x}C^3 + \tfrac{\omega^2}{8s^2x}C^2C' - \tfrac{3\omega^2}{8s^2}CC'^2 - \tfrac{3}{72s}C'^3 \\ &+\left(-\tfrac{3\omega^4}{8s^3x}S^2 - \tfrac{\omega^2}{4s^2x}SS' + \tfrac{3\omega^2}{8s^2}S'^2\right)C + \\ &+\left(-\tfrac{\omega^2}{8s^2x}S^2 + \tfrac{3\omega^2}{4s^2}SS' + \tfrac{1}{8s}S'^2\right)C'\,. \end{aligned} \tag{41}$$

Fourth Harmonic

$$\begin{aligned} S^4_{30} = &\tfrac{\omega^5}{24\,s^4\,x^2}\left(C^4+S^4\right) + \tfrac{\omega}{16\,s^2}\left(C'^4+S'^4\right) \\ &+\left(\tfrac{\omega^3}{24\,s^3\,x^2} - \tfrac{\omega^5}{3\,s^4\,x}\right)\left(C^3C' + S^3S'\right) \\ &+\left(\tfrac{\omega^3}{24\,s^3\,x^2} - \tfrac{\omega^5}{3\,s^4\,x}\right)\left(C^3C' + S^3S'\right) \\ &+\tfrac{3\,\omega^3}{8\,s^3\,x}\left(C^2S'^2 + C'^2S^2 - S^2S'^2 - C^2C'^2\right) \\ &-\tfrac{3\,\omega}{8\,s^2}\,C'^2\,S'^2 + \tfrac{3\,\omega^3}{2\,s^3\,x}CC'\,S\,S' \\ &+\left(\tfrac{3\,\omega}{16\,s^2\,x} - \tfrac{\omega^3}{s^3}\right)\left(CC'\,S'^2 + C'^2\,SS'\right) \\ &+\left(-\tfrac{\omega^3}{8\,s^3\,x^2} + \tfrac{\omega^5}{s^4\,x}\right)\left(C^2S\,S' + C\,C'S^2\right) \\ &-\tfrac{\omega^5}{4\,s^4\,x^2}C^2S^2 \end{aligned} \tag{42}$$

$$\begin{aligned} C^4_{30} = &\left(\tfrac{\omega^3}{24\,s^3\,x^2} - \tfrac{\omega^5}{3\,s^4\,x}\right)\left(C'S^3 - C^3S'\right) \\ &-\tfrac{\omega}{4\,s^2}\left(C'^3\,S' - C'\,S'^3\right) \\ &+\left(-\tfrac{\omega^3}{8\,s^3\,x^2} + \tfrac{\omega^5}{s^4\,x}\right)\left(C^2\,C'S - CS^2\,S'\right) \\ &+\left(-\tfrac{3\,\omega}{16\,s^2\,x} + \tfrac{\omega^3}{s^3}\right)\left(C'\,SS'^2 - C\,C'^2\,S'\right) \\ &+\left(\tfrac{\omega}{16\,s^2\,x} - \tfrac{\omega^3}{3\,s^3}\right)\left(C'^3S - C\,S'^3\right) \\ &+\tfrac{3\,\omega^3}{4\,s^3\,x}\left(C'^2 - S'^2\right)C\,S\,. \end{aligned} \tag{43}$$

Substituting these expressions into (15), we could derive the expressions for $\eta(x,t)$.

$$\begin{aligned}\eta(x,t) &= L^0_{10}(x)\varepsilon + L^0_{12}(x)\varepsilon\mu^2 \\ &+ \left(P^1_{00}(x) + P^1_{20}(x)\varepsilon^2 + P^1_{02}(x)\mu^2 + P^1_{04}(x)\mu^4\right)\sin(\omega t) \\ &+ \left(Q^1_{00}(x) + Q^1_{20}(x)\varepsilon^2 + Q^1_{02}(x)\mu^2 + Q^1_{04}(x)\mu^4\right)\cos(\omega t) \\ &+ \left(P^2_{10}(x)\varepsilon + P^2_{12}(x)\varepsilon\mu^2\right)\sin(2\omega t) \\ &+ \left(Q^2_{10}(x)\varepsilon + Q^2_{12}(x)\varepsilon\mu^2\right)\cos(2\omega t) \\ &+ P^3_{20}(x)\varepsilon^2\sin(3\omega t) + Q^3_{20}(x)\varepsilon^2\cos(3\omega t)\,. \end{aligned} \tag{44}$$

We present here only the lowest terms since the paper is overburden by formulas.

Zero Harmonic

$$L^0_{10} = -\frac{1}{4}S'^2 - \frac{1}{4}C'^2 \tag{45}$$

$$\begin{aligned}L^0_{12} &= \frac{\omega^4}{4}S^2 + \frac{\omega^4}{4}C^2 + \frac{\omega^4 x}{18}SS' \\ &+ \frac{\omega^4 x}{18}CC' - \frac{7s\omega^2 x}{36}S'^2 - \frac{7s\omega^2 x}{36}C'^2\,. \end{aligned} \tag{46}$$

First Harmonic

$$P^1_{00} = \omega C \tag{47}$$

$$\begin{aligned}P^1_{20} &= \frac{a}{x^2}S' + \frac{a\omega^2}{2sx^2}S - \frac{\omega^5}{8s^3x}C^3 + \frac{\omega^3}{8s^2x}C^2C' \\ &+ \left(\frac{\omega}{4sx} - \frac{\omega^3}{8s^2}\right)CC'^2 - \frac{3\omega}{8s}C'^3 - \frac{\omega^5}{8s^3x}CS^2 \\ &- \frac{3\omega^3}{8s^2x}C'S^2 + \left(\frac{\omega}{4sx} - \frac{3\omega^3}{8s^2}\right)CS'^2 \\ &- \frac{3\omega}{8s}C'S'^2 + \frac{\omega^3}{2s^2x}CSS' + \frac{\omega^3}{4s^2}C'SS' \end{aligned} \tag{48}$$

$$P^1_{02} = \frac{5s\omega^3 x}{18}C + \left(\frac{5s^2\omega x}{18} + \frac{s\omega^3x^2}{9}\right)C' \tag{49}$$

$$\begin{aligned}P^1_{04} &= \left(-\frac{283s^3\omega^3x}{2700} + \frac{43s^2\omega^5x^2}{1800} - \frac{s\omega^7x^3}{162}\right)C \\ &+ \left(-\frac{283s^4\omega x}{2700} + \frac{103s^3\omega^3x^2}{1350} + \frac{s^2\omega^5x^3}{25}\right)C' \end{aligned} \tag{50}$$

$$Q^1_{00} = -\omega S \tag{51}$$

$$Q^1_{20} = \frac{a}{x^2}C' + \frac{a\omega^2}{2sx^2}C + \frac{\omega^5}{8s^3x}S^3 - \frac{\omega^3}{8s^2x}S^2S'$$

$$+\left(\frac{\omega^3}{8s^2}-\frac{\omega}{4sx}\right)SS'^2+\frac{3\omega}{8s}S'^3$$
$$+\frac{3\omega^3}{8s^2x}C^2S'-\frac{\omega^3}{4s^2}CC'S'+\frac{3\omega}{8s}S'C'^2+\frac{\omega^5}{8s^3x}C^2S$$
$$-\frac{\omega^3}{2s^2x}SCC'+\left(-\frac{\omega}{4sx}+\frac{3\omega^3}{8s^2}\right)SC'^2 \tag{52}$$

$$Q^1_{02}=-\frac{5s\omega^3x}{18}S+\left(-\frac{5s^2\omega x}{18}-\frac{s\omega^3x^2}{9}\right)S' \tag{53}$$

$$Q^1_{04}=\left(\frac{283s^3\omega^3x}{2700}-\frac{43s^2\omega^5x^2}{1800}+\frac{s\omega^7x^3}{162}\right)S$$
$$+\left(\frac{283s^4\omega x}{2700}-\frac{103s^3\omega^3x^2}{1350}-\frac{s^2\omega^5x^3}{25}\right)S'. \tag{54}$$

Second Harmonic

$$P^2_{10}=-\frac{\omega^2}{s}SC'-\frac{\omega^2}{s}S'C-\frac{1}{2}S'C' \tag{55}$$

$$P^2_{12}=\left(\frac{23\omega^4}{6}+\frac{14\omega^6x}{9s}\right)CS+\frac{49\omega^4x}{18}SC'$$
$$+\frac{49\omega^4x}{18}S'C+\left(-\frac{41s\omega^2x}{18}-\frac{14\omega^4x^2}{9}\right)S'C' \tag{56}$$

$$Q^2_{10}=-\frac{\omega^2}{s}CC'+\frac{\omega^2}{s}SS'-\frac{1}{4}C'^2+\frac{1}{4}S'^2 \tag{57}$$

$$Q^2_{12}=\left(\frac{23\omega^4}{12}+\frac{7\omega^6x}{9s}\right)C^2+\frac{49\omega^4x}{18}CC'$$
$$+\left(-\frac{41s\omega^2x}{36}-\frac{7\omega^4x^2}{9}\right)C'^2$$
$$+\left(-\frac{23\omega^4}{12}-\frac{7\omega^6x}{9s}\right)S^2$$
$$-\frac{49\omega^4x}{18}SS'+\left(\frac{41s\omega^2x}{36}+\frac{7\omega^4x^2}{9}\right)S'^2. \tag{58}$$

Third Harmonic

$$P^3_{20}=\frac{3\omega^5}{8s^3x}C^3+\frac{5\omega^3}{8s^2x}C^2C'+\left(\frac{\omega}{4sx}-\frac{9\omega^3}{8s^2}\right)CC'^2$$
$$-\frac{3\omega}{8s}C'^3-\frac{9\omega^5}{8s^3x}CS^2-\frac{5\omega^3}{8s^2x}C'S^2$$
$$-\frac{5\omega^3}{4s^2x}CSS'+\left(-\frac{\omega}{2sx}+\frac{9\omega^3}{4s^2}\right)SS'C'$$

$$+\left(-\frac{\omega}{4sx}+\frac{9\omega^3}{8s^2}\right)CS'^2+\frac{9\omega}{8s}S'^2C' \tag{59}$$

$$\begin{aligned}Q_{20}^3 = &\frac{3\omega^5}{8s^3x}S^3+\frac{5\omega^3}{8s^2x}S^2S'+\left(\frac{\omega}{4sx}-\frac{9\omega^3}{8s^2}\right)SS'^2\\ &-\frac{3\omega}{8s}S'^3-\frac{5\omega^3}{8s^2x}C^2S'+\left(-\frac{\omega}{2sx}+\frac{9\omega^3}{4s^2}\right)CC'S'\\ &+\frac{9\omega}{8s}C'^2S'-\frac{9\omega^5}{8s^3x}C^2S-\frac{5\omega^3}{4s^2x}CC'S\\ &+\left(-\frac{\omega}{4sx}+\frac{9\omega^3}{8s^2}\right)C'^2S\end{aligned} \tag{60}$$

4 Special Cases

Setting $a_{12}=-\frac{1}{\omega}$, $a_{21}=\frac{1}{\omega}$, $a_{11}=a_{22}=0$, we have $S(x)=-\frac{1}{\omega}Y_0\left(2\omega\sqrt{\frac{x}{s}}\right)$ and $C(x)=\frac{1}{\omega}J_0\left(2\omega\sqrt{\frac{x}{s}}\right)$. Then the main approximation for $\eta(x,t)$ is

$$J_0\left(2\omega\sqrt{x/s}\right)\sin(\omega t)+Y_0\left(2\omega\sqrt{x/s}\right)\cos(\omega t), \tag{61}$$

which is equivalent to

$$\begin{aligned}&\sqrt{\tfrac{2}{\pi x}}\begin{pmatrix}\cos\left(2\omega\sqrt{x/s}-\frac{\pi}{4}\right)\sin(\omega t)\\ +\sin\left(2\omega\sqrt{x/s}-\frac{\pi}{4}\right)\cos(\omega t)\end{pmatrix}\\ &=\sqrt{\tfrac{2}{\pi x}}\sin\left(2\omega\sqrt{x/s}-\tfrac{\pi}{4}+\omega t\right)\end{aligned} \tag{62}$$

for $x\to+\infty$. So, allowing some carelessness, we can consider this case as progressive wave, whereas the case $a_{11}=1$, $a_{22}=a_{12}=a_{21}=0$ is a standing wave studied in [8].

5 Conclusions

Periodic solution with the accuracy of $(\varepsilon^3,\ \varepsilon\mu^2,\ \mu^4)$ to equations (3)–(6) is presented. The intermediate equations are given for illustrating the method of derivation but the expressions (24)–(60) can be proved by substitution into system (3)–(6) (using expression (12) for the potential). Such substitution was performed for validation of the solutions using computer algebra.

We conjecture that these expressions are only the lowest terms of a certain expanded exact solution to the Euler equations (3)–(6). The number of known exact solutions to the Euler equations is very small. This solution can be used for describing water dynamics on beaches.

The derived formulas are obtained by the method of unknown coefficients as solutions of some **overdetermined** systems of linear algebraic equations. The reason for their solvability remains obscure in the moment. Such results could hardly be obtained without use of computer algebra.

The work was supported by the RFBR Project 96-15-96525.

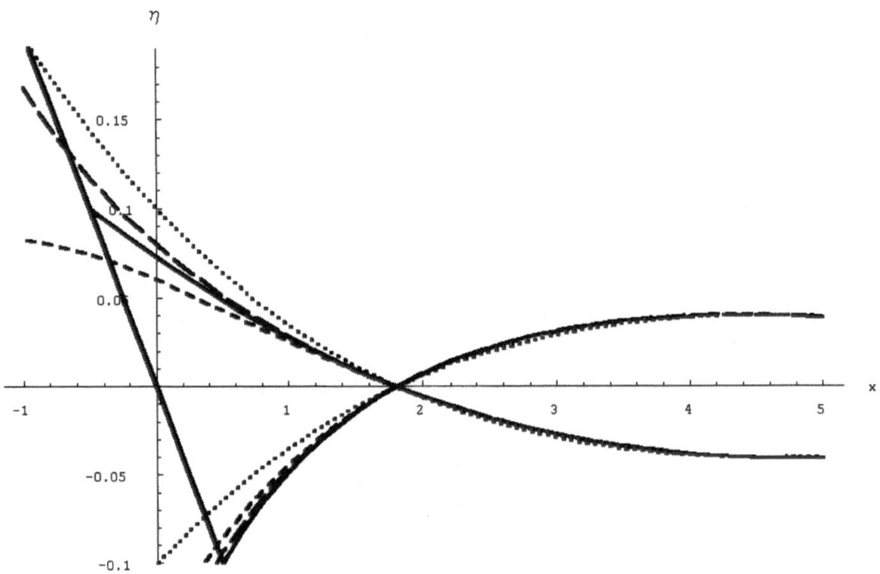

Fig. 1. Standing waves. Successive approximations of two extreme positions (at times $t = 0$ and $t = \frac{\pi}{\omega}$) of the water surface (given by the function $\varepsilon\eta(x,t)$). Parameters of the periodic motion are $\omega = 0.4$, $s = 0.2$, $\varepsilon = 0.25$, and $\mu = 0.4$. Zero-order solutions (first harmonic) are shown by dotted lines while the short- and long-dashed lines are constructed using (ε, μ^2) and $(\varepsilon^2, \varepsilon\mu^2, \mu^4)$ approximations. The solid curves are plotted using $(\varepsilon^3, \varepsilon\mu^2, \mu^4)$ (the four harmonics approximation)

References

[1] Grange, J. L. de la (Lagrange): Mecanique Analitique, v. 2 Paris (1788)

[2] Boussinesq, J.: Theorie des ondes et des remous qui se propagent le long d'un canal rectangulaire horisontal, en communiquant au liquide contenu dans ce canal des vitesses sensiblement pareilles de la surface au fond. *J. Math. Pures Appl. 2nd Series* **17** (1872) 55–108

[3] Friedrichs, K. O.: On the derivation of the shallow water theory. *Comm. Pure Appl. Math.* **1** (1948) 81–85

[4] Madsen, P. A. & Schäffer, H. A.: Higher-order Boussinesq-type equations for surface gravity waves: derivation and analysis. *Phil. Trans. R. Soc. Lond.* A **8** (1998) 441–455

[5] Mei, C. C. & Le Méhauté, B.: Note on equation of long waves over an uneven bottom. *J. Geophys. Res.* **7** (1966) 393–400

[6] Mei, C. C.: The Applied Dynamics of Ocean Surface Waves. Wiley, New York (1983)

[7] Shermenev, A & Shermeneva, M.: Long periodic waves on an even beach. *Physical Review,* **E, 61** (2000) 6000–6002

[8] Shermenev, A.: Nonlinear periodic waves on a beach. *Geophysical and Astrophysical Fluid Dynamics* (2001) (to appear)

[9] Stokes, G. G.: On the theory of oscillatory waves, *Trans. Cambridge Phil. Soc.* (1847) 8

[10] Struik, D. J.: Determination rigoureuse des ondes irrotationelles periodiques dans un canal à profondeur finie, *Math. Ann.* **95** (1926) 595–634

[11] Carrier, G. F. & Greenspan, H. P.: Water waves of finite amplitude on a sloping beach, *J. Fluid Mech.* **4** (1957) 97–109

The Method of Newton Polyhedra for Investigating Singular Positions of Some Mechanisms*

Akhmadjon Soleev and Adizjon S. Barotov

Department of Mathematics, Samarkand State University, Samarkand, 703004, Uzbekistan,
iipp@samarkand.uz and soleev@samarkand.uz

Abstract. In the paper the singular positions of a displacement function of the plane mechanism with three degrees of freedom are investigated. The singularities of this function are subdivided into two types. The singularities of the displacement function as a function of control coordinates belong to the first type. The second type singularities are such singularities, in the neighbourhood of which the coordinates of position and control do not form an n- dimensional manifold (n is the number of degrees of freedom).

1 Introduction

The needs of advanced engineering demand the investigation of displacement functions of mechanisms used in robotics, which have complex structure. The displacement functions of these mechanisms have not been studied at all. This gives rise to difficulties of the solution of such problems such as finding all singularities of the mechanism.

The geometry of the majority of mechanisms is described by a system of non-linear algebraic equations of the form:

$$f_i(x, y) = 0, \qquad i = 1, 2, \ldots, m, \tag{1}$$

where $x = (x_1, x_2, \ldots, x_m)$ and $y = (y_1, y_2, \ldots, y_n)$ are the coordinates of position and control, respectively, f_i are polynomials. There are also many other methods for representing the connection equations, for example, the method of closed loops [1], the theory of screws [2]. But the algebraic form of representation is in our opinion most preferential since the methods for the solution of systems of algebraic equations are most advanced.

From the mathematical viewpoint the special position of the mechanism is characterized by Jacobian $\mathcal{J}(X, Y) \stackrel{\text{def}}{=} \det\left(\frac{\partial f_i}{\partial x_j}\right) \quad (i, j = \overline{1, m})$ of system (1).

In [3] the authors subdivided the set of singular points in which mechanism will be in singular position into two types. The singularities of the displacement

* This work was supported by State Committee for Science and Technology of the Rebublic of Uzbekistan, grant No. 4110.

function as a function of control coordinates belong to the first type. The second type contains such singularities, in the neighbourhood of which the coordinates of position and control do not form an n- dimensional manifold (n is the number of degrees of freedom).

Below we present an algorithm for finding the singularities:

1. Writing the equations of connection in algebraic form:

$$f_i(x,y)=0, \quad i=1,2,\dots,m.$$

2. Compiling a Jacobi matrix on the basis of the system of connection equations:

$$\mathcal{A} \stackrel{\text{def}}{=} \left(\frac{\partial f_i}{\partial z_j}\right), \quad i=\overline{1,m}, \ j=\overline{1,m+n}.$$

3. Allocation of all minors A_I, $I=(i_1...,i_n)$ of the order m of the Jacobi matrix $\mathcal{A}$.

4. Calculation of the allocated minors.

5. Decomposing the obtained determinants in factors and finding such configurations of mechanisms, in which $A_I=0$ at least for one set of indices I.

6. Finding the conditions of a simultaneous vanishing of all minors of the order m of the Jacobi matrix.

7. Finding out the conditions imposed on mechanism parameters at fulfilment of which $A_I=0$ for all sets $I=(i_1...,i_n)$.

To find all the singularities of the first type it is sufficient to apply the above first five steps. The last two steps, namely, **6** and **7** are rather labour-consuming.

The n-dimensional solution of system (1) satisfies the conditions of the Cauchy theorem about an implicit function near a singular point of the first type and, therefore, is unique and is representable in the form of power series convergent in a small neighbourhood of a singular point. Near the second type points, such a representation is impossible, since in this case the conditions of the Cauchy theorem are violated. It is possible to obtain the solution $Z-Z^o$ in the form of power series on additional parameters $T=(t_1,t_2,\dots,t_n)$, where the number of linearly independent parameters t_j is equal to the number of degrees of freedom of the mechanism, by the method of *Newton polyhedra.* This method is universal, since it allows us to obtain the representation of displacement function in both regular and singular points.

Below we present the algorithm for resolving the singularities of the system of equations (1) and find all local branches of its solutions.

2 Plane Mechanism with Three Degrees of Freedom

Consider the mechanism with three degrees of freedom on a plane XOY (Fig. 1). The points O,A,B, and C have the coordinates: $O=O(0,0)$, $A=A(x_0,y_0)$, $B=B(x_1,y_1)$ and $C=C(a,0)$. The lengths $OA=l_0$ and $BC=l_1$ are variable, and $AB=d$ and the base $OC=a$ are constant.

Connection equations for the displacement functions of the mechanism are as follows:

$$\begin{aligned} f_1 &\stackrel{\text{def}}{=} x_0^2+y_0^2-l_0^2=0,\\ f_2 &\stackrel{\text{def}}{=} (x_1-x_0)^2+(y_1-y_0)^2-d^2=0, \end{aligned} \tag{2}$$

$$f_3 \stackrel{\text{def}}{=} (x_1 - a)^2 + y_1^2 - l_1^2 = 0.$$

$$\mathcal{J} \stackrel{\text{def}}{=} 2 \begin{pmatrix} x_0 & y_0 & 0 & 0 & l_0 & 0 \\ u_{01} & v_{01} & u_{10} & v_{10} & 0 & 0 \\ 0 & 0 & x_1 - a & y_1 & 0 & l_1 \end{pmatrix},$$

where $u_{ij} = x_i - x_j$, $v_{ij} = y_i - y_j$ and $u_{ij} = -u_{ji}$, $v_{ij} = -v_{ji}$. In this matrix all nonzero minors of dimension 3 are

$$\begin{array}{llll} A_1 = 2(x_1 - a)M_1, & A_2 = 2y_1 M_1, & A_3 = 2l_1 M_1, & A_4 = 2x_0 M_2, \\ A_5 = 8l_0 u_{01}(x_1 - a), & A_6 = 8l_1 x_0 u_{10}, & A_7 = 8y_1 l_0 u_{01}, & A_8 = 8l_1 x_0 v_{10}, \\ A_9 = 8l_0 l_1 u_{01}, & A_{10} = 2y_0 M_2, & A_{11} = 8l_0 v_{01}(x_1 - a), & A_{12} = 8y_0 l_1 u_{10}, \\ A_{13} = 8l_0 y_1 v_{01}, & A_{14} = 8l_1 y_0 v_{10}, & A_{15} = 8l_0 l_1 v_{01}, & A_{16} = 2l_0 M_2, \\ A_{17} = 8l_0 l_1 u_{10}, & A_{18} = 8l_0 l_1 v_{10}, & & \end{array}$$

where $M_1 = 4(x_0 v_{01} - y_0 u_{01})$, $\quad M_2 = 4(y_1 u_{10} - (x_1 - a)v_{10})$.

$M_1 = 0$ only if three points O, A, B lie on the same straight line. Similarly, $M_2 = 0$ only if three points A, B, C lie on the same straight line. Both cases mean that the mechanism may take the singular configurations, which are singular positions of the first type. The singularities of the second type arise only at $A_k = 0$ $(k = 1, 2, \ldots, 18)$.

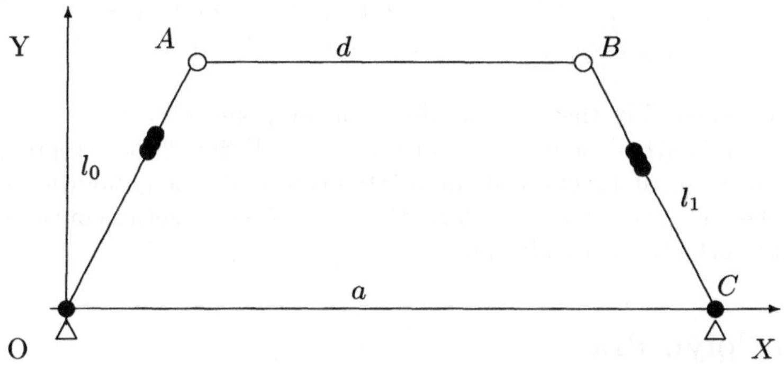

Fig. 1

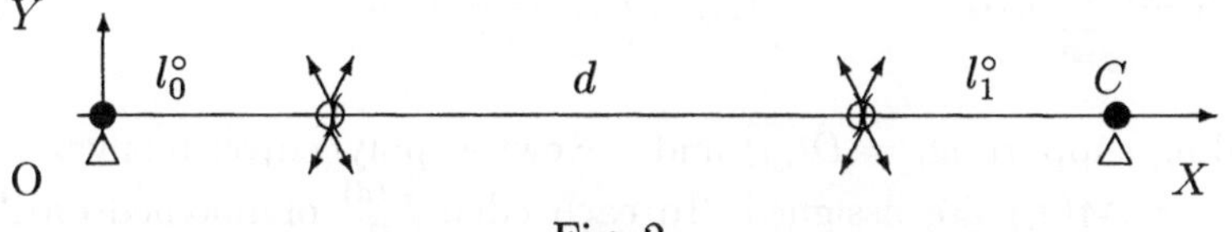

Fig. 2

Theorem. *The plane mechanism with three degrees of freedom (Fig. 1) has no singularities of the second type.*

Proof. Since $l_0 \neq 0$ and $l_1 \neq 0$, all minors A_k $(k \neq 5, 6, 9, 17)$ will vanish simultaneously only at the horizontal position of the mechanism. In this case, four points of the mechanism O, A, B, C lie on a single straight line. Since the points O, C lie on the OX axis, the points A, B should also lie there. It is feasible, when the coordinates y, y_1 are equal to zero. Under such condition the remaining four minors $A_5 = A_6 = A_9 = A_{17}$ are equal to zero only at $x_0 = x_1$. Since $d \neq 0$, it is not feasible. The theorem is proved.

All singular positions of the mechanism will be only of the first type.

Let the coordinates $x_0 = x_0^\circ, y_0 = 0, x_1 = x_1^\circ, y_1 = 0, l_0 = l_0^\circ, l_1 = l_1^\circ$ be the coordinates of the singular position of the mechanism (Fig. 2). From the theorem proved above it follows that $l_0^\circ = \mu_0 x_0^\circ$ and $l_1^\circ = \mu_1(a - x_1^\circ)$, where $\mu_i = \pm 1\,(i = 0, 1)$.

Let us make the following change of variables in system (2):

$$\begin{cases} x_0 = x_0^\circ + z_1, & y_0 = z_2, \\ x_1 = x_1^\circ + z_3, & y_1 = z_4, \\ l_0 = \mu_0 x_0^\circ + z_5, & l_1^\circ = \mu_1(a - x_1^\circ) + z_6. \end{cases} \tag{3}$$

Then we obtain the following system:

$$\begin{cases} g_1 \stackrel{\text{def}}{=} 2x_0^\circ z_1 - 2\mu_0 x_0^\circ z_5 + z_1^2 + z_2^2 - z_5^2 = 0, \\ g_2 \stackrel{\text{def}}{=} 2u_{01}^\circ z_1 - 2u_{10}^\circ z_3 - 2z_1 z_3 - 2z_2 z_4 + z_1^2 + z_2^2 + z_3^2 + z_4^2 = 0, \\ g_3 \stackrel{\text{def}}{=} 2(x_1^\circ - a) z_3 + 2\mu_1(x_1^\circ - a) z_6 + z_3^2 + z_4^2 - z_6^2 = 0. \end{cases} \tag{4}$$

This system is considered in the neighbourhood of the point $Z^0 = 0$.

By means of the method of *Newton polyhedra* we will find a local representation of the displacement function of the mechanism in the neighbourhood of singular point. This representation will have the form of power series convergent in a small neighbourhood of singular point.

3 Newton Polyhedra

The method of *Newton polyhedra* consists of the following [4]: to each of equations

$$g_i \stackrel{\text{def}}{=} \sum_{j=1}^{l_i} b_{ij} Z^{Q_{ij}} = 0, \quad Q_{ij} \in D_i, \quad i = 1, 2, \ldots, m, \tag{5}$$

the corresponding support $D_i = D(g_i)$ and a Newton polyhedron (convex hull of set D_i) $\mathcal{M}_i = \mathcal{M}(g_i)$ are assigned. To each edge $\Gamma_{ij}^{(d)}$ of polyhedron $\mathcal{M}_i$ there corresponds its normal cone $K_{ij}^{(d)}$ in space R_*^n conjugate to R^n. While investigating the g_i in the neighbourhood of $Z^0 = 0$, *the cone of a problem* is $\mathcal{K} = \{P : p_1 < 0, \ldots, p_n < 0\}$ from R_*^n. It is necessary to identify in the D_i sets only those sets $D_{ik}^{(d)}$, for which the normal cones $K_{ik}^{(d)}$ intersect with cone $\mathcal{K}$. The system

$$\hat{g}_i \stackrel{\text{def}}{=} \sum_j b_{ij} Z^{Q_{ij}} = 0, \quad Q_{ij} \in D_{ik}^{(d)}, \quad i = 1, 2, \ldots, m, \tag{6}$$

is called the truncation of system (5) of the order $P = (p_1, \ldots, p_n)$, if for each i the polynomial $\hat{g}_i$ is a truncation of polynomial g_i of the order P [4]. The *cone of the truncated system* is obviously $\Pi = K(g_1, \hat{g}_1) \bigcap \cdots \bigcap K(g_m, \hat{g}_m)$.

Let the solution of system (5) have the form

$$z_i = c_i \tau^{p_i} (1 + o(1)), \ \sum p_i^2 \neq 0, \ i = 1, \ldots, n \ (\tau \to \infty)$$

and $P \in \Pi$. Then according to Theorem 2.1 from [4], $z_i = c_i \tau^{p_i}$, $i = 1, \ldots, n$, is the solution of the truncated system (6).

The above algorithm for determination of all boundary subsets $D_{ik}^{(d)}$ of a finite set of points D_i and their normal cones $K_{ik}^{(d)}$ has been implemented in a program for IBM PC compatible computers (see [5]). Also some calculations were made in MAPLE and *Mathematica 4*.

The results of the program runs are as follows:

1) Dimension of the subspace in which an investigated set of the points lie;

2) The set of the vectors of normals to the hypersurfaces (the coordinates of the vectors have the greatest common divisor 1) marked by "!";

3) The columns of the table of correspondences marked by "V" correspond to the the vertexes, the normal cones of which intersect the cone of the problem, the remaining columns are marked by letter "d".

Below, the listings of the input and output data of program for g_i ($i = 1, 2, 3$) of system (4) are presented.

For g_1 the support is $D_1 = \{Q_{11} = (1,0,0),\ Q_{12} = (0,0,1),\ Q_{13} = (2,0,0),$

$$Q_{14} = (0,2,0),\ Q_{15} = (0,0,2)\}.$$

Here $\mathcal{M}_1 \subset R^3$, $\mathcal{K} \subset R_*^3$. The output of the program run is listed in Table 1.

Table 1

```
      Polyhedron dimension: 3
       Corresponding matrix
                                V  V  V  V  V
 1  !   0   -1    0             +  +  +  -  +
 2  !  -2   -1   -2             +  +  -  +  -
 3  !  -1    0    0             -  +  -  +  +
 4  !   0    0   -1             +  -  +  +  -
=====================================
         Surface dimension: 2
          1    ( 1; 2; 3; 5;)
          2    ( 1; 2; 4;)
          3    ( 2; 4; 5;)
          4    ( 1; 3; 4;)
=====================================
```

For g_2 the support

$$D_2 = \{Q_{21} = (1,0,0,0),\ Q_{22} = (0,0,1,0),\ Q_{23} = (1,0,1,0),\ Q_{24} = (0,1,0,1),$$
$$Q_{25} = (2,0,0,0),\ Q_{26} = (0,2,0,0),\ Q_{27} = (0,0,2,0),\ Q_{28} = (0,0,0,2)\}.$$

Here $\mathcal{M}_2 \subset R^4$, $\mathcal{K} \subset R^4_*$. The output of the program run is listed in Table 2.

Table 2

```
          Polyhedron dimension: 4
            Corresponding matrix
                                 V  V  d  d  V  V  V  V
1 !   0   -1    0    0           +  +  +  -  +  -  +  +
2 !  -2   -1   -2   -1           +  +  -  +  -  +  -  +
3 !   0    0   -1    0           +  -  -  +  +  +  -  +
4 !  -1    0    0    0           -  +  -  +  -  +  +  +
5 !   0    0    0   -1           +  +  +  -  +  +  +  -
=======================================================
            Surface dimension: 3
          1    ( 1; 2; 3; 5; 7; 8;)
          2    ( 1; 2; 4; 6; 8;)
          3    ( 1; 4; 5; 6; 8;)
          4    ( 2; 4; 6; 7; 8;)
          5    ( 1; 2; 3; 5; 6; 7;)
=======================================================
```

For g_3 the support is $D_3 = \{Q_{31} = (1,0,0),\ Q_{32} = (2,0,0),\ Q_{33} = (0,2,0),$

$$Q_{34} = (0,0,1),\ Q_{35} = (0,0,2)\}.$$

Here $\mathcal{M}_3 \subset R^3$, $\mathcal{K} \subset R^3_*$. The output of the program run is listed in Table 3.

Table 3

```
  Polyhedron dimension: 3
    Corresponding matrix
                        V  V  V  V  V
1  !   0   -1    0      +  +  -  +  +
2  !  -2   -1   -2      +  -  +  +  -
3  !  -1    0    0      -  -  +  +  +
4  !   0    0   -1      +  +  +  -  -
=====================================
     Surface dimension: 2
       1   ( 1; 2; 4; 5;)
       2   ( 1; 3; 4;)
       3   ( 3; 4; 5;)
       4   ( 1; 2; 3;)
=====================================
```

The computational results show that for polyhedra $\mathcal{M}_1$ and $\mathcal{M}_3$ the normal cones of 4 faces of dimension 2, 5 faces of dimension 1, and 3 edges intersect with the cone of problem $P \leq 0$. For polyhedron $\mathcal{M}_2$, the normal cones of 5 faces of dimension 3, 9 faces of dimension 2, 8 faces of dimension 1 and 4 edges intersect with the problem cone. The number of points Q_{ij} lying on this face are indicated in brackets, i.e. the numbers j of points $Q_{ij}^{(d)}$ (d is the dimension of the corresponding face) are indicated in brackets. Only such truncations are considered here, which correspond to the faces of maximal dimension (the normal cones of these faces are marked by !) with normal vectors N_i lying inside the problem cone, i.e. N_i with non-zero components. Such a truncation is unique and has the following form (see 2nd row of Table 1, 2nd row of Table 2 and 2nd row of Table 3):

$$\begin{cases} \hat{g}_1 \stackrel{\text{def}}{=} 2x_0^\circ z_1 - 2\mu_0 x_0^\circ z_5 + z_2^2 = 0, \\ \hat{g}_2 \stackrel{\text{def}}{=} 2u_{01}^\circ z_1 - 2u_{10}^\circ z_3 - 2z_2 z_4 + z_2^2 + z_4^2 = 0, \\ \hat{g}_3 \stackrel{\text{def}}{=} 2(x_1^\circ - a)z_3 + 2\mu_1(x_1^\circ - a)z_6 + z_4^2 = 0. \end{cases} \tag{7}$$

From here we obtain the following parametric solution:

$$\begin{aligned} z_1 &= \mu_1 t_3 - \nu_1 t_1 t_2 + \nu_2 t_1^2 + \nu_3 t_2^2, \; z_2 = t_1, \\ z_3 &= -\mu_1 t_3 + \frac{1}{2(a - x_1^\circ)} t_2^2, \; z_4 = t_2, \\ z_5 &= \frac{\mu_1}{\mu_0} t_3 - \lambda_1 t_1 t_2 + \lambda_2 t_1^2 + \lambda_3 t_2^2, \; z_6 = t_3. \end{aligned} \tag{8}$$

Here $t_i (i = 1,2,3)$ are linearly independent parameters,

$$\begin{aligned} \nu_1 &= \frac{1}{x_1^\circ - x_0^\circ}, \; \nu_2 = \frac{\nu_1}{2}, \; \nu_3 = \frac{a - x_0^\circ}{2(x_1^\circ - x_0^\circ)(a - x_1^\circ)}; \\ \lambda_1 &= \frac{\nu_1}{\mu_0}, \; \lambda_2 = \frac{\lambda_1 x_1^\circ}{2x_0^\circ}, \; \lambda_3 = \frac{\nu_3}{\mu_0}. \end{aligned}$$

The first approximation of the solution of system (2) is as follows:

$$\begin{aligned} &x_0 = x_0^\circ + \mu_1 t_3 - \nu_1 t_1 t_2 + \nu_2 t_1^2 + \nu_3 t_2^2 + \dots, \quad y_0 = t_1 + \dots, \\ &x_1 = x_1^\circ - \mu_1 t_3 + \tfrac{1}{2(a - x_1^\circ)} t_2^2 + \dots, \quad y_1 = t_2 + \dots, \\ &l_0 = \mu_0 x_0^\circ + \tfrac{\mu_1}{\mu_0} t_3 - \lambda_1 t_1 t_2 + \lambda_2 t_1^2 + \lambda_3 t_2^2 + \dots, \quad l_1^\circ = \mu_1(a - x_1^\circ) + t_3 + \dots, \end{aligned}$$

where y_0°, y_1° and l_1° are control coordinates.

It is not difficult to see that the mechanism can move from this singular position in four different directions (Fig. 2).

The series expansion of the local representation of system (2) at other singular cases can be obtained in the same manner.

This algorithm is much simpler than the algorithms of other methods.

References

1. Kieffer, J. and Litvin, F.L.: Local parametric representation of displacement functions for linkages and manipulators. *Mech. Mach. Theory* **26** (1991) 41–53
2. Dimentberg, F.M.: *Definition of Positions of Spatial Mechanisms* (in Russian) , Acad. Sci. USSR, Moscow (1950)
3. Soleev, A. and Barotov, A.S.: Local representation in small neighborhood of a singular point of displacement function of robotic mechanism. *Uzb. J. for Mech. Prob.* (in Russian) No. 3 (2000), 5 pp.
4. Bruno, A.D. and Soleev, A.: Local uniformization of branches of space curve and Newton polyhedra. *Algebra and Analysis* **3** (1991) 67–101
5. Soleev, A. and Aranson, A.: Calculation of a polyhedron and normal cones of its faces. Preprint/ Inst. for Appl. Math. of Russian Acad. Sci., Moscow, No. 36 (1994), 25 pp.

Algebraic Predicates for Empirical Data

Hans J. Stetter

Tech.Univ. Vienna, A-1040 Vienna, Austria
stetter@aurora.anum.tuwien.ac.at

Abstract. It is widely assumed that the assignment of truth values to non-trivial algebraic predicates containing numerical data is possible only if the data are exact and if exact computation is employed. But in many *application areas* the answers to questions like "Are all zeros of that (model) polynomial in the left half-plane?" are of principal importance. We develop a framework in which algebraic predicates with *empirical data* are assigned a positive real number in place of a truth value. This *validity value* permits an interpretation which is more informative that the classical "yes – no" answer. It depends *continuously* on the data and can thus be computed approximately by floating-point arithmetic. A number of non-trivial examples support the usefulness of our approach.

1 Introduction

With the increasing refinement of mathematical modelling and simulation, the processing of polynomials in one or several (often many) variables has assumed an important place in scientific computing. Current computer algebra systems (like Maple, Mathematica, and many others) provide powerful tools for that purpose. However, many of these tools assume – explicitly or implicitly – that the numeric data in the polynomials may be considered as *exact*, and they use *exact (rational) arithmetic* for the numerical operations performed. While this second fact will only slow down the processing in many cases, the restriction to exact data prohibits the use of these tools for the majority of problems in scientific computing: Some or all of the real-life numeric data in such problems are of a *limited – often quite low – accuracy*.

If the task under discussion asks for numeric results and if these depend *continuously* on the data, it is generally no principal problem to compute meaningful results also for empirical data, though it may require changes in the algorithmic approach. If, however, a desired result is either *integer* or a *truth value*, and if this result is not invariant in a large neighborhood of the given data, the mapping from the data to this result must be *discontinuous*. Then it seems that meaningful answers can only be given for exact data and that exact computation is required to obtain them.

But tasks of this nature may very well be meaningfully posed for data from real-life situations. E.g., the zeros z_ν of a certain univariate polynomial with *measured* coefficients may be the time constants for a dynamic behavior described by components $\exp(z_\nu t)$; thus it is crucial that all zeros z_ν have a negative real part so that the dynamic is stable. This must be true not only for the measured values of the coefficients but for all possible sets of coefficients within the measuring accuracy. In such situations, it is often suggested that *intervals* should be specified for the data; the mathematical and algorithmic problems originating from interval data have created *Interval Analysis*, a special discipline of computational mathematics. But now the answer to the question "Are all polynomials with coefficients from the specified intervals stable?" depends discontinuously on the *bounds* of the intervals so that nothing has been gained: An exact specification of an interval for an empiric quantitiy is as irrealistic as that of the quantity itself; and, in any case, very precise arithmetic is needed to establish the answer near the discontinuity.

Therefore, in this paper, we suggest a different approach to the assignment of values to algebraic predicates for empirical data: Our previously used concept of interpreting empirical data as *families of neighborhoods* permits the immediate definition of a *backward error* for predicates with an $\exists$ quantifier. This backward error takes positive real values and *smoothes* the dicontinuous transition between "there exists ..." and "there does not exist ..."; as a *validity measure* it permits an intuitive interpretation by the expert from the problem area. Predicates with an $\forall$ quantifier may be turned into $\exists$-predicates by the well-known double negation.

Since the backward error is a continuous and generally smooth function of the data and since its value is only needed with about a 10% accuracy, floating-point computation and other approximate tools may be used for its fast (approximate) computation. This is another important aspect of our approach, in particular when it is compared with the computationally expensive interval methods.

In the following, we will at first recall the concept of empirical quantities. Then we will explain how a backward error may be associated with algebraic predicates in a natural way, and how such backward errors are to be obtained and interpreted. A number of nontrivial examples will show the use of our concepts and conclude the paper.

Throughout the paper, we will (ab)use the O(..) notation in an *order of magnitude* sense as it is commonly done by engineers: O(1) denotes quantities vaguely between .1 and 10, and $O(c) = O(1)\,c$. There is *no asymptotic notion* associated with this usage of O(..). Cf. also the remarks after Definition 1.

2 Empirical Data

Many numeric data in scientific computing have a *limited accuracy*. This may be inherent in their meaning in the mathematical model, it may be due to their origin as measurements or results from preceeding computations, or also to the fact that secondary effects related to that quantity have been neglected. Generally, these indeterminations are, at best, known in terms of *orders of magnitude* or – equivalently – numbers of meaningful digits in the numeric representation. If such data are treated as exact in subsequent manipulations, this can lead to absurd results. Recently, in an exotic country, I saw a sign on a mountain whose top is covered with high trees, specifying its height as 2595.6768 m, which is obviously ridiculous. Most probably, this number originated from the conversion of 8516 feet to metric length, and just as probably, the indetermination in this original height was a few feet. Thus, the metric height has an indetermination of O(1 m).

How should one introduce such *empirical data* into numerical computations? The following concept appears to combine mathematical flexibility with realism; cf., e.g., [7], [9].

Definition 1: A scalar *empirical quantity* $(\bar{\alpha}, \varepsilon)$ consists of a *specified value* $\bar{\alpha} \in \mathbb{R}$ or $\mathbb{C}$ and a *tolerance* $\varepsilon > 0$. It represents the *family* of neighborhoods

$$N_\delta(\bar{\alpha}, \varepsilon) \;:=\; \{\alpha \in \mathbb{C} \,:\, |\alpha - \bar{\alpha}| \leq \varepsilon\,\delta\}\,, \tag{1}$$

parametrized by the real positive parameter δ. (For $\bar{\alpha} \in \mathbb{R}$, it has to be specified or clear from the context whether the N_δ are restricted to the real line or not.) An m-dimensional empirical quantity $(\bar{a}, e)$, with specified value $\bar{a} = (\bar{\alpha}_1, \ldots, \bar{\alpha}_m)$ and tolerance $e = (\varepsilon_1, \ldots, \varepsilon_m)$, represents the family of m-dimensional neighborhoods

$$N_\delta(\bar{a}, \varepsilon) \;:=\; \{a \in \mathbb{C}^m \,:\, \|a - \bar{a}\|_e^* \leq \delta\}\,, \tag{2}$$

where $\|..\|_e^* := \|(.., \frac{|\alpha_\mu - \bar{\alpha}_\mu|}{\varepsilon_\mu}, ..)\|^*$ and $\|..\|^*$ is some absolute vector norm in $\mathbb{R}^m$. In this paper, we will use $\|(.., \gamma_\mu, ..)\|^* := \max |\gamma_\mu|$ throughout[1] so that

$$\|a - \bar{a}\|_e^* \leq \delta \quad \Longleftrightarrow \quad |\alpha_\mu - \bar{\alpha}_\mu| \leq \varepsilon_\mu\,\delta\,,\; \mu = 1(1)m\,.$$

Similarly, the notion may be extended to matrices and other structured quantities. □

Variations in $\mathbb{C}$ of real or complex data may often be specified in terms of real and imaginary parts. In terms of orders of magnitude, the specifications

$$\left.\begin{array}{l} |\mathrm{Re}\;(\alpha_\mu - \bar{\alpha}_\mu)| \\ |\mathrm{Im}\;(\alpha_\mu - \bar{\alpha}_\mu)| \end{array}\right\} \;\leq\; \varepsilon_\mu \quad \text{and} \quad |\alpha_\mu - \bar{\alpha}_\mu| \;\leq\; \varepsilon_\mu \tag{3}$$

may be considered as equivalent. If necessary, problems with complex coefficients may be replaced by equivalent real problems.

At first sight, the presence of the parameter δ in addition to the tolerances ε_μ appears redundant. But the use of a *family of intervals* in place of just one interval reflects the fact that the notion $\bar{\alpha} \pm \varepsilon$ as it is often used in engineering is *not equivalent* to the interval $[\bar{\alpha} - \varepsilon, \bar{\alpha} + \varepsilon]$ but is rather meant to indicate a *cloud* of potential neighborhoods $N_\delta(\bar{\alpha}, \varepsilon)$, with $\delta = \mathrm{O}(1)$, as we have introduced it in Definition 1. This vagueness is also necessary because tolerances can generally be specified, at best, with a one-digit accuracy; usually, they indicate just an

[1] but all our concepts are equally feasible with other norms as well

order of magnitude. Other advantages of that concept will become apparent in the next section.

Intuitively, the parameter δ is to be interpreted thus: Values α in an $N_\delta(\bar{\alpha}, \varepsilon)$ with $\delta < 1$ are considered as *valid values* of the empirical quantity $(\bar{\alpha}, \varepsilon)$. But $\delta = 1$ is no distinguished bound: As δ increases above 1, there is merely a gradual drop in the *validity* of the further values. Beyond $\delta = 5$ or so, we may begin to put less and less trust into the validity of the outlying values. This *smooth decrease of validity* – replacing the discontinuous *loss of validity* in an interval concept – permits a realistic dealing with empirical quantities and is expressed by the use of the O(1) symbol: The values α of the empirical quantity $(\bar{\alpha}, \varepsilon)$ satisfy

$$\frac{|\alpha - \bar{\alpha}|}{\varepsilon} \leq \mathrm{O}(1),$$

with the meaning just explained. In a concrete situation, an expert of the model under consideration may perhaps be able to give a more rigid interpretation of O(1).

Naturally, not all data in a real-life problem are empirical:

Definition 2: A numerical quantity with an *exact fixed value* (like $0, 1, \pi$, etc.) is called *intrinsic*. In our context, intrinsic quantities are generally considered as part of the problem formulation and not regarded as data. □

Our main objects will be *empirical polynomials* in $s \geq 1$ variables. Here, we must accomodate sparsity and the combined occurrence of empirical and intrinsic coefficients. As usual, we let $x^j := x_1^{j_1} \dots x_s^{j_s}$, $j \in \mathbb{N}_0^s$, and $p(x) = \sum_{j \in J} \alpha_j x^j$, with support $J \subset \mathbb{N}_0^s$. The *empirical support* $J' \subset J$ of an empirical polynomial is

$$J' \;:=\; \{j \in J \;:\; \alpha_j \text{ is empirical}\}, \qquad M := |J'| \leq |J|. \tag{4}$$

Thus the associated tolerance vector $e \in \mathbb{R}^M$ has to contain tolerances ε_j for $j \in J'$.

Definition 3: An empirical polynomial $(\bar{p}, e)$, with specified polynomial $\bar{p}(x) = \sum_{j \in J} \bar{\alpha}_j x^j$ and tolerance $e = (\varepsilon_j, j \in J')$ represents the family of polynomial neighborhoods

$$\begin{aligned} N_\delta(\bar{p}, e) \;&:=\; \{p = \textstyle\sum \alpha_j x^j \\ &\text{with } \alpha_j = 0, j \notin J, \alpha_j = \bar{\alpha}_j, j \in J \setminus J', |\alpha_j - \bar{\alpha}_j| \leq \varepsilon_j \, \delta, j \in J'\}. \end{aligned} \tag{5}$$

Note that (5) implies the use of the max-norm for $\|(.., \frac{|\alpha_j - \bar{\alpha}_j|}{\varepsilon_j}, ..)\|^*$; other norms may be used if desired.

Note also that a *vanishing* $\bar{\alpha}_j$ may be intrinsic or empirical: If $j \notin J$, the vanishing is part of the sparsity pattern and extends to all $p \in N_\delta(\bar{p}, e)$; if $j \in J' \subset J$, the neighboring p may have (small) nonzero coefficients for x^j. □

Naturally, a set of empirical polynomials $\{(\bar{p}_\nu, e_\nu)\}$ may form an empirical polynomial system $(\overline{P}, E)$.

An empirical polynomial $(\bar{p}, e)$ or polynomial system defines a linear space $\mathcal{A}$ of dimension M or $\sum M_\nu$ resp. whose components are the α_j or $\alpha_{\nu j}$ with $j \in J'$ or $\nu j \in J'_\nu$ resp., i.e. those coëfficients which vary under the indetermination of the empirical setting. Actually, it is more convenient to consider the differences $\Delta\alpha_j := \alpha_j - \bar{\alpha}_j$ which puts the origin at the point $\bar{a}$ of the specified coefficients.

Definition 4: For an empirical polynomial $(\bar{p}, e)$ or polynomial system $(\bar{P}, E) = \{(\bar{p}_\nu, e_\nu)\}$, the associated *empirical data space* $\Delta\mathcal{A}$ is the normed linear space with components $\Delta\alpha_{(\nu)j}$, $j \in J'_{(\nu)}$, furnished with the norm

$$\|\Delta a\|_e^* := \max_j \frac{|\Delta\alpha_j|}{\varepsilon_j}, \quad \text{or} \quad \|(\Delta a_1, \ldots, \Delta a_n)\|_E^* := \max_\nu \|\Delta a_\nu\|_{e_\nu}^* . \ \square \tag{6}$$

In $\Delta\mathcal{A}$, the neighborhoods $N_\delta(\bar{p}, e)$ appear as norm neighborhoods $\|\Delta a\|_e^* \leq \delta$ of the *origin*, which explains the suitability of this vector space for our investigations.

3 Predicates for Empirical Polynomials

To illustrate the scope of our investigation, we consider a few simple situations: For some *fixed* $z \in \mathbb{C}$, consider the predicate

$$\Pi_z(p) \ := \ 'z \text{ is a zero of } p' .$$

over the ring $\mathbb{P} = \mathbb{C}[x]$ of univariate polynomials. As a map from $\mathbb{P}$ to the set $\{\textbf{true}, \ \textbf{false}\}$, Π_z is necessarily *discontinuous*, which is irrelevant within classical (exact) algebra. A well-known extension of Π to polynomials with coefficients of limited accuracy is

$$\widetilde{\Pi}_z(\tilde{p}) \ := \ 'z \text{ is a pseudo-zero of } \tilde{p}' ,$$

where, generally, the definition of *pseudo-zero* assumes $\tilde{p}$ to be an *interval polyomial*, i.e. to have its coefficients taken from specified (small) intervals; cf., e.g., [1], or the wide literature on interval mathematics. However this is done in detail, it retains $\{\textbf{true}, \ \textbf{false}\}$ as image set of $\widetilde{\Pi}_z$ and the associated discontinuity.

With our definition of an empirical polynomial as a *family* of polynomial intervals, the presence of the real parameter δ permits an extension of the image set for $\widetilde{\Pi}_z$ to the nonnegative reals. We achieve this when we define

$$\widetilde{\Pi}_z((\bar{p}, e)) \ := \ \min \ \{\delta \ : \ \exists p \in N_\delta(\bar{p}, e) \text{ with } \Pi_z(p) = \textbf{true}\} . \tag{7}$$

While this retains the original algebraic meaning of the predicate Π_z, it associates a nonnegative real number with its application to an empirical polynomial. This number has an immediate intuitive meaning in relation with $(\bar{p}, e)$, and it is obvious that only a crude approximation for its mathematically defined value is needed: If $\widetilde{\Pi}_z((\bar{p}, e)) = O(1)$, we will say that z is a valid zero of the empirical polynomial; cf. section 2.

This is nothing new: When we consider Π_z as a predicate on $z \in \mathbb{C}$ for fixed p, (7) is simply the well-known *backward error* of z, with our parametrization of the neighborhoods of $(\bar{p}, e)$. In this simple case, there is even an explicit expression for the image of $\widetilde{\Pi}_z$:

$$\widetilde{\Pi}_z((\bar{p}, e)) = \frac{|\bar{p}(z)|}{\sum_{j \in J'} \varepsilon_j \, |z^j|} ; \tag{8}$$

cf., e.g., [7] or [1]. This indicates that the backward error concept may provide an adequate tool for the extension of algebraic predicates to empirical polynomials.

To test this conjecture, let us consider a predicate which refers to a polynomial:

$$\Pi_2(p) := \text{'}p \text{ has a multiple zero'},$$

or – equivalently –

$$\Pi_2(p) := \text{'}\exists z \in \mathbb{C} : z \text{ is a multiple zero of } p\text{'}.$$

The evaluation of Π_2 is made possible by the fact that polynomials with a multiple zero can be characterized without reference to the location of the multiple zero:

$$\Pi_2(p) = \begin{cases} \textbf{true} & \text{if } \det \operatorname{Sylv}(p, p') = 0 , \\ \textbf{false} & \text{otherwise} . \end{cases}$$

Naturally, this predicate for exact polynomials is discontinuous.

A first, formal extension of this predicate to empirical polynomials is possible just like in (7):

$$\widetilde{\Pi}_2((\bar{p}, e)) := \min \{\delta : \exists p \in N_\delta(\bar{p}, e) \text{ with } \Pi_2(p) = \textbf{true} \} .$$

To turn that into a form which permits evaluation, we observe that the determinant of the Sylvester matrix of p and p' is a polynomial $S(a)$ in the coefficients of p. This provides us with an equivalent definition which utilizes the empirical data space $\Delta\mathcal{A}$ of $(\bar{p}, e)$ introduced at the end of section 2: Let

$$\mathcal{S} := \{\Delta a : S(\bar{a} + \Delta a) = 0 \} \subset \Delta\mathcal{A}; \tag{9}$$

Then the formal definition of $\widetilde{\Pi}_2$ becomes

$$\widetilde{\Pi}_2((\bar{p}, e)) := \min_{\Delta a \in \mathcal{S}} \|\Delta a\|_e^* . \tag{10}$$

While it is not so obvious at first how to evaluate $\widetilde{\Pi}_2$ computationally, it is clear that it is well-defined because the intersection of the algebraic manifold $\mathcal{S}$ with some sufficiently large ball $\|\Delta a\|_e^* \le r$ in $\Delta\mathcal{A}$ is a compact set.

Accidentally, the expression (8) for $\widetilde{\Pi}_z((\bar{p}, e))$ stems from the same translation of the problem into $\Delta\mathcal{A}$: $p(z) = 0$ is equivalent to $\sum_j (\bar{\alpha}_j + \Delta\alpha_j) z^j = 0$ which represents a linear manifold in $\Delta\mathcal{A}$ (for fixed z!). For a linear manifold of codimension 1, the minimal norm distance from the origin can be explicitly stated (cf. (14)), which yields (8).

Obviously, in our extension of algebraic predicates to empirical polynomials, the value of the predicate is the minimal norm of a modification of $\bar{p}$ generating a neighboring polynomial for which the predicate is true in the strict algebraic sense. Thus, this value is nothing but the *backward error of the predicate* with respect to $(\bar{p}, e)$.

But there are also algebraic predicates where we wish to be sure that they are true not only for $\bar{p}$ but also for *all* polynomials in the neighborhoods $N_\delta(\bar{p}, e)$, $\delta \leq O(1)$. Consider, for real polynomials only, the predicate

$$\Pi_r \;:=\; '\,p \text{ has only real zeros}\,',$$

or – equivalently –

$$\Pi_r \;:=\; '\,p \text{ has no non-real zeros}\,'.$$

This negative formulation shows that the extension to empirical polynomials will have to involve an $\forall$ quantifier rather than an $\exists$ quantifier. Our above approach works only for an $\exists$-predicate; therefore we rewrite Π_r once more into

$$\Pi_r \;:=\; \neg\,'\,p \text{ has non-real zeros}\,'.$$

While we can immediately deal with $'\,p$ has non-real zeros$'$, it is not clear how we should treat the negation. One could define the backward error for a negated predicate as the reciprocal of the backward error of the predicate; but then the backward error would no longer display the minimal norm distance from $\bar{p}$ to the polynomials at which the value of the exact predicate jumps.

It is the set of these polynomials in which we are most interested when we deal with algebraic $\forall$-predicates for empirical polynomials. With Π_r, e.g., this set is characterized by $\Pi_2 =$ **true**: Since a zero of a real polynomial can move away from the real axis only after merging with another real zero, we can reach a polynomial with a non-real zero from one with only real zeros only by passing through one which has a double zero. As we can extend Π_2 to empirical polynomials, this settles the problem in this special case; we shall see that this also happens in general.

We conclude this section by general formulations of the insights which we have gained so far:

Definition 5: The mapping $\Pi \; : \; \mathbb{P}^s \to \{\mathbf{true}, \mathbf{false}\}$ is an *algebraic predicate for polynomials* if there exists a semialgebraic set $\mathcal{S}$ such that

$$\Pi(p) = \mathbf{true} \qquad \Longleftrightarrow \qquad a \in \mathcal{S} \subset \mathcal{A}\,, \tag{11}$$

where $a \in \mathcal{A}$ is the coefficient vector of p. We will call Π *critical* if the associated set $\mathcal{S}$ has a dimension[2] which is smaller than M, the dimension of the data space $\mathcal{A}$. □

[2] If $\mathcal{S}$ has components of different dimensions, we consider only the component closest to the coefficients $\bar{a}$ of $\bar{p}$.

When we extend an algebraic predicate to an empirical polynomial $(\bar{p}, e)$, we will consider the associated semialgebraic set (or a suitable component of it) in the empirical data space $\Delta\mathcal{A}$ of $(\bar{p}, e)$. For a critical predicate, $\mathcal{S}$ consists of boundary points and is closed in $\Delta\mathcal{A}$; therefore its intersection with a fixed ball $\|\Delta a\|_e^* \le \bar{\delta}$, $\bar{\delta} > \mathrm{O}(1)$, is either empty or compact. We will use the notation $\mathcal{S}$ for this intersection from now on.

Definition 6: For a critical algebraic predicate Π, its value for an empirical polynomial $(\bar{p}, e)$ is defined as

$$\begin{aligned} \widetilde{\Pi}((\bar{p}, e)) &:= \min\nolimits_{\Delta a \in \mathcal{S}} \|\Delta a\|_e^* \qquad (\text{or } \infty \text{ if } \mathcal{S} = \emptyset)\,. \\ \Pi \text{ is } \textit{valid for } (\bar{p}, e) &\text{ if } \widetilde{\Pi}((\bar{p}, e)) \le \mathrm{O}(1). \end{aligned} \tag{12}$$

For a non-critical algebraic predicate Π, we consider $\partial\mathcal{S}$, the boundary of $\mathcal{S}$, and define

$$\begin{aligned} \partial\widetilde{\Pi}((\bar{p}, e)) &:= \min\nolimits_{\Delta a \in \partial\mathcal{S}} \|\Delta a\|_e^* \qquad (\text{or } \infty \text{ if } \mathcal{S} = \emptyset)\,. \\ \Pi \text{ is } \textit{valid for } (\bar{p}, e) &\text{ if } 0 \in \mathcal{S} \text{ and } \partial\widetilde{\Pi}((\bar{p}, e)) > \mathrm{O}(1). \quad \square \end{aligned} \tag{13}$$

4 Computation of Predicate Values

Definition 6 defines real nonnegative values for certain algebraic predicates Π when they are extended and applied to empirical polynomials $(\bar{p}, e)$. We will now consider the actual computation of these values. (12) and (13) indicate that it is sufficient to generate a 1-digit approximation of $\widetilde{\Pi}((\bar{p}, e))$.

After the relevant empirical data space $\Delta\mathcal{A}$ has been selected (cf. Definition 4), the set $\mathcal{S} \subset \mathcal{A}$ related to Π must be determined (cf. Definition 5); then a representation for $\mathcal{S}$ or $\partial\mathcal{S}$ must be found which is suitable for the minimization task in (12) or (13) resp. Remember that we have restricted the original semi-algebraic set in $\Delta\mathcal{A}$ to its intersection with a closed ball of a radius $\bar{\delta} > \mathrm{O}(1)$ about the origin.

To exclude both unsuitable representations and exotic predicates, we make the following

Assumption: For $\|\Delta a\|_e^* \le \bar{\delta}$, $\mathcal{S}$ or $\partial\mathcal{S}$ may be respresented by one or several polynomial equations in the components $\Delta\alpha_j$ of $\Delta a \in \Delta\mathcal{A}$ or in the real and imaginary parts of the $\Delta\alpha_j$. $\square$

In Section 3, we have found such polynomial representations for the predicates Π_z, Π_2, and $\partial\Pi_r$.

Let us now consider the computational evaluation of (12) or (13) resp. Since there is no difference in the tasks posed, we will restrict our attention to the situation with (12). The dimension of $\Delta\mathcal{A}$ will always be denoted by M.

1) $\mathcal{S}$ is a *linear manifold*:

A representation of $\mathcal{S}$ consists of m linear equations $S_\mu(\Delta a) = 0$, $\mu = 1(1)m$, where $m \le M$ is the *codimension* of $\mathcal{S}$. For $m = 1$, the minimization task (12) may be solved explicitly: Let $S(\Delta a) = \gamma_0 + \sum_{j\in J'} \gamma_j \,\Delta\alpha_j$, then

$$\min_{S(\Delta a)=0} \|\Delta a\|_e^* \;=\; \begin{cases} \dfrac{|\gamma_0|}{\sum_j \varepsilon_j\,|\gamma_j|} & \text{(max-norm)}\,, \\ \dfrac{|\gamma_0|}{(\sum_j \varepsilon_j^2\,|\gamma_j|^2)^{\frac{1}{2}}} & (L_2\text{-norm})\,, \\ \dfrac{|\gamma_0|}{\max_j(\varepsilon_j\,|\gamma_j|)} & (L_1\text{-norm})\,. \end{cases} \tag{14}$$

The associated minimizing Δa may also be expressed explicitly; for the max norm and real γ_j, we have

$$(\Delta\alpha_j)_{\min} \;=\; (\pm\|(\Delta a)_{\min}\|_e^*)\cdot \operatorname{sign}\gamma_j\,; \tag{15}$$

cf. [8]. The various denominators in (14) are the norms of $(\ldots,\gamma_j,\ldots)^T$ dual to the norm $\|..\|_e^*$ in $\Delta\mathcal{A}$; (14) also permits the substitution of $\varepsilon_j = 0$ for j.

For $1 < m < M$, (12) poses a linear minimization problem. For the max-norm and *real* $\Delta\alpha_j$ and γ_j, we may write it as

$$\min\,\delta\,,\ \text{with}\ -\delta\,\varepsilon_j \le \Delta\alpha_j \le \delta\,\varepsilon_j\,,\ j\in J',\ \text{and}\ S_\mu(\Delta a) = 0,\ \mu = 1(1)m,$$

which is a standard "linear program" and permits computational evaluation by relevant minimization software. A crude approximation for the minimum is sufficient, which may shorten the computation considerably. The minimizing Δa is automatically generated.

For complex γ_j and/or $\Delta\alpha_j$, we may introduce real and imaginary parts as separate variables, parameters, and equations. In agreement with (3) we may then pose the minimization in (12) for

$$\|\Delta a\|_e^* \;=\; \max\,(\|\mathrm{Re}\ \Delta a\|_e^*, \|\mathrm{Im}\ \Delta a\|_e^*)\,,$$

which recovers a standard linear minimization problem. With the complex modulus, the minimization is no longer linear; but it remains convex and there is also relevant software available.

2) $\mathcal{S}$ is an *algebraic manifold* :

Consider at first an algebraic manifold $\mathcal{S}$ of codimension $M-1$, i.e. of dimension 1: It can locally be represented in a parametrized form $\Delta a = a(t)$, with a real or complex parameter t, suitably restricted to an interval T corresponding to the compact part of $\mathcal{S}$ under consideration. The minimum of the real function $\|a(t)\|_e^*$ on T can generally be found without great difficulty: For the max-norm, we have to follow the patches of different active $\Delta\alpha_j$; but it often suffices to get to a sufficiently small value (for (12)) or to establish a sufficiently large lower bound bound (for (13)).

For a real $\mathcal{S}$ of codimension 1, i.e. of dimension $M-1$, and the max-norm, we must try to determine the orthants of $\Delta\mathcal{A}$ in which S may attain its shortest distance from the origin; this is often possible through a use of (15) as shown in Example 2. Let one such orthant be characterized by its sequence $\{(-1)^{\sigma_j}, \sigma_j = 0,1\}$ of M + or - signs; then the minimizing Δa must lie on the line $\Delta a(t) = t \cdot ((-1)^{\sigma_1}, \ldots, (-1)^{\sigma_M})$, $t \in \mathbb{R}$. The intersections of this line with $\mathcal{S}$ are found from the univariate polynomial $S(\Delta a(t))$ and we have

$$\widetilde{\Pi}((\bar{p}, e)) \;=\; \min_{t \in \mathbf{R}} \;\{|t| \;:\; S(\Delta a(t)) = 0\}\,.$$

For more general cases, a clever utilization of particular properties of $\mathcal{S}$ will often greatly reduce the computational effort which may be quite large when general minimization software is used to find the minimum. This is particularly true when complex values of the parameters have to be taken into account.

5 Examples

Example 1

We consider the extended predicate Π_2 from section 3

$$\widetilde{\Pi}_2((\bar{p}, e)) \;:=\; \min\;\{\delta \;:\; \exists p \in N_\delta(\bar{p}, e) \text{ with a multiple zero }\}$$

The real polynomial

$$\bar{p} \;:=\; x^4 + 2.326\,x^3 - 7.364\,x^2 - 7\,x + 14$$

has the approximate zeros $-3.532, -1.783, 1.397, 1.592$. Assume that the coefficients of x^3 and x^2 have limited accuracy, with a potential indetermination of a few units in the last digit; can we be sure that this may not permit a merger of the two positive zeros and thus the appearance of a double zero. For an answer, we consider the empirical polynomial $(\bar{p}, e)$, with tolerances (say) $\varepsilon_2 = \varepsilon_3 = .002$. Note that we might just as well have assigned tolerances of .001 or .003 in this situation, which shows the inherent vagueness of a tolerance specification. The dimension of the empirical data space $\Delta\mathcal{A}$ is $M = 2$.

¿From section 3, we know that the set S for the critical algebraic predicate Π_2 may be represented by

$$S(\Delta\alpha_2, \Delta\alpha_3) = \det\;\mathrm{Sylv}(\bar{p} + \Delta p, (\bar{p} + \Delta p)')\,,$$
$$\text{with} \quad \Delta p(x) := (\Delta\alpha_2\,x^2 + \Delta\alpha_3\,x^3)\,,$$

which is a polynomial of degree 4 in $\Delta\alpha_2, \Delta\alpha_3$. We have to determine the approximate minimal norm distance of the manifold $S(\Delta\alpha_2, \Delta\alpha_3) = 0$ from the origin of $\Delta\mathcal{A}$.

For a first orientation, we check for a change of sign of S between the origin and (say) $(\pm.05, \pm.05)$. It turns out that the only change of sign occurs between (0,0) and (.05,.05). With $||\Delta a||_e^* := \max(|\Delta\alpha_2|, |\Delta\alpha_3|)\;/\;.002$, we now look for the zero of S along $\Delta a(t) := (.002, .002) \cdot t$ which is found at $t \approx 14.1$. Thus

$\widetilde{\Pi}_2((\bar{p},e)) \approx 14$ which implies that the assertion $'p$ has a multiple zero$'$ is *not valid* for our empirical polynomial.

As has been indicated in section 3, we have thus also found an answer for the validity of the non-critical predicate Π_r whose algebraic manifold $\partial\mathcal{S}$ coincides with the manifold $\mathcal{S}$ for Π_2: The assertion $'p$ has only real zeros$'$ is *valid* for our empirical polynomial because it would require a perturbation of more than 28 units in the last digits of α_2 and α_3 to arrive at a non-real zero.

Example 2

Now we address the question posed in the abstract and in section 1: Given an *empirical* real polynomial, is it stable? The associated non-critical algebraic predicate is

$$\Pi_s(p) \;:=\; \text{'all zeros of } p \text{ are in the left halfplane } \mathbb{C}_- \text{ ';}$$

the boundary set $\partial\mathcal{S}$ of its semialgebraic set $\mathcal{S}$ in the coefficient space of p may be characterized in various ways:

a) *Routh-Hurwitz criterion*: For $p(x) := \sum_{\nu=0}^{n} \alpha_\nu x^\nu$, with $\alpha_n > 0, \alpha_0 \neq 0$, we form an $n \times n$-matrix $H(a)$ by arranging the coefficients $\alpha_{n-1}, \dots, \alpha_0$ along the main diagonal and filling the rows by further α_ν: to the right with increasing ν, to the left with decreasing ν. It is well-known that the polynomial p is stable iff all principal minors of $H(a)$ are positive. This represents $\mathcal{S}$ by n strict inequalities; the resulting representation of $\partial\mathcal{S}$ is not well manageable.

b) *Direct representation of* $\partial\mathcal{S}$: The domain of stable polynomials with all zeros in $\mathbb{C}_-$ must be bounded by polynomials with a conjugate pair of zeros on the imaginary axis or a zero at 0. With $p(x;a)$ as above and a_0, a_1 denoting the sets of coefficients with even and odd subscripts, a purely imaginary zero $\pm\mathrm{i}\,\eta$ must satisfy

$$\begin{aligned} p(\mathrm{i}\eta;a) &= \sum_{\nu=0}^{\lfloor n/2\rfloor} (-1)^\nu\, \alpha_{2\nu}\, (\eta^2)^\nu + \mathrm{i}\,\eta\, \sum_{\nu=0}^{\lfloor (n-1)/2\rfloor} (-1)^\nu\, \alpha_{2\nu+1}\, (\eta^2)^\nu \\ &=: \hat{p}_r(\eta^2;a_0) + \mathrm{i}\,\eta\, \hat{p}_i(\eta^2;a_1) = 0\,. \end{aligned} \tag{16}$$

Thus, the manifold $\partial\mathcal{S}$ in the empirical data space $\Delta\mathcal{A}$ of $(\bar{p},e)$ is given by

$$\partial\mathcal{S} \;:=\; \{\Delta a \in \Delta\mathcal{A} :\; \exists\hat{\eta} \in \mathbb{R}_+ \;:\; \hat{p}_r(\hat{\eta};\bar{a}_0 + \Delta a_0) = \hat{p}_i(\hat{\eta};\bar{a}_1 + \Delta a_1) = 0\}\,.$$

Via $\mathrm{Sylv}(\hat{p}_r,\hat{p}_i)$, $\partial\mathcal{S}$ may be represented by the polynomial equation $\partial\mathcal{S}(\Delta a) = 0$.

To find $\min_{\Delta a\in\partial\mathcal{S}} ||\Delta a||_e^*$, we observe at first that the coefficient sets of $\hat{p}_r$ and $\hat{p}_i$ are *disjoint*. For fixed $\hat{\eta} > 0$, we can tell from (15) in which orthant of $\Delta\mathcal{A}$ the minimizing Δa_0 and Δa_1 may lie; since this depends only on the *sign* of $\hat{\eta}$, it is independent of the location of $\hat{\eta}$ and holds also for $\partial\mathcal{S}(\Delta a)$. The 4 possible orthants are (cf. (16)) $\pm(+,+,-,-,+,\dots)$, $\pm(+,-,-,+,+,\dots)$. Thus, we can (cf. section 4) introduce the lines $\Delta a(t) := (\varepsilon_0, \varepsilon_1, -\varepsilon_2, -\varepsilon_3, \varepsilon_4, \dots) \cdot t$ and

$(\varepsilon_0, -\varepsilon_1, -\varepsilon_2,$
$\varepsilon_3, \varepsilon_4, \ldots) \cdot t$ resp.; then

$$\partial\, \widetilde{\Pi}_s((\bar{p}, e)) \;:=\; \min\; \{|t| \,:\, \partial S(\Delta a(t)) = 0\}\,,$$

where the minimum is also taken over the two choices of $\Delta a(t)$. Note that $p\,(\mathrm{i}\eta;\, \Delta a(t))$ induces $\hat{p}_r(\hat{\eta}; \Delta a(t)) \;=\; \pm\, t \cdot \sum_\nu \varepsilon_{2\nu}\, \hat{\eta}^\nu$ and a corresponding expression for $\hat{p}_i(\hat{\eta}; \Delta a(t))$.

Remark: Here, the observation about the minimizing orthants has been known as *Kharitonov's Theorem* [6] which reads – in *our notation* – :

If and only if $p(x; \bar{a}+\Delta a(\pm\delta))$ is stable for the two choices of $\Delta a(t)$ introduced above, *all* polynomials in $N_\delta(\bar{p}, e)$ are stable.

c) *Parametrization by the Critical Zero*: The idea of introducing the critical zero as a parameter for the representation of ∂S has been used by Hitz, Kaltofen, Karmarkar, and Lakshman in various papers; cf., e.g., [2], [5]. In [3], it has been used for our present task, with the L_2-norm and in a somewhat different notation. Here, we consider the max-norm backward error $\delta(\eta)$ of $\mathrm{i}\eta$ as a zero of $(\bar{p}, e)$ and find its minimum along $\eta \geq 0$ which equals $\partial\, \widetilde{\Pi}_s((\bar{p}, e)$.

Since we admit only *real* modifications of the empirical coefficients, we cannot apply (8) directly but must consider $\hat{\eta} = \eta^2$ as a *simultaneous* zero of $(\hat{\bar{p}}_r, e_0)$ and $(\hat{\bar{p}}_i, e_1)$; cf. (16). But as the coefficients of $\hat{p}_r$ and $\hat{p}_i$ are disjoint, we can apply (8) to them separately and form

$$\delta(\hat{\eta}) \;=\; \max\; (\delta_r(\hat{\eta}),\, \delta_i(\hat{\eta})) \;=\; \max\; (\frac{|\hat{\bar{p}}_r(\hat{\eta})|}{\sum_\nu \varepsilon_{2\nu}\, \hat{\eta}^\nu},\; \frac{|\hat{\bar{p}}_i(\hat{\eta})|}{\sum_\nu \varepsilon_{2\nu+1}\, \hat{\eta}^\nu})\,.$$

The determination of $\min_{\hat{\eta}} \delta(\hat{\eta})$ requires a piecewise analysis which is, however, greatly simplified by the available contextual information.

For a numerical example, we take $\bar{p}(x) \;=\; x^4 + 3.38\, x^3 + 6.44\, x^2 + 8.19\, x + 7.85$ and assume tolerances .01 for all coefficients except the leading one; the associated real and imaginary parts are $\hat{\bar{p}}_r(\hat{\eta}) = \hat{\eta}^2 - 6.44\, \hat{\eta} + 7.85$ and $\hat{\bar{p}}_i(\hat{\eta}) = -3.38\, \hat{\eta} + 8.19$. $\bar{p}$ has zero pairs with real parts ≈ -1.5704 and -.1196; but can we declare $(\bar{p}, e)$ as stable?

With approach b), when we consider the notice at the end of its description, we obtain

$$\partial S(\Delta a(t)) = \begin{vmatrix} 7.85 + .01\, t & -6.44 + .01\, t & 1 \\ 8.19 \pm .01\, t & -3.38 \pm .01\, t & 0 \\ 0 & 8.19 \pm .01\, t & -3.38 \pm .01\, t \end{vmatrix} \approx \begin{cases} -21.516 + .3340\, t + .00037\, t^2 \\ -21.516 + .4482\, t + .00269\, t^2 \end{cases}$$

The approximate zeros are -960, 60 and -206, 39 resp. which clearly establishes $\widetilde{\Pi}_s((\bar{p}, e))$ as valid.

With approach c), we have

$$\delta(\hat{\eta}) \;=\; \max\; (\frac{|\hat{\eta}^2 - 6.44\, \hat{\eta} + 7.85|}{.01\, (1 + \hat{\eta})},\; \frac{|-3.38\, \hat{\eta} + 8.19|}{.01\, (1 + \hat{\eta})})\,.$$

For small $\hat{\eta}$, the second term dominates; equality occurs at $\hat{\eta} \approx 2.07$ with $\delta \approx 39$ and this value remains the valid minimum for all larger $\hat{\eta}$.

Example 3[3]

The autocalibration of a camera poses *Kruppa's Problem* which requires the solution of a system P of 6 non-sparse quadratic equations in 5 variables. For a generic model, this *overdetermined system* has one solution; with coefficients of limited accuracy, the system is *inconsistent* and we must look for pseudozeros of the empirical system $(\bar{P}, E)$, with tolerances related to measurement accuracies. The associated critical algebraic predicate $\Pi_K(P) :=$ 'P is consistent' may be extended to $(\bar{P}, E)$ as

$$\widetilde{\Pi}_K(\bar{P}, E) \;=\; \min\ \{\delta : \exists p_\nu \in N_\delta(\bar{p}_\nu, e_\nu),\ \nu = 1(1)6, \text{ with a common zero}\}\,.$$

The use of resultants for a representation of the critical manifold in the joint data space $\Delta\mathcal{A}$ of the 6 polynomials is possible but cumbersome. A more pragmatic approach by Mourrain (cf. also [4] for a general discussion of this idea) consists in finding approximations of the 32 zeros of a 5-polynomial subsystem (say $\{\bar{p}_1, \ldots, \bar{p}_5\}$) and to check them against the remaining polynomial $\bar{p}_6$. If one of them is a valid pseudozero $\tilde{z}$ of $(\bar{p}_6, e_6)$, its backward error $\delta_6(\tilde{z})$ is an *upper bound* for $\widetilde{\Pi}_K(\bar{P}, E)$.

If this gives a dubious answer $\gtrapprox O(1)$, we want to refine $\tilde{z}$ such that its backward error δ_6 is reduced while its backward errors δ_ν, $\nu = 1(1)5$, which were negligible at first, remain sufficiently small. With the correction $\Delta z \in \mathbb{R}^5$ for $\tilde{z}$,

$$\delta_\nu(\tilde{z} + \Delta z) \;\approx\; \frac{|\bar{p}_\nu(\tilde{z} + \Delta z)|}{\sum \varepsilon_{\nu j}\, |\tilde{z}^j|}\,, \text{ with } \bar{p}_\nu(\tilde{z} + \Delta z) \;\approx\; \bar{p}_\nu(\tilde{z}) + \text{grad}\, \bar{p}_\nu(\tilde{z})\, \Delta z\,,$$

where we have neglected terms of $O(\|\Delta z\|^2)$ and $O(e\,\|\Delta z\|)$ to *linearize* the situation. We may now determine a suitable Δz and the approximate value of $\widetilde{\Pi}_K(\bar{P}, E)$ from the linear program

$$\min_{\Delta z}\ \delta \ \text{ with } -\delta \sum \varepsilon_{\nu j} |\tilde{z}^j| \;\le\; \bar{p}_\nu(\tilde{z}) + \text{grad}\, \bar{p}_\nu(\tilde{z})\, \Delta z \le \delta \sum \varepsilon_{\nu j} |\tilde{z}^j|\,, \quad \nu = 1(1)6\,.$$

Numerical experiments have confirmed the feasibility and efficiency of this approach.

References

1. F. Chaitin-Chatelin, V. Frayssé: Lectures on Finite Precision Computations, Software - Environments - Tools, SIAM, Philadelphia, 1996

[3] I owe this example to presentations by B.Mourrain; cf. also http://www-sop.inria.fr/saga/POL/BASE/2.multipol/kruppa.html

2. M.A. Hitz, E. Kaltofen: Efficient Algorithms for Computing the Nearest Polynomial with Constrained Roots, in: Proceed. ISSAC'98 (Ed. O.Gloor) (1998) 236-243
3. M.A. Hitz, E. Kaltofen, Y.N. Lakshman: Efficient Algorithms for Computing the Nearest Polynomial with a Real Root and Related Problems, in: Proceed. ISSAC'99 (Ed. S.Dooley) (1999) 205-212
4. Y. Huang, H.J. Stetter, W. Wu, L. Zhi: Pseudofactors of Multivariate Polynomials, in: Proceed. ISSAC 2000 (Ed.: C. Traverso) (2000) 161-168
5. N.K. Karmarkar, Y.N. Lakshman: Approximate Polynomial GCDs and Nearest Singular Polynomials, in: Proceed. ISSAC'96 (Ed. Y.N. Lakshman) (1996) 35-39
6. V.L. Kharitonov: Asymptotic Stability of an Equilibrium Position of a Family of Systems of Linear Differential Equations, Diff. Equations **14** (1979) 1483-1485
7. H.J. Stetter: Polynomials with Coefficients of Limited Accuracy, in: Computer Algebra in Scientific Computing (Eds. V.G.Ganzha, E.W.Mayr, E.V.Vorozhtsov), Springer, 1999, 409-430
8. H.J. Stetter: The Nearest Polynomial with a Given Zero, and Similar Problems, SIGSAM Bull. **33** (1999) no. 4, 2-4.
9. H.J. Stetter: Numerical Polynomial Algebra: Concepts and Algorithms, in: ATCM 2000, Proceed. 5th Asian Technology Conf. in Math. (Eds.: W.-Ch. Yang, S.-Ch. Chu, J.-Ch. Chuan), 22-36, ATCM Inc. USA, 2000.

Fractional Driftless Fokker–Planck Equation with Power Law Diffusion Coefficients

Norbert Südland[2], Gerd Baumann[1,2], and Theo F. Nonnenmacher[1]

[1] Department of Mathematical Physics, University of Ulm, Albert–Einstein–Allee 11, D–89069 Ulm/Donau, Germany
{sued, bau, non}@physik.uni-ulm.de
http://www.physik.uni-ulm.de/math/mathphys.html
[2] Visual Analysis AG, Neumarkter Straße 87, D–81673 München, Germany
{Norbert.Suedland, Gerd.Baumann}@visualanalysis.com
http://www.visualanalysis.com

Abstract. A generalized fractional driftless Fokker–Planck equation is discussed with a diffusion constant depending on the space coordinates by a power law.
The solution of this equation contains Lévy asymptotics for some special power law diffusion constants.

1 Indroduction

The Fokker–Planck equation according to Risken [12] is given by

$$\frac{\partial W(x,t)}{\partial t} = \left(-\frac{\partial}{\partial x} D^{(1)}(x) \;\cdot + \frac{\partial^2}{\partial x^2} D^{(2)}(x) \cdot \right) W(x,t), \tag{1}$$

with $D^{(2)}(x)$ called the diffusion coefficient and $D^{(1)}(x)$ the drift coefficient. We discuss this equation now without the drift coefficient using $d(x)$ instead of $D^{(2)}(x)$ as notation. Equations of this type occur with the velocity selective coherent population trapping (VSCPT) model [15] using a velocity v instead of x.

We generalize this equation using a fractional derivative of time insted of the derivative of first order in (1).

By this the fractional diffusion equation of Schneider and Wyss [17]

$$u(x,t) = \sum_{\nu=0}^{-[-q]-1} u_\nu(x,0)\, \frac{t^\nu}{\nu!} + \frac{d}{\Gamma(q)} \int_0^t (t-\tau)^{q-1} \Delta u(x,t)\; \mathrm{d}\tau, \tag{2}$$

with $\Re(q) > 0$ and $u_\nu(x,0) = \frac{\partial^\nu}{\partial t^\nu} u(x,t)\big|_{t=0}$ is completed by a space dependant diffusion constant $d \to d(x)$.

The diffusion constant is constant in time which means that no interaction by an experimentator is considered. In contrast to Schneider and Wyss we discuss a one-dimensional space coordinate x only. A physical sense is existent for real q with $0 < q \leq 1$ (fractional diffusion) and $1 < q \leq 2$ (fractional wave).

Now a local approximation to the space dependant diffusion constant can be given by a power law $d(x) \to d\,|x|^{2-\mu}$ for some critical points of space where d is a positive constant and μ is an arbitrary real number.

Considering that the equation discussed is linear and thus leading to a fundamental system of solutions we can discuss each single fundamental solution by omitting the sum in (2) and specifying the nonnegative integer parameter ν.

The Laplace convolution integral in (2) is replaced by a Riemann operator of fractional differentiation according to Samko et al. [13].

The equation to be discussed results:

$$u(x,t) = u_\nu(x,0)\,\frac{t^\nu}{\nu!} + d\,\mathcal{D}_{0,t}^{-q}\,\frac{\partial^2}{\partial x^2}\left(|x|^{2-\mu}\,u(x,t)\right). \tag{3}$$

Involving a fractional derivative to this equation (3) we obtain a fractional differential equation which can be modified to a usual differential equation by specifying q to be a positive integer:

$$\mathcal{D}_{0,t}^{q}\,u(x,t) - \frac{t^{\nu-q}\,u_\nu(x,0)}{\Gamma(1+\nu-q)} = \frac{\partial^2}{\partial x^2}\left(d\,|x|^{2-\mu}\,u(x,t)\right). \tag{4}$$

To get a propagator of equation (3) or (4), each initial value problem $u_\nu(x,0)$ is set to be a Dirac's delta function $\delta(x)$.

In section 2 we show the possibilities to solve this equation in the Laplace space. In section 3 we discuss the momenta of the solution. In section 4 we find out the Mellin and the Fourier transform of the solution. In section 5 we discuss the solution and its properties. In section 6 the special case $q \to 1$ and $\nu \to 0$ is discussed in detail. After all we give a conclusion.

2 The Solutions in Laplace Space Leading to a Propagator

2.1 Using the Standard Way

The Laplace transform of equation (4) yields:

$$p^q\,L(x) - p^{q-\nu-1}\,\delta(x) = \frac{\partial^2}{\partial x^2}\left(d\,|x|^{2-\mu}\,L(x)\right), \tag{5}$$

with $L(x) = \int_0^\infty \exp(-p\,t)\,u(x,t)\,\mathrm{d}t$ being the Laplace transform of the solution.

This equation (5) is not easy to be solved at once, but there is a possibility of a substitution $L_h(x) = d\,|x|^{2-\mu}\,L(x)$ leading to the following equation:

$$\frac{p^q\,|x|^{\mu-2}}{d}\,L_h(x) - p^{q-\nu-1}\,\delta(x) = L_h''(x). \tag{6}$$

Indeed, the homogenuous part of equation (6) is a basic equation of the Bessel functions. The results given here are substituted back to $L(x)$.

For $\mu \neq 0$ results:

$$L(x) = \frac{x^{\mu-\frac{3}{2}}}{d} \left(I_{-\frac{1}{\mu}} \left(2\,x^{\frac{\mu}{2}} \sqrt{\frac{p^q}{d\,\mu^2}} \right) C_1(p) + I_{\frac{1}{\mu}} \left(2\,x^{\frac{\mu}{2}} \sqrt{\frac{p^q}{d\,\mu^2}} \right) C_2(p) \right). \tag{7}$$

For $\mu = 0$ the homogenuous solution of (5) is:

$$L(x) = x^{-\frac{\sqrt{d+4\,p^q}}{2\,\sqrt{d}}} \frac{C_1(p)}{d\,\sqrt{x^3}} + x^{\frac{\sqrt{d+4\,p^q}}{2\,\sqrt{d}}} \frac{C_2(p)}{d\,\sqrt{x^3}}\,. \tag{8}$$

Note that the integration constants are optional. This allows to construct a symmetrical propagator like function with $\lim_{x\to\pm\infty} L(x) = 0$ and $L(-x) = L(x)$ according to the symmetry of equation (5). There is one integration constant left which is specified by the transformed normalization conditions to Laplace space:

$$2 \int_0^\infty L(x)\,\mathrm{d}x \stackrel{!}{=} \int_0^\infty \exp(-p\,t)\,\frac{t^\nu}{\nu!}\,\mathrm{d}t = p^{-\nu-1}. \tag{9}$$

For $\mu \neq 0$ we choose $C_2(p) \to -C_1(p)$ in (7) and specify $C_1(p)$ by (9). The result is:

$$L(x) = \frac{p^{q-\nu-1}\,|x|^{\mu-\frac{3}{2}}}{d\,|\mu|\,\Gamma\left(1-\frac{1}{\mu}\right)\left(\frac{p^q}{d\,\mu^2}\right)^{\frac{1}{2\,\mu}}} K_{\frac{1}{\mu}} \left(2\,|x|^{\frac{\mu}{2}} \sqrt{\frac{p^q}{d\,\mu^2}} \right). \tag{10}$$

There is a possibility to reduce $|\mu|\,\Gamma\left(1-\frac{1}{\mu}\right) = -\operatorname{sign}(\mu)\,\Gamma\left(-\frac{1}{\mu}\right)$ in (10) thus there are cases $\mu = \frac{1}{i}$ with integer i which will destroy the propagator. Indeed, the normalization integral diverges in these cases. Discussing a solution without normalization is still possible, but no more unique.

2.2 Problems with $\mu = 0$

Some more problems occur with $\mu = 0$: there seems to be no way to construct a propagator like symmetrical function by using the integration constants in (8), because this function is zero or infinity at $x = 0$ for any integration constant! The limit $\mu \to -2$ of (10) leads to a separation of cases because of the term $p^{\frac{-q}{2\mu}}$.

For this we make use of some functions leading to a Dirac's delta function by a limit $\epsilon \to 0$. The functions we discuss can easily be transformed by Mellin transform:

$$\delta_1^\epsilon(x) = \frac{\exp\left(-\frac{x^2}{\epsilon^2}\right)}{\epsilon\,\sqrt{\pi}}\,, \tag{11}$$

$$\delta_2^\epsilon(x) = \frac{\epsilon}{(\epsilon^2 + x^2)\,\pi}\,. \tag{12}$$

With these functions instead of Dirac's $\delta(x)$ in (5) we get the Mellin transform of the solutions in Laplace space. These solutions lead to a generalization of Fox's

H–function and even of Saxena's I–function [14]. We called it ℵ–function [18] and get now:

$$L_1^\epsilon(x) = \frac{p^{q-\nu-1}}{2\sqrt{\pi}\,\epsilon}\,\aleph^{1,0,0}_{1,1,1}\left[\frac{x}{\epsilon}\,\middle|\,\begin{matrix}(0,1)\\(0,\frac{1}{2})\end{matrix}\,\middle|\,p^q\,\begin{matrix}()\\(0,1)\end{matrix}\,\middle|\,-d\,\begin{matrix}()\\(-2,1)\end{matrix}\right], \tag{13}$$

$$L_2^\epsilon(x) = \frac{p^{q-\nu-1}}{2\,\pi\,\epsilon}\,\aleph^{2,0,0}_{1,1,1}\left[\frac{x}{\epsilon}\,\middle|\,\begin{matrix}(0,1),(0,\frac{1}{2})\\(0,\frac{1}{2})\end{matrix}\,\middle|\,p^q\,\begin{matrix}()\\(0,1)\end{matrix}\,\middle|\,-d\,\begin{matrix}()\\(-2,1)\end{matrix}\right]. \tag{14}$$

Here in this case the ℵ–functions can be reduced to a Meijer's G–function, but the results seem not to be fit for the inverse Laplace transform from p to t:

$$L_1^\epsilon(x) = \frac{p^{q-\nu-1}}{4\,d\sqrt{\pi}\,\epsilon}\,G^{2,1}_{2,3}\left[\frac{x^2}{\epsilon^2}\,\middle|\,\begin{matrix}\left(\frac{1}{4}+\frac{\sqrt{d+4\,p^q}}{4\sqrt{d}}\right) \mid \left(\frac{1}{4}-\frac{\sqrt{d+4\,p^q}}{4\sqrt{d}}\right)\\ \left(0,-\frac{3}{4}-\frac{\sqrt{d+4\,p^q}}{4\sqrt{d}}\right) \mid \left(-\frac{3}{4}+\frac{\sqrt{d+4\,p^q}}{4\sqrt{d}}\right)\end{matrix}\right], \tag{15}$$

$$L_2^\epsilon(x) = \frac{p^{q-\nu-1}}{4\,d\,\pi\,\epsilon}\,G^{2,2}_{3,3}\left[\frac{x^2}{\epsilon^2}\,\middle|\,\begin{matrix}\left(0,\frac{1}{4}+\frac{\sqrt{d+4\,p^q}}{4\sqrt{d}}\right) \mid \left(\frac{1}{4}-\frac{\sqrt{d+4\,p^q}}{4\sqrt{d}}\right)\\ \left(0,-\frac{3}{4}-\frac{\sqrt{d+4\,p^q}}{4\sqrt{d}}\right) \mid \left(-\frac{3}{4}+\frac{\sqrt{d+4\,p^q}}{4\sqrt{d}}\right)\end{matrix}\right]. \tag{16}$$

With the substitutions

$$\frac{x^2}{\epsilon^2} \to z, \qquad \frac{1}{4}+\frac{\sqrt{d+4\,p^q}}{4\sqrt{d}} \to a, \qquad \frac{1}{4}-\frac{\sqrt{d+4\,p^q}}{4\sqrt{d}} \to b, \tag{17}$$

we get some more simplifications for $b \neq 1$ and because of $a \neq b$:

$$L_1^\epsilon(x) = \frac{p^{q-\nu-1}}{4\,d\sqrt{\pi}\,\epsilon\,(a-b)}\left(E(a,z)-E(b,z)-z^{a-1}\,\Gamma(1-a)\right), \tag{18}$$

$$L_2^\epsilon(x) = \frac{p^{q-\nu-1}}{4\,d\,\pi\,\epsilon\,(b-a)}\times$$
$$\left(z^{b-1}\,\pi\,\csc(\pi\,b)+Y(-z,1,1-a)-Y(-z,1,1-b)\right), \tag{19}$$

where the exponential integral $E(n,x) = x^{n-1}\,\Gamma(1-n,x)$ is a special case of an incomplete gamma function which can be expressed by a confluent hypergeometric function $\ldots = \frac{\Gamma(1-n)}{x^{1-n}} - \frac{F_{1,1}(1-n,-n,-x)}{1-n}$ and the Lerch zeta function mentioned here is a special case of the Gauß hypergeometric function $F_{2,1}(1,a,a+1,-z) = a\,Y(-z,1,a)$.

The reduced form (17,18,19) of the result (15,16) is a proper possibility to get some graphics, whereas the form of the ℵ–function (13,14) allows to go on with integral transforms via Mellin transform in an elegant way. The limit $\epsilon \to 0$ is not possible in the Laplace space ad hoc. It can be done later after further transforms.

Another possibility to get analytical results of equation (5) for $\mu = 0$ is the choice of shifted delta functions as an initial value problem in (5). So we choose

$$u_\nu(x,0) = \frac{\delta(x+1)}{2}+\frac{\delta(x-1)}{2}, \tag{20}$$

and get a solution out of (5) by the method (9):

$$L(x) = \frac{p^{q-\nu-1}\,|x|^{\pm\frac{\sqrt{d+4\,p^q}}{2\,\sqrt{d}}}}{2\,\sqrt{d\,|x|^3}\,\sqrt{d+4\,p^q}} \;. \tag{21}$$

The sign plus (+) is correct in (21) for $|x| \le 1$, the sign minus (−) for $|x| \ge 1$. This formula can also be written with $|x|^{\pm c} \to \exp(-c\,|\ln|x||)$ for all real x.

3 The Momenta of the Solution

3.1 Laplace Transform of the Momenta

The results in Laplace space are quite complicated. For this reason we calculate the momenta of the solution before the solution itself.

The result is symmetrical in x, thus we obtain a Mellin integral by calculating the momenta with the following formula:

$$2\int_0^\infty x^M\;L(x)\;\mathrm{d}x = \int_0^\infty \exp(-p\;t)\;\int_{-\infty}^\infty |x|^M\;u(x,t)\;\mathrm{d}x\;\mathrm{d}t\;. \tag{22}$$

After this the inverse Laplace transform is done by the results of Oberhettinger [10] using the Mellin transform $\mathcal{M}_x^z\,f(x)$:

$$\mathcal{M}_x^z\,f(x) := \int_0^\infty f(x)\,x^{z-1}\,\mathrm{d}x\;, \tag{23}$$

$$\mathcal{M}_p^{1-z}\int_0^\infty \exp(-p\;t)\,f(t)\,\mathrm{d}t = \Gamma(1-z)\,\mathcal{M}_t^z\,f(t)\;. \tag{24}$$

Using (22) with (10,13,14,21) leads to the Laplace transform $L(p,M)$ of the momenta:

$$L(p,M) = \frac{d^{\frac{M}{\mu}}\,|\mu|^{\frac{2M}{\mu}}\,\Gamma\left(1+\frac{M-1}{\mu}\right)\,\Gamma\left(1+\frac{M}{\mu}\right)}{p^{\frac{Mq}{\mu}-\nu-1}\,\Gamma\left(1-\frac{1}{\mu}\right)}\;, \tag{25}$$

$$L(p,M) = \frac{p^{q-\nu-1}\,\epsilon^M\,\Gamma\left(\frac{M+1}{2}\right)}{\sqrt{\pi}\,(p^q-(M-1)\,M\,d)}\;, \tag{26}$$

$$L(p,M) = \frac{p^{q-\nu-1}\,\epsilon^M\,\Gamma\left(\frac{M+1}{2}\right)\,\Gamma\left(\frac{1-M}{2}\right)}{\pi\,(p^q-(M-1)\,M\,d)}\;, \tag{27}$$

$$L(p,M) = \frac{p^{q-\nu-1}}{p^q-(M-1)\,M\,d}\;. \tag{28}$$

Now it is obvious that the Laplace transform of the norm ($M=0$) is $p^{-\nu-1}$ for all real μ. The limit $\epsilon \to 0$ leads to a term which is zero for $\Re(M) > 0$ and $p^{-\nu-1}$ for $M=0$:

$$L(p,M) = \frac{0^M}{p^{\nu+1}}\;. \tag{29}$$

3.2 Prolongation of the Mellin Transform

Discussing the inverse Laplace transform of these terms there is a need of a prolongation of the Mellin transform. For example there is a possibility to get the momenta out of (26) and (27) by inverse Laplace transform for $q = 1$ and $\nu = 0$ only. The results which are possible by the inverse Laplace transform without making use of the Mellin transform (24) are:

$$\int_{-\infty}^{\infty} |x|^M\, u(x,t)\, \mathrm{d}x = \frac{d^{\frac{M}{\mu}}\, t^{\frac{Mq}{\mu}+\nu}\, |\mu|^{\frac{2M}{\mu}}\, \Gamma\left(1+\frac{M-1}{\mu}\right)\, \Gamma\left(1+\frac{M}{\mu}\right)}{\Gamma\left(1-\frac{1}{\mu}\right)\, \Gamma\left(\nu+1+\frac{Mq}{\mu}\right)}\;, \tag{30}$$

$$\int_{-\infty}^{\infty} |x|^M\, u(x,t)\, \mathrm{d}x = \frac{\exp\left((M-1)\, M\, d\, t\right)\, \epsilon^M\, \Gamma\left(\frac{M+1}{2}\right)}{\sqrt{\pi}}\;, \tag{31}$$

$$\int_{-\infty}^{\infty} |x|^M\, u(x,t)\, \mathrm{d}x = \frac{\exp\left((M-1)\, M\, d\, t\right)\, \epsilon^M\, \Gamma\left(\frac{M+1}{2}\right)\, \Gamma\left(\frac{1-M}{2}\right)}{\pi}\;, \tag{32}$$

$$\int_{-\infty}^{\infty} |x|^M\, u(x,t)\, \mathrm{d}x = \exp\left((M-1)\, M\, d\, t\right)\;, \tag{33}$$

$$\int_{-\infty}^{\infty} |x|^M\, u(x,t)\, \mathrm{d}x = \frac{0^M\, t^\nu}{\nu!}\;. \tag{34}$$

For $q \neq 1$ there is a need of the Mellin transform of $\mathcal{M}_p^y\, \frac{1}{p+a}$ for any real a. Now there are two different results with the Mellin transform of $\frac{1}{p+a}$ for $a > 0$ according to Oberhettinger [10]:

$$\mathcal{M}_p^y\, \frac{1}{p+a} = a^{y-1}\, \pi\, \csc(\pi\, y)\;, \qquad 0 < \Re(y) < 1\;; \tag{35}$$

$$\mathcal{M}_p^y\, \frac{1}{p-a} = -(a)^{y-1}\, \pi\, \csc(\pi\, y)\, \cos(\pi\, y)\;, \qquad 0 < \Re(y) < 1\;. \tag{36}$$

For $0 < \Re(y) < 1$ and $a = 0$ the formula (35) yields the correct result of a divergent integral, too. If the inverse Mellin transform of (36) is calculated, the residual points of $\csc(\pi\, y)$ are found at integer y only. Thus the replacement $\cos(\pi\, y) \to (-1)^y$ in (36) is true leading to (35) for all real a.

The use of (35) for all real a is needed to get the inverse Laplace transform of (26,27,28) via (24) obtaining the extended Mittag Leffler function $E_{\alpha,\beta}(z)$ [4] and enlarging the possibilities written about by Samko et al. [13] to get out a Mittag Leffler function by inverse Laplace transform:

$$\int_{-\infty}^{\infty} |x|^M\, u(x,t)\, \mathrm{d}x = \frac{t^\nu\, \epsilon^M\, E_{q,\nu+1}\left((M-1)\, M\, d\, t^q\right)\, \Gamma\left(\frac{M+1}{2}\right)}{\sqrt{\pi}}\;, \tag{37}$$

$$\int_{-\infty}^{\infty} |x|^M\, u(x,t)\, \mathrm{d}x = \frac{t^\nu\, \epsilon^M\, E_{q,\nu+1}\left((M-1)\, M\, d\, t^q\right)}{\cos\left(\frac{M\,\pi}{2}\right)}\;, \tag{38}$$

$$\int_{-\infty}^{\infty} |x|^M\, u(x,t)\, \mathrm{d}x = t^\nu\, E_{q,\nu+1}\left((M-1)\, M\, d\, t^q\right)\;. \tag{39}$$

The case $q = 1$ and $\nu = 0$ leads back to the results (31,32,33). The divergence of (32) and (38) for $M \geq 1$ is a property of the Cauchy density (12). The limit $\epsilon \to 0$ in (37,38) leads to (34).

4 The Mellin and Fourier Transform of the Solution

For symmetrical functions the Mellin transform and the momenta are connected:

$$\int_{-\infty}^{\infty} |x|^M \, u(x,t) \, \mathrm{d}x = 2 \, \mathcal{M}_x^{M+1} \, u(x,t) \ . \tag{40}$$

By this relation (40) a Mellin transform of the solution is known. These Mellin transforms are needed to calculate the Fourier (cosine) transform [10, 4] of a symmetrical function:

$$\mathcal{M}_k^z \int_{-\infty}^{\infty} u(x,t) \, \exp(\mathrm{i}\, k\, x) \, \mathrm{d}x = \frac{2^z \, \sqrt{\pi} \, \Gamma\left(\frac{z}{2}\right)}{\Gamma\left(\frac{1-z}{2}\right)} \, \mathcal{M}_x^{1-z} \, u(x,t) \ . \tag{41}$$

With $M \to -z$ we get by (40) the following results of the Fourier transform $u(k,t)$ of the solution $u(x,t)$.

For $\mu > 0$ we obtain:

$$u(k,t) = \frac{\sqrt{\pi}\, t^{\nu}}{\Gamma\left(1 - \frac{1}{\mu}\right)} \, \Psi_{2,2} \left[\begin{matrix} (1, \frac{2}{\mu}), (1 - \frac{1}{\mu}, \frac{2}{\mu}) \\ (\frac{1}{2}, 1), (\nu + 1, \frac{2q}{\mu}) \end{matrix} ; - \frac{k^2 \, (d \, t^q \, \mu^2)^{\frac{2}{\mu}}}{4} \right] . \tag{42}$$

For $\mu < 0$ we obtain:

$$u(k,t) = \frac{\sqrt{\pi}\, t^{\nu}\, |\mu|}{2 \, \Gamma\left(1 + \frac{1}{|\mu|}\right)} \times \tag{43}$$

$$H_{1,4}^{3,0} \left[\frac{2^{-|\mu|} \, k^{|\mu|}}{d \, t^q \, \mu^2} \middle| \begin{matrix} () & | \ (\nu+1, q) \\ \left(0, \frac{|\mu|}{2}\right), (1,1), \left(1 + \frac{1}{|\mu|}, 1\right) & | \ (\frac{1}{2}, \frac{|\mu|}{2}) \end{matrix} \right] .$$

For $\mu = 0$ we discuss the momenta (34) and recognize them as the momenta of Dirac's $\delta(x)$ and a factor depending on ν. By (40) we obtain the Mellin transform of Dirac's $\delta(x)$ for $\Re(z) > 1$ or $z = 1$ with $0^0 = 1$:

$$\mathcal{M}_x^z \, \delta(x) = \frac{0^{z-1}}{2} \ . \tag{44}$$

A very interesting effect of this result is the following property:

$$\mathcal{M}_x^z \, x \, \delta(x) = \frac{0^z}{2} \not\equiv 0 \ . \tag{45}$$

This result proofs that there is no total identity $x \, \delta(x) \equiv 0$. Thus we have a reason to stress the sentence of Mellin [7], which says that the Mellin transform of zero only is zero.

The result (44) leads with (40) and (34) to the solution for $\mu = 0$:

$$u(x,t) = \frac{t^\nu}{\nu!}\,\delta(x)\ . \tag{46}$$

This is the static initial value problem! This solution fits with (4) for $\mu = 0$ according to the following property of Dirac's delta function which can be prooved by partial integration:

$$\delta^{(n)}(x) = -\frac{n\,\delta^{(n-1)}(x)}{x}\ . \tag{47}$$

The proof of (44) is completed by this check.

The Fourier transform of (46) is:

$$u(k,t) = \frac{t^\nu}{\nu!}\ . \tag{48}$$

Because of this static result the other propagator with the initial value problem (20) is an interesting alternative to consider. The Fourier transform of this other case leads by (41) and a power series of (39) to a function which seems to have no classification yet. The power series starts with the following terms:

$$\begin{aligned} u(k,t) = {} & \frac{t^\nu}{\nu!}\cos(k) - \frac{d\,k^2\,t^{q+\nu}\cos(k)}{\Gamma(q+\nu+1)} + \\ & \frac{d^2\,t^{2\,q+\nu}}{\Gamma(2\,q+\nu+1)}\left(-2\,k^2\cos(k) + 4\,\pi\,k^3\sin(k) + k^4\cos(k)\right) + \\ & \frac{d^3\,t^{3\,q+\nu}}{\Gamma(3\,q+\nu+1)}\left(-4\,k^2\cos(k) + 32\,\pi\,k^3\sin(k) + \right. \\ & \left. + 38\,k^4\cos(k) - 12\,\pi\,k^5\sin(k) - k^6\cos(k)\right) + t^\nu\,O\left(t^{4\,q}\right)\ . \end{aligned} \tag{49}$$

For $t \to 0$ (49) leads to the fourier transform of (20). The other fourier transforms according to (37,38) with $\epsilon > 0$ have similar power series to (49) without a systematical form or function name, too.

5 The Solution and its Properties

The inverse Mellin transform of (30) using (40) yields the solution for $\mu \neq 0$:

$$u(x,t) = \frac{t^\nu\,|\mu|\,H^{2,0}_{1,2}\left[\frac{|x|^\mu}{d\,\mu^2\,t^q}\,\middle|\,\begin{matrix} () & |\ (\nu+1,q) \\ (1,1),(1-\frac{1}{\mu},1)\ | & () \end{matrix}\right]}{2\,|x|\,\Gamma\left(1-\frac{1}{\mu}\right)}\ . \tag{50}$$

The asymptotic behaviour of this solution is for $\mu > 0$ and large $|x|$ or for $\mu < 0$ and small $|x|$ according to Schneider and Wyss [17]:

$$u(x,t) \sim \frac{t^\nu\,|\mu|\,\exp\left(-(2-q)\left(\frac{|x|^\mu\,q^q}{d\,t^q\,\mu^2}\right)^{\frac{1}{2-q}}\right)}{2\,\sqrt{2-q}\,|x|\,q^{\nu+\frac{1}{2}}\,\Gamma\left(1-\frac{1}{\mu}\right)}\left(\frac{|x|^\mu\,q^q}{d\,t^q\,\mu^2}\right)^{\frac{1-\frac{1}{\mu}-\nu}{2-q}}\ . \tag{51}$$

Note that all cases $\mu = 2$ discussed so far are equivalent to the results given by Schneider and Wyss [17] for one dimension.

The asymptotic behaviour of (50) is for $\mu > 0$ and small $|x|$ or for $\mu < 0$ and large $|x|$ according to Mathai and Saxena [6]:

$$u(x,t) \sim -\frac{t^\nu \operatorname{sign}(\mu)}{2} \left(\frac{|x|^{\mu-1}}{d\,t^q\,\Gamma(\nu+1-q)} + \frac{|x|^{\mu-2}\,|\mu|^{\frac{2}{\mu}}\,\Gamma\left(\frac{1}{\mu}\right)}{(d\,t^q)^{1-\frac{1}{\mu}}\,\Gamma\left(-\frac{1}{\mu}\right)\Gamma\left(\nu+1-q\,\frac{\mu-1}{\mu}\right)} \right). \tag{52}$$

For $\mu < 0$ and q being no integer this is a Lévy asymptotic with Lévy index $\alpha = -\mu > 0$. For $\mu < 0$ and q being a positive integer relation (52) is another Lévy asymptotic behaviour with Lévy index $\alpha = 1 - \mu > 1$.

For $\mu = 0$ the solution (46) is already known.

The case $\mu = 0$ with the alternative initial value problem (20) leads to a solution which can be expressed by known functions. To get this solution we start with (21) and calculate a Mellin transform from p to y. Formula (24) is also used. The result is the Mellin transform of the solution:

$$\mathcal{M}_t^y\, u(x,t) = \frac{|\ln(|x|)|^{\frac{y+\nu}{q}-\frac{1}{2}}\,K_{\frac{1}{2}-\frac{y+\nu}{q}}\left(\frac{|\ln(|x|)|}{2}\right)\Gamma\left(1-\frac{y+\nu}{q}\right)}{2\sqrt{\pi}\,q\,\sqrt{|x|^3}\,d^{\frac{y+\nu}{q}}\,\Gamma(1-y)}. \tag{53}$$

In this result the factor $\sqrt{|x|^3}$ in the denominator is left out of consideration and the term $|\ln(|x|)| \to g$ is substituted. Then the Mellin transform of (53) from g to z is expressed by gamma functions and exponential factors in y and z only:

$$\frac{4^{\frac{y+\nu}{q}+z-2}\,\Gamma\left(\frac{z}{2}\right)\Gamma\left(1-\frac{y+\nu}{q}\right)\Gamma\left(\frac{z-1}{2}+\frac{y+\nu}{q}\right)}{\sqrt{\pi}\,q\,d^{\frac{y+\nu}{q}}\,\Gamma(1-y)}. \tag{54}$$

This term can be simplified in the case $q = 1$ and $\nu = 0$. Now the inverse Mellin transform of (54) is calculated from y to t. After this the other inverse Mellin transform from z to $|\ln(|x|)|$ results a Fox's H–function in two variables [6]:

$$u(x,t) = \frac{\left(\frac{4}{d}\right)^{\frac{\nu}{q}}}{8\sqrt{\pi\,|x|^3}} \times \tag{55}$$

$$H^{1,0,1,1,0}_{1,(1:1),0,(1:0)}\left[\begin{array}{c|ccc} \frac{4}{d\,t^q} & & (-\frac{1}{2}+\frac{\nu}{q},1) \mid () & \\ & () \mid & (1,q) & ;(1,1)\mid() \\ \frac{16}{\ln^2|x|} & & () & \mid () \\ & (1-\frac{\nu}{q},1)\mid & () & ;\ ()\ \mid() \end{array}\right].$$

The special case $q \to 1$ and $\nu \to 0$ of (55) ends up by the same way of calculation with:

$$u(x,t) = \frac{\exp\left(-\frac{d\,t}{4} - \frac{\ln^2|x|}{4\,d\,t}\right)}{4\sqrt{d\,\pi\,t\,|x|^3}} \,. \tag{56}$$

This last result was also obtained by studies of Lie symmetries. Here, the computer algebra software package *MathLie* was used.

The study of the asymptotic behaviour of (55) may start with the asymptotic behaviour of the Bessel function in (53) and the inverse Mellin transform of the result. The results are Fox's H–functions.

However, if an asymptotic behaviour of these Fox's H-functions is discussed, the results are not consistent to the special case $q \to 1$ and $\nu \to 0$. For this reason it seems to be better to give no asymptotic behaviour of this case of result.

6 The Special Case $q \to 1$ and $\nu \to 0$

In this case a diffusion equation with a spatial dependency of the diffusion constant is discussed. The results are simpler than with arbitrary q and ν. For this reason they are grouped together now in this section. The way of calculation is the same as already shown.

The differential equation according to (4) is:

$$\frac{\partial}{\partial t}\,u(x,t) = \frac{\partial^2}{\partial x^2}\left(d\,|x|^{2-\mu}\,u(x,t)\right) \,. \tag{57}$$

The momenta of the normalized solutions are for $\mu \neq 0$:

$$\int_{-\infty}^{\infty} |x|^M\,u(x,t)\,\mathrm{d}x = \frac{\left(d\,t\,\mu^2\right)^{\frac{M}{\mu}}\,\Gamma\left(1+\frac{M-1}{\mu}\right)}{\Gamma\left(1-\frac{1}{\mu}\right)} \,. \tag{58}$$

The other cases of momenta are already given in (31,32,33), and the moment (34) changes to 0^M.

The Fourier (cosine) transform of the solution now is for $\mu > 0$:

$$u(k,t) = \frac{\sqrt{\pi}}{\Gamma\left(1-\frac{1}{\mu}\right)}\,\Psi_{1,1}\left[\begin{matrix}\left(1-\frac{1}{\mu},\frac{2}{\mu}\right)\\ \left(\frac{1}{2},1\right)\end{matrix};-\frac{k^2\,\left(d\,t\,\mu^2\right)^{\frac{2}{\mu}}}{4}\right] \,. \tag{59}$$

For $\mu < 0$ we obtain:

$$u(k,t) = \frac{\sqrt{\pi}\,|\mu|}{2\,\Gamma\left(1+\frac{1}{|\mu|}\right)} \times \tag{60}$$

$$H^{2,0}_{0,3}\left[\frac{2^{|\mu|}\,|k|^{-|\mu|}}{d\,t\,\mu^2}\,\middle|\,\begin{matrix}() & | & ()\\ \left(0,\frac{|\mu|}{2}\right),\left(1+\frac{1}{|\mu|},1\right) & | & \left(\frac{1}{2},\frac{|\mu|}{2}\right)\end{matrix}\right] \,.$$

For $\mu = 0$ the Fourier transform of the solution is 1 according to (48).

The special initial value problem (20) leads to a power series starting with the following terms:

$$u(k,t) = \cos(k) - d\,t\,k^2 \cos(k) + \frac{(d\,t)^2}{2}\left(-2\,k^2 \cos(k) + 4\,\pi\,k^3 \sin(k) + k^4 \cos(k)\right) + \frac{(d\,t)^3}{6}\left(-4\,k^2 \cos(k) + 32\,\pi\,k^3 \sin(k) + 38\,k^4 \cos(k) - 12\,\pi\,k^5 \sin(k) - k^6 \cos(k)\right) + O\left(t^4\right) . \tag{61}$$

The solution itself is for $\mu \neq 0$:

$$u(x,t) = \frac{\exp\left(-\frac{|x|^\mu}{d\,t\,\mu^2}\right)\left(\frac{|x|^\mu}{d\,t\,\mu^2}\right)^{1-\frac{1}{\mu}} |\mu|}{2\,|x|\,\Gamma\left(1-\frac{1}{\mu}\right)} . \tag{62}$$

This solution can be derived both from (50) and (51) by specifying the parameters $q \to 1$ and $\nu \to 0$.

The power series describing the asymptotic behaviour for $\mu > 0$ and short $|x|$ or for $\mu < 0$ and large $|x|$ now is:

$$u(x,t) \sim -\frac{|x|^{\mu-2}\,|\mu|^{\frac{2}{\mu}}\,\mathrm{sign}(\mu)}{2\,(d\,t)^{1-\frac{1}{\mu}}\,\Gamma\left(-\frac{1}{\mu}\right)} . \tag{63}$$

For $\mu < 0$ this is a Lévy asymptotic with Lévy index $\alpha = 1 + |\mu| > 1$.

Note that T. F. Nonnenmacher in 1990 [9] dicussed a function

$$f(x) = \frac{a^\mu}{\Gamma(\mu)}\,x^{-\mu-1}\exp\left(-\frac{a}{x}\right) , \qquad x > 0\,,\; a > 0\,, \tag{64}$$

of Lévy asymptotic in a different context. Solution (62) with $\mu \to -1$ is almost the same as solution (64) with the parameters $\mu = 2$ and $a = \frac{1}{d\,t}$. The difference is a factor 2 caused by different definitions of the normalization integral in both contexts.

Now we get a solution of the type (64) by discussing a spatial dependency of the diffusion constant. The advantage of (62) is that x is discussed as a real number leading to a unique Fourier transform (60).

For $\mu = 0$ the solution now is a time constant Dirac's $\delta(x)$ function.

For $\mu = 0$ and the spezial initial value problem (20) the solution is given by (56).

7 Conclusions

Schaufler et. al. [16] discuss a special diffusion equation based on the VSCPT theory leading to a time and space dependant diffusion "constant". Their diffusion term is symmetrical in a variable v which is used instead of x in our equation

(4). Equation (57) is discussed explicitly in the PhD thesis of S. Schaufler [15] with a symmetrical space dependant diffusion constant in time.

A simple theory leading to Lévy distributions may be used even in discussing stock prices by a diffusion model [11]. In discussing this we stress that the value of money is not a physical unit given by the SI, but something depending on mans influence. For this a diffusion theory of stock prices may be useful in times of financial stability only. For example, in our theory the normalization integral of all moving substance is a constant of time.

It is possible to modify the equation (4) by another space dependancy of the diffusion constant, e. g. a real power law $x^{2-\mu}$ which will be a restriction to the use of μ. The difference of these models in comparison with the model discussed here is a difference in using and calculating the normalization integral and the momenta.

The use of the ℵ–functions (13,14) is optional, but it enables a systematical and elegant way of calculation. The solutions discussed in this article emphasize a need of more special functions to give a correct analytical result of non–trivial differential equations. Of course there is an analytical continuation of the solutions (49) or (61), but we are not able to give more than a power series only. The advantage of discussing power series as inverse Mellin transforms is the possibility of getting the analytical continuation and the asymptotic behaviour of a function.

All results given here in this article have been checked by computer algebra. An example of this work is given in the appendix B.

Meanwhile, since completion of this work, some other recent works have occured, thus the review article of Metzler and Klafter [8] is to be mentioned, where especially in their §5 a similar Fokker–Planck equation referring to other and some more references than this work is discussed.

Acknowledgement

We thank A. A. Kilbas for discussion. This work was partially supported by Deutsche Forschungsgemeinschaft (SFB 239 B10).

A Definition of the Special Functions

In 1936 Dixon and Ferrar [3] discussed a special function which was named after Ch. Fox [5] later. There are several possibilities of giving a definition to this Fox's H–function. The definition given here is different from the definition given by Mathai and Saxena [6] or Braaksma [2], but enables a simplified understanding of a Fox's H–function to be an inverse Mellin transform [7]:

$$H_{p,q}^{m,n}\left[x\left|\begin{matrix}(a_1,A_1),\ldots,(a_n,A_n) \mid (a_{n+1},A_{n+1}),\ldots,(a_p,A_p)\\(b_1,B_1),\ldots,(b_m,B_m) \mid (b_{m+1},B_{m+1}),\ldots,(b_q,B_q)\end{matrix}\right.\right]:=$$

$$\frac{1}{2\pi\mathrm{i}}\int_{c-\mathrm{i}\infty}^{c+\mathrm{i}\infty}\frac{\prod_{j=1}^{m}\Gamma(b_j+B_jz)\prod_{j=1}^{n}\Gamma(1-a_j-A_jz)}{\prod_{j=m+1}^{q}\Gamma(1-b_j-B_jz)\prod_{j=n+1}^{p}\Gamma(a_j+A_jz)}\,x^{-z}\,\mathrm{d}z\ . \tag{65}$$

If all $\{A_j, B_j\}$ are the same, the function (65) is reduced to a Meijer's G–function [4]. A formula to derive the asymptotic behaviour of (65) is given by Schneider and Wyss [17].

Here and within the other special functions of this article or the software package *FractionalCalculus*, empty brackets () respectively {} show empty products resulting one due to *Mathematica* syntax.

As Barnes [1] pointed out, an analytic continuation of the resulting power series all over the complex area is possible. This is the reason why these Fox's H–functions and some more functions are also called Mellin Barnes integrals.

Now the ℵ–function [18] is a generalization of Fox's H–function with $k \in \{1, 2, \ldots, N\}$:

$$\aleph^{p_0, p_1, \ldots, p_N}_{q_0, q_1, \ldots, q_N} \left[x \left| \begin{matrix} (a_1, A_1), \ldots, (a_{p_0}, A_{p_0}) \\ (b_1, B_1), \ldots, (b_{q_0}, B_{q_0}) \end{matrix} \right| \ldots \right.$$
$$\left. \ldots \left| c_k \begin{matrix} (a_{p_{(k-1)}+1}, A_{p_{(k-1)}+1}) \ldots, (a_{p_k}, A_{p_k}) \\ (b_{q_{(k-1)}+1}, B_{q_{(k-1)}+1}) \ldots, (b_{q_k}, B_{q_k}) \end{matrix} \right| \ldots \right] := \tag{66}$$
$$\frac{1}{2\pi \mathrm{i}} \int_{c-\mathrm{i}\infty}^{c+\mathrm{i}\infty} \frac{\prod_{j=1}^{q_0} \Gamma(b_j + B_j z) \prod_{j=1}^{p_0} \Gamma(1 - a_j - A_j z)\, x^{-z}\, \mathrm{d}z}{\sum_{k=1}^{N} \left(c_k \prod_{j=q_{(k-1)}+1}^{q_k} \Gamma(1 - b_j - B_j z) \prod_{j=p_{(k-1)}+1}^{p_k} \Gamma(a_j + A_j z) \right)} .$$

For $N = 1$ in (66) the definition (65) of the Fox's H–function is recovered. For all $\{c_k\}$ being the same, Saxena's I–function [14] is obtained.

If the sum in the denominator of (66) can be simplified to a polynomial in z, the factors of this polynomial can be expressed by a fraction of Euler's gamma functions leading to a Fox's H–function instead. This possibility is used to get the results (15,16) out of (13,14).

B Examples for Using Computer Algebra

B.1 Type in an Equation

When doing calculations by computer algebra, first of all, it is necessary to type in the equation to the machine. After loading the *Mathematica* package *FractionalCalculus* the following commands (**bold**) give sensible results (plain):

$$\mathbf{Assume[0 < q < 2];}$$

$$\mathbf{gl} = \mathcal{D}^{\mathbf{q}}_{\mathbf{0,t}} \left[\mathbf{u[x, t]} - \frac{\mathbf{t}^{\nu}}{\mathbf{\Gamma[\nu + 1]}} \mathbf{u}_{\nu}\mathbf{[x, 0]} \right] == \partial_{\{\mathbf{x,2}\}} (\mathbf{d[x]\, u[x, t]})$$

RiemannLiouville::conditions :
Conditions to solve the integral:
t > 0 && Re[ν] > −1 && Re[q] < 1

$$\mathcal{D}^{\mathrm{q}}_{0,\mathrm{t}} [\mathrm{u[x, t]}] - \frac{\mathrm{t}^{-\mathrm{q}+\nu} \mathrm{u}_{\nu}[\mathrm{x}, 0]}{\Gamma[1 - \mathrm{q} + \nu]} = \mathrm{u[x, t]\, d''[x]} + 2\, \mathrm{d'[x]\, u^{(1,0)}[x, t]} + \mathrm{d[x]\, u^{(2,0)}[x, t]}$$

This example shows, that even typing in an equation can be improved by making use of computer algebra, because several identities of formulation are shown.

The *Mathematica* package *FractionalCalculus* enables to assume several areas of validity for parameters as needed when handling with fractional calculus and Mellin transforms.

B.2 Simplification of an Equation

The simplification of an equation by substitution is very convenient:

$$\mathbf{glsubst} = \mathbf{gl/.\,u_\nu[x,0] \to dummy/.\,u \to Function}\left[\{\mathbf{x},\mathbf{t}\}, \frac{\mathbf{h[x,t]}}{\mathbf{d[x]}}\right]/.$$

$$\mathbf{dummy \to u_\nu[x,0]\,//\,ExpandAll}$$

$$\frac{\mathcal{D}^{q}_{0,t}\,[h[x,t]]}{d[x]} - \frac{t^{-q+\nu}\,u_\nu[x,0]}{\Gamma[1-q+\nu]} == h^{(2,0)}[x,t]$$

The special case $q = 1$ meaning $\nu = 0$ can be found by the following:

$$\mathbf{glsubst/.\,q \to 1/.\,\nu \to 0}$$

$$\frac{\mathcal{D}^{1}_{0,t}\,[h[x,t]]}{d[x]} == h^{(2,0)}[x,t]$$

These examples stress the substitution used between the equations (5) and (6).

B.3 Simplification of Special Functions

A special function such as Meijer's G–function automatically can be simplified by the standard *Mathematica* program. The examples given here show the simplification of the results (15,16) to (18,19), which can be obtained even without (the input will be different) use of the *Mathematica* package *FractionalCalculus*:

$$\mathbf{lapfunktion3einfach} =$$

$$\frac{\mathbf{p}^{-1+q-\nu}}{4\,\mathbf{d}\sqrt{\pi}\,\epsilon}\,\mathbf{FunctionExpand}\left[\mathbf{G}\left[\mathbf{z}\left|\begin{matrix}\{a\} & | & \{b\}\\ \{0,b-1\} & | & \{a-1\}\end{matrix}\right.\right]\right]\,//$$

$$\mathbf{Simplify\,//\,FullSimplify}$$

$$\frac{p^{-1+q-\nu}\,(z\,\mathrm{ExpIntegralE}[a,z] - z\,\mathrm{ExpIntegralE}[b,z] + a\,z^{a}\,\Gamma[-a])}{4\,(a-b)\,d\sqrt{\pi}\,z\,\epsilon}$$

$$\mathbf{lapfunktion4einfach} =$$

$$\frac{\mathbf{p}^{-1+q-\nu}}{4\,\mathbf{d}\,\pi\,\epsilon}\,\mathbf{FunctionExpand}\left[\mathbf{G}\left[\mathbf{z}\left|\begin{matrix}\{0,a\} & | & \{b\}\\ \{0,b-1\} & | & \{a-1\}\end{matrix}\right.\right]\right]\,//$$

$$\mathbf{Simplify\,//\,FullSimplify}$$

$$\frac{p^{-1+q-\nu}\,\left(\pi\,z^{b}\,\mathrm{Csc}[b\,\pi] + z\,\mathrm{LerchPhi}[-z,1,1-a] - z\,\mathrm{LerchPhi}[-z,1,1-b]\right)}{4\,(-a+b)\,d\,\pi\,z\,\epsilon}$$

B.4 Momenta of Solution

Another quite interesting point of view might be the calculation of the momenta of the solution out of the Laplace transforms of them. Here the input to get the results (39,34) out of (28,29) is shown, thus the possibilities of Mellin transform to enlarge the Laplace transform are demonstrated:

$$\mathbf{mom5} =$$

$$\left(\mathcal{M}^{-1}\right)_{\mathbf{y}}^{\mathbf{t}}\left[\mathbf{Solve}\left[\mathcal{M}_{\mathbf{p}}^{\mathbf{y}}[\mathbf{lapmomente5}]\,/.\,\{\mathbf{Rule} \to \mathbf{Equal}, \mathbf{y} \to \mathbf{1} - \mathbf{y}\},\right.\right.$$

$$\left.\left.\mathcal{M}_{\mathbf{t}}^{\mathbf{y}}[\mathbf{mom}[\mathbf{M}, \mathbf{t}]]\right] \mathbin{//} \mathbf{PowerExpand} \mathbin{//} \mathbf{Simplify}\right] \mathbin{//}$$

$$\mathbf{Flatten} \mathbin{//} \mathbf{PowerExpand} \mathbin{//} \mathbf{Simplify}$$

$$\{\mathrm{mom}[\mathrm{M}, \mathrm{t}] \to \mathrm{t}^{\nu}\, \mathrm{E}_{\mathrm{q},(1+\nu)}\left[(\mathrm{d}\,(-1 + \mathrm{M})\,\mathrm{M}\,\mathrm{t}^{\mathrm{q}})\right]\}$$

$$\mathbf{mom6} = \left(\mathcal{L}^{-1}\right)_{\mathbf{p}}^{\mathbf{t}}[\mathbf{lapmomente6}]$$

$$\left\{\mathrm{mom}[\mathrm{M}, \mathrm{t}] \to \frac{0^{\mathrm{M}}\, \mathrm{t}^{\nu}}{\Gamma[1 + \nu]}\right\}$$

Of course, a lot of more examples could be shown, but it seems to be too boring just to repeat each detail of this article. The examples given here just emphasize the possibilities of the combination of a computer algebra system like *Mathematica* and the Mellin transform calculus. The complete verification of this article fills about 60 pages of a *Mathematica* Notebook where the software package *FractionalCalculus* source needs at about 1 MByte on disk.

References

1. Barnes E W 1908 A New Developement of the Theory of the Hypergeometric Functions *Proc. London Math. Soc.* **(II) 6** 141–77
2. Braaksma B L J 1964 Asymptotic expansions and analytic continuations for a class of Barnes integrals *Comp. Math.* **15** 239–341
3. Dixon A L and Ferrar W L 1936 A Class of Discontinuous Integrals *Quart. J. Math., Oxford Ser.* **7** 81–96
4. Erdélyi A, Magnus W, Oberhettinger F and Tricomi F G 1953 Higher Transcendental Functions *McGraw-Hill Book Company New York Toronto London*
5. Fox C 1961 The G and H Functions as Symmetrical Fourier Kernels *Trans. Amer. Math. Soc.* **98** 395–429
6. Mathai A M and Saxena R K 1978 The H–function with Applications in Statistics and Other Disciplines *Wiley Eastern Limited New Delhi Bangalore Bombay*
7. Mellin H J 1910 Abriß einer einheitlichen Theorie der Gamma– und der hypergeometrischen Funktionen *Math. Ann.* **68** 305–37
8. Metzler R and Klafter Y 2000 *Physics Reports* **239**(1), December 2000, 1-77
9. Nonnenmacher T F 1990 Fractional integral and differential equations for a class of Lévy–type probability densities *J. Phys. A: Math. Gen.* **23** L697S–L700S
10. Oberhettinger F 1974 Tables of Mellin Transforms *Springer–Verlag Berlin Heidelberg New York*

11. Plerou V, Gopikrishnan P, Amaral L A N, Gabaix X and Stanley H E 1999 Economic Fluctuations and Diffusion *cond-mat/9912051 3 Dec 1999*
12. Risken H 1984 The Fokker–Planck Equation *Springer–Verlag Berlin Heidelberg New York Tokio*
13. Samko S G, Kilbas A A and Marichev O I 1993 Fractional Integrals and Derivatives *Gordon and Breach Science Publishers*
14. Saxena V P 1982 Formal Solution of Certain New Pair of Dual Integral Equations Involving H–Functions *Proc. Nat. Acad. Sci. India* **52**(A), III 366-75
15. Schaufler S 1998 Scaling Theory of 1D–VSCPT Laser Cooling *PhD thesis, University of Ulm*
16. Schaufler S, Schleich W P and Yakovlev V P 1997 Scaling and asymptotic laws in subrecoil laser cooling *Europhys. Lett.* **39** 383-8
17. Schneider W R and Wyss W 1989 Fractional diffusion and wave equations *J. Math. Phys.* **30**(1) 134-44
18. Südland N, Baumann G and Nonnenmacher T F 1998 Open Problem: Who knows about the Aleph (ℵ)–Functions? *Fractional Calculus and Applied Analysis* **1**(4) 401-2

Factorization of Overdetermined Systems of Linear Partial Differential Equations with Finite-Dimensional Solution Space

Serguei P. Tsarev*

Department of Mathematics,
Krasnoyarsk State Pedagogical University,
Lebedevoi, 89
660049, Krasnoyarsk, Russia
e-mail: tsarev@edk.krasnoyarsk.su

Abstract. We give an improved algorithm for factorization, i.e. decomposition of such systems with rational function coefficients into lower-order systems. This algorithm is based on the recent algorithm of Z. Li and F. Schwarz for construction of hyperexponential solutions. An analogue of the Loewy-Ore theory of factorization is exposed. We also prove an analogue of a result by E. Landau on uniqueness of the number and orders of irreducible factors.

1 Introduction

Factorization is often used for simplification of solution procedures for polynomials and linear ordinary differential operators (LODO). The theory of factorization of LODO was developed in [11, 12, 15]. From the algorithmic point of view factorization of LODO was addressed for the first time in [1]. In the past decade many improvements of this factorization algorithm and alternative algorithms were proposed (see [2], [4], [6], [25], [26]), for a Galois theory background cf. [20], [21]. Many well-known features of factorization of usual algebraic polynomials carry over to LODO: for example (see [9], [11]), any two different decompositions of a given LODO L into products of irreducible (in the given differential field of coefficients) LODO

$$L = P_1 \circ \cdots \circ P_k = \overline{P}_1 \circ \cdots \circ \overline{P}_p \tag{1}$$

have the same number of factors ($k = p$) and the factors are pairwise similar (in some transposed order). Two (irreducible for simplicity) LODOs L and M are called *similar* if one can find operators A and B such that $\mathrm{ord}\,(A) = \mathrm{ord}\,(B) < \mathrm{ord}\,(L) = \mathrm{ord}\,(M)$ and $A \circ L = M \circ B$ (see e.g. [15], [26]). A more sophisticated theory of factorization of linear partial differential operators (LPDO) was proposed in [27]; in particular, it turns out that in order to prove a Jordan-Hölder

* The research described in this article was partially supported by INTAS grant 99-01782 and Russian Ministry of education grant E00-1.0-57

type analogue of Landau-Loewy theorem (1) for LPDO one shall formulate everything in terms of (special) *ideals* in the ring of LPDO; thus, instead of solutions of the corresponding equation $Lu = 0$ we have to solve the system:

$$\begin{cases} L_1 u = 0, \\ \quad \vdots \\ L_k u = 0, \end{cases} \tag{2}$$

with $L_1, \ldots, L_k$ generating a left ideal $|L_1, \ldots, L_k\rangle$ in the ring of LPDO.

System (2) is often called *overdetermined*: the number of linearly independent equations in it is larger than the number of unknown functions (one function $u(x_1, \ldots, x_n)$ in (2)). Linear overdetermined systems are very common in applications; for example, the problem of finding symmetries or conservation laws for any (systems of) non-linear PDE or ODE is reduced to solution of a system with one or several unknown functions ([14]). As a rule the number of linearly independent symmetries or conservation laws of the given differential order is finite; this means that the solution space of the corresponding system is *finite-dimensional*. Hereafter we will call such systems *fLOS—finite-dimensional linear overdetermined systems*. All steps in the procedure for finding (infinitesimal) symmetries and conservation laws are simple and algorithmic; only finding closed-form solutions of resulting linear overdetermined systems was an art and consisted in many tricks known to experts.

In the pioneering paper [10] an algorithm for computation of rational and hyperexponential solutions of fLOS with rational function coefficients was given; so we are able to find one-dimensional subspaces of solutions of such fLOS which are simultaneously the solution space of the simplest fLOS $\{(D_1 - a_1(x_1, \ldots, x_n))u = 0, \ldots, (D_n - a_n(x_1, \ldots, x_n))u = 0\}$, $a_i(x_1, \ldots, x_n) \in \overline{\mathbf{Q}}(x_1, \ldots, x_n)$ (i.e. they are rational functions with algebraic number coefficients), $D_i = \partial/\partial x_i$. This is an analogue of splitting off a first-order right factor $L = \overline{L} \circ (D - a(x))$ for LODO, this is the basic step in factorization of LODO (cf. [1], [2]).

In this paper we give the theoretical background and describe an algorithm for finding larger "right factors" of fLOS with rational function coefficients. This algorithm may also be used for decomposition of the solution space of a given fLOS into a direct sum of subspaces — solution of "simpler" fLOS if the given fLOS is completely reducible (see Section 2 for the definition).

2 Basic Definitions and Theoretical Results

Suppose we have a system of linear partial differential equations

$$\begin{cases} E_1 = L_{11}u_1 + L_{12}u_2 + \ldots + L_{1k}u_k = 0, \\ \quad \vdots \\ E_m = L_{m1}u_1 + L_{m2}u_2 + \ldots + L_{mk}u_k = 0, \end{cases} \tag{3}$$

$L_* = \sum_{|i| \le N} a^*_{i_1 \cdots i_n}(x) D^{i_1}_{x_1} D^{i_2}_{x_2} \cdots D^{i_n}_{x_n}$ are LPDOs with rational function coefficients, $u_i = u_i(x_1, \ldots, x_n)$ are the unknown functions, $k \ge 1$. The classical Riquier-Janet algorithm (R-JA) — the famous precursor of Buchberger's Gröbner basis algorithm, see [18], [7], [17] [24] — may be used to test if (3) has finite-dimensional solution space. Basically this is done by differentiation of pairs (E_i, E_j) of equations of (3) to equate their leading differential monomials $D_1^{i_1} \cdots D_n^{i_n} u_s$, subtracting them and simplifying ("reducing") the result using the system itself, forming a sort of S-polynomial, we will denote it $S(E_i, E_j)$. After a finite number of steps this process will give an *involutive system* of linear partial differential equations (i.e. every $S(E_i, E_j)$ for this system reduces to 0). For every u_i one shall then build a "Riquier-Janet chart" (*R-J chart*) of leading monomials of respective operators $\widetilde{L}_{ij}$ of this involutive system. In Fig. 1 we give the R-J chart for $L_{11} = D_1^2 D_2^3 + aD_2^2 + bD_1 + c$, $L_{21} = D_1^5 + gD_1^2 + hD_2$, $L_{31} = D_2^4 + pD_1 D_2^2 + qD_1^2 + r$ w.r.t. the degree-lexicographic ordering of differential monomials (the coefficients do not count). The cells **A**, **B**, **C** correspond to

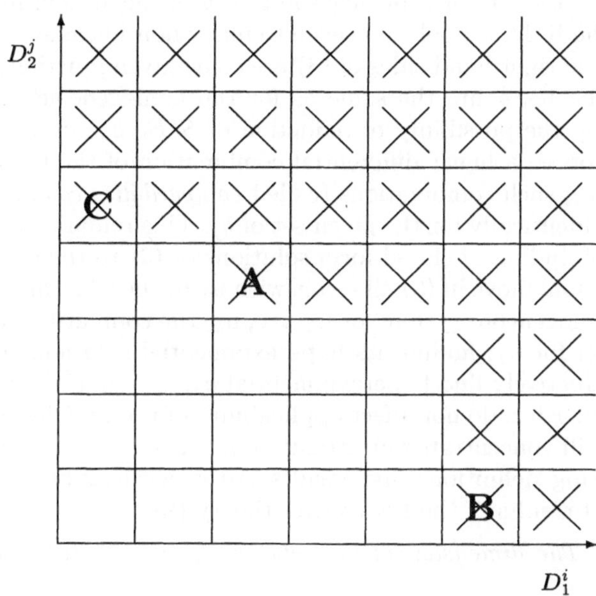

Fig. 1. An R-J chart

the respective leading monomials $D_1^2 D_2^3$, D_1^5, D_2^4 of L_{11}, L_{21}, L_{31}. Every leading monomial **M** on the chart generates a quadrant of its derivatives $D_1^{i_1} \cdots D_n^{i_n} \mathbf{M}$ of all orders (crossed in Fig. 1). The solution space of (3) is finite iff the number of free (non-crossed) cells on the R-J charts of the involutive system for every u_s is finite, the total number of them being the dimension. For this we obviously shall have pure D_i^j as some of the leading monomials for every $i = 1, \ldots, n$ and every u_s, $s = 1, \ldots, k$. The free cells of a R-J chart correspond to the Cauchy data $D_1^{i_1} \cdots D_n^{i_n} u_s(x_1^0, \ldots, x_n^0)$ for solutions of the fLOS.

As in the theory of Gröbner bases one may use the reduced involutive system to check if a LPDE is a member of the left ideal generated by the given LOS (2): for any L we may find its unique reduction modulo the involutive system $\{\widetilde{L}_j u = 0\}$: $L = \sum \widetilde{Q}_i \widetilde{L}_i + R$, where all monomials in R correspond to the free cells of the R-J chart for $\{\widetilde{L}_j u = 0\}$. Since the operators $\widetilde{L}_i$ are linear differential combinations of the operators L of the original LOS, we obtain a representation

$$L = \sum_i Q_i L_i + R \tag{4}$$

with unique R (and non-unique Q_i) for the given differential monomial ordering. We will call (4) the *division* or *reduction* formula of L w.r.t. the system $\{L_j u = 0\}$. Since the free cells correspond to the initial data for the solution of the system and may be chosen arbitrarily we conclude that the necessary and sufficient condition for an operator L to reduce to 0 modulo a given involutive system (i.e. $R \equiv 0$ in (4)) is $Lu = 0$ for arbitrary solution u of the system in question.

Actually the R-JA is applicable even to many non-linear systems of PDEs; in particular, if the right hand sides of (3) are some given functions $b_j(x_1, \ldots, x_n)$, the steps of the R-JA are the same as for the homogeneous system; the only difference being the possibility of reduction of $S(E_i, E_j)$ on intermediate steps to an expression — a linear differential combination of the r.h.s. b_j — without the unknown u_i; such combinations (called *compatibility conditions* or *syzygies*) should vanish identically for the given set of b_j. This remark allows us to reduce the problem of finding of closed-form solutions of (3) to the case with only one unknown function: move in (3) all terms with u_1 to the r.h.s. and apply the R-JA to this non-homogeneous system for $u_2, \ldots, u_k$; the compatibility conditions will give a LOS (2) for u_1; finding its hyperexponential solutions and substituting into (3) we inductively find hyperexponential $u_2, \ldots, u_k$ if they exist. Indefinite constants in the r.h.s. do not affect applicability of the Li-Schwarz algorithm. So hereafter we will concentrate our attention on the case $k = 1$, i.e. (2).

The following definitions and results are generalizations of the respective definitions and results of the Loewy-Ore theory (see e.g. [15], [26]).

Definition 1 *The dimension of the solution space of a fLOS will be called its order.*

This is an obvious generalization of the order of a LODO convenient for our purposes. Using the R-JA we may find it algorithmically. If we have two fLOS $\mathbf{S}_1 = \{\sum_j L_{ij} u_j = 0\}$, $\mathbf{S}_2 = \{\sum_k M_{sk} v_k = 0\}$ and the number of equations of the first is equal to the number of unknown functions of the second we may define their "product" $\mathbf{S}_3 = \mathbf{S}_2 \circ \mathbf{S}_1$ substituting $v_k = \sum_j L_{kj} u_j$: $\mathbf{S}_3 = \{\sum_{jk} M_{sk} L_{kj} u_j = 0\}$. Since the linear dimension of the solution space of a non-homogeneous system $\overline{\mathbf{S}}_1 = \{\sum_j L_{kj} u_j = \overline{v}_k(x)\}$ is either zero (when the compatibility conditions for $\sum_j L_{ij} u_j = \overline{v}_i$ are not satisfied for the given set of v_i) or equals the order of the homogeneous $\mathbf{S}_1$ (when these compatibility conditions are satisfied) obviously we have $\operatorname{ord} \mathbf{S}_3 \leq \operatorname{ord} \mathbf{S}_2 + \operatorname{ord} \mathbf{S}_1$. The equality is achieved when $\mathbf{S}_2$ contains (at least in the left $\mathcal{D}$-module generated by its equations) all syzygies of $\mathbf{S}_1$.

Definition 2 *A system* $\mathbf{S}_1 = \{L_i u = 0\}$ *of type (2) is called right factor of another LOS* $\mathbf{S}_2 = \{M_j u = 0\}$ *of the same type if every solution of* $\mathbf{S}_1$ *is a solution of* $\mathbf{S}_2$.

In such a case using (4) we may find the "quotient system" $\mathbf{Q} = \mathbf{S}_2/\mathbf{S}_1$ that is a fLOS $\mathbf{Q}$ such that $\mathbf{S}_2 = \mathbf{Q}\circ\mathbf{S}_1$ and $\operatorname{ord}\mathbf{Q} = \operatorname{ord}\mathbf{S}_2 - \operatorname{ord}\mathbf{S}_1$: since all R in (4) must vanish we have $M_j = \sum_i Q_{ji}L_i$ so we take $\mathbf{Q} = \{\sum_j Q_{ji}v_i = 0, \sum_j Q^{syz}_{ji}v_i = 0\}$ as such quotient fLOS. We have added the syzygies $\{\sum_j Q^{syz}_{ji}L_i = 0\}$ of $\mathbf{S}_1$ here to guarantee that $\operatorname{ord}\mathbf{Q} = \operatorname{ord}\mathbf{S}_2 - \operatorname{ord}\mathbf{S}_1$.

Below we will also write $\mathbf{S}_1 \subset \mathbf{S}_2$ if $\mathbf{S}_1$ is a right factor of $\mathbf{S}_2$.

Definition 3 *The greatest common divisor of two fLOS is the fLOS which has all the equations of both* $\mathbf{S}_1$ *and* $\mathbf{S}_2$: $GCD(\mathbf{S}_1, \mathbf{S}_2) = \mathbf{S}_1 \bigcup \mathbf{S}_2$.

Obviously the solution space of $GCD(\mathbf{S}_1, \mathbf{S}_2)$ is the intersection of the solution spaces of $\mathbf{S}_1$ and $\mathbf{S}_2$.

Definition 4 *The least common multiple* $LCM(\mathbf{S}_1, \mathbf{S}_2)$ *of two fLOS is the system whose solution space is spanned by the solution spaces of* $\mathbf{S}_1$, $\mathbf{S}_2$, $\operatorname{ord} LCM(\mathbf{S}_1, \mathbf{S}_2) = \operatorname{ord}\mathbf{S}_1 + \operatorname{ord}\mathbf{S}_2 - \operatorname{ord} GCD(\mathbf{S}_1, \mathbf{S}_2)$.

Below we will not distinguish systems whose solution spaces coincide, they give the same reduced involutive systems after the R-JA.

Theorem 1 *The* $LCM(\mathbf{S}_1, \mathbf{S}_2)$ *with coefficients in the same differential field exists for any two fLOS; this* LCM *can be found algorithmically applying only algebraic operations and differentiations to the coefficients of* $\mathbf{S}_1$ *and* $\mathbf{S}_2$.

Proof. Let us write the two given systems

$$\mathbf{S}_1 : \begin{cases} L_1 u = 0, \\ \quad\vdots \\ L_k u = 0, \end{cases} \qquad \mathbf{S}_2 : \begin{cases} M_1 v = 0, \\ \quad\vdots \\ M_p v = 0, \end{cases}$$

with formally different unknown functions u and v and take the union of them with an additional equation for a new unknown function w: $\mathbf{S} = \mathbf{S}_1 \bigcup \mathbf{S}_2 \bigcup \{u + v = w\}$. Treating w as the right-hand (inhomogeneous) side we find the syzygies $Q_j w = 0$ for this united system, obviously this system for w gives us the desired $LCM(\mathbf{S}_1, \mathbf{S}_2)$. □

In order to generalize the Landau-Loewy theorem we have to reformulate (1) in terms of ascending chains of ideals of the ring of LODO (see e.g. [27]):

$$1 \prec L_k \prec L_k \circ L_{k-1} \prec \ldots \prec L_2 \circ \cdots \circ L_k \circ L_{k-1} \prec L,$$

where $A \prec B$ for LODO means that A is a right factor of B. For a given fLOS $\mathbf{S}$ we have the following analogue:

Theorem 2 *Every two ascending maximal chains of right factors for* $\mathbf{S}$

$$\begin{array}{c} \{u = 0\} \subset \mathbf{S}_1 \subset \mathbf{S}_2 \subset \ldots \subset \mathbf{S}_k = \mathbf{S} \\ \{u = 0\} \subset \overline{\mathbf{S}}_1 \subset \overline{\mathbf{S}}_2 \subset \ldots \subset \overline{\mathbf{S}}_m = \mathbf{S} \end{array} \tag{5}$$

have equal lengths: $k = m$. *The quotients* $\mathbf{Q}_i = \mathbf{S}_i/\mathbf{S}_{i-1}$ *and* $\overline{\mathbf{Q}}_j = \overline{\mathbf{S}}_j/\overline{\mathbf{S}}_{j-1}$ *have equal orders after a transposition of the set of indices* $\{i\} \leftrightarrow \{j\}$.

Proof. The partial ordering (w.r.t. the relation $\mathbf{S}_a \subset \mathbf{S}_b$) on the set of fLOS (2) gives it a lattice structure due to existence of GCD and LCM. Since this lattice is isomorphic to a sublattice of the modular lattice of finite-dimensional subspaces (solution spaces of fLOS) of the infinite dimensional universal differential field it is also modular (see e.g. [27] for the definitions and basic facts of the lattice theory). In order to prove the statement about the orders of the quotients we use the notion of similarity of intervals (e.g. Definitions 1, 2 and Proposition 1 in [27]). □

The property of modularity immediately provides for fLOS all the necessary analogues of the basic facts of the Loewy-Ore theory for LODO. In particular we describe the notion of complete reducibility.

Definition 5 *A representation* $\mathbf{S} = LCM(\mathbf{S}_1, \ldots, \mathbf{S}_k)$ *of a given fLOS* $\mathbf{S}$ *is called a direct sum if the set* $\{\mathbf{S}_i\}$ *is* independent *that is* $\forall\ i$, $GCD\big(LCM(\mathbf{S}_1, \ldots, \mathbf{S}_{i-1}, \mathbf{S}_{i+1}, \ldots, \mathbf{S}_k), \mathbf{S}_i\big) = 1$. *Such* $\mathbf{S}$ *are called completely reducible.*

Obviously the solution space of $\mathbf{S}$ is the direct sum of the solution spaces of $\mathbf{S}_i$.

3 Riquier-Janet Wronskians

Wronskians play an important role in the theory of LODO and are essential in Beke's algorithm of LODO factorization [1], [2]. We associate a "Riquier-Janet Wronskian" (called below *R-J Wronskian*) to every R-J chart C with finite number of free cells. Take the differential monomials $\delta_1 = 1$, $\delta_2 = D_1^{j_1} \cdots D_n^{j_n}$, $\ldots \delta_k = D_1^{p_1} \cdots D_n^{p_n}$ corresponding to the free cells of C. Then for every k functions $u_1, \ldots, u_k$ their R-J Wronskian is

$$W_C = \begin{vmatrix} \delta_1 u_1 & \ldots & \delta_1 u_k \\ \delta_2 u_1 & \ldots & \delta_2 u_k \\ & \ldots & \\ \delta_k u_1 & \ldots & \delta_k u_k \end{vmatrix}. \tag{6}$$

We remind here that only fLOS (2) with 1 unknown function are considered for simplicity now; u_i are different solutions of (2).

Proposition 1 *Solutions* u_i *of an involutive fLOS with the R-J chart* C *are linearly independent iff their R-J Wronskian* W_C *does not vanish identically.*

Proof. According to a theorem by Kolchin [8, p. 86] a set of functions $u_1, \ldots, u_k$ is linearly independent iff there exists *some* set of differential monomials $\delta_1, \ldots, \delta_k$ with nonzero Wronskian $\overline{W}$ of type (6). But since u_i are solutions of (2) with the R-J chart C, any their derivative corresponding to a crossed cell of C is a linear differential combination of the derivatives corresponding to the free cells, hence $\overline{W}$ is proportional to W_C so the latter shall be different from 0. □

Using R-J Wronskians we can restore any (reduced involutive) fLOS if we are given its chart and a fundamental set of solutions: from the chart we know the number of equations and their form:

$$E_s = \Delta_s u + \sum_{i=1}^{k} a_{si}\delta_i u = 0, \quad s = 1, \ldots, M, \tag{7}$$

with unknown coefficients $a_{si}(x)$ and the leading differential monomials Δ_s corresponding to the generating cells of the chart (cells **A**, **B**, **C** on Fig. 1), δ_i correspond to the free cells. Fix $s = s_0$ and substitute k given solutions u_i into (7). We get k linear algebraic equations for k unknown coefficients a_{si}; the determinant of this system is $W_C \neq 0$. □

Note that from this proof we obtain the following formula for a_{si}:

$$a_{si} = \widehat{W}_{si}/W_C, \tag{8}$$

where $\widehat{W}_{si}$ is a determinant of the form (6) with Δ_s substituted instead of δ_j in the corresponding row.

4 Li-Schwarz Algorithm

Here we describe in brief for the sake of completeness the basic steps of their algorithm for construction of hyperexponential solutions of fLOS with rational function coefficients ([10]).

Definition 6 *A function $u(x_1, \ldots, x_n)$ is called hyperexponential if all its logarithmic derivatives $l_i(u) = D_i u/u$ are rational functions.*

The computation of hyperexponential solutions consists of four steps. First, for a given fLOS **S** (we follow [10] and suppose for simplicity that $n = 2$) one shall compute two LODOs $L_x \in \mathbf{Q}(x, y)[D_x]$, $L_y \in \mathbf{Q}(x, y)[D_y]$ which belong to the left ideal generated by the operators of the system. Such LODOs exist since we can use the lexicographic ordering of monomials $D_y \succ D_x$ in the R-JA obtaining the respective involutive system with one of the leading monomials being ${D_x}^j$ (it's fLOS!) so all other monomials in that operator also do not involve D_y. Computing with $D_x \succ D_y$ gives L_y. Second, using the available technique [22] of computation of hyperexponential solutions of LODO we compute the set $\mathcal{X}$ of such solutions for L_x (the other variable y being a parameter) and the set $\mathcal{Y}$ of hyperexponential solutions for L_y. Third, from $\mathcal{X}$ and $\mathcal{Y}$ we compute a finite number of inequivalent representatives such that every hyperexponential solution of **S** is equivalent to such representative. Finally, for each representative compute all solutions of **S** that are equivalent to it. Two hyperexponential functions are called equivalent if their ratio is rational.

5 Associated fLOS

The classical algorithm of Beke ([1], see also [2]) reduces the problem of factorization of a LODO L to the problem of finding hyperexponential solutions of the so-called "k-associated" equations of L. We introduce here their natural generalizations. For a given fLOS $\mathbf{S}$, $\operatorname{ord}\mathbf{S} = n$, take an integer k, $0 < k < n$ and a R-J chart C with k free cells. Form the R-J Wronskian W_C of the type (6) for this chart, $u_1, \ldots, u_k$ being (unknown, "dummy") solutions of $\mathbf{S}$.

Lemma 1 *W_C satisfies a fLOS of order not larger than $C_n^k = \frac{n!}{k!(n-k)!}$ with coefficients constructed from the coefficients of $\mathbf{S}$ using rational differential operations.*

Proof. Every derivative of u_i corresponding to a crossed cell in the R-J chart is expressible via the derivatives of the free cells, hence computing any $D_1^{i_1} \cdots D_n^{i_n} W_C$ we obtain a sum of determinants of the same form (of size $k \times k$) with differentiations δ_i corresponding to n free cells. Only C_n^k of these obtained determinants are independent. This is proved along the same lines as for the case of LODO factorization (see [23]): all δ_i corresponding to the free cells in these determinants shall be different; it is possible to number them and make the rows of the obtained determinants increasing w.r.t. this ordering. Computing $D_1 W_C$, $\ldots$, $D_1^{C_n^k} W_C$ (and similar sequential derivatives w.r.t. all other x_i) we prove that W_C satisfy a fLOS and obtain the necessary operators for the Li-Schwarz algorithm. The order of the complete fLOS for W_C does not exceed C_n^k since we can obtain linear differential dependencies between any $N > C_n^k$ derivatives of W_C. □

Hereafter we will call this fLOS *C-associated system for the given system* $\mathbf{S}$.

Lemma 2 *If for a given fLOS $\mathbf{S}$ there exists a fLOS $\mathbf{S}_k \subset \mathbf{S}$ of order k with the chart C and rational coefficients then the corresponding R-J Wronskian W_C (with u_i being the basis of solutions of $\mathbf{S}_k$) is a hyperexponential function.*

Proof. Again computing $D_i W_C$ we obtain a product of W_C and some expression consisting of differential combinations of the coefficients of $\mathbf{S}_k$. □

6 Algorithm for Factorization of fLOS

Now we combine the results of the previous sections: in order to find a right factor of $\mathbf{S}$ of order k one shall:

1. Enumerate all R-J charts $C_1, \ldots, C_N$ of order k.
2. For every C_i form the C_i-associated system $\mathbf{S}_{C_i}$.
3. Using the Li-Schwarz algorithm find the basis W_j of hyperexponential solutions of $\mathbf{S}_{C_i}$.
4. Form the equations (7) for the chosen C_i and find the Wronskian representations (8) for the unknown coefficients.

5. For $\widehat{W}_{si}$ form an auxiliary associated fLOS as we did for W_C in Section 5.
6. Substitute $\widehat{W}_{si} = w_{si}(x)\exp(R(x))$ into this system, here $\exp(R(x))$ is the exponential part in $W_C = r(x)\exp(R(x))$—one of hyperexponential solutions found in step 3; w_{si} will be the new unknown rational functions. Since $D_i W_C / W_C$ are rational the coefficients of the resulting system will remain rational. Note that the number of different exponential parts in W_C found in step 3 is finite (unlike the rational multiplier $r(x)$ which may include additive parameters).
7. Find (using Li-Schwarz algorithm) rational solutions w_{si} for all s, i for the obtained in the step 6 fLOS. If there are several such solutions, form their linear combination with indefinite constant coefficients c_i. Find the coefficients $a_{si} = \widehat{W}_{si}/W_C$ of the hypothetic right factor $\mathbf{S}_{C_i}$ and reduce it to involutive form.
8. Reduce the equations of the original system $\mathbf{S}$ modulo $\mathbf{S}_{C_i}$ and equate the remainders in (4) to zero. This will give a set of algebraic equations for c_i; using the Gröbner technique (or any its alternative at hand) check if there are such c_i and find them. This will give us the final form of the right factor $\mathbf{S}_{C_i}$.

7 Conclusion

Obviously this algorithm is very inefficient and should be improved following the guidelines of improvements given in the last decade for the Beke algorithm for LODO (see e.g. [2]). In particular, one may imitate the approach of [25]) and select among W_C found in step 3 "genuine" solutions and compute a_{si} directly, though this is impossible in some degenerate cases. An alternative very efficient algorithm for LODO factorization [6] (cf. also [4]) requires a deep analogue of singularity theory for fLOS.

Another challenging task would be an algorithm for construction of Liouvillian solutions of fLOS with rational coefficients; a generalization of the differential Galois theory for partial differential equations is being discussed in recent publications [5], [28]. It may be of interest to cast the theoretical results of Section 2 into the modern form of $\mathcal{D}$-modules theory (cf. [19]). Of special interest is an apparently faster algorithm for construction of rational solutions given in [13].

The theoretical results and algorithms described in this paper can easily be carried over to systems of (q)-partial-difference linear equations (see [3]).

Acknowledgements

The author wishes to express his special gratitude to Prof. F. Schwarz (GMD, St. Augustin, Germany) and Dr. Ziming Li who communicated to him their latest results about finding hyperexponential solutions of fLOS and Dr. Yu. Shan'ko (Krasnoyarsk) for his suggestions which greatly simplified the proof of Theorem 1.

References

1. Beke, E.: Die Irreducibilität der homogenen linearen Differentialgleichungen. *Math. Annalen* **45** (1894) 278–300
2. Bronstein, M.: An improved algorithm for factoring linear ordinary differential operators. In *Proceedings of ISSAC'94*, M. Giesbrecht, Ed., ACM Press, New York (1994) 336–340
3. Bronstein, M., and Petkovšek, M.: On Ore rings, linear operators and factorization. *Programming & Computer Software* **20** (1994) 27–44
4. Grigor'ev, D. Yu.: Complexity of factoring and calculating the GCD of linear ordinary differential operators. *J. Symbolic Computation* **10** (1990) 7–37
5. Haraoka, Y.: The Galois theory for linear homogeneous partial differential equations of the first order. *Funkcialaj Ekvacioj* **33** (1990) 79–126
6. van Hoeij, M.: Factorization of differential operators with rational functions coefficients. *J. Symbolic Computation* **24** (1997) 537–562
7. Janet, M.: *Leçons sur les systèmes d'equations aux derivées partielles*, Gauthier-Villars, Paris (1929)
8. Kolchin, E.: *Differential Algebra and Algebraic Groups*, Academic Press, New York (1973)
9. Landau, E.: Ein Satz über die Zerlegung homogener linearer Differentialausdrücke in irreducible Factoren. *J. für die reine und angewandte Math.* **124** (1901/1902) 115–120
10. Li, Z., and Schwarz, F.: Rational Solutions of Riccati-like Partial Differential Equations. *submitted*
11. Loewy, A.: Über reduzible lineare homogene Differentialgleichungen. *Math. Annalen* **56** (1903) 549–584
12. Loewy, A.: Über vollständig reduzible lineare homogene Differentialgleichungen. *Math. Annalen* **62** (1906) 89–117
13. Oaku, T., Takayama, N., and Tsai, H.: Polynomial and rational silutions of holonomic systems. Math.AG e-print (2000); INTERNET URL: http://xxx.lanl.gov/abs/math.AG/0001064
14. Olver, P. J.: *Applications of Lie Groups to Differential Equations*, Springer-Verlag (1986)
15. Ore, O.: Theory of non-commutative polynomials. *Annals of Mathematics* **34** (1933) 480–508
16. Ore, O.: Linear equations in non-commutative fields. *Annals of Mathematics* **32** (1931) 463–477
17. Reid, G. J.: Algorithms for reducing a system of PDEs to standard form, determining the dimension of its solution space and calculating its Taylor series solutions. *Euro. J. Appl. Mech.* **2** (1991) 293–318
18. Riquier, C.: *Les systèmes d'équations aux derivées partielles*, Gauthier-Villars, Paris (1910)
19. Saito, M., Sturmfels, B. and Takayama, N.: Gröbner deformations of hypergeometric differential equations. *Algorithms and Computations in Mathematics* **6**, Springer (1999)
20. Singer, M. F. and Ulmer, F.: Galois groups of second and third order linear differential equations. *Journal of Symbolic Computation* **16** (1993) 9–36
21. Singer, M. F.: Testing reducibility of linear differential operators: a group theoretic perspective. *Applicable Algebra in Engineering, Communication and Computing* **7** (1996) 77–104

22. Singer, M. F.: Liouvillian solution of linear differential equations with Liouvillian coefficients. *Journal of Symbolic Computation* **11** (1991) 251–273
23. Schwarz, F.: A Factorization Algorithm for Linear Ordinary Differential Equations. *Proceedings of ISSAC'89*, ACM Press (1989) 17–25
24. Schwarz, F.: The Riquier-Janet theory and its applications to nonlinear evolution equations. *Physica D* **11** (1984) 243–351
25. Tsarev, S. P.: On some problems in factorization of linear ordinary differential operators. *Programming & Computer Software* **20** (1994) 27–29
26. Tsarev, S. P.: An algorithm for complete enumeration of all factorizations of a linear ordinary differential operator. *Proceedings of ISSAC'96*, ACM Press (1996) 226–231
27. Tsarev, S. P.: Factorization of linear partial differential operators and Darboux integrability of nonlinear PDEs, *SIGSAM Bulletin* **32**, No. 4 (1998) 21–28; also "Computer Science" e-print cs/9811002; INTERNET URL: http://xxx.lanl.gov/abs/cs.SC/9811002
28. Umemura, H.: Differential Galois theory of infinite dimension. *Nagoya Math. Journal* **144** (1996) 59–135

Semilinear Motion Planning Among Moving Objects in REDLOG

Volker Weispfenning
weispfen@uni-passau.de

Fakultät für Mathematik und Informatik
Universität Passau
D-94030 Passau, Germany

Abstract. We study motion planning for rigid objects moving within a non-static, time dependent two or three-dimensional environment. As in a previous paper [19], both the object to be moved and the free space are both semilinear sets with no convexity assumptions. The admissible motions are finite continuous sequences of translations with fixed speeds along finitely many prescribed directions. But now the free space can change in a piecewise linear way with time. So the problem of path finding becomes a problem of finding a piecewise linear time-dependent trajectory of the given semilinear object from a given initial position to a semilinear set of admissible final positions that may change semilinearly with time. Moreover the trajectory is limited to a prescribed time interval. By a strong generalization of the static case we show how to model and solve this problem for a bounded number of translations by linear real quantifier elimination. The worst case complexity of the algorithms is similar as in the static case. Experimental results using the elimination facilities of the REDLOG package of REDUCE show the practical relevance of the approach. This concerns in particular semilinear motion planning within a static environment with additional moving objects. Applications include automatic motion planning of an automobile or aircraft within moving traffic, and automatic motion planning of a carrier for material in a factory with other moving carriers.

1 Introduction

Motion planning for a rigid object in a static environment with obstacles in dimensions two or three is an area of intensive research in computational geometry and computer algebra (compare [1, 10, 7] and the survey article [11]).

By way of contrast the analogous problem for a a dynamic environment with moving obstacles has received much less attention. Important papers in this area are [12, 9]. They show that this problem is much harder than the static one, even if only translational motions are admitted. In particular most motion planning problems in a time-dependent environment are PSPACE-hard. Even the relatively simple one-dimensional problem street crossing problem is NP-hard. Only for constant-speed translations, some of these problems, e. g. the two-dimensional asteroid avoidance problem, are solvable in polynomial time. On the other hand all these problems are of great importance for the motion planning of an automobile, an aircraft, or a space craft moving among other objects of the same kind.

In this paper we study semilinear motion planning within a dynamic time dependent environment. This includes the situation of a static environment with additional moving objects that have to be avoided. The problem is to move a rigid semilinear objects within an semilinear environment that changes dynamically with time on a piecewise linear trajectory from a fixed initial position to a semilinear region of admissible final positions that may also change dynamically with time. Both the environment and the region of admissible final positions may change in a semilinear fashion with time. The solution of the problem is not a *path* in free space as in the static case, but a *trajectory* that is a path in space-time that decribes the actual movement of the object as a function of time. Part of the input data are finitely many vectors $\underline{v}_1, \ldots, \underline{v}_k$ in the ambient space $\mathbb{R}^2$ or $\mathbb{R}^3$, and a natural number n. We admit only semilinear trajectories consisting of n translations with constant directions and speeds specified by some of the vectors $\underline{v}_i$. Morover the whole movement has to take place in a prescribed time interval $[t_{init}, t_{final}]$. We allow, however, idle times before the first translation, between translations, and after the last translation.

In contrast to the problems studied in [12, 9] we prescribe a finite selection of possible constant speed translations that may be used in trajectories. So our problem and result is not directly comparable to the problems and results in these papers. We show that our problem can be solved in polynomial space and time that is polynomial in all data except the maximal number n of translations. The algorithm is exponential in n; so for bounded number of translations it runs in polynomial time. Morover we have implemented the algorithm in the REDLOG package of REDUCE [2, 3].

The corresponding semilinear motion problem for a static environment was studied and solved in [19] by means of extended linear real quantifier elimination [14, 18, 17]. Detailled case studies in the REDLOG package of REDUCE showed that the approach is able to solve significant practical problems both in two- and three-dimensional space.

The purpose of this paper is to extend this systematic approach to the case of the corresponding semilinear motion problem, where the semilinear environment and the final position of the object may both vary dynamically with time again in a semilinear fashion. We show that this is indeed possible and yields algorithms with roughly the same complexity; We describe the implementation of the approach in the REDLOG package of REDUCE 3.7 and present a series of computational examples with timings that show the practical applicability of the approach. Since our model comprises many interesting application cases, e. g. automatic motion planning of an automobile or aircraft within moving traffic, and automatic motion planning of a carrier for material in a factory with other moving carriers, it may be of practical interest.

2 Extended Linear Quantifier Elimination

Linear real quantifier elimination by virtual substitution of test points dates back to the theoretical paper [13]. During the last seven years a lot of theoretical work has been done to improve the method, cf. [8, 16, 5, 4, 17]. The method is implemented in the REDUCE package REDLOG by A. Dolzmann and T. Sturm. REDLOG is in fact a *computer logic system* providing not only quantifier elimination but a sophisticated working environment for first-order logic over various languages and theories, cf. [3]. There are also interfaces to QEPCAD and QERRC available such that these packages can be called from REDLOG and the results are available to be further processed by REDLOG. The REDLOG source code and documentation are freely available on the WWW.[1] The REDLOG package is part of REDUCE 3.7. Here we require only the linear real quantifier elimination and corresponding formula simplification facilities of REDLOG,

We consider first-order formulas for the ordered field of real numbers, an restrict our attention to the elimination of variables that occur linearly in such a formula. In this case quantifier elimination by the methods of REDLOG is guaranteed to succeed in principle. Quantifiers are eliminated one by one. The elimination starts with the innermost quantifier regarding the other quantified variables within ψ as extra parameters. Universal quantifiers are handled by means of the equivalence $\forall x\psi \longleftrightarrow \neg\exists x\neg\psi$. We may thus restrict our attention to a formula beginning with a single existential quantifier $\exists x$.

Roughly speaking, the elimination of the existential quantifier is then achieved by substituting for x a finite selection of suitable test terms depending on $u_1, \dots, u_k$, and by replacing the given formula by the disjunction of the substitution results, see [19] for a concise description of the method.

An extended quantifier elimination with answers can be obtained in a straightforward way from this method by not constructing a disjunction at the end. Instead all the quantifier-free substitution results are kept separately together with the candidate terms yielding them. Then these candidate terms represent sample solutions for the existentially quantifed variables.

The complexity of this method depends on the number of quantified variables and, even more, on the number of quantifier changes. Parameters play a minor role for the complexity. For existential (or universal) formulas with no multiplicative parameters, m quantifiers and at least 2 atomic formulas the number of atomic formulas $\mathrm{at}(\varphi')$ of the quantifier-free output φ' can be bounded in terms of the number of atomic formulas $\mathrm{at}(\varphi)$ of the input formula φ by:

$$\mathrm{at}(\varphi') \leq 2^m \mathrm{at}(\varphi)^{m+1}.$$

(compare [15, 17]).

[1] `http://www.fmi.uni-passau.de/~redlog/`

3 Modeling Time-Dependent Semilinear Motion Planning

In this section we show how to model the time-dependent semilinear motion planning problem addressed in the introduction as an extended linear quantifier elimination problem. This will require a considerable extension of the model for a static environment described in [19].

Let $\varphi(\underline{x}, t)$ be a formula that is linear w. r. t. a specified list $\underline{x} = (x_1, \ldots, x_d)$ of free space variables x_i and a time variable t. Then we denote by $\varphi_t^{\mathbb{R}}$ the set defined by φ in $\mathbb{R}^d$ for a given point t in time, i. e. the set of all $\underline{c} \in \mathbb{R}^d$ such that $\varphi(\underline{c}, t)$ holds in $\mathbb{R}$. We call $\varphi_t^{\mathbb{R}}$ the set defined by φ at time t. A subset S of $\mathbb{R}^{d+1}$ is *semilinear* (compare [19, 6]), if it is defined by a linear formula. As a consequence of linear quantifier elimination, every semilinear set is definable by a quantifier-free linear formula. Every conjunction of atomic linear formulas defines a (not necessarily closed, possibly empty) convex polyhedron. So - via a formation of a disjunctive normal form for quantifier-free linear formulas - every semilinear set is a finite union of (not necessarily closed) convex polyhedrons, i. e. a generalized polyhedron.

In the following we restrict our attention to semilinear sets in $\mathbb{R}^d \times \mathbb{R}$ for $d = 2$ or $d = 3$. Underlined letters will denote vectors in R^d. We consider a rigid semilinear object P in $\mathbb{R}^d$ given in standard position by a quantifier-free linear formula $\varphi(\underline{x})$. We attach to P in standard position the origin as reference point. After translation in $\mathbb{R}^d$ by a vector $\underline{y}$, the reference point has moved to $\underline{y}$, and P is described by the quantifier-free linear formula $\varphi(\underline{x} - \underline{y})$. We let $\psi(\underline{x}, t)$ be a quantifier-free formula that describes the free space S in which the object P may move as semilinear subset of R^d depending also in a semilinear way on time. $\underline{v}_1, \ldots, \underline{v}_k$ are vectors in R^d that define the admissible directions and speed vectors of translations. Note that a translation motion along direction $\underline{v}_i$ is given by an arbitrary scalar multiple of $\underline{v}_i$; so the motion may be of arbitrary length moving forward or backward along $\underline{v}_i$. The constant admissible speed of this motion is also given by the vector $s\underline{v}_i$.

We specify in addition an initial position $(\underline{y}, t_0)$ at starting time t_0, a natural number constant n, and a possibly time-dependent semilinear region of *admissible final positions* of the reference point of P described by quantifier-free linear formula $\delta(\underline{z}, t)$.

Then our **problem** is as follows:

For given time bounds t_{init}, t_{final} find a continuous, piecewise linear trajectory α with constant admissible speed on each piece for the reference point of P, leading from the initial position $(\underline{y}, t_0)$ at some time point $t_0 \geq t_{init}$ to an admissible final position $(\underline{z}, t_1)$ with $t_0 \leq t_1 \leq t_{final}$, such that α consists of at most n successive constant-speed translations with admissible directions and speeds, and such that for every position in space and time $t_0 \leq t \leq t_1$ of the reference point of P on this trajectory the object P is contained in the free space S_t.

We refer to this problem as the *time-dependent semilinear motion planning problem* for the data $t_{init}, t_{final}, \varphi, \psi, \underline{v}_1, \ldots, \underline{v}_k, \underline{y}, \delta$ and n. A trajectory that meets the specifications will be called an *admissible trajectory* for the given data.

As in the static case the problem can be reduced in a preprocessing step to the special case where the object P is a single point:

Let $\gamma(\underline{u}, t)$ be the formula

$$\forall \underline{x}(\varphi(\underline{x} - \underline{u}) \implies \psi(\underline{x}, t)).$$

Notice that γ is linear in $\underline{x}$, $\underline{u}$, and t; hence by linear elimination γ has a quantifier-free equivalent $\gamma'(\underline{u}, t)$ that is linear in $\underline{u}$ and t.

The set in $\gamma_t'^{\mathbb{R}}$ defined by γ' in $\mathbb{R}^d$ at time t is known in the literature as the *free placement space* FP_t of the object P with respect to the free space S_t at time t. FP_t is the set of all positions $\underline{u}$ of the reference point, for which the object P is contained in the free space S_t at time t.

After computation of γ' we are thus reduced to the following **specialized problem**:

For given time bounds t_{init}, t_{final} find a continuous, piecewise linear trajectory α with constant admissible speed on each piece for the reference point of P, leading from the initial position $(\underline{y}, t_0)$ at some time point $t_0 \geq t_{init}$ to an admissible final position $(\underline{z}, t_1)$ with $t_1 \leq t_{final}$, such that α consists of at most n successive constant-speed translations with admissible directions and speeds, and such that for all time points t in $[t_0, t_1]$ every point on this trajectory is contained in the free placement space FP_t.

In order to model this problem as a linear real quantifier elimination problem, we proceed in several steps.

First we describe the possibility of moving with admissible speed from position $(\underline{q}, t_0)$ to position $(\underline{w}, t_1)$ by a translation along a direction specified by the vector $\underline{v}$ within the free placement space FP_t by the formula:

$$(\underline{w} = \underline{q} + (t_1 - t_0)\underline{v} \ \wedge \ \forall s((t_1 - t_0 \geq s \geq 0) \Longrightarrow \gamma'(\underline{q} + s\underline{v}, t_0 + s))) \ \vee$$
$$(\underline{w} = \underline{q} + (t_0 - t_1)\underline{v} \ \wedge \ \forall s((t_0 - t_1 \leq s \leq 0) \Longrightarrow \gamma'(\underline{q} + s\underline{v}, t_0 + s)))$$

This formula $\nu(t_0, t_1, \underline{q}, \underline{w}, \underline{v})$ is linear in $s, t_0, t_1, \underline{q}, \underline{w}$; so again by by linear elimination it has a quantifier-free equivalent $\rho(t_0, t_1, \underline{q}, \underline{w}, \underline{v})$ that is linear in $t_0, t_1, \underline{q}, \underline{w}$.

Next the possibility of moving from position $(\underline{q}, t_0)$ to position $(\underline{w}, t_1)$ by a translation along an admissible direction and speed within time-dependent free space is described by the formula

$$\rho_k(t_0, t_1, \underline{q}, \underline{w}) := \bigvee_{j=1}^{k} \rho(t_0, t_1, \underline{q}, \underline{w}, \underline{v_j})$$

Again ρ_k is quantifier-free and linear in $t_0, t_1, \underline{q}, \underline{w}$.

Finally for specified positive natural number n the following formula describes the possibility of moving within given time bounds t_{init}, t_{final} from initial position $(\underline{y}, t_0)$ with $t_{init} \leq t_0$ to an admissible final position $(\underline{u}_{2n-1}, t_{2n-1})$ with $t_0 \leq t_1 \leq t_{final}$ by a continuous chain of n translations along admissible directions and with admissible speed within time-dependent free space. Notice that we allow idle time intervals $[t_{init}, t_0]$ before the first translation, $[t_{2i-1}, t_{2i}]$ for $1 \leq i < n$ between translations, and $[t_{2n-1}, t_{final}]$ after the last translation.

$$\exists \underline{u}_1 \ldots \underline{u}_n \exists t_0, \ldots t_{2n-1} (\rho_k(t_0, t_1, \underline{y}, \underline{u}_1) \wedge \bigwedge_{i=1}^{n-1} \rho_k(t_{2i}, t_{2i+1}, \underline{u}_i, \underline{u}_{i+1}) \wedge$$

$$t_{init} \leq t_0 \leq t_1 \leq \ldots \leq t_{2n-1} \leq t_{final} \wedge \delta(\underline{u}_n, t_{2n-1}))$$

By definition this formula is linear in $t_{init}, t_{final}, t_0, \ldots t_{2n-1}, \underline{y}, \underline{u}_1, \ldots, \underline{u}_n$. So again by linear elimination it has a quantifier-free equivalent $\sigma_n(t_{init}, t_{final}, \underline{y})$ that is linear in $t_{init}, t_{final}, \underline{y}$. Using extended linear elimination we obtain moreover as answers values for the vectors $\underline{u}_1, \ldots, \underline{u}_n$ and the time points $t_0, \ldots, t_{2n-1}$ that specify the vertices of a continuous admissible piecewise linear trajectory from position $(\underline{y}, t_0)$ to an admissible position $(\underline{u}_n, t_{2n-1})$ that is for all time points $t_0 \leq t \leq t_{2n-1}$ contained in the corresponding free placement space FP_t.

4 Complexity

From the general upper complexity bounds for linear elimination in [15] we can deduce upper complexity bounds for time-dependent semilinear motion planning. Let as before $\mathrm{at}(\varphi)$ denote the number of atomic formulas in the formula φ.

Reviewing the quantifier elimination steps necessary to arrive at a solution of the semilinear motion planning problem, we find the following: The first step is the computation of a quantifier-free formula γ' describing the free placement space from the defining quantifier-free formulas φ and ψ defining the object P and the free space S, respectively. This requires the elimination of a block of d universal quantifiers w. r. t. linear variables in front of a formula with $\mathrm{at}(\varphi) + \mathrm{at}(\psi)$ many atomic formulas. So by the general bound mentioned in the introduction, we get:

$$\mathrm{at}(\gamma') \leq 2^d (\mathrm{at}(\varphi) + \mathrm{at}(\psi))^{d+1}$$

Denote the bound on the right hand side by a. Then the successive elimination of the linear quantifiers $\forall s$ in the computation of ρ leads to the bound

$$\mathrm{at}(\rho) \leq 4a^2.$$

These two elimination steps may be considered as preprocessing steps, since they are independent of the admissible directions, the initial and final position and the number n of translations allowed.

The computation of σ_n and the actual solution path requires the extended elimination of a block of $dn+n+1 = (d+1)n+1$ $dn+2n = (d+2)n$ existential linear quantifiers in front of a formula with at most $k4a^2$ atomic formulas. This leads to an elimination tree with at most

$$2^{(d+2)n}(4ka^2)^{(d+2)n+1} \le (8ka^2)^{(d+3)n}$$

end nodes. By [15] the bit size of the coefficients grows only polynomially during iterated linear elimination. So for $d \le 3$ the whole elimination procedure is polynomial in the data $\underline{v}_1, \ldots, \underline{v}_k$, $\varphi, \psi, \delta, t_{init}, t_{final}$ and exponentially in n.

The same complexity bounds also apply to the case where the initial position is left unspecified. The result will then specify all initial positions for which the time-dependent motion planning problem with the given data is solvable.

5 Computational Examples in REDLOG

For computational examples in the REDLOG package we specify templates for the input of motion semilinear planning problems with fixed external parameters d, k, n.

For dimension $d = 2$, $k = 3$, and $n \le 3$ the *input template* looks e.g. as follows:

```
tinit:=  ;   % lower time bound tinit
tfinal:= ;   % upper time bound tfinal
y1:=   ; y2:=    ;  % Initial position
delta :=  ; % Defining formula for admissible final positions
phi:=    ; % Defining formula for the object in standard position
v11:=   ; v12:=    ; v21:=    ; v22:=    ; v31:=    ; v32:=    ;
% admissible directions and speeds
psi:=    ; % Defining formula for the free space
phip:= sub({x1=x1-u1,x2=x2-u2},phi);
% Defining formula for the object after translation
gamma := all({x1,x2}, phip impl psi);
% Defining formula for the free placement space
gammap:= rlqe(gamma);
% Quantifier-free defining formula for the free placement space
rlatnum(gammap);
% Check for the number of atomic formulas in gammap
gammap1:= rlsimpl(sub({u1=q1+s*v11,u2=q2+s*v12},gammap))$
% Movement from position q in one step in direction v1
gammap2:= rlsimpl(sub({u1=q1+s*v21,u2=q2+s*v22},gammap))$
% Movement from position q in one step in direction v2
gammap3:= rlsimpl(sub({u1=q1+s*v31,u2=q2+s*v32},gammap))$
% Movement from position q in one step in direction v3
rhop1:= rlqe( w1=q1+(t1-t0)*v11 and w2=q2+(t1-t0)*v12 and
      all(s, (t1-t0) >= s >= 0 impl sub({t=t0+s},gammap1))) or
        rlqe( w1=q1+(t0-t1)*v11 and w2=q2+(t0-t1)*v12 and
```

```
      all(s, (t0-t1) <= s <= 0 impl sub({t=t0-s},gammap1)))$
% Movement from position  (q,t0) to position (w,t1) in one translation
%along v1
rhop2:= rlqe( w1=q1+(t1-t0)*v21 and w2=q2+(t1-t0)*v22 and
      all(s, (t1-t0) >= s >= 0 impl sub({t=t0+s},gammap1))) or
        rlqe( w1=q1+(t0-t1)*v21 and w2=q2+(t0-t1)*v22 and
      all(s, (t0-t1) <= s <= 0 impl sub({t=t0-s},gammap1)))$
% Movement from position  (q,t0) to position (w,t1) in one translation
%along v2
rhop3:= rhop1:= rlqe( w1=q1+(t1-t0)*v31 and w2=q2+(t1-t0)*v32 and
      all(s, (t1-t0) >= s >= 0 impl sub({t=t0+s},gammap1))) or
        rlqe( w1=q1+(t0-t1)*v31 and w2=q2+(t0-t1)*v32 and
      all(s, (t0-t1) <= s <= 0 impl sub({t=t0-s},gammap1)))$
% Movement from position  (q,t0) to position (w,t1) in one translation
%along v3
rho:= rhop1 or rhop2 or rho3$
%  Movement from position (q,t0) to position (w,t1) in one translation in
%admissible directions v1 or v2 or v3
rlatnum(rho); %Check for the number of atomic formulas in rho
%Movement in n <=3 translations with three admissible
%directions and speeds v1, v2, v3
sigma1:= rlqea(ex({z1,z2,t0,t1},
         rlsimpl(sub({q1=y1,q2=y2,w1=z1,w2=z2},rho) and
         tinit <= t0 <= t1 <= tfinal and sub({t=t1},delta))));
% Movement in 1 translation in an admissible direction
sigma2:= rlqea(ex({u11,u12,u21,u22,t0,t1,t2,t3},  rlsimpl(
         sub({q1=y1,q2=y2,w1=u11,w2=u12},rho) and
         sub({q1=u11,q2=u12,w1=u21,w2=u22,t0=t2,t1=t3},rho) and
          tinit <= t0 <= t1 <= t2 <= t3 <=tfinal and
         sub({z1=u21,z2=u22,t=t3},delta))));
% Movement in 2 translations in admissible directions
sigma3:= rlqea(ex({u11,u12,u21,u22,u31,u32,t0,t1,t2,t3,t4,t5},  rlsimpl(
         sub({q1=y1,q2=y2,w1=u11,w2=u12},rho) and
         sub({q1=u11,q2=u12,w1=u21,w2=u22,t0=t2,t1=t3},rho) and
         sub({q1=u21,q2=u22,w1=z1,w2=z2,t0=t4,t1=t5},rho) and
         tinit<=t0<=t1<=t2<=t3<=t4<=t5<=tfinal and
         sub({z1=u31,z2=u32,t=t5},delta))));
% Movement in 3 translations in admissible directions
```

Here sigma$n$ has to be selected according to the desired value of n. The reader is invited to test more examples using these templates with the obvious modifications for dimension $d = 3$ and other values of k, n.

The specific examples discussed below use these templates with the specified values of d, k, n. All timings refer to a SUN SPARCstation Ultra I with 140 MHz. The epsilons occuring in some answers stand for positive infinitesimals; in case of two or more epsilons, the second is infinitesimally smaller than the first and so on. In practice the epsilons can be replaced by small enough positive real numbers.

Example 1. **Traffic Light**
In this 1D-example a car travels with constant speed along a straight road furnished with a traffic light. It can pass the traffic light only if the light shows green. The light has two red phases.

```
Data:  d=1, n=1, k=1

tinit:= 0;  tfinal:= 10;
y1:=  a;
delta := z1 = b;
phi:=   x1=0;
v11:= c;
psi:=  not(x1=0 and (2 <= t <= 3  or 5 <= t <= 6));
a:= -3; b:= 5; c:= 2;   %parameter values
phip:= sub({x1=x1-u1,x2=x2-u2},phi);
gamma := all(x1, phip impl psi);
gammap:= rlqe(gamma);
rlatnum(gammap);
gammap1:= rlsimpl(sub({u1=q1+s*v11},gammap))$
rho:= rlqe( w1=q1+(t1-t0)*v11  and
      all(s, (t1-t0) >= s >= 0 impl sub({t=t0+s},gammap1))) or
        rlqe( w1=q1+(t0-t1)*v11 and
      all(s, (t0-t1) <= s <= 0 impl  sub({t=t0-s},gammap1)))$
sigma1:= rlqea(ex({z1,t0,t1},
         rlsimpl(sub({q1=y1,w1=z1},rho) and
         tinit <= t0 <= t1 <= tfinal and sub({t=t1},delta))));

Answer:
sigma1 := {{true,{t0 =  - epsilon1 + 2,t1 =  - epsilon1 + 10,z1 = 5}}}
(Movement in the second green phase immediately before the second red
phase)
Time: ca. 0.15s

The variant with tfinal:= 8.5 yields the answer
sigma1 := {{true,{t0 = 1/2, t1 = 17/2, z1 = 5}}}
(Movement in the second green phase shortly after the first red phase)
Time: ca. 0.15s

The variant with tfinal:= 6 and speed c:= 2 yields the answer: impossible
sigma1 := {},

whereas the variant with tfinal:= 10 and speed c:= 2 yields the answer
sigma1 := {{true,{t0 = 6,t1 = 10,z1 = 5}}}
(Movement in the third green phase at the last possible moment)
Time: ca. 0.15s
```

Example 2. **Street Crossing**
In this 2D-example a person wants to cross a street with two lanes in a perpendicular straight line. On each lane a car is approaching, on the closest lane from

the left, and on the next lane from the right.
The person is modelled by a point, the cars by rectangles. The initial position is one unit away from the curb, the final position across the street also one unit away from the other curb. Between the two lanes there is safe median strip of one unit width.
We consider several variants differing by the number of translations allowed for the person.
In the first, simplest, variant the person is allowed only one fixed speed translation. So it may wait for a while after initial time t_{init}, then move with constant speed to the destination at the other side of the street in order to arrive before final time t_{final}.

```
Data:  d=2, n=1 or n=2, k=1

tinit:= 0;  tfinal:= 6;
y1:=  0; y2:= 0;
delta := z1 = 0 and z2 = 5;
phi:=   x1=0 and x2=0;
v11:= 0; v12:=  1;
psi:=  not(1 <= x2 <= 2 and 0 <= x1 - (c1*t + a1) <= 2) and
       not(3 <= x2 <= 4 and 0 <= x1 - (c2*t + a2) <= 2);
a1:= -5; c1:= 1; a2:= 10; c2:= - 2;  %parameter values
phip:= sub({x1=x1-u1,x2=x2-u2},phi);
gamma := all({x1,x2}, phip impl psi);
gammap:= rlqe(gamma);
rlatnum(gammap);
gammap1:= rlsimpl(sub({u1=q1+s*v11,u2=q2+s*v12},gammap))$
rho:= rlqe( w1=q1+(t1-t0)*v11 and w2=q2+(t1-t0)*v12 and
     all(s, (t1-t0) >= s >= 0 impl sub({t=t0+s},gammap1))) or
       rlqe( w1=q1+(t0-t1)*v11 and w2=q2+(t0-t1)*v12 and
     all(s, (t0-t1) <= s <= 0 impl  sub({t=t0-s},gammap1)))$
sigma1:= rlqea(ex({z1,z2,t0,t1},
         rlsimpl(sub({q1=y1,q2=y2,w1=z1,w2=z2},rho) and
         tinit <= t0 <= t1 <= tfinal and sub({t=t1},delta))));

Answer: (waiting after both cars have passed)
sigma1 := {{true,{t0 = 5,t1 = 10,z1 = 0,z2 = 5}}}
Time: about 0.3 s

Variant with tighter time bound tfinal := 6.

Answer: (passing between both cars)
sigma1 := {{true,{t0 = -epsilon1 + 1,t1 = -epsilon1 + 6,z1 = 0,z2 = 5}}}
Time: about 0.4 s

If the final time is changed to tfinal:= 7 and the starting position of
the second car  to a2:= 6, then the answer is
``sigma1 := {}'', i. e. ``impossible''.
Next we keep these data, but allow the person two moves with the same
```

speed in the same direction with a possible stop in between. This amounts to replacing sigma1 by:

```
sigma2:= rlqea(ex({u11,u12,u21,u22,t0,t1,t2,t3},  rlsimpl(
        sub({q1=y1,q2=y2,w1=u11,w2=u12},rho) and
        sub({q1=u11,q2=u12,w1=u21,w2=u22,t0=t2,t1=t3},rho) and
         tinit <= t0 <= t1 <= t2 <= t3 <=tfinal and
        sub({z1=u21,z2=u22,t=t3},delta))));
```

Then we get within about 1.8 seconds the answer:

```
sigma2 := {{true,
            {t0 = epsilon2,
             t1 =  - epsilon1 + epsilon2 + 3,
             t2 = epsilon1 + epsilon3 + 4,
             t3 = 2*epsilon1 + epsilon3 + 6,
             u11 = 0,
             u12 =  - epsilon1 + 3,
             u21 = 0,
             u22 = 5}}}
```

(After very short waiting first move to the median, then after waiting again for little more than one time unit second move to final position.)

Example 3. **Asteroid Avoidance**

A T-shaped (hence non-convex) 2D-spacecraft can move in the directions of the coordinate axes with different constant speeds. It has to avoid collision with three asteroids moving in different directions each with different constant speed in order to reach a designated target area.

```
Data:  d=2, n=2, k=2

tinit:= 0;  tfinal:= 10;
y1:= 0; y2:=  0;  % Initial position
delta :=  10 <= z1 and 10 <= z2;
phi:=  (0 <=x1<=1 and -2<=x2<=3) or (1<=x1<=3 and 0<=x2<=1);
v11:=  2; v12:=  0; v21:=  0; v22:=  3;
psi:=   not(x1 = 5*t + a1 and x2 = t + b1) and
        not(x1 = 2*t + a2 and x2= -t + b2) and
        not(x1 =  a3 and x2 = -3*t + b3);
a1:= -4; b1:= -2; a2:= 3; b2:= 5; a3:= 7; b3:= 6; % parameter values
phip:= sub({x1=x1-u1,x2=x2-u2},phi);
gamma := all({x1,x2}, phip impl psi);
gammap:= rlqe(gamma);
rlatnum(gammap);
gammap1:= rlsimpl(sub({u1=q1+s*v11,u2=q2+s*v12},gammap))$
gammap2:= rlsimpl(sub({u1=q1+s*v21,u2=q2+s*v22},gammap))$
gammap3:= rlsimpl(sub({u1=q1+s*v31,u2=q2+s*v32},gammap))$
rhop1:= rlqe( w1=q1+(t1-t0)*v11 and w2=q2+(t1-t0)*v12 and
      all(s, (t1-t0) >= s >= 0 impl sub({t=t0+s},gammap1))) or
        rlqe( w1=q1+(t0-t1)*v11 and w2=q2+(t0-t1)*v12 and
```

```
      all(s, (t0-t1) <= s <= 0 impl sub({t=t0-s},gammap1)))$
rhop2:= rlqe( w1=q1+(t1-t0)*v21 and w2=q2+(t1-t0)*v22 and
      all(s, (t1-t0) >= s >= 0 impl sub({t=t0+s},gammap1))) or
        rlqe( w1=q1+(t0-t1)*v21 and w2=q2+(t0-t1)*v22 and
      all(s, (t0-t1) <= s <= 0 impl sub({t=t0-s},gammap1)))$
rho:= rhop1 or rhop2$
rlatnum(rho); %Check for the number of atomic formulas in rho
sigma2:= rlqea(ex({u11,u12,u21,u22,t0,t1,t2,t3},  rlsimpl(
         sub({q1=y1,q2=y2,w1=u11,w2=u12},rho) and
         sub({q1=u11,q2=u12,w1=u21,w2=u22,t0=t2,t1=t3},rho) and
          tinit <= t0 <= t1 <= t2 <= t3 <=tfinal and
         sub({z1=u21,z2=u22,t=t3},delta))));

Answer:
sigma2 := {{true,
          {t0 = epsilon2 + 1,
           t1 = ( - 3*epsilon1 + 6*epsilon2 + 40)/6,
           t2 = 20/3,
           t3 = 10,
           u11 = ( - 3*epsilon1 + 34)/3,
           u12 = 0,
           u21 = ( - 3*epsilon1 + 34)/3,
           u22 = 10}}}$


Time: ca.  135s
(After some waiting first move along the x-axis, then after another
intermission second move along the y-axis; arrival at final position
on the border of the prescribed region exactly at the
final time.
```

6 Conclusions

We have shown that time-dependent admissible semilinear motion planning in a dynamically changing environment can be performed via linear real quantifier elimination with answers. For a bounded number of translations in the trajectories the algorithms work in polynomial time. The approach is implemented in the REDLOG package of REDUCE. It is of practical use for a small number n of edges in admissible paths. Possible industrial applications may include automatic motion planning of an aircraft or automobile within moving traffic, and automatic motion planning of a carrier for material in a factory with other moving carriers.

References

[1] John Canny. *The Complexity of Robot Motion Planning.* ACM Doctoral Dissertation Awards. The MIT Press, Cambridge, MA, 1987.

[2] Andreas Dolzmann and Thomas Sturm. *Redlog User Manual.* FMI, Universität Passau, D-94030 Passau, Germany, October 1996. Edition 1.0 for Version 1.0.

[3] Andreas Dolzmann and Thomas Sturm. Redlog: Computer algebra meets computer logic. *ACM SIGSAM Bulletin*, 31(2):2–9, June 1997.

[4] Andreas Dolzmann and Thomas Sturm. Simplification of quantifier-free formulae over ordered fields. *Journal of Symbolic Computation*, 24(2):209–231, August 1997.

[5] Andreas Dolzmann, Thomas Sturm, and Volker Weispfenning. A new approach for automatic theorem proving in real geometry. *Journal of Automated Reasoning*, 21(3):357–380, 1998.

[6] Lou van den Dries. *Tame Topology and o-minimal structures.* Cambridge University Press, 1998.

[7] Jean-Claude Latombe. *Robot Motion Planning.* Kluwer Academic Publishers, Boston, 1991.

[8] Rüdiger Loos and Volker Weispfenning. Applying linear quantifier elimination. *The Computer Journal*, 36(5):450–462, 1993. Special issue on computational quantifier elimination.

[9] J. Reif and M. Sharir. Motion planning in presence ov moving obstacles. *J. ACM*, 41:764–790, 1994.

[10] J.T. Schwartz, M. Sharir, and Hopcroft J., editors. *Planning, Geometry, and Complexity of Robot Motion.* Ablex Publ. Corporation, Norwood, NJ, USA, 1987.

[11] Micha Sharir. Algorithmic motion planning. In Jacob Goodman and Joseph O'Rourke, editors, *Handbook of Discrete and Computational Geometry*, pages 733–754. CRC Press, 1997.

[12] K. Sutner and W. Maass. Motion planning among time-dependent obstacles. *Acta Informatica*, 26:93–122, 1988.

[13] Volker Weispfenning. The complexity of linear problems in fields. *Journal of Symbolic Computation*, 5(1–2):3–27, February–April 1988.

[14] Volker Weispfenning. Efficient decision procedures for locally finite theories ii. In *ISSAC-88 (Rom 1988)*, volume 358 of *Lecture Notes of Computer Science*, pages 390–401. Springer, 1988.

[15] Volker Weispfenning. Parametric linear and quadratic optimization by elimination. Technical Report MIP-9404, FMI, Universität Passau, D-94030 Passau, Germany, April 1994.

[16] Volker Weispfenning. Quantifier elimination for real algebra—the cubic case. In *Proceedings of the International Symposium on Symbolic and Algebraic Computation (ISSAC 94)*, pages 258–263, Oxford, England, July 1994. ACM Press.

[17] Volker Weispfenning. Quantifier elimination for real algebra—the quadratic case and beyond. *Applicable Algebra in Engineering Communication and Computing*, 8(2):85–101, February 1997.

[18] Volker Weispfenning. Simulation and optimization by quantifier elimination. *Journal of Symbolic Computation*, 24(2):189–208, August 1997. Special issue on applications of quantifier elimination.

[19] Volker Weispfenning. Semilinear motion planning in REDLOG. *AAECC*, 2001. to appear.

Author Index